Mathematics for Engineers

Mathematics for
Engineers

Fourth Edition

Anthony Croft
Loughborough University

Robert Davison

PEARSON
Prentice
Hall

Harlow, England • London • New York • Boston • San Francisco • Toronto • Sydney • Singapore • Hong Kong
Tokyo • Seoul • Taipei • New Delhi • Cape Town • Madrid • Mexico City • Amsterdam • Munich • Paris • Milan

PEARSON EDUCATION LIMITED
Edinburgh Gate
Harlow CM20 2JE
United Kingdom
Tel: +44(0)1279 623623
Web: www.pearson.com.uk

First published 1998 (print)
Second edition published 2004 (print)
Third edition published 2008 (print)
Fourth edition published 2015 (print and electronic)

ISBN: 978-1-292-06593-9 (print)
978-1-292-07775-8 (PDF)
978-1-292-07774-1 (eText)

British Library Cataloguing-in-Publication Data
A catalogue record for the print edition is available from the British Library

Library of Congress Cataloging-in-Publication Data
A catalog record for the print edition is available from the Library of Congress

10 9 8 7 6 5 4 3 2
19 18 17 16 15

Cover: Dubai Meydan bridge, ALMSAEED/Getty Images

Print edition typeset in 10/12 Times by 73
Printed in Italy by L.E.G.O. S.p.A.

NOTE THAT ANY PAGE CROSS REFERENCES REFER TO THE PRINT EDITION

To Kate and Harvey (AC)
To Kathy (RD)

Brief contents

Contents

Companion Website

ON THE
WEBSITE

For open-access student resources specifically
written to complement this textbook and support
your learning, please visit **www.pearsoned.co.uk/croft**

Lecturer Resources

For password-protected online resources tailored to support
the use of this textbook in teaching, please visit
www.pearsoned.co.uk/croft

Publisher's acknowledgements

We are grateful to the following for permission to reproduce copyright material:

Table 9.1 from *Biometrika Tables for Statisticians*, Vol. 1, Oxford University Press, Oxford, (Pearson, E.S. and Hartley, H.O. (eds) 1970), by permission of Oxford University Press; Table 9.1 from STATISTICS: PROBABILITY, INFERENCE AND DECISION: VOL. 1 & 2, 1st edition by Hays. 1970. Reprinted with permission of Brooke/Cole, a division of Thomson Learning: www.thomsonrights.com. Fax 800 730-2215.

The screenshots in this book are Copyright © Parametric Technology Corporation, The MathWorks, Inc., Texas Instruments Incorporated and Waterloo Maple Inc. and reprinted with permission.

Preface

Audience

This book has been written to serve the mathematical needs of students engaged in a first course in engineering or technology at degree level. Students of a very wide range of these programmes will find that the book contains the mathematical methods they will meet in a first-year course in most UK universities. So the book will satisfy the needs of students of aeronautical, automotive, chemical, civil, electronic and electrical, systems, mechanical, manufacturing, and production engineering, and other technological fields. Care has been taken to include illustrative examples from these disciplines where appropriate.

Aims

There are two main aims of this book.

Firstly, we wish to provide a readable, accessible and student-friendly introduction to mathematics for engineers and technologists at degree level. Great care has been taken with explanations of difficult concepts, and wherever possible statements are made in everyday language, as well as symbolically. It is the use of symbolic notation that seems to cause many students problems, and we hope that we have gone a long way to alleviate such problems.

Secondly, we wish to develop in the reader the confidence and competence to handle mathematical methods relevant to engineering and technology through an interactive approach to learning. You will find that the book encourages you to take an active part in the learning process – this is an essential ingredient in the learning of mathematics.

The structure of this book

The book has been divided into 24 chapters. Each chapter is subdivided into a unit called a **block**. A block is intended to be a self-contained unit of study. Each block has a brief introduction to the material in it, followed by explanations, examples and applications. Important results and key points are highlighted. Many of the examples require you to participate in the problem-solving process, and so you will need to have pens or pencils, scrap paper and a scientific calculator to hand. We say more about this aspect below. Solutions to these examples are all given alongside.

Each block also contains a number of practice exercises, and the solutions to these are placed immediately afterwards. This avoids the need for searching at the back of the book for solutions. A further set of exercises appears at the end of each block.

At the end of each chapter you will find end of chapter exercises, which are designed to consolidate and draw together techniques from all the blocks within the chapter.

 Some sections contain computer or calculator exercises. These are denoted by the computer icon. It is not essential that these are attempted, but those of you with access to graphical calculators or computer software can see how these modern technologies can be used to speed up long and complicated calculations.

Learning mathematics

In mathematics almost all early building blocks are required in advanced work. New ideas are usually built upon existing ones. This means that, if some early topics are not adequately mastered, difficulties are almost certain to arise later on. For example, if you have not mastered the arithmetic of fractions, then you will find some aspects of algebra confusing. Without a firm grasp of algebra you will not be able to perform the techniques of calculus, and so on. It is therefore essential to try to master the full range of topics in your mathematics course and to remedy deficiencies in your prior knowledge.

Learning mathematics requires you to participate actively in the learning process. This means that in order to get a sound understanding of any mathematical topic it is essential that you actually perform the calculations yourself. You can't learn mathematics by being a spectator. *You* must use *your* brain to solve the problem, and *you* must write out the solution. These are essential parts of the learning process. It is not sufficient to watch someone else solve a similar problem, or to read a solution in a book, although these things of course can help. The test of real understanding and skill is whether or not you can do the necessary work on your own.

How to use this book

This book contains hundreds of fully worked examples. When studying such an example, read it through carefully and ensure you understand each stage of the calculation.

A central feature of the book is the use of interactive examples that require the reader to participate actively in the learning process. These examples are indicated by the pencil icon. Make sure you have to hand scrap paper, pens or pencils and a calculator. Interactive examples contain 'empty boxes' and 'completed boxes'. An empty box indicates that a calculation needs to be performed by you. The corresponding completed box on the right of the page contains the calculation you should have performed. When working through an interactive example, cover up the completed boxes, perform a calculation when prompted by an empty box, and then compare your work with that contained in the completed box. Continue in this way through the entire example. Interactive examples provide some help and structure while also allowing you to test your understanding.

Sets of exercises are provided regularly throughout most blocks. Try these exercises, always remembering to check your answers with those provided. Practice enhances understanding, reinforces the techniques, and aids memory. Carrying out a large number of exercises allows you to experience a greater variety of problems, thus building your expertise and developing confidence.

Content

The content of the book reflects that taught to first-year engineering and technology students in the majority of UK universities. However, particular care has been taken to develop algebraic skills from first principles and to give students plenty of opportunity to practise using these. It is our firm belief, based on recent experience of teaching engineering undergraduates, that many will benefit from this material because they have had insufficient opportunity in their previous mathematical education to develop such skills fully. Inevitably the choice of contents is a compromise, but the topics covered were chosen after wide consultation coupled with many years of teaching experience. Given the constraint of space we believe our choice is optimal.

Use of modern IT aids

One of the main developments in the teaching of engineering mathematics in recent years has been the widespread availability of sophisticated computer software and its adoption by many educational institutions. Once a firm foundation of techniques has been built, we would encourage its use, and so we have made general references at several points in the text. In addition, in some blocks we focus specifically on two common packages (Matlab and Maple), and these are introduced in the 'Using mathematical software packages' section on page xx. Many features available in software packages can also be found in graphical calculators.

Addition for the fourth edition

We have been delighted with the positive response to *Mathematics for Engineers* since it was first published in 1998. In writing this fourth edition we have been guided and helped by the numerous comments from both staff and students. For these comments, we express our thanks. Our special thanks go to Patrick Bond and Rufus Curnow for the opportunity to write this edition.

This fourth edition has been enhanced by the addition of over 20 extra worked examples from the various fields of engineering. Applicability lies at the heart of engineering mathematics. We believe these additional examples serve to reinforce the crucial role that mathematics plays in engineering. We hope that you agree.

We wish you enjoyment and good luck.

Anthony Croft and Robert Davison
February 2015

Using mathematical software packages

One of the main developments influencing the learning and teaching of engineering mathematics in recent years has been the widespread availability of sophisticated computer software and its adoption by many educational institutions.

As engineering students, you will meet a range of software in your studies. It is also highly likely that you will have access to specialist mathematical software. Two software packages that are particularly useful for engineering mathematics, and which are referred to on occasions throughout this book, are Matlab and Maple. There are others, and you should enquire about the packages that have been made available for your use. A number of these packages come with specialist tools for subjects such as control theory and signal processing, so that you will find them useful in other subjects that you study.

Common features of all these packages include:

- the facility to plot two- and three-dimensional graphs;
- the facility to perform calculations with symbols (e.g. a^2, $x + y$, as opposed to just numbers) including the solution of equations.

In addition, some packages allow you to write computer programs of your own that build upon existing functionality, and enable the experienced user to create powerful tools for the solution of engineering problems.

The facility to work with symbols, as opposed to just numbers, means that these packages are often referred to as **computer algebra systems** or **symbolic processors**. You will be able to enter mathematical expressions, such as $(x + 2)(x - 3)$ or $\dfrac{t - 6}{t^2 + 2t + 1}$, and subject them to all of the common mathematical operations: simplification, factorisation, differentiation, integration, and much more. You will be able to perform calculations with vectors and matrices. With experience you will find that lengthy, laborious work can be performed at the click of a button.

The particular form in which a mathematical problem is entered – that is, the syntax – varies from package to package. Raising to a power is usually performed using the symbol ^. Some packages are menu driven, meaning that you can often select symbols from a menu or toolbar. At various places in the text we have provided examples of this for illustrative purposes. This textbook is not intended to be a manual for any of the packages described. For thorough details you will need to refer to the manual provided with your software or its on-line help.

At first sight you might be tempted to think that the availability of such a package removes the need for you to become fluent in algebraic manipulation and other mathematical techniques. We believe that the converse of this is true. These packages are sophisticated, professional tools, and as such require the user to have a good understanding of the functions they perform, and particularly their limitations. Furthermore, the results provided by the packages can be presented in a variety of forms (as you will see later in the book), and only with a thorough understanding of the mathematics will you be able to appreciate different, yet correct, equivalent forms, and distinguish these from incorrect output.

Figure 1 shows a screenshot from Maple in which we have defined the function $f(x) = x^2 + 3x - 2$ and plotted part of its graph. Note that Maple requires the following particular syntax to define the function: $f: = x \rightarrow x^2 + 3x - 2$. The quantity x^2 is input as x^2.

Finally, Figure 2 shows a screenshot from the package Matlab. Here the package is being used to obtain a three-dimensional plot of the surface $z = \sin(x^2 + y^2)$ as described in Chapter 21. Observe the requirement of Matlab to input x^2 as x $\cdot$ ^2.

Figure 1
A screenshot from Maple showing the package being used to define the function $f(x) = x^2 + 3x - 2$ and plot its graph.

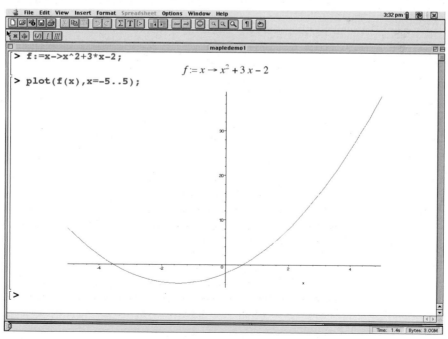

Figure 2
A screenshot from
Matlab showing the
package being used
to plot a three-
dimensional graph.

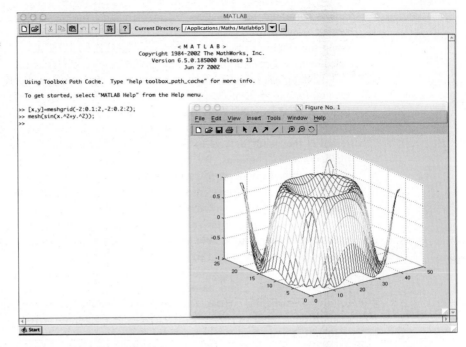

Where appropriate we would encourage you to explore the use of packages such as these. Through them you will find that whole new areas of engineering mathematics become accessible to you, and you will develop skills that will help you to solve engineering problems that you meet in other areas of study and in the workplace.

Arithmetic

Chapter **1**

This chapter reminds the reader of the arithmetic of whole numbers. Arithmetic is the study of numbers. A mastery of numbers and the ways in which we manipulate and operate on them is essential. This mastery forms the bedrock for further study in the field of algebra.

Block 1 introduces some essential terminology and explains rules that determine the order in which operations must be performed. Block 2 focuses on prime numbers. These are numbers that cannot be expressed as the product of two smaller numbers.

Chapter 1 contents

Operations on numbers

1.1 Introduction

Whole numbers are the numbers . . . $-3, -2, -1, 0, 1, 2, 3$ . . . . Whole numbers are also referred to as **integers**. The **positive integers** are $1, 2, 3, 4, \ldots$. The **negative integers** are . . . , $-4, -3, -2, -1$. The . . . indicates that the sequence of numbers continues indefinitely. The number 0 is an integer but it is neither positive nor negative.

Given two or more whole numbers it is possible to perform an **operation** on them. The four arithmetic operations are addition $(+)$, subtraction $(-)$, multiplication $(\times)$ and division $(\div)$.

Addition (+)

We say that $4 + 5$ is the **sum** of 4 and 5. Note that $4 + 5$ is equal to $5 + 4$ so that the order in which we write down the numbers does not matter when we are adding them. Because the order does not matter, addition is said to be **commutative**. When more than two numbers are added, as in $4 + 8 + 9$, it makes no difference whether we add the 4 and 8 first to get $12 + 9$, or whether we add the 8 and 9 first to get $4 + 17$. Whichever way we work we shall obtain the same result, 21. This property of addition is called **associativity**.

Subtraction (−)

We say that $8 - 3$ is the **difference** of 8 and 3. Note that $8 - 3$ is not the same as $3 - 8$ and so the order in which we write down the numbers is important when we are subtracting them. Subtraction is not commutative. Adding a negative number is equivalent to subtracting a positive number; thus $5 + (-2) = 5 - 2 = 3$. Subtracting a negative number is equivalent to adding a positive number: thus $7 - (-3) = 7 + 3 = 10$.

Key point

> Adding a negative number is equivalent to subtracting a positive number.
> Subtracting a negative number is equivalent to adding a positive number.

Multiplication (×)

The instruction to multiply the numbers 6 and 7 is written 6×7. This is known as the **product** of 6 and 7. Sometimes the multiplication sign is missed out altogether and we write $(6)(7)$. An alternative and acceptable notation is to use a dot to represent multiplication and so we could write $6 \cdot 7$, although if we do this care must be taken not to confuse this multiplication dot with a decimal point.

1   **4**   Block 1 Operations on numbers

Note that (6)(7) is the same as (7)(6) so multiplication of numbers is commutative. If we are multiplying three numbers, as in $2 \times 3 \times 4$, we obtain the same result if we multiply the 2 and the 3 first to get 6×4, as if we multiply the 3 and the 4 first to get 2×12. Either way the result is 24. This property of multiplication is known as **associativity**.

Recall that when multiplying positive and negative numbers the sign of the result is given by the following rules:

Key point

(positive) $\times$ (positive) = positive
(positive) $\times$ (negative) = negative
(negative) $\times$ (positive) = negative
(negative) $\times$ (negative) = positive

For example, $(-4) \times 5 = -20$ and $(-3) \times (-6) = 18$.

Division (÷)

The quantity $8 \div 4$ means 8 divided by 4. This is also written as $8/4$ or $\frac{8}{4}$ and is known as the **quotient** of 8 and 4. We refer to a number of the form p/q when p and q are whole numbers as a **fraction**. In the fraction $\frac{8}{4}$ the top line is called the **numerator** and the bottom line is called the **denominator**. Note that $8/4$ is not the same as $4/8$ and so the order in which we write down the numbers is important. Division is not commutative. Division by 0 is never allowed: that is, the denominator of a fraction must never be 0. When dividing positive and negative numbers recall the following rules for determining the sign of the result:

Key point

$$\frac{\text{positive}}{\text{positive}} = \text{positive}$$

$$\frac{\text{positive}}{\text{negative}} = \text{negative}$$

$$\frac{\text{negative}}{\text{positive}} = \text{negative}$$

$$\frac{\text{negative}}{\text{negative}} = \text{positive}$$

Example 1.1
Evaluate
(a) the sum of 9 and 4
(b) the sum of 9 and -4
(c) the difference of 6 and 3

(d) the difference of 6 and −3
(e) the product of 9 and 3
(f) the product of −9 and 3
(g) the product of −9 and −3
(h) the quotient of 10 and 2
(i) the quotient of 10 and −2
(j) the quotient of −10 and −2

Solution
(a) $9 + 4 = 13$
(b) $9 + (-4) = 9 - 4 = 5$
(c) $6 - 3 = 3$
(d) $6 - (-3) = 6 + 3 = 9$
(e) $9 \times 3 = 27$
(f) $(-9) \times 3 = -27$
(g) $(-9) \times (-3) = 27$
(h) $\frac{10}{2} = 5$
(i) $\frac{10}{-2} = -5$
(j) $\frac{-10}{-2} = 5$

Example 1.2 Reliability Engineering – Time between breakdowns
Reliability engineering is concerned with managing the risks associated with break-down of equipment and machinery, particularly when such a breakdown is life-critical or when it can have an adverse effect on business. In Chapter 23 we will discuss the Poisson probability distribution which is used to model the number of breakdowns occurring in a specific time interval. Of interest to the reliability engineer is both the average number of breakdowns in a particular time period and the average time between breakdowns. The **breakdown rate** is the number of breakdowns per unit time.

Suppose a reliability engineer monitors a piece of equipment for a 48-hour period and records the number of times that a safety switch trips. Suppose the engineer found that there were three trips in the 48-hour period.
(a) Assuming that the machine can be restarted instantly, calculate the average time between trips. This is often referred to as the **inter-breakdown** or **inter-arrival time**.
(b) Calculate the breakdown rate per hour.

Solution
(a) With three trips in 48 hours, on average, there will be one trip every 16 hours. Assuming that the machine can be restarted instantly, the average time between trips is 16 hours. This is the inter-breakdown time.
(b) In 16 hours there is one trip. This is equivalent to saying that the breakdown rate is $\frac{1}{16}$ of a trip per hour.
More generally,

$$\text{the breakdown rate} = \frac{1}{\text{the inter-breakdown time}}$$

Exercises

1   Find the sum of -9 and 11.

2   Find the product of 13 and 9.

3   Find the difference of 11 and 4.

4   Find the quotient of 100 and 5.

5   **Manufacturing Engineering – Production of components.** A production line works 14 hours per day and produces 60 electrical components every hour. Find the number of components produced during a working week of 5 days.

6   **Computer Networking – Routing of data.** Computer network traffic can be routed along any of four routes. If a total of 1680 Mbit/s are distributed equally along the four routes, calculate the data rate on each. If one of the routes is disabled, calculate the data rate on each of the three remaining routes.

Solutions to exercises

1   2

2   117

3   7

4   20

5   4200

6   420 Mbit/s, 560 Mbit/s

1.2   Order of operations

The order in which the four operations are carried out is important but may not be obvious when looking at an expression. Consider the following case. Suppose we wish to evaluate $2 \times 3 + 4$. If we carry out the multiplication first the result is $2 \times 3 + 4 = 6 + 4 = 10$. However, if we carry out the addition first the result is $2 \times 3 + 4 = 2 \times 7 = 14$. Clearly we need some rules that specify the order in which the various operations are performed. Fortunately there are rules, called **precedence rules**, that tell us the priority of the various operations – in other words, the order in which they are carried out.

Knowing the order in which operations will be carried out becomes particularly important when programming using software such as Maple, Matlab and Excel if you are to avoid unexpected and erroneous results.

To remind us of the order in which to carry out these operations we can make use of the BODMAS rule. BODMAS stands for:

Key point

Brackets, ( )	first priority
Of, $\times$	
Division, $\div$	second priority
Multiplication, $\times$	
Addition, $+$	third priority
Subtraction, $-$	

Here 'of' means the same as multiply, as in 'a half of 6' means '$\frac{1}{2} \times 6$'.

Later in Chapter 5 we meet a further operation called **exponentiation**. We shall see that exponentiation should be carried out once brackets have been dealt with.

Example 1.3

Evaluate

(a) $2 \times 3 + 4$

(b) $6 \div 2 - 1$

(c) $2 \times (3 + 4)$

(d) $6 \div (2 - 1)$

Solution

(a) There are two operations in the expression: multiplication and addition. Multiplication has a higher priority than addition and so is carried out first.

$$2 \times 3 + 4 = 6 + 4$$
$$= 10$$

(b) There are two operations: division and subtraction. Division is carried out first.

$$6 \div 2 - 1 = 3 - 1$$
$$= 2$$

(c) The bracketed expression, $(3 + 4)$, is evaluated first, even though the addition has a lower priority than multiplication.

$$2 \times (3 + 4) = 2 \times 7$$
$$= 14$$

(d) The bracketed expression is evaluated first.

$$6 \div (2 - 1) = 6 \div 1$$
$$= 6$$

This example illustrates the crucial difference that brackets can make to the value of an expression.

When all the operations in an expression have the same priority, then we simply work from left to right.

Example 1.4

Evaluate

(a) $9 + 3 - 6 + 2 - 4$

(b) $12 \times 2 \div 4 \times 3$

Solution

(a) Noting that the operations addition and subtraction have the same priority we work from left to right thus:

$$9 + 3 - 6 + 2 - 4 = 12 - 6 + 2 - 4$$
$$= 6 + 2 - 4$$
$$= 8 - 4$$
$$= 4$$

(b) Since multiplication and division have equal priority we work from left to right.

$$12 \times 2 \div 4 \times 3 = 24 \div 4 \times 3$$
$$= 6 \times 3$$
$$= 18$$

Example 1.5
Evaluate
(a) $27 \div (7 - 4) + 3 \times 4$
(b) $[(6 \times 2) \div (1 + 2)] \div [3 + 5 - 7 + 3]$

Solution
(a) Evaluation of the expression in brackets is performed first to give

$$27 \div 3 + 3 \times 4$$

The resulting expression contains the operations of division, multiplication and addition. Division and multiplication have higher priority than addition and so are performed first, from left to right. This produces

$$9 + 12$$

Hence the result is 21.
(b) Evaluating the innermost bracketed expressions gives:

$$[12 \div 3] \div [3 + 5 - 7 + 3]$$

Evaluating each of the two remaining bracketed expressions results in

$$4 \div 4$$

and so the final result is 1.

Often a division line replaces bracketed quantities. For example, in the expression
$$\frac{7 + 9}{3 + 1}$$
there is an implied bracketing of the numerator and denominator, meaning
$$\frac{(7 + 9)}{(3 + 1)}$$
The bracketed quantities would be evaluated first, resulting in $\frac{16}{4}$, which simplifies to 4.

Exercises

1 Evaluate the following arithmetic expressions using the rules for priority of operations:
(a) $12 \div 2 \times 6$
(b) $12 \times 6 \div 2$
(c) $12 \div (2 \times 6)$
(d) $(12 \div 2) \times 6$
(e) $12 + 6 \div 2$
(f) $12 \div 6 + 2$
(g) $\dfrac{8 - 2}{2 + 1}$
(h) $\dfrac{-10 \div 2}{-6 + 1}$

2   Insert an appropriate mathematical operation
    as indicated in order to make the given
    expression correct:
    (a) $11 - (5 \, ? \, 3) \times 2 = 7$

(b) $12 \times (15 \, ? \, 5) + 9 = 45$

(c) $\dfrac{5 + (14 \, ? \, 2)}{4} = 3$

Solutions to exercises

1   (a) 36 (b) 36 (c) 1 (d) 36 (e) 15
    (f) 4 (g) 2 (h) 1

2   (a) − (b) ÷ (c) ÷

End of block exercises

1   Find
    (a) the product of 11 and 4
    (b) the sum of −16 and 3
    (c) the difference of 12 and 9
    (d) the quotient of 12 and −2

2   Evaluate the following arithmetical
    expressions:
    (a) $10 - 5 + 2$
    (b) $10 - (5 + 2)$
    (c) $10 \div 5 + 2$
    (d) $10 \times 5 + 2$
    (e) $10 \times (5 + 2)$

3   Evaluate the following expressions:
    (a) $6 - 4 \times 2 + 3$
    (b) $6 - 4 \div 2 + 3$
    (c) $(6 - 4) \times 2 + 3$
    (d) $(6 - 4) \times (2 + 3)$
    (e) $(6 - 4 \div 2) + 3$

4   Evaluate
    $$\dfrac{9 + 3 \times 3}{10 - 3 \div 3}$$

Solutions to exercises

1   (a) 44 (b) −13 (c) 3 (d) −6

3   (a) 1 (b) 7 (c) 7 (d) 10 (e) 7

2   (a) 7 (b) 3 (c) 4 (d) 52 (e) 70

4   2

Prime numbers and prime factorisation

2.1 Introduction

A prime number is a positive integer, larger than 1, which cannot be written as the product of two smaller integers. This means that the only numbers that divide exactly into a prime number are 1 and the prime number itself. Examples of prime numbers are 2, 3, 5, 7, 11, 13, 17 and 19. Clearly 2 is the only even prime. The numbers 4 and 6 are not primes as they can be written as products of smaller integers, namely

$$4 = 2 \times 2, \quad 6 = 2 \times 3$$

When a number has been written as a product we say that the number has been **factorised**. Each part of the product is termed a **factor**. When writing $6 = 2 \times 3$ then both 2 and 3 are factors of 6. Factorisation of a number is not unique. For example, we can write

$$12 = 2 \times 6, \quad 12 = 3 \times 4, \quad 12 = 2 \times 2 \times 3$$

All these are different, but nevertheless correct, factorisations of 12.

When a number is written as the product of prime numbers we say that the number has been **prime factorised**. Prime factorisation is unique.

Prime numbers have a long history, being extensively studied by the ancient Greek mathematicians including Pythagoras. There has been significant renewed interest in prime numbers once it was recognised that they have important applications in cryptography and particularly Internet security. Whilst it is easy to multiply two very large prime numbers together it is very difficult then to factorise the result to obtain the original primes. Prime numbers form the basis of systems such as the RSA cryptosystem in which a message is encoded, but can only be decoded by someone who has knowledge of the original prime numbers.

Example 2.1
Prime factorise the following numbers: (a) 18, (b) 693.

Solution
The technique of prime factorisation entails repeatedly dividing the number and its factors by prime numbers until no further division is possible.

(a) Starting with the first prime, 2, we note that 18 may be written as

$$18 = 2 \times 9$$

We now consider the factor 9. Clearly 2 is not a factor of 9 so we try the next prime number, 3, which is a factor.

$$18 = 2 \times 3 \times 3$$

All the factors are primes: that is, 18 has been prime factorised.

(b) We note that 2 is not a factor of 693 and so try the next prime, 3. We see that 3 is a prime factor and write

$$693 = 3 \times 231$$

Looking at 231, we note that 3 is a factor and write

$$693 = 3 \times 3 \times 77$$

Looking at 77, we note that 3 is not a factor and so try the next prime, 5. Since 5 is not a factor we try the next prime, 7, which is a factor. We write

$$693 = 3 \times 3 \times 7 \times 11$$

All the factors are now prime and so no further factorisation is possible.

Since ancient times methods have been developed to find prime numbers. The interested reader is referred, for example, to the *sieve of Eratosthenes*, which is an efficient method for finding relatively small prime numbers.

Exercises

1   Explain why 2 is the only even prime number.

2   State all prime numbers between 50 and 100.

3   Prime factorise the following numbers:
(a) 30 (b) 96 (c) 500 (d) 589 (e) 3239

Solutions to exercises

2   53, 59, 61, 67, 71, 73, 79, 83, 89, 97

3   (a) $2 \times 3 \times 5$
(b) $2 \times 2 \times 2 \times 2 \times 2 \times 3$
(c) $2 \times 2 \times 5 \times 5 \times 5$
(d) 19×31
(e) 41×79

2.2   Highest common factor

Suppose we prime factorise the numbers 12 and 90. This produces

$$12 = 2 \times 2 \times 3, \quad 90 = 2 \times 3 \times 3 \times 5$$

Some factors are common to both numbers. For example, 2 is such a common factor, as is 3. There are no other common prime factors. Combining these common factors we see that (2×3) is common to both. Thus 6 (i.e. 2×3) is the highest number that is a factor of both 12 and 90. We call 6 the **highest common factor (h.c.f.)** of 12 and 90.

Key point

Given two or more numbers, the highest common factor (h.c.f.) is the largest (highest) number that is a factor of all the given numbers.

To put this another way, the h.c.f. is the highest number that divides exactly into each of the given numbers.

Example 2.2

Find the h.c.f. of 16 and 30.

Solution

We prime factorise each number:

$$16 = 2 \times 2 \times 2 \times 2, \quad 30 = 2 \times 3 \times 5$$

There is only one prime factor common to both: 2. Hence 2 is the h.c.f. of 16 and 30.

Example 2.3

Find the h.c.f. of 30 and 50.

Solution

Prime factorisation yields

$$30 = 2 \times 3 \times 5, \quad 50 = 2 \times 5 \times 5$$

The common prime factors are 2 and 5 and so the h.c.f is $2 \times 5 = 10$.

Example 2.4

Find the h.c.f. of 36, 54 and 126.

Solution

Prime factorisation yields

36 = [   ] $2 \times 2 \times 3 \times 3$

54 = [   ] $2 \times 3 \times 3 \times 3$

126 = [   ] $2 \times 3 \times 3 \times 7$

The common factors are [   ]   2, 3, 3

Hence the h.c.f. is $2 \times 3 \times 3 = 18$.
　　Note that the factor of 3 is included twice because 3×3 is common to all factorisations.

Exercises

1　Calculate the h.c.f. of the following numbers:
(a) 6, 10
(b) 28, 42
(c) 36, 60, 90
(d) 7, 19, 31

Solutions to exercises

1 (a) 2 (b) 14 (c) 6 (d) 1

2.3 Lowest common multiple

Suppose we are given two or more numbers and wish to find numbers into which all the given numbers will divide. For example, given 4 and 6 we see that both divide exactly into 12, 24, 36, 48 and so on. The smallest number into which they both divide is 12. We say that 12 is the **lowest common multiple** of 4 and 6.

Key point

The lowest common multiple (l.c.m.) of a set of numbers is the smallest (lowest) number into which all of the given numbers will divide exactly.

Example 2.5
Find the l.c.m. of 6 and 8.

Solution
There are many numbers into which both 6 and 8 divide exactly: for example, 48, 96 and 120. We seek the smallest number with this property. By inspection and trial and error we see that the smallest number is 24. Hence the l.c.m. of 6 and 8 is 24.

For larger numbers it is not appropriate to use inspection as a means of finding the l.c.m.; a more systematic method is needed, and this is now explained.

The numbers are prime factorised. The l.c.m. is formed by examining the prime factorisations. All the different primes that occur in the prime factorisations are noted. The highest occurrence of each prime is also noted. The l.c.m. is then formed using the highest occurrence of each prime. Consider the following example.

Example 2.6
Find the l.c.m. of 90, 120 and 242.

Solution
Each number is prime factorised to yield

$$90 = 2 \times 3 \times 3 \times 5$$
$$120 = 2 \times 2 \times 2 \times 3 \times 5$$
$$242 = 2 \times 11 \times 11$$

The different primes are noted: these are 2, 3, 5 and 11. The highest occurrence of each prime is noted:

Prime	2	3	5	11
Highest occurrence	3	2	1	2

The highest occurrence of 2 is 3 since 2 occurs three times in the prime factorisation of 120. The highest occurrence of 3 is 2 since 3 occurs twice in the prime factorisation of 90.

The l.c.m. is then $2 \times 2 \times 2 \times 3 \times 3 \times 5 \times 11 \times 11$, which is 43560. Hence 43560 is the smallest number into which 90, 120 and 242 will all divide exactly.

Example 2.7
Find the l.c.m. of 25, 35 and 45.

Solution
Prime factorisation of each number yields

25 = ▨ 5×5

35 = ▨ 5×7

45 = ▨ $3 \times 3 \times 5$

Hence the primes are 3, 5 and 7.

The highest occurrence of 3 is ▨   2

The highest occurrence of 5 is ▨   2

The highest occurrence of 7 is ▨   1

Hence in its prime factorised form the l.c.m. is

▨ $3 \times 3 \times 5 \times 5 \times 7$

The l.c.m. is 1575.

Example 2.8
Find the l.c.m. of 4, 8 and 24.

Solution
Each number is prime factorised: $4 = 2 \times 2$, $8 = 2 \times 2 \times 2$ and $24 = 2 \times 2 \times 2 \times 3$. The l.c.m. is formed using the highest occurrence of each prime and is then $2 \times 2 \times 2 \times 3 = 24$. Note from this example that the l.c.m. of a set of numbers can be one of the numbers in the set.

Exercises

1 Find the l.c.m. of
   (a) 6, 16
   (b) 6, 16, 20
   (c) 20, 30, 75
   (d) 22, 32, 45, 72
   (e) 11, 17, 21, 100

Solutions to exercises

1   (a) 48 (b) 240 (c) 300 (d) 15840 (e) 392700

End of block exercises

1   Find the h.c.f. of the following sets of numbers:
(a) 20, 30, 40, 50 (b) 60, 108, 180 (c) 18, 126, 198, 324 (d) 105, 147, 210, 273

2   Find the l.c.m. of the following:
(a) 5, 8, 12 (b) 6, 8, 9, 10 (c) 15, 18, 20, 25
(d) 20, 22, 30, 35

3   Prime factorise the following numbers:
(a) 315 (b) 2695 (c) 988 (d) 16905

Solutions to exercises

1   (a) 10 (b) 12 (c) 18 (d) 21

2   (a) 120 (b) 360 (c) 900 (d) 4620

3   (a) $3 \times 3 \times 5 \times 7$
(b) $5 \times 7 \times 7 \times 11$
(c) $2 \times 2 \times 13 \times 19$
(d) $3 \times 5 \times 7 \times 7 \times 23$

End of chapter exercises

1   Evaluate
(a) $12 - (3 \div 3)$
(b) $(12 - 3) \div 3$
(c) $12 - 3 \div 3$
(d) $2 \times (6 \div 3)$
(e) $(2 \times 6) \div 3$
(f) $2 \times 6 \div 3$

2   Prime factorise the following numbers:
(a) 15 (b) 16 (c) 50 (d) 91 (e) 119
(f) 323 (g) 682

3   Calculate the h.c.f. of the following sets of numbers:
(a) 6, 21 (b) 16, 24, 72 (c) 30, 45, 60
(d) 18, 30, 42, 100

4   Find the l.c.m. of the following:
(a) 4, 10 (b) 16, 30, 40 (c) 15, 16, 25, 32

Solutions to exercises

1   (a) 11 (b) 3 (c) 11 (d) 4 (e) 4 (f) 4

2   (a) 3×5 (b) $2 \times 2 \times 2 \times 2$
(c) $2 \times 5 \times 5$ (d) 7×13 (e) 7×17
(f) 17×19 (g) $2 \times 11 \times 31$

3   (a) 3 (b) 8 (c) 15 (d) 2

4   (a) 20 (b) 240 (c) 2400

Fractions

Chapter **2**

The methods used to simplify, add, subtract, multiply and divide numerical fractions are exactly the same as those used for algebraic fractions. So it is important to understand and master those methods with numerical fractions before moving on to apply them to algebraic fractions.

Block 1 introduces basic terminology and the idea of equivalent fractions. Fractions that are equivalent have the same value. Using prime factorisation a fraction may be expressed in its simplest form.

Block 2 illustrates how to add and subtract fractions. Key to these operations is the writing of all fractions with a common denominator. Mixed fractions – those fractions that have a whole number part as well as a fractional part – are introduced. Finally multiplication and division of fractions are explained and illustrated.

Chapter 2 contents

Introducing fractions

1.1 Introduction

In this block we introduce some terminology. Fractions can be classed as either proper or improper and this classification is explained. We introduce the idea of equivalent fractions. Using the prime factorisation technique of Chapter 1 we explain how to write a fraction in its simplest form.

A fraction is a number of the form $\frac{p}{q}$, where p and q are whole numbers. So, for example, $\frac{2}{3}, \frac{5}{9}, \frac{11}{3}$ and $\frac{100}{4}$ are all fractions.

The number p (the 'top' of the fraction) is called the **numerator**. The number q (the 'bottom' of the fraction) is called the **denominator**.

Key point

$$\text{Fraction} = \frac{p}{q} = \frac{\text{numerator}}{\text{denominator}}$$

Note that the denominator is never zero because division by 0 is not permissible.

Fractions can be classified as either **proper fractions** or **improper fractions**. When determining whether a fraction is proper or improper we ignore any negative signs in the numerator and denominator.

If the numerator, p, is less than the denominator, q, then the fraction is proper. If p is equal to or greater than q then the fraction is improper. Examples of proper fractions are $\frac{1}{2}, \frac{99}{100}$ and $\frac{2}{5}$. Examples of improper fractions are $\frac{3}{2}, \frac{5}{5}$ and $\frac{100}{99}$.

Exercises

1 Classify each fraction as either proper or improper:

(a) $\frac{7}{9}$ (b) $\frac{5}{2}$ (c) $\frac{5}{5}$ (d) $-\frac{5}{7}$ (e) $\frac{6}{-11}$ (f) $\frac{-3}{2}$

Solutions to exercises

1 (a) proper (b) improper (c) improper
(d) proper (e) proper (f) improper

1.2   Simplest form of a fraction

Every fraction can be expressed in different forms. For example, you may know that $\frac{1}{2}, \frac{2}{4}$ and $\frac{3}{6}$ all have the same value. These are examples of **equivalent fractions**. Equivalent fractions have the same value.

We note that multiplying or dividing both numerator and denominator of a fraction by the same number produces an equivalent fraction: that is, a fraction with the same value as the original fraction. Consider the following example.

Starting with the fraction $\frac{16}{24}$ we

1   multiply both numerator and denominator by 2 to obtain $\frac{32}{48}$

2   multiply both numerator and denominator by 10 to obtain $\frac{160}{240}$

3   divide both numerator and denominator by 4 to obtain $\frac{4}{6}$

4   divide both numerator and denominator by 8 to obtain $\frac{2}{3}$.

Hence $\frac{16}{24}, \frac{32}{48}, \frac{160}{240}, \frac{4}{6}$ and $\frac{2}{3}$ are all equivalent fractions. They all have exactly the same value.

A fraction in its simplest form is one that has no factors common to both numerator and denominator. This means that it is not possible to divide both numerator and denominator exactly. For example, $\frac{17}{49}$ is in its simplest form; there is no number which can divide exactly into both 17 and 49. However, $\frac{18}{48}$ is not in its simplest form: 6 is a factor of both numerator and denominator. It is thus possible to divide both numerator and denominator by 6 to obtain $\frac{3}{8}$. This is an equivalent fraction and is in its simplest form.

Example 1.1

Express $\frac{36}{120}$ in its simplest form.

Solution

We seek factors that are common to both numerator and denominator. By using prime factorisation we can find the h.c.f. of 36 and 120 to be 12. Dividing numerator and denominator by this h.c.f. yields $\frac{3}{10}$. Hence in its simplest form $\frac{36}{120}$ is $\frac{3}{10}$.

An alternative way to find the simplest form is to prime factorise numerator and denominator: thus

$$\frac{36}{120} = \frac{2 \times 2 \times 3 \times 3}{2 \times 2 \times 2 \times 3 \times 5}$$

We note all of the factors common to both numerator and denominator: these are $2 \times 2 \times 3$. These factors are cancelled.

$$\frac{2 \times 2 \times 3 \times 3}{2 \times 2 \times 2 \times 3 \times 5} = \frac{\cancel{2} \times \cancel{2} \times \cancel{3} \times 3}{\cancel{2} \times \cancel{2} \times 2 \times \cancel{3} \times 5} = \frac{3}{2 \times 5} = \frac{3}{10}$$

Cancelling all the common factors is equivalent to dividing numerator and denominator by the h.c.f., 12, so these two approaches, while appearing different, are in fact the same.

Example 1.2

Express $\frac{90}{210}$ in its simplest form.

Solution

Prime factorising numerator and denominator yields

$$\frac{90}{210} = \qquad\qquad\qquad \qquad\qquad \frac{2 \times 3 \times 3 \times 5}{2 \times 3 \times 5 \times 7}$$

The factors common to both numerator and denominator are

$$\qquad\qquad\qquad\qquad\qquad\qquad\qquad\qquad\qquad 2, 3, 5$$

Cancelling the common factors yields

$$\qquad\qquad\qquad\qquad\qquad\qquad\qquad\qquad\qquad\qquad \frac{3}{7}$$

Hence $\frac{90}{210}$ in its simplest form is $\frac{3}{7}$.

Sometimes we need to write a fraction in an equivalent form with a specified denominator. The following example illustrates the technique.

Example 1.3

Express $\frac{3}{4}$ as an equivalent fraction with a denominator of 24.

Solution

The original denominator is 4; the desired denominator is 24. Thus the original denominator must be multiplied by 6. To produce an equivalent fraction, then, both numerator and denominator must be multiplied by 6. Thus

$$\frac{3}{4} = \frac{3 \times 6}{4 \times 6} = \frac{18}{24}$$

Recall that multiplying both numerator and denominator by the same number does not change the value of the fraction.

Example 1.4 Mechanical Engineering – Force on a bar

The force on a metal bar is increased by $\frac{15}{25}$ N. Express the increase in force as a fraction in its simplest form.

Solution

By prime factorising both numerator and denominator and then cancelling common factors we find

$$\frac{15}{25} = \frac{3 \times 5}{5 \times 5} = \frac{3}{5}$$

In its simplest form, the increase in force is $\frac{3}{5}$ N.

Example 1.5 Thermodynamics – Temperature of a cooling liquid

Experiments are conducted to measure the rate at which the temperature of a liquid cools. The reduction in temperature over a measured time is noted. In the first 20 minutes the temperature drops by 6 °C. The average rate of fall of temperature is therefore $\frac{6}{20}$ °C per minute.

Express this rate of decrease in temperature as a fraction in its simplest form.

Solution

$$\frac{6}{20} = \frac{2 \times 3}{2 \times 2 \times 5} = \frac{3}{10}$$

The rate of decrease in temperature is $\frac{3}{10}$ °C per minute.

Example 1.6 Reliability Engineering – Quality control

A quality control officer tests 300 items taken from a production line and finds that 15 of the items are faulty.

(a) Express the number of faulty items as a fraction of those tested.

(b) Express this fraction in its simplest form.

Solution

(a) There are 15 faulty items from a total of 300. So the fraction of items that are faulty is $\frac{15}{300}$.

(b)

$$\frac{15}{300} = \frac{3 \times 5}{2 \times 2 \times 3 \times 5 \times 5} = \frac{1}{20}$$

Exercises

1 Express each fraction in its simplest form:

(a) $\frac{9}{12}$ (b) $\frac{16}{20}$ (c) $\frac{20}{16}$ (d) $\frac{72}{96}$ (e) $-\frac{30}{42}$ (f) $\frac{-20}{-45}$

2 (a) Express $\frac{3}{5}$ as an equivalent fraction with a denominator of 40.

(b) Express $\frac{9}{30}$ as an equivalent fraction with a denominator of 10.

(c) Express 6 as an equivalent fraction with a denominator of 4.

3 Express each fraction as an equivalent fraction with a denominator of 32:

(a) $\frac{1}{16}$ (b) $\frac{3}{8}$ (c) $\frac{1}{4}$

4 Consider the fractions $\frac{2}{3}, \frac{5}{6}$ and $\frac{7}{8}$.

(a) Calculate the l.c.m. of their denominators.

(b) Express each fraction as an equivalent fraction with the l.c.m. found in (a) as the denominator.

5 **Reliability Engineering – Quality control.**

(a) Quality control checks find six faulty items from 200 that were tested. Express the number of faulty items as a fraction, in its simplest form, of the total number tested.

(b) Improvements to the production process mean that the number of faulty items is halved. Express the number of faulty items now as a fraction of the total number tested.

6 **Thermodynamics – Temperature of a metal bar.** A metal bar is heated and its temperature is measured every 5 minutes. The table records the temperature readings.

Time (min)	0	5	10	15	20
Temp (°C)	21	27	31	33	35

Calculate the rate of increase of temperature in units of °C per minute

(a) in the first 5 minutes

(b) in the first 10 minutes

(c) in the first 15 minutes

(d) in the first 20 minutes

(e) in the last 5 minutes

(f) in the last 10 minutes

In each case express your answer as a fraction in its simplest form.

Solutions to exercises

1 (a) $\frac{3}{4}$ (b) $\frac{4}{5}$ (c) $\frac{5}{4}$ (d) $\frac{3}{4}$ (e) $-\frac{5}{7}$ (f) $\frac{4}{9}$

2 (a) $\frac{24}{40}$ (b) $\frac{3}{10}$ (c) $\frac{24}{4}$

3 (a) $\frac{2}{32}$ (b) $\frac{12}{32}$ (c) $\frac{8}{32}$

4 (a) 24 (b) $\frac{16}{24}, \frac{20}{24}, \frac{21}{24}$

5 (a) $\frac{3}{100}$ (b) $\frac{3}{200}$

6 (a) $\frac{6}{5}$ (b) 1 (c) $\frac{4}{5}$ (d) $\frac{7}{10}$ (e) $\frac{2}{5}$ (f) $\frac{2}{5}$

End of block exercises

1 Classify each of the following fractions as proper or improper:
   (a) $\frac{6}{7}$ (b) $\frac{7}{6}$ (c) $\frac{6}{6}$ (d) $\frac{21}{22}$ (e) $\frac{3}{1}$

2 Express the fraction $\frac{2}{7}$ in three different equivalent forms.

3 Express each fraction in its simplest form:
   (a) $\frac{20}{100}$ (b) $\frac{16}{60}$ (c) $\frac{9}{45}$ (d) $\frac{22}{24}$ (e) $\frac{56}{16}$

Solutions to exercises

1 (a) proper (b) improper (c) improper
   (d) proper (e) improper

2 Several possible solutions, for example $\frac{4}{14}, \frac{8}{28}, \frac{20}{70}$

3 (a) $\frac{1}{5}$ (b) $\frac{4}{15}$ (c) $\frac{1}{5}$ (d) $\frac{11}{12}$ (e) $\frac{7}{2}$

Operations on fractions

2.1 Introduction

This block explains and illustrates how to perform the four arithmetic operations on fractions. Writing fractions in equivalent forms so that they all have the same denominator is a crucial first step when adding and subtracting fractions. Mixed fractions comprise a whole number component as well as a fractional component. They can be written as improper fractions and vice versa. The block concludes with multiplication and division of fractions.

2.2 Addition and subtraction of fractions

To add and subtract fractions we must rewrite each fraction so that they have the same denominator. This denominator is called the **common denominator** as it is common to each fraction.

The common denominator is the l.c.m. of the original denominators. Once each fraction has been written in equivalent form with the common denominator, the numerators are then added or subtracted as required. In summary:

1 Calculate the l.c.m. of the original denominators to find the common denominator.
2 Express each fraction in equivalent form with the common denominator.
3 Add/subtract numerators and divide the result by the l.c.m.

Example 2.1

Find $\frac{3}{8} + \frac{5}{6}$.

Solution

The original denominators are 6 and 8. Their l.c.m. is 24. Each fraction is written in equivalent form with 24 as the denominator:

$$\frac{3}{8} \text{ is equivalent to } \frac{9}{24}; \frac{5}{6} \text{ is equivalent to } \frac{20}{24}$$

So

$$\frac{3}{8} + \frac{5}{6} = \frac{9}{24} + \frac{20}{24}$$
$$= \frac{9 + 20}{24}$$
$$= \frac{29}{24}$$

Example 2.2
Find $\frac{1}{3} - \frac{1}{4} + \frac{1}{6}$.

Solution
The original denominators are 3, 4, and 6.

Their l.c.m. is [ ]

12

Each fraction is written in equivalent form with the l.c.m. as denominator:

$\frac{1}{3}$ is equivalent to [ ]

$\frac{4}{12}$

$\frac{1}{4}$ is equivalent to [ ]

$\frac{3}{12}$

$\frac{1}{6}$ is equivalent to [ ]

$\frac{2}{12}$

So

$$\frac{1}{3} - \frac{1}{4} + \frac{1}{6} = \frac{4}{12} - \frac{3}{12} + \frac{2}{12}$$

$$= \boxed{}$$

$\frac{3}{12}$

$$= \frac{1}{4}$$

Exercises

1   Find
   (a) $\frac{2}{3} + \frac{1}{6}$ (b) $\frac{9}{10} - \frac{3}{5}$ (c) $\frac{1}{2} - \frac{2}{5}$ (d) $\frac{5}{6} - \frac{1}{12} + \frac{1}{3}$ (h) $\frac{3}{8} + \frac{2}{5} + \frac{3}{10}$ (i) $\frac{7}{6} - \frac{2}{3} + \frac{5}{9}$

   (e) $\frac{7}{9} - \frac{2}{3} + \frac{5}{6}$ (f) $2 - \frac{3}{4}$ (g) $\frac{1}{4} - \frac{9}{10}$ (j) $\frac{1}{12} + \frac{1}{6} - \frac{1}{3} - \frac{1}{4}$

Solutions to exercises

1   (a) $\frac{5}{6}$ (b) $\frac{3}{10}$ (c) $\frac{1}{10}$ (d) $\frac{13}{12}$ (e) $\frac{17}{18}$ (f) $\frac{5}{4}$ (g) $-\frac{13}{20}$

   (h) $\frac{43}{40}$ (i) $\frac{19}{18}$ (j) $-\frac{1}{3}$

2.3 Mixed fractions

We have already met proper and improper fractions. We now examine **mixed fractions**.

The number $4\frac{2}{3}$ is an example of a mixed fraction. We note that there is a whole number part, 4, and a proper fraction part, $\frac{2}{3}$. The mixed fraction may be written in an equivalent form as an improper fraction:

$$4\frac{2}{3} = 4 + \frac{2}{3}$$

$$= \frac{12}{3} + \frac{2}{3}$$

$$= \frac{14}{3}$$

Hence $4\frac{2}{3}$ is equivalent to $\frac{14}{3}$.

Example 2.3

(a) Express $2\frac{3}{4}$ as an improper fraction.

(b) Find $2\frac{3}{4} + 1\frac{1}{6}$.

Solution

(a) $2\frac{3}{4} = 2 + \frac{3}{4}$

$$= \quad\boxed{}\qquad\qquad\qquad\qquad\qquad\qquad \frac{8}{4} + \frac{3}{4}$$

$$= \quad\boxed{}\qquad\qquad\qquad\qquad\qquad\qquad\qquad \frac{11}{4}$$

(b) $2\frac{3}{4} + 1\frac{1}{6} = \frac{11}{4} + \boxed{}\qquad\qquad\qquad\qquad\qquad \frac{7}{6}$

$$= \quad\boxed{}\qquad\qquad\qquad\qquad \frac{33}{12} + \frac{14}{12}$$

$$= \frac{47}{12}$$

Example 2.4

Express $\frac{47}{12}$ as a mixed fraction.

Solution

Dividing 47 by 12 results in 3, with a remainder of 11. So

$$\frac{47}{12} = 3 + \frac{11}{12}$$

$$= 3\frac{11}{12}$$

Exercises

1 Express the following mixed fractions as improper fractions:

(a) $1\frac{1}{2}$ (b) $2\frac{1}{3}$ (c) $3\frac{1}{4}$ (d) $3\frac{2}{5}$ (e) $-10\frac{2}{5}$

2 Express the following improper fractions as mixed fractions:

(a) $\frac{5}{4}$ (b) $\frac{6}{5}$ (c) $\frac{12}{5}$ (d) $\frac{47}{3}$ (e) $-\frac{62}{7}$

3 Calculate the following, expressing your answer as an improper fraction:

(a) $1\frac{2}{5} + 2\frac{3}{4}$ (b) $4\frac{1}{3} + 6\frac{3}{7}$ (c) $2\frac{4}{5} - 1\frac{2}{3}$

(d) $6\frac{1}{2} - 3\frac{3}{4} + 2\frac{1}{5}$ (e) $3\frac{1}{3} - 4\frac{4}{9}$

4 Express the solutions to question 3 as mixed fractions.

Solutions to exercises

1 (a) $\frac{3}{2}$ (b) $\frac{7}{3}$ (c) $\frac{13}{4}$ (d) $\frac{17}{5}$ (e) $-\frac{52}{5}$

2 (a) $1\frac{1}{4}$ (b) $1\frac{1}{5}$ (c) $2\frac{2}{5}$ (d) $15\frac{2}{3}$ (e) $-8\frac{6}{7}$

3 (a) $\frac{83}{20}$ (b) $\frac{226}{21}$ (c) $\frac{17}{15}$ (d) $\frac{99}{20}$ (e) $-\frac{10}{9}$

4 (a) $4\frac{3}{20}$ (b) $10\frac{16}{21}$ (c) $1\frac{2}{15}$ (d) $4\frac{19}{20}$ (e) $-1\frac{1}{9}$

2.4 Multiplication of fractions

To multiply two fractions we multiply their denominators to form the denominator of the answer, and multiply their numerators to form the numerator of the answer.

Example 2.5
Find $\frac{2}{3} \times \frac{4}{7}$.

Solution

$$\frac{2}{3} \times \frac{4}{7} = \frac{2 \times 4}{3 \times 7} = \frac{8}{21}$$

Note that the new numerator of 8 is formed by multiplying the two original numerators. Similarly the new denominator is the product of the two original denominators. Sometimes it is possible to simplify the result.

Example 2.6
Calculate $\frac{3}{8} \times \frac{14}{27}$.

Solution

$$\frac{3}{8} \times \frac{14}{27} = \frac{3 \times 14}{8 \times 27} = \frac{42}{216}$$

Writing $\frac{42}{216}$ in its simplest form is $\frac{7}{36}$.

The cancelling of common factors can take place at an earlier stage in the calculation. For example, we could have written

$$\frac{3 \times 14}{8 \times 27} = \frac{14}{8 \times 9} \quad \text{by cancelling a factor of 3}$$

$$= \frac{7}{4 \times 9} \quad \text{by cancelling a factor of 2}$$

$$= \frac{7}{36}$$

Example 2.7
Calculate $\frac{9}{12} \times \frac{1}{5} \times \frac{20}{36}$.

Solution
The fraction $\frac{9}{12}$ can be simplified to $\frac{3}{4}$. Similarly $\frac{20}{36}$ can be simplified to $\frac{5}{9}$. Hence

$$\frac{9}{12} \times \frac{1}{5} \times \frac{20}{36} = \frac{3}{4} \times \frac{1}{5} \times \frac{5}{9}$$

$$= \frac{3}{4} \times \frac{1}{1} \times \frac{1}{9} \quad \text{cancelling a factor of 5}$$

$$= \frac{1}{4} \times \frac{1}{1} \times \frac{1}{3} \quad \text{cancelling a factor of 3}$$

$$= \frac{1}{12}$$

Example 2.8
Calculate (a) $\frac{3}{8} \times \frac{9}{10}$, (b) $\frac{3}{8} \times \frac{10}{81}$.

Solution

(a) $\dfrac{3}{8} \times \dfrac{9}{10} = $ ⬜ $\dfrac{27}{80}$

(b) $\dfrac{3}{8} \times \dfrac{10}{81} = $ ⬜ $\dfrac{5}{108}$

We are sometimes asked to calculate quantities such as '$\frac{1}{3}$ of 96' or '$\frac{2}{5}$ of 60'. Note the use of the word 'of' in these expressions. When we are given such an expression we treat the 'of' as we would a multiplication sign. So when calculating, for example, $\frac{1}{3}$ of 96 we calculate

$$\frac{1}{3} \times 96$$

which results in 32.

Example 2.9
Calculate $\frac{2}{5}$ of 60.

Solution

$$\frac{2}{5} \times 60 = 2 \times 12 \quad \text{(cancelling a factor of 5)}$$
$$= 24$$

So $\frac{2}{5}$ of 60 is 24.

Example 2.10
(a) Find $\frac{3}{25}$ of 100.
(b) Given 1 m is 100 cm, express $\frac{3}{25}$ m in cm.

Solution
(a)

$$\frac{3}{25} \times 100 = 12 \quad \text{(cancelling a factor of 25)}$$

(b) $\frac{3}{25}$ of 1 m is the same as $\frac{3}{25}$ of 100 cm which, using the result from (a), is 12 cm.

Example 2.11 Mechanical Engineering – Extension of a metal wire
A metal wire stretches by $\frac{3}{100}$ of its original length when heavy weights are applied. If the original length of the wire is 4 m calculate (a) the length by which the wire stretches, (b) the resulting length of the stretched wire.

Solution
(a) The wire stretches by $\frac{3}{100}$ of 4 m.

$$\frac{3}{100} \times 4 = \frac{3}{25} \quad \text{(cancelling a factor of 25)}$$

The wire stretches by $\frac{3}{25}$ m.

(b) The resulting length of the wire is

$$4 + \frac{3}{25} = 4\frac{3}{25}$$

Using Example 2.10 this may be expressed as 4 m 12 cm.

Example 2.12
Calculate $1\frac{1}{2} \times 3\frac{1}{4}$.

Solution
Mixed fractions are converted into their equivalent improper form. The multiplication is then performed.

$$1\frac{1}{2} \times 3\frac{1}{4} = \frac{3}{2} \times \frac{13}{4}$$
$$= \frac{39}{8}$$
$$= 4\frac{7}{8}$$

Exercises

1 Calculate

(a) $\frac{1}{2} \times \frac{1}{3}$ (b) $\frac{1}{2} \times \frac{2}{3}$ (c) $\frac{3}{4} \times \frac{1}{3}$

(d) $\frac{3}{4} \times \frac{1}{4}$ (e) $\frac{3}{4} \times \frac{2}{3}$

2 Calculate

(a) $\frac{3}{7} \times \frac{14}{15}$ (b) $\frac{4}{5} \times \frac{10}{21}$ (c) $\frac{5}{6} \times \frac{9}{10}$

(d) $\frac{7}{9} \times \frac{12}{35}$ (e) $\frac{5}{8} \times \frac{12}{25}$

3 Calculate

(a) $\frac{1}{2} \times \frac{2}{3} \times \frac{3}{4}$ (b) $\frac{2}{5} \times \frac{9}{20} \times \frac{10}{13}$ (c) $\frac{3}{4} \times \frac{5}{6} \times \frac{10}{11}$

(d) $\frac{6}{7} \times \frac{8}{9} \times \frac{21}{32}$ (e) $\frac{2}{3} \times \frac{5}{6} \times \frac{7}{9} \times \frac{1}{4}$

4 Calculate the following, expressing your answer as a mixed fraction:

(a) $1\frac{1}{2} \times 3\frac{1}{2}$ (b) $4\frac{1}{3} \times 2\frac{1}{2}$ (c) $5\frac{2}{3} \times 1\frac{1}{17}$

(d) $4\frac{3}{4} \times 6\frac{2}{3}$ (e) $3\frac{2}{5} \times 1\frac{2}{3}$

Solutions to exercises

1 (a) $\frac{1}{6}$ (b) $\frac{1}{3}$ (c) $\frac{1}{4}$ (d) $\frac{3}{16}$ (e) $\frac{1}{2}$

2 (a) $\frac{2}{5}$ (b) $\frac{8}{21}$ (c) $\frac{3}{4}$ (d) $\frac{4}{15}$ (e) $\frac{3}{10}$

3 (a) $\frac{1}{4}$ (b) $\frac{9}{65}$ (c) $\frac{25}{44}$ (d) $\frac{1}{2}$ (e) $\frac{35}{324}$

4 (a) $5\frac{1}{4}$ (b) $10\frac{5}{6}$ (c) 6 (d) $31\frac{2}{3}$ (e) $5\frac{2}{3}$

2.5 Division of fractions

Division of one fraction by another is a similar process to multiplication of fractions. When dividing one fraction by another there is one additional step involved: the second fraction is inverted and then the calculation continues as a multiplication. To **invert** a fraction we interchange numerator and denominator.

Example 2.13
Calculate $\frac{3}{8} \div \frac{5}{6}$.

Solution
The second fraction, $\frac{5}{6}$, is inverted to $\frac{6}{5}$. The calculation is now carried out as a multiplication.

$$\frac{3}{8} \div \frac{5}{6} = \frac{3}{8} \times \frac{6}{5}$$
$$= \frac{9}{20}$$

Example 2.14
Calculate $2\frac{2}{3} \div 1\frac{1}{6}$.

Solution

$$2\frac{2}{3} \div 1\frac{1}{6} = \frac{8}{3} \div \frac{7}{6}$$

$$= \frac{8}{3} \times \frac{6}{7}$$

$$= \frac{16}{7}$$

$$= 2\frac{2}{7}$$

Example 2.15
Calculate (a) $\frac{6}{7} \div \frac{2}{21}$, (b) $5 \div 2\frac{1}{4}$.

Solution

(a) $\dfrac{6}{7} \div \dfrac{2}{21} = \dfrac{6}{7} \times$ ▢ $\dfrac{21}{2}$

$=$ ▢ 9

(b) $5 \div 2\dfrac{1}{4} = 5 \div$ ▢ $\dfrac{9}{4}$

$= 5 \times$ ▢ $\dfrac{4}{9}$

$= \dfrac{20}{9}$

$=$ ▢ $2\frac{2}{9}$

Exercises

1 Calculate
   (a) $\frac{8}{9} \div \frac{5}{6}$ (b) $\frac{16}{17} \div \frac{8}{9}$ (c) $\frac{7}{10} \div \frac{14}{15}$
   (d) $\frac{10}{13} \div \frac{5}{39}$ (e) $\frac{7}{12} \div \frac{49}{50}$

2 Calculate the following, expressing your answer as a mixed fraction:
   (a) $5\frac{1}{2} \div 3\frac{1}{2}$ (b) $6\frac{2}{3} \div 1\frac{1}{4}$ (c) $9\frac{1}{5} \div 11\frac{1}{2}$
   (d) $8\frac{2}{5} \div 5\frac{1}{4}$ (e) $14\frac{2}{3} \div 6\frac{2}{7}$

Solutions to exercises

1   (a) $\frac{16}{15}$ (b) $\frac{18}{17}$ (c) $\frac{3}{4}$ (d) 6 (e) $\frac{25}{42}$

2   (a) $1\frac{4}{7}$ (b) $5\frac{1}{3}$ (c) $\frac{4}{5}$ (d) $1\frac{3}{5}$ (e) $2\frac{1}{3}$

End of block exercises

1   Evaluate
(a) $\frac{4}{5} + \frac{3}{10}$ (b) $\frac{7}{8} - \frac{1}{3}$ (c) $\frac{5}{6} \times \frac{8}{15}$ (d) $\frac{3}{4} \div \frac{9}{8}$

2   Evaluate, expressing your answer as a mixed fraction:
(a) $1\frac{3}{4} + 2\frac{1}{2}$ (b) $4\frac{2}{3} - 1\frac{1}{2}$ (c) $2\frac{3}{4} \times 3\frac{1}{3}$

(d) $5\frac{1}{6} \div 2\frac{1}{4}$

3   Evaluate

$$\frac{1\frac{1}{2} + 3\frac{1}{3}}{4\frac{2}{5} - 1\frac{1}{3}}$$

Solutions to exercises

1   (a) $\frac{11}{10}$ (b) $\frac{13}{24}$ (c) $\frac{4}{9}$ (d) $\frac{2}{3}$

2   (a) $4\frac{1}{4}$ (b) $3\frac{1}{6}$ (c) $9\frac{1}{6}$ (d) $2\frac{8}{27}$

3   $1\frac{53}{92}$

End of chapter exercises

1   Express each of the following fractions in their simplest form:
(a) $\frac{12}{60}$ (b) $\frac{9}{36}$ (c) $\frac{27}{81}$ (d) $\frac{6}{92}$ (e) $\frac{377}{390}$

2   Express each of the following mixed fractions as an improper fraction:
(a) $3\frac{2}{3}$ (b) $5\frac{2}{5}$ (c) $7\frac{1}{2}$ (d) $-9\frac{3}{4}$ (e) $10\frac{4}{7}$

3   Express each of the following improper fractions as mixed fractions:
(a) $\frac{20}{3}$ (b) $\frac{32}{7}$ (c) $\frac{60}{9}$ (d) $\frac{102}{50}$ (e) $\frac{120}{11}$

4   Calculate
(a) $\frac{3}{4} + \frac{1}{3}$ (b) $\frac{1}{2} + \frac{3}{5}$ (c) $\frac{5}{6} - \frac{1}{3}$
(d) $\frac{10}{11} - \frac{1}{2}$ (e) $\frac{4}{7} + \frac{1}{2} - \frac{2}{3}$

5   Calculate the following, expressing your answer as a mixed fraction:
(a) $2\frac{1}{4} + 3\frac{1}{3}$ (b) $2\frac{4}{5} - 1\frac{2}{3}$ (c) $5\frac{2}{3} - 1\frac{1}{2} + 2\frac{1}{5}$
(d) $5 - 4\frac{2}{7} + \frac{1}{3}$ (e) $\frac{9}{10} + \frac{6}{7} - 1\frac{2}{5}$

6   Calculate
(a) $\frac{6}{7} \times \frac{14}{27}$ (b) $\frac{7}{10} \times \frac{4}{5} \times \frac{30}{49}$ (c) $\frac{8}{9} \times \frac{18}{25}$
(d) $\left(-\frac{4}{5}\right) \times \left(-\frac{3}{4}\right)$ (e) $\frac{16}{21} \times \left(-\frac{3}{4}\right)$

7   Calculate the following, expressing your answer as a mixed fraction:
(a) $2\frac{3}{4} \times 2\frac{1}{2}$ (b) $6\frac{1}{3} \times 1\frac{2}{5}$ (c) $\left(-3\frac{1}{4}\right) \times 2\frac{5}{6}$
(d) $3\frac{1}{3} \times 3\frac{1}{3}$ (e) $4\frac{3}{4} \times \left(-2\frac{4}{5}\right)$

8 Calculate

(a) $\frac{7}{9} \div \frac{2}{3}$ (b) $\frac{8}{15} \div \frac{4}{5}$ (c) $\frac{9}{10} \div \frac{9}{20}$

(d) $\frac{6}{11} \div \frac{7}{12}$ (e) $\frac{12}{13} \div \frac{6}{7}$

9 Calculate the following, expressing your answer as a mixed fraction:

(a) $3\frac{2}{3} \div 1\frac{1}{2}$ (b) $6\frac{1}{4} \div 2\frac{2}{3}$ (c) $10\frac{3}{4} \div 2\frac{1}{5}$

(d) $-12\frac{2}{5} \div 3\frac{1}{4}$ (e) $10\frac{5}{6} \div (-4\frac{1}{2})$

10 Calculate

(a) $\frac{1}{4}$ of 60 (b) $\frac{2}{3}$ of 75 (c) $\frac{3}{8}$ of 64 (d) $\frac{2}{5}$ of $\frac{15}{16}$

(e) $\frac{3}{4}$ of $\frac{20}{21}$

expressing each answer as a fraction in its simplest form.

11 Evaluate

(a) $6\frac{2}{3} \div 4$ (b) $10 \div 2\frac{1}{3}$

(c) $(\frac{1}{2} + \frac{1}{3}) \div (\frac{2}{3} + \frac{1}{5})$

(d) $(6 - 2\frac{1}{3}) \times (4\frac{1}{2} - 1\frac{3}{4})$ (e) $\dfrac{2\frac{1}{2} + 1\frac{1}{3}}{6\frac{2}{3} - 2\frac{1}{4}}$

12 **Thermodynamics – Cooling of a liquid.**
The temperature of a liquid is measured every 20 minutes and the results recorded in the table below.

Time (min)	0	20	40	60	80	100	120
Temp (°C)	96	88	81	76	72	70	68

Calculate the rate of decrease of temperature in units of °C per minute:
(a) in the first 20 minutes
(b) in the first 40 minutes
(c) in the first 60 minutes
(d) in the last 60 minutes
(e) in the last 20 minutes

In each case express your answer as a fraction in its simplest form.

13 **Dynamics – Pressure in a vessel.**
The pressure in a vessel is 30 N cm^{-2}. If the pressure is reduced by $\frac{7}{100}$ of its original value, calculate (a) the decrease in pressure, (b) the resulting pressure in the vessel.

Solutions to exercises

1 (a) $\frac{1}{5}$ (b) $\frac{1}{4}$ (c) $\frac{1}{3}$ (d) $\frac{3}{46}$ (e) $\frac{29}{30}$

2 (a) $\frac{11}{3}$ (b) $\frac{27}{5}$ (c) $\frac{15}{2}$ (d) $-\frac{39}{4}$ (e) $\frac{74}{7}$

3 (a) $6\frac{2}{3}$ (b) $4\frac{4}{7}$ (c) $6\frac{2}{3}$ (d) $2\frac{1}{25}$ (e) $10\frac{10}{11}$

4 (a) $\frac{13}{12}$ (b) $\frac{11}{10}$ (c) $\frac{1}{2}$ (d) $\frac{9}{22}$ (e) $\frac{17}{42}$

5 (a) $5\frac{7}{12}$ (b) $1\frac{2}{15}$ (c) $6\frac{11}{30}$ (d) $1\frac{1}{21}$ (e) $\frac{5}{14}$

6 (a) $\frac{4}{9}$ (b) $\frac{12}{35}$ (c) $\frac{16}{25}$ (d) $\frac{3}{5}$ (e) $-\frac{4}{7}$

7 (a) $6\frac{7}{8}$ (b) $8\frac{13}{15}$ (c) $-9\frac{5}{24}$ (d) $11\frac{1}{9}$ (e) $-13\frac{3}{10}$

8 (a) $\frac{7}{6}$ (b) $\frac{2}{3}$ (c) 2 (d) $\frac{72}{77}$ (e) $\frac{14}{13}$

9 (a) $2\frac{4}{9}$ (b) $2\frac{11}{32}$ (c) $4\frac{39}{44}$ (d) $-3\frac{53}{65}$ (e) $-2\frac{11}{27}$

10 (a) 15 (b) 50 (c) 24 (d) $\frac{3}{8}$ (e) $\frac{5}{7}$

11 (a) $\frac{5}{3}$ (b) $\frac{30}{7}$ (c) $\frac{25}{26}$ (d) $\frac{121}{12}$ (e) $\frac{46}{53}$

12 (a) $\frac{2}{5}$ (b) $\frac{3}{8}$ (c) $\frac{1}{3}$ (d) $\frac{2}{15}$ (e) $\frac{1}{10}$

13 (a) $2\frac{1}{10}$ N cm^{-2} (b) $27\frac{9}{10}$ N cm^{-2}

Decimal numbers

In Chapter 2 we saw how to write both proper and improper fractions. This chapter introduces an alternative way of writing fractions using decimal notation.

It is often necessary to write a number to a given accuracy. This is usually denoted by stating the number of decimal places to be used or the number of significant figures to be used. Both of these are explained in this chapter.

Chapter 3 contents

Introduction to decimal numbers

This section gives a quick overview of the decimal system by way of revision. Recall that our number system is based on 10. The number 5276 means

$$5000 + 200 + 70 + 6$$

or, to write it another way,

$$(5 \times 1000) + (2 \times 100) + (7 \times 10) + (6 \times 1)$$

This reminds us of the 'thousands', 'hundreds', 'tens' and 'units' from early school days.

To deal with fractions we extend this system to include 'tenths', 'hundredths', 'thousandths' and so on. We separate the whole number part and fractional part by using a **decimal point** '.'.

Consider 37.92. There is a whole number part, 37, and a fractional part, .92. This number represents

$$(3 \times 10) + (7 \times 1) + \left(9 \times \frac{1}{10}\right) + \left(2 \times \frac{1}{100}\right)$$

The 9 is referred to as being in the **first decimal place;** the 2 is in the **second decimal place.** Similarly 2.1006 represents

$$(2 \times 1) + \left(1 \times \frac{1}{10}\right) + \left(0 \times \frac{1}{100}\right) + \left(0 \times \frac{1}{1000}\right) + \left(6 \times \frac{1}{10000}\right)$$

Note that the 6 is in the fourth decimal place.

Example 1.1
Express (a) 0.2, (b) 0.25, (c) 0.624 as fractions in their simplest form.

Solution
(a) 0.2 is equivalent to $2 \times \frac{1}{10}$. So

$$0.2 = 2 \times \left(\frac{1}{10}\right) = \frac{2}{10} = \frac{1}{5}$$

(b)

$$0.25 = 2 \times \left(\frac{1}{10}\right) + 5 \times \left(\frac{1}{100}\right)$$

$$= \frac{2}{10} + \frac{5}{100}$$

$$= \frac{25}{100}$$

$$= \frac{1}{4}$$

(c)

$$0.624 = \frac{6}{10} + \frac{2}{100} + \frac{4}{1000}$$

$$= \frac{624}{1000} = \frac{78}{125}$$

From the above example we note that

$$0.2 = \frac{2}{10}, \quad 0.25 = \frac{25}{100}, \quad 0.624 = \frac{624}{1000}$$

We can similarly deduce, for example, that

$$0.3049 = \frac{3049}{10000}, \quad 0.12348 = \frac{12348}{100000}$$

Example 1.2
Express (a) 5.156, (b) 3.045 as mixed fractions in their simplest form.

Solution
(a)

$$5.156 = 5 + \frac{156}{1000}$$

$$= 5 + \frac{39}{250} = 5\frac{39}{250}$$

(b) $3.045 = 3 + \boxed{}$ $\dfrac{45}{1000}$

$$= 3 + \boxed{} \qquad \frac{9}{200}$$

$$= \boxed{} \qquad 3\frac{9}{200}$$

Example 1.3
Write the following as decimal numbers: (a) $\frac{3}{4}$, (b) $\frac{3}{8}$, (c) $\frac{4}{9}$, (d) $6\frac{2}{5}$.

Solution
The easiest way to convert a fraction to a decimal number is to use a calculator. Check that you can obtain the following using a calculator.
(a) 0.75 (b) 0.375 (c) 0.444 . . . (d) 6.4

Example 1.4 Materials Engineering – Strain experienced by a stressed material

In materials engineering, testing techniques include subjecting a specimen to a force which causes an extension. An important quantity is **strain**, ϵ, defined as

$$\epsilon = \frac{\ell - \ell_0}{\ell_0}$$

where ℓ_0 is the original length of the specimen, and ℓ is the length once the force is applied.

It is important to be able to test materials for their tensile strength – this is the amount of stress the material is able to stand before it breaks. In carrying out such a test an engineer will need to calculate the strain.

Engineers also use the **strain** when mathematically modelling the extension or compression of an elastic material that is subjected to tensile or compressive forces.

Suppose a block of material of length 0.5 m is subject to a force which causes the material to extend to a length of 0.51 m. Calculate the strain in the material and express it as both a proper fraction and a decimal fraction.

Solution

Here ℓ_0 is the original length and so $\ell_0 = 0.5$. Once the force is applied, the length becomes $\ell = 0.51$. Then

$$\ell - \ell_0 = 0.51 - 0.5 = 0.01$$

The strain is then

$$\epsilon = \frac{\ell - \ell_0}{\ell_0} = \frac{0.01}{0.5} = 0.02$$

As a proper fraction this is $\frac{2}{100} = \frac{1}{50}$.

Exercises

1 Express the following decimal numbers as proper fractions in their simplest form:

(a) 0.8 (b) 0.80 (c) 0.08 (d) 0.080
(e) 0.800

2 Express the following decimal numbers as proper fractions in their simplest form:

(a) 0.2 (b) 0.25 (c) 0.225 (d) 0.025
(e) 0.2025

3 Express the following decimal numbers as proper fractions in their simplest form:

(a) 0.16 (b) 0.88 (c) 0.108 (d) 0.555
(e) 0.965

4 Write each of the following as a decimal number:

(a) $\frac{1}{5}$ (b) $\frac{3}{20}$ (c) $\frac{7}{40}$ (d) $\frac{21}{84}$ (e) $\frac{110}{20}$

Solutions to exercises

1 (a) $\frac{4}{5}$ (b) $\frac{4}{5}$ (c) $\frac{2}{25}$ (d) $\frac{2}{25}$ (e) $\frac{4}{5}$

2 (a) $\frac{1}{5}$ (b) $\frac{1}{4}$ (c) $\frac{9}{40}$ (d) $\frac{1}{40}$ (e) $\frac{81}{400}$

3 (a) $\frac{4}{25}$ (b) $\frac{22}{25}$ (c) $\frac{27}{250}$ (d) $\frac{111}{200}$ (e) $\frac{193}{200}$

4 (a) 0.2 (b) 0.15 (c) 0.175 (d) 0.25 (e) 5.5

1.2 Writing to a given number of decimal places

We are often required to write a number to a given number of decimal places (d.p.), usually 2, 3 or 4.

When asked to write to 1 d.p. we need to consider the first two decimal places, when writing to 2 d.p. we need to consider the first three decimal places, and so on.

If the final digit is less than 5, we simply ignore it. This is called **rounding down**. If the final digit is 5 or more we increase the previous digit by 1. This is called **rounding up**.

Example 1.5

Write 6.38623 to (a) 4 d.p., (b) 3 d.p., (c) 2 d.p., (d) 1 d.p.

Solution

(a) When writing to 4 d.p. we need to consider the first five decimal places: in this case that is 6.38623. The final digit is 3. Since this is less than 5 we simply ignore it and are left with 6.3862. Thus to 4 d.p. the number 6.38623 is 6.3862. Note that 6.3862 is less than 6.38623 and so we have rounded down. The number 6.3862 is closer to 6.38623 than any other number with four decimal places.

(b) To write to 3 d.p. we consider the first four d.p.: that is, we consider 6.3862. The final digit is a 2. This is less than 5 and so is simply ignored, leaving 6.386. So, to 3 d.p., the number 6.38623 is 6.386.

Since 6.386 is less than 6.38623 we have rounded down. The number 6.386 is closer to 6.38623 than any other number with 3 d.p.

(c) To write to 2 d.p. we consider the first 3 d.p.: that is, 6.386. The final digit is 6. Since this is greater than or equal to 5 the previous digit, 8, is increased by 1 to 9. So to 2 d.p. we have 6.39.

Since 6.39 is greater than 6.38623 we have rounded up. The number 6.39 is closer to 6.38623 than any other number with 2 d.p.

(d) Writing to 1 d.p. we consider the first 2 d.p.: that is, 6.38. The final digit is 8 and so the previous digit, 3, is rounded up to 4. Thus to 1 d.p. we have 6.4.

Example 1.6

Write 1.9751 to (a) 2 d.p., (b) 1 d.p.

Solution

(a) We consider the first 3 d.p.: that is, 1.975. The final digit is 5 and so the previous digit, 7, is rounded up to 8. This results in 1.98.

(b) To write to 1 d.p. we consider the first 2 d.p.: that is, 1.97. The final digit is 7 and so the 9 must be increased. Clearly the 9 cannot be increased to 10 and so we increase the 1.9 to 2.0. Thus to 1 d.p. we have 2.0. Note that it is important to write 2.0 and not simply 2. The 2.0 signifies that the number is written to 1 d.p.

Exercises

1 Write 2.152637 to (a) 5 d.p., (b) 4 d.p., (c) 3 d.p., (d) 2 d.p., (e) 1 d.p.

2 Write 9.989 to (a) 2 d.p., (b) 1 d.p.

3 Write 9.999 to (a) 2 d.p., (b) 1 d.p.

Solutions to exercises

1   (a) 2.15264 (b) 2.1526 (c) 2.153
    (d) 2.15 (e) 2.2

2   (a) 9.99 (b) 10.0

3   (a) 10.00 (b) 10.0

End of block exercises

1   Write the following numbers to 3 d.p.:
    (a) 7.6931 (b) −2.0456 (c) 0.0004
    (d) 3.9999

2   To 1 d.p. a number, X, is 4.3. State (a) the
    smallest, (b) the largest possible value of X.

3   Convert the following decimal numbers to
    fractions in their simplest form:
    (a) 0.80 (b) 0.55 (c) 0.12 (d) 0.125 (e) 0.625

4   Express the following fractions as decimal
    numbers to 3 d.p.:
    (a) $\frac{4}{7}$ (b) $\frac{2}{11}$ (c) $\frac{3}{5}$ (d) $-\frac{2}{9}$

Solutions to exercises

1   (a) 7.693 (b) −2.046 (c) 0.000 (d) 4.000

2   (a) 4.25 (b) 4.34999 . . .

3   (a) $\frac{4}{5}$ (b) $\frac{11}{20}$ (c) $\frac{3}{25}$ (d) $\frac{1}{8}$ (e) $\frac{5}{8}$

4   (a) 0.571 (b) 0.182 (c) 0.600 (d) −0.222

Significant figures

2.1 Introduction

We have seen how a number can be written to so many decimal places. Similar to, but nevertheless distinct from, writing to so many decimal places is writing to so many **significant figures (s.f.)**.

2.2 Writing to a given number of significant figures

When writing to so many significant figures we consider digits both before and after the decimal point. When writing to 2 s.f. we consider the first three digits, when writing to 3 s.f. we consider the first four digits, and so on. We always consider one more digit than the number of significant figures required.

To write a number to 2 s.f. we can use at most two non-zero digits, to write to 3 s.f. we use at most three non-zero digits, and so on. Rounding up and rounding down follow the same rules as detailed in the previous block.

Example 2.1
Write 86.97529 to (a) 5 s.f., (b) 4 s.f., (c) 3 s.f., (d) 2 s.f., (e) 1 s.f.

Solution
(a) Writing to 5 s.f. we consider the first six digits: that is, 86.9752. The final digit is 2 and so this is ignored: that is, we round down. Hence to 5 s.f. the number is 86.975. Note that there are no more than five non-zero digits in the final answer. The number 86.975 is closer to 86.97529 than any other five-digit number.
(b) We consider the first five digits: that is, 86.975. The final digit is 5 and so the previous digit, 7, is rounded up to 8. Thus to 4 s.f. we have 86.98.
(c) The first four digits are 86.97. The final digit is 7 and so the previous digit is rounded up. We cannot round up a 9 to a 10 and so 86.9 is rounded up to 87.0.
(d) Writing to 2 s.f. we consider the first three digits, namely 86.9. The final digit, 9, means that the 6 is rounded up to 7, producing 87.
(e) Writing to 1 s.f. we consider the first two digits: 86. The final digit, 6, means the 8 is rounded to 9, producing 90. Note that although the number has been written to 1 s.f. there are two digits in the final answer. However, there is only one non-zero digit.

Example 2.2
Write 99.99 to (a) 3 s.f., (b) 2 s.f., (c) 1 s.f.

Solution
(a) The first four digits are considered: that is, 99.99. The final digit is 9 and so there must be rounding up. Here 99.99 is rounded up to 100.0.

(b) The first three digits are considered: that is, 99.9. The final digit is a 9 and so rounding up takes place to produce 100.

(c) The first two digits are considered: that is, 99. When rounded up this becomes 100. Note that, although we are writing to 1 s.f., there are three digits in the final answer. However, there is only one non-zero digit.

Zeros at the beginning of a number are ignored when counting digits. For example, to write 0.000164 to 1 s.f. we consider the number 0.00016. The leading 0s have not been counted. The 1 is rounded up to 2 producing 0.0002 to 1 s.f. Note that there is only one non-zero digit.

End of block exercises

1 Write 28.403951 to (a) 6 s.f., (b) 5 s.f., (c) 4 s.f., (d) 3 s.f., (e) 2 s.f., (f) 1 s.f.

2 Write 9.0046 to (a) 3 s.f., (b) 2 s.f., (c) 1 s.f.

3 Written to 2 s.f. a number is 86. Calculate (a) the maximum value, (b) the minimum value of the original number.

4 Write 0.550 to (a) 2 s.f., (b) 1 s.f.

5 Write 5.5550 to (a) 2 s.f., (b) 1 s.f.

Solutions to exercises

1 (a) 28.4040 (b) 28.404 (c) 28.40 (d) 28.4 (e) 28 (f) 30

2 (a) 9.00 (b) 9.0 (c) 9

3 (a) 86.4999 . . . (b) 85.5

4 (a) 0.55 (b) 0.6

5 (a) 5.6 (b) 6

End of chapter exercises

1 Write each of the following as a proper fraction in its simplest form:
(a) 0.12 (b) 0.125 (c) 0.1250 (d) 0.85 (e) 0.76

2 Write the following fractions as decimals:
(a) $\frac{3}{5}$ (b) $\frac{3}{10}$ (c) $\frac{3}{20}$ (d) $\frac{3}{40}$ (e) $\frac{3}{100}$

3 Write the following fractions as decimals, giving your answer to 3 d.p.:
(a) $\frac{1}{3}$ (b) $\frac{2}{3}$ (c) $\frac{1}{9}$ (d) $\frac{4}{11}$ (e) $\frac{6}{7}$

4 Write 19.919 to (a) 2 d.p., (b) 1 d.p.

5 Write 19.99 to (a) 3 s.f., (b) 2 s.f., (c) 1 s.f.

6   Write 0.0982 to (a) 3 d.p., (b) 2 d.p.,
    (c) 1 d.p., (d) 2 s.f., (e) 1 s.f.

7   To 4 d.p. a number is 9.9190. State
    (a) the maximum possible value,
    (b) the minimum possible value of the original
    number.

8   To three significant figures a number is 3.60.
    State the maximum and minimum possible
    values of the number.

9   To four significant figures a number is 0.3452.
    State the maximum and minimum possible
    values of the number.

10 A block of material of length 0.9 m is subject
    to a force which causes the material to extend
    to a length of 0.93 m. Calculate the strain in
    the material and express it as both a proper
    fraction and a decimal fraction stating your
    answer to four decimal places.

Solutions to exercises

1   (a) $\frac{3}{25}$ (b) $\frac{1}{8}$ (c) $\frac{1}{8}$ (d) $\frac{17}{20}$ (e) $\frac{19}{25}$

2   (a) 0.6 (b) 0.3 (c) 0.15 (d) 0.075 (e) 0.03

3   (a) 0.333 (b) 0.667 (c) 0.111 (d) 0.364
    (e) 0.857

4   (a) 19.92 (b) 19.9

5   (a) 20.0 (b) 20 (c) 20

6   (a) 0.098 (b) 0.10 (c) 0.1 (d) 0.098
    (e) 0.1

7   (a) 9.91904999 . . . (b) 9.91895

8   3.604999 . . . , 3.595

9   0.34524999 . . . , 0.34515

10 $\frac{1}{30}$, 0.0333

Percentage and ratio

This chapter examines percentages and ratio.

Percentages, explained in Block 1, are simply fractions with a denominator of 100. Writing fractions as percentages is a useful way of making comparisons. For example, if a student receives 19 out of a possible 30 marks in a first test, and 27 out of a possible 40 marks in a second test, is the student improving or getting worse? Once the scores are written in their equivalent percentage form, the question is easy to answer.

In Block 2, we look at **ratio**. Ratio provides a concise method of describing the relative sizes of each component when some entity is divided into various parts.

Chapter 4 contents

Percentage

1.1 Introduction

Percentages are simply fractions with a denominator of 100. Writing fractions as percentages is a useful way of making comparisons. In this block we remind you of how to convert a fraction with an arbitrary denominator into a percentage, and how to calculate the percentage of a given quantity. These are basic numerical skills with which all engineers and technologists should be familiar.

1.2 Calculating percentages

A **percentage** is a fraction with a denominator of 100. The symbol % is used to denote a percentage. For example, we may write $\frac{23}{100}$ as 23%.

We often need to convert a fraction into a percentage: that is, write it as an equivalent fraction with a denominator of 100.

Example 1.1
Write $\frac{7}{25}$ as a percentage.

Solution
We express $\frac{7}{25}$ as an equivalent fraction with a denominator of 100.

$$\frac{7}{25} = \frac{7 \times 4}{25 \times 4} = \frac{28}{100} = 28\%$$

An alternative method is to multiply the given fraction by 100 and then label the result as a percentage. For example,

$$\frac{7}{25} \times 100 = 28$$

and so $\frac{7}{25}$ is 28%.

We may be asked to calculate a percentage of a number – for example, calculate 23% of 160. The following examples illustrate how to do this.

Example 1.2
Calculate (a) 23% of 160, (b) 117% of 240, (c) 100% of 20.

Solution
(a)

$$23\% \text{ of } 160 = \frac{23}{100} \times 160 = 36.8$$

(b)

$$117\% \text{ of } 240 = \qquad \tfrac{117}{100} \times 240 = 280.8$$

(c)

$$100\% \text{ of } 20 = \frac{100}{100} \times 20 = 20$$

We note from (c) that 100% of a number is simply the number itself.

Example 1.3 Electrical Engineering – Voltage across a resistor

The voltage across a resistor is 240 volts to within ±5%. Calculate the maximum and minimum possible voltages.

Solution

The maximum voltage is 240 plus 5% of 240: that is, 105% of 240.

$$105\% \text{ of } 240 = \frac{105}{100} \times 240 = 252 \text{ volts}$$

The minimum voltage is 240 less 5% of 240: that is, 95% of 240.

$$95\% \text{ of } 240 = \frac{95}{100} \times 240 = 228 \text{ volts}$$

Example 1.4 Electrical Engineering – Resistor tolerance bands

A resistor often has its resistance value defined by a tolerance band. This is an interval in which the actual value of the resistance is known to lie. For example, a resistor might be described as having a resistance $80 \ \Omega \pm 5\%$. Find the maximum and minimum values of such a resistance.

Solution

The maximum value is 80 plus 5% of 80, that is

$$80 + \left(\frac{5}{100} \times 80 \right) = 80 + 4$$
$$= 84 \ \Omega$$

The minimum value is

$$80 - \left(\frac{5}{100} \times 80 \right) = 80 - 4$$
$$= 76 \ \Omega$$

Example 1.5 Reliability Engineering

Manufactured components are checked by a quality assurance system to ensure their compliance with a manufacturing specification. Out of 450 components tested, 438 were acceptable.

(a) Calculate the percentage of components that were acceptable.
(b) If 6000 components are manufactured, estimate the number that will be rejected (i.e. are not acceptable).

Solution

(a) 438 components out of 450 are acceptable. As a percentage this is

$$\frac{438}{450} \times 100 = 97.33\%$$

Hence 97.33% of the components tested are acceptable.

(b) From (a), 97.33% of the sample tested are acceptable. So the percentage not acceptable is $100 - 97.33$ which is 2.67%. So 2.67% of the sample are rejected.

We assume that 2.67% of the entire production of 6000 components is also rejected.

$$2.67\% \text{ of } 6000 = \frac{2.67}{100} \times 6000 = 160$$

We estimate that 160 components out of the 6000 manufactured will be rejected.

Often a physical characteristic of an item will change. For example, the volume of gas in a cylinder can change as pressure is applied; the length of a metal bar can increase (decrease) when the bar is heated (cooled). Sometimes we are asked to calculate the **percentage change** of such a physical characteristic.

Percentage change is calculated using the formula

$$\text{percentage change} = \frac{\text{new value} - \text{original value}}{\text{original value}} \times 100$$

If the change is positive then there has been an increase in the measured quantity; if the change is negative then there has been a decrease in the quantity.

Key point

$$\text{Percentage change} = \frac{\text{new value} - \text{original value}}{\text{original value}} \times 100$$

Example 1.6 Mechanical Engineering – Extension of a spring

A spring is stretched from 25 cm to 27 cm. Calculate the percentage change in the length of the spring.

Solution

The original length of the spring is 25 cm. The new length of the spring is 27 cm. So

$$\text{percentage change in the length of the spring} = \frac{\text{new value} - \text{original value}}{\text{original value}} \times 100$$

$$= \frac{27 - 25}{25} \times 100$$

$$= 8$$

The length of the spring has been increased by 8%.

End of block exercises

1. Express the following fractions as percentages:
   (a) $\frac{3}{10}$ (b) $\frac{7}{20}$ (c) $\frac{21}{80}$ (d) $\frac{27}{25}$ (e) 3

2. Calculate the following:
   (a) 80% of 75 (b) 37% of 250 (c) 125% of 550

3. The pressure inside a vessel is $17.5 \pm 10\%$ atmospheres. Calculate the maximum and minimum pressures inside the vessel.

4. **Electrical Engineering – Maximum and minimum values of a resistor.** Resistors are manufactured so that their resistance lies within a tolerance band. Calculate the maximum and minimum values of the resistances given by:
   (a) $10\,\Omega \pm 3\%$
   (b) $29\,k\Omega \pm 5\%$
   (c) $3\,M\Omega \pm 0.1\%$

5. **Mechanical Engineering – Expansion of metal with rise in temperature.** A steel track measures 25 m at 20 °C. At 70 °C it measures 25.01 m. Calculate the percentage change in the length of the track as its temperature changes from 20 °C to 70 °C.

6. **Mechanical Engineering – Compression of a gas.** The volume of gas in a cylinder is 1098 cm^3. Pressure is increased and the volume changes to 936 cm^3. Calculate the percentage change in the volume of gas.

7. **Electrical Engineering – Power loss during transmission.** Transmission lines have a nominal power rating of 30000 watts. If there are transmission losses of 9% calculate the actual power transmitted.

8. **Production Engineering – Quality control.** Quality control systems on a production line ensure that 98.5% of components manufactured are of an acceptable standard. During a day's production, 12000 components are made. Calculate the number that are not of an acceptable standard.

Solutions to exercises

1. (a) 30% (b) 35% (c) 26.25% (d) 108% (e) 300%

2. (a) 60 (b) 92.5 (c) 687.5

3. Maximum = 19.25 atmospheres, minimum = 15.75 atmospheres

4. (a) $9.7\,\Omega$, $10.3\,\Omega$ (b) $27.55\,k\Omega$, $30.45\,k\Omega$
   (c) $2.997\,M\Omega$, $3.003\,M\Omega$

5. Increase of 0.04%

6. Decrease of 14.75%

7. 27300 W

8. 180

Ratio

2.1 Introduction

Ratios are often used to describe the relative sizes of parts that result when some quantity is divided up. For example, we might be interested in a current that branches into two parts with one part taking $\frac{3}{10}$ of the initial current and the other part taking $\frac{7}{10}$. This division can be expressed as a ratio, as we shall explain in this block, and we write the ratio using a colon as 3:7. Essentially ratios are a way of describing two (or more) fractions whose sum is 1.

2.2 Calculating ratios

Suppose a number is divided in the ratio 3:2. We note that $3 + 2 = 5$. We can think of the number being divided into two parts. The first part is $\frac{3}{5}$ of the original; the second part is $\frac{2}{5}$ of the original.

As another example, suppose a quantity is divided in the ratio 4:5:3. We note that $4 + 5 + 3 = 12$. We can think of the quantity being divided into three parts. The first part is $\frac{4}{12}$ of the original, the second part is $\frac{5}{12}$ of the original, and the third part is $\frac{3}{12}$ of the original.

The following examples illustrate how calculations with ratio are performed.

Example 2.1 Electrical Engineering – Branch currents
A branch in a circuit divides the current of 30 amps in the ratio of 3:7. Calculate the current in each branch.

Solution
A ratio of 3:7 means that for every 3 amps in one branch there are 7 amps in the other branch. So we can think of every 10 amps as being split into 3 amps in one branch and 7 amps in the other branch. In other words, one branch receives $\frac{3}{10}$ of the current; the other branch receives the remaining $\frac{7}{10}$ of the current. Now

$$\frac{3}{10} \text{ of } 30 = \frac{3}{10} \times 30 = 9; \quad \frac{7}{10} \text{ of } 30 = \frac{7}{10} \times 30 = 21$$

So the current is split into 9 amps in one branch and 21 amps in the other branch.

Note from the above example that a ratio of 3:7 is equivalent to splitting a number into two parts: $\frac{3}{10}$ of the number in one part and $\frac{7}{10}$ of the number in the other part.

In general, we can split a number, x, in the ratio $m:n$. The first part will be $\frac{m}{m + n} \times x$; the second part will be $\frac{n}{m + n} \times x$.

Example 2.2
Divide 108 in the ratio (a) 5:7, (b) 2:3.

Solution
(a) The first number is

$$\frac{5}{5+7} \times 108 = \frac{5}{12} \times 108 = 45$$

The second number is

$$\frac{7}{5+7} \times 108 = \frac{7}{12} \times 108 = 63$$

(b) The first number is

$$\frac{2}{2+3} \times 108 = \frac{2}{5} \times 108 = 43.2$$

The second number is

$$\frac{3}{2+3} \times 108 = \frac{3}{5} \times 108 = 64.8$$

Just as a fraction can be written in many equivalent ways, so can a ratio. The next example illustrates this.

Example 2.3
Divide 90 in the ratio (a) 4:5, (b) 8:10.

Solution
(a) The first part is

$$\frac{4}{4+5} \times 90 = \frac{4}{9} \times 90 = 40$$

The second part is

$$\frac{5}{4+5} \times 90 = \frac{5}{9} \times 90 = 50$$

(b) The first part is

$$\frac{8}{8+10} \times 90 = \frac{8}{18} \times 90 = 40$$

The second part is

$$\frac{10}{8+10} \times 90 = \frac{10}{18} \times 90 = 50$$

We see that dividing 90 in the ratios 4:5 and 8:10 produces the same result. The ratios 4:5 and 8:10 are equivalent.

Key point

The ratios $m{:}n$ and $km{:}kn$ are equivalent. Note that the constant k may be integer or fractional.

From the key point we immediately see that the ratios 4:5 and 8:10 (Example 2.3) are equivalent.

Example 2.4
Divide 132 in the ratio $\frac{2}{3} : \frac{1}{4}$.

Solution
When presented with a ratio involving fractions, it is usually best to write an equivalent ratio which involves only integers. From the key point above we see that $\frac{2}{3} : \frac{1}{4}$ is equivalent to $k\frac{2}{3} : k\frac{1}{4}$. Choosing k to be 12 we see that $\frac{2}{3} : \frac{1}{4}$ is equivalent to 8:3. Hence the question is equivalent to dividing 132 in the ratio 8:3.

The first number is $\frac{8}{8+3}$ of 132: that is, $\frac{8}{11}$ of 132, which is 96. The second number is $\frac{3}{11}$ of 132: that is, 36.

The idea of a ratio can be extended to three or more parts.

Example 2.5
Divide 160 in the ratio 1:2:5.

Solution
The first number is $\frac{1}{1+2+5}$ of 160: that is, $\frac{1}{8}$ of 160, which is 20.
The second number is $\frac{2}{1+2+5}$ of 160. This is $\frac{2}{8}$ of 160, which is 40.
The third number is $\frac{5}{8}$ of 160, which is 100.

Example 2.6 Materials Engineering – Composition of an alloy
A metal alloy is made of copper, zinc and nickel in the ratio 3:5:7. Calculate the weight of copper in 100 kg of the alloy.

Solution
The alloy can be considered as comprising $3 + 5 + 7 = 15$ parts. Three of the 15 parts are copper, that is $\frac{3}{15}$ of the alloy is copper.

$$\frac{3}{15} \text{ of } 100 = \frac{1}{5} \times 100$$
$$= 20$$

So 100 kg of the alloy contains 20 kg of copper.

Example 2.7 Chemical Engineering – Composition of a gas
Gas X comprises oxygen, nitrogen and hydrogen in the ratio 1:2:1. Gas Y comprises oxygen, nitrogen and hydrogen in the ratio 2:1:3.
(a) If equal volumes of gases X and Y are mixed, calculate the ratio of oxygen, nitrogen and hydrogen in the resulting gas.
(b) If gas X and gas Y are mixed in the ratio 3:4 find the ratio of oxygen, nitrogen and hydrogen in the resulting gas.

Solution
(a) In 4 units of gas X, there is 1 unit of oxygen, 2 units of nitrogen and 1 unit of hydrogen.

In 6 units of gas Y, there are 2 units of oxygen, 1 unit of nitrogen and 3 units of hydrogen.

Equal volumes of gases X and Y are mixed together. Noting that the l.c.m. of 4 and 6 is 12, we consider adding 12 units of gas X and 12 units of gas Y:

12 units of gas X comprise 3 units of oxygen, 6 units of nitrogen and 3 units of hydrogen; 12 units of gas Y comprise 4 units of oxygen, 2 units of nitrogen and 6 units of hydrogen.

So adding 12 units of gas X to 12 units of gas Y results in 24 units of the mixture. This mixture comprises

$$3 + 4 = 7 \text{ units of oxygen}$$
$$6 + 2 = 8 \text{ units of nitrogen}$$
$$3 + 6 = 9 \text{ units of hydrogen}$$

The mixture contains oxygen, nitrogen and hydrogen in the ratio 7:8:9.

(b) Gas X and gas Y are mixed in the ratio of 3:4. We will consider mixing $3 \times 12 = 36$ units of gas X with $4 \times 12 = 48$ units of gas Y:

36 units of gas X contains 9 units of oxygen, 18 units of nitrogen and 9 units of hydrogen; 48 units of gas Y contain 16 units of oxygen, 8 units of nitrogen and 24 units of hydrogen.

So the resulting mixture has $9 + 16 = 25$ units of oxygen, $18 + 8 = 26$ units of nitrogen and $9 + 24 = 33$ units of hydrogen.

The resulting mix contains oxygen, nitrogen and hydrogen in the ratio 25:26:33.

Example 2.8 Materials Engineering – Poisson ratio

In Chapter 3, Example 1.4, we introduced the concept of strain ϵ caused when a material is stretched. Strain is defined by $\epsilon = \dfrac{\ell - \ell_0}{\ell_0}$ where ℓ_0 is the original length of the specimen, and ℓ is the length after a tensile or stretching force is applied. The same formula is used when a material is compressed, but this time the final length will be smaller than the original, leading to a negative strain.

Suppose a material such as a metal bar is stretched along its length, the so-called **axial direction**. It will then tend to contract in the sideways or **transverse directions**. The contraction in the transverse directions gives rise to transverse strains. Figure 2.1 shows a plan view of a metal bar before and after stretching it.

Suppose the strain in the axial direction is labelled ϵ_{axial} and that in the transverse direction is labelled ϵ_{trans}. (Here we have assumed that the strain in both transverse directions is the same.) Then an important quantity used in materials engineering is called the **Poisson ratio**, v, defined by

$$v = -\frac{\epsilon_{trans}}{\epsilon_{axial}}$$

Note the minus sign in this definition, which ensures that the Poisson ratio is positive because ϵ_{trans} is negative when ϵ_{axial} is positive since the bar is stretched.

A metal bar of length 1 m is stretched in the axial direction so that it extends by 6 cm. At the same time, its width contracts from 20 mm to 19.5 mm. Calculate ϵ_{axial}, ϵ_{trans} and hence find the Poisson ratio for this metal. Convert any quantities to SI units before performing the calculations.

Solution

$$\epsilon_{axial} = \frac{1.06 - 1}{1} = 0.06, \quad \epsilon_{trans} = \frac{0.0195 - 0.020}{0.020} = -0.025$$

Figure 2.1
A plan view of a
metal bar before
and after
stretching.

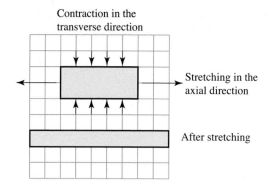

Then

$$\text{Poisson ratio} = v = -\frac{-0.025}{0.06} = 0.42 \quad \text{(to 2 d.p.)}$$

Usually the Poisson ratio of a material lies between 0 and 0.5. Some materials, though, have a negative Poisson ratio. This happens when stretching in the axial direction causes an expansion in the transverse directions too because then both strains are positive. Materials with this property can be naturally occurring or synthetic and are described as **auxetic**. Auxetic materials have many important applications, for example in shock absorbers.

End of block exercises

1   Divide 250 in the ratio 1:9.

2   Divide 1 in the ratio 6:3.

3   Divide 108 in the ratio $\frac{1}{2} : \frac{1}{4} : \frac{3}{5}$.

4   Express the following ratios in their simplest form: (a) 6:3 (b) 5:15 (c) 8:6:10 (d) $\frac{1}{2} : \frac{1}{4}$
    (e) $\frac{1}{2} : \frac{1}{3} : 1$

5   **Materials Engineering – Composition of an alloy.** Brass is a mixture of copper and zinc in

the ratio 8:3. Calculate the weight of zinc in 20 kg of brass.

6   **Materials Engineering – Composition of an alloy.** A metal alloy is made from copper, zinc and steel in the ratio 3:4:1.
    (a) Calculate the amount of copper in a 30 kg block of the alloy.
    (b) 10 kg of copper is added to an existing 40 kg block of the alloy to form a new alloy. Calculate the ratio of copper, zinc and steel in the new alloy.

Solutions to exercises

1   25, 225

2   $\frac{2}{3}, \frac{1}{3}$

3   40, 20, 48

4   (a) 2:1 (b) 1:3 (c) 4:3:5 (d) 2:1 (e) 3:2:6

5   5.45 kg

6   (a) 11.25 kg (b) 5:4:1

End of chapter exercises

1 Calculate 84% of 560.

2 Express $\frac{7}{9}$ as a percentage, giving your answer to 2 d.p.

3 Calculate 220% of −75.

4 The pressure in a vessel is increased by 12% to 130 N m^{-2}. Calculate the original pressure.

5 Divide 315 in the ratio 6:7:8.

6 A mass of 176 kg is divided in the ratio 4:5:7. Calculate the mass of each portion.

7 Express the ratio $1\frac{1}{2} : 3\frac{1}{4}$ using only integers.

8 The temperature of a chemical is reduced by 6% to 130 °C. Calculate the original temperature to 2 d.p.

9 The percentage change is defined as

percentage change

$$= \left(\frac{\text{new value } - \text{ original value}}{\text{original value}} \right) \times 100$$

The voltage in a circuit is increased from 220 volts to 270 volts. Calculate the percentage change.

10 When a number X is increased by 10% its value becomes Y. When a number Z is decreased by 10% its value becomes Y. By what percentage must X be increased so its value equals Z?

11 **Chemical Engineering – Composition of a fluid.** Liquid A is made of water, alcohol and bleach in the ratio of 3:1:2. Liquid B is made of water, alcohol and bleach in the ratio 4:2:1.
  (a) If equal volumes of liquids A and B are mixed, calculate the ratio of water, alcohol and bleach in the new liquid.
  (b) If liquid A and liquid B are mixed in the ratio 2:1, calculate the ratio of water, alcohol and bleach in the resulting mix.

Solutions to exercises

1 470.4

2 77.78%

3 −165

4 116.07 N m^{-2}

5 90, 105, 120

6 44, 55, 77

7 6:13

8 138.30 °C

9 22.73%

10 22.22%

11 (a) 45:19:20 (b) 33:13:17

Basic algebra

Chapter **5**

In order to be able to apply mathematics to solve engineering problems it is necessary to introduce symbols to represent physical quantities. **Algebra** is used to manipulate these symbols. This chapter explains how symbols are used and how they can be combined in various ways. A firm understanding of algebraic techniques is essential for all further mathematical development. This chapter provides the necessary foundation and gives you the opportunity to revise and practise these techniques.

Chapter 5 contents

BLOCK 1   Mathematical notation and symbols

1.1   Introduction

This introductory block reminds you of important notations and conventions used throughout engineering mathematics. We discuss the arithmetic of numbers, the plus or minus sign, $\pm$, the modulus notation $||$, and the factorial notation !. Symbols are introduced to represent physical quantities in formulae and equations. The topic of algebra deals with the manipulation of these symbols. The block closes with an introduction to algebraic conventions. In what follows, a working knowledge of the addition, subtraction, multiplication and division of numerical fractions is essential.

1.2   Numbers and common notations

A knowledge of the properties of numbers is fundamental to the study of engineering mathematics. These properties are an essential foundation. Students who master them will be well prepared for the study of algebra. Much of the terminology used throughout the rest of the book can be most easily illustrated by applying it to numbers. It is for this reason that we strongly recommend that you work through this block even if you think that the material is familiar.

The number line

A useful way of picturing numbers is to use a **number line**. Figure 1.1 shows part of this line. Positive numbers are on the right-hand side of this line; negative numbers are on the left-hand side. Any whole or fractional number can be represented by a point on this line. It is also called the **real number line**, or simply the **real line**. Study Figure 1.1 and note that a minus sign is always used to indicate that a number is negative, whereas the use of a plus sign is optional when describing positive numbers.

Figure 1.1
Numbers can be represented on a number line.

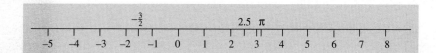

The line extends indefinitely both to the left and to the right. Mathematically we say that the line extends from minus infinity to plus infinity. The symbol for **infinity** is ∞.

The symbol $>$ means 'greater than': for example, $6 > 4$. Given any number, all numbers to the right of it on the number line are greater than the given number. The symbol $<$ means 'less than': for example, $-3 < 19$. We also use the symbols $\geq$ meaning 'greater than or equal to' and $\leq$ meaning 'less than or equal to'. For example, $7 \leq 10$ and $7 \leq 7$ are both true statements.

Sometimes we are interested in only a small section, or **interval**, of the real line. We write $[1, 3]$ to denote all the real numbers between 1 and 3 inclusive: that is, 1 and 3 are included in the interval. Thus the interval $[1, 3]$ consists of all real numbers x, such that $1 \leq x \leq 3$. The square brackets [,] mean that the end-points are included in the interval, and such an interval is said to be **closed**. We write $(1, 3)$ to represent all real numbers between 1 and 3, but not including the end-points. Thus $(1, 3)$ means all real numbers x such that $1 < x < 3$, and such an interval is said to be **open**. An interval may be closed at one end and open at the other. For example, $(1, 3]$ consists of all numbers x such that $1 < x \leq 3$. Intervals can be represented on a number line. A closed end-point is denoted by ●; an open end-point is denoted by ○. The intervals $(-6, -4)$, $[-1, 2]$ and $(3, 4]$ are illustrated in Figure 1.2.

Figure 1.2
The intervals $(-6, -4)$, $[-1, 2]$ and $(3, 4]$ are depicted on the real line.

The plus or minus sign ±

In engineering calculations we often use the notation **plus or minus**, $\pm$. For example, we write 12 ± 8 to mean the two numbers $12 + 8$ and $12 - 8$: that is, 20 and 4. If we say a number lies in the range 12 ± 8 we mean that the number can lie between 4 and 20 inclusive.

The reciprocal of a number

If the number $\frac{2}{3}$ is **inverted** we get $\frac{3}{2}$. The **reciprocal** of a number is found by inverting it. So, for example, the reciprocal of $\frac{2}{7}$ is $\frac{7}{2}$. Note that the old denominator has become the new numerator, and the old numerator has become the new denominator. Because we can write 4 as $\frac{4}{1}$, the reciprocal of 4 is $\frac{1}{4}$.

Example 1.1
State the reciprocal of (a) $\frac{6}{11}$, (b) $\frac{1}{5}$, (c) 7.

Solution

(a) The reciprocal of a number is found by inverting it. Thus the reciprocal of $\dfrac{6}{11}$ is

$$\dfrac{11}{6}$$

(b) The reciprocal of $\dfrac{1}{5}$ is

$\dfrac{5}{1}$ or simply 5

(c) The reciprocal of 7 is

$\dfrac{1}{7}$

The modulus notation | |

We shall make frequent use of the modulus notation | |. The **modulus** of a number is the size of the number regardless of its sign. It is denoted by vertical lines around the number. For example, $|4|$ is equal to 4, and $|-3|$ is equal to 3. The modulus of a number is never negative.

Example 1.2

State the modulus of (a) -17, (b) $\dfrac{1}{5}$, (c) $-\dfrac{1}{7}$.

Solution

(a) The modulus of a number is found by ignoring its sign. Thus the modulus of -17 is

17

(b) The modulus of $\dfrac{1}{5}$ is

$\dfrac{1}{5}$

(c) The modulus of $-\dfrac{1}{7}$ is

$\dfrac{1}{7}$

Factorials !

Another commonly used notation is the **factorial**,!. The number 5!, read 'five factorial', or 'factorial five', means $5 \times 4 \times 3 \times 2 \times 1$, and the number 7! means $7 \times 6 \times 5 \times 4 \times 3 \times 2 \times 1$. Note that 1! equals 1, and by convention 0! is defined to equal 1 as well. Your scientific calculator is probably able to evaluate factorials.

Key point

Factorial notation
If n is a positive whole number then

$$n! = n \times (n-1) \times (n-2) \times \cdots \times 5 \times 4 \times 3 \times 2 \times 1$$

For example,

$$6! = 6 \times 5 \times 4 \times 3 \times 2 \times 1$$
$$= 720$$

Example 1.3
(a) Evaluate without using a calculator 4! and 5!.
(b) Use your calculator to find 10!.

Solution
(a) $4! = 4 \times 3 \times 2 \times 1 = 24$. Similarly, $5! = 5 \times 4 \times 3 \times 2 \times 1 = 120$.
   Note that $5! = 5 \times 4!$.
(b) From your calculator check that $10! = 3628800$.

Example 1.4
Find the factorial button on your calculator and hence state the value of 11!.

Solution
The button may be marked !. Refer to the manual if necessary. Check that

$11! = $ 　　　　　　　　　　　　　　　　　　　　　 39916800

Example 1.5 Coding Theory – Arrangements
In coding theory the letters of a word can be arranged in different ways in order to disguise their meaning. The number of different arrangements of n *different* letters is $n!$. This is because there are n choices available for the first position, $n - 1$ for the second position, and so on.

   For example, the four-letter word NATO has $4! = 24$ different rearrangements:

NATO, NTOA, NOAT, NAOT, NOTA, NTAO;

ANTO, TNOA, ONAT, ANOT, ONTA, TNAO;

ATNO, TONA, OANT, AONT, OTNA, TANO;

ATON, TOAN, OATN, AOTN, OTAN, TAON.

If some of the letters in the word are the same we divide the letters into groups of identical letters and count the number of letters in each group. For example, suppose there are k distinct groups of letters with n_1 identical letters in the first group, n_2 in the second, n_3 in the third, and so on, with n_k in the kth group. The total number of letters is still n so that $n_1 + n_2 + n_3 + \cdots + n_k = n$. The number of different arrangements of the n letters is

$$\frac{n!}{n_1! \times n_2! \times n_3! \times \cdots \times n_k!}$$

For example, the four-letter word NASA can be divided into three groups of identical letters, N, A and S and the number of letters in these groups is 1, 2, 1 respectively. The number of arrangements of the word NASA is

$$\frac{4!}{1!2!1!} = 12$$

These are:

NASA, NSAA, NAAS;
ANSA, SNAA, ANAS;
ASNA, SANA, AASN;
ASAN, SAAN, AASN.

In this context the word **permutation** is often used as an alternative to the word 'arrangement'.

Exercises

1 Draw a number line and on it label points to represent $-5, -3.8, -\pi, -\frac{5}{6}, -\frac{1}{2}, 0, \sqrt{2}, \pi$ and 5.

2 Evaluate (a) $|-18|$, (b) $|4|$, (c) $|-0.001|$, (d) $|0.25|$, (e) $|0.01 - 0.001|$, (f) 2!, (g) $8! - 3!$, (h) $\dfrac{9!}{8!}$.

3 State the reciprocal of (a) 8, (b) $\dfrac{9}{13}$.

4 Evaluate (a) 7 ± 3, (b) 16 ± 7, (c) $-15 \pm \frac{1}{2}$, (d) -16 ± 0.05, (e) $|-8| \pm 13$, (f) $|-2| \pm 8$.

5 Which of the following statements are true?
(a) $-8 \le 8$ (b) $-8 \le -8$ (c) $-8 \le |8|$
(d) $|-8| < 8$ (e) $|-8| \le -8$
(f) $9! \le 8!$ (g) $8! \le 10!$

6 Calculate the number of permutations of the letters in the words:
(a) submarine (b) satellite

Solutions to exercises

1

2 (a) 18 (b) 4 (c) 0.001 (d) 0.25 (e) 0.009 (f) 2 (g) 40314 (h) 9

3 (a) $\dfrac{1}{8}$ (b) $\dfrac{13}{9}$

4 (a) 4, 10 (b) 9, 23 (c) $-15\frac{1}{2}, -14\frac{1}{2}$ (d) $-16.05, -15.95$ (e) $-5, 21$ (f) $-6, 10$

5 (a), (b), (c), (g) are true.

6 (a) $9! = 362880$ (b) $\dfrac{9!}{2!2!2!} = 45360$

1.3 Using symbols

Mathematics provides a very rich language for the communication of engineering concepts and ideas, and a set of powerful tools for the solution of engineering problems. In order to use this language it is essential to appreciate how **symbols** are used to represent physical quantities, and to understand the rules and conventions that have been developed to manipulate them.

The choice of which letters or other symbols to use is largely up to the user, although it is helpful to choose letters that have some meaning in any particular context. For instance, if we wish to choose a symbol to represent the temperature in a room we might choose the capital letter T. Usually the lowercase letter t is used to represent time. Because both time and temperature can vary we refer to T and t as **variables**. In a particular calculation some symbols represent fixed and unchanging quantities and we call these **constants**, although constants can vary from one calculation to the next. Often we reserve the letters x, y and z to stand for variables and use the earlier letters of the alphabet, such as a, b and c, to represent constants. The Greek letter pi, written π, is used to represent the constant $3.14159\ldots$ which appears in the formula for the area of a circle. Other Greek letters are frequently used, and for reference the Greek alphabet is given in Table 1.1.

Table 1.1
The Greek alphabet.

A	α	alpha	I	ι	iota	P	ρ	rho
B	β	beta	K	κ	kappa	Σ	σ	sigma
Γ	γ	gamma	Λ	λ	lambda	T	τ	tau
Δ	δ	delta	M	μ	mu	Υ	υ	upsilon
E	ε	epsilon	N	ν	nu	Φ	ϕ	phi
Z	ζ	zeta	Ξ	ξ	xi	X	χ	chi
H	η	eta	O	o	omicron	Ψ	ψ	psi
Θ	θ	theta	Π	π	pi	Ω	ω	omega

Mathematics is a very precise language and care must be taken to note the exact position of any symbol in relation to any other. If x and y are two symbols, then the quantities xy, x^y, x_y can all mean different things. In the expression x^y you will note that the symbol y is placed to the right of and slightly higher than the symbol x. In this context y is called a **superscript**. In the expression x_y, y is placed lower than and to the right of x, and is called a **subscript**.

Example 1.6 The temperature in a room

The temperature in a room is measured at four points as shown in Figure 1.3. Rather than use different letters to represent the four measurements we can use one symbol, T, together with four subscripts to represent the temperature. Thus the four measurements are T_1, T_2, T_3 and T_4.

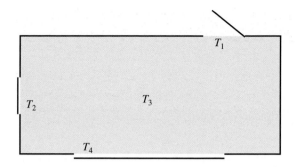

Figure 1.3
The temperature is measured at four points.

Addition (+)

If the letters x and y represent two numbers, then their **sum** is written as $x + y$. Note that $x + y$ is the same as $y + x$ just as $4 + 7$ is equal to $7 + 4$.

Subtraction (−)

Subtracting y from x yields $x - y$. This quantity is also called the **difference** of x and y. Note that $x - y$ is not the same as $y - x$ just as $11 - 7$ is not the same as $7 - 11$.

Multiplication (×)

The instruction to multiply x and y together is written as $x \times y$. Usually the multiplication sign is missed out altogether and we write simply xy. An alternative and acceptable notation is to use a dot to represent multiplication, and so we could write $x.y$. The quantity xy is called the **product** of x and y. Note that xy is the same as yx just as 4×3 is the same as 3×4. Because of this we say that multiplication is **commutative**. Multiplication is also **associative**. This means that when we multiply three quantities together, such as $x \times y \times z$, it doesn't matter whether we evaluate $x \times y$ first and then multiply the result by z, or evaluate $y \times z$ first and then multiply the result by x. Thus

$$(x \times y) \times z = x \times (y \times z)$$
$$= x \times y \times z$$

When mixing numbers and symbols it is usual to write the numbers first. Thus $3 \times x \times y \times 4 = 3 \times 4 \times x \times y = 12xy$.

Example 1.7
Simplify (a) $9(2y)$, (b) $-3(5z)$, (c) $4(2a)$, (d) $2x \times (2y)$.

Solution
(a) Note that $9(2y)$ means $9 \times (2 \times y)$. Because of the associativity of multiplication $9 \times (2 \times y)$ means the same as $(9 \times 2) \times y$, that is $18y$.
(b) $-3(5z)$ means $-3 \times (5 \times z)$. Because of associativity this is the same as $(-3 \times 5) \times z$, that is $-15z$.
(c) $4(2a)$ means $4 \times (2 \times a)$. We can write this as $(4 \times 2) \times a$, that is $8a$.

(d) Because of the associativity of multiplication, the brackets are not needed and we can write $2x \times (2y) = 2x \times 2y$, which equals

$$2 \times x \times 2 \times y = 2 \times 2 \times x \times y$$
$$= 4xy$$

Example 1.8

What is the distinction between $9(-2y)$ and $9 - 2y$?

Solution

The expression $9(-2y)$ means $9 \times (-2y)$. Because of associativity of multiplication we can write this as $9 \times (-2) \times y$ which equals $-18y$. On the other hand $9 - 2y$ means subtract $2y$ from 9. This cannot be simplified.

Division (÷)

The quantity $x \div y$ means x divided by y. This is also written as x/y or $\dfrac{x}{y}$ and is known as the **quotient** of x and y. In the expression $\dfrac{x}{y}$ the top line is called the **numerator** and the bottom line is called the **denominator**. Note that x/y is not the same as y/x. Division by 1 leaves a quantity unchanged so that $\dfrac{x}{1}$ is simply x. Division by 0 is never allowed.

Algebraic expressions

A quantity made up of symbols and the operations $+$, $-$, $\times$ and $/$ is called an **algebraic expression**. One algebraic expression divided by another is called an **algebraic fraction**. Thus

$$\frac{x + 7}{x - 3} \quad \text{and} \quad \frac{3x - y}{2x + z}$$

are algebraic fractions. The **reciprocal** of an algebraic fraction is found by inverting it so that the old numerator becomes the new denominator, and the old denominator becomes the new numerator. Thus the reciprocal of $\dfrac{2}{x}$ is $\dfrac{x}{2}$. The reciprocal of $\dfrac{x + 7}{x - 3}$ is $\dfrac{x - 3}{x + 7}$.

Example 1.9

State the reciprocal of each of the following expressions:

(a) $\dfrac{y}{z}$ (b) $\dfrac{x + z}{a - b}$ (c) $3y$ (d) $\dfrac{1}{a + 2b}$ (e) $\dfrac{1}{y}$

Solution

(a) The reciprocal of $\dfrac{y}{z}$ is $\dfrac{z}{y}$.

(b) The reciprocal of $\dfrac{x + z}{a - b}$ is $\dfrac{a - b}{x + z}$.

(c) $3y$ is the same as $\dfrac{3y}{1}$. The reciprocal of $3y$ is $\dfrac{1}{3y}$.

(d) The reciprocal of $\dfrac{1}{a + 2b}$ is $\dfrac{a + 2b}{1}$ or simply $a + 2b$.

(e) The reciprocal of $\dfrac{1}{y}$ is $\dfrac{y}{1}$ or simply y.

Finding the reciprocal of a complicated expression can cause confusion. Study the following example carefully.

Example 1.10

State the reciprocal of:

(a) $p + q$ (b) $\dfrac{1}{R_1} + \dfrac{1}{R_2}$

Solution

(a) Because $p + q$ can be thought of as $\dfrac{p + q}{1}$ its reciprocal is

$$\frac{1}{p + q}$$

Note in particular that the reciprocal of $p + q$ is not $\dfrac{1}{p} + \dfrac{1}{q}$. This distinction is important and a common cause of error. To avoid an error carefully identify the numerator and denominator in the original expression before inverting.

(b) The reciprocal of $\dfrac{1}{R_1} + \dfrac{1}{R_2}$ is

$$\frac{1}{\dfrac{1}{R_1} + \dfrac{1}{R_2}}$$

To simplify this further requires knowledge of the addition of algebraic fractions, which is dealt with in Block 6. It is important to note that the reciprocal of $\dfrac{1}{R_1} + \dfrac{1}{R_2}$ is not $R_1 + R_2$.

Example 1.11 Electrical Engineering – Conductance and reactance

Reciprocals of physical quantities occur in several engineering contexts. For example, if R is the resistance value of a resistor, then its reciprocal, $\dfrac{1}{R}$, is known as the **conductance**, G, so that

$$G = \frac{1}{R}$$

Clearly, when the resistance of a component is large, its conductance will be small, and vice versa.

In a.c. circuit theory, if C is the capacitance of a capacitor, the reciprocal of C appears in the definition of **capacitive reactance**, X_C. This is defined as

$$X_C = \frac{1}{\omega} \times \frac{1}{C} = \frac{1}{\omega C}$$ where ω is the angular frequency of the applied alternating voltage (see Chapter 9, Block 7, Section 7.4).

Example 1.12 Electrical Engineering – Transfer function

A control engineer often makes use of a mathematical expression called a **transfer function** which tells the engineer how the output of a complicated system is related to the input or driving force. The transfer function often takes the form of an algebraic fraction, where the variable used is the letter s. Typical transfer functions may look like

$$\frac{1}{s^2 + 3s + 4} \quad \text{or} \quad \frac{1}{s^2 - 1}$$

The equals sign, $=$

The equals sign, $=$, is used in several different ways. Firstly, an equals sign is used in **equations**. The left-hand side and right-hand side of an equation are equal only when the variable involved takes specific values known as **solutions** of the equation. For example, in the equation $x - 8 = 0$, the variable is x. The left-hand side and right-hand side are equal only when x has the value 8. If x has any other value the two sides are not equal.

Secondly, the equals sign is used in **formulae**. Physical quantities are often related through a formula. For example, the formula for the length, C, of the circumference of a circle expresses the relationship between the circumference of the circle and its radius, r. This formula states $C = 2\pi r$. When used in this way the equals sign expresses the fact that the quantity on the left is found by evaluating the expression on the right.

Finally, an equals sign is used in **identities**. At first sight an identity looks rather like an equation, except that it is true for *all* values of the variable. We shall see shortly that $(x - 1)(x + 1) = x^2 - 1$ for any value of x whatsoever. This means that the quantity on the left means exactly the same as that on the right whatever the value of x. To distinguish this usage from other uses of the equals symbol it is more correct to write $(x - 1)(x + 1) \equiv x^2 - 1$, where $\equiv$ means 'is identically equal to', although in practice the equals sign is often used.

The 'not equals' sign, $\neq$

The sign $\neq$ means 'is not equal to'. For example, it is correct to write $5 \neq 6$, $7 \neq -7$.

The δ notation for the change in a variable

The **change** in the value of a quantity is found by subtracting its initial value from its final value. For example, if the temperature of a mixture is initially $13\,°C$ and after some time is found to be $17\,°C$, the change in temperature is $17 - 13 = 4\,°C$. The Greek letter δ is often used to indicate such a change. If x is a variable we write δx to stand for a change in the value of x. We sometimes refer to δx as an **increment** in x. For example, if the value of x changes from 3 to 3.01 we could write $\delta x = 3.01 - 3 = 0.01$. It is important to note that this is not the product of δ and x, rather the whole quantity δx means the increment in x.

Sigma notation, Σ

Sigma notation provides a concise and convenient way of writing long sums. The sum

$$x_1 + x_2 + x_3 + x_4 + \cdots + x_{11} + x_{12}$$

is written using the capital Greek letter sigma, Σ, as

$$\sum_{k=1}^{k=12} x_k \quad \text{or} \quad \sum_{k=1}^{k=12} x_k$$

The Σ stands for the sum of all the values of x_k as k ranges from 1 to 12. Note that the lowermost and uppermost values of k are written at the bottom and top of the sigma sign respectively.

Example 1.13
Write out explicitly what is meant by

$$\sum_{k=1}^{k=5} k^3$$

Solution
We must let k range from 1 to 5:

$$\sum_{k=1}^{k=5} k^3 = 1^3 + 2^3 + 3^3 + 4^3 + 5^3$$

Example 1.14
Express $\dfrac{1}{1} + \dfrac{1}{2} + \dfrac{1}{3} + \dfrac{1}{4}$ concisely using sigma notation.

Solution
Each term takes the form $\dfrac{1}{k}$ where k varies from 1 to 4. Write down the sigma notation:

$$\sum_{k=1}^{k=4} \frac{1}{k}$$

Example 1.15
Write out explicitly $\sum_{k=1}^{k=3} 1$.

Solution
Here k does not appear explicitly in the terms to be added. This means add the number 1, three times:

$$\sum_{k=1}^{k=3} 1 = 1 + 1 + 1$$
$$= 3$$

In general $\sum_{k=1}^{k=n} 1 = n$.

Exercises

1 State the reciprocal of (a) x, (b) $\dfrac{1}{z}$, (c) xy, (d) $\dfrac{1}{xy}$, (e) $a + b$, (f) $\dfrac{2}{a + b}$.

2 The pressure p in a reaction vessel changes from 35 pascals to 38 pascals. Write down the value of δp.

3 Express as simply as possible:
(a) $(-3) \times x \times (-2) \times y$
(b) $9 \times x \times z \times (-5)$

4 Simplify (a) $8(2y)$, (b) $17x(-2y)$, (c) $5x(8y)$, (d) $5x(-8y)$.

5 What is the distinction between $5x(2y)$ and $5x - 2y$?

6 The value of x is 100 ± 3. The value of y is 120 ± 5. Find the maximum and minimum values of
(a) $x + y$ (b) xy (c) $\dfrac{x}{y}$ (d) $\dfrac{y}{x}$

7 Write out explicitly (a) $\sum_{i=1}^{N} f_i$,
(b) $\sum_{i=1}^{N} f_i x_i$.

8 By writing out the terms explicitly show that
$$\sum_{k=1}^{k=5} 3k = 3 \sum_{k=1}^{k=5} k$$

9 Write out explicitly $\sum_{k=1}^{3} y(x_k)\delta x$.

Solutions to exercises

1 (a) $\dfrac{1}{x}$ (b) z (c) $\dfrac{1}{xy}$ (d) xy (e) $\dfrac{1}{a + b}$ (f) $\dfrac{a + b}{2}$

2 $\delta p = 3$ pascals

3 (a) $6xy$ (b) $-45xz$

4 (a) $16y$ (b) $-34xy$ (c) $40xy$ (d) $-40xy$

5 $5x(2y) = 10xy$, $5x - 2y$ cannot be simplified.

6 (a) max 228, min 212 (b) 12875, 11155
(c) 0.8957, 0.7760 (d) 1.2887, 1.1165

7 (a) $\sum_{i=1}^{N} f_i = f_1 + f_2 + \cdots + f_{N-1} + f_N$

(b) $\sum_{i=1}^{N} f_i x_i = f_1 x_1 + f_2 x_2 + \cdots$
$$+ f_{N-1} x_{N-1} + f_N x_N$$

9 $y(x_1)\delta x + y(x_2)\delta x + y(x_3)\delta x$

1.4 Computer algebra packages

Computer algebra packages are specially designed computer programs with the ability to manipulate symbols, solve equations, etc. There are several different packages available but the common ones are Maple and Matlab. You should enquire about the availability of this software in your college or university.

End of block exercises

1   On a number line indicate all numbers greater than or equal to $-\frac{1}{2}$ but less than $\frac{3}{4}$.

2   Evaluate (a) $3 - 2 + |5 - 7|$,
(b) $3 - |2 + 5| - 7$, (c) $|3 - 2| + 5 - 7$,
(d) $|3 - 2| + |5 - 7|$.

3   Evaluate (a) $3! - 2!$, (b) $3! - 2$,
(c) $2! - 3!$, (d) $4 \times 3!$.

4   State the reciprocal of (a) 9, (b) $\dfrac{4}{3}$, (c) $\dfrac{4x}{3y}$.

5   Calculate the maximum and minimum values of the following resistances defined by their tolerance bands:
(a) $35\ \Omega \pm 5\%$ (b) $470\ \Omega \pm 10\%$

6   The speed, v, of a vehicle changes from 40 metres per second to 38 metres per second. Write down the value of δv.

7   Express as simply as possible:
(a) $7 \times x \times 4 \times 2 \times y$
(b) $-8 \times 2 \times a \times b$
(c) $2 \times x \times 3 \times y \times 4z$

8   Simplify (a) $5(3x)$, (b) $4(2x)$, (c) $8(-7x)$.

9   What is the distinction between $x(-y)$ and $x - y$?

10 Show that (a) $\dfrac{n!}{(n-1)!} = n$,

(b) $\dfrac{(n+1)!}{(n-1)!} = n(n+1)$,

(c) $\dfrac{n!}{(n+1)!} = \dfrac{1}{n+1}$.

11 Write out explicitly $\sum_{i=1}^{3} f_i(x_i - \bar{x})^2$.

Solutions to exercises

1

2   (a) 3 (b) -11 (c) -1 (d) 3

3   (a) 4 (b) 4 (c) -4 (d) 24

4   (a) $\dfrac{1}{9}$ (b) $\dfrac{3}{4}$ (c) $\dfrac{3y}{4x}$

5   (a) 33.25, 36.75 (b) 423, 517

6   -2 metres per second

7   (a) $56xy$ (b) $-16ab$ (c) $24xyz$

8   (a) $15x$ (b) $8x$ (c) $-56x$

9   $x(-y)$ is the product of x and $-y$. The expression $x - y$ means subtract y from x.

11 $\sum_{i=1}^{3} f_i(x_i - \bar{x})^2 = f_1(x_1 - \bar{x})^2$
$+ f_2(x_2 - \bar{x})^2 + f_3(x_3 - \bar{x})^2$

Indices

2.1 Introduction

Indices, or powers, provide a convenient notation when we need to multiply a number by itself several times. In this block we explain how indices are written, and state the rules or laws that are used for manipulating them.

Expressions built up using non-negative whole number powers of a variable and known as **polynomials** occur frequently in engineering mathematics. We introduce some common polynomials in this block.

Scientific notation is used to express very large or very small numbers concisely. This requires the use of indices. We explain how to use scientific notation towards the end of the block.

2.2 Index notation

When we wish to multiply a number by itself several times we make use of **index** or **power** notation. The number $4 \times 4 \times 4$ is written as 4^3 and read '4 raised to the power 3' or '4 cubed'. Note that the number of times '4' occurs in the product is written as a superscript. In this context we call the superscript 3 an **index**. Similarly we could write

$$5 \times 5 = 5^2, \text{ read '5 to the power 2'} \quad \text{or} \quad \text{'5 squared'}$$

and

$$7 \times 7 \times 7 \times 7 \times 7 = 7^5, \quad a \times a \times a = a^3, \quad m \times m \times m \times m = m^4$$

More generally, in the expression x^y, x is called the **base** and y is called the index or power. The plural of index is **indices**. The process of raising to a power is also known as **exponentiation** because yet another name for a power is an **exponent**. When dealing with numbers your calculator is able to evaluate expressions involving powers, probably using the x^y button.

Example 2.1
Use a calculator to evaluate 3^{12}.

Solution
Using the x^y button on the calculator check that you obtain $3^{12} = 531441$.

Example 2.2

Identify the index and base in the following expressions:
(a) 8^{11} (b) $(-2)^5$ (c) z^m (d) p^{-q}

Solution

(a) In the expression 8^{11}, 8 is the base and 11 is the index.
(b) In the expression $(-2)^5$, -2 is the base and 5 is the index.
(c) In the expression z^m, z is the base and m is the index.
(d) In the expression p^{-q}, p is the base and $-q$ is the index. The interpretation of a negative index will be given in Section 2.5.

Recall from Chapter 1 that when several operations are involved we can make use of the BODMAS rule for deciding the order in which operations must be carried out. The BODMAS rule makes no mention of exponentiation. Exponentiation should be carried out once any brackets have been dealt with. Consider the following examples.

Example 2.3

Evaluate 7×3^2.

Solution

There are two operations involved here, exponentiation and multiplication. The exponentiation should be carried out before the multiplication, so $7 \times 3^2 = 7 \times 9 = 63$.

Example 2.4

Write out fully (a) $3m^4$, (b) $(3m)^4$.

Solution

(a) In the expression $3m^4$ the exponentiation is carried out before the multiplication by 3, so

$$3m^4 \text{ means } 3 \times (m \times m \times m \times m)$$

that is

$$3 \times m \times m \times m \times m$$

(b) Here the bracketed expression is raised to the power 4 and so should be multiplied by itself four times:

$$(3m)^4 = (3m) \times (3m) \times (3m) \times (3m)$$

Because of the associativity and commutativity of multiplication we can write this as

$$3 \times 3 \times 3 \times 3 \times m \times m \times m \times m$$

or simply $81m^4$.

Note the important distinction between $(3m)^4$ and $3m^4$ and in particular the way that the power is applied.

Care must be taken when using your calculator to find powers of negative numbers. For example, $(-2)^2$ means $(-2) \times (-2) = +4$. Check that your calculator gives the correct answer. It may be necessary to enclose the -2 in brackets.

Exercises

1 Evaluate, without using a calculator:
(a) 3^3 (b) 3^4 (c) 2^5

2 Evaluate, without using a calculator:
(a) 0.2^2 (b) 15^2

3 Evaluate using a calculator:
(a) 7^3 (b) $(14)^{3.2}$

4 Write each of the following using index notation:
(a) $7 \times 7 \times 7 \times 7 \times 7$ (b) $t \times t \times t \times t$

(c) $\dfrac{1}{2} \times \dfrac{1}{2} \times \dfrac{1}{7} \times \dfrac{1}{7} \times \dfrac{1}{7}$

5 Evaluate without using a calculator:
(a) $\left(\dfrac{2}{3}\right)^2$ (b) $\left(\dfrac{2}{5}\right)^3$ (c) $\left(\dfrac{1}{2}\right)^2$ (d) $\left(\dfrac{1}{2}\right)^3$

(e) 0.1^3

6 Explain what is meant by 5×3^4. How would you write the expression if you wanted the multiplication by 5 carried out first?

Solutions to exercises

1 (a) 27 (b) 81 (c) 32

2 (a) 0.04 (b) 225

3 (a) 343 (b) 4651.7

4 (a) 7^5 (b) t^4 (c) $\left(\dfrac{1}{2}\right)^2 \left(\dfrac{1}{7}\right)^3$

5 (a) $\dfrac{4}{9}$ (b) $\dfrac{8}{125}$ (c) $\dfrac{1}{4}$ (d) $\dfrac{1}{8}$ (e) 0.001

6 5×3^4 means $5 \times 81 = 405$. If the intention is to carry out the multiplication first we would write $(5 \times 3)^4$.

2.3 Laws of indices

There are a number of rules that enable us to manipulate expressions involving indices. These rules are known as the **laws of indices**, and they occur so commonly that it will be worthwhile to memorise them.

Key point

The **laws of indices** state:

first law: $a^m \times a^n = a^{m+n}$

second law: $\dfrac{a^m}{a^n} = a^{m-n}$

third law: $(a^m)^n = a^{mn}$

In each case note that the base must be the same throughout.

In words, these state that to multiply two numbers having the same base and possibly different indices, the indices are added. To divide, the indices are subtracted. If a number is raised to a power and the result itself is raised to a power, the two powers are multiplied together. Consider the following examples.

Example 2.5
Simplify (a) $a^5 \times a^4$, (b) $2x^5(x^3)$.

Solution
In each case we are required to multiply expressions involving indices. The bases are the same and we can use the first of the given laws.
(a) The indices must be added: thus $a^5 \times a^4 = a^{5+4} = a^9$.
(b) Because of the associativity of multiplication we can write

$$2x^5(x^3) = 2(x^5 x^3)$$
$$= 2x^{5+3}$$
$$= 2x^8$$

The first law extends in an obvious way when more terms are involved.

Example 2.6
Simplify $b^5 \times b^4 \times b^7$.

Solution
The indices are added. Thus $b^5 \times b^4 \times b^7 = b^{5+4+7} = b^{16}$.

Example 2.7
Simplify $y^4 y^2 y^3$.

Solution
All quantities have the same base. To multiply the quantities together, the indices are added:

$$y^4 y^2 y^3 = \boxed{} \qquad\qquad y^9$$

Example 2.8
Simplify (a) $\dfrac{8^4}{8^2}$, (b) $x^{18} \div x^7$.

Solution
In each case we are required to divide expressions involving indices. The bases are the same and we can use the second of the given laws.

(a) The indices must be subtracted: thus $\dfrac{8^4}{8^2} = 8^{4-2} = 8^2$.

(b) Again the indices are subtracted, and so $x^{18} \div x^7 = x^{18-7} = x^{11}$.

Example 2.9
Simplify (a) $\dfrac{5^9}{5^7}$, (b) $\dfrac{y^5}{y^2}$.

Solution

(a) The bases are the same, and the division is carried out by subtracting the indices:

$$\frac{5^9}{5^7} = \qquad$$

$$5^{9-7} = 5^2$$

(b) $\dfrac{y^5}{y^2} = \qquad$

$$y^{5-2} = y^3$$

Example 2.10

Simplify (a) $(8^2)^3$, (b) $(z^3)^4$.

Solution

We use the third of the given laws.

(a) $(8^2)^3 = 8^{2\times3} = 8^6$

(b) $(z^3)^4 = z^{3\times4} = z^{12}$

Example 2.11

Simplify $(x^2)^5$.

Solution

Apply the third law.

$$(x^2)^5 = \qquad$$

$$x^{2\times5} = x^{10}$$

Example 2.12

Simplify $(e^x)^y$.

Solution

Again, using the third law, the two powers are multiplied:

$$(e^x)^y = \qquad$$

$$e^{x\times y} = e^{xy}$$

Two important results that can be derived from the laws state:

> **Key point**
>
> Any number raised to the power 0 is 1, that is $a^0 = 1$.
>
> Any number raised to the power 1 is itself, that is $a^1 = a$.

A generalisation of the third law states:

> **Key point**
>
> $$(a^m b^n)^k = a^{mk} b^{nk}$$

Example 2.13

Remove the brackets from (a) $(3x)^2$, (b) $(x^3 y^7)^4$.

Solution

(a) Noting that $3 = 3^1$ and $x = x^1$ then

$$
\begin{aligned}
(3x)^2 &= (3^1 x^1)^2 \\
&= 3^{1\times2} x^{1\times2} \\
&= 3^2 x^2 \\
&= 9x^2
\end{aligned}
$$

Alternatively

$$(3x)^2 = (3x) \times (3x)$$
$$= 9x^2$$

(b) $(x^3y^7)^4 = x^{3 \times 4}y^{7 \times 4}$
$$= x^{12}y^{28}$$

Exercises

1   Show that $(-xy)^2$ is equivalent to x^2y^2.

2   Show that $(-xy)^3$ is equivalent to $-x^3y^3$.

3   Write each of the following expressions with a single index:

   (a) $6^7 6^9$ (b) $\dfrac{6^7}{6^{19}}$ (c) $(x^4)^3$

4   Remove the brackets from (a) $(8a)^2$, (b) $(7ab)^3$, (c) $7(ab)^3$, (d) $(6xy)^4$, (e) $(6x)^4y$.

5   Simplify (a) $15x^2(x^3)$, (b) $3x^2(5x)$, (c) $18x^{-1}(3x^4)$.

6   Simplify (a) $5x(x^3)$, (b) $4x^2(x^3)$, (c) $3x^7(x^4)$, (d) $2x^8(x^{11})$, (e) $5x^2(3x^9)$.

Solutions to exercises

3   (a) 6^{16} (b) 6^{-12} (c) x^{12}

4   (a) $64a^2$ (b) $343a^3b^3$ (c) $7a^3b^3$
   (d) $1296x^4y^4$ (e) $1296x^4y$

5   (a) $15x^5$ (b) $15x^3$ (c) $54x^3$

6   (a) $5x^4$ (b) $4x^5$ (c) $3x^{11}$ (d) $2x^{19}$ (e) $15x^{11}$

2.4   Polynomial expressions

An important group of mathematical expressions that use indices are known as **polynomials**. Examples of polynomials are

$$4x^3 + 2x^2 + 3x - 7, \quad x^2 + x, \quad 17 - 2t + 7t^4, \quad z - z^3$$

Notice that they are all constructed using non-negative whole number powers of the variable. Recall that $x^0 = 1$ and so the number -7 appearing in the first example can be thought of as $-7x^0$. Similarly the 17 appearing in the third example can be read $17t^0$.

Key point

A **polynomial expression** takes the form

$$a_0 + a_1x + a_2x^2 + a_3x^3 + \dots$$

where $a_0, a_1, a_2, a_3, \dots$ are all constants called the **coefficients** of the polynomial. The number a_0 is also called the **constant term**. The highest power in a polynomial is called the **degree** of the polynomial. Polynomials with low degrees have special names:

Polynomial	Degree	Name
$ax^3 + bx^2 + cx + d$	3	cubic
$ax^2 + bx + c$	2	quadratic
$ax + b$	1	linear
a	0	constant

Example 2.14

Which of the following expressions are polynomials? Give the degree of those that are.

(a) $3x^2 + 4x + 2$ (b) $\dfrac{1}{x+1}$ (c) $\sqrt{x}$ (d) $2t + 4$

Solution

Recall that a polynomial expression must contain only terms involving whole number powers of the variable.

Give your answers:

(a) $3x^2 + 4x + 2$,   polynomial of degree 2

(b) $\dfrac{1}{x+1}$,   not a polynomial

(c) $\sqrt{x}$,   not a polynomial

(d) $2t + 4$,   polynomial of degree 1

Exercises

1   State which of the following are linear polynomials, which are quadratic polynomials, and which are constants:
(a) x (b) $x^2 + x + 3$ (c) $x^2 - 1$ (d) $3 - x$
(e) $7x - 2$ (f) $\frac{1}{2}$ (g) $\frac{1}{2}x + \frac{3}{4}$ (h) $3 - \frac{1}{2}x^2$

2   State which of the following are polynomials:
(a) $-\alpha^2 - \alpha - 1$ (b) $x^{1/2} - 7x^2$ (c) $\dfrac{1}{x}$ (d) 19

3   Which of the following are polynomials?
(a) $4t + 17$ (b) $\frac{1}{2} - \frac{1}{2}t$ (c) 15
(d) $t^2 - 3t + 7$ (e) $\dfrac{1}{t^2} + \dfrac{1}{t} + 7$

4   State the degree of each of the following polynomials. For those of low degree, give their name.
(a) $2t^3 + 7t^2$ (b) $7t^7 + 14t^3 - 2t^2$ (c) $7x + 2$
(d) $x^2 + 3x + 2$ (e) $2 - 3x - x^2$ (f) 42

Solutions to exercises

1   (a), (d), (e) and (g) are linear; (b), (c) and (h) are quadratic; (f) is a constant.

2   (a) is a polynomial, (d) is a polynomial of degree 0, (b) and (c) are not polynomials.

3   (a), (b), (c) and (d) are polynomials.

4   (a) 3, cubic, (b) 7, (c) 1, linear, (d) 2, quadratic, (e) 2, quadratic, (f) 0, constant.

2.5 Negative indices

Sometimes a number is raised to a negative power. This is interpreted as follows:

Key point

$$\text{negative powers: } a^{-m} = \frac{1}{a^m}, \quad a^m = \frac{1}{a^{-m}}$$

Thus a negative index can be used to indicate a reciprocal.

Example 2.15
Write each of the following expressions using a positive index and simplify if possible.

(a) 2^{-3} (b) $\dfrac{1}{4^{-3}}$ (c) x^{-1} (d) x^{-2} (e) 10^{-1}

Solution

(a) $2^{-3} = \dfrac{1}{2^3} = \dfrac{1}{8}$

(b) $\dfrac{1}{4^{-3}} = 4^3 = 64$

(c) $x^{-1} = \dfrac{1}{x^1} = \dfrac{1}{x}$

(d) $x^{-2} = \dfrac{1}{x^2}$

(e) $10^{-1} = \dfrac{1}{10^1} = \dfrac{1}{10}$ or 0.1

Example 2.16
Write each of the following using a positive index:

(a) $\dfrac{1}{t^{-4}}$ (b) 17^{-3} (c) y^{-1} (d) 10^{-2}

Solution
Use the previous key point.

(a) $\dfrac{1}{t^{-4}} = \boxed{}$ t^4

(b) $17^{-3} = \boxed{}$ $\dfrac{1}{17^3}$

(c) $y^{-1} = \boxed{}$ $\dfrac{1}{y}$

(d) $10^{-2} = \boxed{}$ $\dfrac{1}{10^2}$ which equals $\dfrac{1}{100}$ or 0.01

Example 2.17

Simplify (a) $\dfrac{a^8 \times a^7}{a^4}$, (b) $\dfrac{m^9 \times m^{-2}}{m^{-3}}$.

Solution

(a) Use the first law of indices to simplify the numerator:

$$\frac{a^8 \times a^7}{a^4} = \boxed{} \qquad\qquad \frac{a^{15}}{a^4}$$

Then use the second law to simplify the result:

$$\boxed{} \qquad\qquad a^{11}$$

(b) First simplify the numerator using the first law of indices:

$$\frac{m^9 \times m^{-2}}{m^{-3}} = \boxed{} \qquad\qquad \frac{m^7}{m^{-3}}$$

Then use the second law to simplify the result:

$$\boxed{} \qquad\qquad m^{7-(-3)} = m^{10}$$

Exercises

1   Write the following numbers using a positive index and also express your answers as decimal fractions:
(a) 10^{-1} (b) 10^{-3} (c) 10^{-4}

2   Simplify as much as possible:
(a) $x^3 x^{-2}$ (b) $\dfrac{t^4}{t^{-3}}$ (c) $\dfrac{y^{-2}}{y^{-6}}$

Solutions to exercises

1   (a) $\dfrac{1}{10} = 0.1$ (b) $\dfrac{1}{10^3} = 0.001$

(c) $\dfrac{1}{10^4} = 0.0001$

2   (a) $x^1 = x$ (b) $t^{4+3} = t^7$ (c) $y^{-2+6} = y^4$

2.6   Fractional indices

So far we have used indices that are whole numbers. We now consider those that are fractions. Consider the expression $(16^{1/2})^2$. Using the third law of indices, $(a^m)^n = a^{mn}$, we can write

$$(16^{1/2})^2 = 16^{1/2 \times 2}$$
$$= 16^1$$
$$= 16$$

So $16^{1/2}$ is a number that when squared equals 16, that is 4 or -4. In other words $16^{1/2}$ is a square root of 16, that is $\sqrt{16}$. There are always two square roots of a non-zero number, and we write

$$16^{1/2} = \pm 4$$

In general

Key point

$a^{1/2}$ is a square root of a

Similarly

$$(8^{1/3})^3 = 8^{1/3 \times 3}$$
$$= 8^1$$
$$= 8$$

so that $8^{1/3}$ is a number that when cubed equals 8. Thus $8^{1/3}$ is the cube root of 8, that is $\sqrt[3]{8}$, namely 2. Each number has only one cube root, and so

$$8^{1/3} = 2$$

Key point

$a^{1/3}$ is the cube root of a

In general we have

Key point

$x^{1/n}$ is the nth root of x, that is $\sqrt[n]{x}$

Your calculator will be able to evaluate fractional powers, and roots of numbers. Check that you can obtain the results of the following examples.

Example 2.18
Evaluate (a) $144^{1/2}$, (b) $125^{1/3}$.

Solution
(a) $144^{1/2}$ is a square root of 144, that is ± 12.
(b) Noting that $5^3 = 125$, we see that $125^{1/3} = \sqrt[3]{125} = 5$.

Example 2.19
Evaluate (a) $32^{1/5}$, (b) $32^{2/5}$, (c) $8^{2/3}$.

Solution
(a) $32^{1/5}$ is the fifth root of 32, that is $\sqrt[5]{32}$. Now $2^5 = 32$ and so $\sqrt[5]{32} = 2$.
(b) Using the third law of indices we can write $32^{2/5} = 32^{2 \times 1/5} = (32^{1/5})^2$. Thus

$$32^{2/5} = (32^{1/5})^2$$
$$= 2^2$$
$$= 4$$

(c) Note that $8^{1/3} = 2$. Then

$$8^{2/3} = 8^{2 \times 1/3}$$
$$= (8^{1/3})^2$$
$$= 2^2$$
$$= 4$$

Note the following alternatives:

$$8^{2/3} = (8^{1/3})^2$$
$$= (8^2)^{1/3}$$

Example 2.20
Write the following with a single index:
(a) $\sqrt{x^5}$ (b) $\sqrt[4]{x^3}$

Solution
(a) $\sqrt{x^5} = (x^5)^{1/2}$. Then using the third law of indices we can write this as
$x^{5 \times 1/2} = x^{5/2}$.
(b) $\sqrt[4]{x^3} = (x^3)^{1/4}$. Using the third law we can write this as $x^{3 \times 1/4} = x^{3/4}$.

Example 2.21
Show that $z^{-1/2} = \dfrac{1}{\sqrt{z}}$.

Solution

$$z^{-1/2} = \frac{1}{z^{1/2}}$$

$$= \frac{1}{\sqrt{z}}$$

Example 2.22
Simplify $\dfrac{\sqrt{z}}{z^3 z^{-1/2}}$.

Solution
Rewrite $\sqrt{z}$ using an index and simplify the denominator using the first law of indices:

$$\frac{\sqrt{z}}{z^3 z^{-1/2}} = \boxed{} \qquad \qquad \frac{z^{1/2}}{z^{2.5}}$$

Finally, use the second law to simplify the result:

$$\boxed{} \qquad \qquad z^{0.5-2.5} = z^{-2} \text{ or } \frac{1}{z^2}$$

Example 2.23
The generalisation of the third law of indices states that $(a^m b^n)^k = a^{mk} b^{nk}$. By taking $m = 1$, $n = 1$ and $k = \frac{1}{2}$ show that $\sqrt{ab} = \sqrt{a}\sqrt{b}$.

Solution

Taking $m = 1$, $n = 1$ and $k = \frac{1}{2}$ gives $(ab)^{1/2} = a^{1/2}b^{1/2}$, and the required result follows immediately.

Key point

$$\sqrt{ab} = \sqrt{a}\,\sqrt{b}$$

This result often allows answers to be written in alternative forms. For example, we may write $\sqrt{48}$ as $\sqrt{3 \times 16} = \sqrt{3}\,\sqrt{16} = 4\sqrt{3}$.

Exercises

1   Evaluate using a calculator (a) $3^{1/2}$, (b) $15^{-1/3}$, (c) 85^3, (d) $81^{1/4}$.

2   Evaluate using a calculator (a) 15^{-5}, (b) $15^{-2/7}$.

3   Simplify (a) $\dfrac{a^{11}\,a^{3/4}}{a^{-1/2}}$, (b) $\dfrac{\sqrt{z}}{z^{3/2}}$.

4   Simplify (a) $\dfrac{z^{-5/2}}{\sqrt{z}}$, (b) $\dfrac{\sqrt[3]{a}}{\sqrt[2]{a}}$, (c) $\dfrac{\sqrt[5]{z}}{z^{1/2}}$.

5   From the third law of indices show that $(ab)^{1/2} = a^{1/2}\,b^{1/2}$. Deduce that the square root of a product is equal to the product of the individual square roots.

6   Write each of the following expressions with a single index:

   (a) $(x^{-4})^3$ (b) $x^{1/2}x^{1/4}$ (c) $\dfrac{x^{1/2}}{x^{1/4}}$

Solutions to exercises

1   (a) 1.7321 (b) 0.4055 (c) 614125 (d) 3

2   (a) 1.317×10^{-6} (4 s.f.) (b) 0.4613 (4 s.f.)

3   (a) $a^{12.25}$ (b) z^{-1}

4   (a) z^{-3} (b) $a^{-1/6}$ (c) $z^{-3/10}$

6   (a) x^{-12} (b) $x^{3/4}$ (c) $x^{1/4}$

2.7   Scientific notation

It is often necessary to use very large or very small numbers such as 78000000 or 0.00000034. **Scientific notation** can be used to express such numbers in a more concise form. Each number is written in the form

$$a \times 10^n$$

where a is usually a number between 1 and 10. We can make use of the following facts:

$$10 = 10^1, \quad 100 = 10^2, \quad 1000 = 10^3 \text{ and so on}$$

and

$$0.1 = 10^{-1}, \quad 0.01 = 10^{-2}, \quad 0.001 = 10^{-3} \text{ and so on}$$

Furthermore, to multiply a number by 10^n the decimal point is moved n places to the right if n is a positive integer, and n places to the left if n is a negative integer. If necessary, additional zeros are inserted to make up the required number of decimal places. Then, for example,

the number 5000 can be written $5 \times 1000 = 5 \times 10^3$

the number 403 can be written $4.03 \times 100 = 4.03 \times 10^2$

the number 0.009 can be written $9 \times 0.001 = 9 \times 10^{-3}$

Example 2.24

Write the number 0.00678 in scientific notation.

Solution

$$0.00678 = 6.78 \times 10^{-3}$$

Example 2.25 Engineering constants

Many constants appearing in engineering calculations are expressed in scientific notation. For example, the charge on an electron equals 1.6×10^{-19} coulombs. Avogadro's constant is equal to 6.022×10^{26} and is the number of atoms in 1 kilo-mole of an element. Clearly the use of scientific notation avoids writing lengthy strings of zeros.

Your scientific calculator will be able to accept numbers in scientific notation. Often the E button is used, and a number such as 4.2×10^7 will be entered as 4.2E7. Note that 10E4 means 10×10^4, that is 10^5. To enter the number 10^3, say, you would key in 1E3. Entering powers of 10 incorrectly is a common cause of error. You must check how your particular calculator accepts numbers in scientific notation.

Example 2.26

Use your calculator to find $4.2 \times 10^{-3} \times 3.6 \times 10^{-4}$.

Solution

This exercise is designed to check that you can enter numbers given in scientific notation into your calculator. Check that

$$4.2 \times 10^{-3} \times 3.6 \times 10^{-4} = \qquad\qquad 1.512 \times 10^{-6}$$

Exercises

1 Express each of the following numbers in scientific notation:
(a) 45 (b) 456 (c) 2079 (d) 7000000 (e) 0.1
(f) 0.034 (g) 0.09856

2 Simplify $6 \times 10^{24} \times 1.3 \times 10^{-16}$.

Solutions to exercises

1 (a) 4.5×10^1 (b) 4.56×10^2
(c) 2.079×10^3 (d) 7.0×10^6
(e) 1.0×10^{-1} (f) 3.4×10^{-2}
(g) 9.856×10^{-2}

2 7.8×10^8

2.8 Powers and number bases

We are used to counting in the **base 10** or **decimal** system in which we use the 10 digits 0,1,2,3,4,5,6,7,8 and 9. In Chapter 3 Block 1 we reminded you that the number 5276 means

$$5000 + 200 + 70 + 6$$

or, to write it another way,

$$(5 \times 1000) + (2 \times 100) + (7 \times 10) + (6 \times 1)$$

This reminds us of the 'thousands', 'hundreds', 'tens' and 'units' from early school days. It is helpful in what follows to note that we can also think of this representation as

10^3	10^2	10^1	10^0
5	2	7	6

Note that the 'thousands', 'hundreds', 'tens' and 'units' are simply powers of the number base 10. In the remainder of this section we shall indicate the number base being used by a subscript, as in 5276_{10}.

In several applications, particularly in digital computing, it is essential to use bases other than 10. In base 2 we use only the two digits 0 and 1. Numbers in the base 2 system are called **binary** numbers, and we call 0 and 1 **binary digits** or simply **bits**.

To evaluate the decimal equivalent of a binary number such as 1101_2 we note that powers of 2 are now used to determine the place value:

2^3	2^2	2^1	2^0
1	1	0	1

So

$$1101_2 = (1 \times 2^3) + (1 \times 2^2) + (0 \times 2^1) + (1 \times 2^0)$$
$$= 8 + 4 + 0 + 1$$
$$= 13_{10}$$

Note that using two binary digits we can represent the four (i.e. 2^2) decimal numbers, 0, 1, 2, 3:

$$00_2 = 0_{10} \quad 01_2 = 1_{10} \quad 10_2 = 2_{10} \quad 11_2 = 3_{10}$$

With three binary digits we can represent eight (i.e. 2^3) decimal numbers, 0, 1, 2, 3, 4, 5, 6, 7:

$$000_2 = 0_{10} \quad 001_2 = 1_{10} \quad 010_2 = 2_{10} \quad 011_2 = 3_{10}$$
$$100_2 = 4_{10} \quad 101_2 = 5_{10} \quad 110_2 = 6_{10} \quad 111_2 = 7_{10}$$

In general, with n binary digits we can represent the 2^n decimal numbers 0, 1, 2, . . . , $2^n - 1$.

Example 2.27
Find which decimal numbers can be represented using (a) 8 binary digits, (b) 16 binary digits.

Solution
(a) With 8 binary digits we can represent the 256 ($=2^8$) decimal numbers 0, 1, 2, . . . 255. Note that a base 2 number having 8 binary digits is often referred to as a **byte.**
(b) With 16 binary digits we can represent the 65536 ($=2^{16}$) decimal numbers 0, 1, 2, . . . , 65535.

Example 2.28 Music Technology – Powers of 2 and compact disc technology
In digital audio technology an analogue signal (e.g. a voltage from a microphone) is sampled at 44100 times each second. The value of each of these samples must be recorded. To do this digitally it is necessary to **quantise** the sample. This means to approximate its value by one of a set of predetermined values. Compact discs usually use 16-bit technology, which means that each sample value is recorded using a 16-bit number. In turn, this means that we can store whole numbers in the range 0 to 65535 to represent the sample values. So, we need 2 bytes of storage for each sample, and 44100 samples each second. A stereo signal will require twice as much storage. This means we need 176400 bytes for each second of music stored on a CD. In fact, additional storage is required because other quantities are built in to reduce errors, and additional data is stored (track length, title, etc.). A simple calculation will show that a 4-minute audio track will require in excess of 42336000 bytes of storage – that is, a massive 42 megabytes. You will see why a standard CD-R (recordable CD) that has 700 megabytes of storage capacity will hold well under 16 tracks. To try to download a 42 MB file over the Internet using a domestic modem (56 kilobytes per second) would take over 12 minutes. Whilst a broadband connection would do this much more quickly, this is not a practical way of obtaining music files. Mathematicians have developed compression techniques (e.g. MP3) which can vastly reduce the size of the audio file and thus make Internet transmission a realistic possibility.

Exercises

1 Find the decimal equivalent of the binary number 110011001.

2 Find which decimal numbers can be represented using a 6-bit binary number.

3   Base 8, or **octal**, is used by computer scientists. Here the place values are powers of 8, and we can use the digits 0, 1, 2, 3, 4, 5, 6, 7. Find the decimal equivalent of the octal number 756_8.

4   Estimate the capacity needed to store a 20-minute stereo CD recording of a piece of classical music.

Solutions to exercises

1   409

2   $0, 1, 2, \ldots, 63.$

3   494

4   Well in excess of 200 MB.

End of block exercises

1   Write down the three laws of indices and give a numerical example illustrating each.

2   In the expression 5^4 state which number is the index and which is the base.

3   Simplify each of the following:
(a) $\dfrac{a^{11}}{a^4}$ (b) $(a^b)^4$

4   Remove the brackets from the expression $(4x^3)^5$.

5   Write $\dfrac{1}{x^{-1/2}}$ using a positive index.

6   Simplify
$$\frac{a^7 \times a^{-13}}{a^{-5}}$$

7   Rewrite $\sqrt{a^5}$ using a single index.

8   Express the numbers 4320 and 0.0065 in scientific notation.

9   Remove the brackets from $(7x^2)^{-3}$.

10   Simplify
$$\frac{y^6 y^{-3} y^{0.5}}{y^{-2} y^7}$$

11   Simplify
(a) $(3a^2b)(2a^3b^2)$ (b) $(a^4b^3)(7a^{-2}b^{-1})$
(c) xx^2x^3 (d) $(-2y)(-3y^2)(-4y^{-2})$
(e) $(e^{2x})^y$

Solutions to exercises

1   For example, $2^3 \times 2^7 = 2^{10}$, $\dfrac{3^6}{3^2} = 3^4$, $(5^7)^2 = 5^{14}$.

2   index 4, base 5

3   (a) a^7 (b) a^{4b}

4   $4^5 x^{15}$

5   $x^{1/2}$

6   a^{-1}

7   $a^{5/2}$

8   $4.320 \times 10^3, 6.5 \times 10^{-3}$

9   $7^{-3} x^{-6}$

10   $y^{-1.5}$

11   (a) $6a^5b^3$ (b) $7a^2b^2$ (c) x^6 (d) $-24y$ (e) e^{2xy}

Simplification by collecting like terms

3.1 Introduction

In this block we explain what is meant by the phrase 'like terms' and show how like terms are collected together and simplified.

3.2 Addition and subtraction of like terms

Like terms are multiples of the same quantity. For example, $5y$, $17y$ and $\frac{1}{2}y$ are all multiples of y and so are like terms. Similarly, $3x^2$, $-5x^2$ and $\frac{1}{4}x^2$ are all multiples of x^2 and so are like terms.

Further examples of like terms are:

- kx and lx, which are both multiples of x
- x^2y, $6x^2y$, $-13x^2y$, $-2yx^2$, which are all multiples of x^2y
- abc^2, $-7abc^2$, $kabc^2$, which are all multiples of abc^2

Like terms can be collected together and added or subtracted in order to simplify them.

Example 3.1
Simplify $5x - 13x + 22x$.

Solution
All three terms are multiples of x and so are like terms. The expression can be simplified to $14x$.

Example 3.2
Simplify $5z + 2x$.

Solution
$5z$ and $2x$ are not like terms. They are not multiples of the same quantity. This expression cannot be simplified.

Example 3.3
Simplify $5a + 2b - 7a - 9b$.

Solution
$5a + 2b - 7a - 9b = $ 　　　　　　　　 $-2a - 7b$

Example 3.4
Simplify $2x^2 - 7x + 11x^2 + x$.

Solution
$2x^2$ and $11x^2$, both being multiples of x^2, can be collected together and added to give $13x^2$. Similarly, $-7x$ and x are like terms and these can be added to give $-6x$. We find
$$2x^2 - 7x + 11x^2 + x = 13x^2 - 6x$$
which cannot be simplified further.

Example 3.5
Simplify $\frac{1}{2}x + \frac{3}{4}x - 2y$.

Solution
$\frac{1}{2}x + \frac{3}{4}x - 2y = \qquad\qquad\qquad\qquad\qquad \frac{5}{4}x - 2y$

Example 3.6
Simplify $3a^2b - 7a^2b - 2b^2 + a^2$.

Solution
Note that $3a^2b$ and $7a^2b$ are both multiples of a^2b and so are like terms. There are no other like terms. Therefore
$$3a^2b - 7a^2b - 2b^2 + a^2 = -4a^2b - 2b^2 + a^2$$

Exercises

1   Simplify, if possible:
(a) $5x + 2x + 3x$ (b) $3q - 2q + 11q$
(c) $7x^2 + 11x^2$ (d) $-11v^2 + 2v^2$
(e) $5p + 3q$

2   Simplify, if possible:
(a) $5w + 3r - 2w + r$ (b) $5w^2 + w + 1$
(c) $6w^2 + w^2 - 3w^2$

3   Simplify, if possible:
(a) $7x + 2 + 3x + 8x - 11$
(b) $2x^2 - 3x + 6x - 2$
(c) $-5x^2 - 3x^2 + 11x + 11$
(d) $4q^2 - 4r^2 + 11r + 6q$
(e) $a^2 + ba + ab + b^2$
(f) $3x^2 + 4x + 6x + 8$
(g) $s^3 + 3s^2 + 2s^2 + 6s + 4s + 12$

4   Explain the distinction, if any, between each of the following expressions, and simplify if possible:
(a) $18x - 9x$ (b) $18x(9x)$ (c) $18x(-9x)$
(d) $-18x - 9x$ (e) $-18x(9x)$

5   Explain the distinction, if any, between each of the following expressions, and simplify if possible:
(a) $4x - 2x$ (b) $4x(-2x)$ (c) $4x(2x)$
(d) $-4x(2x)$ (e) $-4x - 2x$ (f) $(4x)(2x)$

6   Simplify, if possible:
(a) $\frac{2}{3}x^2 + \frac{x^2}{3}$ (b) $0.5x^2 + \frac{3}{4}x^2 - \frac{11}{2}x$
(c) $3x^3 - 11x + 3yx + 11$ (d) $-4\alpha x^2 + \beta x^2$
where α and β are constants

Solutions to exercises

1   (a) $10x$ (b) $12q$ (c) $18x^2$ (d) $-9v^2$
    (e) cannot be simplified

2   (a) $3w + 4r$ (b) cannot be simplified
    (c) $4w^2$

3   (a) $18x - 9$ (b) $2x^2 + 3x - 2$
    (c) $-8x^2 + 11x + 11$ (d) cannot be simplified
    (e) $a^2 + 2ab + b^2$ (f) $3x^2 + 10x + 8$
    (g) $s^3 + 5s^2 + 10s + 12$

4   (a) $9x$ (b) $162x^2$ (c) $-162x^2$ (d) $-27x$
    (e) $-162x^2$

5   (a) $4x - 2x = 2x$ (b) $4x(-2x) = -8x^2$
    (c) $4x(2x) = 8x^2$ (d) $-4x(2x) = -8x^2$
    (e) $-4x - 2x = -6x$ (f) $(4x)(2x) = 8x^2$

6   (a) x^2 (b) $1.25x^2 - \frac{11}{2}x$
    (c) cannot be simplified (d) $(\beta - 4\alpha)x^2$

End of block exercises

In each case, simplify the given expression, if possible.

1   $3x - 2y + 7x - 11y$

2   $5x^2 + 3x + 1$

3   $\dfrac{1}{3}x + \dfrac{1}{2}x$

4   $ab + ba$

5   $ab - ba$

6   $8pq + 11pq - 9pq$

7   $ab + abc$

8   $0.01x + 0.35x$

9   $4y + 2x - 3xy$

10   $7\alpha - 3\beta + 2\gamma - 7\alpha + 11\beta$

Solutions to exercises

1   $10x - 13y$

2   cannot be simplified

3   $\dfrac{5}{6}x$

4   $2ab$

5   0

6   $10pq$

7   cannot be simplified

8   $0.36x$

9   cannot be simplified

10   $8\beta + 2\gamma$

Removing brackets

4.1 Introduction

In order to simplify an expression that contains brackets it is often necessary to rewrite the expression in an equivalent form but without any brackets. This process of removing brackets must be carried out according to particular rules, which are described in this block.

4.2 Removing brackets from expressions of the form $a(b + c)$ and $a(b - c)$

In an expression such as $5(x + y)$ it is intended that the 5 multiplies both x and y to produce $5x + 5y$. Thus the expressions $5(x + y)$ and $5x + 5y$ are equivalent. In general we have the following rules known as **distributive laws**:

Key point

$$a(b + c) = ab + ac$$
$$a(b - c) = ab - ac$$

Note that when the brackets are removed both terms in the brackets are multiplied by a. If you insert numbers instead of letters into these expressions you will see that both left- and right-hand sides are equivalent. For example,

$$4(3 + 5) \text{ has the same value as } 4(3) + 4(5), \text{ that is } 32$$

and

$$7(8 - 3) \text{ has the same value as } 7(8) - 7(3), \text{ that is } 35$$

Example 4.1
Remove the brackets from: (a) $9(2 + y)$, (b) $9(2y)$.

Solution
(a) In the expression $9(2 + y)$ the 9 must multiply both terms in the brackets:

$$9(2 + y) = 9(2) + 9(y)$$
$$= 18 + 9y$$

(b) Recall that $9(2y)$ means $9 \times (2 \times y)$ and that when multiplying numbers together the presence of brackets is irrelevant. Thus $9(2y) = 9 \times 2 \times y = 18y$

The crucial distinction between the role of the factor 9 in the two expressions $9(2 + y)$ and $9(2y)$ should be noted.

Example 4.2
Remove the brackets from $9(x + 2y)$.

Solution
In the expression $9(x + 2y)$ the 9 must multiply both the x and the $2y$ in the brackets. Thus

$$9(x + 2y) = 9x + 9(2y)$$
$$= 9x + 18y$$

Example 4.3
Remove the brackets from $-3(5x - z)$.

Solution
The number -3 must multiply both the $5x$ and the z. Thus

$$-3(5x - z) = (-3)(5x) - (-3)(z)$$
$$= -15x + 3z$$

Example 4.4
Remove the brackets from $6x(3x - 2y)$.

Solution

$$6x(3x - 2y) = 6x(3x) - (6x)(2y)$$
$$= 18x^2 - 12xy$$

Example 4.5
Remove the brackets from $-(3x + 1)$.

Solution
Although unwritten, the minus sign outside the brackets stands for -1. We must consider the expression $-1(3x + 1)$.

$$-1(3x + 1) = (-1)(3x) + (-1)(1)$$
$$= -3x + (-1)$$
$$= -3x - 1$$

Example 4.6
Remove the brackets from $-(5x - 3y)$.

Solution
$-(5x - 3y)$ means $-1(5x - 3y)$. Thus

$$-1(5x - 3y) = (-1)(5x) - (-1)(3y)$$
$$= -5x + 3y$$

Example 4.7
Remove the brackets from
(a) $9(2x + 3y)$ (b) $m(m + n)$

Solution
(a) The 9 must multiply both the term $2x$ and the term $3y$. Thus

$$9(2x + 3y) = \qquad\qquad\qquad\qquad 18x + 27y$$

(b) In the expression $m(m + n)$ the first m must multiply both terms in the brackets. Thus

$$m(m + n) = \qquad\qquad\qquad\qquad m^2 + mn$$

Example 4.8
Remove the brackets from the expression $5x - (3x + 1)$ and simplify the result by collecting like terms.

Solution
The brackets in $-(3x + 1)$ were removed in Example 4.5. Thus

$$\begin{aligned} 5x - (3x + 1) &= 5x - 1(3x + 1) \\ &= 5x - 3x - 1 \\ &= 2x - 1 \end{aligned}$$

Example 4.9
Show that $\dfrac{-x - 1}{4}$, $\dfrac{-(x + 1)}{4}$ and $-\dfrac{x + 1}{4}$ are all equivalent.

Solution

Consider $-(x + 1)$. Removing the brackets we obtain $-x - 1$ and so $\dfrac{-x - 1}{4}$ is equivalent to $\dfrac{-(x + 1)}{4}$

A negative quantity divided by a positive quantity will be negative. Hence $\dfrac{-(x + 1)}{4}$ is equivalent to $-\dfrac{x + 1}{4}$

Study all three expressions carefully to recognise the variety of equivalent ways in which we can write an algebraic expression.

Sometimes the bracketed expression can appear on the left, as in $(a + b)c$. To remove the brackets here we use the following rules:

Key point

$$(a + b)c = ac + bc$$
$$(a - b)c = ac - bc$$

Note that when the brackets are removed both the terms in the brackets multiply c.

Example 4.10
Remove the brackets from $(2x + 3y)x$.

Solution
Both terms in the brackets multiply the x outside. Thus

$$(2x + 3y)x = 2x(x) + 3y(x)$$
$$= 2x^2 + 3xy$$

Example 4.11
Remove the brackets from
(a) $(x + 3)\,(-2)$ (b) $(x - 3)\,(-2)$

Solution
(a) Both terms in the brackets must multiply the -2.
$(x + 3)\,(-2) = $ ⬚ $-2x - 6$

(b) $(x - 3)\,(-2) = $ ⬚ $-2x + 6$

Exercises

1   Remove the brackets from each of the
following expressions:
(a) $2(mn)$ (b) $2(m + n)$ (c) $a(mn)$
(d) $a(m + n)$ (e) $a(m - n)$ (f) $(am)n$
(g) $(a + m)n$ (h) $(a - m)n$ (i) $5(pq)$
(j) $5(p + q)$ (k) $5(p - q)$ (l) $7(xy)$
(m) $7(x + y)$ (n) $7(x - y)$ (o) $8(2p + q)$

(p) $8(2pq)$ (q) $8(2p - q)$ (r) $5(p - 3q)$
(s) $5(p + 3q)$ (t) $5(3pq)$

2   Remove the brackets from each of the
following expressions:
(a) $4(a + b)$ (b) $2(m - n)$ (c) $9(x - y)$

Solutions to exercises

1   (a) $2mn$ (b) $2m + 2n$ (c) amn
(d) $am + an$ (e) $am - an$ (f) amn
(g) $an + mn$ (h) $an - mn$ (i) $5pq$
(j) $5p + 5q$ (k) $5p - 5q$ (l) $7xy$
(m) $7x + 7y$ (n) $7x - 7y$ (o) $16p + 8q$

(p) $16pq$ (q) $16p - 8q$ (r) $5p - 15q$
(s) $5p + 15q$ (t) $15pq$

2   (a) $4a + 4b$ (b) $2m - 2n$ (c) $9x - 9y$

4.3 Removing brackets from expressions of the form $(a + b)(c + d)$

Sometimes it is necessary to consider two bracketed terms multiplied together. In an expression such as $(a + b)(c + d)$, by regarding the first bracket as a single term we can use the result in Section 4.2 to write $(a + b)c + (a + b)d$. Removing the brackets from each of these terms produces

Key point

$$(a + b)(c + d) = (a + b)c + (a + b)d$$
$$= ac + bc + ad + bd$$

Alternatively, we see that each term in the first bracket multiplies each term in the second. To remind us of this we can use the picture in Figure 4.1.

Figure 4.1

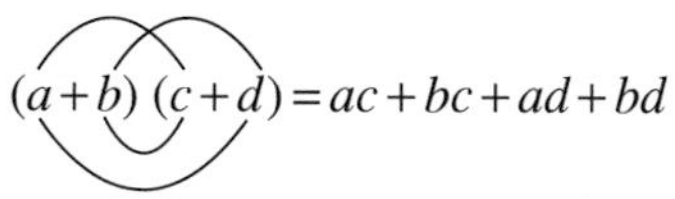

$$(a + b)\,(c + d) = ac + bc + ad + bd$$

Example 4.12
Remove the brackets from $(3 + x)(2 + y)$.

Solution
We find

$$(3 + x)(2 + y) = (3 + x)(2) + (3 + x)(y)$$
$$= (3)(2) + (x)(2) + (3)(y) + (x)(y)$$
$$= 6 + 2x + 3y + xy$$

Example 4.13
Remove the brackets from $(3x + 4)(x + 2)$ and simplify your result.

Solution

$$(3x + 4)(x + 2) = (3x + 4)(x) + (3x + 4)(2)$$
$$= 3x^2 + 4x + 6x + 8$$
$$= 3x^2 + 10x + 8$$

Example 4.14
Remove the brackets from $(a + b)^2$ and simplify your result.

Solution
When a quantity is squared it must be multiplied by itself. Thus

$$(a + b)^2 = (a + b)(a + b)$$
$$= (a + b)a + (a + b)b$$
$$= a^2 + ba + ab + b^2$$
$$= a^2 + 2ab + b^2$$

Example 4.15

Remove the brackets from the following expressions and simplify the results:

(a) $(x + 7)(x + 3)$ (b) $(x + 3)(x - 2)$ (c) $(3 - x)(x + 2)$

Solution

(a) Remove the brackets to obtain

$(x + 7)(x + 3) =$ ▨▨▨▨▨ $x^2 + 7x + 3x + 21$

Simplify the result to obtain

$(x + 7)(x + 3) =$ ▨▨▨▨▨ $x^2 + 10x + 21$

(b) Remove the brackets to obtain

$(x + 3)(x - 2) = (x + 3)(x) + (x + 3)(-2)$

$=$ ▨▨▨▨▨ $x^2 + 3x - 2x - 6$

Simplify the result to obtain

$(x + 3)(x - 2) =$ ▨▨▨▨▨ $x^2 + x - 6$

(c) Remove the brackets and simplify to find

$(3 - x)(x + 2) = (3 - x)x + (3 - x)2$

$=$ ▨▨▨▨▨ $6 + x - x^2$

Example 4.16

Explain the distinction between $(x + 3)(x + 2)$ and $x + 3(x + 2)$.

Solution

In the first case, on removing the brackets we find

$$(x + 3)(x + 2) = x^2 + 3x + 2x + 6$$
$$= x^2 + 5x + 6$$

In the second case we have

$$x + 3(x + 2) = x + 3x + 6 = 4x + 6$$

Note that in the second case the term $(x + 2)$ is only multiplied by 3 and not by x.

Example 4.17

Remove the brackets from $(s^2 + 2s + 4)(s + 3)$.

Solution

Each term in the first set of brackets must multiply each term in the second. Working through all combinations systematically we have

$$(s^2 + 2s + 4)(s + 3) = (s^2 + 2s + 4)(s) + (s^2 + 2s + 4)(3)$$
$$= s^3 + 2s^2 + 4s + 3s^2 + 6s + 12$$
$$= s^3 + 5s^2 + 10s + 12$$

**Example 4.18 Reliability Engineering – Reliability in communication
         networks**

Some communication networks are designed with built-in redundancy, so that in the event of certain components failing, the network can still function. For example, consider the parallel network shown in Figure 4.2. Communication traffic can pass along either route from left to right, so that in the event that one of the components fails the network can still function. Clearly the network will fail to function if both components fail. The **reliability** of a component, or collection of components, is the probability or likelihood that it will function normally during a given period of time. It is a number between 0 and 1 where 0 represents sure failure, and 1 represents guaranteed success.

Figure 4.2
The network will
fail only if both
components fail.

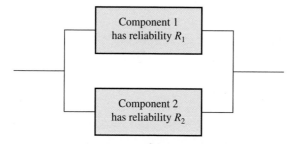

If the two components in Figure 4.2 have reliability R_1 and R_2, respectively, it can be shown that the reliability, R, of the combined system is

$$R = 1 - (1 - R_1)(1 - R_2)$$

Remove the brackets from the right-hand side of this expression and simplify the result.

Solution
Consider first the bracketed terms on the right:

$$(1 - R_1)(1 - R_2) = (1 - R_1) \times 1 - (1 - R_1) \times R_2$$
$$= 1 - R_1 - R_2 + R_1R_2$$

This expression must be subtracted from 1 to give R:

$$R = 1 - (1 - R_1 - R_2 + R_1R_2)$$
$$= R_1 + R_2 - R_1R_2$$

We shall see the significance of this result in Block 7 – *Formulae and transposition*.

Exercises

1 Remove the brackets from each of the following expressions and simplify where possible:
(a) $(2 + a)(3 + b)$ (b) $(x + 1)(x + 2)$
(c) $(x + 3)(x + 3)$ (d) $(x + 5)(x - 3)$

2 Remove the brackets from each of the following expressions:
(a) $(7 + x)(2 + x)$ (b) $(9 + x)(2 + x)$
(c) $(x + 9)(x - 2)$ (d) $(x + 11)(x - 7)$
(e) $(x + 2)x$ (f) $(3x + 1)x$

(g) $(3x + 1)(x + 1)$ (h) $(3x + 1)(2x + 1)$
(i) $(3x + 5)(2x + 7)$ (j) $(3x + 5)(2x - 1)$
(k) $(5 - 3x)(x + 1)$ (l) $(2 - x)(1 - x)$

3   Remove the brackets from
$(s + 1)(s + 5)(s - 3)$.

Solutions to exercises

1   (a) $6 + 3a + 2b + ab$ (b) $x^2 + 3x + 2$
   (c) $x^2 + 6x + 9$ (d) $x^2 + 2x - 15$

2   (a) $14 + 9x + x^2$ (b) $18 + 11x + x^2$
   (c) $x^2 + 7x - 18$ (d) $x^2 + 4x - 77$

(e) $x^2 + 2x$ (f) $3x^2 + x$
(g) $3x^2 + 4x + 1$ (h) $6x^2 + 5x + 1$
(i) $6x^2 + 31x + 35$ (j) $6x^2 + 7x - 5$
(k) $-3x^2 + 2x + 5$ (l) $x^2 - 3x + 2$

3   $s^3 + 3s^2 - 13s - 15$

End of block exercises

In questions 1–12 remove the brackets from the given expression:

1   $15(x + y)$

2   $7(2x + y)$

3   $(-2)(a + b)$

4   $(2 + x)(4 + x)$

5   $(x - 1)(x + 2)$

6   $(x + 2)(x - 2)$

7   $(x^2 + 2)(3x)$

8   $(x + 1)(x - 3)$

9   $(x + 1)(x - 3)x$

10   $(x + 1)(x - 3)(x - 1)$

11   $\frac{3}{4}\left(\frac{1}{2}x + 7\right)$

12   $15(x + 3)xy$

Solutions to exercises

1   $15x + 15y$

2   $14x + 7y$

3   $-2a - 2b$

4   $8 + 6x + x^2$

5   $x^2 + x - 2$

6   $x^2 - 4$

7   $3x^3 + 6x$

8   $x^2 - 2x - 3$

9   $x^3 - 2x^2 - 3x$

10   $x^3 - 3x^2 - x + 3$

11   $\frac{3}{8}x + \frac{21}{4}$

12   $15x^2y + 45xy$

Factorisation

In Block 4 we showed the way in which brackets were removed from algebraic expressions. Factorisation, which can be considered as the reverse of this process, is dealt with in this block. It is essential that you have had a lot of practice removing brackets before you attempt this block.

A number is said to be **factorised** when it is written as a product. For example, 21 can be factorised into 7×3. We say that 7 and 3 are **factors** of 21. Always remember that the factors of a number are *multiplied* together.

Algebraic expressions can also be factorised. Consider the expression $7(2x + 1)$. Removing the brackets we can rewrite this as

$$7(2x + 1) = 7(2x) + (7)(1)$$
$$= 14x + 7$$

Thus $14x + 7$ is equivalent to $7(2x + 1)$. We see that $14x + 7$ has factors 7 and $(2x + 1)$. The factors 7 and $(2x + 1)$ *multiply* together to give $14x + 7$. The process of writing an expression as a product of its factors is called **factorisation**. When asked to factorise $14x + 7$ we write

$$14x + 7 = 7(2x + 1)$$

and so we see that factorisation can be regarded as reversing the process of removing brackets in that we are now inserting them.

Always remember that the factors of an algebraic expression are *multiplied* together.

Example 5.1
Factorise the expression $4x + 20$.

Solution
Both terms in the expression $4x + 20$ are examined to see if they have any factors in common. Clearly 20 can be factorised as $(4)(5)$ and so we can write

$$4x + 20 = 4x + (4)(5)$$

The factor 4 is common to both terms on the right; it is called a **common factor**. The common factor is placed at the front and outside the brackets to give

$$4x + 20 = 4(x + 5)$$

Note that the solution can and should be checked by removing the brackets again.

Example 5.2
Factorise $z^2 - 5z$.

Solution
Note that since $z^2 = z \times z$ we can write

$$z^2 - 5z = z(z) - 5z$$

so that there is a common factor of z. Hence

$$z^2 - 5z = z(z) - 5z$$
$$= z(z - 5)$$

Example 5.3
Factorise $6x - 9y$.

Solution
By observation we note that there is a common factor of 3. Thus

$$6x - 9y = 3(2x) - 3(3y)$$
$$= 3(2x - 3y)$$

Example 5.4
Identify the factor common to both $14z$ and $21w$. Hence factorise $14z + 21w$.

Solution
The factor common to both $14z$ and $21w$ is

7

We can then write

$$14z + 21w =$$ $7(2z + 3w)$

Example 5.5
Factorise $6x - 12xy$.

Solution
First identify any common factors. In this case there are two,

and 6 and x

Then we can write

$$6x - 12xy =$$ $6x(1 - 2y)$

If there is any doubt, check your answer by removing the brackets again.

Exercises

1   Factorise (a) $5x + 15y$, (b) $3x - 9y$,
    (c) $2x + 12y$, (d) $4x + 32z + 16y$,
    (e) $\frac{1}{2}x + \frac{1}{4}y$.
    In each case check your answer by removing
    the brackets again.

2   Factorise (a) $a^2 + 3ab$, (b) $xy + xyz$,
    (c) $9x^2 - 12x$.

3   Explain why a is a factor of $a + ab$ but b is
    not. Factorise $a + ab$.

4   Explain why x^2 is a factor of $4x^2 + 3yx^3 + 5yx^4$
    but y is not. Factorise $4x^2 + 3yx^3 + 5yx^4$.

Solutions to exercises

1   (a) $5(x + 3y)$ (b) $3(x - 3y)$ (c) $2(x + 6y)$
    (d) $4(x + 8z + 4y)$ (e) $\frac{1}{2}(x + \frac{1}{2}y)$

2   (a) $a(a + 3b)$ (b) $xy(1 + z)$ (c) $3x(3x - 4)$

3   $a(1 + b)$

4   $x^2(4 + 3yx + 5yx^2)$

5.3   Factorising quadratic expressions

Key point

An expression of the form $ax^2 + bx + c$ where a, b and c are numbers is called a
quadratic expression.

The numbers b or c may be zero but a must not be zero. The number a is called the
coefficient of x^2, b is the coefficient of x and c is called the **constant term**.
    Consider the product $(x + 1)(x + 2)$. Removing brackets yields $x^2 + 3x + 2$.
We see that the factors of $x^2 + 3x + 2$ are $(x + 1)$ and $(x + 2)$. However, if we
were given the quadratic expression first, how would we factorise it? The following
examples show how to do this but note that not all quadratic expressions can be
factorised.
    To enable us to factorise a quadratic expression in which the coefficient of x^2
equals 1, note the following expansion:

$$(x + m)(x + n) = x^2 + mx + nx + mn$$
$$= x^2 + (m + n)x + mn$$

So, given a quadratic expression we can think of the coefficient of x as $m + n$ and
the constant term as mn. Once the values of m and n have been found the factors can
be easily stated.

Example 5.6
Factorise $x^2 + 4x - 5$.

Solution
Writing $x^2 + 4x - 5 = (x + m)(x + n) = x^2 + (m + n)x + mn$ we seek numbers m and n so that $m + n = 4$ and $mn = -5$. By trial and error it is not difficult to find that $m = 5$ and $n = -1$. So we can write

$$x^2 + 4x - 5 = (x + 5)(x - 1)$$

The answer can be checked easily by removing brackets.

Example 5.7
Factorise $x^2 + 6x + 8$.

Solution
The coefficient of x^2 is 1. We can write

$$x^2 + 6x + 8 = (x + m)(x + n)$$
$$= x^2 + (m + n)x + mn$$

so that $m + n = 6$ and $mn = 8$. Try various possibilities for m and n until you find values that satisfy both of these equations.

$m = \rule{2cm}{0.4pt}$ $n = \rule{2cm}{0.4pt}$ 4 and 2, or 2 and 4

Finally factorise the quadratic:

$x^2 + 6x + 8 = \rule{3cm}{0.4pt}$ $(x + 4)(x + 2)$

When the coefficient of x^2 is not equal to 1 it may be possible to extract a numerical factor. For example, note that $3x^2 + 18x + 24$ can be written as $3(x^2 + 6x + 8)$ and then factorised as in Example 5.7. Sometimes no numerical factor can be found and a slightly different approach may be taken. We shall demonstrate a technique that can always be used to transform the given expression into one in which the coefficient of the squared variable equals 1.

Example 5.8
Factorise $2x^2 + 5x + 3$.

Solution
First note the coefficient of x^2, in this case 2. Multiply the whole expression by this number and rearrange as follows:

$$2(2x^2 + 5x + 3) = 2(2x^2) + 2(5x) + 2(3)$$
$$= (2x)^2 + 5(2x) + 6$$

If we now introduce a new variable such that $z = 2x$ we find that the coefficient of the squared term equals 1. Thus we can write

$$(2x)^2 + 5(2x) + 6 \text{ as } z^2 + 5z + 6$$

This can be factorised to give $(z + 3)(z + 2)$. Returning to the original variable by writing $z = 2x$ we find

$$2(2x^2 + 5x + 3) = (2x + 3)(2x + 2)$$

A factor of 2 can be extracted from the second bracket on the right so that

$$2(2x^2 + 5x + 3) = 2(2x + 3)(x + 1)$$

so that

$$2x^2 + 5x + 3 = (2x + 3)(x + 1)$$

As an alternative to the technique of Example 5.8, experience and practice can often help us to identify factors. For example, suppose we wish to factorise $3x^2 + 7x + 2$. We write

$$3x^2 + 7x + 2 = (\quad)(\quad)$$

In order to obtain the term $3x^2$ we can place terms $3x$ and x in the brackets to give

$$3x^2 + 7x + 2 = (3x + ?)(x + ?)$$

In order to obtain the constant 2, we consider the factors of 2. These are 1, 2 or -1, -2. By placing these factors in the brackets we can factorise the quadratic expression. Various possibilities exist: we could write $(3x + 2)(x + 1)$, $(3x + 1)(x + 2)$, $(3x - 2)(x - 1)$ or $(3x - 1)(x - 2)$, only one of which is correct. By removing brackets from each in turn we look for the factorisation that produces the correct middle term, $7x$. The correct factorisation is found to be

$$3x^2 + 7x + 2 = (3x + 1)(x + 2)$$

With practice you will be able to carry out this process quite easily.

Example 5.9
Factorise the quadratic expression $5x^2 - 7x - 6$.

Solution
Write

$$5x^2 - 7x - 6 = (\quad)(\quad)$$

To obtain the quadratic term $5x^2$, insert $5x$ and x in the brackets:

$$5x^2 - 7x - 6 = (5x + ?)(x + ?)$$

Now examine the factors of -6. These are

$$3, -2; -3, 2; -6, 1; 6, -1$$

Use these factors to find which pair, if any, gives rise to the middle term, $-7x$, and complete the factorisation.

$$5x^2 - 7x - 6 = \qquad\qquad (5x + 3)(x - 2)$$

Example 5.10
On occasions you will meet expressions of the form $x^2 - y^2$. Such an expression is known as the **difference of two squares**. Note that here we are finding the difference between two squared terms. It is easy to verify by removing brackets that this factorises as

$$x^2 - y^2 = (x + y)(x - y)$$

So, if you can learn to recognise such expressions it is an easy matter to factorise them. Factorise

(a) $x^2 - 36z^2$ (b) $25x^2 - 9z^2$ (c) $\alpha^2 - 1$

Solution

In each case we are required to find the difference of two squared terms.

(a) Note that $x^2 - 36z^2 = x^2 - (6z)^2$. This factorises as $(x + 6z)(x - 6z)$.

(b) Here $25x^2 - 9z^2 = (5x)^2 - (3z)^2$. This factorises as $(5x + 3z)(5x - 3z)$.

(c) $\alpha^2 - 1 = (\alpha + 1)(\alpha - 1)$.

Exercises

1 Factorise

(a) $x^2 + 8x + 7$ (b) $x^2 + 6x - 7$

(c) $x^2 + 7x + 10$ (d) $x^2 - 6x + 9$

(e) $x^2 + 5x + 6$

2 Factorise

(a) $2x^2 + 3x + 1$ (b) $2x^2 + 4x + 2$

(c) $3x^2 - 3x - 6$ (d) $5x^2 - 4x - 1$

(e) $16x^2 - 1$ (f) $-x^2 + 1$ (g) $-2x^2 + x + 3$

3 Factorise

(a) $x^2 + 9x + 14$ (b) $x^2 + 11x + 18$

(c) $x^2 + 7x - 18$ (d) $x^2 + 4x - 77$

(e) $x^2 + 2x$ (f) $3x^2 + x$

(g) $3x^2 + 4x + 1$ (h) $6x^2 + 5x + 1$

(i) $6x^2 + 31x + 35$ (j) $6x^2 + 7x - 5$

(k) $-3x^2 + 2x + 5$ (l) $x^2 - 3x + 2$

4 Factorise

(a) $z^2 - 144$ (b) $z^2 - \frac{1}{4}$ (c) $s^2 - \frac{1}{9}$

Solutions to exercises

1 (a) $(x + 7)(x + 1)$ (b) $(x + 7)(x - 1)$

(c) $(x + 2)(x + 5)$ (d) $(x - 3)(x - 3)$

(e) $(x + 3)(x + 2)$

2 (a) $(2x + 1)(x + 1)$ (b) $2(x + 1)^2$

(c) $3(x + 1)(x - 2)$ (d) $(5x + 1)(x - 1)$

(e) $(4x + 1)(4x - 1)$ (f) $(x + 1)(1 - x)$

(g) $(x + 1)(3 - 2x)$

3 (a) $(7 + x)(2 + x)$ (b) $(9 + x)(2 + x)$

(c) $(x + 9)(x - 2)$ (d) $(x + 11)(x - 7)$

(e) $(x + 2)x$ (f) $(3x + 1)x$

(g) $(3x + 1)(x + 1)$ (h) $(3x + 1)(2x + 1)$

(i) $(3x + 5)(2x + 7)$ (j) $(3x + 5)(2x - 1)$

(k) $(5 - 3x)(x + 1)$ (l) $(2 - x)(1 - x)$

4 (a) $(z + 12)(z - 12)$ (b) $(z + \frac{1}{2})(z - \frac{1}{2})$

(c) $(s + \frac{1}{3})(s - \frac{1}{3})$

End of block exercises

1 Factorise (a) $3z - 12$, (b) $-18 + 3t$,

(c) $\frac{1}{2}x + \frac{1}{4}$.

2 Factorise (a) $mn + 11m$, (b) $3pq - 2p$,

(c) $11pq - 3qr$, (d) $4zx + 20yz$.

3 Factorise (a) $x^2 + x$, (b) $3x^2 + 6x$,

(c) $9x^2 - 12x$.

4 Factorise (a) $x^2 + 8x - 9$, (b) $x^2 + 9x - 22$,

(c) $x^2 + 10x + 9$, (d) $x^2 + 7x + 12$,

(e) $x^2 - 7x + 12$.

5 Factorise (a) $14x^2 - 127x - 57$,
(b) $45x^2 + 44x + 7$, (c) $6x^2 + 19x - 11$.

6 Factorise $3x^3 + 17x^2 + 11x$.

7 Factorise $\alpha^2 - \beta^2$.

8 Factorise $144\alpha^2 - 49\beta^2$.

Solutions to exercises

1 (a) $3(z - 4)$ (b) $3(-6 + t)$ (c) $\frac{1}{2}(x + \frac{1}{2})$

2 (a) $m(n + 11)$ (b) $p(3q - 2)$
(c) $q(11p - 3r)$ (d) $4z(x + 5y)$

3 (a) $x(x + 1)$ (b) $3x(x + 2)$ (c) $3x(3x - 4)$

4 (a) $(x + 9)(x - 1)$ (b) $(x + 11)(x - 2)$
(c) $(x + 9)(x + 1)$ (d) $(x + 4)(x + 3)$
(e) $(x - 4)(x - 3)$

5 (a) $(2x - 19)(7x + 3)$
(b) $(5x + 1)(9x + 7)$ (c) $(3x + 11)(2x - 1)$

6 $x(3x^2 + 17x + 11)$

7 $(\alpha + \beta)(\alpha - \beta)$

8 $(12\alpha - 7\beta)(12\alpha + 7\beta)$

Arithmetic of algebraic fractions

Just as one whole number divided by another is a numerical fraction, one algebraic expression divided by another is called an **algebraic fraction**. Examples are

$$\frac{x}{y}, \quad \frac{3x + 2y}{x - y} \quad \text{and} \quad \frac{x^2 + 3x + 1}{x - 4}$$

The top line is called the **numerator** of the fraction and the bottom line is called the **denominator:**

Key point

$$\text{algebraic fraction} = \frac{\text{numerator}}{\text{denominator}}$$

In this block we explain how algebraic fractions can be simplified, added, subtracted, multiplied and divided.

6.2 Cancelling common factors

Consider the fraction $\frac{10}{35}$. To simplify this we can factorise the numerator and the denominator and then cancel any common factors. Common factors are those factors that occur in both the numerator and the denominator. For instance,

$$\frac{10}{35} = \frac{\cancel{5} \times 2}{7 \times \cancel{5}}$$
$$= \frac{2}{7}$$

Note that the common factor of 5 has been cancelled. It is important to remember that only *common factors* can be cancelled. The fractions $\frac{10}{35}$ and $\frac{2}{7}$ have identical values – they are equivalent fractions – but $\frac{2}{7}$ is in a simpler form than $\frac{10}{35}$. We apply the same process when simplifying algebraic fractions.

Example 6.1

Simplify, if possible, (a) $\dfrac{yx}{2x}$, (b) $\dfrac{x}{xy}$, (c) $\dfrac{x}{x+y}$.

Solution

(a) In the expression $\dfrac{yx}{2x}$, x is a factor common to both numerator and denominator. This common factor can be cancelled to give

$$\frac{y\cancel{x}}{2\cancel{x}} = \frac{y}{2}$$

(b) Note that $\dfrac{x}{xy}$ can be written $\dfrac{1x}{xy}$. The common factor of x can be cancelled to give

$$\frac{1\cancel{x}}{\cancel{x}y} = \frac{1}{y}$$

(c) In the expression $\dfrac{x}{x+y}$ notice that an x appears in both numerator and denominator. However, x is not a common factor. Recall that factors of an expression are multiplied together whereas in the denominator x is added to y. This expression cannot be simplified.

Example 6.2

Simplify, if possible, (a) $\dfrac{abc}{3ac}$, (b) $\dfrac{3ab}{b+a}$.

Solution

When simplifying remember that only common factors can be cancelled.

(a) $\dfrac{abc}{3ac} =$ $\qquad\qquad\qquad\qquad\qquad\qquad\qquad\qquad\qquad\qquad \dfrac{b}{3}$

(b) $\dfrac{3ab}{b+a} =$ $\qquad\qquad\qquad\qquad\qquad\qquad$ This cannot be simplified.

Example 6.3

Simplify (a) $\dfrac{21x^3}{14x}$, (b) $\dfrac{36x}{12x^3}$.

Solution

Factorising and cancelling common factors gives:

(a)

$$\frac{21x^3}{14x} = \frac{\cancel{7} \times 3 \times \cancel{x} \times x^2}{\cancel{7} \times 2 \times \cancel{x}}$$

$$= \frac{3x^2}{2}$$

(b)
$$\frac{36x}{12x^3} = \frac{12 \times 3 \times x}{12 \times x \times x^2}$$
$$= \frac{3}{x^2}$$

Example 6.4

Simplify $\dfrac{3x + 6}{6x + 12}$.

Solution

First we factorise the numerator and the denominator to see if there are any common factors.

$$\frac{3x + 6}{6x + 12} = \frac{3(x + 2)}{6(x + 2)}$$

The factor $x + 2$ is common and can be cancelled. There is also a common factor of 3. Thus

$$\frac{3x + 6}{6x + 12} = \frac{3(x + 2)}{6(x + 2)}$$
$$= \frac{1}{2}$$

Example 6.5

Simplify $\dfrac{12}{2x + 8}$.

Solution

Factorise the numerator and denominator, and cancel any common factors.

$$\frac{12}{2x + 8} =$$
$$\frac{6 \times 2}{2(x + 4)} = \frac{6}{x + 4}$$

Example 6.6

Show that $\dfrac{3}{x + 1}$ and $\dfrac{3(x + 4)}{x^2 + 5x + 4}$ are equivalent.

Solution

The denominator, $x^2 + 5x + 4$, can be factorised as $(x + 1)(x + 4)$ so that we can consider

$$\frac{3(x + 4)}{(x + 1)(x + 4)}$$

Note that $(x + 4)$ is a factor common to both the numerator and the denominator. In this form we see that $(x + 4)$ is a common factor which can be cancelled to leave $\dfrac{3}{x + 1}$. Thus $\dfrac{3}{x + 1}$ and $\dfrac{3(x + 4)}{x^2 + 5x + 4}$ are equivalent fractions.

Example 6.7

Show that $\dfrac{1}{x-1}$ and $\dfrac{x-1}{x^2-2x+1}$ are equivalent.

Solution

First factorise the denominator $x^2 - 2x + 1$:

$$(x-1)(x-1)$$

Identify the factor that is common to both numerator and denominator, and cancel this common factor.

$$\frac{x-1}{(x-1)(x-1)} = \qquad\qquad \frac{1}{x-1}$$

Hence the two given fractions are equivalent.

Example 6.8

Simplify $\dfrac{6(4-8x)(x-2)}{1-2x}$.

Solution

The factor $4 - 8x$ can be factorised to $4(1 - 2x)$. Thus

$$\frac{6(4-8x)(x-2)}{1-2x} = \frac{(6)(4)(1-2x)(x-2)}{(1-2x)} = 24(x-2)$$

Example 6.9

Simplify $\dfrac{x^2 + 2x - 15}{2x^2 - 5x - 3}$.

Solution

First factorise the numerator and the denominator.

$$\frac{x^2+2x-15}{2x^2-5x-3} = \qquad\qquad \frac{(x+5)(x-3)}{(2x+1)(x-3)}$$

Finally cancel any common factors to leave

$$\frac{x+5}{2x+1}$$

Exercises

1   Simplify, if possible, (a) $\frac{19}{38}$, (b) $\frac{14}{28}$, (c) $\frac{35}{40}$, (d) $\frac{7}{11}$, (e) $\frac{14}{56}$.

2   Simplify, if possible, (a) $\frac{14}{21}$, (b) $\frac{36}{96}$, (c) $\frac{13}{52}$, (d) $\frac{52}{13}$.

3   Simplify (a) $\dfrac{5z}{z}$, (b) $\dfrac{25z}{5z}$, (c) $\dfrac{5}{25z^2}$, (d) $\dfrac{5z}{25z^2}$.

4   Simplify (a) $\dfrac{4x}{3x}$, (b) $\dfrac{15x}{x^2}$, (c) $\dfrac{4s}{s^3}$, (d) $\dfrac{21x^4}{7x^3}$.

5 Simplify, if possible,

(a) $\dfrac{x+1}{2(x+1)}$ (b) $\dfrac{x+1}{2x+2}$

(c) $\dfrac{2(x+1)}{x+1}$ (d) $\dfrac{3x+3}{x+1}$

(e) $\dfrac{5x-15}{5}$ (f) $\dfrac{5x-15}{x-3}$

6 Simplify, if possible,

(a) $\dfrac{5x+15}{25x+5}$ (b) $\dfrac{5x+15}{25x}$

(c) $\dfrac{5x+15}{25}$ (d) $\dfrac{5x+15}{25x+1}$

7 Simplify

(a) $\dfrac{x^2+10x+9}{x^2+8x-9}$ (b) $\dfrac{x^2-9}{x^2+4x-21}$

(c) $\dfrac{2x^2-x-1}{2x^2+5x+2}$ (d) $\dfrac{3x^2-4x+1}{x^2-x}$

(e) $\dfrac{5z^2-20z}{2z-8}$

8 Simplify

(a) $\dfrac{6}{3x+9}$ (b) $\dfrac{2x}{4x^2+2x}$ (c) $\dfrac{3x^2}{15x^3+10x^2}$

9 Simplify

(a) $\dfrac{x^2-1}{x^2+5x+4}$ (b) $\dfrac{x^2+5x+6}{x^2+x-6}$

Solutions to exercises

1 (a) $\frac{1}{2}$ (b) $\frac{1}{2}$ (c) $\frac{7}{8}$ (d) $\frac{7}{11}$ (e) $\frac{1}{4}$

2 (a) $\frac{2}{3}$ (b) $\frac{3}{8}$ (c) $\frac{1}{4}$ (d) 4

3 (a) 5 (b) 5 (c) $\dfrac{1}{5z^2}$ (d) $\dfrac{1}{5z}$

4 (a) $\dfrac{4}{3}$ (b) $\dfrac{15}{x}$ (c) $\dfrac{4}{s^2}$ (d) $3x$

5 (a) $\dfrac{1}{2}$ (b) $\dfrac{1}{2}$ (c) 2 (d) 3 (e) $x-3$ (f) 5

6 (a) $\dfrac{x+3}{5x+1}$ (b) $\dfrac{x+3}{5x}$ (c) $\dfrac{x+3}{5}$ (d) $\dfrac{5x+15}{25x+1}$

7 (a) $\dfrac{x+1}{x-1}$ (b) $\dfrac{x+3}{x+7}$ (c) $\dfrac{x-1}{x+2}$ (d) $\dfrac{3x-1}{x}$

(e) $\dfrac{5z}{2}$

8 (a) $\dfrac{2}{x+3}$ (b) $\dfrac{1}{2x+1}$ (c) $\dfrac{3}{5(3x+2)}$

9 (a) $\dfrac{x-1}{x+4}$ (b) $\dfrac{x+2}{x-2}$

6.3 Multiplication and division of algebraic fractions

To multiply two fractions we multiply their numerators together and then multiply their denominators together. That is:

Key point

Multiplication

$$\frac{a}{b} \times \frac{c}{d} = \frac{ac}{bd}$$

Any factors common to both numerator and denominator can be cancelled. This cancellation can be performed before or after the multiplication.

Division is performed by inverting the second fraction and then multiplying.

Key point

Division

$$\frac{a}{b} \div \frac{c}{d} = \frac{a}{b} \times \frac{d}{c} = \frac{ad}{bc}$$

Example 6.10

Simplify (a) $\dfrac{2a}{c} \times \dfrac{4}{c}$, (b) $\dfrac{2a}{c} \times \dfrac{c}{4}$, (c) $\dfrac{2a}{c} \div \dfrac{4}{c}$.

Solution

(a)
$$\frac{2a}{c} \times \frac{4}{c} = \frac{8a}{c^2}$$

(b)
$$\frac{2a}{c} \times \frac{c}{4} = \frac{2ac}{4c}$$
$$= \frac{2a}{4} = \frac{a}{2}$$

(c) Division is performed by inverting the second fraction and then multiplying.

$$\frac{2a}{c} \div \frac{4}{c} = \frac{2a}{c} \times \frac{c}{4}$$
$$= \frac{a}{2} \text{ from the result in (b)}$$

Example 6.11

Simplify (a) $\dfrac{1}{5x} \times 3x$, (b) $\dfrac{1}{x} \times x$, (c) $\dfrac{1}{y} \times x$, (d) $y \times \dfrac{1}{x}$, (e) $\dfrac{y}{x} \times x$.

Solution
(a) Note that $3x = \dfrac{3x}{1}$. Then

$$\frac{1}{5x} \times 3x = \frac{1}{5x} \times \frac{3x}{1}$$
$$= \frac{3x}{5x}$$
$$= \frac{3}{5}$$

(b) x can be written as $\dfrac{x}{1}$. Then

$$\frac{1}{x} \times x = \frac{1}{x} \times \frac{x}{1}$$
$$= \frac{x}{x}$$
$$= 1$$

(c)
$$\frac{1}{y} \times x = \frac{1}{y} \times \frac{x}{1}$$
$$= \frac{x}{y}$$

(d)
$$y \times \frac{1}{x} = \frac{y}{1} \times \frac{1}{x}$$
$$= \frac{y}{x}$$

(e)
$$\frac{y}{x} \times x = \frac{y}{x} \times \frac{x}{1}$$
$$= \frac{yx}{x}$$
$$= y$$

Example 6.12
Simplify

$$\frac{\dfrac{2x}{y}}{\dfrac{3x}{2y}}$$

Solution
We can write the fraction as $\dfrac{2x}{y} \div \dfrac{3x}{2y}$. Inverting the second fraction and multiplying we find

$$\frac{2x}{y} \times \frac{2y}{3x} = \frac{4xy}{3xy}$$
$$= \frac{4}{3}$$

Example 6.13
Simplify $\dfrac{4x + 2}{x^2 + 4x + 3} \times \dfrac{x + 3}{7x + 5}$.

Solution
Factorising the numerator and denominator we find

$$\frac{4x + 2}{x^2 + 4x + 3} \times \frac{x + 3}{7x + 5} = \frac{2(2x + 1)}{(x + 1)(x + 3)} \times \frac{x + 3}{7x + 5}$$
$$= \frac{2(2x + 1)(x + 3)}{(x + 1)(x + 3)(7x + 5)}$$
$$= \frac{2(2x + 1)}{(x + 1)(7x + 5)}$$

It is usually better to factorise first and cancel any common factors before multiplying. Don't remove any brackets unnecessarily otherwise common factors will be difficult to spot.

Example 6.14

Simplify $\dfrac{15}{3x - 1} \div \dfrac{3}{2x + 1}$.

Solution

To divide we invert the second fraction and multiply:

$$\frac{15}{3x - 1} \div \frac{3}{2x + 1} = \frac{15}{3x - 1} \times \frac{2x + 1}{3}$$

$$= \frac{(5)(3)(2x + 1)}{3(3x - 1)}$$

$$= \frac{5(2x + 1)}{3x - 1}$$

Example 6.15 Control Engineering—Multiplying algebraic fractions

When control engineers analyse engineering systems they often represent different parts of the system using algebraic fractions in which the variable used is s. This will become apparent particularly once you have a knowledge of Laplace transforms (Chapter 22). It is often necessary to multiply such fractions together. So fluency with manipulation of algebraic fractions becomes important.

Find the product of the fractions $\dfrac{3}{s + 2}$ and $\dfrac{\omega}{s^2 + \omega^2}$.

Solution

$$\frac{3}{s + 2} \times \frac{\omega}{s^2 + \omega^2} = \frac{3\omega}{(s + 2)(s^2 + \omega^2)}$$

Exercises

1   Simplify (a) $\frac{5}{9} \times \frac{3}{2}$, (b) $\frac{14}{3} \times \frac{3}{9}$, (c) $\frac{6}{11} \times \frac{3}{4}$,
   (d) $\frac{4}{7} \times \frac{28}{3}$.

2   Simplify (a) $\frac{5}{9} \div \frac{3}{2}$, (b) $\frac{14}{3} \div \frac{3}{9}$, (c) $\frac{6}{11} \div \frac{3}{4}$,
   (d) $\frac{4}{7} \div \frac{28}{3}$.

3   Simplify

   (a) $2 \times \dfrac{x + y}{3}$ (b) $\frac{1}{3} \times 2(x + y)$
   (c) $\frac{2}{3} \times (x + y)$

4   Simplify
   (a) $3 \times \dfrac{x + 4}{7}$ (b) $\frac{1}{7} \times 3(x + 4)$

   (c) $\frac{3}{7} \times (x + 4)$ (d) $\dfrac{x}{y} \times \dfrac{x + 1}{y + 1}$

   (e) $\dfrac{1}{y} \times \dfrac{x^2 + x}{y + 1}$ (f) $\dfrac{\pi d^2}{4} \times \dfrac{Q}{\pi d^2}$ (g) $\dfrac{Q}{\pi d^2/4}$

5   Find (a) $\dfrac{6/7}{s + 3}$, (b) $\dfrac{3/4}{x - 1}$.

6   Find $\dfrac{3}{x + 2} \div \dfrac{x}{2x + 4}$.

7   Find $\dfrac{5}{2x + 1} \div \dfrac{x}{3x - 1}$.

Solutions to exercises

1 (a) $\frac{5}{6}$ (b) $\frac{14}{9}$ (c) $\frac{9}{22}$ (d) $\frac{16}{3}$

2 (a) $\frac{10}{27}$ (b) 14 (c) $\frac{8}{11}$ (d) $\frac{3}{49}$

3 (a) $\dfrac{2(x+y)}{3}$ (b) $\dfrac{2(x+y)}{3}$ (c) $\dfrac{2(x+y)}{3}$

4 (a) $\dfrac{3(x+4)}{7}$ (b) $\dfrac{3(x+4)}{7}$ (c) $\dfrac{3(x+4)}{7}$

  (d) $\dfrac{x(x+1)}{y(y+1)}$ (e) $\dfrac{x(x+1)}{y(y+1)}$ (f) $\dfrac{Q}{4}$ (g) $\dfrac{4Q}{\pi d^2}$

5 (a) $\dfrac{6}{7(s+3)}$ (b) $\dfrac{3}{4(x-1)}$

6 $\dfrac{6}{x}$

7 $\dfrac{5(3x-1)}{x(2x+1)}$

6.4 Addition and subtraction of algebraic fractions

To add two algebraic fractions the **lowest common denominator** must be found first. This is the simplest algebraic expression that has the given denominators as its factors. All fractions must be written with this lowest common denominator. Their sum is found by adding the numerators and dividing the result by the lowest common denominator.

Key point

Addition
To add two fractions:

1 Find the lowest common denominator.
2 Express each fraction with this denominator.
3 Add the numerators and divide the result by the lowest common denominator.

To subtract two fractions the process is similar. The fractions are written with the lowest common denominator. The difference is found by subtracting the numerators and dividing the result by the lowest common denominator.

Example 6.16
State the simplest expression that has $x+1$ and $x+4$ as its factors.

Solution
The simplest expression is $(x+1)(x+4)$. Note that both $x+1$ and $x+4$ are factors.

Example 6.17
State the simplest expression that has $x-1$ and $(x-1)^2$ as its factors.

Solution

The simplest expression is $(x - 1)^2$. Clearly $(x - 1)^2$ must be a factor of this expression. Because we can write $(x - 1)^2 = (x - 1)(x - 1)$ it follows that $x - 1$ is a factor too.

Example 6.18

Express as a single fraction

$$\frac{3}{x + 1} + \frac{2}{x + 4}$$

Solution

The simplest expression that has both denominators as its factors is $(x + 1)(x + 4)$. This is the lowest common denominator. Both fractions must be written using this denominator.

Note that $\dfrac{3}{x + 1}$ is equivalent to $\dfrac{3(x + 4)}{(x + 1)(x + 4)}$ and also $\dfrac{2}{x + 4}$ is equivalent to $\dfrac{2(x + 1)}{(x + 1)(x + 4)}$. Thus writing both fractions with the same denominator we have

$$\frac{3}{x + 1} + \frac{2}{x + 4} = \frac{3(x + 4)}{(x + 1)(x + 4)} + \frac{2(x + 1)}{(x + 1)(x + 4)}$$

The sum is found by adding the numerators and dividing the result by the lowest common denominator.

$$\frac{3(x + 4)}{(x + 1)(x + 4)} + \frac{2(x + 1)}{(x + 1)(x + 4)} = \frac{3(x + 4) + 2(x + 1)}{(x + 1)(x + 4)}$$

$$= \frac{5x + 14}{(x + 1)(x + 4)}$$

Example 6.19

Express $\dfrac{1}{x - 1} + \dfrac{5}{(x - 1)^2}$ as a single fraction.

Solution

The simplest expression having both denominators as its factors is $(x - 1)^2$. We write both fractions with this denominator. Note that $\dfrac{1}{x - 1}$ is equivalent to $\dfrac{x - 1}{(x - 1)^2}$.

$$\frac{1}{x - 1} + \frac{5}{(x - 1)^2} = \frac{x - 1}{(x - 1)^2} + \frac{5}{(x - 1)^2}$$

$$= \frac{x - 1 + 5}{(x - 1)^2} = \frac{x + 4}{(x - 1)^2}$$

Example 6.20

Find $\dfrac{3}{x + 7} + \dfrac{5}{x + 2}$.

Solution

First find the lowest common denominator:

$(x + 7)(x + 2)$

Both fractions are rewritten using this lowest common denominator:

$$\frac{3}{x + 7} + \frac{5}{x + 2} =$$

$$\frac{3(x + 2)}{(x + 7)(x + 2)} + \frac{5(x + 7)}{(x + 7)(x + 2)}$$

Add the numerators and simplify the result to find the sum of the given fractions:

$$\frac{3}{x + 7} + \frac{5}{x + 2} =$$

$$\frac{8x + 41}{(x + 7)(x + 2)}$$

Example 6.21

Find $\dfrac{5x}{7} - \dfrac{3x - 4}{2}$.

Solution

In this example both denominators are simply numbers. The lowest common denominator is 14, and both fractions are rewritten with this denominator. Thus

$$\frac{10x}{14} - \frac{7(3x - 4)}{14} = \frac{10x - 7(3x - 4)}{14}$$

$$= \frac{28 - 11x}{14}$$

Example 6.22

Find $\dfrac{1}{x} + \dfrac{1}{y}$.

Solution

The simplest expression that has x and y as its factors is xy. This is the lowest common denominator. Both fractions are written using this denominator. Noting that $\dfrac{1}{x} = \dfrac{y}{xy}$ and that $\dfrac{1}{y} = \dfrac{x}{xy}$ we find

$$\frac{1}{x} + \frac{1}{y} = \frac{y}{xy} + \frac{x}{xy}$$

$$= \frac{y + x}{xy}$$

No cancellation is possible because neither x nor y is a factor of the numerator.

Exercises

1 Find (a) $\dfrac{x}{4} + \dfrac{x}{7}$, (b) $\dfrac{2x}{5} + \dfrac{x}{9}$,

(c) $\dfrac{2x}{3} - \dfrac{3x}{4}$, (d) $\dfrac{x}{x+1} - \dfrac{2}{x+2}$,

(e) $\dfrac{x+1}{x} + \dfrac{3}{x+2}$, (f) $\dfrac{2x+1}{3} - \dfrac{x}{2}$,

(g) $\dfrac{x+3}{2x+1} - \dfrac{x}{3}$, (h) $\dfrac{x}{4} - \dfrac{x}{5}$.

2 Find (a) $\dfrac{1}{x+2} + \dfrac{2}{x+3}$,

(b) $\dfrac{2}{x+3} + \dfrac{5}{x+1}$, (c) $\dfrac{2}{2x+1} - \dfrac{3}{3x+2}$,

(d) $\dfrac{x+1}{x+3} + \dfrac{x+4}{x+2}$, (e) $\dfrac{x-1}{x-3} + \dfrac{x-1}{(x-3)^2}$.

3 Find $\dfrac{5}{2x+3} + \dfrac{4}{(2x+3)^2}$.

4 Find $\dfrac{1}{7}s + \dfrac{11}{21}$.

5 Express $\dfrac{A}{2x+3} + \dfrac{B}{x+1}$ as a single fraction.

6 Express $\dfrac{A}{2x+5} + \dfrac{B}{(x-1)} + \dfrac{C}{(x-1)^2}$ as a single fraction.

7 Express $\dfrac{A}{x+1} + \dfrac{B}{(x+1)^2}$ as a single fraction.

8 Express $\dfrac{Ax+B}{x^2+x+10} + \dfrac{C}{x-1}$ as a single fraction.

9 Express $Ax + B + \dfrac{C}{x+1}$ as a single fraction.

10 Show that

$$\dfrac{x_1}{\dfrac{1}{x_3} - \dfrac{1}{x_2}}$$

is equal to $\dfrac{x_1 x_2 x_3}{x_2 - x_3}$.

11 Find (a) $\dfrac{3x}{4} - \dfrac{x}{5} + \dfrac{x}{3}$, (b) $\dfrac{3x}{4} - \left(\dfrac{x}{5} + \dfrac{x}{3}\right)$.

Solutions to exercises

1 (a) $\dfrac{11x}{28}$ (b) $\dfrac{23x}{45}$ (c) $-\dfrac{x}{12}$

(d) $\dfrac{x^2 - 2}{(x+1)(x+2)}$ (e) $\dfrac{x^2 + 6x + 2}{x(x+2)}$

(f) $\dfrac{x+2}{6}$ (g) $\dfrac{9 + 2x - 2x^2}{3(2x+1)}$ (h) $\dfrac{x}{20}$

2 (a) $\dfrac{3x+7}{(x+2)(x+3)}$ (b) $\dfrac{7x+17}{(x+3)(x+1)}$

(c) $\dfrac{1}{(2x+1)(3x+2)}$ (d) $\dfrac{2x^2 + 10x + 14}{(x+3)(x+2)}$

(e) $\dfrac{x^2 - 3x + 2}{(x-3)^2}$

3 $\dfrac{10x + 19}{(2x+3)^2}$

4 $\dfrac{3s + 11}{21}$

5 $\dfrac{A(x+1) + B(2x+3)}{(2x+3)(x+1)}$

6

$\dfrac{A(x-1)^2 + B(x-1)(2x+5) + C(2x+5)}{(2x+5)(x-1)^2}$

7 $\dfrac{A(x+1) + B}{(x+1)^2}$

8 $\dfrac{(Ax+B)(x-1) + C(x^2 + x + 10)}{(x-1)(x^2 + x + 10)}$

9 $\dfrac{(Ax+B)(x+1) + C}{x+1}$

11 (a) $\dfrac{53x}{60}$ (b) $\dfrac{13x}{60}$

End of block exercises

1   Simplify $\dfrac{5x}{25x + 10y}$.

2   Simplify $\dfrac{8r^3}{4\pi r^2}$.

3   Simplify $\dfrac{x + 2}{x^2 + 3x + 2}$.

4   Explain why no cancellation is possible in the expression $\dfrac{a + 2b}{a - 2b}$.

5   Find $\dfrac{3}{11} \times \dfrac{x}{y}$.

6   Find $3 \times \dfrac{x}{11(x + y)}$.

7   Simplify $\dfrac{x + 2}{x^2 + 9x + 20} \times \dfrac{x + 5}{x + 2}$.

8   Find $\dfrac{5}{7y} + \dfrac{2x}{3}$.

9   Express as a single fraction
$$\dfrac{3}{x - 4} - \dfrac{2}{(x - 4)^2}.$$

10   Express as a single fraction
$$2x - 1 + \dfrac{4}{x} + \dfrac{3}{2x + 1}.$$

11   (a) Express $\dfrac{1}{u} + \dfrac{1}{v}$ as a single fraction.
(b) Hence find the reciprocal of $\dfrac{1}{u} + \dfrac{1}{v}$.

12   Express $\dfrac{1}{s} + \dfrac{1}{s^2}$ as a single fraction.

13   Express $-\dfrac{6}{s + 3} - \dfrac{4}{s + 2} + \dfrac{3}{s + 1} + 2$ as a single fraction.

Solutions to exercises

1   $\dfrac{x}{5x + 2y}$

2   $\dfrac{2r}{\pi}$

3   $\dfrac{1}{x + 1}$

5   $\dfrac{3x}{11y}$

6   $\dfrac{3x}{11(x + y)}$

7   $\dfrac{1}{x + 4}$

8   $\dfrac{15 + 14xy}{21y}$

9   $\dfrac{3x - 14}{(x - 4)^2}$

10   $\dfrac{4x^3 + 10x + 4}{2x^2 + x}$

11   (a) $\dfrac{v + u}{uv}$ (b) $\dfrac{uv}{v + u}$

12   $\dfrac{s + 1}{s^2}$

13   $\dfrac{2s^3 + 5s^2 + 3s + 6}{s^3 + 6s^2 + 11s + 6}$

Formulae and transposition

7.1 Introduction

Formulae are used frequently in almost all aspects of engineering in order to relate a physical quantity to one or more others. Many well-known physical laws are described using formulae. For example, you may have already seen Ohm's law, $v = iR$, or Newton's second law of motion, $F = ma$. In this block we describe the process of evaluating a formula, explain what is meant by the subject of a formula, and show how a formula is rearranged or transposed. These are basic skills required in all aspects of engineering.

7.2 Using formulae and substitution

In the study of engineering, physical quantities can be related to each other using a formula. The formula will contain variables and constants that represent the physical quantities. To evaluate a formula we must **substitute** numbers in place of the variables.

Example 7.1 Electrical Engineering – Ohm's law

Figure 7.1
v is given by the
formula $v = iR$.

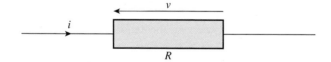

Ohm's law provides a formula for relating the voltage, v, across a resistor to the current through it, i, and the resistance value, R (Figure 7.1). It states

$$v = iR$$

We can use this formula to calculate v if we know values for i and R. For example, if $i = 13$ A, and $R = 5\ \Omega$, then

$$
\begin{aligned}
v &= iR \\
&= (13)(5) \\
&= 65
\end{aligned}
$$

The voltage is 65 V.

Note from this example that it is important to pay attention to the units of any physical quantities involved. Unless a consistent set of units is used a formula is not valid. The most common set of units is the SI system given in Appendix 1 at the end of the book. A knowledge of prefix conventions is also useful, and details are also given in Appendix 1.

Example 7.2 Mechanical Engineering – The kinetic energy of an object

In Figure 7.2 the kinetic energy, E, of an object of mass M moving with speed v can be calculated from the formula $E = \frac{1}{2}Mv^2$.

Figure 7.2
Kinetic energy, $E = \frac{1}{2}Mv^2$.

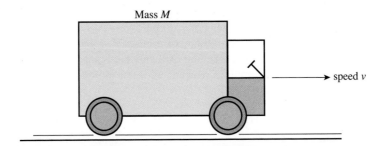

Mass M

speed v

Calculate the kinetic energy of an object of mass 5000 kg moving with a speed of 20 m s^{-1}.

Solution

In this example $M = 5000$ and $v = 20$. Substituting these into the formula we find

$$E = \frac{1}{2}Mv^2$$
$$= \frac{1}{2}(5000)(20^2)$$
$$= 1000000$$

In the SI system the unit of energy is the joule. Hence the kinetic energy of the object is 1000000 joules.

Example 7.3 The area of a circle

The area, A, of the circle of radius r shown in Figure 7.3 can be calculated from the formula $A = \pi r^2$. Equivalently, if we know the diameter of the circle, d, we can use the formula $A = \frac{1}{4}\pi d^2$. Your calculator will be preprogrammed with the value of π. Find the area of a circle having diameter 0.1 m.

Figure 7.3
The area, A, is given by $A = \pi r^2$ or $A = \frac{1}{4}\pi d^2$.

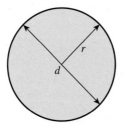

Solution

The area of the circle is

$$A = \qquad\qquad\qquad \frac{\pi(0.1)^2}{4} = 0.0079 \text{ m}^2$$

Example 7.4 The volume of a cylinder

The volume, V, of the cylinder in Figure 7.4 is equal to its cross-sectional area, A, times its length, h.

Figure 7.4
The volume, V, is
given by $V = Ah$.

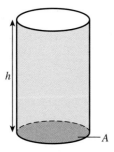

Find the volume of a cylinder having diameter 0.1 m and length 0.3 m.

Solution

We can use the result of Example 7.3 to obtain the cross-sectional area. Then

$$\begin{aligned} V &= Ah \\ &= \frac{\pi(0.1)^2}{4} \times 0.3 \\ &= 0.0024 \end{aligned}$$

The volume is 0.0024 m^3.

Example 7.5 Mechanical Engineering – Positive displacement pump

The pump outlined in Figure 7.5 is known as a **positive displacement pump**. As the piston is being withdrawn, fluid is sucked into the pump chamber. As the piston moves forwards, fluid is ejected through the delivery duct. Valves are used to ensure the correct directions of these flows. The cross-section of the pump chamber is circular with diameter d. As the piston moves through a distance L, known as its **stroke**, the volume of fluid that is pumped through the delivery duct is

$$\text{cross-sectional area of chamber} \times \text{stroke} = \frac{\pi d^2}{4} \times L$$

This is the volume of fluid ejected in one revolution of the pump. If the pump rotates at n revolutions per second, the volume of fluid ejected per second, known as the **theoretical pump delivery**, Q_T, can be found from the formula

$$Q_T = \frac{\pi d^2 L n}{4}$$

Figure 7.5
A positive
displacement
pump.

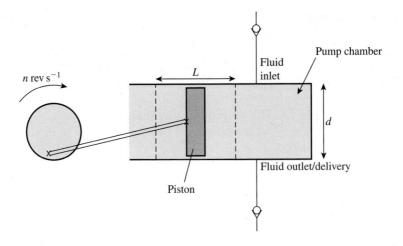

The actual delivery will be less than this owing to internal leakage. Calculate the theoretical pump delivery of such a pump when $n = 0.5$ revolutions per second, $L = 0.3$ m, $d = 0.1$ m.

Solution
Substituting the given values we find

$$Q_T = \frac{\pi(0.1)^2(0.3)(0.5)}{4}$$
$$= 0.0012$$

The theoretical pump delivery is therefore $0.0012 \text{ m}^3 \text{ s}^{-1}$.

Example 7.6 Reliability Engineering – Reliability in communication networks

In Example 4.18 of this chapter we met the formula

$$R = R_1 + R_2 - R_1 R_2$$

for calculating the reliability of two components in parallel when the individual components have reliabilities R_1 and R_2. Suppose each component has the same reliability, $R_1 = R_2 = 0.8$. Find the reliability of the whole system.

Solution
Substituting $R_1 = R_2 = 0.8$ in the given formula:

$$R = 0.8 + 0.8 - (0.8)(0.8) = 0.96$$

So the system as a whole has a 96% chance of functioning normally, whereas the individual components had only an 80% chance of functioning normally. We see the great improvement in reliability achieved by including a second component in parallel.

Example 7.7 Materials Engineering – Measuring the hardness of a material

One of the ways that materials engineers measure the hardness of a material is by an indentation test. Typically a tungsten carbide ball of diameter D is pressed for several seconds onto the material through the action of a known force F.

Figure 7.6
Diameter of indentation is used to measure hardness.

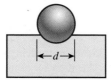

The diameter of the resulting indentation is measured after the force is removed. There are several test standards but one of the most commonly used is called a Brinell test and the material's hardness (written HBW) is calculated from the formula

$$\text{Brinell hardness} = HBW = 0.102\,\frac{2F}{\pi D(D - \sqrt{D^2 - d^2})}$$

where D is the diameter of the tungsten carbide ball (in mm), d is the diameter of the indentation (in mm) at the surface of the material, and F is the applied force (in newtons). This formula can be expressed in the equivalent form

$$\text{Brinell hardness} = HBW = 0.102\,\frac{F}{\dfrac{\pi D(D - \sqrt{D^2 - d^2})}{2}}$$

It can be shown that the quantity $\dfrac{\pi D(D - \sqrt{D^2 - d^2})}{2}$ is the curved surface area of the spherical indentation. Consequently, the Brinell number is proportional to the ratio of the applied force to the curved surface area produced by the indenter. For a given force, a small Brinell number corresponds to a large surface area and hence a relatively soft material. Conversely, a large Brinell number corresponds to a small surface area and hence a material which is relatively hard.

In an indentation test a tungsten carbide ball of diameter 10 mm is impressed upon a sheet of steel with a force of 3000g newtons. (Here g is the constant acceleration due to gravity, 9.81 m s^{-2}.) When the force is removed the diameter of the indentation is measured as 4 mm. Calculate the Brinell hardness.

Solution

We substitute the given values into the formula:

$$HBW = 0.102\,\frac{2F}{\pi D(D - \sqrt{D^2 - d^2})}$$

$$= 0.102\,\frac{2 \times 3000 \times 9.81}{\pi \times 10(10 - \sqrt{10^2 - 4^2})}$$

$$= 229 \ (3 \text{ s.f.})$$

Compare this value with a typical hardness value of a much softer metal such as brass that has a Brinell number of around 60.

Exercises

1   The formula for the volume of a cylinder is $V = \pi r^2 h$. Find V when $r = 5$ cm and $h = 15$ cm.

2   If $R = 5p^2$, find R when (a) $p = 10$, (b) $p = 16$.

3   For the following formulae, find y at the given values of x:
(a) $y = 3x + 2$, $x = -1$, $x = 0$, $x = 1$
(b) $y = -4x + 7$, $x = -2$, $x = 0$, $x = 1$

4   If $P = \dfrac{3}{QR}$ find P if $Q = 15$ and $R = 0.3$.

5   If $y = \sqrt{x/z}$ find y if $x = 13.2$ and $z = 15.6$.

6   Evaluate $M = \dfrac{\pi}{2r + s}$ when $r = 23.7$ and $s = -0.2$.

7   To convert a length measured in metres to one measured in centimetres, the length in metres is multiplied by 100. Convert the following lengths to cm: (a) 5 m, (b) 0.5 m, (c) 56.2 m.

8   To convert an area measured in m² to one measured in cm², the area in m² is multiplied by 10^4. Convert the following areas to cm²: (a) 5 m², (b) 0.33 m², (c) 6.2 m².

9   To convert a volume measured in m³ to one measured in cm³, the volume in m³ is multiplied by 10^6. Convert the following volumes to cm³: (a) 15 m³, (b) 0.25 m³, (c) 8.2 m³.

10   If $\eta = \dfrac{4Q_P}{\pi d^2 Ln}$ evaluate η when $Q_P = 0.0003$, $d = 0.05$, $L = 0.1$ and $n = 2$.

11   For the following formulae, find y at the given values of x:
(a) $y = 2 - x$, $x = -3$, $x = -1$, $x = 1$, $x = 2$
(b) $y = x^2$, $x = -2$, $x = -1$, $x = 0$, $x = 1$, $x = 2$

12   **Mechanical Engineering – Moment of inertia.** The moment of inertia of an object is a measure of its resistance to rotation. It depends upon both the mass of the object and the distribution of mass about the axis of rotation. It can be shown that the moment of inertia, J, of a solid disc rotating about an axis through its centre and perpendicular to the plane of the disc is given by the formula

$$J = \frac{1}{2}Ma^2$$

where M is the mass of the disc and a is its radius. Find the moment of inertia of a disc of mass 12 kg and diameter 10 m. The SI unit of moment of inertia is kg m².

13   **Reliability Engineering.** Refer to Example 7.6. Calculate the percentage improvement achieved in the reliability of the whole system if the individual component reliabilities can be increased to 0.9.

14   **Materials Engineering – Brinell hardness.** The formula $H = 0.102\dfrac{2F}{\pi D(D - \sqrt{D^2 - d^2})}$ has been discussed above in connection with the calculation of the hardness of a material. (For simplicity, we refer to the Brinell hardness here as H.) Transpose this formula to make d^2 the subject.

Solutions to exercises

1   1178.1 cm³

2   (a) 500 (b) 1280

3   (a) $-1, 2, 5$ (b) 15, 7, 3

4   $P = 0.667$

5   $y = 0.920$

6   $M = 0.067$

7   (a) 500 cm (b) 50 cm (c) 5620 cm

8   (a) 50000 cm^2 (b) 3300 cm^2 (c) 62000 cm^2

9   (a) 15000000 cm^3 (b) 250000 cm^3
     (c) 8200000 cm^3

10 $\eta = 0.764$

11 (a) 5, 3, 1, 0 (b) 4, 1, 0, 1, 4

12 150 kg m^2

13 The system reliability increases from 0.96 to
     0.99. The percentage improvement is 3.125%.

14 $d^2 = \dfrac{0.408F}{\pi H} - \left(\dfrac{0.204F}{\pi DH}\right)^2.$

7.3   Rearranging a formula

In the formula for the area of a circle, $A = \pi r^2$, we say that A is the **subject** of the formula. A variable is the subject of the formula if it appears by itself on one side of the formula, usually the left-hand side, and nowhere else in the formula. If we are asked to **transpose** the formula for r, or **solve** for r, then we have to make r the subject of the formula. When transposing a formula there are five rules that must be adhered to. You may

Key point

1   Add the same quantity to both sides of the formula.
2   Subtract the same quantity from both sides of the formula.
3   Multiply both sides of the formula by the same quantity.
4   Divide both sides of the formula by the same quantity.
5   Take 'functions' of both sides of the formula: for example, square both sides, square root both sides, find the reciprocal of both sides.

In summary, and loosely speaking, we must do precisely the same to both sides.

Example 7.8
Transpose the formula $p = 5t - 17$ for t.

Solution
We must try to obtain t on its own on the left-hand side. We do this in stages by using one or more of the five rules. For example, by adding 17 to both sides of $p = 5t - 17$ we find

$$p + 17 = 5t - 17 + 17$$

so that

$$p + 17 = 5t$$

By dividing both sides by 5 we obtain t on its own:

$$\frac{p + 17}{5} = t$$

so that

$$t = \frac{p + 17}{5}$$

Example 7.9

Find the result of squaring both sides of the formula $z = a + b$.

Solution

To square the right-hand side we must find $(a + b)^2$. We find

$$z^2 = (a + b)^2$$
$$= a^2 + 2ab + b^2$$

Note that if $z = a + b$, then $z^2 \neq a^2 + b^2$.

Example 7.10

Transpose the formula $\sqrt{2q} = p$ for q.

Solution

First of all square both sides to remove the square root around $2q$. Note that $(\sqrt{2q})^2 = 2q$. This gives

$$2q = p^2$$

Dividing both sides by 2 gives $q = \dfrac{p^2}{2}$, which may be written as $q = \dfrac{1}{2}p^2$.

Example 7.11

Transpose the formula $v = \sqrt{t^2 + w}$ for w.

Solution

We must try to obtain w on its own on the left-hand side. We do this in several stages. First of all square both sides to remove the square root around $t^2 + w$. This gives

$$v^2 = t^2 + w$$

Then subtract t^2 from both sides to obtain an expression for w:

$$v^2 - t^2 = w$$

Finally, we can write down the formula for w:

$$w = v^2 - t^2$$

Example 7.12

Transpose $x = \dfrac{1}{y}$ for y.

Solution

We must try to obtain an expression for y. In the given formula y appears in the form of a fraction. Multiplying both sides by y has the effect of removing this fraction:

multiply both sides of $\qquad x = \dfrac{1}{y} \qquad$ by y to get

$$yx = y \times \dfrac{1}{y}$$

so that $\qquad yx = 1$

Dividing both sides by x leaves y on its own, $y = \dfrac{1}{x}$.

Example 7.13
Make R the subject of the formula

$$\frac{2}{R} = \frac{3}{x + y}$$

Solution
In the given formula R appears in a fraction. Multiplying both sides by $(x + y)$ we find

$$\frac{2(x + y)}{R} = 3$$

Multiplying both sides by R we find

$$2(x + y) = 3R$$

Finally, dividing both sides by 3 gives

$$R = \frac{2(x + y)}{3}$$

Example 7.14 Electrical Engineering – Resistors in parallel
The net resistance, R, of two resistors, R_1 and R_2, connected in parallel, can be obtained from the formula

$$\frac{1}{R} = \frac{1}{R_1} + \frac{1}{R_2}$$

Make R the subject of the formula.

Solution
The two terms on the right can be added to give

$$\frac{1}{R_1} + \frac{1}{R_2} = \frac{R_2 + R_1}{R_1 R_2}$$

The given formula becomes

$$\frac{1}{R} = \frac{R_2 + R_1}{R_1 R_2}$$

Multiplying both sides by $R_1 R_2$ we find

$$\frac{R_1 R_2}{R} = R_2 + R_1$$

and multiplying by R gives

$$R_1 R_2 = R(R_2 + R_1)$$

Finally dividing by $(R_2 + R_1)$ we find

$$R = \frac{R_1 R_2}{R_2 + R_1}$$

Example 7.15 Mechanical Engineering – Positive displacement pump

Recall the positive displacement pump described in Example 7.5. The actual delivery, Q_P, of such a pump is less than the theoretical value, Q_T, owing to leakage. A relevant quantity is the **volumetric efficiency**, η, defined as

$$\eta = \frac{\text{actual delivery}}{\text{theoretical delivery}}$$

$$= \frac{Q_P}{Q_T}$$

If the actual and theoretical deliveries are equal the volumetric efficiency will be 1. A designer will try to ensure that the value of η is as close to 1 as is possible in practice. The theoretical delivery was given in Example 7.5 as $\frac{1}{4}\pi d^2 Ln$. Hence

$$\eta = \frac{Q_P}{\frac{1}{4}\pi d^2 Ln}$$

$$= \frac{4Q_P}{\pi d^2 Ln}$$

Transpose this formula to make the actual delivery, Q_P, the subject of the formula.

Solution

From

$$\eta = \frac{4Q_P}{\pi d^2 Ln}$$

multiplying both sides of the formula by $\pi d^2 Ln$ we find

$$\pi d^2 Ln\eta = 4Q_P$$

Dividing both sides by 4 we find

$$Q_P = \frac{\pi d^2 Ln\eta}{4}$$

In this form it is possible to calculate the actual delivery of the pump from knowledge of its volumetric efficiency, the physical dimensions of the pump and the pump speed.

Example 7.16 Mechanical Engineering – The coefficient of restitution

Consider Figure 7.7, which shows two moving masses before and after they have collided. The separation speed $w_2 - w_1$ is proportional to the approach speed $v_1 - v_2$, and the constant of proportionality is called the **coefficient of restitution**, e. That is,

$$w_2 - w_1 = e(v_1 - v_2)$$

Transpose this formula to make v_1 the subject.

Figure 7.7
Two masses before
and after collision.

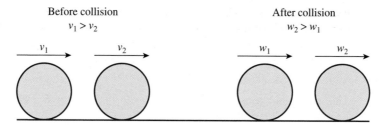

Before collision
$v_1 > v_2$

After collision
$w_2 > w_1$

Solution

We must try to obtain v_1 on its own. Divide both sides by e to obtain an expression for $v_1 - v_2$.

$$v_1 - v_2 = \qquad\qquad\qquad \frac{w_2 - w_1}{e}$$

Finally add v_2 to both sides to leave v_1 on the left.

$$v_1 = v_2 + \frac{w_2 - w_1}{e}$$

Example 7.17 Electrical Engineering – Equivalent resistance of a ladder of resistors

Figure 7.8(a) shows two resistors in series. Figure 7.8(b) shows two resistors in parallel.

Figure 7.8
(a) Two resistors in
series;
(b) two resistors in
parallel.

When two resistors, R_1 and R_2, are placed in series in a circuit the equivalent resistance, R, is $R_1 + R_2$. When the two are placed in parallel the equivalent resistance, R, is given by

$$\frac{1}{R} = \frac{1}{R_1} + \frac{1}{R_2}$$

So, using Example 7.14 we can write

$$R = \frac{R_1 R_2}{R_1 + R_2}$$

Consider the circuit shown in Figure 7.9. Find the equivalent resistance of the circuit.

Figure 7.9

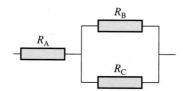

Solution

The equivalent resistance of the parallel part of the circuit is

$$\frac{R_B R_C}{R_C + R_B}$$

This equivalent resistance is in series with the resistor R_A. Hence the equivalent resistance of the whole circuit is

$$R = R_A + \frac{R_B R_C}{R_C + R_B}$$

This can be written as follows:

$$R = R_A + \frac{R_B R_C}{R_C + R_B}$$

$$= \frac{R_A(R_C + R_B) + R_B R_C}{R_C + R_B}$$

$$= \frac{R_A R_C + R_A R_B + R_B R_C}{R_C + R_B}$$

Example 7.18 Mechanical Engineering – Heat flow in an insulated pipe

With rising energy costs and increasing awareness of global warming, much attention is given to insulation and heat flow in domestic and industrial appliances. Figure 7.10 depicts a situation in which heat can flow along a metal rod. The rod is insulated along its length so that no heat can enter or leave the rod through the insulated parts. The two ends of the rod are maintained at temperatures T_2 and T_1, respectively, where $T_2 > T_1$ so that heat flows from left to right as shown.

Figure 7.10
Heat flows along the rod from the warmer to the cooler end.

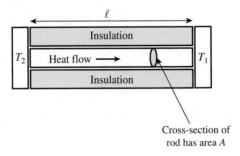

Cross-section of
rod has area A

If the cross-sectional area of the rod is A and its length is ℓ the amount of heat energy crossing any cross-section parallel to the heated ends, per unit time, is given by

$$Q = \lambda A\left(\frac{T_2 - T_1}{\ell}\right)$$

where λ is a constant called the thermal conductivity of the metal. Observe from this formula that if λ is large, then the rate of heat flow, Q, will be relatively large.

Similarly if the temperature gradient between the two ends $\left(\dfrac{T_2 - T_1}{\ell}\right)$ is large, then so too will be the rate of heat flow. Transpose this formula to make T_2 the subject.

Solution
From

$$Q = \lambda A \left(\frac{T_2 - T_1}{\ell}\right)$$

multiplying both sides by ℓ and dividing both sides by λA gives

$$\frac{Q\ell}{\lambda A} = T_2 - T_1$$

Adding T_1 to both sides gives

$$T_2 = \frac{Q\ell}{\lambda A} + T_1$$

This formula enables an engineer to calculate the temperature, for a given heat flow rate, at which the left-hand end needs to be maintained in order to achieve a temperature of T_1 on the right-hand end.

Exercises

1 Transpose the given formulae to make the given variable the subject:
(a) $y = 3x - 7$, for x
(b) $8y + 3x = 4$, for x
(c) $8x + 3y = 4$, for y
(d) $13 - 2x - 7y = 0$, for x

2 Transpose the formula $PV = RT$ for (a) V, (b) P, (c) R, (d) T.

3 Transpose $v = \sqrt{x + 2y}$, (a) for x, (b) for y.

4 Transpose $8u + 4v - 3w = 17$ for (a) u, (b) v and (c) w.

5 **Mechanical Engineering – Coefficient of restitution.** When a ball is dropped from rest onto a horizontal surface it will bounce before eventually coming to rest after a time T where

$$T = \frac{2v}{g}\left(\frac{1}{1 - e}\right)$$

where v is the speed immediately after the first impact, and g is a constant called the acceleration due to gravity. Transpose this formula to make e, the coefficient of restitution, the subject.

6 Transpose

$$q = A_1 \sqrt{\frac{2gh}{(A_1/A_2)^2 - 1}}$$

for A_2.

7 Make x the subject of
(a) $y = \dfrac{r + x}{1 - rx}$

(b) $y = \sqrt{\dfrac{x - 1}{x + 1}}$

Solutions to exercises

1 (a) $x = \dfrac{y + 7}{3}$ (b) $x = \dfrac{4 - 8y}{3}$

(c) $y = \dfrac{4 - 8x}{3}$ (d) $x = \dfrac{13 - 7y}{2}$

2 (a) $V = \dfrac{RT}{P}$ (b) $P = \dfrac{RT}{V}$

(c) $R = \dfrac{PV}{T}$ (d) $T = \dfrac{PV}{R}$

3 (a) $x = v^2 - 2y$ (b) $y = \dfrac{v^2 - x}{2}$

4 (a) $u = \dfrac{17 - 4v + 3w}{8}$

(b) $v = \dfrac{17 - 8u + 3w}{4}$

(c) $w = \dfrac{8u + 4v - 17}{3}$

5 $e = 1 - \dfrac{2v}{gT}$

6 $A_2 = \sqrt{\dfrac{A_1^2 q^2}{2A_1^2 gh + q^2}}$

7 (a) $x = \dfrac{y - r}{1 + yr}$

(b) $x = \dfrac{1 + y^2}{1 - y^2}$

End of block exercises

1 Evaluate $V = r_1^2 + r_2^2$ when $r_1 = 0.1$ and $r_2 = 0.05$.

2 If $A + B + C = 180$, find C when $A = 30$ and $B = 60$.

3 For the following formulae, find y at the given values of x:

(a) $y = \dfrac{1}{2}x + \dfrac{1}{3}$, $x = -2$, $x = 0$, $x = 1$

(b) $y = 2x^2 + 3x + 1$, $x = -2$, $x = -1$, $x = 0$, $x = 1$, $x = 2$

(c) $y = 2x^3 + 3x^2$, $x = -4$, $x = 0$, $x = 4$

4 Transpose

$$p = \dfrac{c}{\sqrt{t}}$$

for t.

5 Transpose

$$L = \dfrac{\mu N^2 A}{l}$$

for N.

6 Make the specified variable the subject of the formula:

(a) $h = c + d + 2e$, for e

(b) $S = 2\pi r^2 + 2\pi rh$, for h

(c) $Q = \sqrt{\dfrac{c + d}{c - d}}$, for c

(d) $\dfrac{x + y}{3} = \dfrac{x - y}{7} + 2$, for x

7 Make n the subject of the formula

$$J = \dfrac{nE}{nL + m}$$

8 Transpose the formula

$$Q = \lambda A \left(\dfrac{T_2 - T_1}{\ell} \right)$$

to make ℓ the subject.

Solutions to exercises

1   0.0125

2   90

3   (a) $-\frac{2}{3}, \frac{1}{3}, \frac{5}{6}$ (b) 3, 0, 1, 6, 15
     (c) $-80, 0, 176$

4   $t = \left(\dfrac{c}{p}\right)^2$

5   $N = \sqrt{\dfrac{Ll}{\mu A}}$

6   (a) $e = \dfrac{h - c - d}{2}$ (b) $h = \dfrac{S - 2\pi r^2}{2\pi r}$

     (c) $c = \dfrac{d(1 + Q^2)}{Q^2 - 1}$ (d) $x = \dfrac{21 - 5y}{2}$

7   $n = \dfrac{Jm}{E - JL}$

8   $\ell = \lambda A\left(\dfrac{T_2 - T_1}{Q}\right)$ or $\ell = \dfrac{\lambda A}{Q}(T_2 - T_1)$

End of chapter exercises

1   Find the sum and product of 7, 11 and 91.

2   Simplify (a) $9 - (-18)$, (b) $11 - (-11)$,
     (c) $-3 - 14$, (d) $(-3)(-14)$.

3   Find the value of $5(8 + 3) - 2(-3 - 6)$.

4   Find the value of $\dfrac{9!}{4!}$.

5   Reduce each of the following fractions to its
     simplest form:
     (a) $\frac{4}{13}$ (b) $\frac{15}{20}$ (c) $\frac{15}{35}$ (d) $\frac{11}{12}$

6   Add $\frac{2}{7}$ and $\frac{1}{5}$ without using a calculator.

7   Find $\frac{3}{8} \times \frac{5}{7}$ without using a calculator.

8   Find $\frac{8}{5}$ of 120.

9   Simplify, if possible,
     (a) $(9x)(-3y)$ (b) $9(4x + 5x)$
     (c) $-2x + 11x$ (d) $(-2x)(11x)$
     (e) $3x - 2y + 4z - 2x - 3y + 9z$
     (f) $(8m)(-3q)$ (g) $8m(-3q)$ (h) $8m - 3q$
     (i) $8(m - 3q)$ (j) $12x \div 3$ (k) $\dfrac{8ab^2d}{4abc}$
     (l) $x - x$ (m) $(x)(-x)$ (n) $x(-x)$
     (o) $-x(x)$

10 Remove the brackets from the following
     expressions and simplify the result if possible:
     (a) $7(x + 2y) + 8(2x - y)$
     (b) $7(p + q) - 6(p + q)$
     (c) $2(x + y) + 3(x + 2y)$

11 Write down the reciprocal of
     (a) 18 (b) $\dfrac{1}{11}$ (c) $\dfrac{3}{8}$ (d) $\dfrac{2x}{3y}$

12 Simplify (a) $x^2x^5x^9$, (b) $\dfrac{x^9}{x^4}$, (c) $\dfrac{x^4}{x^5}$.

13 If $V = Ah$ find the value of V when $A = 3$ and
     $h = 26$.

14 Factorise (a) $16x^2 - 4x$, (b) $11y + 121$,
     (c) $ax^2 - bx$, (d) $c^2 - 2cs$, (e) $s^2 - 2cs$.

15 Remove the brackets from the following and
     simplify the result:
     (a) $(a + 3)(a - 5)$ (b) $(2b + 6)(b - 7)$
     (c) $(3c + 4)(2c - 1)$ (d) $(5d + 8)(-2d - 1)$

16 Factorise (a) $2x^2 - 11x + 5$,
     (b) $2x^2 + 13x + 15$.

17 Simplify (a) $\dfrac{x}{4} + \dfrac{x}{5}$, (b) $\dfrac{x}{4} + \dfrac{x}{5} + \dfrac{x}{6}$.

18　Simplify (a) $\dfrac{2}{x} - \dfrac{3}{2x}$, (b) $\dfrac{2}{x} + \dfrac{3}{2x}$,

(c) $\dfrac{2}{x} \times \dfrac{3}{2x}$.

19　Transpose each of the following formulae to make the given variable the subject:

(a) $x = \dfrac{c}{y}$, for y

(b) $x = \dfrac{c}{y}$, for c

(c) $k = \dfrac{2n + 5}{n + 3}$, for n

(d) $T = 2\pi\sqrt{\dfrac{R - L}{g}}$, for R

20　Simplify each of the following:

(a) $\dfrac{m^{-7}}{m^{-4}}$　(b) $(3ab^2c)^3$

(c) $y^3 \times y^{-2} \times x^7 \times x^5 \times x^{-3}$

21　Write down the reciprocal of the following:

(a) $\dfrac{1}{3} + \dfrac{1}{2}$　(b) $\dfrac{x + y}{13}$　(c) $\dfrac{2R + 1}{R - 1}$　(d) $4!$

22　Simplify $\dfrac{2x - 5}{10} - \dfrac{3x - 2}{15}$.

23　Simplify $\dfrac{3}{2x - 5} \div \dfrac{4}{x - 3}$.

24　Without using a calculator find the value of

$$3\dfrac{13}{17} + \dfrac{4 - \dfrac{1}{3}}{3/7}$$

25　Express the following using single powers:

(a) $\dfrac{4^{1/3}\,4^8}{4^{-1/2}}$　(b) $\sqrt{y}\ \sqrt[3]{y}\sqrt[4]{y}$

26　Express as a single fraction

$$\dfrac{u}{(u - v)(u - w)} + \dfrac{v}{(u - v)(u - w)} + \dfrac{w}{u - v}$$

27　Simplify

$$\dfrac{\dfrac{5}{x} - \dfrac{2}{x^2}}{x + 1}$$

28　Simplify $\dfrac{A}{2z\ s - \omega} \cdot \dfrac{1}{} - \dfrac{A}{2z\ s + \omega} \cdot \dfrac{1}{}$.

29　Factorise $\dfrac{A}{t_0 s} - \dfrac{A}{t_0 s^2}$.

30　Remove the square brackets from

$$\dfrac{1}{\omega}\left[\dfrac{s\omega}{s^2 + \omega^2} - s\right]$$

31　Express as a single fraction

$$\dfrac{a_1}{s + p_1} + \dfrac{a_2}{s + p_2}$$

32　Express as a single fraction

$$s + 2 + \dfrac{s + 3}{(s + 1)(s + 2)}$$

33　Express as a single fraction

$$\dfrac{1}{\omega^2 s} - \dfrac{s}{\omega^2(s^2 + \omega^2)}$$

34　By multiplying both numerator and denominator of $\dfrac{1}{a + b\sqrt{c}}$ by $a - b\sqrt{c}$ show that

$$\dfrac{1}{a + b\sqrt{c}} = \dfrac{a - b\sqrt{c}}{a^2 - b^2\,c}$$

Use this approach to show that

$$\dfrac{1}{2 + \sqrt{3}} = 2 - \sqrt{3}$$

Solutions to exercises

1   109, 7007

2   (a) 27 (b) 22 (c) −17 (d) 42

3   73

4   15120

5   (a) $\dfrac{4}{13}$ (b) $\dfrac{3}{4}$ (c) $\dfrac{3}{7}$ (d) $\dfrac{11}{12}$

6   $\dfrac{17}{35}$

7   $\dfrac{15}{56}$

8   192

9   (a) $-27xy$ (b) $81x$ (c) $9x$ (d) $-22x^2$
    (e) $x - 5y + 13z$ (f) $-24mq$ (g) $-24mq$
    (h) cannot be simplified
    (i) cannot be simplified (j) $4x$ (k) $\dfrac{2bd}{c}$ (l) 0
    (m) $-x^2$ (n) $-x^2$ (o) $-x^2$

10 (a) $23x + 6y$ (b) $p + q$ (c) $5x + 8y$

11 (a) $\dfrac{1}{18}$ (b) 11 (c) $\dfrac{8}{3}$ (d) $\dfrac{3y}{2x}$

12 (a) x^{16} (b) x^5 (c) x^{-1}

13 78

14 (a) $4x(4x - 1)$ (b) $11(y + 11)$ (c) $x(ax - b)$
    (d) $c(c - 2s)$ (e) $s(s - 2c)$

15 (a) $a^2 - 2a - 15$ (b) $2b^2 - 8b - 42$
    (c) $6c^2 + 5c - 4$ (d) $-10d^2 - 21d - 8$

16 (a) $(2x - 1)(x - 5)$ (b) $(x + 5)(2x + 3)$

17 (a) $\dfrac{9x}{20}$ (b) $\dfrac{37x}{60}$

18 (a) $\dfrac{1}{2x}$ (b) $\dfrac{7}{2x}$ (c) $\dfrac{3}{x^2}$

19 (a) $y = \dfrac{c}{x}$ (b) $c = xy$
    (c) $n = \dfrac{5 - 3k}{k - 2}$ (d) $R = L + g\left(\dfrac{T}{2\pi}\right)^2$

20 (a) m^{-3} (b) $27a^3b^6c^3$ (c) yx^9

21 (a) $\dfrac{6}{5}$ (b) $\dfrac{13}{x + y}$ (c) $\dfrac{R - 1}{2R + 1}$ (d) $\dfrac{1}{24}$

22 $-\dfrac{11}{30}$

23 $\dfrac{3(x - 3)}{4(2x - 5)}$

24 $\dfrac{1885}{153}$

25 (a) $4^{53/6}$ (b) $y^{13/12}$

26 $\dfrac{uw + u + v - w^2}{(u - v)(u - w)}$

27 $\dfrac{5x - 2}{x^2(x + 1)}$

28 $\dfrac{A\omega}{z(s + \omega)(s - \omega)}$

29 $\dfrac{A}{t_0 s}\left(1 - \dfrac{1}{s}\right)$

30 $\dfrac{s}{s^2 + \omega^2} - \dfrac{s}{\omega}$

31 $\dfrac{a_1 s + a_2 s + a_1 p_2 + a_2 p_1}{(s + p_1)(s + p_2)}$

32 $\dfrac{s^3 + 5s^2 + 9s + 7}{(s + 2)(s + 1)}$

33 $\dfrac{1}{s(s^2 + \omega^2)}$

Functions and mathematical models

Chapter **6**

The study of functions is fundamental to engineering mathematics because functions provide a way of mathematically describing relationships between physical variables.

In this chapter we show how to represent functions and describe many of their properties. Many functions that arise commonly in engineering applications are introduced. Confidence and competence in the algebra of functions are essential for further study, and you will have the opportunity to develop these. Other functions such as trigonometrical, exponential and logarithmic functions are introduced in later blocks once the initial foundations have been laid.

Graphs, block diagrams and sets provide visual ways of representing functions. You will plot and interpret some simple graphs in the following blocks. If you have access to a graphical calculator or computer software you will be able to experiment with more complicated ones. Straight line graphs are particularly important, and a block is devoted to the study of these.

Chapter 6 contents

Basic concepts of functions

In engineering there are many quantities that change their value as time changes. For example, the temperature of a furnace may change with time as it is heated. Similarly, there are many quantities that change their value as the location of a point of interest changes. For example, the shear stress in a bridge girder will vary from point to point across the bridge. A quantity whose value can change is known as a **variable**. We use **functions** to describe how a variable changes as a consequence of another changing. There are many different types of function that are used by engineers. We shall be examining some of these in later blocks.

The introduction of functions leads naturally to the concept of mathematical modelling. A mathematical model is a mathematical description of a real-life situation. The mathematical description usually involves several functions and takes the form of a set of interrelated equations, although a single equation can represent some simple phenomena.

Consider, for example, the combustion of gas in a piston cylinder. There are several quantities of interest. These include the volume, temperature and pressure of the gas, the velocity of the piston, and so on. As another example, consider an electric circuit. The relevant quantities could include voltages, currents, resistances, conductivity, cross-sectional area of the conductor, and so on. Each quantity is assigned a symbol: R for resistance, v for velocity, p for pressure, V for volume, and so on. Sometimes the symbols used are Greek letters.

In any particular situation or experiment, the values of some quantities remain fixed. These are referred to as constants. The cross-sectional area of a wire may be constant for a particular circuit, or maybe the resistance of a component is constant. The values of other quantities may change. The volume and pressure of gas in a cylinder vary as the piston moves up and down. As we have already mentioned, quantities whose values change are referred to as variables.

Sometimes there are physical laws which relate the various quantities to each other. In a simple circuit, as we have seen already in Section 7.2 of Chapter 5, if V denotes voltage, I denotes current and R denotes resistance (a constant) then we may write

$$V = IR$$

This is known as Ohm's law. We can say that V is a function of I. As another example, if u denotes the initial velocity of a body, v denotes velocity at any point in time, a

denotes acceleration and t denotes the time that has elapsed, then we may establish that

$$v = u + at$$

This is a mathematical equation for modelling the velocity of a body that is changing due to acceleration. In words we would say 'The velocity (v) at any particular instant is its initial velocity (u) plus the product of its acceleration (a) and the time elapsed (t).' Note that velocity, v, is changing with time and so is a variable whereas initial velocity, u, is a fixed quantity and so is a constant.

However, some caution is needed. Both of these equations apply only under very particular circumstances which are not apparent simply by examining the equations. These particular circumstances may rarely if ever be achieved in reality and so the equation is then an approximation to what is really happening.

So mathematical models are usually an idealised description of the real situation. In the above example, the equation $v = u + at$ requires that acceleration, a, is constant. If the acceleration of a body was *almost* constant, then we may decide that this equation, which requires constant acceleration, is an adequate model of reality. As another example, consider again Ohm's law which describes the relationship between voltage across a resistor, V, current, I and resistance, R:

$$V = IR$$

This is, however, based on some assumptions that simplify the situation. The equation assumes that there is no variation in the current density across the cross-section of the resistor. For large voltages, the resistor breaks down and so it is assumed that the voltage is within limits such that the resistance remains constant. So it must be remembered that a mathematical model is an approximation to real life.

Sometimes there is no known physical law that relates particular variables. In such cases, measurements are sometimes made and from this experimental data it may be possible to construct an equation (or equations) that adequately captures the relationship between the measured quantities. Any assumptions which are made in constructing such equations must be clearly stated.

Many situations require more than one equation to describe them. Consider an electric circuit comprising several meshes (see Chapter 13 Block 6 for details). The currents in the meshes are modelled by a system of simultaneous equations. These simultaneous equations need to be solved to determine each individual mesh current.

We now return to the concept of a variable. It is often useful to distinguish between independent and dependent variables. Dependent variables are dependent upon the values of the independent variables. The purpose of a mathematical model is to have a clear mathematical understanding, in the form of a set of equations, of how the dependent variables change in response to changes in the independent variables.

Even though mathematical models are approximations of reality, they have many advantages. By modelling a situation accurately, it is possible to calculate the effects of changing the values of the physical quantities, without actually conducting real-life experiments. For example, by modelling the strength of a metal beam it is possible to see the effects of changing its cross-sectional area or length without having to manufacture and test a set of actual metal beams. So, although mathematical models are theoretical constructs of an idealised situation, their ability to predict the effects of changing values of key characteristics of a situation is widely used.

So how do we know if a mathematical model is a 'good enough' description of reality? Predictions from the model can be compared with actual measurements made during the experiment. Modelled results are compared with real results and a judgement is made on whether the fit is good enough. If not, then the model must be improved. This may necessitate additional equations being added to the model, or existing equations having terms modified, added or removed. The cycle is then repeated and the new model tested against reality.

1.3   The function rule

A function can be thought of as a **rule** that operates on an **input** and produces an **output**. This can be illustrated pictorially in two ways, as shown in Figure 1.1. The first way is by using a **block diagram**, which consists of a box showing the input and output, and the rule. We often write the rule inside the box. The second way is to use two sets, one to represent the input and one to represent the output. An arrow shows the relationship between them.

Figure 1.1
A function is a rule operating on an input to produce an output.

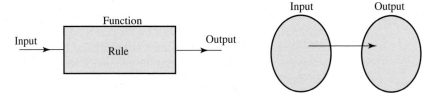

In order for a rule to be a function it must produce only a *single* output for any given input. The function with the rule 'treble the input' is shown in Figure 1.2.

Note that with an input of 4 the function would produce an output of 12. With a more general input, x say, the output will be $3x$. It is usual to assign a letter to a function in order to label it. The trebling function in Figure 1.2 has been given the symbol f.

> **Key point**
>
> A **function** is a rule that operates on an input and produces a single output from that input.

Note that the set diagram in Figure 1.2 illustrates that for any input we choose we obtain just one output, because just one arrow leaves each input value.

Figure 1.2
The function with rule 'treble the input'.

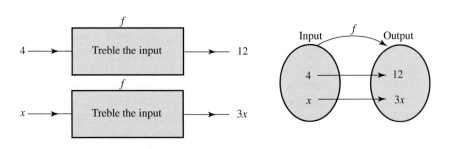

Example 1.1
Write down the output from the function shown in Figure 1.3 when the input is
(a) 4, (b) −3, (c) x.

Figure 1.3

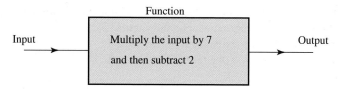

Solution
In each case the function rule instructs us to multiply the input by 7 and then subtract 2.

(a) When the input is 4 the output is   26

(b) When the input is −3 the output is   −23

(c) When the input is x the output is $7x - 2$

Several different notations are used by engineers to describe functions. For the trebling function in Figure 1.2 it is common to write

$$f(x) = 3x$$

This indicates that with an input x, the function, f, produces an output of $3x$. The input to the function is placed in the brackets after the 'f'. Do not interpret this as multiplication. $f(x)$ is read as 'f is a function of x', or simply 'f of x', meaning that the value of the output from the function depends upon the value of the input x. The value of the output is often called the **value of the function**.

Sometimes $f(x) = 3x$ is abbreviated to simply $f = 3x$, and it is then left to the reader to note that f is a function of x.

Example 1.2 Thermodynamics – Converting centigrade to kelvin
Temperatures can be recorded in either centigrade (°C) or kelvin (K). Temperatures in centigrade can be converted to their equivalent in kelvin by adding 273. So, for example, 70 °C is equivalent to 70 + 273 = 343 K.

This may be expressed diagrammatically as shown in Figure 1.4 where the input, T, is the temperature in °C and the output is the temperature in K or algebraically as

$$f(T) = T + 273$$

Figure 1.4
Block diagram
showing the
function
$f(T) = T + 273$.

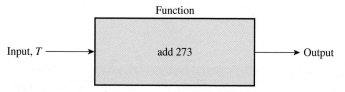

Example 1.3
State the rule of each of the following functions:
(a) $f(x) = 6x$ (b) $f(t) = 6t - 1$ (c) $g(z) = z^2 - 7$ (d) $h(t) = t^3 + 5$
(e) $p(x) = x^3 + 5$

Solution

(a) The rule for f is 'multiply the input by 6'.

(b) Here the input has been labelled t. The rule for f is 'multiply the input by 6 and subtract 1'.

(c) Here the function has been labelled g and the input has been labelled z. The rule for g is 'square the input and subtract 7'.

(d) The rule for h is 'cube the input and add 5'.

(e) The rule for p is 'cube the input and add 5'.

Note from Example 1.3 parts (d) and (e) that it is the rule that is important when describing a function and not the letters being used. Both $h(t)$ and $p(x)$ instruct us to 'cube the input and add 5'.

Example 1.4

Write down a mathematical function that can be used to describe the following rules:

(a) 'square the input and divide the result by 2'

(b) 'divide the input by 3 and then add 7'

Solution

(a) Use the letter x for input and the letter f to represent the function. Then

$$f(x) = \qquad \qquad\qquad f(x) = \frac{x^2}{2}$$

(b) Label the function g and call the input t:

$$g(t) = \qquad \qquad\qquad g(t) = \frac{t}{3} + 7$$

Exercises

1 Explain what is meant by a function.

2 State the rule of each of the following functions:

(a) $f(x) = 5x$ (b) $f(t) = 5t$

(c) $f(x) = 8x + 10$ (d) $f(t) = 7t - 27$

(e) $f(t) = 1 - t$ (f) $h(t) = \dfrac{t}{3} + \dfrac{2}{3}$

(g) $f(x) = \dfrac{1}{1 + x}$

Solutions to exercises

2 (a) multiply the input by 5, (b) same as (a), (c) multiply the input by 8 and then add 10, (d) multiply the input by 7 and then subtract 27, (e) subtract the input from 1, (f) divide the input by 3 and then add $\frac{2}{3}$, (g) add 1 to the input and then find the reciprocal of the result.

1.4 The argument of a function

The input to a function is called its **argument**. It is often necessary to obtain the output from a function if we are given its argument. For example, given the function $g(t) = 3t + 2$ we may require the value of the output when the argument is 2. We write this as $g(t = 2)$ or more usually and compactly as $g(2)$. In this case the value of $g(2)$ is $3 \times 2 + 2 = 8$.

Example 1.5
Given the function $f(x) = 3x + 1$ find
(a) $f(2)$
(b) $f(-1)$
(c) $f(6)$

Solution
(a) State the function rule:

	multiply input by 3 and add 1

When the argument is 2, we find
$$f(2) = 3 \times 2 + 1$$
$$= 7$$

(b) Here the argument is -1.

$$f(-1) = \qquad \qquad 3 \times (-1) + 1 = -2$$

(c) $f(6) = \qquad \qquad 3(6) + 1 = 19$

Example 1.6 Dynamics – Constant acceleration
A body has an initial velocity of u m s^{-1} and a constant acceleration of a m s^{-2}. Then, after t seconds, the velocity, v, of the body is given by

$$v(t) = u + at$$

Note that velocity, v, is a function of time t. The argument of the function v is t.

Example 1.7 Dynamics – Constant acceleration
A body has an initial velocity of 6 m s^{-1} and a constant acceleration of 2 m s^{-2}. Calculate the velocity after 14 seconds.

Solution
$$v(t) = u + at$$
$$= 6 + 2t$$
$$v(14) = 6 + 2(14) = 34$$

The velocity after 14 seconds is 34 m s^{-1}.

It is possible to obtain the value of a function when the argument is an algebraic expression. Consider the following example.

Example 1.8

Given the function $y(x) = 3x + 2$ find

(a) $y(t)$ (b) $y(2t)$ (c) $y(z + 2)$ (d) $y(5x)$ (e) $y\left(\dfrac{t}{\alpha}\right)$

Solution

State the rule for this function:

multiply the input by 3 and then add 2

We can apply this rule whatever the argument.

(a) In this case the argument is t. Multiplying this by 3 and adding 2 we find $y(t) = 3t + 2$. Equivalently we can replace x by t in the expression for the function, so $y(t) = 3t + 2$.

(b) In this case the argument is $2t$. We need to replace x by $2t$ in the expression for the function. So

$y(2t) = $ 　　　　　　　　　　　　　　　　　$3(2t) + 2 = 6t + 2$

(c) In this case the argument is $z + 2$.

$y(z + 2) = $ 　　　　　　　　　　　　　　$3(z + 2) + 2 = 3z + 8$

(d) The argument is $5x$ and so there appears to be a clash of notation with the original expression for the function. There is no problem if we remember that the rule is to multiply the input by 3 and then add 2. The input now is $5x$, so

$$y(5x) = 3(5x) + 2$$
$$= 15x + 2$$

(e) $y\left(\dfrac{t}{\alpha}\right) = $ 　　　　　　　　　　　　　　　　$3\dfrac{t}{\alpha} + 2$

Exercises

1　Explain what is meant by the 'argument' of a function.

2　Given the function $g(t) = 8t + 3$ find
(a) $g(7)$ (b) $g(2)$ (c) $g(-0.5)$
(d) $g(-0.11)$

3　Given the function $f(t) = 2t^2 + 4$ find
(a) $f(x)$ (b) $f(2x)$ (c) $f(-x)$ (d) $f(4x + 2)$

(e) $f(3t + 5)$ (f) $f(\lambda)$ (g) $f(t - \lambda)$ (h) $f\left(\dfrac{t}{\alpha}\right)$

4　Given $g(x) = 3x^2 - 7$ find
(a) $g(3t)$ (b) $g(t + 5)$ (c) $g(6t - 4)$
(d) $g(4x + 9)$

5　Calculate $f(x + h)$ when
(a) $f(x) = x^2$ (b) $f(x) = x^3$ (c) $f(x) = \dfrac{1}{x}$
In each case write down the corresponding expression for $f(x + h) - f(x)$.

6　If $f(x) = \dfrac{1}{(1 - x)^2}$ find $f\left(\dfrac{x}{\ell}\right)$.

Solutions to exercises

1   The argument is the input.

2   (a) 59 (b) 19 (c) −1 (d) 2.12

3   (a) $2x^2 + 4$ (b) $8x^2 + 4$ (c) $2x^2 + 4$
    (d) $32x^2 + 32x + 12$ (e) $18t^2 + 60t + 54$
    (f) $2\lambda^2 + 4$ (g) $2(t - \lambda)^2 + 4$ (h) $\dfrac{2t^2}{\alpha^2} + 4$

4   (a) $27t^2 - 7$ (b) $3t^2 + 30t + 68$
    (c) $108t^2 - 144t + 41$
    (d) $48x^2 + 216x + 236$

5   (a) $x^2 + 2xh + h^2$
    (b) $x^3 + 3x^2h + 3xh^2 + h^3$

(c) $\dfrac{1}{x + h}$

The corresponding expressions are (a) $2xh + h^2$
(b) $3x^2h + 3xh^2 + h^3$

(c) $\dfrac{1}{x + h} - \dfrac{1}{x} = -\dfrac{h}{x(x + h)}$

6   $\dfrac{1}{\left(1 - \dfrac{x}{\ell}\right)^2}$

End of block exercises

1   State the rule of each of the following
    functions:
    (a) $f(v) = 9v$ (b) $f(t) = 12t$
    (c) $f(x) = 208x + 36$ (d) $f(t) = 25t - 18$

    (e) $f(x) = \frac{1}{2}x + \frac{3}{4}$ (f) $g(x) = \dfrac{10}{x}$

2   Given the function $g(x) = 6 - 12x$ find
    (a) $g(1)$ (b) $g(7)$ (c) $g(-2)$ (d) $g(\frac{1}{2})$
    (e) $g(0.01)$ (f) $g(-0.5)$

3   Given the function $f(x) = \dfrac{1}{x}$ find
    (a) $f(\lambda)$ (b) $f(t)$ (c) $f(t - \lambda)$ (d) $f(\omega - x)$
    (e) $f\left(\dfrac{t}{\alpha}\right)$

4   If $F(s) = \dfrac{1}{s + 1}$ find (a) $F(s - 1)$,
    (b) $F(s + 1)$, (c) $F(s^2 + \omega^2)$.

5   **Area of a circle.** The area, A, of a circle
    depends upon the radius, r according to

    $$A(r) = \pi r^2$$

    Calculate $A(2r)$ and hence show that when the
    radius of a circle is doubled, its area is
    increased by a factor of 4.

6   **Volume of a sphere.** The volume, V, of a
    sphere depends upon the radius, r, according to

    $$V(r) = \dfrac{4\pi r^3}{3}$$

    (a) Calculate $V(2r)$. Hence determine how the
    volume of a sphere changes when the
    radius is doubled.
    (b) By what factor does the volume of a
    sphere change when the radius is halved?

Solutions to exercises

1   (a) multiply the input by 9 (b) multiply the
    input by 12 (c) multiply the input by 208 and
    add 36 (d) multiply the input by 25 and subtract

18 (e) find one-half of the input and then add $\frac{3}{4}$
    (f) divide the number 10 by the input.

2 (a) -6 (b) -78 (c) 30 (d) 0 (e) 5.88 (f) 12

3 (a) $\dfrac{1}{\lambda}$ (b) $\dfrac{1}{t}$ (c) $\dfrac{1}{t-\lambda}$ (d) $\dfrac{1}{\omega-x}$ (e) $\dfrac{\alpha}{t}$

4 (a) $\dfrac{1}{s}$ (b) $\dfrac{1}{s+2}$ (c) $\dfrac{1}{s^2+\omega^2+1}$

5 $A(2r) = 4\pi r^2 = 4A(r)$

6 (a) $V(2r) = \dfrac{32\pi r^3}{3} = 8V(r)$. So when the radius is doubled the volume increases by a factor of 8.

(b) $V\left(\dfrac{r}{2}\right) = \dfrac{4\pi}{3}\left(\dfrac{r}{2}\right)^3 = \dfrac{1}{8}\left(\dfrac{4\pi r^3}{3}\right) = \dfrac{V(r)}{8}$. When the radius is halved the volume decreases by a factor of 8.

The graph of a function

2.1 Introduction

Engineers often find mathematical ideas easier to understand when these are portrayed visually as opposed to algebraically. Graphs are a convenient and widely used way of portraying functions. By inspecting a graph it is easy to describe a number of properties of the function being considered. For example, where is the function positive, and where is it negative? Where is it increasing and where is it decreasing? Do function values repeat? Questions such as these can be answered once the graph of a function has been drawn. In this block we shall describe how the graph of a function is obtained and introduce various terminology associated with graphs.

2.2 The graph of a function

Consider the function $f(x) = 2x$. The output from this function is obtained by multiplying the input by 2. We can choose several values for the input to this function and calculate the corresponding outputs. We have done this for integer values of x between -2 and 2, and the results are shown in Table 2.1.

Table 2.1

input, x	-2	-1	0	1	2
output, $f(x)$	-4	-2	0	2	4

To construct the graph of this function we first draw a pair of **axes** – a vertical axis and a horizontal axis. These are drawn at right angles to each other and intersect at the **origin O** as shown in Figure 2.1.

Figure 2.1
The two axes intersect at the origin. Each pair of values, x and $f(x)$, gives a point on the graph.

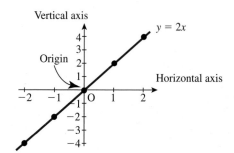

Each pair of input and output values can be represented on a graph by a single point. The input values are measured along the horizontal axis and the output values are measured along the vertical axis. The horizontal axis is often called the **x axis**. The vertical axis is commonly referred to as the **y axis**, so that we often write the function as

$$y = f(x)$$
$$= 2x$$

or simply

$$y = 2x$$

Each pair of x and y values in the table is plotted as a single point shown as $\bullet$ in Figure 2.1. The point is often labelled as (x, y). The values of x and y are said to be the **coordinates** of the point. The points are then joined with a smooth curve to produce the required graph as shown in Figure 2.1. Note that in this case the graph is a straight line.

Dependent and independent variables

Since x and y can have a number of different values they are variables. Here x is called the **independent variable** and y is called the **dependent variable**. By knowing or choosing a value of the independent variable x, the function rule enables us to calculate the corresponding value of the dependent variable y. To show this dependence we often write $y(x)$. This is read as 'y is a function of x' or 'y depends upon x', or simply 'y of x'. Note that it is the independent variable that is the input to the function and the dependent variable that is the output.

The domain and range of a function

The set of values that we allow the independent variable to take is called the **domain** of the function. A domain is often an interval on the x axis. For example, the function

$$y = g(x)$$
$$= 5x + 2 \quad -5 \le x \le 20$$

has any value of x between -5 and 20 inclusive as its domain because it has been stated as this. If the domain of a function is not stated then it is taken to be the largest set possible. For example,

$$h(t) = t^2 + 1$$

has domain $(-\infty, \infty)$ since h is defined for every value of t and the domain has not been stated otherwise.

The set of values of the function for a given domain, that is the set of y values, is called the **range** of the function. The range of $g(x)$ is $[-23, 102]$. The range of $h(t)$ is $[1, \infty)$ although this may not be apparent to you at this stage. The range can usually be identified quite easily once a graph has been drawn.

Later, you will meet some functions for which certain values of the independent variable must be excluded from the domain because at these values the function would be undefined. One such example is $f(x) = \dfrac{1}{x}$ for which we must exclude the value $x = 0$, since $\frac{1}{0}$ is meaningless.

In the function $y = f(x)$, x is called the **independent variable** and y is called the **dependent variable** because the value of y depends upon the value chosen for x.
The set of x values used as input to the function is called the **domain** of the function.
The set of values that y takes as x is varied is called the **range** of the function.

Example 2.1
Consider the function given by $g(t) = 2t^2 + 1$, $-2 \le t \le 2$.
(a) State the domain of the function.
(b) Plot a graph of the function.
(c) Deduce the range of the function from the graph.

Solution
(a) The domain is given as the closed interval $[-2, 2]$, that is any value of t between -2 and 2 inclusive.
(b) To construct the graph a table of input and output values must be constructed first. Such a table is shown in Table 2.2.

Table 2.2

t	-2	-1	0	1	2
$y = g(t)$	9	3	1	3	9

Each pair of t and y values in the table is plotted as a single point shown as •. The points are then joined with a smooth curve to produce the required graph as shown in Figure 2.2.

Figure 2.2
Graph of
$g(t) = 2t^2 + 1$.

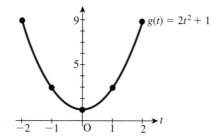

$g(t) = 2t^2 + 1$

(c) The range is the set of values that the function takes as x is varied. By inspecting the graph we see that the range of g is the closed interval $[1, 9]$.

Example 2.2

Consider the function given by $f(x) = x^2 + 2, \quad -3 \le x \le 3$.

(a) State the domain of the function.

(b) Draw up a table of input and output values for this function.

(c) Plot a graph of the function.

(d) Deduce the range of the function by inspecting the graph.

Solution

(a) Recall that the domain of a function $f(x)$ is the set of values that x is allowed to take. Write this set of values as an interval:

$[-3, 3]$

(b) The table of values has been partially calculated. Complete this now:

x	-3	-2	-1	0	1	2	3
$x^2 + 2$		6		2			

x	-3	-2	-1	0	1	2	3
$x^2 + 2$	11	6	3	2	3	6	11

(c) The graph is shown in Figure 2.3.

Figure 2.3
Graph of
$f(x) = x^2 + 2$.

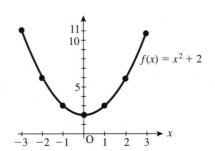

(d) Recall that the range of the function is the set of values that the function takes as x is varied. It is possible to deduce this from the graph. Write this set as an interval.

$[2, 11]$

Example 2.3

Explain why the value $x = -4$ must be excluded from the domain of the function

$$f(x) = \frac{3}{(x + 4)^2}$$

Solution

When $x = -4$ the denominator is zero, so f is undefined here.

Example 2.4 Chemical Engineering – Discharge of a liquid from a tank

Figure 2.4
A liquid storage
system.

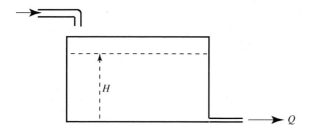

Figure 2.4 shows a tank that is used to store a liquid. Liquid can be let into the tank through an inlet pipe at the top, and it discharges from the tank through a spout in the side near its base. Such a situation occurs frequently in chemical engineering applications. Under certain conditions the flow through the spout will be *laminar* or smooth, and the rate of outflow, Q, is proportional to the depth, or head, H, of liquid in the tank. This is expressed mathematically as

$$Q = KH$$

where K is a constant of proportionality called the **discharge coefficient**. The dependent variable Q is a function of the independent variable H. Here the input to the function is the head H, the function rule is 'multiply the input by K', and the resulting output is the flow rate Q. A graph of Q against H is shown in Figure 2.5(a).

Figure 2.5
(a) Laminar flow
characteristic;
(b) turbulent flow
characteristic.

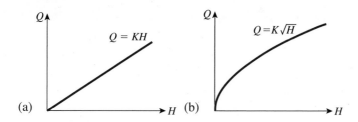

If the flow through the spout is turbulent then a different functional relationship exists between Q and H:

$$Q = K\sqrt{H}$$

Here the function rule is 'take the positive square root of the input, H, and multiply this by the discharge coefficient, K'. The output is the flow rate Q. A graph of Q against H is shown in Figure 2.5(b).

Exercises

1. Explain the meaning of the terms 'dependent variable' and 'independent variable'. When plotting a graph, which variables are plotted on which axes?

2. When stating the coordinates of a point, which coordinate is given first?

3   Explain the meaning of an expression such as $y(x)$ in the context of functions. What is the interpretation of $x(t)$?

4   Explain the meaning of the terms 'domain' and 'range' when applied to functions.

5   Plot a graph of the following functions. In each case state the domain and the range of the function.
(a) $f(x) = 3x + 2, -2 \leq x \leq 5$
(b) $g(x) = x^2 + 4, -2 \leq x \leq 3$
(c) $p(t) = 2t^2 + 8, -2 \leq t \leq 4$
(d) $f(t) = 6 - t^2, 1 \leq t \leq 5$

Solutions to exercises

1   The independent variable is plotted on the horizontal axis. The dependent variable is plotted on the vertical axis.

2   The independent variable is given first, as in (x, y).

3   $x(t)$ means that the dependent variable x is a function of the independent variable t.

5   (a) domain $[-2, 5]$, range $[-4, 17]$
(b) $[-2, 3], [4, 13]$ (c) $[-2, 4], [8, 40]$
(d) $[1, 5], [-19, 5]$.

2.3   Using computer software to plot graphs of functions

Computer software packages make it very easy to plot graphs of functions. Not only can they produce graphs accurately and quickly, but also they generally have other useful facilities. For example, it is possible to zoom in on particular parts of the graph, and to redraw a graph using different axes. It is straightforward to produce multiple plots with several graphs in one figure. Furthermore, packages can be used to plot other forms of graph such as polar plots, parametric plots (see Block 5), etc. When you have managed to produce some simple graphs you will find on-line help facilities that will enable you to explore this topic further.

Example 2.5
Use a computer package to plot a graph of the function $f(x) = 2x^3 - 3x^2 - 39x + 20$ for values of x between -6 and 6. By inspecting the graph locate the values of x where the graph cuts the horizontal axis.

Solution

Maple
In Maple the command to plot a graph of this function is

```
plot(2*x^ 3 − 3*x^ 2 − 39*x + 20,x = −6..6);
```

Note that Maple requires the multiplication symbol $*$ to be inserted, and uses the symbol $^$ to denote a power. Note that the domain of interest is entered in the form $x = -6..6$. Maple produces the output shown in Figure 2.6, which can be customised as required by the user. By inspection the graph cuts the horizontal axis at $x = -4$, $x = 5$ and $x = \frac{1}{2}$.

Figure 2.6

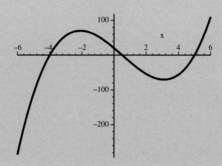

Matlab
In Matlab it is first necessary to specify the coordinates of all the required points on the graph. Matlab will then plot the points and join them with a smooth curve. Clearly we want points for which the value of x lies between -6 and 6. Suppose we want to plot points at intervals of 0.1. The appropriate values of x are defined using the command

```
x = -6:0.1:6;
```

Then at each of these values of x, the value of y is calculated using the command

```
y = 2*x.^ 3 - 3*x.^ 2 - 39*x + 20;
```

Note the requirement to input x^3 as `x.^3`, and so on. Finally the command to plot the graph is

```
plot(x,y);
```

Matlab produces the graph, similar to that produced by Maple, in a new window. You should explore the commands to add labels, title, reposition the axes, etc.

End of block exercises

1   Plot graphs of $y = x^2$ and $y = -x^2$ for $-4 \le x \le 4$. In each case state the domain and range of the function.

2   On the same diagram draw graphs of $y = 7x + 1$ and $y = 7x + 2$. Comment on any similarities between the two graphs.

3   On the same diagram draw graphs of $y = 3x$ and $y = 4x$. Comment on any similarities between the two graphs.

4   The relationship between a temperature, T_F, measured in degrees Fahrenheit (°F) and a temperature T_C measured in degrees Celsius (°C) is given by the function $T_F = \frac{9}{5}T_C + 32$. Plot a graph of this function for the domain

$0 \le T_C \le 100$. What is the range of this function?

5   Plot a graph of the function

$$f(x) = (x - 2)(x + 4)$$

What is the range of this function?

6   Explain why the value $x = 0$ must be excluded from the domain of

$$g(x) = \frac{1}{x(x - 7)}$$

What other value must be excluded and why?

Solutions to exercises

1   Domain $[-4, 4]$ in both cases; range of x^2 is
    $[0, 16]$, range of $-x^2$ is $[-16, 0]$.

2   Both graphs have the same slope.

3   Both pass through the origin.

4   $[32, 212]$

5   $[-9, \infty)$

6   When $x = 0$, $g(x)$ is not defined. The value
    $x = 7$ must also be excluded because once
    again this value makes the denominator zero.

Composition of functions

3.1 Introduction

When the output from one function is used as the input to another function we form what is known as a **composite function**. We study composite functions in this block.

3.2 Composition of functions

Consider the two functions $g(x) = x^2$ and $h(x) = 3x + 5$. Block diagrams showing the rules for these functions are shown in Figure 3.1.

Figure 3.1
Block diagrams of two functions g and h.

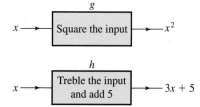

Suppose we place these block diagrams together in series as shown in Figure 3.2, so that the output from function g is used as the input to function h.

Figure 3.2
The composition of the two functions to give $h(g(x))$.

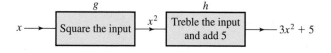

Study Figure 3.2 and deduce that when the input to g is x the output from the two functions in series is $3x^2 + 5$. Since the output from g is used as input to h we write

$$h(g(x)) = h(x^2)$$
$$= 3x^2 + 5$$

The form $h(g(x))$ is known as the **composition** of the functions g and h.

Suppose we interchange the two functions so that h is applied first, as shown in Figure 3.3.

Figure 3.3
The composition of the two functions to give $g(h(x))$.

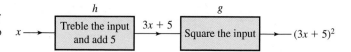

Study Figure 3.3 and note that when the input to h is x the final output is $(3x + 5)^2$. We write

$$g(h(x)) = (3x + 5)^2$$

Note that the function $h(g(x))$ is different from $g(h(x))$.

Example 3.1
Given two functions $g(t) = 3t + 2$ and $h(t) = t + 3$ obtain expressions for the compositions
(a) $g(h(t))$ (b) $h(g(t))$

Solution
(a) We have

$$g(h(t)) = g(t + 3)$$

Now the rule for g is 'triple the input and add 2', and so we can write

$$g(t + 3) = 3(t + 3) + 2$$
$$= 3t + 11$$

So, $g(h(t)) = 3t + 11$.
(b) We have

$$h(g(t)) = h(3t + 2)$$

State the rule for h:

'add 3 to the input'

$$h(g(t)) = \qquad\qquad\qquad h(3t + 2) = 3t + 5$$

We note that $h(g(t)) \neq g(h(t))$.

Example 3.2
Find $f(f(x))$ when $f(x) = 3x + 2$.

Solution
Here the function rule is 'multiply the input by 3 and then add 2'. The composite function $f(f(x))$ is illustrated in Figure 3.4.

$$f(f(x)) = f(3x + 2)$$
$$= 3(3x + 2) + 2$$
$$= 9x + 8$$

Figure 3.4
The composition
of $f(x)$ with itself.

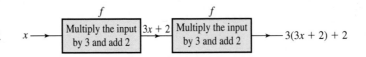

Exercises

1 Find $f(g(x))$ when $f(x) = x - 7$ and $g(x) = x^2$.

2 If $f(x) = 8x + 2$ find $f(f(x))$.

3 If $f(x) = x + 6$ and $g(x) = x^2 - 5$ find
   (a) $f(g(0))$, (b) $g(f(0))$, (c) $g(g(2))$, (d) $f(g(7))$.

4 If $f(x) = \dfrac{x - 3}{x + 1}$ and $g(x) = \dfrac{1}{x}$ find $g(f(x))$.

Solutions to exercises

1 $x^2 - 7$

2 $8(8x + 2) + 2 = 64x + 18$

3 (a) 1 (b) 31 (c) -4 (d) 50

4 $\dfrac{x + 1}{x - 3}$

End of block exercises

1 If $f(x) = x + 9$, $g(x) = 3 - x^2$ and
   $h(x) = 1 + 2x$ find (a) $g(f(x))$, (b) $h(g(f(x)))$,
   (c) $f(h(g(x)))$, (d) $f(f(x))$.

2 If $f(x) = \dfrac{1}{x}$ find $f(f(x))$.

3 If $y(x) = x^2 + 3$ and $z(x) = \sqrt{x}$ find $y(z(x))$
   and $z(y(x))$.

4 Express the function $f(x) = 7(x - 2)$ as the
   composition of two simpler functions.

5 Express the function $f(x) = 7x - 2$ as the
   composition of two simpler functions.

6 If $r(t) = 2t + 3$ and $s(t) = 7t^2 - t$ find
   (a) $r(s(t))$, (b) $s(r(t))$.

7 If $f(x) = \dfrac{1}{x - 3}$, and $g(x) = \dfrac{1}{x + 3}$, find
   (a) $f(g(x))$, (b) $g(f(x))$, (c) $f(f(x))$.

8 (a) If $f(x) = \dfrac{2}{x}$ and $g(x) = 7 - x$ find $g(f(x))$.
   (b) What is the domain of $g(f(x))$?

Solutions to exercises

1 (a) $-x^2 - 18x - 78$ (b) $-2x^2 - 36x - 155$
   (c) $16 - 2x^2$ (d) $x + 18$

2 x

3 $x + 3, \sqrt{x^2 + 3}$

4 If $g(x) = 7x$, and $h(x) = x - 2$, then
   $g(h(x)) = 7(x - 2)$.

5 If $g(x) = 7x$, and $h(x) = x - 2$, then
   $h(g(x)) = 7x - 2$.

6 (a) $14t^2 - 2t + 3$ (b) $28t^2 + 82t + 60$

7 (a) $-\dfrac{x + 3}{3x + 8}$ (b) $\dfrac{x - 3}{3x - 8}$ (c) $\dfrac{3 - x}{3x - 10}$

8 (a) $7 - \dfrac{2}{x}$ (b) $x \neq 0$

One-to-one functions and inverse functions

In this block we examine more terminology associated with functions. We explain one-to-one and many-to-one functions and show how the rule associated with certain functions can be reversed to give so-called inverse functions. These ideas will be needed when we deal with particular functions in later blocks.

One-to-many rules

Recall from Block 1 that a rule for a function must produce a single output for a given input. Not all rules satisfy this criterion. For example, the rule 'take the square root of the input' cannot be a rule for a function because for a given input, other than zero, there are two outputs: an input of 4 produces outputs of 2 and -2. Figure 4.1 shows two ways in which we can picture this situation, the first being a block diagram, and the second using two sets representing input and output values and the relationship between them.

Figure 4.1
This rule cannot be a function – it is a one-to-many rule.

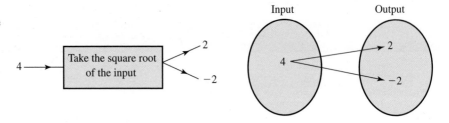

Such a rule is described as a **one-to-many rule**. This means that one input produces more than one output. This is obvious from inspecting the sets in Figure 4.1.

The graph of the rule 'take $\pm\sqrt{x}$' can be drawn by constructing a table of values:

x	0	1	2	3	4
$y = \pm\sqrt{x}$	0	± 1	$\pm\sqrt{2}$	$\pm\sqrt{3}$	± 2

The graph is shown in Figure 4.2. Plotting a graph of a one-to-many rule will result in a curve through which a vertical line can be drawn that cuts the curve more than once as shown.

Figure 4.2
A vertical line cuts the graph of a one-to-many rule more than once.

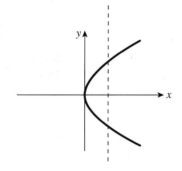

By describing a rule more specifically it is possible to make it a valid rule for a function. For example, the rule 'take the positive square root of the input' is a valid function rule because a given input produces a single output. The graph of this function is the upper branch in Figure 4.2.

Many-to-one and one-to-one functions

Consider the function $y(x) = x^2$. An input of $x = 3$ produces an output of 9. Similarly, an input of -3 *also* produces an output of 9. In general, a function for which *different* inputs can produce the *same* output is called a **many-to-one function**. This is represented pictorially in Figure 4.3, from which it is clear why we call this a many-to-one function.

Figure 4.3
This represents a many-to-one function.

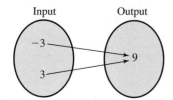

It is possible to decide whether a function is many-to-one by examining its graph. Consider the graph of $y = x^2$ shown in Figure 4.4.

Figure 4.4
The function
$y = x^2$ is a many-to-one function.

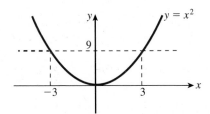

We see that a horizontal line drawn on the graph cuts it more than once. This means that two different inputs have yielded the same output and so the function is many-to-one.

If a function is not many-to-one then it is said to be **one-to-one**. This means that each different input to the function yields a different output. Consider the function $y(x) = x^3$, which is shown in Figure 4.5.

Figure 4.5
The function
$y(x) = x^3$ is a one-to-one function.

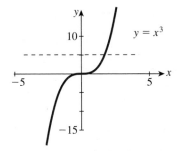

A horizontal line drawn on this graph will intersect the curve only once. This means that each input value of x yields a different output value for y.

Example 4.1 Electrical Engineering – Voltage in a circuit

The function shown in Figure 4.6 is often used to model voltage, V, in electric circuits. It is known as a sine function – this function is described in detail in Chapter 9.

Figure 4.6
A sine function used to model voltage.

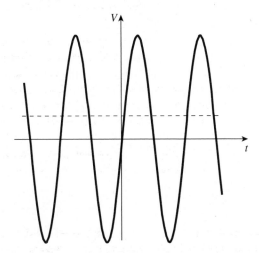

State whether this is a one-to-one function or a many-to-one function.

Solution
This is a many-to-one function as there are many values of t that correspond to the same value of voltage, V as indicated by the dashed line in Figure 4.6.

Example 4.2
Study the graphs shown in Figure 4.7. Decide which, if any, are graphs of functions. For those which are, state whether the function is one-to-one or many-to-one.

Solution

not a function

Figure 4.7(a)

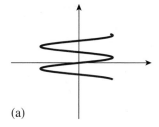

(a)

Figure 4.7(b)

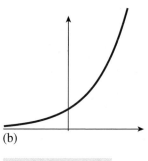

(b)

one-to-one function

Exercises

1   Explain why a one-to-many rule cannot be a function.

2   Illustrate why $y = x^4$ is a many-to-one function by providing a suitable numerical example.

3   By sketching a graph of $y = 3x - 1$ show that this is a one-to-one function.

4.3 Inverse of a function

We have seen that a function can be regarded as taking an input, x, and processing it in some way to produce a single output $f(x)$ as shown in Figure 4.8.

A natural question to ask is whether we can find a function that will reverse the process. In other words, can we find a function that will start with $f(x)$ and process it to produce x? This idea is also shown in Figure 4.8. If we can find such a function it

Figure 4.8
The second block
reverses the
process in the first.

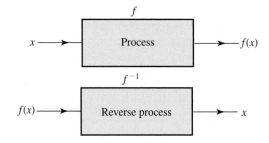

is called the **inverse function** to $f(x)$ and is given the symbol $f^{-1}(x)$. Do not confuse the '−1' with an index, or power. Here the superscript is used purely as the notation for the inverse function. Note that $f^{-1}(f(x)) = x$, as shown in Figure 4.9.

Figure 4.9
f^{-1}, if it exists,
reverses the
process in f.

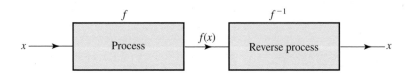

Key point

$f^{-1}(x)$ is the notation used to denote the **inverse function** of $f(x)$. The inverse function, if it exists, reverses the process in $f(x)$.

Example 4.3
Find the inverse function for $f(t) = 3t - 8$.

Solution
The function $f(t)$ takes an input, t and produces an output, $3t - 8$. The inverse function, f^{-1}, must take an input $3t - 8$ and give an output t. That is,

$$f^{-1}(3t - 8) = t$$

If we introduce a new variable $z = 3t - 8$, and transpose this to give $t = \dfrac{z + 8}{3}$, then

$$f^{-1}(z) = \frac{z + 8}{3}$$

So the rule for f^{-1} is add 8 to the input and divide the result by 3. Writing f^{-1} with t as its argument instead of z gives

$$f^{-1}(t) = \frac{t + 8}{3}$$

This is the inverse function of $f(t)$.

Example 4.4

Find the inverse function of $g(x) = 8 - 7x$.

Solution

This function takes an input x and produces an output $8 - 7x$. The inverse function g^{-1} must take an input $8 - 7x$ and produce an output x. That is,

$$g^{-1}(8 - 7x) = x$$

Introduce a new variable $z = 8 - 7x$ and transpose this for x in order to find the inverse function.

$$g^{-1}(z) = \frac{8 - z}{7}$$

Write the inverse function with an argument of x.

$$g^{-1}(x) = \frac{8 - x}{7}$$

Example 4.5

Find the inverse function of $f(x) = \dfrac{3x - 2}{x}$.

Solution

The inverse function must be such that $f^{-1}\left(\dfrac{3x - 2}{x}\right) = x$. By letting $z = \dfrac{3x - 2}{x}$ find the inverse function.

$$f^{-1}(z) = \frac{2}{3 - z}$$

Write this with x as its argument.

$$f^{-1}(x) = \frac{2}{3 - x}$$

Not all functions possess an inverse function. In fact, only one-to-one functions do so. If a function is many-to-one the process to reverse it would require many outputs from one input, contradicting the definition of a function.

Exercises

1 Explain what is meant by the inverse of a function.

2 Explain why a many-to-one function does not have an inverse function. Give an example.

3 Find the inverse of each of the following functions:
(a) $f(x) = 4x + 7$ (b) $f(x) = x$

(c) $f(x) = -23x$ (d) $f(x) = \dfrac{1}{x + 1}$

Solutions to exercises

3 (a) $f^{-1}(x) = \dfrac{x - 7}{4}$ (b) $f^{-1}(x) = x$ (c) $f^{-1}(x) = -\dfrac{x}{23}$ (d) $f^{-1}(x) = \dfrac{1 - x}{x}$

End of End of block exercises

1 By sketching a suitable graph, or otherwise, determine which of the following are one-to-one functions:
(a) $f(x) = -x$ (b) $f(x) = -3x + 7$
(c) $f(x) = x^4$ (d) $f(x) = \dfrac{1}{x}$

2 Find the inverse of the function
$f(x) = \frac{1}{7}(4x - 3)$.

3 Find the inverse of the function
$f(x) = \frac{1}{2}x + 1$.

4 If $f(x) = 5 - 4x$ find $f^{-1}(x)$. Show that $f(f^{-1}(x)) = x$.

5 If $f(t) = \dfrac{4t - 3}{5}$ show that $f^{-1}(t) = \dfrac{5t + 3}{4}$.

6 Find the inverse function of $g(x) = \dfrac{x - 1}{x + 1}$, $x \neq -1$.

Solutions to exercises

1 (a), (b) and (d) are one-to-one.

2 $f^{-1}(x) = \dfrac{7x + 3}{4}$

3 $2x - 2$

4 $f^{-1}(x) = \dfrac{5 - x}{4}$

6 $g^{-1}(x) = \dfrac{x + 1}{1 - x}$

Parametric representation of a function

We have already seen that it is possible to represent a function using the form $y = f(x)$. This is sometimes called the **cartesian form**. An alternative representation is to write expressions for y and x in terms of a third variable known as a **parameter**. Commonly the variables t or θ are used to denote the parameter.

For example, when a projectile such as a ball or rocket is thrown or launched, the x and y coordinates of its path can be described by a function in the form $y = f(x)$. However, it is often useful also to give its x coordinate as a function of the time after launch, that is $x(t)$, and its y coordinate in the same way as $y(t)$. Here time t is the parameter.

5.2 Parametric representation of a function

Suppose we write x and y in terms of t in the form

$$x = 4t \quad y = 2t^2, \quad \text{for} \quad -1 \le t \le 1 \tag{1}$$

For different values of t between -1 and 1, we can calculate pairs of values of x and y. For example, when $t = 1$ we see that $x = 4(1) = 4$ and $y = 2 \times 1^2 = 2$. That is, $t = 1$ corresponds to the point with xy coordinates $(4, 2)$.

A complete table of values is given in Table 5.1.

Table 5.1

t	-1	-0.5	0	0.5	1
x	-4	-2	0	2	4
y	2	0.5	0	0.5	2

If the resulting points are plotted on a graph then different values of t correspond to different points on the graph. The graph of (1) is plotted in Figure 5.1. The arrow on the graph shows the direction of increasing t.

It is sometimes possible to convert a parametric representation of a function into the more usual form by combining the two expressions to eliminate the parameter.

Thus if $x = 4t$ and $y = 2t^2$ we can write $t = \dfrac{x}{4}$ and so

$$y = 2t^2$$
$$= 2\left(\frac{x}{4}\right)^2$$
$$= \frac{2x^2}{16}$$
$$= \frac{x^2}{8}$$

So $y = \dfrac{x^2}{8}$. Using $y = \dfrac{x^2}{8}$ and giving values to x we can find corresponding values of y. Plotting (x, y) values gives exactly the same curve as in Figure 5.1.

Figure 5.1
Graph of the function defined parametrically by $x = 4t$, $y = 2t^2$, $-1 \le t \le 1$.

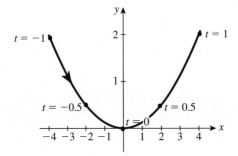

Example 5.1

Consider the function $x = \dfrac{1}{2}\left(t + \dfrac{1}{t}\right)$, $y = \dfrac{1}{2}\left(t - \dfrac{1}{t}\right)$, $1 \le t \le 8$.

(a) Draw up a table of values of this function.
(b) Plot a graph of the function.

Solution

(a) A partially completed table of values has been prepared. Complete the table.

t	1	2	3	4	5	6	7	8
x	1	1.25	1.67					4.06
y	0	0.75						3.94

(b) The graph is shown in Figure 5.2. Add your points to those already marked on the graph. The arrow on the graph shows the direction of increasing t.

It is possible to eliminate t between the two equations so that the original parametric form can be expressed as $x^2 - y^2 = 1$.

Figure 5.2
Graph of
$$x = \frac{1}{2}\left(t + \frac{1}{t}\right),$$
$$y = \frac{1}{2}\left(t - \frac{1}{t}\right),$$
$$1 \leq t \leq 8.$$

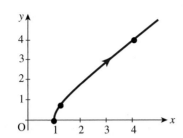

Exercises

1 Explain what is meant by the term 'parameter'.

2 Consider the parametric equations
$x = +\sqrt{t}, y = t$, for $0 \leq t \leq 10$.
(a) Draw up a table of values of t, x and y for values of t between 0 and 10.

(b) Plot a graph of this function.
(c) Obtain an explicit equation for y in terms of x.

Solution to exercise

2 (c) $y = x^2, 0 \leq x \leq \sqrt{10}$.

End of block exercises

1 Consider the parametric equations $x = 3t$,
$y = 9t^2$.
(a) Plot a graph of this function.
(b) Find an explicit expression for y in terms of x.

2 Given the parametric equations
$x = 3t + 2 \quad y = -3t + 5$
plot a graph of y against x.

3 Obtain the cartesian equation of the function defined parametrically by $x = t, y = 1 - t$, for $0 \leq t \leq 1$.

4 Plot a graph of the function defined by
$x = -\sqrt{t}, y = t$, for $t \geq 0$.

Solutions to exercises

1 $y = x^2$

3 $y = 1 - x, 0 \leq x \leq 1$

4 This is the graph of $y = x^2$ but only for $x \leq 0$.

Describing functions

6.1 Introduction

There are a number of different terms used to describe the ways in which functions behave. Some graphs have gaps or jumps known as discontinuities, others possess various symmetries. Some graphs have a pattern that repeats at regular intervals. In this block we explain some of these terms and give examples.

6.2 Continuous and discontinuous functions and limits

Figure 6.1
(a) A continuous function;
(b) a discontinuous function.

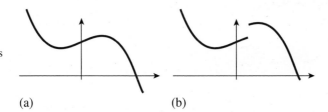

(a)   (b)

Consider the graph shown in Figure 6.1(a). The curve can be traced out from left to right without moving pen from paper. Such a function is called **continuous**. If we try to trace out the curve in Figure 6.1(b), the presence of the jump in the graph means that a pen must be lifted from the paper and moved in order to trace the graph. Such a function is said to be **discontinuous**. A jump is known as a **discontinuity**.

Example 6.1
Sketch a graph of a function that has two discontinuities.

   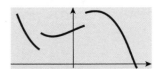

When defining a discontinuous function algebraically it is often necessary to give different function rules for different intervals of the x axis. Consider the following example.

Example 6.2

A function is defined as

$$f(x) = \begin{cases} 3 & x < 0 \\ x^2 & x \geq 0 \end{cases}$$

Notice that there is one rule for when x is less than 0 and another rule for when x is greater than or equal to 0. Sketch a graph of this function.

Solution

A graph of this function is shown in Figure 6.2. Notice the convention that ● is used to show an included end-point, and ○ is used to show an excluded end-point. This convention was introduced in Chapter 5 Block 1.

Figure 6.2
A discontinuous function.

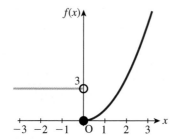

The limit of a function

Refer back to Figure 6.2. Suppose we ask 'to what value does $f(x)$ approach as x approaches 0?' Another way of saying this is 'what value is $f(x)$ close to when x is close to 0?' Note that this is not the same as asking 'what is the value of $f(x)$ when x equals 0?' If we approach from the right-hand side, then from the graph we see that as x gets nearer and nearer to 0 the value of $f(x)$ gets nearer to 0. We write this as

$$\lim_{x \to 0^+} f(x) = 0$$

and say 'the limit of $f(x)$ as x tends to 0 from above is 0'. The small superscript $+$ is used to indicate that 0 is approached from the right-hand side.

On the other hand if x gets closer to 0 from the left-hand side, the value of $f(x)$ remains at 3. We write

$$\lim_{x \to 0^-} f(x) = 3$$

and say 'the limit of $f(x)$ as x tends to 0 from below is 3'. The small superscript $-$ is used to indicate that 0 is approached from the left-hand side.

In this example the right-hand limit and the left-hand limit are not equal, and this is indicative of the fact that the function is discontinuous.

Example 6.3

Consider again the function

$$f(x) = \begin{cases} 3 & x < 0 \\ x^2 & x \geq 0 \end{cases}$$

shown in Figure 6.2.

(a) State $f(3)$.

(b) State $\lim_{x \to 3^+} f(x)$.

(c) State $\lim_{x \to 3^-} f(x)$.

Solution

(a) $f(3) = $ ▨ 9

(b) Ask 'what value does y approach when x approaches 3 from the right?'

▨ 9

(c) Ask 'what value does y approach when x approaches 3 from the left?'

▨ 9

Note that in this case the right- and left-hand limits are the same, and equal to the value of the function at $x = 3$. In such a case we say the limit of the function exists at the point $x = 3$ and write

$$\lim_{x \to 3} f(x) = 9$$

The limit exists because the function is continuous at the point where $x = 3$.

In general a function is continuous at a point $x = a$ if the left-hand and right-hand limits are the same there, and if both of these are equal to the value of the function at that point. That is,

$$\lim_{x \to a^+} f(x) = \lim_{x \to a^-} f(x) = f(a)$$

If the right-hand and left-hand limits are the same, we can simply describe this common limit as $\lim_{x \to a} f(x)$. If the limits are not the same we say the limit of the function does not exist at $x = a$.

Key point

The function $f(x)$ is continuous at $x = a$ if

$$\lim_{x \to a^+} f(x) = \lim_{x \to a^-} f(x) = f(a)$$

Exercises

1 Explain the distinction between a continuous and a discontinuous function. Draw a graph showing an example of each type of function.

2 Study graphs of the functions $y = x^2$ and $y = -x^2$. Are these continuous functions?

3 Study graphs of $y = 3x - 2$ and $y = -7x + 1$. Are these continuous functions?

4 Draw a graph of the function

$$f(x) = \begin{cases} 2x + 1 & x < 3 \\ 5 & x = 3 \\ 6 & x > 3 \end{cases}$$

Find

(a) $\lim_{x \to 0^+} f(x)$

(b) $\lim_{x \to 0^-} f(x)$

(c) $\lim_{x \to 0} f(x)$

(d) $\lim_{x \to 3^+} f(x)$

(e) $\lim_{x \to 3^-} f(x)$

(f) $\lim_{x \to 3} f(x)$

Solutions to exercises

2 yes

3 yes

4 (a) 1 (b) 1 (c) 1 (d) 6 (e) 7
(f) limit does not exist

6.3 Periodic functions

Any function that has a definite pattern repeated at regular intervals is said to be **periodic**. Each complete pattern is known as a **cycle**. The interval over which the repetition takes place is called the **period** of the function, and is usually given the symbol T. Each cycle has length T. The pattern of the function is repeated every cycle. Hence increasing or decreasing the value of t by T will not change the value of the function. Example 6.4 illustrates this behaviour. This is stated mathematically as

$$f(t) = f(t + T) \quad \text{for all values of } t \text{ in the domain of } f(t)$$

Key point

Let $f(t)$ be a periodic function with period T. Then for all values of t in the domain of $f(t)$:

$$f(t) = f(t + T)$$

Example 6.4

A periodic function is shown in Figure 6.3. The horizontal axis is labelled t, and t is measured in milliseconds.

(a) State the period, T, of this function.

(b) How many complete cycles will take place in 1 second?

Figure 6.3
A periodic
function.

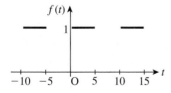

Solution

(a)   10 milliseconds

(b)   100

Exercises

1 Explain what is meant by a periodic function.

2 Sketch a graph of a periodic function that has no discontinuities.

3 Sketch a graph of a periodic function that has discontinuities.

4 A periodic function has period 0.01 seconds. How many times will the pattern in the graph repeat over an interval of 10 seconds?

Solution to exercise

4 1000

6.4   Odd and even functions

Example 6.5
Figure 6.4 shows graphs of several functions. They share a common property. Study the graphs and comment on any symmetry.

Figure 6.4

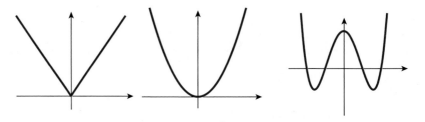

Solution

The graphs are symmetrical about the vertical axis

Any function that is symmetrical about the vertical axis – that is, the graph on the right is a mirror image of that on the left – is said to be an **even** function. Even functions have the following property:

Key point

An **even function** is such that $f(-x) = f(x)$ for all x.
The graph of an even function is symmetrical about the vertical axis.

This is saying that the function value at a negative value of x is the same as the function value at the corresponding positive value of x.

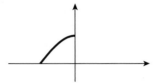

Example 6.6
Extend the graph in Figure 6.5 in order to produce a graph of an even function.

Figure 6.5
Extend this graph to produce an even function.

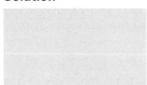

Solution

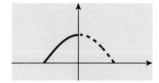

Example 6.7 Mechanical Engineering – Catenary
Figure 6.6 represents a heavy cable hanging under gravity from two points at the same height. Such a curve, known as a **catenary**, is described by a mathematical function known as a hyperbolic cosine, $f(x) = \cosh x$. The hyperbolic functions are discussed in detail in Chapter 8.

Figure 6.6
The catenary.

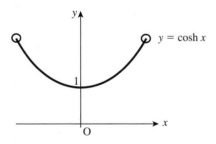

(a) Comment upon any symmetry.
(b) Is this function one-to-one or many-to-one?
(c) Is this a continuous or discontinuous function?
(d) State $\lim_{x \to 0} \cosh x$.

Solution

(a)   function is even

(b)   many-to-one

(c)   continuous

(d)   1

Example 6.8
Show algebraically that the function $f(x) = 2x^4 - 3$ is an even function.

Solution
To show that f is even we must show that $f(-x) = f(x)$. Now

$$f(-x) = 2(-x)^4 - 3$$
$$= 2x^4 - 3 \text{ since } (-x)^4 = x^4$$

But this is the same as $f(x)$. We have shown $f(-x) = f(x)$ and so f is an even function. Check for yourself that $f(-2) = f(2)$.

Imagine pivoting a graph at the origin and rotating it through $180°$. If the result is the same as the original graph there is said to be rotational symmetry of order 2 about the origin.

Example 6.9
Figure 6.7 shows graphs of several functions. They share a common property. Study the graphs and comment on any symmetry.

Figure 6.7

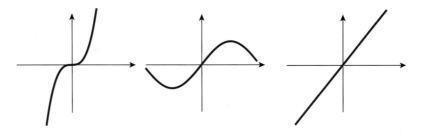

There is rotational symmetry of order 2 about the origin.

Any function that possesses rotational symmetry of order 2 – that is, the graph on the right can be obtained by rotating the curve on the left through $180°$ about the origin – is said to be an **odd** function. Odd functions have the following property:

Key point

An **odd function** is such that $f(-x) = -f(x)$ for all x.
The graph of an odd function possesses rotational symmetry of order 2 about the origin.

This is saying that the function value at a negative value of x is the negative of the function value at the corresponding positive value of x.

Example 6.10

Extend the graph in Figure 6.8 in order to produce a graph of an odd function.

Figure 6.8
Extend this graph
to produce an odd
function.

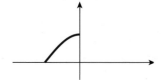

Solution

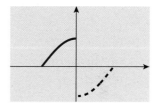

Example 6.11

Show algebraically that the function $f(x) = 6x^3 - 5x$ is an odd function.

Solution

To show that f is odd we must show that $f(-x) = -f(x)$. Now

$$f(-x) = 6(-x)^3 - 5(-x)$$
$$= -6x^3 + 5x \quad \text{since } (-x)^3 = -x^3$$
$$= -(6x^3 - 5x)$$

But this is $-f(x)$. We have shown $f(-x) = -f(x)$ and so f is an odd function. Check for yourself that $f(-2) = -f(2)$.

Exercises

1   Classify the following functions as odd, even or neither. If necessary sketch a graph to help you decide.
   (a) $f(x) = 6$ (b) $f(x) = x^2$
   (c) $f(x) = 2x + 1$ (d) $f(x) = x$
   (e) $f(x) = 2x$

2   Sketch a graph of a function that is neither odd nor even.

Solutions to exercises

1   (a) even (b) even (c) neither (d) odd (e) odd

End of block exercises

1   Sketch a graph of the function

$$f(x) = \begin{cases} x^2 & x < 0 \\ x + 2 & 0 \le x \le 2 \\ 4 & x > 2 \end{cases}$$

Find
(a) $\lim_{x \to 0^-} f(x)$
(b) $\lim_{x \to 0^+} f(x)$
(c) $\lim_{x \to 0} f(x)$

(d) $\lim_{x \to 2^-} f(x)$
(e) $\lim_{x \to 2^+} f(x)$
(f) $\lim_{x \to 2} f(x)$

2   A function is periodic with period 2 and is even. Sketch a possible form of this function.

3   A function is periodic with period 1 and is odd. Sketch a possible form of this function.

Solutions to exercises

1   (a) 0 (b) 2 (c) does not exist (d) 4 (e) 4 (f) 4

The straight line

7.1 Introduction

Probably the most important function and graph that you will use are those associated with the straight line. A large number of relationships between engineering variables can be described using a straight line or **linear** graph. Even when this is not strictly the case it is often possible to approximate a relationship by a straight line. In this block we study the equation of a straight line, its properties and graph.

7.2 Linear functions

Any function of the form $y = f(x) = ax + b$ where a and b are constants is called a **linear function**. The constant a is called the **coefficient of** x, and b is referred to as the **constant term**.

Key point

All linear functions can be written in the form

$$f(x) = ax + b$$

or simply

$$y = ax + b$$

For example,

$$f(x) = 3x + 2, \quad g(x) = \frac{1}{2}x - 7 \quad \text{and} \quad h(x) = -3x + \frac{2}{3}$$

are all linear functions. The graph of a linear function is always a straight line. Such a graph can be plotted by finding just two distinct points and joining these with a straight line.

Example 7.1
Plot the graph of the linear function $y = f(x) = 4x + 3$.

Solution
We start by finding two points. Suppose we choose $x = 0$; then $y = f(0) = 3$. So the first point has coordinates $(0, 3)$. Secondly, suppose we choose $x = 5$; then

$y = f(5) = 23$. The second point is (5, 23). These two points are then plotted and joined with a straight line as shown in Figure 7.1.

Figure 7.1
A straight line
graph can be
drawn by finding
two distinct points.

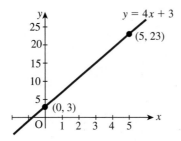

Example 7.2
Plot graphs of the three linear functions $y = 4x - 3$, $y = 4x$ and $y = 4x + 5$.

Solution
For each function it is necessary to find two points on the line.

For $y = 4x - 3$, suppose for the first point we choose $x = 0$, so that $y = -3$. For the second point, let $x = 2$ so that $y = 5$. So, the points $(0, -3)$ and $(2, 5)$ can be plotted and joined. This is shown in Figure 7.2.

Figure 7.2
These graphs show
the effect of
varying the
constant term.

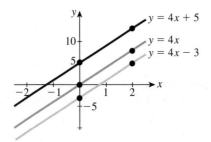

For $y = 4x$ we find the points $(0, 0)$ and $(2, 8)$. Similarly for $y = 4x + 5$ we find points $(0, 5)$ and $(2, 13)$. The corresponding lines are shown in Figure 7.2.

Example 7.3
Refer to Example 7.2. Comment upon the effect of varying the constant term of the linear function.

As different constant terms are used a series of parallel lines is generated. The value of the constant term is also known as the **vertical intercept** because this is the value of y where the line cuts the y axis.

Example 7.4
State the vertical intercept of each of the following lines:
(a) $y = 3x + 3$
(b) $y = \frac{1}{2}x - \frac{1}{3}$
(c) $y = 1 - 3x$
(d) $y = -5x$

Solution
In each case identify the constant term.

(a) For $y = 3x + 3$ the constant term is

3

(b) For $y = \frac{1}{2}x - \frac{1}{3}$ the constant term is

$-\frac{1}{3}$

(c) For $y = 1 - 3x$ the constant term is

1

(d) For $y = -5x$ the constant term is

0

Example 7.5
Plot graphs of the lines $y = 3x + 3$, $y = 5x + 3$ and $y = -2x + 3$.

Solution
Note that all three lines have the same constant term, that is 3. So all three lines pass through (0, 3), the vertical intercept. A further point has been calculated for each of the lines, and their graphs are shown in Figure 7.3.

Figure 7.3
All these graphs have the same vertical intercept but different gradients.

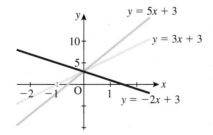

Note from the graphs in Example 7.5 that as the coefficient of x is changed the gradient of the graph changes. The coefficient of x gives the slope or **gradient** of the line. In general, for the line $y = ax + b$, a positive value of a produces a graph that slopes upwards from left to right. The larger the value of a, the steeper is the graph. This can be seen by comparing the graphs of $y = 5x + 3$ and $y = 3x + 3$.

A negative value of a produces a graph that slopes downwards from left to right. If a is zero the line is horizontal, that is its gradient is zero. These properties are summarised in Figure 7.4. Lines with the same gradient are known as **parallel lines**.

Figure 7.4
The gradient of a line $y = ax + b$ is given by the value of a.

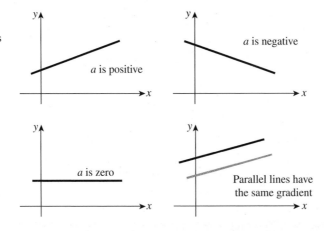

a is positive

a is negative

a is zero

Parallel lines have the same gradient

Key point

In the linear function $y = ax + b$, a is the **gradient** of the graph and b is its **vertical intercept**.

You should note that it is also common practice to write the equation of a straight line as $y = mx + c$, in which case m is the gradient of the line and c is the vertical intercept.

Example 7.6
State the gradients of the following lines:
(a) $y = 7x + 2$

(b) $y = -\dfrac{1}{3}x + 4$

(c) $y = \dfrac{x + 2}{3}$

Solution
In each case the coefficient of x must be examined.
(a) The gradient of $y = 7x + 2$ is

7

(b) The gradient of $y = -\dfrac{1}{3}x + 4$ is

$-\dfrac{1}{3}$

(c) The gradient of $y = \dfrac{x + 2}{3}$ is

$\dfrac{1}{3}$

Example 7.7
Which of the following lines has the steepest gradient?

(a) $y = \dfrac{17x + 4}{5}$ (b) $y = 9x - 2$ (c) $y = \dfrac{1}{3}x + 4$

Solution
The line with the largest gradient is

(b)

Example 7.8 Structural Engineering – Shear force in a beam
The calculation of forces in beams is important in the design of structures such as bridges and buildings. Beams are classified according to how they are supported and/or fixed at their ends. Here we consider a **simply supported beam** of length L which carries a load uniformly distributed along its length as shown in Figure 7.5. A beam is said to be simply supported if, at each end, vertical motion is restricted by the supports but the beam is free to rotate. Placing one of the supports on a roller as indicated ensures that discussion of horizontal reaction forces can be avoided. By symmetry, the vertical reaction force at each support will be the same and is denoted by R. The weight of the load being supported is usually quoted as a 'weight per unit length' w, say, which means that the total weight supported is wL.

Figure 7.5
A simply supported beam with a uniform load.

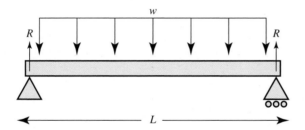

Within the beam itself there are internal forces. Imagine a vertical cut in the beam at a point x units from the left-hand end as shown in Figure 7.6. The part of the beam on the right exerts a force known as a shear force, having magnitude V, on the part that is on the left, as shown. An equal and opposite force is exerted by the left part on the right. Structural engineers need to calculate the shear force in order to assess whether the beam is capable of withstanding the load.

Figure 7.6
Shear force V.

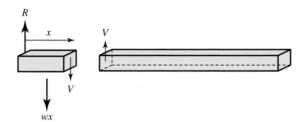

(a) By considering vertical forces for the entire beam (Figure 7.5) show that $R = \dfrac{wL}{2}$ where R is the reaction at each support point.

(b) By considering vertical forces for the short section shown in Figure 7.6 show that

$$V(x) = \frac{w}{2}(L - 2x).$$

(c) Show that $V(x)$ takes the form of an equation of a straight line and state the gradient and the vertical intercept.

(d) Sketch a graph showing the shear force at any point x.

Solution

(a) Considering forces vertically for the entire beam (Figure 7.5), the reaction R at the support points is found from

$$2R = wL$$

so that $R = \dfrac{wL}{2}$.

(b) We consider forces vertically for the short section (Figure 7.6), noting that the weight of the section is wx:

$$R = wx + V$$

Then, using the result of part (a),

$$\frac{wL}{2} = wx + V$$

so that

$$V = \frac{wL}{2} - wx = \frac{w}{2}(L - 2x)$$

(c) Comparing the equation $V = \dfrac{w}{2}(L - 2x) = \dfrac{wL}{2} - wx$ with the general equation of a straight line $y = ax + b$, we see that the gradient, a, is $-w$. The vertical intercept, b, is $\dfrac{wL}{2}$.

(d) It is straightforward to plot the graph by noting that when $x = 0$, $V = \dfrac{wL}{2}$. Similarly, when $x = L$, $V = -\dfrac{wL}{2}$. The graph is shown in Figure 7.7. The magnitude of the shear force is zero half-way along the beam, and achieves a maximum value of $\dfrac{wL}{2}$ at each end.

Figure 7.7
The shear force varies linearly along the length of the beam.

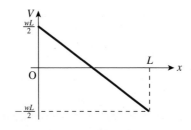

Exercises

1   State the general form of the equation of a straight line, explaining the role of each of the terms in your answer.

2   State which of the following functions will have straight line graphs:
(a) $y(x) = 3x - 3$ (b) $f(t) = t^{1/2}$
(c) $f(x) = \dfrac{1}{x}$ (d) $g(x) = 13$
(e) $f(t) = -2 - t$

3   For each of the following, identify the gradient and vertical intercept:
(a) $f(x) = 2x + 1$ (b) $f(t) = 3$
(c) $g(t) = -2t$ (d) $y(x) = -7 - 17x$
(e) $f(x) = mx + c$.

Solutions to exercises

1   For example, $y = ax + b$. x is the independent variable, y is the dependent variable, a is the gradient and b is the vertical intercept.

2   (a), (d) and (e) will have straight line graphs.

3   (a) gradient $= 2$, vertical intercept $= 1$
(b) 0, 3   (c) $-2, 0$ (d) $-17, -7$ (e) m, c

7.3   Finding the gradient of a line given two points on the line

A common requirement is to find the gradient of the line when we know two points on the line. Suppose the two points are $A(x_1, y_1)$, $B(x_2, y_2)$, as shown in Figure 7.8. The vertical distance between A and B is $y_2 - y_1$. The horizontal distance between A and B is $x_2 - x_1$. The gradient of the line is given by

$$\text{gradient} = \frac{\text{vertical distance}}{\text{horizontal distance}}$$

Figure 7.8
The coordinates of points A and B are (x_1, y_1) and (x_2, y_2) respectively.

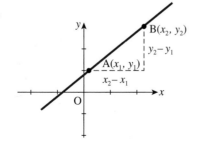

The gradient of the line joining A and B can then be calculated from the following formula.

Key point

The **gradient** of the line joining $A(x_1, y_1)$ and $B(x_2, y_2)$ is given by

$$\text{gradient} = \frac{\text{vertical distance}}{\text{horizontal distance}}$$

$$= \frac{y_2 - y_1}{x_2 - x_1}$$

Example 7.9 Electrical Engineering – Sawtooth wave

A sawtooth waveform, $f(t)$, is shown in Figure 7.9.

Figure 7.9
A sawtooth waveform.

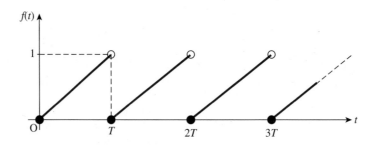

Note that $f(t)$ is a periodic function with period T. State $f(t)$ in mathematical form.

Solution
Consider $0 \leq t < T$. Here, $f(t)$ is a linear function passing through the origin and so the vertical intercept is 0. We now calculate the gradient of the line. The horizontal distance is T, the vertical distance is 1 and so the gradient is $1/T$. Hence on $0 \leq t < T, f(t)$ has the form

$$f(t) = \frac{t}{T}$$

Since $f(t)$ is periodic with period T then

$$f(t) = f(t + T) \quad \text{for all } t \geq 0$$

So $f(t)$ can be expressed as

$$f(t) = \frac{t}{T} \quad \text{for } 0 \leq t < T$$

and $f(t) = f(t + T)$ for all $t \geq 0$.

Example 7.10

Find the gradient of the line joining each of the following pairs of points:
(a) $A(0, 3)$ and $B(4, 5)$
(b) $A(-1, 4)$ and $B(2, 1)$

Solution

(a) We calculate the gradient as follows:

$$\text{gradient} = \frac{y_2 - y_1}{x_2 - x_1}$$

$$= \frac{5 - 3}{4 - 0}$$

$$= \frac{2}{4}$$

$$= \frac{1}{2}$$

Thus the gradient of the line is $\frac{1}{2}$. Graphically, this means that if x increases by 1 unit, the value of y increases by $\frac{1}{2}$.

(b) $\text{gradient} = \frac{y_2 - y_1}{x_2 - x_1} = \qquad\qquad \frac{1 - 4}{2 - (-1)} = -1$

Thus the gradient of the line is -1. Graphically, this means that for every unit increase in x, the value of y decreases by 1 unit.

Example 7.11

A straight line has equation $y = 3x + 7$. State, without calculation, the increase in y obtained from a unit increase in x.

Solution

The gradient of this line is 3. This means that y increases by 3 units for every unit increase in x.

Example 7.12 Mechanical Engineering – Tension in a spring

The tension T in a spring when it is extended by a distance e is given by $T = 0.25e$. Calculate the increase in tension that follows from a unit increase in extension.

Solution

$T = 0.25e$ is a linear function with gradient 0.25. Thus a unit increase in extension results in an increase in tension of 0.25 units.

This is an example of Hooke's law, which states that the tension is proportional to the extension. Hooke's law is a mathematical model which assumes the extended spring remains within its limits of elasticity.

Exercises

1 Calculate the gradient of the line joining $(1, 0)$ and $(15, -3)$.

2 Calculate the gradient of the line joining $(10, -3)$ and $(15, -3)$.

3 State the increase or decrease in y which follows from increasing x by 1 unit in each of the following cases:
(a) $y = -17x + 2$ (b) $y = \frac{1}{3}x + 2$
(c) $y = 3$

Solutions to exercises

1 $-\dfrac{3}{14}$

2 0

3 (a) decrease of 17 (b) increase of $\frac{1}{3}$ (c) y is constant so there is no change in y

7.4   Finding the equation of a line

There are a number of ways in which the equation of a line can be found. Much depends upon the information that is available. We may know

- the gradient of the line and the coordinates of one point through which the line passes
- the coordinates of two points on the line.

In what follows remember that any straight line has the form $y = ax + b$, where a is the gradient of the line and b is the vertical intercept.

Finding the equation of a line given its gradient and the coordinates of one point on the line

Consider the following example.

Example 7.13
Find the equation of the straight line that has gradient 11 and passes through the point with coordinates (2, 3).

Solution
In this example let us suppose that the equation of the line has the form $y = mx + c$, in which case m is the gradient and c is the vertical intercept. We are told the gradient is 11 and so $m = 11$. The equation is therefore

$$y = 11x + c$$

We are told that the line passes through the point (2, 3). Thus when $x = 2$, y must equal 3. Therefore, substituting these values gives

$$3 = 11(2) + c$$

That is,

$$3 = 22 + c$$

It follows, by solving this equation, that c equals -19. A thorough treatment of equations like this is given in Chapter 7. The equation of the line is therefore $y = 11x - 19$.

Example 7.14
Find the equation of the line with gradient 7 and which passes through $(5, 6)$.

Solution
Let the equation be $y = mx + c$.
   Use the information given about the gradient of the line.

$$y = 7x + c$$

Use the fact that the line passes through $(5, 6)$ to find c.

$$y = 7x - 29$$

Finding the equation of a line given two points on the line

Consider the following example.

Example 7.15
Find the equation of the line passing through $(2, 1)$ and $(-4, 2)$.

Solution
The equation of the line must take the form $y = ax + b$. We are told that $(2, 1)$ lies on the line, so when $x = 2, y = 1$. Similarly $(-4, 2)$ lies on the line, so when $x = -4, y = 2$. Substituting these pairs of values into $y = ax + b$ produces the two equations

$$1 = 2a + b \quad \text{and} \quad 2 = -4a + b$$

These are known as **simultaneous equations**. Solution of simultaneous equations is described in detail in Chapter 7. It can be shown that values of a and b which satisfy both equations are

$$a = -\frac{1}{6}, \quad b = \frac{4}{3}$$

So the equation of the line is $y = -\frac{1}{6}x + \frac{4}{3}$.

An alternative way of finding the equation of the line is by use of a formula.
   The equation of the line passing through the points with coordinates $A(x_1, y_1)$ and $B(x_2, y_2)$ is given by the following.

Key point

The line passing through points $A(x_1, y_1)$ and $B(x_2, y_2)$ is given by

$$\frac{y - y_1}{y_2 - y_1} = \frac{x - x_1}{x_2 - x_1}$$

Example 7.16
Find the equation of the line passing through A$(-7, 11)$ and B$(1, 3)$.

Solution
Here $x_1 = -7$ and $x_2 = 1$. Write down y_1 and y_2.

$$y_1 = 11, y_2 = 3$$

Apply the formula:

$$\frac{y - y_1}{y_2 - y_1} = \frac{x - x_1}{x_2 - x_1} =$$ $$\frac{y - 11}{3 - 11} = \frac{x + 7}{1 + 7}$$

Simplify this to obtain the required equation.

$$y = 4 - x$$

Example 7.17 Mechanical Engineering – Extension of a spring
A spring has a natural (unstretched) length of 60 cm. The length is extended by adding a mass to the spring. Every 1 kg added extends the spring by 2 cm. If l cm is the length of the spring when m kg is applied, determine an equation for l in terms of m.

Solution
When 1 kg is applied the extension is 2 cm. So when m kg is applied the extension is $2m$ cm. The length of the spring is made up of its natural length plus the extension

$$l = \text{natural length} + \text{extension}$$
$$= 60 + 2m$$

So $l = 60 + 2m$. This is the equation of a straight line.

Exercises

1   Find the equation of the line joining $(1, 5)$ and $(-9, 2)$.

2   Find the gradient and vertical intercept of the line joining $(8, 1)$ and $(-2, -3)$.

Solutions to exercises

1   $y = \dfrac{3}{10}x + \dfrac{47}{10}$

2   $0.4, -2.2$

7.5   Finding the distance between two points on a line

Referring again to Figure 7.8, the distance between the points A(x_1, y_1) and B(x_2, y_2) is given by the following formula, which follows immediately from Pythagoras's theorem. This theorem is described in detail in Chapter 10.

Key point	The distance between points $A(x_1, y_1)$ and $B(x_2, y_2)$ is given by

$$\text{distance} = \sqrt{(x_2 - x_1)^2 + (y_2 - y_1)^2}$$

Example 7.18
Find the distance between $A(-7, 11)$ and $B(1, 3)$.

Solution
Apply the formula:

$$\text{distance} = \sqrt{(x_2 - x_1)^2 + (y_2 - y_1)^2} = \qquad\qquad \sqrt{128}$$

Exercise

1   Find the distance between the points $(4, 5)$ and $(-17, 1)$.

Solution to exercise

1   $\sqrt{457}$

End of block exercises

1   State the gradient and vertical intercept of
(a) $y = 8x - 3$ (b) $y = 3t - 2$
(c) $y = 9$ (d) $y = -3t$ (e) $f(x) = -3 - 4x$

2   An alternative way of finding the equation of a straight line through the points $(x_1, f(x_1))$, $(x_2, f(x_2))$ is by use of Lagrange's formula. This states that

$$f(x) = \frac{x - x_2}{x_1 - x_2} f(x_1) + \frac{x - x_1}{x_2 - x_1} f(x_2)$$

Use this formula to find the equation of the line through $(4, 1)$ and $(7, -5)$.

3   Find the equation of the line that passes through $A(0, 3)$ and $B(11, -1)$.

4   Find the gradient of the line that passes through $A(-9, 1)$ and $B(2, 16)$.

5   Find the distance between the points with coordinates $(9, 1)$ and $(12, 1)$.

6   Find the distance between the points with coordinates $(19, -2)$ and $(-12, 1)$.

7   The point $A(x_1, y_1)$ lies on the line $y = -2x + 3$. If the value of x_1 is increased by 7, what is the resulting change in the value of y_1?

8   Find the equation of the line passing through $A(2, -1)$ and $B(5, 8)$.

9   A spring has length 90 cm when a mass of 10 kg is applied and a length of 81 cm when a 4 kg mass is applied. If l cm is the length of the spring when mass m kg is applied
(a) find an equation expressing l in terms of m
(b) calculate the length of the spring when a mass of 7 kg is applied.

Solutions to exercises

1 (a) gradient 8, vertical intercept -3
(b) 3, -2 (c) 0, 9 (d) $-3, 0$
(e) $-4, -3$

2 $f(x) = -2x + 9$

3 $y = -\dfrac{4}{11}x + 3$

4 $\dfrac{15}{11}$

5 3

6 $\sqrt{970} = 31.14$

7 -14

8 $y = 3x - 7$

9 (a) $l = 75 + 1.5m$ (b) 85.5 cm

Common engineering functions

8.1 Introduction

This block provides a catalogue of some functions commonly met in engineering. These include polynomials, rational functions, the modulus function, the unit step function and the unit impulse or delta function. Important properties and definitions are stated. This block can be used as a reference when the need arises throughout the rest of the book. There are, of course, other types of function that arise in engineering applications, such as trigonometrical, exponential and logarithmic functions. These are dealt with in later chapters.

8.2 Polynomial functions

A **polynomial expression** is one made up of multiples of non-negative whole number powers of a variable, such as $3x^2$, $-7x^3$, and so on. You are already familiar with many such expressions:

$$x^2 + 3x + 9, \quad 3t^4 - 2t^3 + 7t^2 - 11t, \quad 8x + 9$$

are all polynomial expressions.

Polynomial expressions are used to define polynomial functions. Polynomial functions include

$$P_1(x) = 3x^2 - x + 2$$
$$P_2(z) = 7z^4 + z^2 - 1$$
$$P_3(t) = 3t + 9$$
$$P_4(t) = 6$$

where x, z and t are independent variables. Purely for convenience, we have labelled these polynomials P_1, P_2, P_3 and P_4.

When the context is clear, both polynomial functions and polynomial expressions are loosely called **polynomials**.

Note that fractional and negative powers of the independent variable are not allowed, so that $f(x) = x^{-1}$ and $g(x) = x^{3/2}$ are not polynomials. The function $P_4(t) = 6$ is a polynomial – we can regard it as $6t^0$.

By convention a polynomial is written with the powers either increasing or decreasing. For example,

$$3x + 9x^2 - x^3 + 2$$

would be written as either

$$-x^3 + 9x^2 + 3x + 2 \quad \text{or} \quad 2 + 3x + 9x^2 - x^3$$

In general we have the following definition.

Key point

A **polynomial expression** has the form

$$a_n x^n + a_{n-1} x^{n-1} + a_{n-2} x^{n-2} + \cdots + a_2 x^2 + a_1 x + a_0$$

where n is a non-negative integer, $a_n, a_{n-1}, \ldots, a_1, a_0$ are constants and x is a variable.
A **polynomial function** $P(x)$ has the form

$$P(x) = a_n x^n + a_{n-1} x^{n-1} + a_{n-2} x^{n-2} + \cdots + a_2 x^2 + a_1 x + a_0$$

The **degree** of a polynomial is the value of the highest power. Referring to the examples listed above, polynomial P_1 has degree 2, because the term with the highest power is $3x^2$, P_2 has degree 4, P_3 has degree 1 and P_4 has degree 0. Polynomials with low degrees have the special names given in Table 8.1.

Table 8.1

	Degree	Name
$ax^4 + bx^3 + cx^2 + dx + e$	4	Quartic
$ax^3 + bx^2 + cx + d$	3	Cubic
$ax^2 + bx + c$	2	Quadratic
$ax + b$	1	Linear
a	0	Constant

Typical graphs of some polynomial functions are shown in Figure 8.1. In particular, observe that the graphs of the linear polynomials P_1 and P_2 are straight lines.

Figure 8.1
Graphs of some typical linear, quadratic and cubic polynomials.

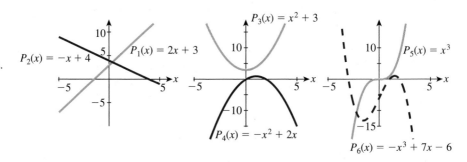

Another important feature of all polynomial graphs is that they are continuous.

Example 8.1
Which of the polynomial graphs in Figure 8.1 are odd and which are even? Are any periodic?

Solution

P_3 is even, P_5 is odd. None is periodic.

Example 8.2
State which of the following are polynomial functions. For those that are, give the degree and name.

(a) $f(x) = 6x^2 + 7x^3 - 2x^4$

(b) $f(t) = t^3 - 3t^2 + 7$

(c) $g(x) = \dfrac{1}{x^2} + \dfrac{3}{x}$

(d) $f(x) = 16$

(e) $g(x) = \frac{1}{6}$

Solution
Remember that a polynomial is built up by adding terms involving non-negative whole number powers of the independent variable, and that the degree is the value of the highest power.

(a)   polynomial of degree 4 (quartic)

(b)   polynomial of degree 3 (cubic)

(c)   not a polynomial

(d)   polynomial of degree 0 (constant)

(e)   polynomial of degree 0 (constant)

Example 8.3 Dynamics – Motion with constant acceleration
Time, t, distance, s, initial velocity, u, final velocity, v, and constant acceleration, a, are related by the following equations:

$$s = ut + \frac{1}{2}at^2 \tag{1}$$

$$v = u + at \tag{2}$$

Note that (1) is a polynomial of degree 2 in t; equation (2) is a polynomial of degree 1 (i.e. linear) in t.

Example 8.4 Structural engineering – Bending moment in a beam
In Example 7.8 we introduced the shear force in a simply supported beam of length L carrying a uniform load, w per unit length (Figure 8.2).

Figure 8.2
A simply supported beam with a uniform load.

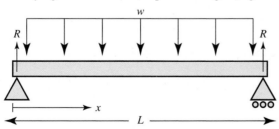

Another quantity of importance to a structural engineer is the **bending moment**, which is a measure of the internal stress caused when the beam is loaded. A bending moment is not a force but a turning effect, or a tendency to cause a rotation about an axis, and arises through the interaction of internal forces in the beam. Generally a moment is calculated as the product of the magnitude of a force and its perpendicular distance from the axis of rotation and hence the units of bending moment are N m. Like the shear force, the bending moment, M, varies with position x along the beam. It is given by the quadratic polynomial

$$M(x) = \frac{w}{2}(Lx - x^2)$$

(a) Evaluate $M(0)$ and $M(L)$ and deduce that the bending moment is zero at both ends of the beam.

(b) Sketch a graph of this quadratic function for $0 \leq x \leq L$.

Solution

(a) Substituting $x = 0$ into $M(x) = \frac{w}{2}(Lx - x^2)$ immediately shows that $M(0) = 0$. Similarly, substituting $x = L$ we find $M(L) = \frac{w}{2}(L^2 - L^2) = 0$. We deduce that the bending moment is zero at both ends of the beam, as required.

(b) Observe that the form of this quadratic function corresponds closely to the function $P_4(x)$ illustrated in Figure 8.1 in that the coefficient of x^2 is negative and the constant term is zero. To sketch the graph it is helpful to locate the values of x where $M(x) = 0$, that is the values of x where the graph crosses the horizontal axis. We have already noted that $M(0) = 0$ and $M(L) = 0$ so the graph crosses the horizontal axis at $x = 0$ and $x = L$. This information is sufficient to sketch the graph which is shown in Figure 8.3.

Figure 8.3
A bending moment diagram for a simply supported beam with a uniform load.

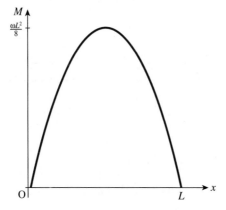

From the symmetry of the graph we see that $M(x)$ has its maximum value when $x = \frac{L}{2}$ (i.e. at the middle of the beam). This maximum value is calculated thus:

$$M\left(\frac{L}{2}\right) = \frac{w}{2}\left(L \cdot \frac{L}{2} - \left(\frac{L}{2}\right)^2\right)$$

$$= \frac{w}{2}\left(\frac{L^2}{2} - \frac{L^2}{4}\right)$$

$$= \frac{w}{2}\left(\frac{L^2}{4}\right)$$

$$= \frac{wL^2}{8}$$

This allows the value $\dfrac{wL^2}{8}$ to be indicated on the vertical axis.

An alternative approach to finding the points where the graph crosses the x axis requires the solution of the equation

$$\frac{w}{2}(Lx - x^2) = 0$$

which is the topic of Chapter 7 Block 2.

Exercises

1 Write down a polynomial of degree 3 with independent variable t.

2 Write down a function that is not a polynomial.

3 Explain why $y = 1 + x + x^{1/2}$ is not a polynomial.

4 State the degree of the following polynomials:
(a) $P(t) = t^4 + 7$ (b) $P(t) = -t^3 + 3$
(c) $P(t) = 11$ (d) $P(t) = t$

5 Write down a polynomial of degree 0 with independent variable z.

6 Referring to Figure 8.1, state which functions are one-to-one and which are many-to-one.

Solutions to exercises

1 For example, $f(t) = 1 + t + 3t^2 - t^3$

2 For example, $y = \dfrac{1}{x}$

3 A term such as $x^{1/2}$, with a fractional index, is not allowed in a polynomial.

4 (a) 4 (b) 3 (c) 0 (d) 1

5 $P(z) = 13$, for example

6 P_1, P_2 and P_5 are one-to-one. The rest are many-to-one.

8.3 Rational functions

A rational function is formed by dividing one polynomial by another. Examples include

$$R_1(x) = \frac{x + 2}{x^2 + 1}, \quad R_2(t) = \frac{t^3 - 1}{2t + 3}, \quad R_3(z) = \frac{2z^2 + z - 1}{z^2 + z - 2}$$

For convenience we have labelled these rational functions R_1, R_2 and R_3.

Key point

A **rational function** has the form

$$R(x) = \frac{P(x)}{Q(x)}$$

where P and Q are polynomial expressions; P is called the **numerator** and Q is called the **denominator.**

The graphs of rational functions can take a variety of different forms and can be difficult to plot by hand. Use of a graphics calculator or computer software can help. If you have access to a plotting package or calculator it would be useful to obtain graphs of these functions for yourself. The next three examples allow you to explore some of the features of the graphs.

Example 8.5

Study the graph in Figure 8.4 and the algebraic form of the function $R_1(x) = \dfrac{x + 2}{x^2 + 1}$ carefully and try to answer the following questions.

Figure 8.4
Graph of
$$R_1(x) = \frac{x + 2}{x^2 + 1}.$$

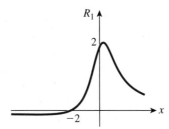

(a) For what values of x, if any, is the denominator zero?
(b) For what values of x, if any, is the denominator negative?
(c) For what values of x is the function negative?
(d) What is the value of the function when x is zero?
(e) What happens to the function if x gets very large (say 10, 100, . . .)? Substitute some values to see.

Solution

(a) $x^2 + 1$ is never zero

(b) $x^2 + 1$ is never negative

(c)   only when x is less than -2

(d)   2

(e) R_1 approaches zero because the x^2 term in the denominator becomes very large.

Note that for large x values the graph gets closer and closer to the x axis. We say that the x axis is a **horizontal asymptote** of this graph.

Asking yourself, and answering, questions such as these will help you to sketch graphs of rational functions.

Example 8.6

Study the graph in Figure 8.5 and the algebraic form of the function $R_2(t) = \dfrac{t^3 - 1}{2t + 3}$ carefully and try to answer the following questions.

Figure 8.5
Graph of
$R_2(t) = \dfrac{t^3 - 1}{2t + 3}$.

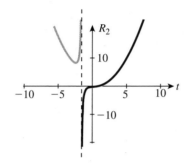

(a) What is the function value when $t = 1$?
(b) What is the value of the denominator when $t = -\frac{3}{2}$?
(c) What do you think happens when $t = -\frac{3}{2}$?

Solution

(a)   0

(b)   0

(c)   When $t = -\frac{3}{2}$ the function is not defined. An explanation is given below.

Note from parts (b) and (c) that we must exclude the value $t = -\frac{3}{2}$ from the domain of this function because division by 0 is not defined. When t approaches $-\frac{3}{2}$ from below, R_2 approaches positive infinity. When t approaches $-\frac{3}{2}$ from above, R_2 approaches negative infinity. Thus there is a discontinuity at $t = -\frac{3}{2}$. The dashed line in Figure 8.5 is $t = -\frac{3}{2}$. This line is approached by the curve as t approaches $-\frac{3}{2}$. It is known as a **vertical asymptote**.

Example 8.7

Study the graph in Figure 8.6 and the algebraic form of the function $R_3(z) = \dfrac{2z^2 + z - 1}{(z - 1)(z + 2)}$ carefully and try to answer the following questions.

Figure 8.6
Graph of
$R_3(z)$
$$= \frac{2z^2 + z - 1}{(z - 1)(z + 2)}.$$

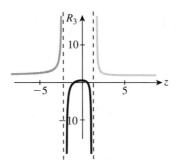

(a) What is happening to the graph when z approaches -2 or 1?
(b) Which values should be excluded from the domain of this function?
(c) Try substituting some large values for z (e.g. 10, 100, . . .). What happens to R_3 as z gets large?
(d) Is there a horizontal asymptote?
(e) What is the name given to the vertical lines $z = 1$ and $z = -2$?

Solution

(a)    denominator is zero, R^3 tends to $+$infinity as z approaches -2 from below or 1 from above. It tends to $-$infinity as z approaches -2 from above or 1 from below.

(b) $z = -2$ and $z = 1$

(c) R_3 approaches the value 2

(d) $y = 2$ is a horizontal asymptote

(e)   vertical asymptotes

Examples 8.5–8.7 are intended to give you some guidance so that you will be able to sketch rational functions of your own. Each function must be looked at individually, but some general guidelines are given below:

1 Find the value of the function when the independent variable is zero. This is generally easy to evaluate and gives you a point on the graph.
2 Find values of the independent variable that make the denominator zero. These values must be excluded from the domain of the function and give rise to vertical asymptotes.
3 Find values of the independent variable that make the dependent variable zero. This gives you points where the graph cuts the horizontal axis (if at all).
4 Study the behaviour of the function when x is positive and large, and negative and large.
5 Are there any vertical or horizontal asymptotes? Oblique asymptotes can also occur but these are beyond the scope of this book.

6 When you have studied differentiation you will know a further technique that will enable you to locate maximum and minimum points of a function.

It is particularly important for engineers to find values for which the denominator is zero. These values are known as the **poles** of the rational function.

Example 8.8
State the poles of the following rational functions:

(a) $f(t) = \dfrac{t - 3}{t + 7}$

(b) $F(s) = \dfrac{s + 7}{(s + 3)(s - 3)}$

(c) $r(x) = \dfrac{2x + 5}{(x + 1)(x + 2)}$

Solution
In each case we locate the poles by seeking values of the independent variable that make the denominator zero.

(a) The denominator of $f(t) = \dfrac{t - 3}{t + 7}$ is zero when

   $t = $ -7

(b) The denominator of $F(s) = \dfrac{s + 7}{(s + 3)(s - 3)}$ is zero when

   $s = $ $3 \text{ or } -3$

(c) The denominator of $r(x) = \dfrac{2x + 5}{(x + 1)(x + 2)}$ is zero when

   $x = $ $-1 \text{ or } -2$

In each case the calculated values are the poles of the rational function. If you have access to a plotting package, plot these functions now.

Exercises

1 Explain what is meant by a rational function.

2 State the degree of the numerator and the degree of the denominator of the rational function

$$R(x) = \frac{3x^2 + x + 1}{x - 1}$$

3 Explain the term 'pole' of a rational function.

4 Referring to Figures 8.4, 8.5 and 8.6, state which functions, if any, are one-to-one and which are many-to-one.

5 Without using a graphical calculator plot graphs of $y = \dfrac{1}{x}$ and $y = \dfrac{1}{x^2}$. Comment upon whether these graphs are odd, even or neither, whether they are continuous or discontinuous, and state the position of any poles.

Solutions to exercises

1 $R(x) = \dfrac{P(x)}{Q(x)}$

where P and Q are polynomials.

2   2, 1, respectively

3   The pole is a value of the independent variable that makes the denominator zero.

4   All are many-to-one.

5 $\dfrac{1}{x}$ is odd, and discontinuous. Pole at $x = 0$.

$\dfrac{1}{x^2}$ is even and discontinuous. Pole at $x = 0$.
Neither of these functions is defined at $x = 0$, so this value should be excluded from the domain.

8.4   The modulus function

The modulus of a number is the size of that number with no regard paid to its sign. For example, the modulus of -7 is 7. The modulus of $+7$ is also 7. We can write $|-7| = 7$ and $|+7| = 7$. The modulus function is defined as follows:

Key point

> The **modulus function** is defined as
>
> $$f(x) = |x|$$

The output from this function is simply the modulus of the input.

A graph of this function is shown in Figure 8.7. Note that this graph is continuous. It is smooth everywhere except at the origin where there is a corner.

Figure 8.7
Graph of the modulus function.

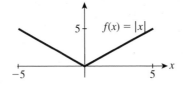

Study Figure 8.7 and note that when x is positive the graph is the same as that of $y = x$. When x is negative the graph is the same as that of $y = -x$. Consequently we can also write the modulus function in the form

$$f(x) = |x| = \begin{cases} x & x \geq 0 \\ -x & x < 0 \end{cases}$$

Example 8.9
Draw up a table of values of the function $f(x) = |x - 2|$ for values of x between -3 and 5. Sketch a graph of this function.

Solution

The table has been started. Complete it for yourself.

x	-3	-2	-1	0	1	2	3	4	5
$f(x)$	5		3	2		0			

The graph is shown in Figure 8.8. Plot your calculated points on the graph.

Figure 8.8
Graph for
Example 8.9.

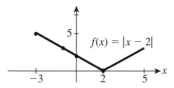

Plot: $f(x) = |x - 2|$

Exercises

1   Sketch a graph of the following functions:
    (a) $f(x) = 3|x|$ (b) $f(x) = |x + 1|$
    (c) $f(x) = 7|x - 3|$

2   Is the modulus function one-to-one or many-to-one?

Solutions to exercises

2   Many-to-one

8.5   The unit step function

The unit step function is defined as follows:

Key point

$$u(t) = \begin{cases} 1 & t \geq 0 \\ 0 & t < 0 \end{cases}$$

Study this definition carefully. You will see that it is defined in two parts, with one expression to be used when t is greater than or equal to 0, and another expression to be used when t is less than 0. The graph of this function is shown in Figure 8.9, from which it is obvious why it is called a step function. Such functions are useful in engineering applications when we wish to model a quantity that is 'off' but which is 'switched on' at $t = 0$.

Figure 8.9
Graph of the unit
step function.

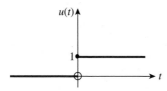

There is a discontinuity in the graph when $t = 0$. That is why we need to define the function in two parts: one part for when t is negative, and one part for when t is non-negative. The point with coordinates $(0, 1)$ is part of the function defined on $t \geq 0$.

The position of the discontinuity may be shifted to the left or right. The graph of $u(t - d)$ is shown in Figure 8.10.

Figure 8.10
Graph of $u(t - d)$.

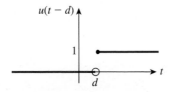

In Figures 8.9 and 8.10 the function takes the value 0 or 1. We can adjust the value 1 by multiplying the function by any other number we choose. The graph of $2u(t - 3)$ is shown in Figure 8.11.

Figure 8.11
Graph of
$2u(t - 3)$.

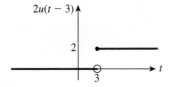

Example 8.10
Sketch the functions
(a) $A(t) = u(t) - u(t - 1)$
(b) $B(t) = u(t) - u(t - 1) + u(t - 2)$
(c) $C(t) = 2u(t) - u(t - 3)$

Solution
(a) Figure 8.9 above shows $u(t)$. By considering Figure 8.10 with $d = 1$ we can sketch $u(t - 1)$ and this is shown in Figure 8.12.

Figure 8.12
The function
$u(t - 1)$.

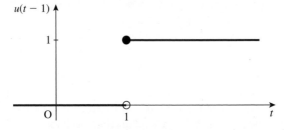

Looking at the two graphs referred to, we see that $u(t) - u(t - 1)$ is 0 everywhere except for the interval $0 \leq t < 1$ where the function has a value of 1. This is illustrated in Figure 8.13.

Figure 8.13
The function $A(t) = u(t) - u(t-1)$.

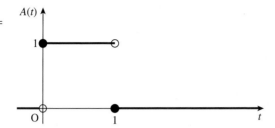

Note that in engineering terms this is a unit pulse of duration 1.

(b) The function $B(t) = u(t) - u(t-1) + u(t-2)$ can be thought of as $B(t) = [u(t) - u(t-1)] + u(t-2) = A(t) + u(t-2)$. We note that $u(t-2)$ is 0 for $t < 2$ and 1 for $t \geq 2$. Adding $u(t-2)$ to $A(t)$ yields the graph shown in Figure 8.14.

Figure 8.14
The function $u(t) - u(t-1) + u(t-2)$.

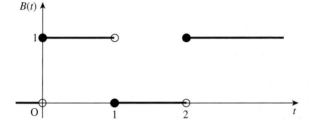

(c) Figure 8.15 (a), (b) shows the functions $f(t) = 2u(t)$ and $g(t) = u(t-3)$ respectively.

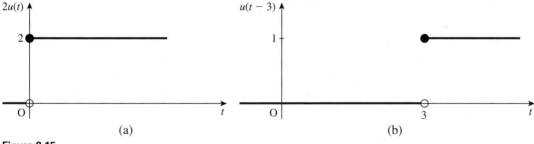

Figure 8.15
(a) The function $f(t) = 2u(t)$; (b) the function $g(t) = u(t-3)$.

By considering these two graphs the graph of $C(t) = f(t) - g(t) = 2u(t) - u(t-3)$ is then evident. The value of $2u(t)$ is reduced by 1 for $t \geq 3$ owing to subtracting the function $u(t-3)$. This results in the graph shown in Figure 8.16.

Figure 8.16
The function $2u(t) - u(t-3)$.

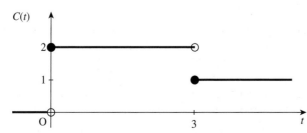

Example 8.11 Electronic Engineering – Pulse wave
A pulse wave, $f(t)$, of magnitude M and period T is illustrated in Figure 8.17.

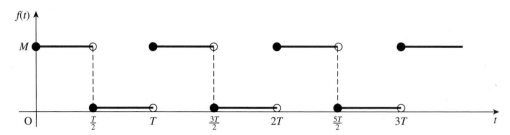

Figure 8.17

Express $f(t)$ using the unit step function, $u(t)$.

Solution
By drawing upon Example 8.10 we can state that

$$f(t) = M\left[u(t) - u\left(t - \frac{T}{2}\right) + u(t - T) - u\left(t - \frac{3T}{2}\right) + u(t - 2T) - \cdots\right]$$

Note that this can be expressed as an infinite sum using the sigma notation:

$$f(t) = M\sum_{0}^{\infty}(-1)^i u\left(t - \frac{iT}{2}\right)$$

Example 8.12 Electronic Engineering – Sawtooth pulse
A sawtooth pulse, $f(t)$, of duration T, starting at $t = 0$, is illustrated in Figure 8.18.

Figure 8.18
Graph showing
$f = 0$ for $t < 0$,
$f = t$ for
$0 < t < T$ and
$f = 0$ for $t \geq T$.

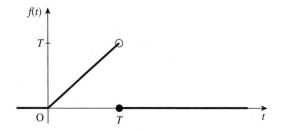

Note that for $t \geq T$ the sawtooth pulse is 0. Express $f(t)$ in terms of the unit step function.

Solution
Consider the function $g(t) = tu(t)$. Since $u(t) = 0$ for $t < 0$ and $u(t) = 1$ for $t \geq T$, then $g(t)$ may be expressed as

$$g(t) = \begin{cases} 0 & \text{for} \quad t < 0 \\ t & \text{for} \quad t \geq 0 \end{cases}$$

A graph of $g(t)$ is shown in Figure 8.19(a).

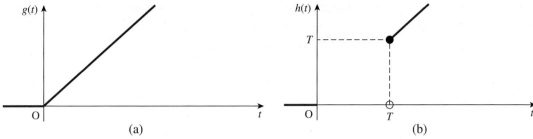

Figure 8.19

Consider now the function $h(t) = tu(t - T)$. Noting that $u(t - T) = 0$ for $t < T$, and $u(t) = 1$ for $t \geq T$, we can express $h(t)$ as

$$h(t) = \begin{cases} 0 & \text{for} \quad t < T \\ t & \text{for} \quad t \geq T \end{cases}$$

A graph of $h(t)$ is shown in Figure 8.19(b).

Finally we consider $g(t) - h(t)$. By looking at the graphs of $g(t)$ and $h(t)$ we can see that $g(t) - h(t)$ is simply the function shown in Figure 8.18, that is

$$f(t) = g(t) - h(t) = tu(t) - tu(t - T) = t[u(t) - u(t - T)]$$

Exercises

1 Sketch graphs of the following functions:
(a) $u(t)$ (b) $-u(t)$ (c) $u(t - 1)$ (d) $u(t + 1)$
(e) $u(t - 3) - u(t - 2)$ (f) $3u(t)$
(g) $-2u(t - 3)$

2 Sketch a graph of the function $t^2 u(t - 1)$.

3 A **periodic square wave**, $f(t)$, is illustrated in Figure 8.20.
   Note that $f(t)$ has period T. State $f(t)$ in terms of the unit step function.

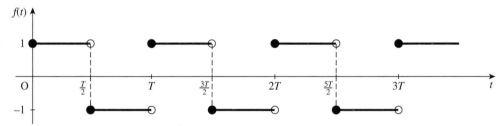

Figure 8.20

Solutions to exercises

3 $f(t) = u(t) - 2u\left(t - \dfrac{T}{2}\right) + 2u(t - T) - 2u\left(t - \dfrac{3T}{2}\right) + 2u(t - 2T) - \cdots$

This can be expressed as

$$u(t) + 2\sum_{1}^{\infty} (-1)^i u\left(t - \dfrac{iT}{2}\right)$$

8.6 The delta function, or unit impulse function, $\delta(t)$

Consider the rectangular-shaped function $R(t)$ shown in Figure 8.21. The base of the rectangle has width h, and its height is $\dfrac{1}{h}$. This means that the area of the rectangle is 1.

A function of particular interest to control engineers is the **impulse function** or **delta function**, written $\delta(t)$, which is obtained by allowing h to become smaller and

Figure 8.21
A rectangular-shaped function of width h and height $\dfrac{1}{h}$.

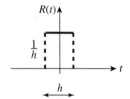

smaller whilst the height of the rectangle is allowed to grow, so that the total area enclosed remains constant at 1. Thus

$$\delta(t) = \text{a rectangle function for which } h \text{ is allowed to approach zero}$$

So this function can be thought of as being zero everywhere except at the origin, when it is infinitely large. In reality a function that has infinite size and which lasts for zero time is impossible. However, this function can be used as a model of a situation in which, for example, a large voltage pulse is applied to an electrical system for a very short period of time.

Because the area under the graph is 1 we say that $\delta(t)$ represents an impulse of strength 1. The function $k\delta(t)$ encloses an area k and is referred to as an impulse of strength k.

Because an infinitely large value cannot be drawn on a graph, the delta function is often illustrated as in Figure 8.22 where the height of the arrow represents the strength of the impulse.

The position of the delta function can be shifted to the left or the right. $\delta(t - d)$ is an impulse occurring at $t = d$.

Figure 8.22
The height of the arrow represents the strength of the impulse.

End of block exercises

1   State the degree of the following polynomials:
    (a) $(t - 1)(t - 2)t$ (b) $7x^2(x^3 + 3x^2)$

2   Use a graphics calculator, or computer
    software, to plot $y = x^5 - x^2 + 2$ for x
    between -2 and 2.

3   Without using a graphics calculator or
    software plot

    $$y = \frac{x - 1}{x + 2}$$

    State which, if any, values should be excluded
    from the domain of the function. State the
    position of any poles. State any asymptotes.

4   Plot a graph of

    $$f(R) = \frac{R}{1 + R} \text{ for } R \geq 0$$

    Are there any poles of this function lying
    within the given domain?

5   Sketch a graph of $u(t - 6)$, $u(t - 5)$ and
    $u(t - 6) - u(t - 5)$.

6   Sketch a graph of $-\frac{1}{2}u(t + 1)$.

7   Sketch $t^3 u(t - 1)$.

Solutions to exercises

1   (a) 3 (b) 5

3   Exclude $x = -2$. Pole at $x = -2$. Vertical
    asymptote $x = -2$. Horizontal asymptote
    $y = 1$.

4   No

End of chapter exercises

1   If $G(s) = \dfrac{s}{s^2 + \omega^2}$ find $G(s + \alpha)$.

2   If $F(s) = \dfrac{\omega}{s^2 + \omega^2}$ find $F(s + \alpha)$.

3   If $F(s) = \dfrac{1}{s + 1}$ find $5F(5s)$.

4   If $F(t) = \dfrac{1}{t + 1}$ find $sF(s)$.

5   If $f(x) = 3x^2$ find $f(t - \tau)$.

6   If $G(s) = \dfrac{1}{s + 1}$ find $G(j\omega)$.

7   If $f(x) = 3x^4$ find $f\left(\dfrac{\sqrt{15}}{2}t\right)$.

8   The **signum function** is defined as

    $$f(x) = \text{sgn}(x) = \begin{cases} +1 & x > 0 \\ -1 & x < 0 \\ 0 & (x = 0) \end{cases}$$

    (a) Sketch a graph of this function.
    (b) Is this function discontinuous or
        continuous?
    (c) Is this function odd, even or neither?
    (d) Is this function periodic?
    (e) Is this function many-to-one or one-to-one?

9   (a) Sketch a graph of the function
$$u(t - 1) - u(t - 2).$$
(b) State the position of any discontinuities.

10   The **ramp** function is defined by
$$f(x) = \begin{cases} 0 & x < 0 \\ kx & x \geq 0 \end{cases}$$
(a) Sketch a graph of this function.
(b) State the position of any discontinuities.
(c) Find $\lim_{x \to 0} f(x)$.

11   Sketch a graph of $y = x + |x|$.

12   Sketch a graph of $f(x) = u(x - 1)|x|$.

13   State the poles of the following rational functions:
(a) $F(s) = \dfrac{1}{s}$ (b) $F(s) = \dfrac{s^2 + 2s + 3}{(s + 1)^3}$

(c) $F(s) = \dfrac{1}{3s - 2}$

14   Consider the function $f(x) = 5 - 4x$. Find $f^{-1}(x)$. Show that $f(f^{-1}(x)) = f^{-1}(f(x)) = x$.

15   Find the inverse of the function
$$f(x) = \frac{1}{7}(4x - 3)$$

16   If $f(x) = x^2 + 3x$ and $g(x) = x + 2$ find
(a) $f(f(x))$, (b) $f(g(x))$, (c) $g(f(x))$, (d) $g(g(x))$.

17   State the rule that describes the function $y = 3(x^2 - 1)$.

18   Write a formula for the function given by the rule 'subtract the cube of the input from the square of the input'.

19   State the domain and range of the function $y = 3t - 17, 0 \leq t \leq 9$.

20   The **maximal domain** of a function is the largest possible domain that can be defined for that function. Find the maximal domain of the function
$$f(t) = \frac{5}{t}$$

21   Find the inverse of the function
$$f(x) = -\frac{3}{x}$$

22   Find the equation of the straight line passing through $(-1, 4)$ and $(-4, 1)$. Does the line pass through $(-2, 3)$?

23   **Electrical Engineering and Electronics – Reactance of a capacitor.** The **reactance of a capacitor** is its resistance to the passage of alternating current. Reactance, X, measured in ohms, is given by
$$X = \frac{1}{2\pi f C}$$
where f is the frequency of the current in hertz and C is the capacitance, measured in farads. Note that X is a function of f.
  Calculate the reactance when the frequency is 50 hertz and the capacitance is 10^{-6} farads.

24   **Volume of a cone.** The volume, V, of a cone with base radius, r, and vertical height, h, is given by
$$V = \frac{1}{3}\pi r^2 h$$
Note that V is a function of two variables: r and h.
(a) If the radius is doubled and the height remains constant, describe the effect on the volume.
(b) If the height is doubled and the radius is constant, describe the effect on the volume.
(c) If both radius and height are doubled, describe the effect on the volume.

25   **Dynamics – Constant acceleration.** The distance travelled, s, by a particle with initial velocity, u, and constant acceleration, a, after t seconds is
$$s(t) = ut + \frac{1}{2}at^2$$
Note that s is a function of t. If the initial velocity is 6 m s^{-1} and the acceleration is 0.5 m s^{-2} calculate the distance travelled
(a) in the first 10 seconds
(b) from $t = 10$ to $t = 20$.

26 **Extension of a spring.** A spring has a natural length of 90 cm. When a 1.5 kg mass is suspended from the spring, the length extends to 115 cm. Calculate the length when a 2.5 kg mass is suspended from the spring.

27 A curve is defined parametrically by

$$x = t + 1, \quad y = t^2 + 1$$

Obtain y explicitly in terms of x.

Solutions to exercises

1   $\dfrac{s + \alpha}{(s + \alpha)^2 + \omega^2}$

2   $\dfrac{\omega}{(s + \alpha)^2 + \omega^2}$

3   $\dfrac{5}{5s + 1}$

4   $\dfrac{s}{s + 1}$

5   $3(t - \tau)^2$

6   $\dfrac{1}{j\omega + 1}$

7   $3\left(\dfrac{\sqrt{15}}{2} t\right)^4 = \dfrac{675}{16} t^4$

8   (b) discontinuous (c) odd (d) no (e) many-to-one

9   discontinuities at $t = 1$ and $t = 2$

10   (b) none (c) 0

13   (a) $s = 0$ (b) $s = -1$ (c) $s = \dfrac{2}{3}$

14   $f^{-1}(x) = \dfrac{5 - x}{4}$

15   $f^{-1}(x) = \dfrac{7x + 3}{4}$

16   (a) $x^4 + 6x^3 + 12x^2 + 9x$ (b) $x^2 + 7x + 10$ (c) $x^2 + 3x + 2$ (d) $x + 4$

17   Square the input, subtract 1, and multiply the result by 3.

18   $f(x) = x^2 - x^3$

19   domain $[0, 9]$, range $[-17, 10]$

20   all t except $t = 0$

21   $f^{-1}(x) = -\dfrac{3}{x}$

22   $y = x + 5$
     $(-2, 3)$ lies on this line.

23   3183 Ω

24   (a) volume increased by a factor of 4 (b) volume is doubled (c) volume increases by a factor of 8

25   (a) 85 m (b) 135 m

26   131.7 cm

27   $y = x^2 - 2x + 2$

Polynomial equations, inequalities, partial fractions and proportionality

Chapter **7**

This chapter is concerned with developing algebraic techniques.

Equations are made up of mathematical expressions in which there are one or more unknown quantities. Equations arise in engineering problems when the underlying laws or physical principles are applied to model the problem.

To solve an equation means to find these unknown quantities. There are many different types of equation, and the solution of these different types is tackled in different ways. In the first instance it is necessary to recognise the sort of equation that you are dealing with. Having done this it is necessary to select an appropriate method of solution.

In this chapter several sorts of equations and methods for solving them will be described.

Inequalities are made up of mathematical expressions connected through relationships of the form 'is less than', 'is less than or equal to', 'is greater than' or 'is greater than or equal to'. Inequalities, like equations, contain unknowns which must be found. Block 5 describes two ways in which inequalities can be solved.

The chapter continues with a block on **partial fractions**. This technique allows some complicated algebraic fractions to be expressed as the sum of simpler fractions. There are many situations when this is useful, especially when performing calculations with Laplace transforms, for example in control theory.

The chapter concludes with a treatment of proportionality. The concepts of direct and inverse proportion are explained and illustrated.

Chapter 7 contents

Solving linear equations

1.1 Introduction

Many problems in engineering reduce to the solution of an equation or a set of equations. An equation is a type of mathematical expression that contains one or more unknown quantities which you will be required to find. In this block we consider a particular type of equation that contains a single unknown quantity, known as a linear equation. Later blocks will describe techniques for solving other types of equation.

1.2 Linear equations

Key point

> A **linear equation** is an equation of the form
>
> $$ax + b = 0$$
>
> where a and b are known numbers, and x represents an unknown quantity that we must find.

In the equation $ax + b = 0$, the number a is called the **coefficient of** x, and the number b is called the **constant term**.

The following are examples of linear equations:

$$3x + 4 = 0, \; -2x + 3 = 0, \; -\frac{1}{2}x - 3 = 0$$

Note that the unknown, x, appears only to the first power, that is as x, and not as x^2, $\sqrt{x}$, $x^{1/2}$, etc. Linear equations often appear in a non-standard form, and different letters are often used for the unknown quantity. For example,

$$3i - 7 = 17, \; 13 = 3z + 1 \text{ and } 1 - \frac{1}{2}y = 3$$

are all linear equations. Where necessary they can be rearranged and written in the form $ax + b = 0$. We shall explain how to do this later in this block.

Example 1.1
Which of the following are linear equations and which are not linear?
(a) $3x + 7 = 0$ (b) $-3t + 17 = 0$ (c) $3x^2 + 7 = 0$ (d) $5x = 0$

Solution
The equations that can be written in the form $ax + b = 0$ are linear.

(a) $3x + 7 = 0$ is   linear

(b) $-3t + 17 = 0$ is   linear; the unknown is t

(c) $3x^2 + 7 = 0$ is   not linear because of the term x^2

(d) $5x = 0$ is   linear; here the constant term is zero

To **solve** a linear equation means to find the value of x that can be substituted into the equation so that the left-hand side equals the right-hand side. Any such value is known as a **solution** or **root** of the equation. If a number is a root, we say that it **satisfies** the equation.

Example 1.2
Consider the linear equation $3x - 2 = 10$.
(a) Check that $x = 4$ is a solution.
(b) Check that $x = 2$ is not a solution.

Solution
(a) To check that $x = 4$ is a solution we substitute the value 4 for x and see whether both sides of the equation are equal. Evaluating the left-hand side we find $3(4) - 2$, which equals 10, the same as the right-hand side. So, $x = 4$ is a solution. We say that $x = 4$ satisfies the equation.
(b) Substituting $x = 2$ into the left-hand side we find $3(2) - 2$, which equals 4. Clearly the left-hand side is not equal to 10, and so $x = 2$ is not a solution. The number $x = 2$ does not satisfy the equation.

Example 1.3
Test which of the following values are solutions of the equation

$$18 - 4x = 26$$

(a) $x = 2$, (b) $x = -2$, (c) $x = 8$.

Solution
(a) Substituting $x = 2$, the left-hand side equals

                                                             10

But $10 \neq 26$ so $x = 2$ is not a solution.
(b) Substituting $x = -2$, the left-hand side equals

                                              $18 - 4(-2) = 26$

This is the same as the right-hand side, so $x = -2$ is a solution.

(c) Substituting $x = 8$, the left-hand side equals

$$18 - 4(8) = -14$$

But $-14 \neq 26$ and so $x = 8$ is not a solution.

Exercises

1. (a) Write down the general form of a linear equation.
   (b) Explain what is meant by the root of a linear equation.

In questions 2–8 verify that the given value is a solution of the given equation.

2. $3x - 7 = -28, x = -7$

3. $8x - 3 = -11, x = -1$

4. $2x + 3 = 4, x = \dfrac{1}{2}$

5. $\dfrac{1}{3}x + \dfrac{4}{3} = 2, x = 2$

6. $7x + 7 = 7, x = 0$

7. $11x - 1 = 10, x = 1$

8. $0.01x - 1 = 0, x = 100$

Solutions to exercises

1. (a) The general form is $ax + b = 0$ where a and b are known numbers and x represents the unknown quantity.

   (b) A root is a number that satisfies the equation.

1.3 Solving a linear equation

To solve a linear equation we try to make the unknown quantity the **subject** of the equation. This means we attempt to obtain the unknown quantity on its own on the left-hand side. To do this we may apply the same five rules used for transposing formulae given in Chapter 5 Block 7. These are given again here.

Key point

Operations that can be used in the process of solving a linear equation:

1. Add the same quantity to both sides.
2. Subtract the same quantity from both sides.
3. Multiply both sides by the same quantity.
4. Divide both sides by the same quantity.
5. Take functions of both sides: for example, square both sides.

A useful summary of these rules is 'whatever we do to one side of an equation we must also do to the other'.

Example 1.4
Solve the equation $x + 14 = 5$.

Solution
Note that by subtracting 14 from both sides, we leave x on its own on the left. Thus

$$x + 14 - 14 = 5 - 14$$
$$x = -9$$

Hence the solution of the equation is $x = -9$. It is easy to check that this solution is correct by substituting $x = -9$ into the original equation and checking that both sides are indeed the same. You should get into the habit of doing this.

Example 1.5
Solve the equation $19y = 38$.

Solution
In order to make y the subject of the equation we can divide both sides by 19:

$$19y = 38$$
$$\frac{19y}{19} = \frac{38}{19}$$

Cancelling the 19s gives $\qquad y = \dfrac{38}{19}$

so $\qquad y = 2$

Hence the solution of the equation is $y = 2$.

Example 1.6
Solve the equation $4x + 12 = 0$.

Solution
Starting from $4x + 12 = 0$ we can subtract 12 from both sides to obtain

$$4x + 12 - 12 = 0 - 12$$
so that $\qquad 4x = -12$

If we now divide both sides by 4 we find

$$\frac{4x}{4} = \frac{-12}{4}$$

Cancelling the 4s gives $\quad x = -3$

So the solution is $x = -3$.

Example 1.7
Solve the linear equation $14t - 56 = 0$.

Solution
First add 56 to both sides and obtain an expression for $14t$:

$14t = \rule{1.5em}{1.2em}$ $\qquad\qquad\qquad\qquad\qquad$ 56

Finally divide both sides by 14 to find

$$t = \boxed{} \qquad\qquad \frac{56}{14} = 4$$

Example 1.8
Solve the following equations:

(a) $x + 3 = \sqrt{7}$ (b) $x + 3 = -\sqrt{7}$

Solution
(a) Subtracting 3 from both sides gives $x = \sqrt{7} - 3$.
(b) Subtracting 3 from both sides gives $x = -\sqrt{7} - 3$.

Note that when asked to solve $x + 3 = \pm\sqrt{7}$ we can write the two solutions as $x = -3 \pm \sqrt{7}$. It is usually acceptable to leave the solutions in this form rather than work out their decimal approximations. This form is known as the **surd form**.

Example 1.9
Solve the equation $\frac{2}{3}(t + 7) = 5$.

Solution
There are a number of ways in which the solution can be attempted. The idea is to remove unwanted terms on the left-hand side gradually to leave t on its own. By multiplying both sides by $\frac{3}{2}$ we find

$$\frac{3}{2} \times \frac{2}{3}(t + 7) = \frac{3}{2} \times 5$$

$$= \frac{3}{2} \times \frac{5}{1}$$

and after simplifying and cancelling, $t + 7 = \dfrac{15}{2}$

Finally, subtracting 7 from both sides gives

$$t = \frac{15}{2} - 7$$

$$= \frac{15}{2} - \frac{14}{2}$$

$$= \frac{1}{2}$$

So the solution is $t = \frac{1}{2}$.

Example 1.10
Solve the equation $3(p - 2) + 2(p + 4) = 5$.

Solution
At first sight this may not appear to be in the form of a linear equation. Some preliminary work is necessary. Removing the brackets and collecting like terms we find the left-hand side can be written

$$3(p - 2) + 2(p + 4) = 3p - 6 + 2p + 8$$
$$= 5p + 2$$

So the equation becomes $5p + 2 = 5$. We can then proceed to find p. Subtracting 2 from both sides yields $5p = 3$ so that, finally, $p = \frac{3}{5}$.

Example 1.11
Solve the equation $2(x - 5) = 3 - (x + 6)$.

Solution
First remove the brackets on both sides:

$$2x - 10 = 3 - x - 6$$

We may write this as

$$2x - 10 = -x - 3$$

We shall try to rearrange this equation so that terms involving x appear only on the left-hand side, and constants on the right. Start by adding 10 to both sides:

$$2x = -x + 7$$

Now add x to both sides:

$$3x = 7$$

Finally solve this to find x:

$$x = \qquad \frac{7}{3}$$

Example 1.12
Solve the equation

$$\frac{6}{1 - 2x} = \frac{7}{x - 2}$$

Solution
This equation appears in an unfamiliar form but can be rearranged into the standard form of a linear equation. By multiplying both sides by $(1 - 2x)$ and $(x - 2)$ we find

$$(1 - 2x)(x - 2) \times \frac{6}{1 - 2x} = (1 - 2x)(x - 2) \times \frac{7}{x - 2}$$

Consider each side in turn and cancel any common factors:

$$6(x - 2) = 7(1 - 2x)$$

Removing the brackets and rearranging to find x we have

$$6x - 12 = 7 - 14x$$

and further rearrangement gives $\quad 20x = 19$

$$x = \frac{19}{20}$$

The solution is therefore $x = \frac{19}{20}$.

Example 1.13 Electrical Engineering – Kirchhoff's current law

Figure 1.1
The current entering the node equals the total current leaving the node: $I = I_1 + I_2$.

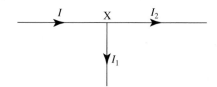

Consider Figure 1.1, which shows part of an electric circuit. At point X the circuit divides into two **branches**. Point X is known as a **node**. The current in each of the branches is denoted by I, I_1 and I_2. Kirchhoff's current law states that the current entering any node must equal the current leaving that node. Thus we have the equation

$$I = I_1 + I_2$$

(a) If $I_2 = 10$ A and $I = 18$ A calculate I_1.
(b) Suppose $I = 36$ A and it is known that current I_2 is five times as great as I_1. Find the branch currents.

Solution
(a) Substituting the given values into the equation we find $18 = I_1 + 10$. Solving for I_1 we find

$$I_1 = 18 - 10$$
$$= 8$$

Thus I_1 equals 8 A.
(b) We are given that, from Kirchhoff's law, $I = I_1 + I_2$. We are told that I_2 is five times as great as I_1, and so we can write $I_2 = 5I_1$. Since $I = 36$ we have

$$36 = I_1 + 5I_1$$
$$= 6I_1$$

Solving the linear equation $36 = 6I_1$ gives $I_1 = 6$ A. Finally, since I_2 is five times as great as I_1, $I_2 = 5I_1 = 30$ A.

Exercises

In questions 1–24 solve each equation:

1   $7x = 14$

2   $-3x = 6$

3   $\frac{1}{2}x = 7$

4   $3x = \frac{1}{2}$

5   $4t = -2$

6   $2t = 4$

7   $4t = 2$

8   $2t = -4$

9   $\dfrac{x}{6} = 3$

10   $\dfrac{x}{6} = -3$

11   $7x + 2 = 9$

12   $7x + 2 = 23$

13   $-7x + 1 = -6$

14   $-7x + 1 = -13$

15   $\dfrac{17}{3}t = -2$

16   $3 - x = 2x + 8$

17   $x - 3 = 8 + 3x$

18   $\dfrac{x}{4} = 16$

19   $\dfrac{x}{9} = -2$

20   $-\dfrac{13}{2}x = 14$

21   $-2y = -6$

22   $-7y = 11$

23   $-69y = -690$

24   $-8 = -4\gamma$

In questions 25–47 solve each equation:

25   $3y - 8 = \dfrac{1}{2}y$

26   $7t - 5 = 4t + 7$

27   $3x + 4 = 4x + 3$

28   $4 - 3x = 4x + 3$

29   $3x + 7 = 7x + 2$

30   $3(x + 7) = 7(x + 2)$

31   $2x - 1 = x - 3$

32   $2(x + 4) = 8$

33   $-2(x - 3) = 6$

34   $-2(x - 3) = -6$

35   $-3(3x - 1) = 2$

36   $2 - (2t + 1) = 4(t + 2)$

37   $5(m - 3) = 8$

38   $5m - 3 = 5(m - 3) + 2m$

39   $2(y + 1) = -8$

40   $17(x - 2) + 3(x - 1) = x$

41   $\dfrac{1}{3}(x + 3) = -9$

42   $\dfrac{3}{m} = 4$

43   $\dfrac{5}{m} = \dfrac{2}{m + 1}$

44   $-3x + 3 = 18$

45   $3x + 10 = 31$

46   $x + 4 = \sqrt{8}$

47   $x - 4 = \sqrt{23}$

48   If $y = 2$ find x if $4x + 3y = 9$.

49   If $y = -2$ find x if $4x + 5y = 3$.

50   If $y = 0$ find x if $-4x + 10y = -8$.

51   If $x = -3$ find y if $2x + y = 8$.

52   If $y = 10$ find x when $10x + 55y = 530$.

53   If $\gamma = 2$ find β if $54 = \gamma - 4\beta$.

In questions 54–63 solve each equation:

54   $\dfrac{x - 5}{2} - \dfrac{2x - 1}{3} = 6$

55   $\dfrac{x}{4} + \dfrac{3x}{2} - \dfrac{x}{6} = 1$

56   $\dfrac{x}{2} + \dfrac{4x}{3} = 2x - 7$

57   $\dfrac{5}{3m + 2} = \dfrac{2}{m + 1}$

58   $\dfrac{2}{3x - 2} = \dfrac{5}{x - 1}$

59   $\dfrac{x - 3}{x + 1} = 4$

60   $\dfrac{x + 1}{x - 3} = 4$

61 $\dfrac{y-3}{y+3} = \dfrac{2}{3}$

64 Solve the linear equation $ax + b = 0$ to find x.

62 $\dfrac{4x+5}{6} - \dfrac{2x-1}{3} = x$

65 Solve the linear equation $\dfrac{1}{ax+b} = \dfrac{1}{cx+d}$ to find x.

63 $\dfrac{3}{2s-1} + \dfrac{1}{s+1} = 0$

Solutions to exercises

1	2		23	10
2	-2		24	2
3	14		25	$\dfrac{16}{5}$
4	$\dfrac{1}{6}$		26	4
5	$-\dfrac{1}{2}$		27	1
6	2		28	$\dfrac{1}{7}$
7	$\dfrac{1}{2}$		29	$\dfrac{5}{4}$
8	-2		30	$\dfrac{7}{4}$
9	18			
10	-18		31	-2
11	1		32	0
12	3		33	0
13	1		34	6
14	2		35	$\dfrac{1}{9}$
15	$-\dfrac{6}{17}$		36	$-\dfrac{7}{6}$
16	$-\dfrac{5}{3}$		37	$\dfrac{23}{5}$
17	$-\dfrac{11}{2}$		38	6
18	64		39	-5
19	-18		40	$\dfrac{37}{19}$
20	$-\dfrac{28}{13}$		41	-30
21	3		42	$\dfrac{3}{4}$
22	$-\dfrac{11}{7}$			

43 $-\dfrac{5}{3}$

44 -5

45 7

46 $\sqrt{8} - 4$

47 $\sqrt{23} + 4$

48 $\dfrac{3}{4}$

49 $\dfrac{13}{4}$

50 2

51 14

52 -2

53 -13

54 -49

55 $\dfrac{12}{19}$

56 42

57 1

58 $\dfrac{8}{13}$

59 $-\dfrac{7}{3}$

60 $\dfrac{13}{3}$

61 15

62 $\dfrac{7}{6}$

63 $-\dfrac{2}{5}$

64 $-\dfrac{b}{a}$

65 $\dfrac{d - b}{a - c}$

End of block exercises

1 Solve the equation $\dfrac{x}{9} = 5$.

2 Solve the equation $7x = 63$.

3 Solve the equation $7I = -28$.

4 Solve the equation $3x + 11 = 20$.

5 Solve the equation $3 - 11x = 25$.

6 Solve the equation $7R + 13 = 2R - 7$.

7 Solve the equation $3(4x + 11) = 8$.

8 Solve the equation $5(x + 3) + 2(2x - 1) = 9$.

9 Solve the equation $\dfrac{1}{5x} + \dfrac{1}{4x} = 10$.

10 Solve the equation $\dfrac{3}{s - 1} = \dfrac{2}{s - 5}$.

Solutions to exercises

1 $x = 45$

2 $x = 9$

3 $I = -4$

4 $x = 3$

5 $x = -2$

6 $R = -4$

7 $x = -\dfrac{25}{12}$

8 $x = -\dfrac{4}{9}$

9 $x = \dfrac{9}{200}$

10 $s = 13$

Solving quadratic equations

A **quadratic equation** is one that can be written in the form $ax^2 + bx + c = 0$, where a, b and c are numbers and x is the unknown whose value(s) we wish to find. In this block we shall see that a quadratic equation can possess

- two solutions,
- a single solution, or
- no solutions at all.

We describe several ways in which quadratic equations can be solved. Familiarity with all the techniques is important.

Key point

A **quadratic equation** is one that can be written in the form

$$ax^2 + bx + c = 0$$

where a, b and c are numbers and x is the unknown whose value(s) we wish to find.

For example,

$$2x^2 + 7x - 3 = 0, \quad x^2 + x + 1 = 0, \quad 0.5x^2 + 3x + 9 = 0$$

are all quadratic equations. To ensure the presence of the x^2 term the number a cannot be zero. However, b and c may be zero, so that

$$4x^2 + 3x = 0, \quad 2x^2 - 3 = 0 \quad \text{and} \quad 6x^2 = 0$$

are all quadratic equations. Frequently quadratic equations occur in non-standard form but where necessary they can be rearranged into standard form using the rules for rearranging linear equations given in Block 1 of this chapter.

For example,

$$3x^2 + 5x = 8 \quad \text{and} \quad 2x^2 = 8x - 9$$

can be rewritten as

$$3x^2 + 5x - 8 = 0 \quad \text{and} \quad 2x^2 - 8x + 9 = 0$$

respectively.

To solve a quadratic equation we must find values of the unknown x that make the left-hand and right-hand sides equal. Such values are known as **solutions** or **roots** of the quadratic equation. We shall now describe four techniques for solving quadratic equations:

1 factorisation
2 using a formula
3 completing the square
4 solution using graphs.

Exercises

1 Verify that $x = 2$ and $x = 3$ are both solutions of $x^2 - 5x + 6 = 0$.

2 Verify that $x = -2$ and $x = -3$ are both solutions of $x^2 + 5x + 6 = 0$.

2.3 Solution by factorisation

It may be possible to solve a quadratic equation by factorisation using the method described for factorising quadratic expressions in Chapter 5 Block 5, although you should be aware that not all quadratic equations can be factorised.

Example 2.1
Solve the quadratic equation $x^2 + x - 6 = 0$.

Solution
Factorising we find

$$x^2 + x - 6 = (x + 3)(x - 2) = 0$$

When the product of two quantities equals zero, at least one of the two must equal zero. In this case either $(x + 3)$ is zero or $(x - 2)$ is zero. It follows that

$$x + 3 = 0, \text{ giving } x = -3$$

or

$$x - 2 = 0, \text{ giving } x = 2$$

There are two solutions, $x = -3$ and $x = 2$. These solutions can be checked quite easily by substitution back into the given equation.

Example 2.2
Solve the quadratic equation $2x^2 - 7x - 4 = 0$.

Solution
Factorising we find

$$2x^2 - 7x - 4 = (2x + 1)(x - 4) = 0$$

In this case either $(2x + 1)$ is zero or $(x - 4)$ is zero. It follows that

$$2x + 1 = 0, \text{ giving } x = -\frac{1}{2}$$

or

$$x - 4 = 0, \text{ giving } x = 4$$

There are two solutions, $x = -\frac{1}{2}$ and $x = 4$.

Example 2.3
Solve the equation $x^2 + 5x = 0$.

Solution
Factorising and equating each factor to zero we find

$$x^2 + 5x = x(x + 5) = 0$$

so that $x = 0$ and $x = -5$ are the two solutions.

Example 2.4 Structural Engineering—Bending moment in a beam
In Example 8.4 in Chapter 6 we considered a simply supported beam of length L carrying a uniform load w, per unit length (Figure 2.1), and introduced the **bending moment**, M, which is a measure of the internal stress caused when the beam is loaded.

Figure 2.1
A simply supported beam with a uniform load.

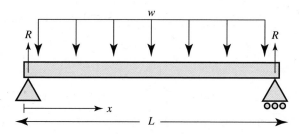

M varies with position x along the beam. It can be shown (see Example 1.18 in Chapter 17) that M is given by the quadratic polynomial

$$M(x) = \frac{w}{2}(Lx - x^2)$$

Determine the values of x for which the bending moment is zero.

Solution
The values of x for which the bending moment is zero are found by solving

$$\frac{w}{2}(Lx - x^2) = 0$$

which is a quadratic equation. We can proceed in exactly the same way as in Example 2.3 and factorise the equation noting the common factor x in both terms in the brackets:

$$\frac{w}{2}(Lx - x^2) = \frac{w}{2}x(L - x) = 0$$

Noting that $\frac{w}{2}$ cannot be zero, we have

$$x(L - x) = 0$$

Each factor is then equated to zero to give $x = L$ and $x = 0$ as the two solutions. We conclude that the bending moment is zero at both ends of the beam.

Example 2.5
Solve the equation $4x^2 + 12x + 9 = 0$.

Solution
Factorising we find

$$4x^2 + 12x + 9 = (2x + 3)(2x + 3) = (2x + 3)^2$$

This time the factor $(2x + 3)$ occurs twice. The equation becomes

$$(2x + 3)^2 = 0$$

so that

$$2x + 3 = 0$$

and we obtain the root $x = -\frac{3}{2}$. Because the factor $2x + 3$ appears twice in the equation $(2x + 3)^2 = 0$ we say that this root is a **repeated** or **double root**.

Example 2.6
Solve the quadratic equation $7x^2 - 20x - 3 = 0$.

Solution
First factorise the left-hand side:

$$7x^2 - 20x - 3 = \boxed{} \qquad\qquad (7x + 1)(x - 3)$$

Each factor is then equated to zero to obtain the two solutions:

$$x = \boxed{} \text{ and } \boxed{} \qquad\qquad -\frac{1}{7} \text{ and } 3$$

Exercises

Solve the following equations by factorisation:

1 $x^2 - 3x + 2 = 0$

2 $x^2 - x - 2 = 0$

3 $x^2 + x - 2 = 0$

4 $x^2 + 3x + 2 = 0$

5   $x^2 + 8x + 7 = 0$

6   $x^2 - 7x + 12 = 0$

7   $x^2 - x - 20 = 0$

8   $x^2 - 1 = 0$

9   $x^2 - 2x + 1 = 0$

10   $x^2 + 2x + 1 = 0$

11   $x^2 + 11x = 0$

12   $2x^2 + 2x = 0$

13   $x^2 - 3x = 0$

14   $x^2 + 9x = 0$

Solve the following quadratic equations by factorisation:

15   $2x^2 - 5x + 2 = 0$

16   $6x^2 - x - 1 = 0$

17   $-5x^2 + 6x - 1 = 0$

18   $-x^2 + 4x - 3 = 0$

Solutions to exercises

1   1, 2

2   $-1, 2$

3   $-2, 1$

4   $-1, -2$

5   $-7, -1$

6   4, 3

7   $-4, 5$

8   $1, -1$

9   1 twice

10   -1 twice

11   $-11, 0$

12   $0, -1$

13   0, 3

14   $0, -9$

15   $2, \dfrac{1}{2}$

16   $\dfrac{1}{2}, -\dfrac{1}{3}$

17   $\dfrac{1}{5}, 1$

18   1, 3

2.4   Solution by formula

When it is difficult or impossible to factorise a quadratic equation, it may be possible to solve it using a formula that is used to calculate the roots.

Key point

If $ax^2 + bx + c = 0$ then

$$x = \frac{-b \pm \sqrt{b^2 - 4ac}}{2a}$$

To apply the formula it is necessary to identify carefully the values of a, b and c, paying particular attention to the signs of these numbers. Substitution into the formula gives the desired solutions.

Note that if the quantity $b^2 - 4ac$ is a positive number we can take its square root and the formula will produce two solutions known as **distinct real roots**. The first root is found by using the positive square root in the formula, and the second is found using the negative square root. If $b^2 - 4ac = 0$ there will be a single root known as a **repeated root**. The value of this root is $x = -\dfrac{b}{2a}$. Finally if $b^2 - 4ac$ is negative we say the equation possesses **complex roots**. These require special treatment and are described in Chapter 11.

Key point

The quadratic equation $ax^2 + bx + c = 0$ has

- distinct real roots if $b^2 - 4ac > 0$
- a repeated root if $b^2 - 4ac = 0$
- complex roots if $b^2 - 4ac < 0$.

Example 2.7

Compare the given equation with the standard form $ax^2 + bx + c = 0$ and identify a, b and c. Calculate $b^2 - 4ac$ and use this information to state the nature of the roots.

(a) $3x^2 + 2x - 7 = 0$

(b) $3x^2 + 2x + 7 = 0$

(c) $3x^2 - 2x + 7 = 0$

(d) $x^2 + x + 2 = 0$

(e) $-x^2 + 3x - \frac{1}{2} = 0$

(f) $5x^2 - 3 = 0$

(g) $x^2 - 2x + 1 = 0$

Solution

(a) $a = 3, b = 2$ and $c = -7$. So

$$b^2 - 4ac = (2)^2 - 4(3)(-7)$$
$$= 88$$

and the roots will be real and distinct.

(b) $a = 3, b = 2$ and $c = 7$. So

$$b^2 - 4ac = (2)^2 - 4(3)(7)$$
$$= -80$$

The roots will be complex.

(c) $a = 3, b = -2$ and $c = 7$. So

$$b^2 - 4ac = (-2)^2 - 4(3)(7)$$
$$= -80$$

Again the roots will be complex.

(d) $a = 1, b = 1$ and $c = 2$. So

$$b^2 - 4ac = (1)^2 - 4(1)(2)$$
$$= -7$$

and the roots will be complex.

(e) $a = -1, b = 3$ and $c = -\frac{1}{2}$. So

$$b^2 - 4ac = 3^2 - 4(-1)(-\frac{1}{2})$$
$$= 7$$

The roots will be real and distinct.

(f) $a = 5, b = 0$ and $c = -3$. So

$$b^2 - 4ac = 0 - 4(5)(-3)$$
$$= 60$$

The roots will be real and distinct.

(g) $a = 1, b = -2$ and $c = 1$. So

$$b^2 - 4ac = (-2)^2 - 4(1)(1)$$
$$= 0$$

There will be a single repeated root.

Example 2.8

Solve the quadratic equation $2x^2 + 3x - 6 = 0$ using the formula.

Solution

We compare the given equation with the standard form $ax^2 + bx + c = 0$ in order to identify a, b and c. We see that $a = 2$, $b = 3$ and $c = -6$. Note particularly the sign of c. Substituting these values into the formula we find

$$x = \frac{-b \pm \sqrt{b^2 - 4ac}}{2a}$$
$$= \frac{-3 \pm \sqrt{3^2 - 4(2)(-6)}}{(2)(2)}$$
$$= \frac{-3 \pm \sqrt{9 + 48}}{4}$$
$$= \frac{-3 \pm \sqrt{57}}{4}$$
$$= \frac{-3 \pm 7.5498}{4}$$
$$= 1.1375 \text{ and } -2.6375$$

The two solutions are $x = 1.1375$ and $x = -2.6375$. However, it is often sufficient to leave your answer in the so-called **surd form** $x = \dfrac{-3 \pm \sqrt{57}}{4}$.

Example 2.9

Solve the equation $3x^2 - x - 6 = 0$.

Solution

First identify a, b and c.

$a =$ ☐ $b =$ ☐ $c =$ ☐ $a = \boxed{3}$, $b = \boxed{-1}$, $c = \boxed{-6}$

Substitute these values into the formula

$$x = \frac{-b \pm \sqrt{b^2 - 4ac}}{2a}$$

$=$ ☐

$\dfrac{-(-1) \pm \sqrt{(-1)^2 - (4)(3)(-6)}}{(2)(3)}$

$= \dfrac{1 \pm \sqrt{73}}{6}$

Finally calculate the values of x:

$x =$ ☐ $1.5907, -1.2573$

Exercises

Solve the following quadratic equations by using the formula:

1 $x^2 + 8x + 1 = 0$

2 $x^2 + 7x - 2 = 0$

3 $x^2 + 6x - 2 = 0$

4 $4x^2 + 3x - 2 = 0$

5 $2x^2 + 3x - 1 = 0$

6 $x^2 + x - 1 = 0$

7 $-x^2 + 3x + 1 = 0$

8 $-2x^2 - 3x + 1 = 0$

9 $2x^2 + 5x - 3 = 0$

10 $-2s^2 - s + 3 = 0$

11 $9x^2 + 6x + 1 = 0$

12 $x^2 + 6x + 9 = 0$

13 $3x^2 - 6x + 3 = 0$

Solutions to exercises

1 $-0.1270, -7.8730$

2 $-7.2749, 0.2749$

3 $0.3166, -6.3166$

4 $-1.1754, 0.4254$

5 $0.2808, -1.7808$

6 $-1.6180, 0.6180$

7 $3.3028, -0.3028$

8 $-1.7808, 0.2808$

9 $\dfrac{1}{2}, -3$

10    $-\dfrac{3}{2}, 1$

11    $-\dfrac{1}{3}$

12   -3

13   1

2.5   Completing the square

The technique known as completing the square can be used to solve quadratic equations, although it is applicable in other circumstances as well. We shall develop this technique in the following examples.

Example 2.10
(a) Show that $(x + 3)^2 = x^2 + 6x + 9$.
(b) Hence show that $x^2 + 6x$ can be written as $(x + 3)^2 - 9$.

Solution
(a) Removing the brackets we find

$$\begin{aligned}(x + 3)^2 &= (x + 3)(x + 3) \\ &= x^2 + 3x + 3x + 9 \\ &= x^2 + 6x + 9\end{aligned}$$

Thus

$$(x + 3)^2 = x^2 + 6x + 9$$

(b) By subtracting 9 from both sides of the previous equation it follows that

$$(x + 3)^2 - 9 = x^2 + 6x$$

Example 2.11
(a) Show that $(x - 4)^2 = x^2 - 8x + 16$.
(b) Hence show that $x^2 - 8x$ can be written as $(x - 4)^2 - 16$.

Solution
(a) Removing the brackets we find

$$\begin{aligned}(x - 4)^2 &= (x - 4)(x - 4) \\ &= x^2 - 4x - 4x + 16 \\ &= x^2 - 8x + 16\end{aligned}$$

(b) Subtracting 16 from both sides we can write

$$(x - 4)^2 - 16 = x^2 - 8x$$

We shall now generalise the results of Examples 2.10 and 2.11. Noting that

$$(x + k)^2 = x^2 + 2kx + k^2$$

we can write

$$x^2 + 2kx = (x + k)^2 - k^2$$

Note that the constant term in the brackets on the right is always half the coefficient of x on the left. This process is called **completing the square**.

Key point	**Completing the square** $x^2 + 2kx = (x + k)^2 - k^2$

Example 2.12
Complete the square for the expression $x^2 + 16x$.

Solution
Comparing $x^2 + 16x$ with the general form $x^2 + 2kx$ we see that $k = 8$. Hence

$$x^2 + 16x = (x + 8)^2 - 8^2$$
$$= (x + 8)^2 - 64$$

Note that the constant term in the brackets on the right, that is 8, is half the coefficient of x on the left, which is 16.

Example 2.13
Complete the square for the expression $5x^2 + 4x$.

Solution
Consider $5x^2 + 4x$. First of all the coefficient 5 is removed outside a bracket as follows

$$5x^2 + 4x = 5(x^2 + \tfrac{4}{5}x)$$

We can now complete the square for the quadratic expression in the brackets:

$$x^2 + \tfrac{4}{5}x = (x + \tfrac{2}{5})^2 - (\tfrac{2}{5})^2$$
$$= (x + \tfrac{2}{5})^2 - \tfrac{4}{25}$$

Hence, multiplying both sides by 5 we find

$$5x^2 + 4x = 5\left[(x + \tfrac{2}{5})^2 - \tfrac{4}{25}\right]$$

Completing the square can be used to solve quadratic equations as shown in the following examples.

Example 2.14
Solve the equation $x^2 + 6x + 2 = 0$ by completing the square.

Solution
First of all just consider $x^2 + 6x$, and note from Example 2.10 that we can write this as

$$x^2 + 6x = (x + 3)^2 - 9$$

Then the quadratic equation can be written as

$$x^2 + 6x + 2 = (x + 3)^2 - 9 + 2 = 0$$

that is

$$(x + 3)^2 = 7$$

Taking the square root of both sides gives

$$x + 3 = \pm\sqrt{7}$$
$$x = -3 \pm \sqrt{7}$$

The two solutions are $x = -3 + \sqrt{7} = -0.3542$ and $x = -3 - \sqrt{7} = -5.6458$.

Example 2.15

Solve the equation $x^2 - 8x + 5 = 0$.

Solution

First consider $x^2 - 8x$. Using Example 2.11 we can write

$$x^2 - 8x = (x - 4)^2 - 16$$

so that the equation becomes

$$x^2 - 8x + 5 = (x - 4)^2 - 16 + 5 = 0$$
$$(x - 4)^2 = 11$$
$$x - 4 = \pm\sqrt{11}$$
$$x = 4 \pm\sqrt{11}$$
$$= 7.3166, 0.6834$$

Example 2.16

Solve the equation $x^2 - 4x + 1 = 0$ by completing the square.

Solution

First examine the two leftmost terms in the equation, $x^2 - 4x$. Complete the square for these terms:

$x^2 - 4x = $ ▨ $(x - 2)^2 - 4$

The equation $x^2 - 4x + 1 = 0$ can then be written

$x^2 - 4x + 1 = $ ▨

                                                        $(x - 2)^2 - 4 + 1 = (x - 2)^2 - 3$

$\qquad\qquad = 0$

From which

$(x - 2)^2 = $ ▨ 3

$x - 2 = $ ▨ $\pm\sqrt{3}$

$x = $ ▨ $2 \pm \sqrt{3} = 3.7321, 0.2679$

Exercise

1   Solve the quadratic equations at the end of Section 2.4 by completing the square.

2.6 Solution of quadratic equations using graphs

Equations can be solved quite simply using graphs. For example, to solve the equation $x^2 - 4x + 1 = 0$ we can plot a graph of the function $y = x^2 - 4x + 1$. If the graph crosses the horizontal axis it will do so when $y = 0$, and so the x coordinates at such points are solutions of $x^2 - 4x + 1 = 0$. This is illustrated in the following example.

Example 2.17

Solve the equation $x^2 - 4x + 1 = 0$ by plotting a graph of the function

$$y = x^2 - 4x + 1$$

Solution

By constructing a table of function values we can plot the graph as shown in Figure 2.2.

Figure 2.2
The graph of
$y = x^2 - 4x + 1$
cuts the x axis at
C and D.

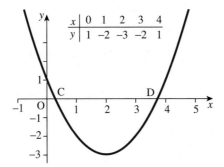

x	0	1	2	3	4
y	1	-2	-3	-2	1

The solutions of the equation $x^2 - 4x + 1 = 0$ are found by looking for points where the graph crosses the horizontal axis. The two points are approximately $x = 0.27$ and $x = 3.73$, marked C and D on the figure.

If you have access to a graphics calculator or computer software for graph plotting check that you can use it to plot quadratic graphs and locate solutions of quadratic equations.

Computer and calculator exercises

Solve the following quadratic equations
by plotting graphs:

1   $7x^2 - 17x - 9 = 0$

2   $0.56x^2 + 0.05x - 3 = 0$

3   $-0.25x^2 - 0.05x + 0.01 = 0$

End of block exercises

1 Solve the following quadratic equations:
(a) $x^2 - 9 = 0$ (b) $s^2 - 25 = 0$
(c) $3x^2 - 12 = 0$

2 Solve the equation $x^2 - 5x + 6 = 0$.

3 Solve the equation $6s^2 + s - 15 = 0$.

4 Solve the equation $x^2 + 7x = 0$.

5 Solve the equation $2x^2 - 3x - 7 = 0$.

Solutions to exercises

1 (a) $x = 3, -3$ (b) $s = 5, -5$ (c) $x = 2, -2$

2 $x = 3, 2$

3 $s = \frac{3}{2}, -\frac{5}{3}$

4 $x = 0, -7$

5 $2.766, -1.266$

Factorising polynomial expressions and solving polynomial equations

A **polynomial expression** is one of the form

$$a_n x^n + a_{n-1} x^{n-1} + \cdots + a_2 x^2 + a_1 x + a_0$$

where $a_0, a_1, \ldots$ are known coefficients (or numbers) and x is a variable.

For example, $x^3 - 17x^2 + 54x - 8$ is a polynomial expression. Note that only non-negative whole number powers of the variable x are allowed in a polynomial expression. When a polynomial expression is equated to zero, a **polynomial equation** is obtained. Thus

$$3x^2 + 6x - 7 = 0, \quad x^3 - 2x^2 + 8x - 3 = 0$$

are examples of polynomial equations. The linear and quadratic equations that you have already met are particular types of polynomial equation. In this block you will learn how to factorise some polynomial expressions and how to solve some polynomial equations. You will also learn the technique of **equating coefficients**, which is very important when you need to perform calculations involving partial fractions in Block 6.

3.2 Multiplying polynomials together

The **degree** of a polynomial is the highest power to which the variable is raised. Thus $x^3 + 6x + 2$ has degree 3, $t^6 - 6t^4 + 2t$ has degree 6, and $5x + 2$ has degree 1.

Let us consider what happens when two polynomials are multiplied together. For example,

$$(x + 1)(3x - 2)$$

is the product of two first-degree polynomials. Expanding the brackets we obtain

$$(x + 1)(3x - 2) = 3x^2 + x - 2$$

which is a second-degree polynomial.

In general we can regard a second-degree polynomial, or quadratic, as the product of two first-degree polynomials, provided that the quadratic can be factorised.

On the other hand

$$(x - 1)(x^2 + 3x - 7) = x^3 + 2x^2 - 10x + 7$$

is a third-degree, or cubic, polynomial that is also the product of a linear polynomial and a quadratic polynomial.

In general we can regard a cubic polynomial as the product of a linear polynomial and a quadratic polynomial. This fact will be important in the following section when we come to factorise cubics.

Key point	cubic = linear × quadratic

Exercises

1 If $x^3 - 17x^2 + 54x - 8 = (x - 4)$ × a polynomial, state the degree of the polynomial.

2 (a) If $3x^2 + 13x + 4 = (x + 4)$ × a polynomial, state the degree of the polynomial.
(b) What is the coefficient of x in this unknown polynomial?

3 If $2x^2 + 5x + 2 = (x + 2)$ × a polynomial, what must be the coefficient of x in this unknown polynomial?

4 Two quadratic polynomials are multiplied together. What is the degree of the resulting polynomial?

Solutions to exercises

1 two

2 (a) one (b) it must be three in order to generate the term $3x^2$ when the brackets are removed.

3 two

4 four

3.3 Factorising polynomial expressions and equating coefficients

Factorisation of polynomial expressions can sometimes be achieved if one or more of the factors is already known. This requires a knowledge of the technique of **equating coefficients**, which is illustrated in the following example.

Example 3.1
Factorise the expression $x^3 - 17x^2 + 54x - 8$ given that one of the factors is $(x - 4)$.

Solution
Given that $x - 4$ is a factor we can write

$$x^3 - 17x^2 + 54x - 8 = (x - 4) \times \text{an unknown quadratic polynomial}$$

The unknown polynomial must be quadratic because the expression on the left is cubic. Suppose we write this quadratic as $\alpha x^2 + \beta x + \gamma$ where α, β and γ are unknown numbers, which we shall try to find. Then

$$x^3 - 17x^2 + 54x - 8 = (x - 4)(\alpha x^2 + \beta x + \gamma)$$

Removing the brackets on the right and collecting like terms together we have

$$x^3 - 17x^2 + 54x - 8 = \alpha x^3 + (\beta - 4\alpha)x^2 + (\gamma - 4\beta)x - 4\gamma$$

Equating coefficients means that we compare the coefficients of each term on the left with the corresponding term on the right. Thus if we look at the x^3 terms on each side we see that

$$x^3 = \alpha x^3$$

that is, α must equal 1 for both sides to be the same. Similarly by equating coefficients of x^2 we find

$$-17 = \beta - 4\alpha$$

With $\alpha = 1$ we have $-17 = \beta - 4$ so that β must equal -13. Finally, equating constant terms we find

$$-8 = -4\gamma$$

so that $\gamma = 2$. Check for yourself that, with these values of γ and β, the coefficient of x is the same on both sides. We can now write the polynomial expression as

$$x^3 - 17x^2 + 54x - 8 = (x - 4)(x^2 - 13x + 2)$$

The quadratic expression on the right cannot be factorised further.

Exercises

In questions 1–4 factorise the given polynomial expressions into three linear factors:

1 $x^3 - 6x^2 + 11x - 6$, given $x - 1$ is a factor.

2 $x^3 - 7x - 6$, given $x + 2$ is a factor.

3 $2x^3 + 7x^2 + 7x + 2$, given $x + 1$ is a factor.

4 $3x^3 + 7x^2 - 22x - 8$, given $x + 4$ is a factor.

Solutions to exercises

1 $(x - 1)(x - 2)(x - 3)$

2 $(x - 3)(x + 1)(x + 2)$

3 $(x + 1)(2x + 1)(x + 2)$

4 $(x - 2)(x + 4)(3x + 1)$

3.4 Polynomial equations

Key point

A **polynomial equation** has the form

$$a_n x^n + a_{n-1} x^{n-1} + \cdots + a_2 x^2 + a_1 x + a_0 = 0$$

where $a_0, a_1, \ldots$ are known coefficients, and x represents an unknown whose value(s) are to be found.

The following are examples of polynomial equations:

$$5x^6 - 3x^4 + x^2 + 7 = 0, \quad -7x^4 + x^2 + 9 = 0$$

Recall that the **degree** of the equation is the highest power of x occurring. The **solutions** or **roots** of the equation are those values of x that satisfy the equation. A polynomial equation of degree n has n roots.

Example 3.2
Verify that $x = -1$, $x = 1$ and $x = 0$ are roots of the equation

$$x^3 - x = 0$$

Solution
We substitute each value in turn into $x^3 - x = 0$.

$$(-1)^3 - (-1) = -1 + 1$$
$$= 0$$

so $x = -1$ is clearly a root. It is easy to verify that $x = 1$ and $x = 0$ are also solutions.

Polynomial equations of low degree have special names. A polynomial equation of degree 1 is a linear equation and such equations have been solved in Block 1. A polynomial equation of degree 2 is a quadratic equation and such equations have been solved in Block 2. We now consider ways in which polynomial equations of higher degree can be solved.

Exercises

In questions 1 and 2 verify that the given values are solutions of the given equations:

1 $x^2 - 5x + 6 = 0$, $x = 3$, $x = 2$

2 $2t^3 + t^2 - t = 0$, $t = 0$, $t = -1$, $t = \frac{1}{2}$

3.5 Solving polynomial equations when one solution is known

In Block 2 we gave a formula that can be used to solve quadratic equations. Unfortunately, when dealing with equations of higher degree no such simple formulae exist. If one of the roots can be spotted we can sometimes find the others by the method shown in Example 3.3.

Example 3.3

Verify that $x = 4$ is a solution of the equation $P(x) = x^3 - 17x^2 + 54x - 8 = 0$. Hence find the other solutions.

Solution

We substitute $x = 4$ into the left-hand side of the given equation:

$$4^3 - 17(4^2) + 54(4) - 8 = 64 - 272 + 216 - 8$$
$$= 0$$

So, when $x = 4$ the left-hand side equals zero. Hence $x = 4$ is indeed a solution. Knowing that $x = 4$ is a root we can state that $(x - 4)$ must be a factor of $P(x)$. Therefore $P(x)$ can be written

$$P(x) = x^3 - 17x^2 + 54x - 8$$
$$= (x - 4) \times \text{quadratic polynomial}$$

The quadratic polynomial has already been found in Example 3.1. So the given equation can be written

$$P(x) = x^3 - 17x^2 + 54x - 8$$
$$= (x - 4)(x^2 - 13x + 2) = 0$$

In this form we see that

$$x - 4 = 0 \quad \text{or} \quad x^2 - 13x + 2 = 0$$

The first equation gives $x = 4$, which we already knew. The second must be solved using one of the methods for solving quadratic equations given in Block 2. For example, using the formula we find

$$x = \frac{-b \pm \sqrt{b^2 - 4ac}}{2a}$$
$$= \frac{13 \pm \sqrt{(-13)^2 - 4.1.2}}{2}$$
$$= \frac{13 \pm \sqrt{161}}{2}$$
$$= \frac{13 \pm 12.6886}{2}$$
$$= 12.844, 0.156$$

Hence the solutions are $x = 4$, $x = 12.844$ and $x = 0.156$.

Example 3.4
Solve the equation $x^3 + 8x^2 + 16x + 3 = 0$ given that $x = -3$ is a root.

Solution
Consider the equation $x^3 + 8x^2 + 16x + 3 = 0$. If $x = -3$ is a root then $x + 3$ must be a factor of the left-hand side. We can therefore write the left-hand side as

$$x^3 + 8x^2 + 16x + 3 = (x + 3)(\alpha x^2 + \beta x + \gamma)$$

Expanding the right-hand side we have

$$x^3 + 8x^2 + 16x + 3 = \alpha x^3 + (3\alpha + \beta)x^2 + (3\beta + \gamma)x + 3\gamma$$

Equating coefficients of x^3 we find

$$\alpha = \boxed{} \qquad\qquad 1$$

Equating constant terms we find

$$\boxed{} \qquad\qquad 3 = 3\gamma \text{ so that } \gamma = 1$$

Finally equating coefficients of x^2 gives

$$\boxed{} \qquad\qquad 8 = 3\alpha + \beta$$

from which $\beta = 5$.

This enables us to write the equation as

$$(x + 3)(x^2 + 5x + 1) = 0$$

Thus $x + 3 = 0$ or $x^2 + 5x + 1 = 0$. The quadratic equation can be solved using the formula to obtain $x = -4.791$ and $x = -0.209$.

Example 3.5 Control Engineering – Poles of a transfer function
We have already noted in Example 1.11 (Chapter 5 Block 1) that algebraic fractions can appear in control engineering applications in the form of **transfer functions** which relate the output of a system to the input. **Poles** of the transfer function are those values which make the denominator of the fraction zero and knowledge of these is particularly important when examining the stability of a control system.

Calculate the poles of the system which has transfer function

$$\frac{500}{s^3 + 15s^2 + 50s}$$

Solution
We examine the denominator and seek those values of s for which

$$s^3 + 15s^2 + 50s = 0$$

Observe that this is a polynomial equation of degree 3, that is a cubic equation.

Note that there is a common factor in the terms of the equation, namely s, and so the equation can be factorised into

$$s(s^2 + 15s + 50) = 0$$

In turn, the quadratic term can be factorised:

$$s(s + 10)(s + 5) = 0$$

from which $s = 0$, $s = -10$ and $s = -5$. There are three poles.

Exercises

In questions 1–3 verify that the given value is a solution of the equation and hence find all solutions:

1 $x^3 + 7x^2 + 11x + 2 = 0$, $x = -2$

2 $2x^3 + 11x^2 - 2x - 35 = 0$, $x = -5$

3   Verify that $x = 1$ and $x = 2$ are solutions of $x^4 + 4x^3 - 17x^2 + 8x + 4 = 0$ and hence find all solutions.

Solutions to exercises

1 $-2, -0.2087, -4.7913$

2 $-5, -2.1375, 1.6375$

3 $1, 2, -0.2984, -6.7016$

3.6   Solving a polynomial equation graphically

The real roots of the polynomial equation $P(x) = 0$ are given by the values of the intercepts of the function $y = P(x)$ and the x axis because on the x axis, y, and hence $P(x)$, is zero. Many excellent computer software packages and graphics calculators exist that can be used for plotting graphs and hence solving polynomial equations. Suppose the graph of $y = P(x)$ is plotted and takes a form similar to that shown in Figure 3.1.

The graph intersects the x axis at $x = x_1$, $x = x_2$ and $x = x_3$ and so the equation $P(x) = 0$ has real roots x_1, x_2 and x_3.

Figure 3.1
A polynomial function that cuts the x axis at points x_1, x_2 and x_3.

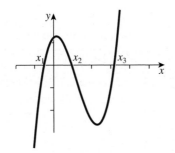

Example 3.6
Plot a graph of the function $y = 4x^4 - 15x^2 + 5x + 6$ and hence solve the equation $4x^4 - 15x^2 + 5x + 6 = 0$.

Solution
The graph has been plotted with the aid of a computer graph-plotting package and is shown in Figure 3.2.

Figure 3.2
Graph of
$y = 4x^4 -$
$15x^2 + 5x + 6$.

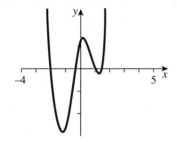

The solutions of the equation are found by looking for where the graph crosses the horizontal axis. We see that the solutions are $x = 1$, $x = 1.5$, $x = -0.5$ and $x = -2$.

An important feature of the graph of a polynomial is that it is continuous. By studying the graph in Figure 3.2 you will see that if we choose any two values of x, say a and b, such that $y(a)$ and $y(b)$ have opposite signs, then at least one root lies between $x = a$ and $x = b$.

Exercise

1   Consider the polynomial
$P(x) = 5x^3 - 47x^2 + 84x$. By evaluating $P(2)$ and $P(3)$ show that at least one root of $P(x) = 0$ lies between $x = 2$ and $x = 3$.

 ## Computer and calculator exercises

By plotting graphs estimate the roots of each of the following equations:

1   $x^3 + 2x^2 - 7x + 3 = 0$

2   $2x^3 + 3x + 7 = 0$

3   $x^4 + 2x^3 - 7x^2 + 11x + 3 = 0$

Solutions to exercises

1   $0.53, -3.96, 1.43$

2   -1.20

3   $-4.2, -0.25$

End of block exercises

1   Factorise $x^3 - x^2 - 65x - 63$ given that $(x + 7)$ is a factor.

2   Show that $x = -1$ is a root of $x^3 + 11x^2 + 31x + 21 = 0$ and locate the other roots algebraically.

3   Show that $x = 2$ is a root of $x^3 - 3x - 2 = 0$ and locate the other roots.

4   Solve the equation $x^4 - 2x^2 + 1 = 0$.

5   Factorise $x^4 - 7x^3 + 3x^2 + 31x + 20$ given that $(x + 1)$ is a factor.

6   Given that two of the roots of $x^4 + 3x^3 - 7x^2 - 27x - 18 = 0$ have the same modulus but different sign, solve the equation. (Hint: let two of the roots be α and $-\alpha$ and use the technique of equating coefficients.)

7   Without solving the equation, or using a graphical calculator, show that

$$x^4 + 4x - 1 = 0$$

has a root between $x = 0$ and $x = 1$.

Solutions to exercises

1   $(x + 7)(x + 1)(x - 9)$

2   $x = -1, -3, -7$

3   $x = 2, -1$

4   $x = -1, 1$

5   $(x + 1)^2(x - 4)(x - 5)$

6   $x = -3, 3, -1, -2$

Solving simultaneous equations

4.1 Introduction

Equations often arise in which there is more than one unknown quantity. When this is the case there will usually be more than one equation involved. For example, in the two equations

$$7x + y = 9, \quad -3x + 2y = 1$$

there are two unknowns: x and y. In order to solve the equations we must find values for x and y that satisfy both of the equations. The two equations are called **simultaneous equations**. You should verify that the solution of these equations is $x = 1, y = 2$ because by substituting these values into both equations, the left-hand and right-hand sides are equal.

In this block we shall show how two simultaneous equations can be solved either by a method known as **elimination** or by drawing graphs. If more than two equations are involved you should refer to the techniques described in Chapter 13 Blocks 1, 2 and 3.

Exercises

1 In each case verify that the given values satisfy the given simultaneous equations:
   (a) $x = 2, y = -2$ satisfy $7x + y = 12$ and $-3x - y = -4$.
   (b) $x = 4, y = 3$ satisfy $x + y = 7$ and $x - y = 1$.
   (c) $x = -3, y = 2$ satisfy $8x - y = -26$ and $9x + 2y = -23$.

4.2 Solving simultaneous equations by elimination

One way of solving simultaneous equations is by **elimination**. Elimination, as the name implies, involves removing one of the unknowns. Note that if both sides of an equation are multiplied or divided by a non-zero number an equivalent equation results. For example, if we are given the equation

$$x + 4y = 5$$

then by multiplying both sides by 7, say, we find

$$7x + 28y = 35$$

and this modified equation is equivalent to the original one.

Given two simultaneous equations, elimination of one unknown can be achieved by modifying the equations so that the coefficients of that unknown in each equation are the same. By subtracting one equation from the other that unknown is eliminated.

If the coefficients of one unknown differ only in sign, that unknown can be eliminated by adding the equations together.

Consider the following example.

Example 4.1
Solve the simultaneous equations

$$3x + 5y = 31 \tag{1}$$
$$2x + 3y = 20 \tag{2}$$

Solution
We first try to modify each equation so that the coefficient of x is the same in both equations. This can be achieved if equation (1) is multiplied by 2 and equation (2) is multiplied by 3. This gives

$$6x + 10y = 62$$
$$6x + 9y = 60$$

Note that in this form the unknown x can be removed or eliminated if the second equation is subtracted from the first:

$$
\begin{array}{r}
6x + 10y = 62\ - \\
\underline{6x + \ 9y = 60} \\
0x + \ 1y = 2
\end{array}
$$

The result implies that $1y = 2$, and we see immediately that y must equal 2. To find x we substitute the value found for y into either of the given equations. For example, using equation (1),

$$3x + 5(2) = 31$$
$$3x = 21$$
$$x = 7$$

Thus the solution of the given equations is $x = 7$, $y = 2$. You should always check your solution by substituting back into both of the given equations.

If the coefficients of x in the two equations differ only in sign, elimination can be achieved by adding the two equations:

Example 4.2
Solve the equations

$$-3x + \ y = \ 18 \tag{3}$$
$$7x - 3y = -44 \tag{4}$$

Solution

We modify the equations so that x can be eliminated by adding. For example, by multiplying equation (3) by 7 and equation (4) by 3 we find

$$-21x + 7y = 126$$
$$21x - 9y = -132$$

If both equations are now added we can eliminate x. Therefore

$$-21x + 7y = 126 +$$
$$\underline{21x - 9y = -132}$$
$$0x - 2y = -6$$

from which $-2y = -6$, so that $y = 3$. Substituting this value into equation (3) we can find x:

$$-3x + 3 = 18$$

so that

$$-3x = 15$$

that is

$$x = -5$$

Sometimes it will be easier to eliminate y rather than x:

Example 4.3
Solve the equations

$$5x + 3y = -74$$
$$-2x - 3y = 26$$

Solution

Note that the coefficients of y differ only in sign. By adding the two equations we find $3x = -48$ so that $x = -16$. It follows that $y = 2$.

Example 4.4
Solve the equations

$$5x - 7y = -80 \tag{5}$$
$$2x + 11y = 106 \tag{6}$$

Solution

First modify the equations so that the coefficient of x is the same in both. This means that if equation (5) is multiplied by 2 then equation (6) must be multiplied by

5

Write down the resulting equations:

$$10x - 14y = -160, \quad 10x + 55y = 530$$

Subtract these to eliminate x and hence find y.

$$y = 10$$

Finally verify that $x = -2$.

Equations with no solution

On occasions you may come across a pair of simultaneous equations that have no solution. Consider the following example.

Example 4.5

Solve the equations

$$10x - 2y = -3 \qquad (7)$$
$$-5x + \ y = \ \ \ 1 \qquad (8)$$

Solution

Leaving equation (7) unaltered and multiplying equation (8) by 2 we find

$$10x - 2y = -3$$
$$-10x + 2y = \ \ \ 2$$

Adding these equations to eliminate x we find y is eliminated as well:

$$\begin{array}{r} 10x - 2y = -3 \ + \\ -10x + 2y = \ \ \ 2 \\ \hline 0x + 0y = -1 \end{array}$$

The last line is clearly nonsense. We say that the equations are **inconsistent** and they have no solution.

Equations with an infinite number of solutions

Simultaneous equations can also possess an infinite number of solutions. Consider the following example.

Example 4.6

Solve the equations

$$2x + \ y = 8 \qquad (9)$$
$$4x + 2y = 16 \qquad (10)$$

If equation (9) is multiplied by 2 we find

$$4x + 2y = 16$$
$$4x + 2y = 16$$

Note that both equations are now identical. This means that one of them is redundant. We need only consider the single equation

$$2x + y = 8$$

There are infinitely many pairs of values of x and y that satisfy this equation. For example, if $x = 0$, $y = 8$. Similarly, if $x = 1$, $y = 6$, and if $x = -3$, $y = 14$. We could continue like this, producing more and more solutions. Suppose we choose a value for x. Because we are not restricted in any way we call x a **free variable**. Let us call our choice λ. We can then write

$$2\lambda + y = 8 \text{ so that } y = 8 - 2\lambda$$

The solution is therefore $x = \lambda$, $y = 8 - 2\lambda$ for any value of λ whatsoever. There are an infinite number of such solutions.

Example 4.7 Civil Engineering – Curve fitting

In many engineering applications physical quantities are specified at a few distinct points rather than everywhere in a given domain. For example, if a civil engineer wishes to analyse the strain in a building, strain gauges may be attached at several specific points in the structure. Interpolation techniques can be used to find the strain at other points. Curve fitting is used to achieve this, by fitting a curve whose mathematical equation can be calculated through the known points. Curve fitting is particularly important in computer-aided design (CAD) and computer-aided manufacture (CAM) because the geometry or shape of a solid can be described by specifying a number of points on the surface and interpolating using an appropriate curve or line.

Suppose two measured points have (x, y) coordinates $(5, 8)$ and $(6, 5)$ as shown in Figure 4.1.
(a) Find the equation of the straight line that passes through or interpolates these points.
(b) Use this linear approximation to estimate $y(5.3)$.

Figure 4.1
The straight line interpolates the two given points.

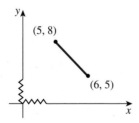

Solution
(a) Suppose the straight line has equation $y = mx + c$. Because the line passes through $(5, 8)$ and $(6, 5)$ we know that

$$8 = 5m + c \qquad (11)$$
$$\text{and} \quad 5 = 6m + c \qquad (12)$$

These are two simultaneous equations, which must be solved in order to find m and c. Subtracting equation (12) from equation (11) we find

$$3 = -m$$

so that $m = -3$. Substituting this value for m into equation (11) we find

$$8 = 5(-3) + c$$

so that $c = 23$. Then the linear approximation has equation $y = -3x + 23$.
(b) We can use the linear approximation to estimate y when $x = 5.3$:

$$y(5.3) = -3(5.3) + 23 = 7.1$$

In practice, higher-degree polynomials or other curves are usually used for interpolation purposes rather than simple straight lines.

Exercises

In questions 1–3 solve the given simultaneous equations by elimination:

1   $5x + y = 8, -3x + 2y = -10$

2   $2x + 3y = -2, 5x - 5y = 20$

3   $7x + 11y = -24, -9x + y = 46$

4   A straight line has an equation of the form $y = ax + b$. The line passes through the points with coordinates $(2, 4)$ and $(-1, 3)$. Write down the simultaneous equations that

must be satisfied by a and b. Solve them and hence find the equation of the line.

5   A quadratic function $y = ax^2 + bx + c$ is used in signal processing to approximate a more complicated signal. If this function must pass through the points with coordinates $(0, 0)$, $(1, 3)$ and $(5, -11)$ write down the simultaneous equations satisfied by a, b and c. Solve these to find the quadratic function.

Solutions to exercises

1   $x = 2, y = -2$

2   $x = 2, y = -2$

3   $x = -5, y = 1$

4   $y = \dfrac{1}{3}x + \dfrac{10}{3}$

5   $y = -\dfrac{13}{10}x^2 + \dfrac{43}{10}x$

4.3   Solving simultaneous equations using graphs

An alternative way of solving two simultaneous equations in two unknowns is to draw a graph. Note that each of the equations we have studied is a linear equation and plotting its graph will produce a straight line. We can plot a line for each equation and locate the point of intersection of the two lines. This point represents the solution of the equations. If the two lines do not intersect then the equations have no solution. If the two lines are identical, there are an infinite number of solutions. Consider the following examples.

Example 4.8

Solve the simultaneous equations

$$4x + y = 9 \tag{13}$$
$$-x + y = -1 \tag{14}$$

by plotting two straight line graphs.

Solution

Equation (13) is rearranged into the standard form of the equation of a straight line: $y = -4x + 9$. By selecting two points on the line a graph can be drawn as shown in Figure 4.2.

Similarly, equation (14) can be rearranged as $y = x - 1$ and its graph drawn. This is also shown in Figure 4.2. The coordinates of any point on line I satisfy $4x + y = 9$. The coordinates of any point on line II satisfy $-x + y = -1$. At the point where the two lines intersect the x and y coordinates must satisfy both equations and so the point of intersection represents the solution. We see from the graph that the point of intersection is $(2, 1)$. The solution of the given equations is therefore $x = 2, y = 1$.

Figure 4.2
The coordinates of the point of intersection give the required solution.

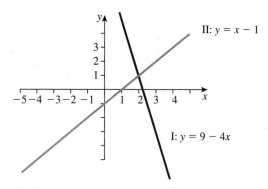

Example 4.9

Solve the equations: $10x - 2y = -3, 5x - y = -1$.

Solution

Rewriting the equations in standard form we find

$$y = 5x + \frac{3}{2} \quad \text{and} \quad y = 5x + 1$$

Graphs of these lines are shown in Figure 4.3. Note that these lines do not intersect because they are parallel. This means that the given simultaneous equations do not have a solution; they are inconsistent.

Figure 4.3
The graphs do not intersect – the equations have no solution.

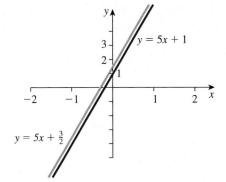

$y = 5x + 1$

$y = 5x + \frac{3}{2}$

Exercises

In questions 1–4 solve the given equations graphically:

1 $5x - y = 7, 2x + y = 7$

2 $2x - 2y = -2, 5x + y = -9$

3 $7x + 3y = 25, -2x + y = 4$

4 $4x + 4y = -4, x + 7y = -19$

Solutions to exercises

1 $x = 2, y = 3$

2 $x = -\frac{5}{3}, y = -\frac{2}{3}$

3 $x = 1, y = 6$

4 $x = 2, y = -3$

Computer and calculator exercise

1 Use your calculator or computer software to plot graphs and hence estimate the solution of
$$0.5x - 0.25y = -0.6$$
$$0.33x + 0.8y = -0.15$$

Solution to exercise

1 approx. $x = -1.07, y = 0.25$

End of block exercises

1 Solve the following simultaneous equations:
(a) $3x + 2y = 7, x + y = 3$
(b) $3s + 4t = 0, 2s - 2t = 7$

2 Show that the equations $11x + y = 7$ and $7y = 7 - 77x$ have no solution.

3 Solve the equations

$$\frac{3}{s} + \frac{2}{t} = 11, \quad -\frac{7}{s} - \frac{5}{t} = -26$$

(Hint: let $x = \frac{1}{s}, y = \frac{1}{t}$.)

4 Find the equation of the straight line passing

through $\left(4, 13\frac{3}{4}\right)$ and $\left(1, \frac{13}{4}\right)$.

5 A graph has equation $y = ax^2 + bx$. Find a and b if the graph passes through the points $(1, 2)$ and $(3, 5)$.

Solutions to exercises

1 (a) 1, 2 (b) 2, −1.5

3 $\frac{1}{3}$, 1

4 $y = \frac{7}{2}x - \frac{1}{4}$

5 $-\frac{1}{6}, \frac{13}{6}$

Solution of inequalities

5.1 Introduction

An inequality is an expression involving one of the symbols $\geq$, $\leq$, $>$ or $<$. This block will first show how to manipulate inequalities correctly. Then the solution of inequalities, both algebraically and graphically, will be described.

5.2 The inequality symbols

Recall the meaning of the following symbols:

Key point

> means: 'is greater than', $\geq$ means: 'is greater than or equal to'
< means: 'is less than', $\leq$ means: 'is less than or equal to'

So, we may state, for example, $8 > 7, 9 \geq 2, -2 < 3$ and $7 \leq 7$. The symbols $>$, $<$, $\geq$, $\leq$ are called **inequalities**.

A number line is often a helpful way of picturing inequalities. Given two numbers a and b, if $b > a$ then b will be to the right of a on the number line, as shown in Figure 5.1.

Figure 5.1
If $b > a$, b will be to the right of a on the number line.

Note from Figure 5.2 that $-3 > -5, 4 > -2$ and $8 > 5$.

Figure 5.2

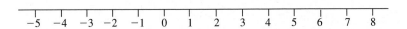

Inequalities can always be written in two ways. For example, in English we can state that 8 is greater than 7, or equivalently, 7 is less than 8. Mathematically we write $8 > 7$ or $7 < 8$. Similarly if $b > a$ then $a < b$. If $a < b$ then a will be to the left of b on the number line.

Example 5.1

Rewrite the inequality $-\frac{2}{5} < x$ using only the 'greater than' sign, $>$.

Solution

$-\frac{2}{5} < x$ can be written as $x > -\frac{2}{5}$.

Example 5.2

Rewrite the inequality $5 > x$ using only the 'less than' sign, $<$.

Solution

$5 > x$ can be written as $x < 5$.

Sometimes two inequalities are combined into a single statement. Consider for example $3 < x < 6$. This is a compact way of writing $3 < x$ and $x < 6$. Now $3 < x$ is equivalent to $x > 3$ and so $3 < x < 6$ means x is greater than 3 but less than 6.

It would be incorrect to write $5 > x > 9$ because 5 is not greater than 9. Instead two inequalities would be used and we would write: $x < 5$ or $x > 9$, that is x is less than 5 or greater than 9.

Inequalities obey simple rules when used in conjunction with arithmetical operations.

Key point

1   Adding or subtracting the same quantity from both sides of an inequality leaves the inequality sign unchanged.
2   Multiplying or dividing both sides by a **positive** number leaves the inequality sign unchanged.
3   Multiplying or dividing both sides by a **negative** number reverses the inequality.

For example, since $8 > 5$, by adding k to both sides we can state

$$8 + k > 5 + k$$

for any value of k.

Further, by multiplying both sides of $8 > 5$ by k we can state $8k > 5k$ provided k is positive. However,

$$8k < 5k$$

if k is negative. Note that the inequality sign is reversed when multiplying both sides by a negative number. A common mistake is to forget to reverse the inequality sign when multiplying inequalities by a negative number. For example, $8 > 5$, but multiplying both sides by -1 gives $-8 < -5$.

Example 5.3

Find the result of multiplying both sides of the inequality $-18 < 9$ by the number -3.

Solution

$54 > -27$

The **modulus sign** is sometimes used with inequalities. For example, $|x| < 1$ means all numbers whose actual size, irrespective of sign, is less than 1. This means any value between -1 and 1. Thus

$$|x| < 1 \quad \text{means} \quad -1 < x < 1$$

In general, if k is a positive number,

Key point

$$|x| < k \quad \text{means} \quad -k < x < k$$

Similarly $|x| > 4$ means all numbers whose size, irrespective of sign, is greater than 4. This means any value greater than 4 or less than -4. Thus

$$|x| > 4 \quad \text{means} \quad x > 4 \text{ or } x < -4$$

In general, if k is a positive number,

Key point

$$|x| > k \quad \text{means} \quad x > k \text{ or } x < -k$$

Exercises

1 State whether each of the following statements is true or false.
(a) $4 > 9$ (b) $4 > 4$ (c) $4 \geq 4$
(d) $0.001 < 10^{-5}$ (e) $|-19| < 100$
(f) $|-19| > -20$ (g) $0.001 \leq 10^{-3}$

In questions 2–9 rewrite each of the statements without using a modulus sign:

2 $|x| < 2$

3 $|x| < 5$

4 $|x| \leq 7.5$

5 $|x - 3| < 2$

6 $|x - a| < 1$

7 $|x| > 2$

8 $|x| > 7.5$

9 $|x| \geq 0$

Solutions to exercises

1 (a) F (b) F (c) T (d) F (e) T (f) T (g) T

2 $-2 < x < 2$

3 $-5 < x < 5$

4 $-7.5 \leq x \leq 7.5$

5 $-2 < x - 3 < 2$

6 $-1 < x - a < 1$

7 $x > 2$ or $x < -2$

8 $x > 7.5$ or $x < -7.5$

9 $x \geq 0$ or $x \leq 0$, in fact any x

5.3 Solving inequalities algebraically

When we are asked to **solve** an inequality, the inequality will contain an unknown variable, x say. To solve means to find all values of x for which the inequality is true. In a **linear inequality** the unknown appears only to the first power, that is as x, and not as x^2, x^3, $x^{1/2}$ and so on. It is possible to solve a linear inequality by making the unknown the subject. Consider the following examples.

Example 5.4
Solve the inequality $4x + 3 > 0$.

Solution

$$4x + 3 > 0$$
$$4x > -3 \quad \text{by subtracting 3 from both sides}$$
$$x > -\tfrac{3}{4} \quad \text{by dividing both sides by 4}$$

Hence all values of x greater than $-\tfrac{3}{4}$ satisfy $4x + 3 > 0$.

Example 5.5
Solve the inequality $-3x - 7 \le 0$.

Solution

$$-3x - 7 \le 0$$
$$-3x \le 7 \quad \text{by adding 7 to both sides}$$
$$x \ge -\tfrac{7}{3} \quad \text{dividing both sides by } -3$$
$$\text{and reversing the inequality}$$

Hence all values of x greater than or equal to $-\tfrac{7}{3}$ satisfy $-3x - 7 \le 0$.

Example 5.6
Solve the inequality $17x + 2 < 4x + 1$.

Solution
We try to make x the subject and obtain it on its own on the left-hand side. Start by subtracting $4x$ from both sides to remove quantities involving x from the right.

$$13x + 2 < 1$$

Then subtract 2 from both sides to remove the 2 on the left:

$$13x < -1$$

Finally find the range of values satisfied by x:

$$x < -\tfrac{1}{13}$$

Example 5.7

Solve the inequality $|5x - 2| < 4$ and depict the solution graphically.

Solution

$$|5x - 2| < 4$$

is equivalent to

$$-4 < 5x - 2 < 4$$

We shall treat both parts of the inequality separately:

$$
\begin{aligned}
-4 &< 5x - 2 \\
-2 &< 5x &&\text{by adding 2 to both sides} \\
-\tfrac{2}{5} &< x &&\text{by dividing both sides by 5}
\end{aligned}
$$

So $x > -\tfrac{2}{5}$. Now consider the second part: $5x - 2 < 4$.

$$
\begin{aligned}
5x - 2 &< 4 \\
5x &< 6 &&\text{by adding 2 to both sides} \\
x &< \tfrac{6}{5} &&\text{by dividing both sides by 5}
\end{aligned}
$$

Putting both parts of the solution together we see that the inequality is satisfied when $-\tfrac{2}{5} < x < \tfrac{6}{5}$. This range of values is shown in Figure 5.3.

Figure 5.3
$|5x - 2| < 4$
when
$-\tfrac{2}{5} < x < \tfrac{6}{5}$.

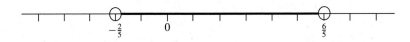

Example 5.8

Solve the inequality $|1 - 2x| < 5$.

Solution

First of all rewrite the inequality without using the modulus sign.

$|1 - 2x| < 5$ is equivalent to $-5 < 1 - 2x < 5$

Then treat each part separately. First of all consider $-5 < 1 - 2x$. Solve this.

$$x < 3$$

The second part is $1 - 2x < 5$. Solve this.

$$x > -2$$

Finally, confirm that the solution is $-2 < x < 3$.

Exercises

In questions 1–16 solve the given inequality algebraically:

1 $4x > 8$ **2** $5x > 8$

3 $8x > 5$

4 $8x \leq 5$

5 $2x > 1$

6 $3x < -1$

7 $5x > 2$

8 $2x > 0$

9 $8x < 0$

10 $3x \geq 0$

11 $3x > 4$

12 $\frac{3}{4}x > 1$

13 $4x \leq -3$

14 $3x \leq -4$

15 $5x \geq 0$

16 $4x \leq 0$

In questions 17–26 solve the given inequality algebraically:

17 $5x + 1 < 8$

18 $5x + 1 \leq 8$

19 $7x + 3 \geq 0$

20 $18x + 2 > 9$

21 $14x + 11 > 22$

22 $1 - 5x \leq 0$

23 $2 + 5x \geq 1$

24 $11 - 7x < 2$

25 $5 + 4x > 2x + 1$

26 $7 - 3x > x - 5$

In questions 27–33 solve the inequality:

27 $|7x - 3| > 1$

28 $|2x + 1| \geq 3$

29 $|5x| < 1$

30 $|5x| \leq 0$

31 $|1 - 5x| > 2$

32 $|2 - 5x| \geq 3$

33 $|2x - 1| < 1$

Solutions to exercises

1 $x > 2$

2 $x > \frac{8}{5}$

3 $x > \frac{5}{8}$

4 $x \leq \frac{5}{8}$

5 $x > \frac{1}{2}$

6 $x < -\frac{1}{3}$

7 $x > \frac{2}{5}$

8 $x > 0$

9 $x < 0$

10 $x \geq 0$

11 $x > \frac{4}{3}$

12 $x > \frac{4}{3}$

13 $x \leq -\frac{3}{4}$

14 $x \leq -\frac{4}{3}$

15 $x \geq 0$

16 $x \leq 0$

17 $x < \frac{7}{5}$

18 $x \leq \frac{7}{5}$

19 $x \geq -\frac{3}{7}$

20 $x > \frac{7}{18}$

21 $x > \frac{11}{14}$

22 $x \geq \frac{1}{5}$

23 $x \geq -\frac{1}{5}$

24 $x > \frac{9}{7}$

25 $x > -2$

26 $x < 3$

27 $x > \frac{4}{7}$ or $x < \frac{2}{7}$

28 $x \geq 1$ or $x \leq -2$

29 $-\frac{1}{5} < x < \frac{1}{5}$

30 $x = 0$

31 $x < -\frac{1}{5}, x > \frac{3}{5}$

32 $x \leq -\frac{1}{5}, x \geq 1$

33 $0 < x < 1$

5.4 Solving inequalities using graphs

Graphs can be used to solve inequalities, particularly if the inequality is not linear. Graphics calculators or software can help save a lot of time and effort here.

Example 5.9
Solve graphically the inequality $5x + 2 < 0$.

Solution
We consider the function $y = 5x + 2$ whose graph is shown in Figure 5.4.

Figure 5.4
Graph of
$y = 5x + 2$.

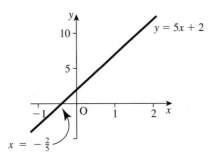

The values of x that make $5x + 2$ negative are those for which y is negative. We see directly from the graph that y is negative when $x < -\frac{2}{5}$.

Example 5.10
Find the range of values of x for which $x^2 - x - 6 < 0$.

Solution
We consider the graph of $y = x^2 - x - 6$, which is shown in Figure 5.5.

Figure 5.5
Graph of
$y = x^2 - x - 6$.

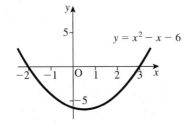

Note that the graph crosses the x axis when $x = -2$ and when $x = 3$. Now $x^2 - x - 6$ will be negative when y is negative. Directly from the graph we see that y is negative when $-2 < x < 3$.

Example 5.11

Find the range of values of x for which $x^2 - x - 6 > 0$.

Solution

The graph of $y = x^2 - x - 6$ has been drawn in Figure 5.5. We require $y = x^2 - x - 6$ to be positive. Use the graph to solve the problem.

$x < -2$ or $x > 3$

Example 5.12

By plotting a graph of $y = 20x^4 - 4x^3 - 143x^2 + 46x + 165$ find the range of values of x for which

$$20x^4 - 4x^3 - 143x^2 + 46x + 165 < 0$$

Solution

A software package has been used to plot the graph, which is shown in Figure 5.6. We see that y is negative when $-2.5 < x < -1$ and when $1.5 < x < 2.2$.

Figure 5.6
Graph of
$y = 20x^4 - 4x^3 - 143x^2 + 46x + 165$.

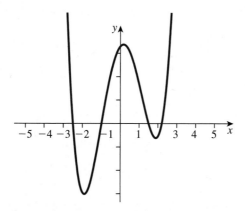

Exercises

In questions 1–4 solve the given inequality graphically:

1 $3x + 1 < 0$

2 $2x - 7 < 0$

3 $6x + 9 > 0$

4 $5x - 3 > 0$

Solutions to exercises

1 $x < -\dfrac{1}{3}$

2 $x < \dfrac{7}{2}$

3 $x > -\dfrac{3}{2}$

4 $x > \dfrac{3}{5}$

Computer and calculator exercises

Use a graphics calculator or software to solve the following inequalities:

1 $6x^3 + x^2 - 4x + 1 < 0$

2 $4x^4 + 9x^3 - 3x^2 - 10x > 0$

3 $\dfrac{x + 2}{x - 3} < 0$

4 $\dfrac{x + 1}{x + 2} > 0$

5 $\dfrac{(x - 1)(x + 1)}{x - 3} < 0$

End of block exercises

1 Express in words, the meaning of
(a) $N \leq -21$ (b) $0 \leq p \leq 1$

2 Express in symbols the statement 'a can take any value between -3 and 9 inclusive'.

3 Express in symbols 'x is non-negative'.

4 Solve the inequality $5 - 4x \leq -2$.

5 Solve the given inequalities:
(a) $|2x - \frac{1}{2}| < 2$ (b) $|3x + 4| > 8$
(c) $|4x + 3| \geq 1$ (d) $|5x - 2| < 10$

6 Find the range of values satisfied by X if

$$\left|\frac{X - \mu}{\sigma}\right| < 1 \text{ and } \sigma > 0.$$

7 Sketch a graph of the rational function

$$R(x) = \frac{x + 1}{2x - 6}$$

Hence state the range of values for which

$$\frac{x + 1}{2x - 6} > 0$$

Solutions to exercises

1 (a) N is any number less than or equal to -21.
(b) p lies between 0 and 1, inclusive.

2 $-3 \leq a \leq 9$

3 $x \geq 0$

4 $x \geq \dfrac{7}{4}$

5 (a) $-\dfrac{3}{4} < x < \dfrac{5}{4}$

(b) $x < -4$ or $x > \dfrac{4}{3}$

(c) $x \leq -1$ or $x \geq -\dfrac{1}{2}$

(d) $-\dfrac{8}{5} < x < \dfrac{12}{5}$

6 $-\sigma + \mu < X < \sigma + \mu$

7 $x > 3$ or $x < -1$

Partial fractions

6.1 Introduction

It is often helpful to break down a complicated fraction into the sum of simpler fractions. For example, it can be shown that $\dfrac{4x + 7}{x^2 + 3x + 2}$ has the same value as $\dfrac{1}{x + 2} + \dfrac{3}{x + 1}$ for any value of x. We say that

$$\frac{4x + 7}{x^2 + 3x + 2} \text{ is identically equal to } \frac{1}{x + 2} + \frac{3}{x + 1}$$

and that the **partial fractions** of $\dfrac{4x + 7}{x^2 + 3x + 2}$ are $\dfrac{1}{x + 2}$ and $\dfrac{3}{x + 1}$. The ability to express a fraction as its partial fractions is particularly useful in the study of Laplace transforms, and in control theory. In this block we explain how partial fractions are found.

6.2 Proper and improper fractions

Frequently we find that an algebraic fraction appears in the form

$$\text{algebraic fraction} = \frac{\text{numerator}}{\text{denominator}} = \frac{\text{polynomial}}{\text{polynomial}}$$

that is as the ratio of two polynomials. For example,

$$\frac{x^3 + x^2 + 3x + 7}{x^2 + 1}, \quad \frac{3x^2 - 2x + 5}{x^2 - 7x + 2} \quad \text{and} \quad \frac{x}{x^4 + 1}$$

The procedure for finding partial fractions depends critically upon the form of the denominator, as we shall see. The **degree** of the numerator, n say, is the highest power occurring in the numerator. The degree of the denominator, d say, is the highest power occurring in the denominator. If $d > n$ the fraction is said to be **proper**. If $d \le n$ the fraction is said to be **improper**. Before calculating the partial fractions of an algebraic fraction it is important to decide whether the fraction is proper or improper.

Example 6.1
For each of the following fractions state the degrees of the numerator and denominator. Hence classify the fractions as proper or improper.

(a) $\dfrac{x^3 + x^2 + 3x + 7}{x^2 + 1}$ (b) $\dfrac{3x^2 - 2x + 5}{x^2 - 7x + 2}$ (c) $\dfrac{x}{x^4 + 1}$ (d) $\dfrac{s^2 + 4s + 5}{(s^2 + 2s + 4)(s + 3)}$

Solution

(a) The degree of the numerator, n, is 3. The degree of the denominator, d, is 2. Because $d \leq n$ the fraction is improper.

(b) Here $n = 2$ and $d = 2$. State whether this fraction is proper or improper.

$d \leq n$; the fraction is improper

(c) Noting that $x = x^1$ we see that $n = 1$ and $d = 4$. State whether this fraction is proper or improper.

$d > n$; the fraction is proper

(d) Removing the brackets in the denominator we see that it has degree 3. The degree of the numerator is 2 and so this fraction is proper.

The denominator of an algebraic fraction can be factorised into a product of linear and quadratic factors. Linear factors are those of the form $ax + b$, for example $2x + 7$, $3x - 2$ and $4 - x$. Quadratic factors are those of the form $ax^2 + bx + c$ such as $x^2 + x + 1$, and $4x^2 - 2x + 3$, which cannot be factorised into linear factors.

Exercises

1 For each fraction state the degrees of the numerator and denominator, and hence determine which are proper and which are improper:

(a) $\dfrac{x + 1}{x}$ (b) $\dfrac{x^2}{x^3 - x}$

(c) $\dfrac{(x - 1)(x - 2)(x - 3)}{x - 5}$

Solutions to exercises

1 (a) $n = 1$, $d = 1$, improper
  (b) $n = 2$, $d = 3$, proper
  (c) $n = 3$, $d = 1$, improper

6.3 Proper fractions with linear factors

First we describe how to calculate partial fractions of proper fractions where the denominator may be written as a product of linear factors. The steps needed to calculate the partial fractions are as follows:

1 Factorise the denominator.
2 Each factor will produce a partial fraction. A factor such as $3x + 2$ will produce a partial fraction of the form $\dfrac{A}{3x + 2}$ where A is an unknown constant. In general

a linear factor $ax + b$ will produce a partial fraction $\dfrac{A}{ax + b}$. The unknown constants for each partial fraction may be different and so we shall call them A, B, C and so on.

3 Evaluate the unknown constants by equating coefficients or substituting specific values of x.

The sum of the partial fractions is identical in value to the original algebraic fraction for any value of x.

Key point

> A linear factor $ax + b$ produces a partial fraction of the form $\dfrac{A}{ax + b}$.

The steps involved are illustrated in the following example.

Example 6.2

Express $\dfrac{7x + 10}{2x^2 + 5x + 3}$ as its partial fractions.

Solution

Note that this fraction is proper. The denominator is factorised to give $(2x + 3)(x + 1)$. It has been written as the product of two linear factors. Each factor produces a partial fraction. The factor $2x + 3$ produces a partial fraction of the form $\dfrac{A}{2x + 3}$. The factor $x + 1$ produces a partial fraction $\dfrac{B}{x + 1}$, where A and B are unknown constants which we now try to find. We write

$$\frac{7x + 10}{(2x + 3)(x + 1)} = \frac{A}{2x + 3} + \frac{B}{x + 1}$$

By multiplying both sides by $(2x + 3)(x + 1)$ we obtain

$$7x + 10 = A(x + 1) + B(2x + 3) \tag{1}$$

We may now let x take any value we choose. By an appropriate choice we can simplify the right-hand side. Let $x = -1$ because this choice eliminates A. We find

$$7(-1) + 10 = A(0) + B(-2 + 3)$$
$$3 = B$$

so that the constant B must equal 3. The constant A can be found by substituting other values for x or alternatively by equating coefficients. Observe that, by rearranging the right-hand side, equation (1) can be written as

$$7x + 10 = (A + 2B)x + (A + 3B)$$

Comparing the coefficients of x on both sides we see that $7 = A + 2B$. We already know $B = 3$ and so

$$7 = A + 2(3)$$
$$= A + 6$$

from which $A = 1$. We can therefore write

$$\frac{7x + 10}{2x^2 + 5x + 3} = \frac{1}{2x + 3} + \frac{3}{x + 1}$$

We have succeeded in expressing the given fraction as the sum of its partial fractions. The result can always be checked by adding the fractions on the right.

Example 6.3

Express $\dfrac{9 - 4x}{3x^2 - x - 2}$ in partial fractions.

Solution

First factorise the denominator:

$$3x^2 - x - 2 = \boxed{} \qquad\qquad (3x + 2)(x - 1)$$

Because there are two linear factors we write

$$\frac{9 - 4x}{3x^2 - x - 2} = \frac{9 - 4x}{(3x + 2)(x - 1)}$$

$$= \frac{A}{3x + 2} + \frac{B}{x - 1}$$

Multiply both sides by $(3x + 2)(x - 1)$ to obtain the equation from which we can find values for A and B.

$$9 - 4x = \boxed{} \qquad\qquad A(x - 1) + B(3x + 2)$$

By substituting an appropriate value for x obtain B.

$$\boxed{} \qquad\qquad \text{substitute } x = 1 \text{ and get } B = 1$$

Finally by equating coefficients of x obtain the value of A.

$$\boxed{} \qquad\qquad -4 = A + 3B, A = -7 \text{ since } B = 1$$

Finally, write down the partial fractions:

$$\frac{9 - 4x}{3x^2 - x - 2} = \boxed{} \qquad\qquad \frac{-7}{3x + 2} + \frac{1}{x - 1}$$

Exercises

1   (a) Find the partial fractions of

$$\frac{5x - 1}{(x + 1)(x - 2)}$$

(b) Check your answer by adding the partial fractions together again.

(c) Express in partial fractions

$$\frac{7x + 25}{(x + 4)(x + 3)}$$

(d) Check your answer by adding the partial fractions together again.

2 Find the partial fractions of $\dfrac{11x + 1}{(x - 1)(2x + 1)}$.

Express each of the following as the sum of partial fractions:

3 $\dfrac{3}{(x + 1)(x + 2)}$

4 $\dfrac{5}{x^2 + 7x + 12}$

5 $\dfrac{-3}{(2x + 1)(x - 3)}$

Solutions to exercises

1 (a) $\dfrac{2}{x + 1} + \dfrac{3}{x - 2}$ (c) $\dfrac{3}{x + 4} + \dfrac{4}{x + 3}$

2 $\dfrac{4}{x - 1} + \dfrac{3}{2x + 1}$

3 $\dfrac{3}{x + 1} - \dfrac{3}{x + 2}$

4 $\dfrac{5}{x + 3} - \dfrac{5}{x + 4}$

5 $\dfrac{6}{7(2x + 1)} - \dfrac{3}{7(x - 3)}$

6.4 Proper fractions with repeated linear factors

As before, the denominator is factorised first. Sometimes a linear factor appears more than once. For example, in

$$\frac{1}{(x + 1)(x + 1)} \quad \text{which equals} \quad \frac{1}{(x + 1)^2}$$

the factor $(x + 1)$ occurs twice. We call it a **repeated linear factor**. The repeated linear factor $(x + 1)^2$ produces two partial fractions of the form $\dfrac{A}{x + 1} + \dfrac{B}{(x + 1)^2}$. In general, a repeated linear factor of the form $(ax + b)^2$ generates two partial fractions of the form

$$\frac{A}{ax + b} + \frac{B}{(ax + b)^2}$$

Key point

Repeated linear factors $(ax + b)^2$ produce partial fractions of the form

$$\frac{A}{ax + b} + \frac{B}{(ax + b)^2}$$

Once again the unknown constants are found by a mixture of equating coefficients and substituting specific values for x.

Example 6.4

Express

$$\frac{10x + 18}{4x^2 + 12x + 9}$$

in partial fractions.

Solution

First the denominator is factorised.

$$4x^2 + 12x + 9 = \boxed{}$$

$$(2x + 3)(2x + 3) = (2x + 3)^2$$

You should have found a repeated linear factor. The repeated linear factor $(2x + 3)^2$ gives rise to two partial fractions of the form

$$\frac{10x + 18}{(2x + 3)^2} = \frac{A}{2x + 3} + \frac{B}{(2x + 3)^2}$$

Multiply both sides through by $(2x + 3)^2$ to obtain the equation that must be solved to find A and B.

$$10x + 18 = A(2x + 3) + B$$

Now evaluate the constants A and B by equating coefficients. Equating coefficients of x gives

$$10 = 2A, A = 5$$

Equating constant terms gives $18 = 3A + B$, from which $B = 3$. So, finally, we may write

$$\frac{10x + 18}{(2x + 3)^2} = \frac{5}{2x + 3} + \frac{3}{(2x + 3)^2}$$

Exercises

Express the following in partial fractions:

1. $\dfrac{3 - x}{x^2 - 2x + 1}$

2. $-\dfrac{7x - 15}{(x - 1)^2}$

3. $\dfrac{3x + 14}{x^2 + 8x + 16}$

4. $\dfrac{5x + 18}{(x + 4)^2}$

5. $\dfrac{2x^2 - x + 1}{(x + 1)(x - 1)^2}$

6. $\dfrac{5x^2 + 23x + 24}{(2x + 3)(x + 2)^2}$

7. $\dfrac{6x^2 - 30x + 25}{(3x - 2)^2 (x + 7)}$

8. $\dfrac{s + 2}{(s + 1)^2}$

9. $\dfrac{2s + 3}{s^2}$

Solutions to exercises

1 $\quad -\dfrac{1}{x-1} + \dfrac{2}{(x-1)^2}$

2 $\quad -\dfrac{7}{x-1} + \dfrac{8}{(x-1)^2}$

3 $\quad \dfrac{3}{x+4} + \dfrac{2}{(x+4)^2}$

4 $\quad \dfrac{5}{x+4} - \dfrac{2}{(x+4)^2}$

5 $\quad \dfrac{1}{x+1} + \dfrac{1}{x-1} + \dfrac{1}{(x-1)^2}$

6 $\quad \dfrac{3}{2x+3} + \dfrac{1}{x+2} + \dfrac{2}{(x+2)^2}$

7 $\quad -\dfrac{1}{3x-2} + \dfrac{1}{(3x-2)^2} + \dfrac{1}{x+7}$

8 $\quad \dfrac{1}{s+1} + \dfrac{1}{(s+1)^2}$

9 $\quad \dfrac{2}{s} + \dfrac{3}{s^2}$

6.5 Proper fractions with quadratic factors

Sometimes a denominator is factorised producing a quadratic factor that cannot be factorised into linear factors. One such quadratic factor is $x^2 + x + 1$. This factor would produce a partial fraction of the form $\dfrac{Ax + B}{x^2 + x + 1}$. In general a quadratic factor of the form $ax^2 + bx + c$ produces a single partial fraction of the form $\dfrac{Ax + B}{ax^2 + bx + c}$.

Key point

A quadratic factor of the form $ax^2 + bx + c$ produces a partial fraction of the form

$$\frac{Ax + B}{ax^2 + bx + c}$$

Example 6.5

Express as partial fractions

$$\frac{3x + 1}{(x^2 + x + 10)(x - 1)}$$

Solution

Note that the quadratic factor cannot be factorised further. We have

$$\frac{3x + 1}{(x^2 + x + 10)(x - 1)} = \frac{Ax + B}{x^2 + x + 10} + \frac{C}{x - 1}$$

Multiplying both sides by $(x^2 + x + 10)(x - 1)$ gives

$$3x + 1 = $$

$$(Ax + B)(x - 1) + C(x^2 + x + 10)$$

To evaluate C we can let $x = 1$, which eliminates the first term on the right. This gives

$$4 = 12C, \text{ so that } C = \frac{1}{3}$$

Equate coefficients of x^2 and hence find A. Finally substitute any other value for x or equate coefficients of x to find B.

$$A = -\frac{1}{3}, B = \frac{7}{3}$$

Finally

$$\frac{3x + 1}{(x^2 + x + 10)(x - 1)} = \frac{-\frac{1}{3}x + \frac{7}{3}}{x^2 + x + 10} + \frac{\frac{1}{3}}{x - 1}$$

$$= \frac{1}{3}\frac{7 - x}{x^2 + x + 10} + \frac{1}{3(x - 1)}$$

Example 6.6 Electrical Engineering – Admittance

Admittance, Y, is a quantity that is used in analysing electronic circuits. A typical expression for admittance might take the form

$$Y(s) = \frac{s^2 + 4s + 5}{(s^2 + 2s + 4)(s + 3)}$$

where s can be thought of as representing frequency. To calculate the behaviour of the circuit it is often necessary to express the admittance as the sum of its partial fractions and find the effect of each part separately. Express $Y(s)$ in partial fractions.

Solution

The fraction is proper. The denominator contains a quadratic factor that cannot be factorised further, and also a linear factor. Thus

$$\frac{s^2 + 4s + 5}{(s^2 + 2s + 4)(s + 3)} = \frac{As + B}{s^2 + 2s + 4} + \frac{C}{s + 3}$$

Multiplying both sides by $(s^2 + 2s + 4)(s + 3)$ we obtain

$$s^2 + 4s + 5 = (As + B)(s + 3) + C(s^2 + 2s + 4)$$

To find the constant C we can let $s = -3$ to eliminate A and B. Thus

$$(-3)^2 + 4(-3) + 5 = C[(-3)^2 + 2(-3) + 4]$$

so that

$$2 = 7C$$

and so $C = \frac{2}{7}$.

Equating coefficients of s^2 we find

$$1 = A + C$$

so that $A = 1 - C = 1 - \frac{2}{7} = \frac{5}{7}$.

Equating constant terms gives

$$5 = 3B + 4C$$

so that

$$3B = 5 - 4C$$
$$= 5 - 4\left(\frac{2}{7}\right)$$
$$= 5 - \frac{8}{7}$$
$$= \frac{27}{7}$$
$$\text{so } B = \frac{9}{7}$$

Finally

$$Y(s) = \frac{s^2 + 4s + 5}{(s^2 + 2s + 4)(s + 3)} = \frac{\frac{5}{7}s + \frac{9}{7}}{s^2 + 2s + 4} + \frac{\frac{2}{7}}{s + 3}$$

which can be written as

$$Y(s) = \frac{5s + 9}{7(s^2 + 2s + 4)} + \frac{2}{7(s + 3)}$$

Exercises

Express each of the following as the sum of its partial fractions:

1 $\dfrac{3}{(x^2 + x + 1)(x - 2)}$

2 $\dfrac{27x^2 - 4x + 5}{(6x^2 + x + 2)(x - 3)}$

3 $\dfrac{2x + 4}{4x^2 + 12x + 9}$

4 $\dfrac{6x^2 + 13x + 2}{(x^2 + 5x + 1)(x - 1)}$

Solutions to exercises

1 $\dfrac{3}{7(x - 2)} - \dfrac{3(x + 3)}{7(x^2 + x + 1)}$

2 $\dfrac{3x + 1}{6x^2 + x + 2} + \dfrac{4}{x - 3}$

3 $\dfrac{1}{2x + 3} + \dfrac{1}{(2x + 3)^2}$

4 $\dfrac{3x + 1}{x^2 + 5x + 1} + \dfrac{3}{x - 1}$

6.6 Improper fractions

When calculating the partial fractions of improper fractions an extra term needs to be included. The extra term is a polynomial of degree $n - d$ where d is the degree of the denominator and n is the degree of the numerator. Recall that

- a polynomial of degree 0 is a constant, A say,
- a polynomial of degree 1 has the form $Ax + B$,
- a polynomial of degree 2 has the form $Ax^2 + Bx + C$,

and so on. For example, if the numerator has degree 5 and the denominator has degree 3, then $n - d = 2$, the fraction is improper, and we need to include an extra term of the form $Ax^2 + Bx + C$.

Key point

> If a fraction is improper an additional term is included taking the form of a polynomial of degree $n - d$, where n is the degree of the numerator and d is the degree of the denominator.

Example 6.7

Express as partial fractions

$$\frac{2x^2 - x - 2}{x + 1}$$

Solution

The fraction is improper because $n = 2$, $d = 1$ and so $d \leq n$. Further, note that $n - d = 1$. We therefore need to include an extra term: a polynomial of the form $Ax + B$, in addition to the usual partial fractions. So

$$\frac{2x^2 - x - 2}{x + 1} = Ax + B + \frac{C}{x + 1}$$

where the final term arises from the linear factor in the denominator. Multiplying both sides by $x + 1$ we find

$$2x^2 - x - 2 = (Ax + B)(x + 1) + C$$
$$= Ax^2 + (A + B)x + (B + C)$$

Equating coefficients of x^2 gives $A = 2$. Equating coefficients of x gives $-1 = A + B$ and so $B = -1 - A = -3$. Equating the constant terms gives $-2 = B + C$ and so $C = -2 - B = -2 - (-3) = 1$. Finally we have

$$\frac{2x^2 - x - 2}{x + 1} = 2x - 3 + \frac{1}{x + 1}$$

Exercises

Express the following in partial fractions:

1. $\dfrac{x + 3}{x + 2}$

2. $\dfrac{3x - 7}{x - 3}$

3. $\dfrac{x^2 + 2x + 2}{x + 1}$

4. $\dfrac{2x^2 + 7x + 7}{x + 2}$

5. $\dfrac{3x^5 + 4x^4 - 21x^3 - 40x^2 - 24x - 29}{(x + 2)^2 (x - 3)}$

6. $\dfrac{4x^5 + 8x^4 + 23x^3 + 27x^2 + 25x + 9}{(x^2 + x + 1)(2x + 1)}$

Solutions to exercises

1 $1 + \dfrac{1}{x + 2}$

2 $3 + \dfrac{2}{x - 3}$

3 $1 + x + \dfrac{1}{x + 1}$

4 $2x + 3 + \dfrac{1}{x + 2}$

5 $\dfrac{1}{(x + 2)^2} + \dfrac{1}{x + 2} + \dfrac{1}{x - 3} + 3x^2 + x + 2$

6 $2x^2 + x + 7 + \dfrac{1}{2x + 1} + \dfrac{1}{x^2 + x + 1}$

End of block exercises

1 Express in partial fractions:

(a) $\dfrac{2x - 4}{x(x - 1)(x - 3)}$

(b) $\dfrac{1 + x}{(x + 3)^2 (x + 1)}$

(c) $\dfrac{x^2 + 1}{(2x + 1)(x - 1)(x - 3)}$

(d) $\dfrac{4s - 3}{2s + 1}$

(e) $\dfrac{3s + 1}{s(s - 2)}$

2 Express in partial fractions

$$\dfrac{K(1 + \alpha s)}{(1 + \tau s)s}$$

where K, α and τ are constants.

3 Express in partial fractions

(a) $\dfrac{2s + 1}{s^5(s + 1)}$ (b) $\dfrac{2s^3 + 6s^2 + 6s + 3}{s + 1}$

4 Express in partial fractions

$$\dfrac{2x + 1}{(x - 2)(x + 1)(x - 3)}$$

Solutions to exercises

1 (a) $-\dfrac{4}{3x} + \dfrac{1}{x - 1} + \dfrac{1}{3(x - 3)}$

(b) $\dfrac{1}{(x + 3)^2}$

(c) $\dfrac{5}{21(2x + 1)} - \dfrac{1}{3(x - 1)} + \dfrac{5}{7(x - 3)}$

(d) $2 - \dfrac{5}{2s + 1}$

(e) $\dfrac{7}{2(s - 2)} - \dfrac{1}{2s}$

2 $\dfrac{K}{s} + \dfrac{K(\alpha - \tau)}{1 + \tau s}$

3 (a) $\dfrac{1}{s^5} + \dfrac{1}{s^4} - \dfrac{1}{s^3} + \dfrac{1}{s^2} - \dfrac{1}{s} + \dfrac{1}{s + 1}$

(b) $\dfrac{1}{s + 1} + 2s^2 + 4s + 2$

4 $-\dfrac{5}{3(x - 2)} - \dfrac{1}{12(x + 1)} + \dfrac{7}{4(x - 3)}$

Proportionality

Consider two variables, a and b. We say that a is **proportional to** b if

$$a = kb \qquad (1)$$

where k is a constant. The constant k is called the **constant of proportionality**. We may also write

$$a \propto b$$

to show that a is proportional to b. Note that if a is proportional to b, then it also follows that b is proportional to a.

You will notice that (1) is the equation of a straight line passing through the origin; that is, the vertical intercept is 0. The gradient of this line is k. This means that when the value of a doubles (or trebles) then likewise the value of b doubles (or trebles). If the value of a halves, then the value of b halves too.

Example 7.1 Mechanical Engineering – Extension of a spring

When a force, F, is applied to the end of a spring, it produces an extension, e. Hooke's law states that the extension is proportional to the force applied. So we may write

$$e \propto F$$

or equivalently

$$e = kF$$

where k is a constant of proportionality.

Example 7.2 Mechanical Engineering – Extension of a spring

When a force of 10 N is applied to the end of a spring it produces an extension of 12 cm. Given that the extension of the spring is proportional to the applied force, calculate
(a) the extension produced when a force of 17 N is applied
(b) the force required to produce an extension of 15 cm.

Solution
Extension, e, is proportional to the applied force, F, so

$$e = kF$$

We are given that $e = 12$ when $F = 10$ so

$$12 = k(10)$$

from which $k = \frac{6}{5}$, so

$$e = \frac{6F}{5}$$

(a) A force of 17 N is applied, that is $F = 17$. Then the extension produced is

$$e = \frac{6F}{5} = \frac{6(17)}{5} = 20.4$$

The extension produced is 20.4 cm.

(b) The extension, e, is 15. So

$$15 = \frac{6F}{5}$$

$$F = \frac{15 \times 5}{6} = 12.5$$

The force required is 12.5 N

Sometimes it is useful to introduce a new variable when doing calculations involving proportion. Consider the following illustration. The area, A, of a circle is given by

$$A = \pi r^2$$

where r is the radius. If we introduce a new variable, x, where x is the square of the radius, that is $x = r^2$, then

$$A = \pi x$$

So, area, A, is proportional to x, the square of the radius. The constant of proportionality is π. Note that A is *not* proportional to r. A graph of A against x would be a straight line through the origin; a graph of A against r, however, would be a curve.

Example 7.3 Electronic Engineering – Resistance

Resistivity is a measure of how difficult it is for electrons to flow through a material. If the material has length, L, cross-sectional area, A, and resistance, R, then

$$R = \frac{\rho L}{A}$$

where ρ is a constant called the **resistivity** of the material. If the cross-sectional area is fixed (constant) then

$$R = kL$$

where $k = \dfrac{\rho}{A} = $ constant. So resistance, R, is proportional to length, L.

Example 7.4 Electrical Engineering – Power in a resistor

In an electric circuit the power, P, current, I, and resistance, R, are related by

$$P = I^2 R \qquad (2)$$

If the resistance is constant, k_1 say, then

$$P = k_1 I^2$$

So for constant resistance, the power is proportional to the square of the current. Note that power is not proportional to current, but is proportional to the square of the current.

Referring again to equation (2), if the current, I, is constant, then I^2 is also constant; call it k_2, giving

$$P = k_2 R$$

So for constant current, the power is proportional to resistance.

Example 7.5 Fluid Flow

The rate of flow, Q, of a fluid through a valve depends upon the pressure, P, across the valve, according to the law

$$Q = k\sqrt{P}$$

where k is a constant. Here we see that flow rate, Q, is proportional to $\sqrt{P}$, the square root of the pressure. Note that Q is not proportional to P.

7.2 Inverse proportion

A variable, a, is inversely proportional to a variable, b, if a is proportional to $\dfrac{1}{b}$. In other words

$$a \propto \frac{1}{b}$$

which can be expressed as

$$a = \frac{k}{b}$$

where k is the constant of proportionality. Note that when b doubles, then a is halved; when b is halved, then a doubles.

Example 7.6 Electrical Engineering – Voltage, current and resistance

Suppose that in an electric circuit

$$R = \frac{V}{I}$$

where R = resistance, V = voltage and I = current. For a fixed voltage, that is $V = k$ (a constant), then

$$R = \frac{k}{I}$$

and so resistance is inversely proportional to current.

Example 7.7 Mechanical Engineering – Vibration

The resonant frequency of a body, ω, is given by

$$\omega = \sqrt{\frac{k}{m}} \tag{3}$$

where k is the stiffness of the body and m is its mass. We may write (3) as

$$\omega = \frac{K}{\sqrt{m}}$$

where $K = \sqrt{k}$. We see that the resonant frequency is inversely proportional to the square root of the mass.

Example 7.8

It is known that a is inversely proportional to b. Measurements taken during an experiment are recorded in Table 7.1.

Table 7.1

a	5	10	15	20
b	6	3	2	1.5

(a) Find the equation connecting a and b.
(b) Calculate the value of a when $b = 30$.

Solution

(a) We are told that a is inversely proportional to b and so

$$a = \frac{k}{b} \qquad (4)$$

where k is a constant. From the measurements given in Table 7.1, $a = 5$ when $b = 6$. Substituting these values into (1) gives

$$5 = \frac{k}{6}$$

from which $k = 30$. Hence (4) becomes

$$a = \frac{30}{b} \qquad (5)$$

Note that all the pairs of values from Table 7.1 fit this equation. For example, when $b = 3$, equation (5) gives

$$a = \frac{30}{3} = 10$$

which concurs with the value in Table 7.1.

(b) Using (5), with $b = 30$, gives

$$a = \frac{30}{30} = 1$$

So when $b = 30$, then $a = 1$.

End of block exercises

1 Explain what is meant by the phrase 'a is proportional to b'.

2 Explain what is meant by the phrase 'a is inversely proportional to b'.

3 Given a is proportional to b, state which of the following are true and which are false:
(a) when a doubles, then b also doubles
(b) when a is halved, then b is doubled
(c) a graph of a against b is a straight line graph
(d) a divided by b is a constant

4   It is known that y is proportional to x.
Experimental measurements are recorded in
Table 7.2.

Table 7.2

y	30	40	50	60
x	5	6.67	8.33	10

(a) Determine the equation connecting y and x.
(b) Calculate y when $x = 2$.

5   Given that y is inversely proportional to x, state
which of the following are true and which are
false:
(a) when x is doubled, y is doubled also
(b) x is inversely proportional to y
(c) when x is halved, y is doubled
(d) a graph of y against x is a straight line with
a negative gradient

Solutions to exercises

3   (a) T (b) F (c) T (d) T

4   (a) $y = 6x$ (b) $y(2) = 12$

5   (a) F (b) T (c) T (d) F

End of chapter exercises

1   Solve the equation $4x - 20 = 9$.

2   Solve the equation $15 - 3x = 3(x - 7) + 11$.

3   Solve the equation $\dfrac{x + 2}{5} + 3 = \dfrac{x}{7}$.

4   Solve the simultaneous equations
$3x - 2y = 11, 5x + 7y = 39$.

5   On a number line show the numbers $-\pi$, 0,
$\sqrt{2}$, $-\sqrt{3}$, $|-0.5|$, $-(3!)$ and $\frac{11}{19}$.

6   Is the statement $\left(\frac{2}{3}\right)^{1/2} \leq \left(\frac{1}{2}\right)^{2/3}$ true or false?

7   Draw an x-y coordinate frame and shade the
region for which $x < 3$ and $y > -2$.

8   Use the method of completing the square to
derive the formula for solving a quadratic
equation.

9   By sketching an appropriate graph, or
otherwise, solve the inequality $\dfrac{1}{2 - x} < 5$.

10   Express $\dfrac{x^2 + 2x - 1}{x^3 - x}$ in partial fractions.

11   Express in partial fractions $\dfrac{x - \gamma}{(x - \alpha)(x - \beta)}$.

12   Express in partial fractions $\dfrac{x^3 + x + 1}{x^2 + 7x + 12}$.

13   Express in partial fractions

$$G(s) = \dfrac{6(s + 2)}{(s + 1)(s + 3)}.$$

14   Calculate the poles of the rational function

$$G(s) = \dfrac{s + 5}{s^2 + 3s + 2}.$$

15   Express in partial fractions

$$C(s) = \dfrac{K}{s(1 + \tau s)}$$

where K and τ are constants.

16 Express in partial fractions

$$C(s) = \frac{K}{(1 + \tau s)s^2}$$

where K and τ are constants.

17 Factorise $v^3 - 3v^2 - 33v + 35$ given that $(v - 7)$ is a factor.

18 Factorise $x^3 + 6x^2 + 6x + 5$ given that $x + 5$ is a factor.

19 Solve the following quadratic equations by an appropriate method.
(a) $x^2 + 16x + 64 = 0$ (b) $x^2 - 6x + 3 = 0$
(c) $2x^2 - 6x - 3 = 0$ (d) $x^2 - 4x + 1 = 0$
(e) $x^2 - 22x + 121 = 0$ (f) $x^2 - 8 = 0$

20 Solve the equation $x^3 - 5x^2 + 2x + 8 = 0$.

21 Solve the inequality $|3x + 2| \leq 4$.

22 Factorise $t^3 + 3t^2 + 2t$.

23 Show that $3 - 2t - t^2 = -(t + 3)(t - 1)$.

24 Table 7.3 shows the values of x and y. Given that y is proportional to x
(a) find an equation connecting y and x
(b) calculate the value of y when $x = 36$

(c) calculate the value of x when $y = 200$

Table 7.3

x	5	10	15	20	25
y	22.5	45	67.5	90	112.5

25 If a is proportional to b state which of the following are true and which are false:
(a) a multiplied by b is a constant
(b) a divided by b is a constant
(c) $\sqrt{a}$ is proportional to $\sqrt{b}$

26 A variable P is proportional to I^2.
(a) Use the measurements in Table 7.4 to determine an equation connecting P and I.

Table 7.4

P	24	54	96	150	216
I	2	3	4	5	6

(b) Calculate P when $I = 10$.

27 If a is inversely proportional to b state which of the following are true and which are false:
(a) a multiplied by b is a constant
(b) a divided by b is a constant
(c) a^2 is inversely proportional to b^2

Solutions to exercises

1 $x = \dfrac{29}{4}$

2 $x = \dfrac{25}{6}$

3 $-\dfrac{119}{2}$

4 $x = 5, y = 2$

6 False

9 $x > 2$ and $x < \frac{9}{5}$

10 $-\dfrac{1}{x + 1} + \dfrac{1}{x - 1} + \dfrac{1}{x}$

11 $\dfrac{\alpha - \gamma}{(\alpha - \beta)(x - \alpha)} + \dfrac{\beta - \gamma}{(\beta - \alpha)(x - \beta)}$

12 $\dfrac{67}{x + 4} - \dfrac{29}{x + 3} + x - 7$

13 $\dfrac{3}{s + 3} + \dfrac{3}{s + 1}$

14 $s = -1, s = -2$

15 $C(s) = \dfrac{K}{s} - \dfrac{K\tau}{1 + \tau s}$

16 $C(s) = \dfrac{K\tau^2}{1 + \tau s} + \dfrac{K}{s^2} - \dfrac{K\tau}{s}$

17 $(v - 7)(v + 5)(v - 1)$

18 $(x + 5)(x^2 + x + 1)$

19 (a) $x = -8$ twice (b) $x = 3 \pm \sqrt{6}$

(c) $x = \dfrac{3}{2} \pm \dfrac{\sqrt{15}}{2}$ (d) $x = 2 \pm \sqrt{3}$

(e) $x = 11$ twice (f) $x = \pm 2\sqrt{2}$

20 $x = 4, 2, -1$

21 $-2 \le x \le \frac{2}{3}$

22 $t(t^2 + 3t + 2) = t(t + 2)(t + 1)$

24 (a) $y = 4.5x$ (b) $y(36) = 162$ (c) $x = 44.4$

25 (a) F (b) T (c) T

26 (a) $P = 6l^2$ (b) $P(10) = 600$

27 (a) T (b) F (c) T

Logarithms and exponentials

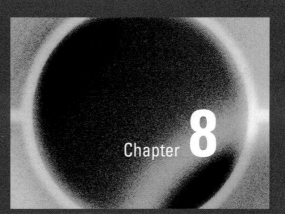

This chapter examines two important functions: the exponential function and the logarithmic function. Each function is the inverse of the other.

The **exponential function** has been found to have widespread application in many areas of science and engineering, whereas the gain of an amplifier can be modelled using the logarithmic function. **Logarithms** are used extensively in the study of sound, and the decibel, used in defining the intensity of sound, is based on a logarithmic scale.

The properties and laws of both functions are covered in this chapter, and methods of solving equations involving exponentials and logarithms are explained.

Chapter 8 contents

The exponential function

1.1 Introduction

The exponential function has widespread application in many areas of science and engineering. Areas that utilise the exponential function include expansion of materials, laws of cooling, radioactive decay, and the discharge of a capacitor. This block details some properties of the exponential function and includes some of its applications.

The hyperbolic functions are also introduced. These functions are defined in terms of the exponential function. Identities involving the hyperbolic functions are tabulated.

1.2 Exponential expressions

An **exponent** is another name for a power or index. Expressions involving exponents are called **exponential expressions**. For example, 3^4 and a^b are exponential expressions. In the exponential expression a^x, a is called the **base** and x is the exponent. Exponential expressions can be simplified and manipulated using the laws of indices. These laws were given in Chapter 5 Block 2 but are stated again here for reference.

Laws of indices

$$a^m a^n = a^{m+n}, \quad \frac{a^m}{a^n} = a^{m-n}, \quad (a^m)^n = a^{mn}$$

In this chapter we shall be dealing with exponential expressions in which the base will always be a particular constant called the **exponential constant**. This has the symbol e and is approximately equal to 2.718. Exponential expressions with this base dominate engineering applications.

Example 1.1
Most calculators have powers of the exponential constant preprogrammed; these can be calculated, probably with a button marked e^x. Check that you can use your calculator by calculating the following:

(a) $e^{3.7}$ (b) $e^{-1.6}$ (c) $2e^{1.5}$

Solution

(a) $e^{3.7} =$ [blank]   40.4473

(b) $e^{-1.6} =$ [blank]   0.2019

(c) $2e^{1.5} =$ [blank]   8.9634

1.3   Simplifying exponential expressions

The laws of indices and the rules of algebra apply to exponential expressions. The following examples illustrate this.

Example 1.2

Simplify (a) $e^{2x}e^{3x}$, (b) $\dfrac{e^{4x}}{e^{7x}}$, (c) $(e^{2x})^3$.

Solution

(a) $e^{2x}e^{3x} = e^{2x+3x} = e^{5x}$.

(b) $\dfrac{e^{4x}}{e^{7x}} = e^{4x-7x} = e^{-3x}$. This may be written as $\dfrac{1}{e^{3x}}$.

(c) $(e^{2x})^3 = e^{2x \times 3} = e^{6x}$.

Example 1.3

Expand the brackets and simplify where possible:
(a) $e^x(e^{-x} + 1)$
(b) $(e^x + 1)^2 - 1$
(c) $(e^x + e^{-x})^2 - 2$

Solution

(a) $e^x(e^{-x} + 1) = e^x e^{-x} + e^x$

$\qquad\qquad\qquad = e^0 + e^x$

$\qquad\qquad\qquad = 1 + e^x$

(b) $(e^x + 1)^2 - 1 =$ [blank] $e^x e^x + 2e^x + 1 - 1$

$\qquad\qquad = e^{2x} + 2e^x$

(c) $(e^x + e^{-x})^2 - 2 =$ [blank] $e^x e^x + 2e^x e^{-x} + e^{-x} e^{-x} - 2$

$\qquad\qquad\qquad = $ [blank] $e^{2x} + 2 + e^{-2x} - 2$

$\qquad\qquad\qquad = e^{2x} + e^{-2x}$

Exercises

1 Simplify each expression as far as possible:

(a) $e^{2x}e^{7x}$

(b) $(3e^x)(2e^{-x})$

(c) $e^{2x}(e^{-2x} + e^{-x} + 1) - e^x(1 + e^x)$

(d) $\dfrac{e^{-3x}}{2e^{-x}}$

2 Expand the brackets of the following expressions:

(a) $(e^x + 2)^2$

(b) $(e^x + 1)(e^{-x} - 1)$

(c) $(e^{2x} + e^x)(e^{-2x} + e^{-x})$

(d) $(1 + e^{2x} + e^{-2x})(1 - e^x)$

3 Simplify as far as possible:

(a) $\dfrac{e^x + e^{-x}}{2} + \dfrac{e^x - e^{-x}}{2}$

(b) $e^x + \dfrac{1}{e^x} - e^{-x}$

(c) $\dfrac{e^{2x} + e^x}{e^x} - 1$

(d) $e^{3x}(e^{-2x} - e^{-3x}) + 1$

Solutions to exercises

1 (a) e^{9x} (b) 6 (c) 1 (d) $\dfrac{e^{-2x}}{2}$

2 (a) $e^{2x} + 4e^x + 4$ (b) $e^{-x} - e^x$

(c) $2 + e^x + e^{-x}$

(d) $1 - e^x + e^{2x} - e^{3x} + e^{-2x} - e^{-x}$

3 (a) e^x (b) e^x (c) e^x (d) e^x

1.4 The exponential function and its graph

Key point

An exponential function has the form

$$y = a^x$$

where a is a positive constant called the **base**.

Hence $y = (0.4)^x$, $y = 2.5^x$ and $y = 10^x$ are all exponential functions. Note that, in an exponential function, the independent variable appears as a power.

Key point

The most commonly used exponential function, commonly called the **exponential function**, is

$$y = e^x$$

where the base e is the exponential constant whose value is e = 2.71828182 . . .

Table 1.1 gives values of the exponential function e^x for various x values and Figure 1.1 illustrates a graph of $y = e^x$ for $-3 \le x \le 3$.

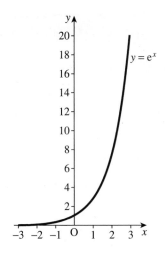

Figure 1.1
The exponential
function, $y = e^x$.

Table 1.1
Values of e^x.

x	e^x	x	e^x
-3	0.05	0.5	1.65
-2.5	0.08	1.0	2.72
-2.0	0.14	1.5	4.48
-1.5	0.22	2.0	7.39
-1.0	0.37	2.5	12.18
-0.5	0.61	3.0	20.09
0	1.00		

From Figure 1.1 we note some properties of the exponential function:

1   As x becomes large and positive, e^x increases without bound. We express this mathematically as $e^x \to \infty$ as $x \to \infty$.
2   As x becomes large and negative, e^x approaches 0. We write $e^x \to 0$ as $x \to -\infty$.
3   e^x is never negative.

The property that e^x increases as x increases is referred to as **exponential growth**.
    Figure 1.2 shows a graph of the related function $y = e^{-x}$ for $-3 \le x \le 3$ and Table 1.2 lists appropriate values of e^{-x}.

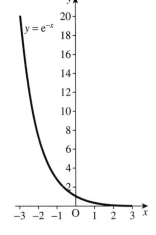

Figure 1.2
The function
$y = e^{-x}$ exhibits
exponential decay.

Table 1.2
Values of e^{-x}.

x	e^{-x}	x	e^{-x}
-3	20.09	0.5	0.61
-2.5	12.18	1.0	0.37
-2.0	7.39	1.5	0.22
-1.5	4.48	2.0	0.14
-1.0	2.72	2.5	0.08
-0.5	1.65	3.0	0.05
0	1.00		

From Figure 1.2 we see that e^{-x} decreases as x increases: this is referred to as **exponential decay**.

Example 1.4

Plot $y = e^{0.5x}$ and $y = e^{0.7x}$ for $0 \le x \le 3$. Comment on your graphs.

Solution

Table 1.3 shows values of $e^{0.5x}$ and $e^{0.7x}$ for $0 \le x \le 3$. Figure 1.3 illustrates the graphs.

From Figure 1.3 we see that $y = e^{0.7x}$ grows more rapidly than $y = e^{0.5x}$.

Figure 1.3
$y = e^{0.7x}$ grows more rapidly than $y = e^{0.5x}$.

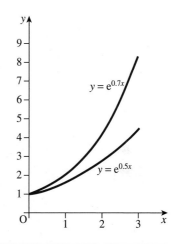

Table 1.3
Values of $e^{0.5x}$ and $e^{0.7x}$.

x	$e^{0.5x}$	$e^{0.7x}$
0	1.00	1.00
0.5	1.28	1.42
1.0	1.65	2.01
1.5	2.12	2.86
2.0	2.72	4.06
2.5	3.49	5.75
3.0	4.48	8.17

Exercises

1   Evaluate (a) $e^{2.7}$, (b) $e^{-1.6}$, (c) $3e^{0.7}$, (d) $-1.9e^{0.5}$.

2   Sketch

$$x = 3 + 2e^{-t}$$

Which value does x approach as t increases?

3   Sketch

$$R(t) = 4 - 2e^{-1.5t}$$

Which value does R approach as t increases?

4   State the domain and range of (a) $y = e^x$, (b) $y = e^{-x}$.

Solutions to exercises

1   (a) 14.8797 (b) 0.2019
    (c) 6.0413 (d) -3.1326

2   See Figure 1.4. As t increases, x approaches 3.

3   See Figure 1.5. As t increases R approaches 4.

4   (a) $(-\infty, \infty)$, $(0, \infty)$ (b) $(-\infty, \infty)$, $(0, \infty)$

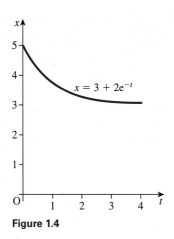

Figure 1.4

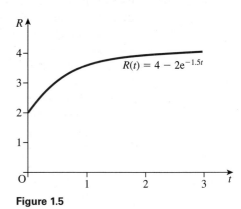

Figure 1.5

1.5 Applications of the exponential function

The following examples illustrate some of the applications of the exponential function.

Example 1.5 Electrical Engineering – Discharge of a capacitor

The charge, $q(t)$, stored on a capacitor having a capacitance C, discharging through a resistor of resistance R, is given by

$$q(t) = Qe^{-\frac{t}{CR}}$$

where Q is the initial charge.
(a) Find the value of the charge when $t = 1$ given $Q = 50$, $C = 0.25$ and $R = 2$.
(b) If the value of R is doubled calculate the new value of the charge when $t = 1$.

Solution
(a)

$$q(1) = 50e^{-\frac{1}{(0.25)(2)}}$$
$$= 50e^{-2}$$
$$= 6.77$$

(b)

$$q(1) = 50e^{-\frac{1}{(0.25)(4)}}$$
$$= 50e^{-1}$$
$$= 18.39$$

Example 1.6 Electrical Engineering – Decay of a current in a circuit

Consider a circuit with resistance R, inductance L and an initial current of i_0. The current $i(t)$ will decay with time, t, according to the law

$$i(t) = i_0 e^{-\frac{Rt}{L}}$$

(a) Calculate the current when $t = 1.5$ given $i_0 = 12$, $R = 2$ and $L = 6$.
(b) If L is increased from 6 to 9 calculate the new value of $i(1.5)$.

Solution

(a) $i(1.5) = 12e^{-\frac{2(1.5)}{6}}$
$= 12e^{-0.5}$
$= 7.28$

(b) $i(1.5) = $ $12e^{-\frac{2(1.5)}{9}} = 12e^{-\frac{1}{3}} = 8.60$

Example 1.7 Chemical Engineering – Newton's law of cooling

Newton's law of cooling states that the rate at which a body cools is proportional to the excess of its temperature above the temperature of the environment in which it is placed. Let $\Theta(t)$ be the temperature of a body at time t, Θ_0 its initial temperature and Θ_e the temperature of the environment. Newton's law of cooling can be stated mathematically as

$$\Theta(t) = \Theta_e + (\Theta_0 - \Theta_e)e^{-kt}$$

where k is a positive constant that depends upon the material of the body. From this equation we see that the temperature is dropping exponentially owing to the term e^{-kt}: that is, the temperature follows an exponential decay curve. In this example we take $k = 1$, $\Theta_e = 20$ and $\Theta_0 = 120$, so

$$\Theta(t) = 20 + 100e^{-t}$$

(a) Plot $\Theta(t)$ for $t = 0$ to $t = 6$.
(b) Use your graph to find the temperature that the body approaches as t increases.

Solution

(a) Table 1.4 gives values of $\Theta(t)$ for various values of t from 0 to 6. Figure 1.6 shows the graph of $\Theta(t)$.

Table 1.4

t	$\Theta(t) = 20 + 100e^{-t}$	t	$\Theta(t) = 20 + 100e^{-t}$
0	120.0	3.5	23.0
0.5	80.7	4.0	21.8
1.0	56.8	4.5	21.1
1.5	42.3	5.0	20.7
2.0	33.5	5.5	20.4
2.5	28.2	6.0	20.2
3.0	25.0		

Figure 1.6
Temperature decays exponentially with time.

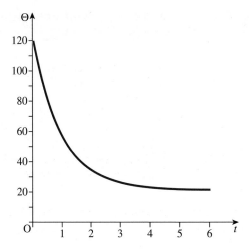

(b) From Figure 1.6 we see that the temperature, $\Theta(t)$, approaches 20 as t increases. This is to be expected: the temperature of the body approaches the temperature of the environment.

The result stated in Example 1.7(b) can be seen for the general case by considering the equation

$$\Theta(t) = \Theta_e + (\Theta_0 - \Theta_e)e^{-kt}$$

We have already noted that e^{-x} approaches 0 as x increases; clearly e^{-t} and e^{-kt} approach 0 as t increases since k is positive. Hence $(\Theta_0 - \Theta_e)e^{-kt}$ approaches 0 as t increases. Finally $\Theta_e + (\Theta_0 - \Theta_e)e^{-kt}$ approaches Θ_e as t increases.

Exercises

1 The number of particles, $N(t)$, emitted by a radioactive substance varies with time, t, according to the law

$$N(t) = 9.63 \times 10^{17}e^{-0.6t} \quad t \geq 0$$

(a) Calculate the number of particles emitted when $t = 0$.

(b) Calculate the number of particles emitted when $t = 2$.

2 The length, $l(T)$, of a bar depends upon the temperature, T, according to the law

$$l(T) = l_0e^{\alpha T}$$

where l_0 is the length of the bar when $T = 0$, and α is a positive constant.

(a) Calculate the percentage change in length when T increases from $T = 20$ to $T = 100$, given $\alpha = 0.001$.

(b) Calculate the percentage change in length when T decreases from $T = 150$ to $T = 50$.

3 Atmospheric pressure, $P(h)$ atmospheres, varies according to the height, h metres, above the surface of the Earth. Given

$$P(h) = e^{-0.000016h}$$

calculate the pressure at a height of

(a) 2 km
(b) 10 km

above the Earth.

Solutions to exercises

1   (a) 9.63×10^{17} (b) 2.90×10^{17}

2   (a) 8.33% increase (b) 9.52% decrease

3   (a) 0.9685 atmospheres
     (b) 0.8521 atmospheres

1.6   Hyperbolic functions

Closely associated to the exponential function are the **hyperbolic functions**. The hyperbolic functions are defined as follows:

Key point

$$\text{hyperbolic sine: } \sinh x = \frac{e^x - e^{-x}}{2}$$

$$\text{hyperbolic cosine: } \cosh x = \frac{e^x + e^{-x}}{2}$$

$$\text{hyperbolic tangent: } \tanh x = \frac{\sinh x}{\cosh x} = \frac{e^x - e^{-x}}{e^x + e^{-x}}$$

$$\text{cosech } x = \frac{1}{\sinh x} = \frac{2}{e^x - e^{-x}}$$

$$\text{sech } x = \frac{1}{\cosh x} = \frac{2}{e^x + e^{-x}}$$

$$\coth x = \frac{1}{\tanh x} = \frac{e^x + e^{-x}}{e^x - e^{-x}}$$

Readers already familiar with the trigonometrical functions will note the similarity of the names. Most scientific calculators have values of the hyperbolic functions pre-programmed. Use Example 1.8 to check that you can use your calculator to find the values of the hyperbolic functions.

Example 1.8

Use a calculator to evaluate
(a) sinh 2.1 (b) cosh (−1.3) (c) tanh 1.4
(d) cosech 2.2 (e) sech (−1.6) (f) coth (−2)

Solution

(a) $\sinh 2.1 = 4.0219$

(b) $\cosh (-1.3) = 1.9709$

(c) $\tanh 1.4 = 0.8854$

(d) $\text{cosech } 2.2 = \dfrac{1}{\sinh 2.2} = 0.2244$

(e) $\text{sech}(-1.6) = \dfrac{1}{\cosh(-1.6)} = 0.3880$

(f) $\coth(-2) = \dfrac{1}{\tanh(-2)} = -1.0373$

Figure 1.7 illustrates graphs of $y = \sinh x$, $y = \cosh x$ and $y = \tanh x$.

Figure 1.7
(a) $y = \sinh x$,
(b) $y = \cosh x$,
(c) $y = \tanh x$.

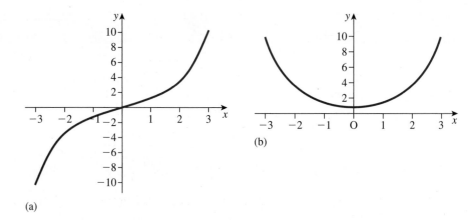

(a)

(b)

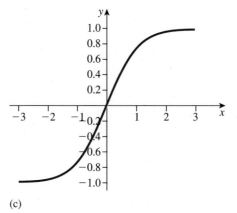

(c)

Example 1.9

Express $7e^x + 3e^{-x}$ in terms of $\sinh x$ and $\cosh x$.

Solution

We note that

$$\cosh x + \sinh x = \frac{e^x + e^{-x}}{2} + \frac{e^x - e^{-x}}{2}$$

$$= e^x$$

$$\cosh x - \sinh x = \frac{e^x + e^{-x}}{2} - \frac{e^x - e^{-x}}{2}$$

$$= e^{-x}$$

So

$$7e^x + 3e^{-x} = 7(\cosh x + \sinh x) + 3(\cosh x - \sinh x)$$
$$= 10\cosh x + 4\sinh x$$

Example 1.10 Mechanical Engineering – The catenary

A heavy chain, suspended at both ends, hangs in the shape defined by

$$y = a\cosh\left(\frac{x}{b}\right)$$

Note that when $x = 0$ we have

$$y(0) = a\cosh 0$$
$$= a$$

and that when $x = b$ we have

$$y(b) = a\cosh\left(\frac{b}{b}\right)$$

$$= a\cosh 1$$
$$= 1.54a$$

The points $(0, a)$ and $(b, 1.54a)$ are marked on Figure 1.8.

Figure 1.8
A catenary is
defined by
$$y = a\cosh\left(\frac{x}{b}\right).$$

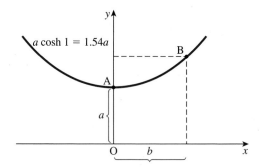

Such a curve is known as a **catenary**.

1.7 Hyperbolic identities

If two expressions have the same value for all values of x, we say that they are identical. For example, e^x and $\cosh x + \sinh x$ have the same value for any value of x. A statement of the form $e^x = \cosh x + \sinh x$ is called an **identity**. There are many identities that relate different hyperbolic functions. Table 1.5 lists some of the common hyperbolic identities.

Table 1.5
Hyperbolic
identities.

$$e^x = \cosh x + \sinh x$$
$$e^{-x} = \cosh x - \sinh x$$
$$\cosh^2 x - \sinh^2 x = 1$$
$$1 - \tanh^2 x = \text{sech}^2 x$$
$$\coth^2 x - 1 = \text{cosech}^2 x$$
$$\sinh (x \pm y) = \sinh x \cosh y \pm \cosh x \sinh y$$
$$\cosh (x \pm y) = \cosh x \cosh y \pm \sinh x \sinh y$$
$$\sinh 2x = 2 \sinh x \cosh x$$
$$\cosh 2x = \cosh^2 x + \sinh^2 x$$
$$\cosh^2 x = \frac{\cosh 2x + 1}{2}$$
$$\sinh^2 x = \frac{\cosh 2x - 1}{2}$$

Note that $\cosh^2 x$ means $(\cosh x)^2$ and $\sinh^2 x$ means $(\sinh x)^2$.

Example 1.11

Prove the following identities:
(a) $\cosh^2 x - \sinh^2 x = 1$
(b) $\sinh 2x = 2 \sinh x \cosh x$

Solution

(a) Note that from Example 1.9

$$\cosh x + \sinh x = e^x, \quad \cosh x - \sinh x = e^{-x}$$

Now

$$\cosh^2 x - \sinh^2 x = (\cosh x + \sinh x)(\cosh x - \sinh x)$$
$$= (e^x)(e^{-x})$$
$$= 1$$

(b) We have

$$\sinh 2x = \frac{e^{2x} - e^{-2x}}{2}$$

Also

$$2 \sinh x \cosh x = 2 \left(\frac{e^x - e^{-x}}{2} \right) \left(\frac{e^x + e^{-x}}{2} \right)$$
$$= \frac{(e^x - e^{-x})(e^x + e^{-x})}{2}$$
$$= \frac{(e^x)^2 + e^x e^{-x} - e^{-x} e^x - (e^{-x})^2}{2}$$
$$= \frac{e^{2x} + 1 - 1 - e^{-2x}}{2}$$
$$= \frac{e^{2x} - e^{-2x}}{2}$$

Hence

$$\sinh 2x = 2 \sinh x \cosh x$$

Example 1.12 Electrical Engineering – Voltage in a transmission line

A transmission line is an arrangement of electrical conductors for transporting electromagnetic waves. The coaxial cable used to carry a signal from a TV aerial to a TV set is a common example.

The voltage in a line depends upon both distance along the line, z, and time, t. For any fixed time, the voltage at distance z along the line has the form

$$v(z) = Ae^{-\alpha z} + Be^{\alpha z}$$

where A and B are constants. For a particular transmission line $A = 1$ and $B = 4$ giving

$$v = e^{-\alpha z} + 4e^{\alpha z}$$

Show that v may be written as

$$v = 3 \sinh \alpha z + 5 \cosh \alpha z$$

Solution

$$
\begin{aligned}
v &= e^{-\alpha z} + 4e^{\alpha z} \\
&= (\cosh \alpha z - \sinh \alpha z) + 4(\cosh \alpha z + \sinh \alpha z) \quad \text{(using Table 1.5)} \\
&= 5 \cosh \alpha z + 3 \sinh \alpha z
\end{aligned}
$$

Exercises

1 Evaluate
(a) $\sinh(-3)$ (b) $\cosh 1.9$ (c) $\tanh(-3.2)$
(d) $\operatorname{cosech} 2$ (e) $\operatorname{sech}(-2.5)$ (f) $\coth 0.6$

2 Sketch $y = \operatorname{cosech} x$ for $-3 \leq x \leq 3$.

3 Sketch $y = \operatorname{sech} x$ for $-3 \leq x \leq 3$.

4 Prove the identities
(a) $1 - \tanh^2 x = \operatorname{sech}^2 x$
(b) $\sinh^2 x = \dfrac{\cosh 2x - 1}{2}$

5 Express $4e^{2x} - 6e^{-2x}$ in terms of $\sinh 2x$ and $\cosh 2x$.

6 **Electrical Engineering – Transmission line.**
The voltage in a transmission line, v, is given by

$$v = 7e^{-\alpha z} + 3e^{\alpha z}$$

Express the voltage, v, using hyperbolic functions.

Solutions to exercises

1. (a) -10.02 (b) 3.42 (c) -1.00 (d) 0.28
   (e) 0.16 (f) 1.86

2. See Figure 1.9.

3. See Figure 1.10.

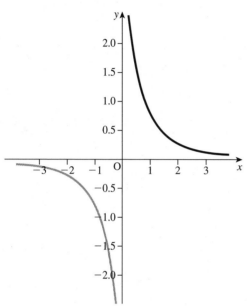

Figure 1.9
$y = \text{cosech}\, x.$

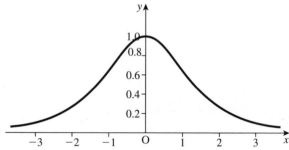

Figure 1.10
$y = \text{sech}\, x.$

5. $10 \sinh 2x - 2 \cosh 2x$

6. $10 \cosh \alpha z - 4 \sinh \alpha z$

End of block exercises

1. Sketch $y = \coth x$ for $-3 \leq x \leq 3$.

2. Simplify each of the following as much as possible:
   (a) $e^{-7x}e^{4x}$ (b) $2e^{3x}e^{5x}$ (c) $(3e^{3x})(2e^{2x})(-e^{-x})$
   (d) $\dfrac{8e^{6x}}{2e^{-2x}}$

3. Prove the identities
   (a) $\cosh 2x = \cosh^2 x + \sinh^2 x$
   (b) $\sinh(x + y) = \sinh x \cosh y + \cosh x \sinh y$

4. The charge on a capacitor, $C(t)$, decays according to the law
   $$C(t) = 100e^{-0.2t} \quad t \geq 0$$

   (a) Calculate the charge when $t = 3$.
   (b) Calculate the charge when $t = 5$.

5. (a) Plot $y = 6 + 4e^{-2t}$, for $0 \leq t \leq 3$.
   (b) What value does y approach as t increases?

6. Given
   $$y = a + be^{-t}$$
   where a and b are constants, state the value that y approaches as t increases.

7. (a) Show that
   $$\frac{e^x + 6}{2e^x + 5}$$

may be expressed as

$$\frac{1 + 6e^{-x}}{2 + 5e^{-x}}$$

(b) What value does

$$\frac{e^x + 6}{2e^x + 5}$$

approach as x becomes large and positive?

8 **Electrical Engineering – Decay of a current in a circuit.** The current, $i(t)$, in a circuit changes with time, t, according to

$$i(t) = i_0 e^{-\frac{Rt}{L}}$$

(a) Calculate the current when $t = 0.5$ given $i_0 = 10$, $R = 3$ and $L = 5$.
(b) Describe the effect on $i(t)$ if the value of R is increased, all other values remaining constant.
(c) Describe the effect on $i(t)$ if the value of L is increased, all other values remaining constant.

Solutions to exercises

1   See Figure 1.11.

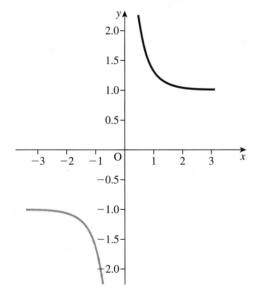

Figure 1.11
$y = \coth x$.

2   (a) e^{-3x} (b) $2e^{8x}$ (c) $-6e^{4x}$ (d) $4e^{8x}$

4   (a) 54.88 (b) 36.79

5   (a) See Figure 1.12 (b) 6

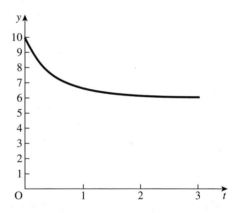

Figure 1.12
$y = 6 + 4e^{-2t}$.

6   a

7   (b) 0.5

8   (a) 7.41
(b) The current decreases as R increases.
(c) The current increases as L increases.

Logarithms and their laws

Logarithms are an alternative way of writing expressions that involve powers, or indices. They are used extensively in the study of sound. The decibel, used in defining the intensity of sound, is based on a logarithmic scale. In this block we define the logarithm and give the laws that are used to simplify and manipulate expressions involving logarithms.

2.2 Definition of the logarithm

Consider the statement that

$$16 = 2^4$$

Here 2 is the base and 4 is the power. This may be expressed in an alternative way using logarithms as

$$\log_2 16 = 4$$

We read this as 'log 16, to the base 2, equals 4'. Some more examples follow:

$8 = 2^3$ may be written as $\log_2 8 = 3$

$25 = 5^2$ may be written as $\log_5 25 = 2$

$81 = 3^4$ may be written as　　　　　　　　　　　　　　　　　$\log_3 81 = 4$

$0.125 = (0.5)^3$ may be written as　　　　　　　　　　　$\log_{0.5} 0.125 = 3$

$10000 = 10^4$ may be written as $\log_{10} 10000 = 4$

$10^{-1} = 0.1$ may be written as　　　　　　　　　$\log_{10} 0.1 = -1$

In general we have

Key point

If a is a positive constant and

$$N = a^x$$

then

$$\log_a N = x$$

The number a is called the **base** of the logarithm. In practice, most logarithms are to the base 10 or e. Logarithms to base e are called **natural logarithms**. Natural logarithms are often denoted by ln rather than $\log_e$, whereas logarithms to base 10 are denoted simply by log rather than $\log_{10}$.

Key point

If $N = 10^x$ then $x = \log_{10} N = \log N$.

If $N = e^x$ then $x = \log_e N = \ln N$.

Scientific calculators are preprogrammed with both natural logarithms and logarithms to base 10. Use Example 2.1 to check that you can use your calculator.

Example 2.1

Use a scientific calculator to find:

(a) log 79 (b) ln 79 (c) log 0.21 (d) ln 0.036

Solution

(a) log 79 = 1.8976

(b) ln 79 = 4.3694

(c) log 0.21 =   −0.6778

(d) ln 0.036 =   −3.3242

Example 2.2

Express each of the following statements using logarithms:

(a) $10^3 = 1000$ (b) $10^{-2} = 0.01$ (c) $12.1825 = e^{2.5}$ (d) $e^{-2.3026} = 0.1$

Solution

(a) log 1000 = 3 (b) log 0.01 = −2 (c) ln 12.1825 = 2.5

(d) ln 0.1 = −2.3026

Exercises

1   Write the following using logarithms:

(a) $32 = 2^5$ (b) $125 = 5^3$ (c) $243 = 3^5$

(d) $4^3 = 64$ (e) $6^2 = 36$

2   Write the following using logarithms:

(a) $10^2 = 100$ (b) $0.001 = 10^{-3}$

(c) $e^{-1.3} = 0.2725$ (d) $e^{1.5} = 4.4817$

3   Evaluate

(a) log 250 (b) ln 250 (c) log 0.46

(d) ln 0.46

4   Write the following using indices:

(a) $\log_5 625 = 4$

(b) $\log_2 256 = 8$

(c) $\log 0.0251 = -1.6$

(d) $\ln 17 = 2.8332$

Solutions to exercises

1 (a) $\log_2 32 = 5$ (b) $\log_5 125 = 3$
   (c) $\log_3 243 = 5$ (d) $\log_4 64 = 3$
   (e) $\log_6 36 = 2$

2 (a) $\log 100 = 2$ (b) $\log 0.001 = -3$
   (c) $\ln 0.2725 = -1.3$ (d) $\ln 4.4817 = 1.5$

3 (a) 2.3979 (b) 5.5215
   (c) -0.3372 (d) -0.7765

4 (a) $5^4 = 625$ (b) $2^8 = 256$
   (c) $10^{-1.6} = 0.0251$ (d) $e^{2.8332} = 17$

2.3 Logarithms to any base

To calculate a logarithm to base a, knowing logarithms to base b, we use

Key point

$$\log_a N = \frac{\log_b N}{\log_b a}$$

Example 2.3
Evaluate $\log_3 40$.

Solution

$$\log_3 40 = \frac{\log 40}{\log 3}$$
$$= \frac{1.6021}{0.4771}$$
$$= 3.358$$

Example 2.4
Evaluate $\log_7 36$.

Solution

$$\log_7 36 = \qquad\qquad\qquad\qquad \frac{\log 36}{\log 7} = 1.842$$

Exercises

1 Evaluate
   (a) $\log_4 36$ (b) $\log_2 9$ (c) $\log_{12} 10$

Solutions to exercises

1 (a) 2.585 (b) 3.170 (c) 0.927

2.4 Laws of logarithms

Just as expressions involving indices can be simplified using appropriate laws, so expressions involving logarithms can be simplified using the laws of logarithms. The laws of logarithms hold true for any base. However, it is essential that the same base is used throughout an expression before the laws can be applied.

The first law of logarithms is

Key point

$$\log A + \log B = \log AB$$

Example 2.5

Simplify each of the following to a single logarithmic expression:
(a) $\log 7 + \log 2$
(b) $\ln 9 + \ln 10$
(c) $\log x + \log x^2$
(d) $\ln 2y + \ln 3y$

Solution

(a) $\log 7 + \log 2 = \log (7 \times 2) = \log 14$
(b) $\ln 9 + \ln 10 = \ln 90$

(c) $\log x + \log x^2 = $ \qquad $\log(xx^2) = \log x^3$

(d) $\ln 2y + \ln 3y = $ \qquad $\ln(2y3y) = \ln 6y^2$

The second law of logarithms states

Key point

$$\log A - \log B = \log \left(\frac{A}{B}\right)$$

Example 2.6

Express as a single logarithm:
(a) $\log 15 - \log 3$
(b) $\ln 6 - \ln (0.5)$
(c) $\ln 6x^2 - \ln 2x$
(d) $\log 1 - \log x$

Solution

(a) $\log 15 - \log 3 = \log \left(\dfrac{15}{3}\right) = \log 5$

(b) $\ln 6 - \ln (0.5) = \ln \left(\dfrac{6}{0.5}\right) = \ln 12$

(c) $\ln 6x^2 - \ln 2x =$ ⬚ $\quad\quad\quad\quad\quad\quad\quad\quad \ln\left(\dfrac{6x^2}{2x}\right) = \ln 3x$

(d) $\log 1 - \log x =$ ⬚ $\quad\quad\quad\quad\quad\quad\quad\quad\quad\quad \log\dfrac{1}{x}$

We use the second law of logarithms to develop another result. Clearly $\log A - \log A = 0$. Using the second law we may also write

$$\log A - \log A = \log\left(\frac{A}{A}\right)$$
$$= \log 1$$

Hence we see that

Key point

$$\log 1 = 0$$

The third law of logarithms states

Key point

$$\log A^n = n \log A$$

Example 2.7

Rewrite the following in an alternative form without using a power:
(a) $\log 2^3$ (b) $\log 2^2$ (c) $\log x^2$ (d) $\ln t^5$

Solution

(a) $\log 2^3 =$ ⬚ $\quad\quad\quad\quad\quad\quad\quad\quad\quad\quad\quad\quad\quad$ 3 log 2

(b) $\log 2^2 =$ ⬚ $\quad\quad\quad\quad\quad\quad\quad\quad\quad\quad\quad\quad\quad$ 2 log 2

(c) $\log x^2 =$ ⬚ $\quad\quad\quad\quad\quad\quad\quad\quad\quad\quad\quad\quad\quad$ 2 log x

(d) $\ln t^5 =$ ⬚ $\quad\quad\quad\quad\quad\quad\quad\quad\quad\quad\quad\quad\quad\quad$ 5 ln t

Example 2.8

Simplify each of the following as much as possible:
(a) $3 \log 2 - \log 6$
(b) $\ln 250 + 2 \ln 2$
(c) $\log 3x^3 + 2 \log x$
(d) $\ln t^5 - 3 \ln t^2$

Solution

(a)
$$3 \log 2 - \log 6 = \log 2^3 - \log 6$$
$$= \log 8 - \log 6$$
$$= \log\left(\frac{8}{6}\right)$$
$$= \log\left(\frac{4}{3}\right)$$

(b)
$$\ln 250 + 2 \ln 2 = \ln 250 + \ln 2^2$$
$$= \ln 250 + \ln 4$$
$$= \ln (250 \times 4)$$
$$= \ln 1000$$

(c)
$$\log 3x^3 + 2 \log x = \log 3x^3 + \log x^2$$
$$= \log (3x^3 \cdot x^2) = \log 3x^5$$

(d)
$$\ln t^5 - 3 \ln t^2 = \ln t^5 - \ln (t^2)^3$$
$$= \ln t^5 - \ln t^6$$
$$= \ln \frac{t^5}{t^6}$$
$$= \ln \frac{1}{t}$$

Note that

$$\ln \frac{1}{t} = \ln t^{-1}$$
$$= -\ln t$$

Example 2.9
Simplify

(a) $\dfrac{1}{2} \log 25$ (b) $\dfrac{1}{3} \log 8$ (c) $\dfrac{1}{2} \log x^2$ (d) $\dfrac{1}{2} \ln 9t^2$

Solution

(a) $\dfrac{1}{2} \log 25 = \log 25^{1/2} = \log 5$

(b) $\dfrac{1}{3} \log 8 = \log 8^{1/3} = \log 2$

(c) $\dfrac{1}{2} \log x^2 =$ ⬚ $\log (x^2)^{1/2} = \log x$

(d) $\dfrac{1}{2} \ln 9t^2 =$ ⬚ $\ln (9t^2)^{1/2} = \ln 3t$

(Hint: $(9t^2)^{1/2} = 9^{1/2}(t^2)^{1/2} = 3t$ using the generalised third law of indices.)

Example 2.10
Simplify

(a) $3 \log x - 2 \log \left(\dfrac{1}{x} \right) - \dfrac{1}{2} \log 9$

(b) $4 \log t^2 - 2 \log t^3 + \log \left(\dfrac{1}{t^2} \right)$

Solution

(a) $3 \log x = \log x^3$, $2 \log \dfrac{1}{x} = \log \dfrac{1}{x^2} = \log x^{-2}$, and $\dfrac{1}{2} \log 9 = \log 9^{1/2} = \log 3$, so

$$3 \log x - 2 \log\left(\frac{1}{x}\right) - \frac{1}{2} \log 9 = \log x^3 - \log x^{-2} - \log 3$$

$$= \log\left(\frac{x^3}{3x^{-2}}\right)$$

$$= \log\left(\frac{x^5}{3}\right)$$

(b) $4 \log t^2 = \log (t^2)^4 = \log t^8$

$2 \log t^3 = $ [blank] $\log(t^3)^2 = \log t^6$

$\log \dfrac{1}{t^2} = \log t^{-2}$

So

$$4 \log t^2 - 2 \log t^3 + \log \frac{1}{t^2} = \log t^8 - \log t^6 + \log t^{-2}$$

$$= \text{[blank]} \qquad\qquad \log\left(\frac{t^8 \, t^{-2}}{t^6}\right)$$

$$= \log 1$$

$$= 0$$

Exercises

1 Simplify to a single log term:
(a) $\log 10 + \log 20$
(b) $\log 30 - \log 6$
(c) $\log 1 + \log 2 + \log 3 + \log 4$
(d) $\log 4 - \log 3$
(e) $\log 6 - \log 5 + \log 4 - \log 3$

2 Simplify to a single log term:
(a) $2 \log 3 + 3 \log 2$
(b) $4 \log 3 - \log 27$
(c) $2 \log 8 - 4 \log 2 + \log 3$
(d) $2 \log 12 - 3 \log 6 + 2 \log 2$
(e) $5 \log 3 - 2 \log 9 + 2 \log 6 - 3 \log 2$

3 Simplify to a single log term:
(a) $\dfrac{1}{2} \log 16 + \dfrac{1}{3} \log 8$

(b) $\dfrac{2}{3} \log 27 - 2 \log 3$

(c) $\dfrac{1}{4} \log 1 + \dfrac{3}{4} \log 16 + \dfrac{1}{2} \log 9 - 2 \log 2$

(d) $\dfrac{3}{2} \log 4 - \dfrac{2}{3} \log 8$

(e) $\dfrac{4}{3} \log 64 + \dfrac{2}{3} \log 27 - \log 24$

4 Simplify to a single log term:
(a) $\log x + \log 2x$
(b) $\log x + \log 3x + \log 5x$
(c) $\log x^2 y - \log y^2$
(d) $\log x^2 + \log y^2 - \log z^2$
(e) $\log A + \log B - \log \dfrac{1}{C}$

5 Simplify to a single log term:
(a) $2 \log x^2 - 3 \log x$
(b) $3 \log x^2 y + 2 \log xy$
(c) $4 \log AB^2 - 3 \log A^2 B$
(d) $2 \log rs + 3 \log rs^2 - 4 \log r^2 s$
(e) $4 \log abc - 2 \log a^2 b - 3 \log bc$

6 Simplify to a single log term:

(a) $2 \log \dfrac{x}{y} + 3 \log \dfrac{y}{x}$

(b) $2 \log \dfrac{x^2}{y} - 3 \log \dfrac{y^2}{x}$

(c) $4 \log rs^2 t - 2 \log r^2 st^2 + 3 \log \dfrac{r^2}{st}$

(d) $\dfrac{1}{2} \log \dfrac{x^2}{y^2} - \dfrac{1}{3} \log \dfrac{x^3}{y^6}$

(e) $\dfrac{2}{3} \log \dfrac{a^2 b}{c^2} + \dfrac{3}{2} \log \dfrac{c}{ab} - \dfrac{1}{2} \log abc$

Solutions to exercises

1 (a) $\log 200$ (b) $\log 5$ (c) $\log 24$

(d) $\log \dfrac{4}{3}$ (e) $\log \dfrac{8}{5}$

2 (a) $\log 72$ (b) $\log 3$ (c) $\log 12$

(d) $\log \dfrac{8}{3}$ (e) $\log \dfrac{27}{2}$

3 (a) $\log 8$ (b) $\log 1 = 0$ (c) $\log 6$
(d) $\log 2$ (e) $\log 96$

4 (a) $\log 2x^2$ (b) $\log 15x^3$ (c) $\log \dfrac{x^2}{y}$

(d) $\log \dfrac{x^2 y^2}{z^2}$ (e) $\log ABC$

5 (a) $\log x$ (b) $\log x^8 y^5$ (c) $\log \dfrac{B^5}{A^2}$

(d) $\log \dfrac{s^4}{r^3}$ (e) $\log \dfrac{c}{b}$

6 (a) $\log \dfrac{y}{x}$ (b) $\log \dfrac{x^7}{y^8}$ (c) $\log \dfrac{r^6 s^3}{t^3}$

(d) $\log y$ (e) $\log \dfrac{1}{a^{2/3} b^{4/3} c^{1/3}}$

2.5 The logarithmic functions

We are now ready to introduce the logarithmic functions. They are defined as follows:

Key point

$$y = \log x \text{ and } y = \ln x \text{ for } x > 0$$

Table 2.1 shows values of x, $\log x$ and $\ln x$. Their graphs are shown in Figure 2.1. We note the following common properties:

1 As x increases, both $\log x$ and $\ln x$ increase indefinitely. We write this mathematically as $\log x \to \infty$ as $x \to \infty$, $\ln x \to \infty$ as $x \to \infty$.
2 As x approaches 0, both $\log x$ and $\ln x$ approach minus infinity. We express this as $\log x \to -\infty$ as $x \to 0$, $\ln x \to -\infty$ as $x \to 0$.
3 $\log 1 = \ln 1 = 0$.

4 $\log x$ and $\ln x$ are not defined when x is negative or zero. Thus the domain of the logarithm functions is $x > 0$. The range of these functions is $(-\infty, \infty)$.

Table 2.1

x	$\log x$	$\ln x$
0.1	-1	-2.30
0.2	-0.70	-1.61
0.5	-0.30	-0.69
1	0	0
2	0.30	0.69
3	0.48	1.10
4	0.60	1.39
5	0.70	1.61
6	0.78	1.79
7	0.85	1.95
8	0.90	2.08
9	0.95	2.20
10	1	2.30

Figure 2.1
The graphs of
$y = \log x$ and
$y = \ln x$.

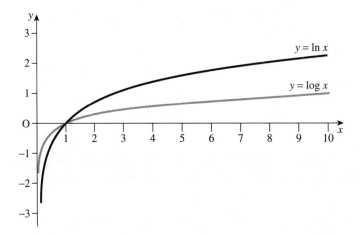

End of block exercises

1 Express the following statements using logarithms:
   (a) $2^7 = 128$
   (b) $4^5 = 1024$
   (c) $243 = 3^5$
   (d) $0.125 = 2^{-3}$
   (e) $4^{-1/2} = 0.5$

2 Express the following using indices:
   (a) $\log_3 81 = 4$
   (b) $\log_{10} 0.0001 = -4$
   (c) $\log_{16} 8 = 0.75$
   (d) $\log_4 8 = 1.5$
   (e) $\log_4 0.125 = -1.5$

3 Express as a single log term in its simplest form:

(a) $2 \log y + 3 \log x$

(b) $3 \log AB - 2 \log B - \log A$

(c) $\frac{1}{2} \log 4x^2 - \frac{1}{3} \log x$

(d) $\frac{1}{2} \log 8x + \frac{3}{2} \log 2x$

(e) $\log 2x - 2 \log x + x \log 2$

4 Simplify as far as possible

(a) $\log(x^2 + 4x + 3) - \log(x + 1)$

(b) $2 \log(x - 1) - \log(x^2 - 1)$

5 Evaluate (a) $\log_4 9$, (b) $\log_{12} 6$.

Solutions to exercises

1 (a) $\log_2 128 = 7$ (b) $\log_4 1024 = 5$
(c) $\log_3 243 = 5$ (d) $\log_2 0.125 = -3$
(e) $\log_4 0.5 = -0.5$

2 (a) $3^4 = 81$ (b) $10^{-4} = 0.0001$
(c) $16^{0.75} = 8$ (d) $4^{1.5} = 8$
(e) $4^{-1.5} = 0.125$

3 (a) $\log x^3 y^2$ (b) $\log A^2 B$ (c) $\log 2x^{2/3}$
(d) $\log 8x^2$ (e) $\log \left(\dfrac{2^{x+1}}{x} \right)$

4 (a) $\log(x + 3)$ (b) $\log \left(\dfrac{x - 1}{x + 1} \right)$

5 (a) 1.585 (b) 0.721

Solving equations involving logarithms and exponentials

In this block we examine the techniques used in solving equations involving logarithmic and exponential terms. We recall the connection between logarithmic and exponential expressions:

Key point

$$\text{if } N = a^x \text{ then } x = \log_a N$$

In particular,

Key point

$$\text{if } N = 10^x \text{ then } x = \log N$$

and

Key point

$$\text{if } N = e^x \text{ then } x = \ln N$$

The process of finding x, having been given 10^x or e^x, is known as **taking logs**.

The following examples illustrate the technique for solution.

Example 3.1
Solve
(a) $10^x = 17.4$ (b) $e^x = 5.25$

Solution
(a) We have

$$10^x = 17.4$$

so taking logs gives

$$x = \log 17.4$$
$$= 1.2405$$

(b) We have $e^x = 5.25$. Taking logs gives

$$x = \ln 5.25$$
$$= 1.6582$$

Example 3.2

Solve

(a) $10^{2x} = 39$ (b) $e^{3x-1} = 75$

Solution

(a) We have $10^{2x} = 39$. Taking logs gives

$$2x = \log 39$$
$$x = \frac{\log 39}{2}$$
$$= 0.7955$$

(b)
$$e^{3x-1} = 75$$
$$3x - 1 = \ln 75$$
$$3x = \ln 75 + 1$$
$$x = \frac{\ln 75 + 1}{3}$$
$$= 1.7725$$

Example 3.3

Solve $3(10^{0.5x-2}) = 96$.

Solution

$$3(10^{0.5x-2}) = 96$$
$$10^{0.5x-2} = \frac{96}{3}$$
$$= 32$$
$$0.5x - 2 = \boxed{} \qquad\qquad \log 32$$
$$0.5x = \boxed{} \qquad\qquad \log 32 + 2$$
$$x = \boxed{} \qquad\qquad 2(\log 32 + 2)$$
$$= 7.0103$$

Example 3.4

Solve

(a) $\log x = 1.4$ (b) $\ln x = -0.5$

Solution

(a)
$$\log x = 1.4$$
$$x = 10^{1.4}$$
$$= 25.12$$

(b)
$$\ln x = -0.5$$
$$x = e^{-0.5}$$
$$= 0.61$$

Example 3.5
Solve
(a) $\log(x^2 + 2) = 2.6$ (b) $3 \ln (5x + 1) = 9.3$

Solution
(a) $\log(x^2 + 2) = 2.6$

$$x^2 + 2 = \boxed{} \qquad\qquad 10^{2.6}$$

$$x^2 = \boxed{} \qquad\qquad 10^{2.6} - 2$$

$$x = \boxed{} \qquad\qquad \sqrt{10^{2.6} - 2}$$

$$= \pm 19.90$$

(b) $3 \ln (5x + 1) = 9.3$

$$\ln (5x + 1) = \frac{9.3}{3}$$

$$= 3.1$$

$$5x + 1 = \boxed{} \qquad\qquad e^{3.1}$$

$$5x = \boxed{} \qquad\qquad e^{3.1} - 1$$

$$x = \boxed{} \qquad\qquad x = \frac{e^{3.1} - 1}{5}$$

$$= 4.24$$

Example 3.6 Electronic Engineering – Current through a diode
The current through a diode, I, is given by

$$I = I_s (e^{40V} - 1)$$

where I_s is the reverse saturation current and V is the voltage across the diode.
(a) Express V as the subject of the equation.
(b) Evaluate V when $I = 3 \times 10^{-2}, I_s = 1.5 \times 10^{-4}$.

Solution
(a)
$$I = I_s (e^{40V} - 1)$$

$$e^{40V} - 1 = \frac{I}{I_s}$$

$$e^{40V} = \frac{I}{I_s} + 1$$

Now, taking logs,

$$40V = \ln \left(\frac{I}{I_s} + 1\right)$$

$$V = \frac{1}{40} \ln \left(\frac{I}{I_s} + 1\right)$$

(b) Substituting the given values of I and I_s into the expression for V results in

$$V = \frac{1}{40} \ln \left(\frac{3 \times 10^{-2}}{1.5 \times 10^{-4}} + 1\right)$$

$$= \frac{1}{40} \ln (200 + 1)$$

$$= \frac{1}{40} \ln 201$$

$$= 0.133$$

Example 3.7 Electronic Engineering – Gain of an amplifier

The voltage gain, measured in decibels (dB), of an amplifier is given by

$$\text{gain} = 20 \log \left(\frac{V_o}{V_i} \right)$$

where V_i is the input voltage and V_o is the output voltage.

The output voltage from an amplifier is 250 mV. If the amplifier has a gain of 17 dB calculate the input voltage.

Solution

We have

$$\text{gain} = 20 \log \left(\frac{V_o}{V_i} \right)$$

where the gain is 17 dB and $V_o = 250$ mV. So

$$17 = 20 \log \left(\frac{250}{V_i} \right)$$

$$\log \left(\frac{250}{V_i} \right) = \frac{17}{20}$$

$$= 0.85$$

$$\frac{250}{V_i} = 10^{0.85}$$

$$V_i = \frac{250}{10^{0.85}}$$

$$= 35.31$$

The input voltage is 35.3 mV.

Exercises

1   Solve the following equations:
    (a) $\log x = 0.7531$
    (b) $\log x = 1.6431$
    (c) $\log x = -0.4213$
    (d) $\log x = -2.3500$

2   Solve
    (a) $\ln x = 2.4050$
    (b) $\ln x = 0.9611$
    (c) $\ln x = -0.9611$
    (d) $\ln x = -2.0000$

3   Solve
    (a) $10^x = 7$
    (b) $10^x = 70$
    (c) $10^x = 17$
    (d) $10^x = 0.7000$

4   Solve
    (a) $e^x = 5$
    (b) $e^x = 0.5$
    (c) $e^x = 25$
    (d) $e^x = 0.001761$

5 Solve
   (a) $\log 2x = 1.5$
   (b) $\log (3x + 1) = 2.1500$
   (c) $\log (x^2 + 3) = 2.3671$
   (d) $4 \log (5x - 6) = -0.8000$

6 Solve
   (a) $\ln (2x^2) = 3$
   (b) $2 \ln (x^2) = 3$
   (c) $\ln (x^2 + 2) = 1.3$
   (d) $3 \ln (x^2 + 1) = 3.9$

7 An amplifier has a gain of 25 dB. If the input voltage is 15 mV calculate the output voltage.

8 The current, I, through a diode is given by

   $$I = I_s(e^{40V} - 1)$$

   where I_s is the reverse saturation current and V is the voltage across the diode. If the current is 300 times greater than the reverse saturation current, calculate the voltage across the diode.

Solutions to exercises

1 (a) 5.6637 (b) 43.9643 (c) 0.3791
   (d) 0.004467

2 (a) 11.0784 (b) 2.6146 (c) 0.3825
   (d) 0.1353

3 (a) 0.8451 (b) 1.8451 (c) 1.2304
   (d) −0.1549

4 (a) 1.6094 (b) −0.6931 (c) 3.2189
   (d) −6.3419

5 (a) 15.8114 (b) 46.7513 (c) ±15.1612
   (d) 1.3262

6 (a) ±3.1690 (b) ±2.1170 (c) ±1.2920
   (d) ±1.6338

7 267 mV

8 143 mV

End of block exercises

1 Solve
   (a) $\log (3x - 7) = 2.6500$
   (b) $2 \ln (x^2 + 4) = 3.9$

2 Solve
   (a) $e^{-2x} = 4$ (b) $10^{3-x} = 20$

3 Solve
   (a) $\dfrac{10^{3x}}{10^x} = 30$ (b) $\log x + \log (2x) = 3$

4 Solve
   (a) $3e^{4x} = 2.7$ (b) $3 \ln 4x = -2.7$

5 Solve
   (a) $\log (5 - x^2) = \dfrac{1}{2}$ (b) $e^{\frac{3}{x}} = \dfrac{1}{3}$

Solutions to exercises

1 (a) 151.2 (b) ±1.7403

2 (a) −0.6931 (b) 1.6990

3 (a) 0.7386 (b) 22.36

4 (a) −0.02634 (b) 0.1016

5 (a) ±1.3556 (b) −2.7307

Applications of logarithms

4.1 Introduction

Block 2 looked at the definition and laws of logarithms. This block concentrates on some of the applications of logarithms.

4.2 Signal ratio and decibels

The ratio between two signal levels is often of interest to engineers. For example, the output and input signals of an electronic system can be compared to see whether the system has increased the level of a signal. A common case is an amplifier, where the output signal is usually much larger than the input signal. This signal ratio is often expressed in decibels (dB), given by

$$\text{power gain (dB)} = 10 \log \left(\frac{P_o}{P_i} \right)$$

where P_o is the power of the output signal and P_i is the power of the input signal. The term **gain** is used because if $P_o > P_i$ then the logarithm function is positive, corresponding to an increase in power. If $P_o < P_i$ then the gain is negative, corresponding to a decrease in power. A negative gain is often termed an **attenuation**.

The advantage of using decibels as a measure of gain is that if several electronic systems are connected together then it is possible to obtain the overall system gain in decibels by adding together the individual system gains. We shall show this for three systems connected together, but the development is easily generalised to more systems. Let the power input to the first system be P_{i1}, and the power output from the third system be P_{o3}. Suppose the three are connected so that the power output from system 1, P_{o1}, is used as input to system 2: that is, $P_{i2} = P_{o1}$. The power output from system 2, P_{o2}, is then used as input to system 3: that is, $P_{i3} = P_{o2}$. We wish to find the overall power gain, $10 \log \left(\dfrac{P_{o3}}{P_{i1}} \right)$. Now

$$\frac{P_{o3}}{P_{i1}} = \frac{P_{o3}P_{o2}P_{o1}}{P_{i3}P_{i2}P_{i1}}$$

because $P_{i3} = P_{o2}$ and $P_{i2} = P_{o1}$. Therefore

$$10 \log\left(\frac{P_{o3}}{P_{i1}}\right) = 10 \log\left(\frac{P_{o3}P_{o2}P_{o1}}{P_{i3}P_{i2}P_{i1}}\right)$$

That is,

$$10 \log\left(\frac{P_{o3}}{P_{i1}}\right) = 10 \log\left(\frac{P_{o3}}{P_{i3}}\right) + 10 \log\left(\frac{P_{o2}}{P_{i2}}\right) + 10 \log\left(\frac{P_{o1}}{P_{i1}}\right)$$

using the laws of logarithms.

It follows that the overall power gain is equal to the sum of the individual power gains. Often engineers are more interested in voltage gain rather than power gain. The power of a signal is proportional to the square of its voltage. We define **voltage gain** (dB) by

$$\text{voltage gain (dB)} = 10 \log\left(\frac{V_o^2}{V_i^2}\right) = 20 \log\left(\frac{V_o}{V_i}\right)$$

Example 4.1 Electronic Engineering – Gain of an amplifier

Calculate the voltage gain in decibels of an amplifier where the input signal is 0.8 V and the output signal is 1.2 V.

Solution

We have $V_o = 1.2$ and $V_i = 0.8$. Then

$$\text{voltage gain (dB)} = 20 \log\frac{V_o}{V_i}$$

$$= 20 \log\frac{1.2}{0.8}$$

$$= 3.52$$

The voltage gain is 3.52 decibels.

Exercises

1 Calculate the voltage gain in decibels of an amplifier where the input signal is 0.5 V and the output signal is 2.2 V.

2 Calculate the voltage gain in decibels of an amplifier where the input signal is 0.15 V and the output signal is 1.9 V.

Solutions to exercises

1 12.87 dB

2 22.05 dB

4.3 Use of log–linear and log–log scales

We look at each kind of scale separately.

Log–linear scales

Suppose we wish to plot

$$y(x) = x^6 \quad 1 \le x \le 10$$

This may appear a straightforward exercise, but consider the variation in the x and y values. As x varies from 1 to 10, then y varies from 1 to 1000000, as tabulated in Table 4.1.

Table 4.1

x	y
1	1
2	64
3	729
4	4096
5	15625
6	46656
7	117649
8	262144
9	531441
10	1000000

Several of these points would not be discernible on a graph, and so information would be lost. This can be overcome by using a **log scale**, which accommodates the large variation in y. Thus log y is plotted against x, rather than y against x.

Table 4.2 shows values of x and log y, and the corresponding graph is illustrated in Figure 4.1.

Table 4.2

x	log y
1	0
2	1.81
3	2.86
4	3.61
5	4.19
6	4.67
7	5.07
8	5.42
9	5.73
10	6

Figure 4.1
The function $y = x^6$ is plotted on a log–linear graph.

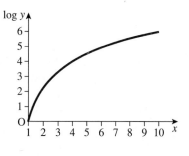

As x varies from 1 to 10, log y varies from 0 to 6. A plot in which one scale is logarithmic and the other is linear is known as a **log–linear** graph. In effect, use of the log scale has compressed a large variation into one that is much smaller and easier to observe.

Example 4.2

Consider $y = 7^x$ for $-3 \leq x \leq 3$. Plot a log–linear graph of this function.

Solution

We have

$$y = 7^x$$

and so

$$\log y = \log (7^x)$$
$$= x \log 7$$
$$= 0.8451x$$

Putting $Y = \log y$ we have $Y = 0.8451x$, which is the equation of a straight line passing through the origin with gradient log 7. Hence when log y is plotted against x a straight line graph is produced. This is shown in Figure 4.2. Note that, by taking logs, the range on the vertical axis has been greatly reduced.

Figure 4.2
A log–linear plot
of $y = 7^x$ pro-
duces a straight
line graph.

x	y	$Y = \log y$
-3	0.003	-2.54
-2	0.020	-1.69
-1	0.143	-0.85
0	1	0
1	7	0.85
2	49	1.69
3	343	2.54

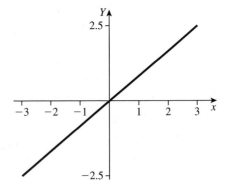

Log–log scales

A plot in which both scales are logarithmic is known as a **log–log plot**. Here log y is plotted against log x.

Example 4.3

Consider $y = x^7$ for $1 \leq x \leq 10$. Plot a log–log graph of this function.

Solution

We have

$$y = x^7$$

and so

$$\log y = \log (x^7)$$
$$= 7 \log x$$

We plot log y against log x for a log–log plot. Putting $Y = \log y$ and $X = \log x$ we have $Y = 7X$, which is a straight line through the origin with gradient 7, as shown in Figure 4.3.

Figure 4.3
A log–log plot of $y = x^7$ produces a straight line graph.

x	y	$X = \log x$	$Y = \log y$
1	1	0	0
2	128	0.301	2.107
3	2187	0.477	3.340
4	16384	0.602	4.214
5	78125	0.699	4.893
6	279936	0.778	5.447
7	823543	0.845	5.916
8	2097152	0.903	6.322
9	4782969	0.954	6.680
10	10000000	1	7

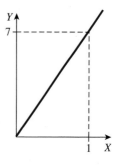

We have seen how a log scale has the effect of compressing a very large range of values into a more meaningful size. Similarly a log scale expands a very small range into a more meaningful size. Consider the following example.

Example 4.4
Plot a log–log graph of

$$y = x^3 \quad 10^{-6} \leq x \leq 10^{-1}$$

Solution
The small range of x values would almost be impossible to plot. The plot is more meaningful when a log–log scale is used. Table 4.3 shows values of x, y, log x and log y.

Table 4.3

x	y	$\log x$	$\log y$
10^{-6}	10^{-18}	-6	-18
10^{-5}	10^{-15}	-5	-15
10^{-4}	10^{-12}	-4	-12
10^{-3}	10^{-9}	-3	-9
10^{-2}	10^{-6}	-2	-6
10^{-1}	10^{-3}	-1	-3

From Table 4.3 we see that both x and y have a very small range but that log x and log y have a much more meaningful range. Figure 4.4 shows a plot of log y against log x.

Figure 4.4
A plot of log y
against log x.

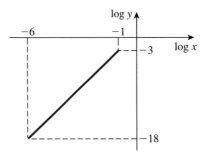

Figure 4.4 shows a straight line graph with gradient 3. This can be seen from

$$y = x^3$$
$$\log y = \log x^3$$
$$= 3 \log x$$

4.4 Use of log–linear and log–log paper

The requirement to take logarithms is a tedious process, which can be avoided by using special graph papers called log–linear graph paper and log–log graph paper. An example of log–linear graph paper is shown in Figure 4.5.

Note that on one axis the scale is uniform; this is the linear scale. On the other, the scale is not uniform and is marked in **cycles** from 1 to 9. This is the logarithmic scale. On this scale values of y are plotted directly, without first taking logarithms. On the graph paper shown in Figure 4.5 there are two cycles but papers are also available with three or more cycles. To decide which sort of graph paper is appropriate it is necessary to examine the variation in size of the variable to be plotted measured in powers of 10. If, for example, y varies from 1 to 10, then paper with one cycle is appropriate. If y varies from 1 to 10^2, two-cycle paper is necessary. If y varies from 10^{-1} to 10^4, then paper with $4 - (-1) = 5$ cycles would be appropriate. To see how log–linear paper is used in practice, consider the following example.

Example 4.5
Following an experiment the following pairs of data values were recorded:

	A	B	C	D
x	0	1	5	12
y	4.00	5.20	14.85	93.19

It is believed that y and x are related by the equation $y = ab^x$. By plotting a log–linear graph verify the relationship is of this form and determine a and b.

Figure 4.5
Two-cycle
log–linear
graph paper.

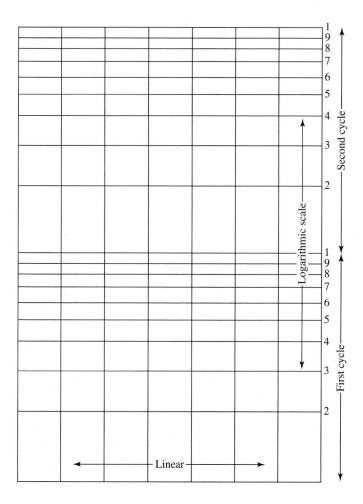

Solution

If the relationship is given by $y = ab^x$, then taking logarithms yields

$$\log y = \log a + x \log b$$

So, plotting $\log y$ against x should produce a straight line graph with gradient $\log b$ and vertical intercept $\log a$. The need to find $\log y$ is eliminated by plotting the y values directly on a logarithmic scale. Examining the table of data we see that y varies from approximately 10^0 to 10^2 so that two-cycle paper is appropriate. Values of y between 1 and 10 are plotted on the first cycle, and those between 10 and 100 are plotted on the second. The points are plotted in Figure 4.6. Note in particular that in this example the '1' at the start of the second cycle represents the value 10, the '2' represents the value 20, and so on. From the graph, the straight line relationship between $\log y$ and x is evident. It is therefore reasonable to assume that the relationship between y and x is of the form $y = ab^x$.

Figure 4.6
The log–linear
graph is a
straight line.

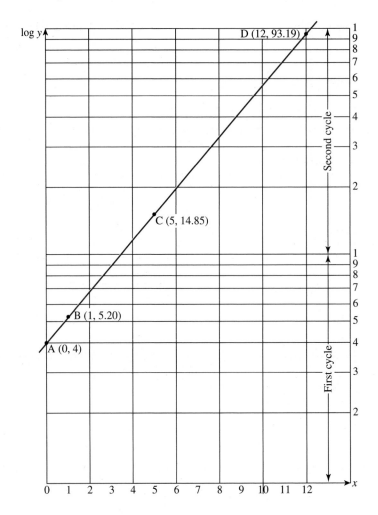

To find the gradient of the graph we can choose any two points on the line, for example C and B. The gradient is then

$$\frac{\log 14.85 - \log 5.20}{5 - 1} = \frac{\log\left(\dfrac{14.85}{5.20}\right)}{4}$$

$$= 0.1139$$

Recall that log b is the gradient of the line and so

$$\log b = 0.1139$$

that is

$$b = 10^{0.1139}$$
$$= 1.2999$$

The vertical intercept is log a. From the graph the vertical intercept is log 4 so that

$$\log a = \log 4$$

that is

$$a = 4$$

We conclude that the relationship between y and x is given by $y = 4 \, (1.3)^x$.

Example 4.6

The variables x and y are thought to be connected by an equation of the form

$$y = ax^n$$

Some experimental values are tabulated below:

	A	B	C	D
x	1	3	7	9
y	2	31	255	500

By plotting the data on appropriate graph paper, determine the law connecting x and y.

Solution

If $y = ax^n$ then

$$\log y = \log (ax^n)$$
$$= \log a + n \log x$$

Letting $Y = \log y$ and $X = \log x$ we have

$$Y = nX + \log a$$

This is a straight line equation with gradient n and vertical intercept log a. Since both log x and log y are used then a log scale on both axes is needed: that is, log–log paper is used. One cycle is needed to accommodate the x variation; three cycles are needed to accommodate the y variation. Log–log paper comes in 1 cycle $\times$ 1 cycle, 2 cycle $\times$ 2 cycle and 3 cycle $\times$ 3 cycle. Consequently, for the example in question, 3 cycle $\times$ 3 cycle paper is needed. Figure 4.7 shows the log–log paper with the data points plotted. The straight line fit tells us that x and y are connected by a law of the form $y = ax^n$.

Using points A and D to calculate the gradient we have

$$\text{gradient} = \frac{\log 500 - \log 2}{\log 9 - \log 1}$$
$$= \frac{\log 250}{\log 9}$$
$$= 2.5$$

Hence we see that $n = 2.5$.

Figure 4.7
If $y = ax^n$, then log–log paper produces a straight line.

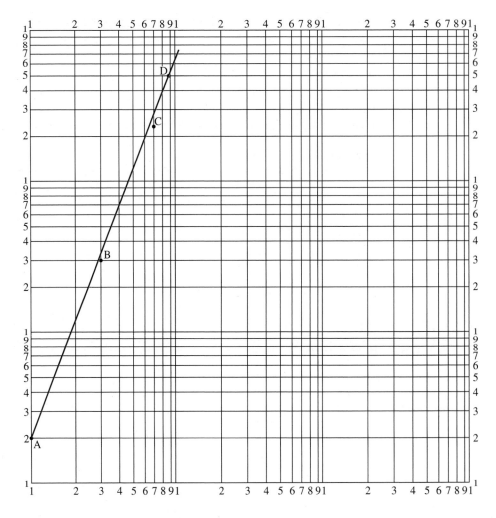

From the graph the vertical intercept is log 2 and so

$$\log a = \log 2$$
$$a = 2$$

Hence x and y are connected by $y = 2x^{2.5}$.

Examples 4.5 and 4.6 illustrate the following points.

Key point

If $y = ax^n$ then a log–log plot produces a straight line with gradient $= n$ and vertical intercept $= \log a$.

If $y = ab^x$, then a log–linear plot produces a straight line with gradient $= \log b$ and vertical intercept $= \log a$.

Example 4.7 Electrical Engineering – Bode plot of a linear circuit

Engineers are often interested in how a circuit will respond to a sinusoidal signal. A Bode plot helps in the analysis.

A **Bode plot** consists of two components:

1 The ratio of the amplitudes of the output signal and the input signal is plotted against frequency.
2 The phase shift between the input and output signals is plotted against frequency.

A log scale is used for the frequency in order to compress its length: for example, a typical frequency range is 0.1 Hz to 10^6 Hz, which corresponds to a range of -1 to 6 on a log scale. A log scale is also used for the ratio of the signal amplitudes as this is calculated in decibels. Phase shift is plotted on a linear scale. So the signal amplitude ratio against frequency is a log–log graph and the phase shift against frequency is a linear–log graph.

Exercises

1 It is thought that x and y are connected by a law of the form $y = ax^n$.
Measurements of x and y are

x	1.5	3	5	10	20	25
y	7.2	23	55	180	600	860

(a) By using appropriate paper find the law connecting x and y.
(b) Predict the y measurement when $x = 30$.

2 It is thought that x and y are connected by a law of the form $y = ax^n$.
Measurements of x and y are

x	0.25	0.5	0.8	1.3	4	7
y	0.18	0.30	0.42	0.61	1.41	2.2

(a) By using appropriate paper find the law connecting x and y.
(b) Predict the y measurement when $x = 10$.

3 It is thought that x and y are connected by a law of the form $y = ab^x$.
Measurements of x and y are

x	2	3	4	4.5	5	5.5
y	18	54	160	280	500	840

(a) By using appropriate paper find the law connecting x and y.
(b) Predict the y measurement when $x = 6$.

4 It is thought that x and y are connected by a law of the form $y = ab^x$.
Measurements of x and y are

x	2	5	10	20	30	35	38
y	1	1.75	4.33	27	170	400	700

(a) By using appropriate paper find the law connecting x and y.
(b) Predict the y measurement when $x = 40$.

5 The time of swing of a pendulum, T seconds, is measured for different lengths of the pendulum, l metres. The results are

l	0.2	0.4	0.6	0.8	1.0	1.2
T	0.9	1.3	1.5	1.8	2.0	2.2

If $T = kl^n$ use appropriate graph paper to find the values of k and n.

Solutions to exercises

1   (a) $y = 3.6x^{1.7}$ (b) 1168

2   (a) $y = 0.5x^{0.75}$ (b) 2.81

3   (a) $y = 2(3^x)$ (b) 1458

4   (a) $y = 0.7(1.2^x)$ (b) 1029

5    $k = 2$, $n = 0.5$

End of block exercises

1   Measurements of the variables x and y are recorded as follows

x	1.5	2.0	2.5	3.0	3.5
y	1.76	1.47	1.23	1.03	0.86

   (a) Given $y = ab^x$, use appropriate graph paper to find a and b.
   (b) Predict y when $x = 4$.

2   It is believed that in a particular circuit the power, P, and voltage, V, are related by a law of the form $P = kV^n$. Measurements of P and V are

P	11	45	125	245	405
V	1.5	3	5	7	9

   (a) By plotting the data on appropriate paper, find the values of k and n.

(b) If the voltage is increased to 12 V, calculate the power.
(c) Calculate the minimum voltage required if the power must exceed 1000 W.

3   Calculate the voltage gain (dB) of an amplifier that produces an output signal of 2.35 V when the input signal is 40 mV.

4   An amplifier consists of a preamplifier and a main amplifier. The input signal to the preamplifier is 5 mV and the output signal is 80 mV. The input to the main amplifier is 80 mV and the output signal is 3 V.
   (a) Calculate the voltage gain of the preamplifier.
   (b) Calculate the voltage gain of the main amplifier.
   (c) Calculate the overall gain of the amplifier.

Solutions to exercises

1   (a) $a = 3$, $b = 0.7$ (b) 0.72

2   (a) $k = 5$, $n = 2$ (b) 720 (c) 14.2

3   35.4

4   (a) 24.08 (b) 31.48 (c) 55.6

End of chapter exercises

1   Evaluate (a) $e^{1.6}$, (b) $e^{-1.6}$, (c) $\dfrac{1}{e^{1.6}}$.

2   The current in a circuit, $i(t)$, is given by

$$i(t) = 25e^{-0.2t} \quad t \geq 0$$

(a) State the current when $t = 0$.
(b) Calculate the value of the current when $t = 2$.
(c) Calculate the time when the value of the current is 12.5.

3   Simplify as far as possible:

(a) $\dfrac{e^{2x}e^{x}}{e^{-3x}}$ (b) $(4e^{2x})(3e^{-x})$

(c) $\dfrac{2e^{x}+1}{2}+\dfrac{2-e^{x}}{3}$ (d) $e^{4x}-(e^{2x}+1)^{2}$

4   Simplify as far as possible:

(a) $e^{2t}e^{3t}$ (b) $3e^{2t}e^{-t}$ (c) $\dfrac{(e^{x})^{2}}{e^{2x}}$ (d) $\sqrt{\dfrac{e^{4x}}{9}}$

5   Evaluate (a) $\sinh 4.7$, (b) $\cosh(-1.6)$, (c) $\tanh 1.2$.

6   Express $6e^{x}+3e^{-x}$ in terms of the hyperbolic functions $\sinh x$ and $\cosh x$.

7   Prove the hyperbolic identity

$$\cosh 2x = \sinh^{2} x + \cosh^{2} x$$

8   Express the following statements using logarithms:
(a) $8^{2}=64$ (b) $4^{3}=64$ (c) $2^{6}=64$

9   Express the following using indices:
(a) $\log 35 = 1.5441$ (b) $\log_{6} 1296 = 4$
(c) $\ln 50 = 3.9120$ (d) $\log_{9} 3 = 0.5$

10 Evaluate (a) $\log_{2} 20$, (b) $\log_{7} 2$.

11 Simplify to a single logarithmic expression:
(a) $\ln 4y + \ln x$
(b) $3 \ln t^{2} - 2 \ln t$
(c) $3 \log t - \log 3t$
(d) $\log 2x + \log 5x - 1$

12 Solve
(a) $2 \ln (3x - 10) = 8.5$
(b) $\log (x^{3} + 1) = 2.4$
(c) $3 \log 4x - 8 = 0$
(d) $\dfrac{\ln 5x}{2} = 1.6$

13 Solve
(a) $e^{4x} = 90$
(b) $10^{x/2} - 20 = 0$
(c) $3e^{-x} = 20$
(d) $10^{3x-6} = 40$

14 Solve
(a) $10^{\log x} = 17$
(b) $10^{2 \log x} = 17$
(c) $10^{x}10^{2x} = 90$
(d) $10^{2x} = 30(10^{x})$

15 Solve
(a) $\ln (e^{x}) = 5000$
(b) $\ln (e^{x} + 10) = 5$
(c) $\dfrac{2}{3} \ln (x^{2} + 9) = 3$
(d) $\log \left(\dfrac{x}{2} + 1 \right) = 1.5$

16 Simplify (a) $\sqrt{e^{2\ln x}}$, (b) $\log (100^{x})$.

17 The temperature, T, of a chemical reaction is given by

$$T = 120e^{0.02t} \qquad t \geq 0$$

Calculate the time needed for the temperature to (a) double its initial value, (b) treble its initial value.

18 Calculate the voltage gain in decibels of an amplifier where the input voltage is 17 mV and the output voltage is 300 mV.

19 The voltage input to an amplifier is 30 mV.
(a) Calculate the output voltage if the amplifier has a gain of 16 dB.
(b) Calculate the output voltage if the amplifier has a gain of 32 dB.

20 The variables x and y are believed to be connected by a law of the form $y = ax^{n}$. Measurements of x and y are

x	2	5	7	10	15	20
y	7.9	28.6	46.0	74.7	134.1	200.0

(a) By drawing an appropriate graph, find the law connecting x and y.
(b) Predict y when $x = 17$.

21 Variables y and t are thought to be connected by a law of the form

$$y = \dfrac{a^{t}}{k}$$

Measurements of t and y are

t	2	4	6	8	10
y	0.80	2.62	8.50	27.8	90.0

(a) By using appropriate graph paper find the law connecting y and t.
(b) Predict y when $t = 12$.
(c) Predict the value of t when y first exceeds 1000.

Solutions to exercises

1   (a) 4.9530 (b) 0.2019 (c) 0.2019

2   (a) 25 (b) 16.76 (c) 3.4657

3   (a) e^{6x} (b) $12e^x$ (c) $\dfrac{2e^x}{3} + \dfrac{7}{6}$ (d) $-(2e^{2x} + 1)$

4   (a) e^{5t} (b) $3e^t$ (c) 1 (d) $\dfrac{e^{2x}}{3}$

5   (a) 54.9690 (b) 2.5775 (c) 0.8337

6   $9\cosh x + 3\sinh x$

8   (a) $\log_8 64 = 2$ (b) $\log_4 64 = 3$
    (c) $\log_2 64 = 6$

9   (a) $10^{1.5441} = 35$ (b) $6^4 = 1296$
    (c) $e^{3.9120} = 50$ (d) $9^{0.5} = 3$

10 (a) 4.3219 (b) 0.3562

11 (a) $\ln 4xy$ (b) $\ln t^4$ (c) $\log\left(\dfrac{t^2}{3}\right)$ (d) $\log x^2$

12 (a) 26.7018 (b) 6.3012 (c) 116.0397
    (d) 4.9065

13 (a) 1.1250 (b) 2.6021 (c) -1.8971
    (d) 2.5340

14 (a) 17 (b) $\sqrt{17}$ (c) 0.6514 (d) 1.4771

15 (a) 5000 (b) 4.9302 (c) 9.0010 (d) 61.2456

16 (a) x (b) $2x$

17 (a) 34.66 (b) 54.93

18 24.93 dB

19 (a) 189.3 (b) 1194.3

20 (a) $y = 3x^{1.4}$ (b) 158

21 (a) $y = \dfrac{1.8^t}{4}$ (b) 289 (c) 14.11

Trigonometry

Chapter **9**

This chapter opens with a treatment of the two common units used for measuring angles: degrees and radians. The trigonometrical ratios of sine, cosine and tangent are then introduced. Initially the ratios of angles between 0° and 90° are dealt with and then the ratios of angles of any size are incorporated. The common identities involving trigonometrical ratios are studied together with the solution of trigonometrical equations. The application of trigonometry to combining two waves into a single wave concludes the chapter.

Chapter 9 contents

Angles

1.1 Introduction

Angles measure the amount through which a line or object has been turned. The Greek letters α, θ and ϕ are commonly used to denote angles. In Figure 1.1 the angle between lines AB and AC is θ.

Figure 1.1
The angle between AB and AC is θ.

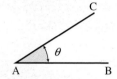

We can think of the line AB as being turned through or rotated an angle θ to the new position AC.

1.2 Units

There are two main units used to measure angles: the **degree** and the **radian**. Both units are defined with reference to a circle.

Degree

Consider a circle, centre O, as shown in Figure 1.2.

A typical radius, OA, is shown. If the radius OA is rotated as indicated so that it ends up in its original position we say it has been turned through a complete revolution. The angle that is equivalent to a complete revolution is 360 degrees, denoted 360°.

Figure 1.2
One complete revolution is 360°.

Key point

1 complete revolution = 360°

Radian

Consider a circle of radius r, centre O. An arc AB of length r is shown. We say that the arc AB **subtends** an angle at the centre O. This is angle AOB. The situation is illustrated in Figure 1.3. Note that there is no symbol to denote that an angle is being measured in radians. Hence if an angle is given and no symbol is present then you must assume the angle is measured in radians.

Then the angle AOB is defined to be 1 radian.

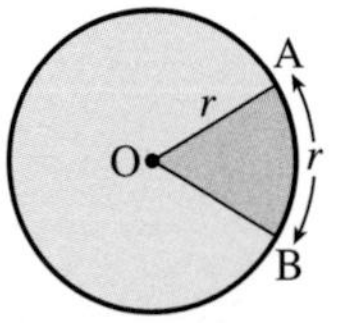

Figure 1.3
The arc AB has length r.

Key point

1 radian = angle subtended at centre by an arc whose length is one radius

An arc length r subtends an angle of 1 radian. Then an arc of length $2r$ subtends an angle of 2 radians and in general an arc of length αr subtends an angle of α radians. Let us examine the case where $\alpha = 2\pi$. An arc length of $2\pi r$ is the entire circumference of the circle, and this subtends an angle of 2π radians. But the circumference subtends a complete revolution at the centre, that is 360°, and so

$$2\pi \text{ radians} = 360°$$

Hence we have

Key point

π radians = 180°

Some common angles, marked in both degrees and radians, are shown in Figure 1.4.

Example 1.1
Convert 37° to radians.

Solution
We have

$$180° = \pi \text{ radians}$$

and so

$$1° = \frac{\pi}{180} \text{ radians}$$
$$37° = 37 \times \frac{\pi}{180} \text{ radians}$$
$$= 0.6458 \text{ radians}$$

Figure 1.4
Some common angles.

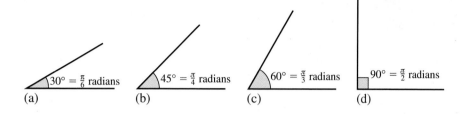

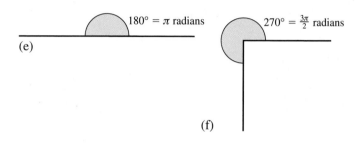

Example 1.2
Convert 1.2 radians to degrees.

Solution

$$\pi \text{ radians} = 180°$$
$$1 \text{ radian} = \frac{180°}{\pi}$$
$$1.2 \text{ radians} = 1.2 \times \frac{180°}{\pi}$$
$$= 68.75°$$

Commonly, angles measured in radians are expressed as multiples of π: for example, 3π radians, $\frac{\pi}{2}$ radians and $\frac{2\pi}{3}$ radians.

Example 1.3
Express 72° in the form $\alpha\pi$ radians.

Solution
We have

$$180° = \pi \text{ radians}$$
$$1° = \frac{1}{180} \times \pi \text{ radians}$$
$$72° = \frac{72}{180} \times \pi \text{ radians}$$
$$= \frac{2\pi}{5} \text{ radians}$$

Example 1.4
Express 117° in radians.

Solution
We have
$180° = \pi$ radians

$1° = $ [ ] $\dfrac{\pi}{180}$ radians

$117° = $ [ ] $117 \times \dfrac{\pi}{180} = 2.0420$ radians

Example 1.5
Express 3.12 radians in degrees.

Solution
π radians $= 180°$

1 radian $= $ [ ] $\dfrac{180°}{\pi}$

3.12 radians $= $ [ ] $3.12 \times \dfrac{180°}{\pi} = 178.8°$

End of block exercises

1 Convert the following angles in radians to degrees:
(a) 0.3609 (b) 0.4771 (c) 1.3692 (d) $\frac{\pi}{3}$
(e) $\frac{2\pi}{3}$ (f) 6π (g) $\frac{\pi}{5}$ (h) $\frac{3\pi}{2}$

2 Convert the following angles in degrees to radians:
(a) 12° (b) 65° (c) 200° (d) 340° (e) 1000°

3 Express the following angles in the form $\alpha\pi$ radians:
(a) 90° (b) 45° (c) 60° (d) 120° (e) 240°
(f) 72° (g) 216° (h) 135° (i) 108° (j) 270°

Solutions to exercises

1 (a) 20.68° (b) 27.34° (c) 78.45° (d) 60°
(e) 120° (f) 1080° (g) 36° (h) 270°

2 (a) 0.2094 (b) 1.1345 (c) 3.4907 (d) 5.9341
(e) 17.4533

3 (a) $\frac{\pi}{2}$ (b) $\frac{\pi}{4}$ (c) $\frac{\pi}{3}$ (d) $\frac{2\pi}{3}$ (e) $\frac{4\pi}{3}$ (f) $\frac{2\pi}{5}$
(g) $\frac{6\pi}{5}$ (h) $\frac{3\pi}{4}$ (i) $\frac{3\pi}{5}$ (j) $\frac{3\pi}{2}$

The trigonometrical ratios

2.1 Introduction

The three common trigonometrical ratios of sine, cosine and tangent are defined with reference to a right-angled triangle. Some simple properties of the ratios are developed. The use of scientific calculators to find the trigonometrical ratios and their inverses is explained.

2.2 Some terms associated with a right-angled triangle

A **right angle** is an angle of $90°$. In Figure 2.1, $\triangle ABC$ has a right angle at C. The side opposite a right angle is called the **hypotenuse**. In Figure 2.1, AB is the hypotenuse.

Figure 2.1
The hypotenuse is AB. The hypotenuse is always opposite the right angle.

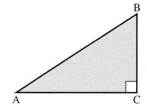

Consider now the angle at A. We often write just A when referring to the angle at A; similarly with B. The **side opposite** A is BC. The **side adjacent** to A is AC. Similarly the side opposite B is AC; the side adjacent to B is BC.

2.3 Definition of the trigonometrical ratios

We refer to the right-angled triangle in Figure 2.1 to define the trigonometrical ratios, sine, cosine and tangent. The sine of A is written $\sin A$, the cosine of A is written $\cos A$, and the tangent of A is written $\tan A$. We define:

Key point

$$\sin A = \frac{\text{length of side opposite}}{\text{length of hypotenuse}} = \frac{\text{BC}}{\text{AB}}$$

$$\cos A = \frac{\text{length of side adjacent}}{\text{length of hypotenuse}} = \frac{AC}{AB}$$

$$\tan A = \frac{\text{length of side opposite}}{\text{length of side adjacent}} = \frac{BC}{AC}$$

The sine, cosine and tangent of B are defined in exactly the same way, leading to

$$\sin B = \frac{AC}{AB}, \quad \cos B = \frac{BC}{AB}, \quad \tan B = \frac{AC}{BC}$$

2.4 Properties of the trigonometrical ratios

We note some properties of the trigonometrical ratios:

1 Since all are defined as the ratio of two lengths, none of the trigonometrical ratios has any units.
2 Since the hypotenuse is always the longest side of a right-angled triangle, the sine and cosine ratios can never be greater than 1.
3 The tangent ratio does not involve the hypotenuse and so this ratio can be greater than 1.

4 $\tan A = \dfrac{BC}{AC}$

$ = \dfrac{BC}{AB} \dfrac{AB}{AC}$

$ = \dfrac{BC/AB}{AC/AB}$

$ = \dfrac{\sin A}{\cos A}$

Similarly we see that

$$\tan B = \frac{\sin B}{\cos B}$$

Indeed, for any angle, θ, we have

Key point

$$\tan \theta = \frac{\sin \theta}{\cos \theta}$$

5 $\sin A = \dfrac{BC}{AB}$

$ = \cos B$

$$\cos A = \frac{AC}{AB}$$
$$= \sin B$$

We note that $B = 90° - A$. Hence we have for any angle A

Key point

$$\sin A = \cos(90° - A)$$
$$\cos A = \sin(90° - A)$$

6 We have

$$\tan A = \frac{BC}{AC}$$
$$= \frac{1}{AC/BC}$$
$$= \frac{1}{\tan B}$$

We also note that $B = 90° - A$ and so

Key point

$$\tan A = \frac{1}{\tan(90° - A)}$$

We now look at some examples.

Example 2.1
Figure 2.2 shows a right-angled triangle with the lengths of the sides labelled. Calculate
(a) $\sin A$ (b) $\cos A$ (c) $\tan A$ (d) $\sin B$ (e) $\cos B$ (f) $\tan B$

Figure 2.2

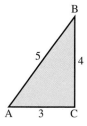

Solution
(a) $\sin A = \dfrac{BC}{AB} = \dfrac{4}{5} = 0.8$

(b) $\cos A = \dfrac{AC}{AB} = \dfrac{3}{5} = 0.6$

(c) $\tan A = \dfrac{BC}{AC} = \dfrac{4}{3} = 1.3333$

(d) $\sin B = \dfrac{AC}{AB} = 0.6$

(e) $\cos B = \dfrac{BC}{AB} = 0.8$

(f) $\tan B = \dfrac{AC}{BC} = \dfrac{3}{4} = 0.75$

Example 2.2

Figure 2.3 shows $\triangle XYZ$. The angle at X is 90°. Calculate
(a) $\sin Y$ (b) $\cos Y$ (c) $\sin Z$ (d) $\tan Y$ (e) $\tan Z$ (f) $\cos Z$

Figure 2.3

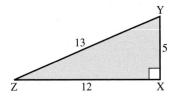

Solution

(a) $\sin Y = \dfrac{\text{opposite}}{\text{hypotenuse}} = $ $\dfrac{XZ}{YZ} = \dfrac{12}{13}$

(b) $\cos Y = \dfrac{\text{adjacent}}{\text{hypotenuse}} = $ $\dfrac{XY}{YZ} = \dfrac{5}{13}$

(c) $\sin Z = $ $\dfrac{XY}{YZ} = \dfrac{5}{13}$

(d) $\tan Y = \dfrac{\text{opposite}}{\text{adjacent}} = $ $\dfrac{XZ}{XY} = \dfrac{12}{5}$

(e) $\tan Z = $ $\dfrac{XY}{XZ} = \dfrac{5}{12}$

(f) $\cos Z = $ $\dfrac{XZ}{YZ} = \dfrac{12}{13}$

When an angle is known, a scientific calculator can be used to find its trigonometrical ratios. The angle may be expressed in degrees or radians.

Example 2.3
Use a scientific calculator to evaluate the following:
(a) sin 37° (b) cos 80° (c) tan 53° (d) sin 1.1 (e) cos 0.6321 (f) tan 0.5016

Solution
Using a scientific calculator we find:
(a) sin 37° = 0.6018
(b) cos 80° = 0.1736
(c) tan 53° = [   ]   1.3270

For (d), (e) and (f) the angles are given in radians. Make sure your calculator is set to RADIAN mode.
(d) sin 1.1 = 0.8912
(e) cos 0.6321 = [   ]   0.8068
(f) tan 0.5016 = 0.5484

Example 2.4 Mechanics – The tension in a cable
Traffic lights are suspended from a gantry above a carriageway as shown in Figure 2.4. The cable anchor points are 12 metres apart and the traffic light unit must hang 1 m below the gantry. By resolving forces vertically it can be shown that the tension, T, in each of the cables suspending the lights is given by

$$T = \frac{mg}{2 \cos \theta}$$

where θ is the angle between a cable and the upward vertical, m is the mass of the traffic light unit and g is a constant – the acceleration due to gravity – nominally $9.81 \ \mathrm{m \ s^{-2}}$. In this model, the mass of the cables has been ignored on the assumption that they are light compared with the mass of the traffic light unit. From this formula it can be deduced that as the angle θ increases – as will happen if the anchor points are placed further apart – then the tension in the cable will also increase.

Figure 2.4
Using trigonometric ratios to calculate the tension in a cable.

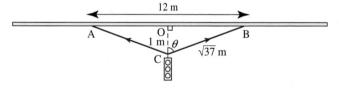

(a) Find the tension T when the mass of the traffic light unit is 18 kg.
(b) If the maximum permitted tension in the cable is 400 N and the traffic light unit must remain 1 m below the gantry, calculate how far apart the suspension points, A and B, must be.

Solution

(a) By inspection of the right-angled triangle OBC we see that $\cos \theta = \dfrac{1}{\sqrt{37}}$. Then

$$T = \frac{18 \times 9.81}{2 \times \dfrac{1}{\sqrt{37}}} = 537 \ \mathrm{N} \quad (3 \ \mathrm{s.f.})$$

Note that when the traffic light unit is in this position,

$$\theta = \cos^{-1}\frac{1}{\sqrt{37}} = 80.5°$$

(b) Let us calculate the corresponding angle when $T = 400$ N:

$$\cos\theta = \frac{mg}{2T} = \frac{18 \times 9.81}{2 \times 400} = 0.2207$$

$$\theta = \cos^{-1}0.2207 = 77.25°$$

Here $\cos^{-1}$ is the inverse cosine function described in detail on p339, and which can be found using a calculator. Therefore, referring to Figure 2.5,

Figure 2.5

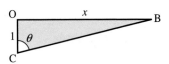

$$\tan\theta = \tan 77.25 = \frac{x}{1}$$

so that

$$x = \tan 77.25 = 4.42 \text{ m}$$

We deduce that the suspension points must be no more than 8.84 metres apart.

Example 2.5 Structural Engineering – Trigonometry for the method of sections

A truss is a structure consisting of straight members connected at joints. An example is shown in Figure 2.6. In the study of structural mechanics, the *method of sections* is a technique used to find the forces in the different members of the truss.

Figure 2.6
A truss consisting of several members.

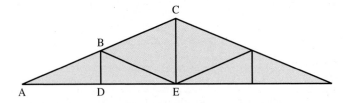

The method consists of isolating a particular part of the truss and considering only those forces which act on that isolated part. Figure 2.7 shows the part which could be isolated in order to find the forces in members BC, BE and DE.

Figure 2.7
A section, or cut, through the truss.

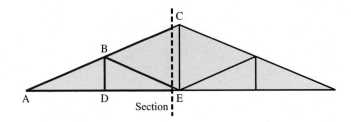

To calculate the forces it is necessary to use trigonometry to find angles and lengths which are not immediately available. In particular it is necessary to find the perpendicular distance of the member BC from the point E if an engineer wants to calculate the moment of the force in BC about the point E. Use the information provided in Figure 2.8 to find this distance (EF).

Figure 2.8

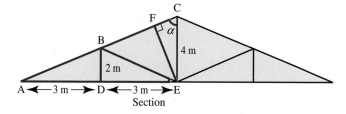

Section

Solution

Note that the required length EF is one side of the right-angled triangle EFC another side of which is already known (CE = 4 m). The angle at C, labelled α, in this triangle can be found by considering the larger triangle ACE.

Note that

$$\tan \alpha = \frac{6}{4}, \quad \alpha = \tan^{-1}\left(\frac{6}{4}\right) = 56.3° \quad (1 \text{ d.p.})$$

Then, in triangle EFC,

$$\sin \alpha = \frac{EF}{4}$$
$$EF = 4 \sin \alpha$$
$$= 4 \sin 56.3°$$
$$= 3.3 \ (1 \text{ d.p.})$$

that is EF = 3.3 m. Knowing this distance enables a structural engineer to write down an expression for the moment about E of the force in member BC. In turn, and with knowledge of other moments, the force in BC can be calculated.

Exercises

1 In $\triangle$CDE, D is a right angle. The lengths of CD, DE and CE are α, β and γ respectively. State
   (a) sin C (b) cos C (c) tan C (d) sin E
   (e) tan E (f) cos E

2 Use a scientific calculator to evaluate
   (a) cos 61° (b) tan 0.4 (c) sin 70°
   (d) cos 0.7613 (e) tan 51° (f) sin 1.2

Solutions to exercises

1 (a) $\dfrac{\beta}{\gamma}$ (b) $\dfrac{\alpha}{\gamma}$ (c) $\dfrac{\beta}{\alpha}$ (d) $\dfrac{\alpha}{\gamma}$ (e) $\dfrac{\alpha}{\beta}$ (f) $\dfrac{\beta}{\gamma}$

2 (a) 0.4848 (b) 0.4228 (c) 0.9397 (d) 0.7239
   (e) 1.2349 (f) 0.9320

2.5 Secant, cosecant and cotangent ratios

These ratios are the reciprocals of the cosine, sine and tangent ratios. Secant, cosecant and cotangent are usually abbreviated to sec, cosec and cot, respectively.

Key point

$$\sec A = \frac{1}{\cos A}$$

$$\operatorname{cosec} A = \frac{1}{\sin A}$$

$$\cot A = \frac{1}{\tan A}$$

Example 2.6
Evaluate
(a) cosec 50° (b) cot 20° (c) sec 70° (d) cot 0.7

Solution

(a)
$$\operatorname{cosec} 50° = \frac{1}{\sin 50°}$$
$$= 1.3054$$

(b)
$$\cot 20° = \frac{1}{\tan 20°}$$
$$= 2.7475$$

(c) sec 70° =

$$\frac{1}{\cos 70°} = 2.9238$$

(d) Note that the angle is in radians.

$$\cot 0.7 = \frac{1}{\tan 0.7}$$
$$= 1.1872$$

Exercise

1   Evaluate the following:
(a) cot 1.2 (b) sec 45° (c) cosec 0.6391
(d) cot 57° (e) sec 0.9600

Solution to exercise

1   (a) 0.3888 (b) 1.4142 (c) 1.6765 (d) 0.6494
(e) 1.7436

2.6 The inverse trigonometrical ratios

Suppose we know the value of sin A, but not the value of A. For example, let sin $A = 0.6513$, and we wish to find the value of A. Given

$$\sin A = 0.6513$$

we write

$$A = \sin^{-1}(0.6513)$$

This states that A is the angle whose sine is 0.6513. We read this as A is the **inverse sine** of 0.6513. So the notation $\sin^{-1}$ means 'the angle whose sine is . . .'. The -1 should not be interpreted as a power. Other notations are sometimes used, namely

$$A = \text{inv} \sin(0.6513) \quad \text{and} \quad A = \arcsin(0.6513)$$

Similarly if cos $B = 0.3619$ we write $B = \cos^{-1}(0.3619)$; if tan $C = 1.4703$ then $C = \tan^{-1}(1.4703)$.

Key point

> $\sin^{-1}$ means 'the angle whose sine is . . .'
> $\cos^{-1}$ means 'the angle whose cosine is . . .'
> $\tan^{-1}$ means 'the angle whose tangent is . . .'

We use a scientific calculator to find the inverse sine, inverse cosine and inverse tangent of a number.

Example 2.7
Find A given
(a) sin $A = 0.4213$ (b) cos $A = 0.5316$ (c) tan $A = 1.7503$

Solution
(a) We have

$$\sin A = 0.4213$$

and so

$$A = \sin^{-1}(0.4213)$$

Using a scientific calculator we see

$$A = 24.92°$$

If your calculator is in radian mode you will obtain the equivalent answer in radians, that is 0.4349 radians.

(b)
$$\cos A = 0.5316$$
$$A = \cos^{-1}(0.5316)$$
$$= 57.89°$$

(c)
$$\tan A = 1.7503$$
$$A = \tan^{-1}(1.7503)$$
$$= 60.26°$$

Example 2.8

Find B given

(a) $\cos B = 0.8061$ (b) $\sin B = 0.4611$ (c) $\tan B = 1.2500$

Solution

(a) $\cos B = 0.8061$

$\qquad B = \cos^{-1}(0.8061)$

$\qquad\quad = \boxed{}$ 36.28°

(b) $\sin B = 0.4611$

$\qquad B = \boxed{}$ 27.46°

(c) $\tan B = 1.2500$

$\qquad B = \boxed{}$ 51.34°

End of block exercises

1 θ is an angle between 0° and 90°. In each case find θ given:
   (a) $\sin \theta = 0.3467$
   (b) $\cos \theta = 0.6419$
   (c) $\tan \theta = 1.7500$
   (d) $\sin \theta = 0.7396$
   (e) $\tan \theta = 0.5050$
   (f) $\cos \theta = 0.3507$

2 Evaluate sec 37°.

3 Evaluate cot 75°.

4 Evaluate cosec 17°.

5 Evaluate cosec 1.

6 θ is an angle between 0° and 90°. In each case, find θ.
   (a) $\sin \theta = \cos \theta$
   (b) $\sin \theta = 2 \cos \theta$

7 $\triangle ABC$ has a right angle at C, AC = 6 cm, BC = 10 cm and AB = 11.66 cm. Find
   (a) $\sin A$
   (b) $\cos A$
   (c) $\tan A$
   (d) $\sin B$
   (e) $\cos B$
   (f) $\tan B$
   (g) A
   (h) B

8 We have seen that $\sin^{-1} x$ means 'the angle whose sine is x' and not the reciprocal $\dfrac{1}{\sin x}$. How would you write the reciprocal using a negative power?

Solutions to exercises

1 (a) 20.29° (b) 50.07° (c) 60.26° (d) 47.70°
   (e) 26.79° (f) 69.47°

2 1.2521

3 0.2679

4 3.4203

5 1.1884

6 (a) 45° (b) 63.43°

7 (a) 0.8576 (b) 0.5146 (c) 1.6667 (d) 0.5146
(e) 0.8576 (f) 0.6 (g) 59.04° (h) 30.96°

8 Brackets would be inserted to show the intention, that is

$$\frac{1}{\sin x} = (\sin x)^{-1}$$

The trigonometrical ratios in all quadrants

3.1 Introduction

Block 2 defined the trigonometrical ratios sine, cosine and tangent with reference to the sides of a right-angled triangle. No angle in a right-angled triangle is greater than 90°. So, if we wish to define the trigonometrical ratios of angles greater than 90° we need a method that does not use right-angled triangles. This block looks at how this is achieved.

3.2 The four quadrants

Figure 3.1 shows the x and y axes intersecting at the origin O. The axes divide the x–y plane into four sections, called **quadrants**. These are numbered 1 to 4 as indicated in Figure 3.1.

Figure 3.1
The x and y axes divide the plane into four quadrants.

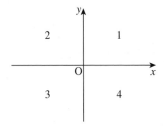

We now consider a rotating arm, OC. The arm is fixed at the origin, O. We measure the angle θ from the positive x axis to the arm, measuring in an anticlockwise direction. Figure 3.2 shows the arm in each of the four quadrants.

Note that in quadrant 1, θ lies between 0° and 90°; in quadrant 2, θ is between 90° and 180°; in quadrant 3, θ is between 180° and 270°; and in quadrant 4, θ is between 270° and 360°. Figure 3.3 illustrates this.

Figure 3.2
The arm OC
rotates
anticlockwise into
each of the four
quadrants.

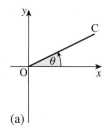
(a)

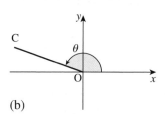
(b)

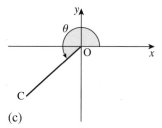
(c)

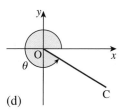
(d)

Figure 3.3
(a) In quadrant 1, θ
lies between 0°
and 90°; (b) in
quadrant 2, θ lies
between 90° and
180°; (c) in
quadrant 3, θ lies
between 180° and
270°; (d) in
quadrant 4, θ lies
between 270° and
360°.

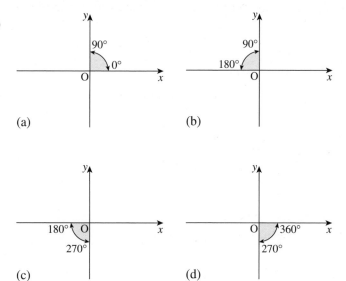
(a)   (b)

(c)   (d)

3.3   Projections onto the x and y axes

We now introduce **projections** of the arm OC onto the x and y axes. The projection
of OC onto the x axis is OA; the projection onto the y axis is OB. Figure 3.4 shows
the x and y projections as the arm rotates into the four quadrants.

Note that the projections may be positive or negative. For example, when OC is
in the second quadrant, the x projection, OA, is on the negative x axis and so is

Figure 3.4
The x projection is OA, the y projection is OB.

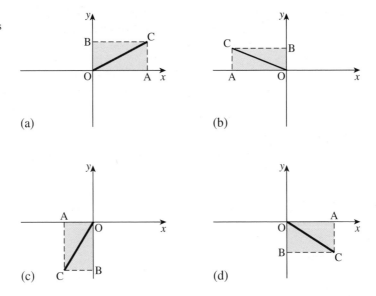

negative. The y projection, OB, is on the positive y axis and so is positive. The arm OC is considered to be always positive.

Table 3.1 gives the signs of the x and y projections of OC as it rotates through the four quadrants.

Table 3.1

	First quadrant	Second quadrant	Third quadrant	Fourth quadrant
x projection	+	−	−	+
y projection	+	+	−	−

3.4 Extended definition of the trigonometrical ratios

We define the trigonometrical ratios in terms of the x and y projections of a rotating arm, OC.

Key point

$$\sin \theta = \frac{y \text{ projection of OC}}{\text{OC}}$$

$$\cos \theta = \frac{x \text{ projection of OC}}{\text{OC}}$$

$$\tan \theta = \frac{y \text{ projection of OC}}{x \text{ projection of OC}}$$

By examining the signs of the x and y projections in Table 3.1 it is easy to determine the signs of $\sin\theta$, $\cos\theta$ and $\tan\theta$ for the four quadrants. For example, consider θ in the second quadrant. Since the y projection is positive, then $\sin\theta$ is positive. Similarly, as the x projection is negative, then $\cos\theta$ is negative. The sign of $\tan\theta$ is negative since it is given by the ratio of a positive number and a negative number. Table 3.2 shows the sign of $\sin\theta$, $\cos\theta$ and $\tan\theta$ for θ in the four quadrants.

Table 3.2

	First quadrant	Second quadrant	Third quadrant	Fourth quadrant
$\sin\theta$	$+$	$+$	$-$	$-$
$\cos\theta$	$+$	$-$	$-$	$+$
$\tan\theta$	$+$	$-$	$+$	$-$

Example 3.1
An angle θ is such that $\sin\theta < 0$ and $\cos\theta > 0$. In which quadrant does θ lie?

Solution
Referring to Table 3.2 we see that $\sin\theta < 0$ when θ lies in the third or fourth quadrants. When $\cos\theta > 0$, θ lies in the first or fourth quadrants. Hence for both conditions to be satisfied simultaneously θ must be in the fourth quadrant.

Example 3.2
Show $\sin 17° = \sin 163°$.

Solution
Although $\sin 17°$ and $\sin 163°$ can easily be evaluated using a calculator and hence shown to be equal, it is instructive to show their equality using the definition of $\sin\theta$.

Figure 3.5
OC and OC′ have the same y projection and so $\sin 17° = \sin 163°$.

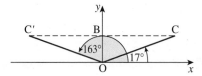

Consider two arms, OC and OC′, of equal length. Figure 3.5 shows the arm OC forming an angle of 17° with the positive x axis, and the arm OC′ forming an angle of 163° with the positive x axis. Noting that $163° = 180° - 17°$, we can see that OC′ is the reflection of OC in the y axis. By symmetry both the y projections of OC and OC′ will be OB. Since the y projections are equal and the arms are of equal length, then from the definition of $\sin\theta$ we have $\sin 17° = \sin 163°$.

Example 3.2 illustrates a general rule that is true for any value of θ:

Key point

$$\sin(180° - \theta) = \sin\theta$$

Example 3.3

An angle θ is such that $\tan \theta > 0$ and $\sin \theta < 0$. In which quadrant does θ lie?

Solution

Using Table 3.2, we see that $\tan \theta > 0$ when θ lies in the

    [   ] and [   ] quadrants   first and third

Also, $\sin \theta < 0$ when θ lies in the

    [   ] and [   ] quadrants   third and fourth

Hence θ lies in the [   ] quadrant.   third

Exercises

1   An angle θ is such that $\sin \theta < 0$ and $\cos \theta < 0$. In which quadrant does θ lie?

2   An angle β is such that $\cos \beta > 0$ and $\tan \beta < 0$. State the range of possible values of β.

Solutions to exercises

1   third

2   $270° < \beta < 360°$

3.5   Adding and subtracting multiples of 360°

The sine of an angle is governed by the position of the rotating arm, OC. We note that 360° corresponds to exactly one complete revolution. Thus rotating the arm, either clockwise or anticlockwise, through multiples of 360° will leave it in exactly the same position, and consequently the value of the sine remains unchanged. This may be stated mathematically as:

$$\begin{aligned}
\sin \theta &= \sin(\theta + 360°) \\
&= \sin(\theta + 720°) \\
&= \sin(\theta + 1080°) \\
&= \cdots
\end{aligned}$$

and also

$$\begin{aligned}
\sin \theta &= \sin(\theta - 360°) \\
&= \sin(\theta - 720°)
\end{aligned}$$

$$= \sin(\theta - 1080°)$$
$$= \cdots$$

Written compactly we have

Key point

$$\sin \theta = \sin(\theta \pm n\,360°) \quad n = 1, 2, 3, \ldots$$

An exactly similar argument can be applied to $\cos \theta$. Thus

Key point

$$\cos \theta = \cos(\theta \pm n\,360°) \quad n = 1, 2, 3, \ldots$$

We now consider $\tan \theta$. Recall that $\tan \theta$ is defined by

$$\tan \theta = \frac{y \text{ projection}}{x \text{ projection}}$$

The corresponding result for $\tan \theta$ is

Key point

$$\tan \theta = \tan(\theta \pm n\,180°) \quad n = 1, 2, 3, \ldots$$

Hence, adding 180° to, or subtracting 180° from, an angle leaves the tangent unaltered. Note that with sine and cosine we can add or subtract multiples of 360°, but with the tangent we can add and subtract multiples of 180°.

Expressions that repeat their values at regular intervals are called **periodic**. Hence $\sin \theta$, $\cos \theta$ and $\tan \theta$ are all periodic.

Example 3.4
Simplify $\sin(\theta + 1000°)$.

Solution
Subtracting multiples of 360° from $\theta + 1000°$ leaves the sine unchanged. Hence

$$\sin(\theta + 1000°) = \sin[\theta + 1000° - 2(360°)]$$
$$= \sin(\theta + 280°)$$

Example 3.5
Simplify $\cos(x - 900°)$.

Solution
Adding multiples of 360° to $x - 900°$ leaves the cosine unchanged. Hence

$$\cos(x - 900°) = \cos[x - 900° + 3(360°)]$$
$$= \cos(x + 180°)$$

Example 3.6
Simplify $\tan(z - 540°)$.

Solution
Adding multiples of $180°$ to $z - 540°$ leaves the tangent unchanged. So we have

$$\tan(z - 540°) = \tan[z - 540° + 3(180°)]$$
$$= \tan z$$

Example 3.7
Simplify $\sin(-180° - \theta)$.

Solution
Adding $360°$ to an angle leaves the sine unchanged. Hence
$\sin(-180° - \theta) = \sin(-180° - \theta + 360°)$

$$= \boxed{} \qquad\qquad \sin(180° - \theta)$$

Using the result following Example 3.2, we can further simplify this to

$$\boxed{} \qquad\qquad \sin\theta$$

Hence

$$\sin(-180° - \theta) = \sin\theta$$

Exercises

1 Simplify
(a) $\sin(\alpha + 400°)$
(b) $\sin(\beta - 500°)$
(c) $\sin(\alpha - 450°)$
(d) $\sin(\theta + 1000°)$
(e) $\sin(2\theta + 1080°)$

2 Simplify
(a) $\cos(x - 300°)$
(b) $\cos(\theta - 810°)$
(c) $\tan(x + 200°)$
(d) $\tan(x - 540°)$
(e) $\cos(\beta + 500°)$

Solutions to exercises

1 (a) $\sin(\alpha + 40°)$ (b) $\sin(\beta - 140°)$
(c) $\sin(\alpha - 90°)$ (d) $\sin(\theta - 80°)$
(e) $\sin 2\theta$

2 (a) $\cos(x + 60°)$ (b) $\cos(\theta - 90°) = \sin\theta$
(c) $\tan(x + 20°)$ (d) $\tan x$ (e) $\cos(\beta + 140°)$

End of block exercises

1 The angles are given in radians. Evaluate the following:
(a) sin 30
(b) cos 27
(c) tan 31

2 An angle α is such that $\sin \alpha < 0$ and $\cos \alpha > 0$. In which quadrant does α lie?

3 Simplify $\cos(450° - \theta)$.

4 Simplify $\tan(y + 1260°)$.

5 Simplify $\sin(-1260° - \theta)$.

6 An angle β is such that $\tan \beta > 0$ and $\sin \beta < 0$. In which quadrant does β lie?

7 An angle α is such that $\sin \alpha \geq 0$ and $\sin 2\alpha \geq 0$. State the range of possible values of α.

8 An angle θ is such that $\cos \theta < 0$ and $\cos 2\theta > 0$. State the range of possible values for θ.

9 An angle θ is such that $\tan \theta > 0$ and $\tan \frac{\theta}{2} < 0$. State the range of possible values of θ.

10 An angle ϕ is such that $\sin \phi > 0$ and $\cos 2\phi < 0$. State the range of possible values of ϕ.

Solutions to exercises

1 (a) -0.9880 (b) -0.2921 (c) -0.4417

2 fourth quadrant

3 $\sin \theta$

4 $\tan y$

5 $\sin \theta$

6 third quadrant

7 $0° \leq \alpha \leq 90°$

8 $135° < \theta < 180°$

9 $180° < \theta < 270°$

10 $45° < \phi < 135°$

Trigonometrical functions and their graphs

4.1 Introduction

Having introduced the trigonometrical ratios of sine, cosine and tangent we are ready to consider the three trigonometrical functions $y = \sin x$, $y = \cos x$ and $y = \tan x$. A number of properties and graphs of these functions are considered. Extensions are made to include the functions $y = \sin kx$, $y = \cos kx$ and $y = \tan kx$ for various values of k.

4.2 The function $y = \sin x$

Table 4.1 gives values of x in degrees and the corresponding values of $\sin x$ found using a scientific calculator.

Table 4.1

x	0	30	60	90	120	150	180
$\sin x$	0	0.500	0.866	1	0.866	0.500	0

x	210	240	270	300	330	360
$\sin x$	−0.500	−0.866	−1	−0.866	−0.500	0

Plotting these values produces the graph shown in Figure 4.1.
Note that the maximum value of $\sin x$ is 1; the minimum value of $\sin x$ is −1.

Figure 4.1
A graph of
$y = \sin x$ for
$0° \leq x \leq 360°$.

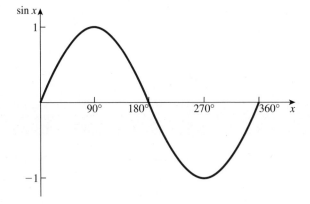

If we had been working in radians, then $0° \leq x \leq 360°$ would be replaced by $0 \leq x \leq 2\pi$. The shape of the sine function is often referred to as a **sine wave** or a **cycle**.

We saw in Block 3 that $\sin x = \sin(x + 360°) = \sin(x + 720°)$ and $\sin x = \sin(x - 360°) = \sin(x - 720°)$ and so on. In other words, adding or subtracting multiples of 360° to an angle does not alter the sine of the angle. Hence the graph in Figure 4.1 can be extended to the left and to the right by repeating cycles of the same shape every 360°. Figure 4.2 illustrates this.

Figure 4.2
Cycles are repeated every 360°.

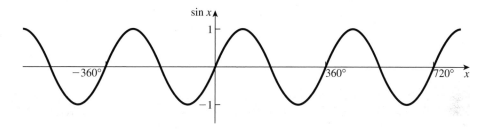

When using radian measure, cycles are repeated every 2π radians.

Exercises

Use the graphs in Figures 4.1 and 4.2 to answer the following questions:

1   What is the maximum possible domain of the function $y = \sin x$?

2   What is the range of $y = \sin x$?

3   Is the function $y = \sin x$ one-to-one or many-to-one?

Solutions to exercises

1   all x

2   $[-1, 1]$

3   many-to-one

4.3 The function $y = \cos x$

Using a scientific calculator values of cos x are found for various values of x measured in degrees. These are recorded in Table 4.2.

Table 4.2

x	0	30	60	90	120	150	180
cos x	1	0.866	0.500	0	−0.500	−0.866	−1

x	210	240	270	300	330	360
cos x	−0.866	−0.500	0	0.500	0.866	1

The values are graphed in Figure 4.3.

Figure 4.3
The function
$y = \cos x$ for
$0° \leq x \leq 360°$.

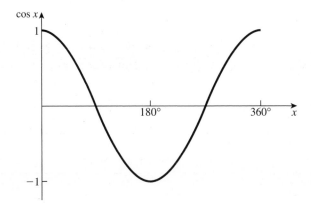

As with $y = \sin x$, the function $y = \cos x$ may be extended to the left and to the right. Since adding or subtracting multiples of 360° to an angle leaves its cosine unchanged, then full cycles will be repeated every 360°. Figure 4.4 illustrates this.

Figure 4.4
The function
$y = \cos x$
completes a
full cycle
every 360°.

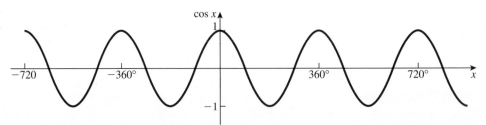

Note that $y = \cos x$ completes a full cycle every 360° or 2π radians. The maximum value of cos x is 1; the minimum value is −1.

Exercises

Use the graphs in Figures 4.3 and 4.4 to answer the following questions:

1 What is the maximum possible domain of the function $y = \cos x$?

2 What is the range of $y = \cos x$?

3 Is the function $y = \cos x$ one-to-one or many-to-one?

Solutions to exercises

1 all x

2 $[-1, 1]$

3 many-to-one

4.4   The function $y = \tan x$

Table 4.3 gives values of x in degrees and corresponding values of $\tan x$.

Table 4.3

x	0	30	60	90	120	150	180
$\tan x$	0	0.577	1.732	–	−1.732	−0.577	0

x	210	240	270	300	330	360
$\tan x$	0.577	1.732	–	−1.732	−0.577	0

Figure 4.5 shows a graph of $y = \tan x$ for $0° \leq x \leq 360°$.

Figure 4.5
The function
$y = \tan x$ for
$0° \leq x \leq 360°$.

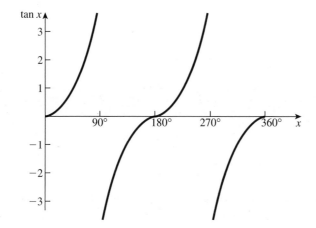

Note that the graph is radically different from those of $y = \sin x$ and $y = \cos x$. The function $y = \tan x$ has no maximum or minimum value.

Recalling from Block 3 that adding or subtracting multiples of 180° to an angle leaves its tangent unchanged, we see that extending the graph in Figure 4.5 to the left and right produces that shown in Figure 4.6.

Figure 4.6
For $y = \tan x$, cycles are repeated every 180° or π radians.

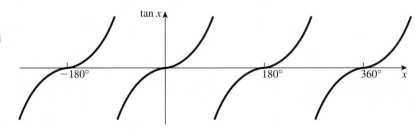

Note that the function is discontinuous at $x = \cdots -270°, -90°, 90°, 270°, \ldots$.

Exercises

Use the graphs in Figures 4.5 and 4.6 to answer the following questions:

1 What is the maximum possible domain of the function $y = \tan x$?

2 What is the range of $y = \tan x$?

3 Is the function $y = \tan x$ one-to-one or many-to-one?

Solutions to exercises

1 All values except $x = \cdots -270°, -90°, 90°, 270°, \ldots$. These values must be excluded from the domain.

2 $(-\infty, \infty)$

3 many-to-one

4.5 The amplitude of $y = A \sin x$ and $y = A \cos x$

Consider the function $y = A \sin x$, where A is a positive constant, that is $A > 0$. The number A is called the **amplitude**. It is the maximum value of y. The minimum value of y is $-A$. Thus $y = 3 \sin x$ has an amplitude of 3. The maximum value of $3 \sin x$ is 3; the minimum value is -3. Note that the amplitude of $y = \sin x$ is 1. Figure 4.7 shows graphs of $y = \sin x$ and $y = 3 \sin x$ for $0° \leq x \leq 360°$.

Figure 4.7
The amplitude of
$y = \sin x$ is 1; the
amplitude of
$y = 3 \sin x$ is 3.

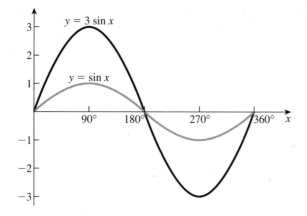

Note that a full cycle of $y = 3 \sin x$ is completed in 360°. The amplitude does not affect the periodic properties of sin x: that is, it takes 360° for both $y = \sin x$ and $y = 3 \sin x$ to complete a full cycle. In general, $y = A \sin x$ completes a cycle every 360°.

Similar comments apply to $y = A \cos x$. The amplitude of $A \cos x$ is A. It takes 360° for a full cycle of $A \cos x$ to be completed.

Key point	The amplitude of $y = A \sin x$ and $y = A \cos x$ is A.

Example 4.1
State the amplitude of each of the following functions:
(a) $y = 2 \sin x$
(b) $y = 4.7 \cos x$
(c) $y = \dfrac{2 \sin x}{3}$
(d) $y = 0.8 \cos x$

Solution
(a) 2 (b) 4.7 (c) $\frac{2}{3}$ (d) 0.8

Exercise

1 State the amplitude of
(a) $10 \sin x$ (b) $7.3 \cos x$ (c) $0.01 \sin t$
(d) $1.2 \cos \theta$

Solution to exercise

1 (a) 10 (b) 7.3 (c) 0.01 (d) 1.2

4.6 The functions $y = A \sin kx$ and $y = A \cos kx$

We have already seen that A is the amplitude and this is the maximum value of y. We now consider the effect of the parameter k in $A \sin kx$. To allow us to focus on k we take A to be 1. Thus we examine the function $y = \sin kx$ for various values of k, for example $y = \sin 2x$, $y = \sin 3x$, $y = \sin \dfrac{x}{2}$ and $y = \sin \dfrac{3x}{2}$. We can deduce the graphs of these functions by reference to the graph of $y = \sin x$.

We begin by looking at $y = \sin 2x$. Recall that $y = \sin x$ completes one full cycle as x varies from $0°$ to $360°$. Then $y = \sin 2x$ completes one full cycle as $2x$ varies from $0°$ to $360°$, that is as x varies from $0°$ to $\frac{360°}{2} = 180°$. If x is measured in radians then $y = \sin 2x$ completes a full cycle as x varies from 0 to π radians. A graph of the function is illustrated in Figure 4.8.

Figure 4.8
The function
$y = \sin 2x$
completes a full
cycle every $180°$ or
π radians.

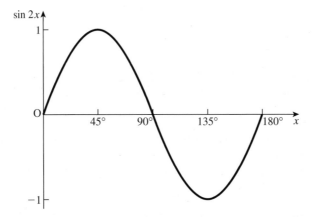

We have noted previously that the amplitude does not affect the periodic property of $y = \sin x$. Since $y = \sin 2x$ completes a cycle every $180°$, then in general $y = A \sin 2x$ completes a cycle every $180°$ also. Note that the amplitude of $y = \sin 2x$ is 1.

As another example consider $y = \sin \dfrac{3x}{2}$. A full cycle is completed as $\dfrac{3x}{2}$ varies from $0°$ to $360°$: that is, as x varies from $0°$ to $\frac{360°}{3/2} = 240°$. A graph of $y = \sin \dfrac{3x}{2}$ is illustrated in Figure 4.9.

Similarly, $y = A \sin \dfrac{3x}{2}$ completes a cycle every $240°$.

In general, $y = A \sin kx$ completes a full cycle in $\dfrac{360°}{k}$ or $\dfrac{2\pi}{k}$ radians.

The function $y = A \cos kx$ also has these properties. We have

Key point

The functions $y = A \sin kx$ and $y = A \cos kx$ complete a cycle every $\dfrac{360°}{k}$ or $\dfrac{2\pi}{k}$ radians.

Figure 4.9
$y = \sin\dfrac{3x}{2}$ completes a cycle every $240°$ or $\dfrac{4\pi}{3}$ radians.

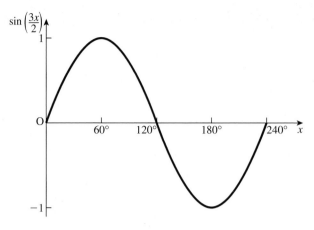

Example 4.2

State the number of cycles of y in $360°$ given

(a) $y = 3 \sin 4x$

(b) $y = 4 \cos 3x$

(c) $y = \dfrac{1}{2} \cos \dfrac{x}{2}$

(d) $y = 5 \sin \dfrac{3x}{4}$

Solution

(a) $y = 3 \sin 4x$ completes a cycle every $\dfrac{360°}{4} = 90°$. In $360°$, four cycles are completed.

(b) $y = 4 \cos 3x$ completes a cycle every $\dfrac{360°}{3} = 120°$. In $360°$ three cycles are completed.

(c) $y = \dfrac{1}{2} \cos \dfrac{x}{2}$ completes a cycle every $\dfrac{360°}{1/2} = 720°$. In $360°$, 0.5 of a cycle is completed.

(d) $y = 5 \sin \dfrac{3x}{4}$ completes a cycle every $\dfrac{360°}{3/4} = 480°$. In $360°$, 0.75 of a cycle is completed.

Example 4.2 illustrates the following general statement.

Key point

The functions $y = A \sin kx$ and $y = A \cos kx$ complete k cycles every $360°$ or every 2π radians.

Example 4.3

For each function, state (i) the minimum value and (ii) the increase in x required for y to complete a full cycle.

(a) $y = 3 \sin 5x$

(b) $y = 0.65 \cos 6x$

(c) $y = \dfrac{4}{5} \sin \dfrac{x}{3}$

(d) $y = -2 \cos \dfrac{5x}{3}$

Solution

(a) The amplitude is 3 and so the minimum value is -3. A full cycle requires x to increase by $\dfrac{360°}{5} = 72°$.

(b) The minimum value is -0.65. A full cycle requires x to increase by $\dfrac{360°}{6} = 60°$.

(c) The minimum value is $-\dfrac{4}{5}$. A full cycle requires x to increase by $\dfrac{360°}{1/3} = 1080°$.

(d) The minimum value is -2. A full cycle requires x to increase by $\dfrac{360°}{5/3} = 216°$.

Example 4.4

State the maximum value, the minimum value and the number of cycles completed in $720°$ for the function $y = \dfrac{3}{2} \cos \dfrac{5x}{2}$.

Solution

The amplitude of $y = \dfrac{3}{2} \cos \dfrac{5x}{2}$ is $\dfrac{3}{2}$. Hence the maximum value is ▮ $\dfrac{3}{2}$

and the minimum value is ▮ $-\dfrac{3}{2}$

The function $y = \dfrac{3}{2} \cos \dfrac{5x}{2}$ completes ▮ cycles every $360°$. $\dfrac{5}{2}$

Hence in $720°$ the function completes ▮ cycles. 5

Example 4.5 Electronic Engineering – Rectified half sine wave

A rectified half sine wave, $f(t)$, is illustrated in Figure 4.10. Recall (from Chapter 6 Block 6) that any function that has a pattern which repeats at regular intervals is said to be periodic and the interval over which the repetition takes place is called the period. Hence $f(t)$ is a periodic function with period T. Write a mathematical expression for this rectified half sine wave. Note that, in this example, the independent variable is t, not x, and that angles are measured in radians.

Figure 4.10
A rectified half
sine wave.

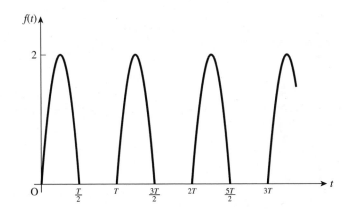

Solution

For $0 \leq t < T/2$ the rectified half-wave is a sine wave of amplitude 2 and period T. Note again the key point above which states that sin kt completes a cycle every $\frac{2\pi}{k}$, that is, the period is $\frac{2\pi}{k}$. Applying this to the current function we see that $T = \frac{2\pi}{k}$ from which $k = \frac{2\pi}{T}$. So $f(t)$ can be expressed as

$$f(t) = 2 \sin\left(\frac{2\pi t}{T}\right)$$

For $\frac{T}{2} \leq t \leq T$ then $f(t) = 0$. Note also that the rectified half-wave has period T. So

$$f(t) = \begin{cases} 2 \sin\left(\dfrac{2\pi t}{T}\right) & \text{for } 0 \leq t < \dfrac{T}{2} \\[2ex] 0 & \text{for } \dfrac{T}{2} \leq t < T \end{cases}$$

and $f(t) = f(t + T)$ for all values of t.

4.7 The function $y = A$ tan kx

We consider the function $y = A$ tan kx. As with $y = A$ sin kx and $y = A$ cos kx, the factor A does not affect the periodic properties of the tangent function. Recall that $y = \tan x$ completes a cycle as x increases by 180° or π radians. Hence tan kx completes a cycle as kx increases through 180°, that is as x increases through $\frac{180°}{k}$. So, for example, $y = A$ tan $2x$ completes a cycle every $\frac{180°}{2} = 90°$, $y = A$ tan $3x$ completes a cycle every $\frac{180°}{3} = 60°$, and $y = A$ tan $\frac{2x}{3}$ completes a cycle every $\frac{180°}{2/3} = 270°$. A graph of $y = 2$ tan $3x$ for $-30° \leq x \leq 150°$ is shown in Figure 4.11.

Figure 4.11
$y = 2$ tan $3x$
completes a cycle
every 60°.

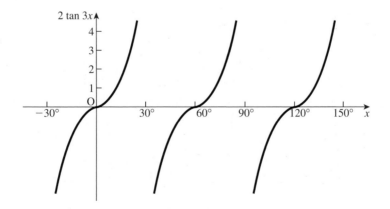

In general $y = A \tan kx$ completes a cycle every $\dfrac{180°}{k}$. In other words, k cycles are completed every 180°.

Example 4.6

State the number of cycles completed every 360° for

(a) $y = 3 \tan x$

(b) $y = 7 \tan 2x$

(c) $y = -\dfrac{1}{2} \tan \dfrac{3x}{2}$

Solution

(a) For $y = 3 \tan x$, the '3' does not affect the periodic properties.

Now $y = 3 \tan x$ completes ▢ cycles every 180° and so in 360°,

▢ cycles are completed.

(b) $y = 7 \tan 2x$ completes ▢ cycles every 180° and hence completes

▢ cycles every 360°.

(c) $y = -\dfrac{1}{2} \tan \dfrac{3x}{2}$ completes ▢ cycles every 180° and so completes

▢ cycles every 360°.

1

2

2

4

$\dfrac{3}{2}$

3

End of block exercises

State the amplitude of the functions given in questions 1–6.

1 $y = 3 \sin 4x$

2 $y = \frac{3}{7} \cos 2x$

3 $y = 0.96 \sin x$

4 $y = -2 \cos 7x$

5 $y = -0.6 \sin 3x$

6 $y = \dfrac{4 \cos x}{7}$

State the number of cycles completed every 360° for the functions given in questions 7–12.

7 $y = 2 \sin 3x$

8 $y = -2 \sin 3x$

9 $y = 4 \cos \dfrac{5x}{6}$

10 $y = 0.5 \cos 0.5x$

11 $y = 6 \tan 6x$

12 $y = -3 \tan 9x$

13 Sketch $y = \sin(-x)$ for $0° \le x \le 360°$.

14 Sketch $y = \tan \dfrac{x}{2}$ for $0° \le x \le 360°$.

15 State two properties that are common to both $y = A \sin kx$ and $y = A \cos kx$.

Solutions to exercises

1	3	7	3
2	$\frac{3}{7}$	8	3
3	0.96	9	$\frac{5}{6}$
4	2	10	0.5
5	0.6	11	12
6	$\frac{4}{7}$	12	18

Trigonometrical identities

Consider the two expressions $\sin 2x$ and $2 \sin x \cos x$. These two expressions have the same value for every value of x. Try evaluating the expressions for, say, $x = 25°$, $60°$, 1.2 radians.

If two expressions are equal for all values of the variables used, then we say the expressions are **identical**. Hence $\sin 2x$ and $2 \sin x \cos x$ are identical, that is equal for all values of x. A statement such as $\sin 2x = 2 \sin x \cos x$ is known as an **identity**. In this block we provide a table of important identities and show how expressions involving the trigonometrical ratios can be simplified using them.

There are several commonly used trigonometrical identities. These are listed in Table 5.1.

Table 5.1

$$\sin^2 A + \cos^2 A = 1$$

$$\frac{\sin A}{\cos A} = \tan A$$

$$\sin(A + B) = \sin A \cos B + \sin B \cos A$$

$$\sin(A - B) = \sin A \cos B - \sin B \cos A$$

$$\cos(A + B) = \cos A \cos B - \sin A \sin B$$

$$\cos(A - B) = \cos A \cos B + \sin A \sin B$$

$$\sin 2A = 2 \sin A \cos A$$

$$\cos 2A = 1 - 2 \sin^2 A = 2 \cos^2 A - 1 = \cos^2 A - \sin^2 A$$

$$\tan(A + B) = \frac{\tan A + \tan B}{1 - \tan A \tan B}$$

$$\tan(A - B) = \frac{\tan A - \tan B}{1 + \tan A \tan B}$$

$$2 \sin A \cos B = \sin(A + B) + \sin(A - B)$$

$$2 \cos A \cos B = \cos(A + B) + \cos(A - B)$$

$$2 \sin A \sin B = \cos(A - B) - \cos(A + B)$$

$$\sin^2 A = \tfrac{1}{2}(1 - \cos 2A)$$

$$\cos^2 A = \tfrac{1}{2}(1 + \cos 2A)$$

Note that we write $\sin^2 A$ to mean $(\sin A)^2$. Similarly, $\cos^2 A$ is the accepted notation for $(\cos A)^2$. The identities can be used to simplify trigonometrical expressions. The first entry in the table is particularly important and should be remembered.

Key point

$$\sin^2 A + \cos^2 A = 1$$

Choose any angle A for yourself and verify that this identity is true.

Example 5.1
Use trigonometrical identities to simplify $\cos A \tan A$.

Solution
We note that $\tan A = \dfrac{\sin A}{\cos A}$ and so

$$\cos A \tan A = \cos A \frac{\sin A}{\cos A}$$
$$= \sin A$$

Hence $\cos A \tan A$ is identical to $\sin A$.

Example 5.2
(a) Use the identity for $\sin (A + B)$ to show that $\sin 2A = 2 \sin A \cos A$.
(b) Use the identity for $\cos (A + B)$ to show that $\cos 2A = \cos^2 A - \sin^2 A$.

Solution
(a) We use the identity

$$\sin(A + B) = \sin A \cos B + \sin B \cos A$$

A special case of this identity occurs when $B = A$. We then have

$$\sin 2A = \sin A \cos A + \sin A \cos A$$

that is

$$\sin 2A = 2 \sin A \cos A$$

(b) We use the identity

$$\cos(A + B) = \cos A \cos B - \sin A \sin B$$

A special case of this identity occurs when $B = A$. We then have

$$\cos 2A = \cos A \cos A - \sin A \sin A$$

that is

$$\cos 2A = \cos^2 A - \sin^2 A$$

Example 5.3
Show that $\cos(-\theta) = \cos\theta$.

Solution
We use the identity for $\cos(A - B)$ with $A = 0$ and $B = \theta$.
   Now from Table 5.1

$$\cos(A - B) = \cos A \cos B + \sin A \sin B$$

With $A = 0$ and $B = \theta$ this becomes

$$\cos(-\theta) = \qquad \cos 0 \cos\theta + \sin 0 \sin\theta$$

Noting that $\sin 0 = 0$ and $\cos 0 = 1$ this simplifies to

$$ \qquad\qquad\qquad\qquad\qquad\qquad \cos\theta$$

Hence $\cos(-\theta) = \cos\theta$ as required.

Example 5.4
Show that

$$\tan(\theta + 45°) = \frac{1 + \tan\theta}{1 - \tan\theta}$$

Solution
We use the identity

$$\tan(A + B) = \frac{\tan A + \tan B}{1 - \tan A \tan B}$$

Putting $A = \theta$ and $B = 45°$ gives

$$\tan(\theta + 45°) = \frac{\tan\theta + \tan 45°}{1 - \tan\theta \tan 45°}$$

Now $\tan 45° = 1$ and so

$$\tan(\theta + 45°) = \frac{\tan\theta + 1}{1 - \tan\theta}$$
$$= \frac{1 + \tan\theta}{1 - \tan\theta}$$

Example 5.5
Simplify

$$\frac{\sin^3 A}{\cos A} + \sin A \cos A$$

Solution
We write the expression with a common denominator of $\cos A$:

$$\frac{\sin^3 A}{\cos A} + \sin A \cos A = \frac{\sin^3 A + \sin A \cos^2 A}{\cos A}$$

Now

$$\frac{\sin^3 A + \sin A \cos^2 A}{\cos A} = \frac{\sin A(\sin^2 A + \cos^2 A)}{\cos A}$$

$$= \frac{\sin A}{\cos A} \quad \text{since } \sin^2 A + \cos^2 A = 1$$

$$= \tan A$$

Hence $\dfrac{\sin^3 A}{\cos A} + \sin A \cos A$ simplifies to tan A.

Example 5.6

Simplify

$$\frac{\sin 2A \sin A}{2 \cos A} + \cos 2A$$

Solution

From Table 5.1 we have $\sin 2A = 2 \sin A \cos A$ and so

$$\frac{\sin 2A \sin A}{2 \cos A} = \qquad\qquad\qquad \frac{2 \sin A \cos A \sin A}{2 \cos A} = \sin^2 A$$

Also from Table 5.1, $\cos 2A = \cos^2 A - \sin^2 A$. Hence

$$\frac{\sin 2A \sin A}{2 \cos A} + \cos 2A$$

$$= \qquad\qquad\qquad\qquad\qquad \sin^2 A + \cos^2 A - \sin^2 A = \cos^2 A$$

Example 5.7 Electrical Engineering – Amplitude modulation

Amplitude modulation is a technique which allows a signal of a certain frequency (the signal frequency) to be transmitted at a different frequency (the carrier frequency). (Note that the term **frequency** is explained in Block 7.)

The signal can be represented by the cosine wave $y_s = S \cos \omega_s t$. The carrier can be represented by $y_c = C \cos \omega_c t$. The modulated signal is given by the product $y_c y_s$. Use a trigonometrical identity to show that the modulated signal can be written as the sum of two cosine waves.

Solution

The modulated signal is

$$y_c y_s = (C \cos \omega_c t) \times (S \cos \omega_s t)$$

$$= CS(\cos \omega_c t \cos \omega_s t)$$

Using the identity $2 \cos A \cos B = \cos(A + B) + \cos(A - B)$ (Table 5.1), we can write

$$y_c y_s = \frac{1}{2} CS(\cos(\omega_c + \omega_s)t + \cos(\omega_c - \omega_s)t)$$

which is the sum of two cosine waves, oscillating at different frequencies from the original signal. This result enables the design of an antenna which can be used to transmit the original signal efficiently.

End of block exercises

1 From Table 5.1 we have

$$\cos(A - B) = \cos A \cos B + \sin A \sin B$$

Verify this identity when $A = 80°$ and $B = 30°$.

2 Verify the identity

$$2 \sin A \sin B = \cos(A - B) - \cos(A + B)$$

with $A = 50°$ and $B = 15°$.

3 Verify

$$\tan(A - B) = \frac{\tan A - \tan B}{1 + \tan A \tan B}$$

with $A = 65°$ and $B = 30°$.

4 Show $\sin(-\theta) = -\sin \theta$.

5 Show $\sin\left(\frac{\pi}{2} - \theta\right) = \cos \theta$.

6 Show $\cos\left(\frac{\pi}{2} - \theta\right) = \sin \theta$.

7 Show $\sin\left(\theta + \frac{\pi}{2}\right) = \cos \theta$.

8 Show $\cos\left(\theta + \frac{\pi}{2}\right) = -\sin \theta$.

9 Show $\sin(180° - \theta) = \sin \theta$.

10 Show $\cos(180° - \theta) = -\cos \theta$.

11 Show $\tan(180° - \theta) = -\tan \theta$.

12 Show $\sin(180° + \theta) = -\sin \theta$.

13 Show $\cos(180° + \theta) = -\cos \theta$.

14 Show $\tan(180° + \theta) = \tan \theta$.

15 Show $\sin(360° - \theta) = -\sin \theta$.

16 Show $\cos(360° - \theta) = \cos \theta$.

17 Show $\tan(360° - \theta) = -\tan \theta$.

18 Show $\sin 3A = 3 \sin A \cos^2 A - \sin^3 A$.

19 Show $\cos 3A = 4 \cos^3 A - 3 \cos A$.

20 Show $\sin 4A = 4 \sin A \cos A(\cos^2 A - \sin^2 A)$.

21 Show $\cos 4A = 8 \cos^4 A - 8 \cos^2 A + 1$.

22 Simplify

$$\sin A \cos A \tan A + \frac{2 \sin A \cos^3 A}{\sin 2A}$$

23 Simplify

$$\tan A + \frac{1}{\tan A}$$

24 Show

$$\frac{\sin 3A}{\sin 2A} = 2 \cos A - \frac{1}{2 \cos A}$$

Solutions to exercises

22 1

23 $\dfrac{1}{\cos A \sin A}$ which may be written as $\dfrac{2}{\sin 2A}$

Trigonometrical equations

6.1 Introduction

We examine the method of solving equations of the form $\sin\theta = k$, $\cos\theta = k$ and $\tan\theta = k$, where k is a constant. Because $\sin\theta$, $\cos\theta$ and $\tan\theta$ are periodic functions, then there are an infinite number of solutions of trigonometrical equations. Often the values of θ are restricted to a limited range. This restriction means there are then only a finite number of solutions.

6.2 Notation

If

$$\sin\theta = k$$

then we write

$$\theta = \sin^{-1}k$$

This is read as 'θ equals the inverse sine of k'. Similarly if $\cos\theta = k$ then $\theta = \cos^{-1}k$ and if $\tan\theta = k$ then $\theta = \tan^{-1}k$. The inverse functions $\sin^{-1}$, $\cos^{-1}$ and $\tan^{-1}$ are available on scientific calculators. Sometimes $\sin^{-1}$ is written as 'inv sin' or 'arcsin'.

Example 6.1
Given $\sin\theta = 0.3214$ find θ using a scientific calculator.

Solution

$$\sin\theta = 0.3214$$

and so

$$\theta = \sin^{-1}(0.3214)$$
$$= 18.75°$$

using a scientific calculator.

Example 6.2
Given $\sin\theta = -0.2000$ find θ using a scientific calculator.

Solution
$$\sin\theta = -0.2000$$

$$\theta = \qquad\qquad\qquad\qquad\qquad\qquad \sin^{-1}(-0.2000) = -11.54°$$

Example 6.3
Given $\cos \theta = -0.6132$ find θ using a scientific calculator.

Solution

$\cos \theta = -0.6132$

$$\theta = \qquad \cos^{-1}(-0.6132) = 127.82°$$

A scientific calculator returns only one solution to a trigonometrical equation. As $\sin \theta$, $\cos \theta$ and $\tan \theta$ are periodic functions, then trigonometrical equations involving these functions have many solutions. The following section illustrates how these equations are solved.

6.3   Solving trigonometrical equations

The method of solution is illustrated by examples.

Example 6.4
Solve

$$\sin \theta = 0.7215 \quad 0° \leq \theta \leq 360°$$

Solution

Figure 6.1 illustrates a graph of $y = \sin \theta$ for $0° \leq \theta \leq 360°$. We require solutions between $0°$ and $360°$. The values of θ such that $\sin \theta = 0.7215$ are marked as A and B.

From Figure 6.1 we see there are two solutions, one between $0°$ and $90°$, that is the first quadrant, and one between $90°$ and $180°$, that is the second quadrant. Using a scientific calculator gives

$$\theta = \sin^{-1}(0.7215) = 46.18°$$

Figure 6.1
$\sin \theta = 0.7215$ has a solution in the first quadrant and a solution in the second quadrant.

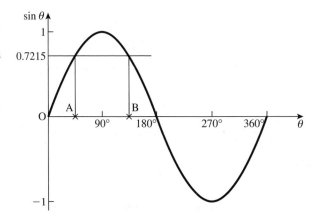

This is the solution represented by point A. By symmetry, the solution at B is given by

$$\theta = 180° - 46.18° = 133.82°$$

The required solutions are $\theta = 46.18°, 133.82°$.

Example 6.5
Solve

$$\cos x = 0.3456 \quad 0 \leq x \leq 2\pi$$

Solution
A graph of $y = \cos x$ for $0 \leq x \leq 2\pi$ is shown in Figure 6.2. Note that in this example angles are measured in radians.

Figure 6.2
$\cos x = 0.3456$
has a solution in
the first quadrant
and a solution in
the fourth
quadrant.

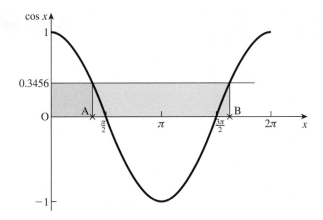

From Figure 6.2 there are two solutions, A and B. Solution A is in the first quadrant, that is between 0 and $\frac{\pi}{2}$, and solution B is in the fourth quadrant, that is between $\frac{3\pi}{2}$ and 2π.

We have

$$\cos x = 0.3456$$

Using a scientific calculator we find

$$x = \cos^{-1}(0.3456) = 1.2179$$

This is the solution represented by A. Using symmetry the solution in the fourth quadrant is found from

$$x = 2\pi - 1.2179 = 5.0653$$

The required solutions are $x = 1.2179, 5.0653$.

Example 6.6
Solve

$$\tan t = -1.3000 \quad 0 \leq t \leq 2\pi$$

Solution

Figure 6.3 illustrates $y = \tan t$ and marks the solution points A and B.

Point A lies between ░░░░░░░░░░ $\dfrac{\pi}{2}$ and π

Point B lies between ░░░░░░░░░░ $\dfrac{3\pi}{2}$ and 2π

Using a scientific calculator we have

$$\tan t = -1.3000$$

$$t = \;\;░░░░░░░░░░░░░░░\;\; \tan^{-1}(-1.3000) = -0.9151$$

Clearly the value of t returned by a calculator is not within the range of values of interest, namely 0 to 2π.

Recall from Block 3 that adding π radians to an angle does not change its tangent, and so another solution is given by $-0.9151 + \pi = 2.2265$. This is the solution represented in the second quadrant by point A.

Figure 6.3
$\tan t = -1.3000$
has solutions at
A and B.

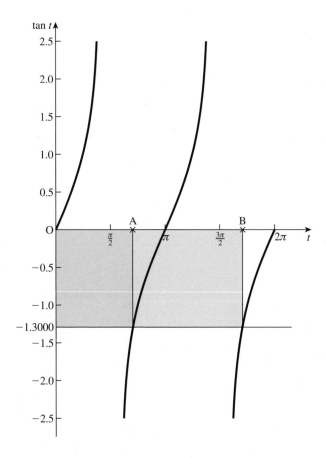

By examining the graph in Figure 6.3, we see the solution at point B is π radians above point A. So the solution at point B is given by $2.2265 + \pi = 5.3681$.

The required solutions are $t = 2.2265, 5.3681$.

Example 6.7

Solve

$$\sin \theta = -0.6500 \quad 0 \le \theta \le 2\pi$$

Solution

From Figure 6.4, $\sin \theta = -0.6500$ has solutions at A and B.

Figure 6.4
$\sin \theta = -0.6500$
has solutions at
A and B.

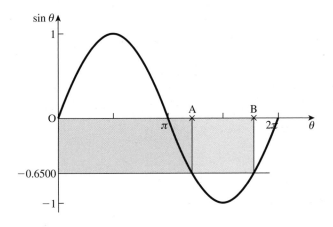

Point A is between π and $\dfrac{3\pi}{2}$

Point B is between $\dfrac{3\pi}{2}$ and 2π

We have

$$\sin \theta = -0.6500$$

and so

$\theta = $ $\sin^{-1}(-0.6500) = -0.7076$

Clearly this solution is not within the range of interest.

Recall from Block 3 that adding $2\pi(360°)$ to an angle does not change its sine. Hence $-0.7076 + 2\pi = 5.5756$ is a solution.

This solution is represented by point   B

Using the symmetry of Figure 6.4 we see that point A is above π by the same amount that B is below 2π.

Hence the solution at point A is given by

$$\pi + 0.7076 = 3.8492$$

The required solutions are $\theta = 3.8492, 5.5756$.

Example 6.8

Solve

$$\cos \theta = -0.4215 \quad 0° \leq \theta \leq 360°$$

Solution

Sketch $y = \cos \theta$ and mark the points where $\cos \theta = -0.4215$.

 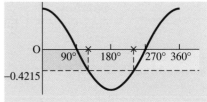

There are two solutions.

One is between   and the other is between $90°$ and $180°$

                                                                                                                                            $180°$ and $270°$

We have

$$\cos \theta = -0.4215$$

and so

$\theta = $ $\cos^{-1}(-0.4215) = 114.93°$

Using the symmetry of the cosine curve, the other solution is

$\theta = $ $360° - 114.93° = 245.07°$

Exercises

1   Solve
    (a) $\sin \theta = 0.3510, 0° \leq \theta \leq 360°$
    (b) $\sin \theta = 0.4161, 0 \leq \theta \leq 2\pi$
    (c) $\cos t = -0.3778, 0 \leq t \leq 2\pi$
    (d) $\cos x = 0.7654, 0° \leq x \leq 360°$
    (e) $\tan y = 1.7136, 0° \leq y \leq 360°$
    (f) $\tan y = -0.3006, 0° \leq y \leq 360°$

Solutions to exercises

1   (a) $20.55°$, $159.45°$ (b) $3.5707, 5.8540$ (e) $59.73°$, $239.73°$ (f) $163.27°$, $343.27°$
     (c) $1.9582, 4.3250$ (d) $40.06°$, $319.94°$

6.4 Further trigonometrical equations

We now consider more complex trigonometrical equations.

Example 6.9
Solve

$$\sin 3\theta = 0.7215 \qquad 0° \leq \theta \leq 360°$$

Solution
We introduce a new variable, z, defined by $z = 3\theta$. As θ moves from $0°$ to $360°$ then z moves through $0°$ to $1080°$. Hence the problem may be written as one of finding z given

$$\sin z = 0.7215 \qquad 0° \leq z \leq 1080°$$

Using Example 6.4, the solutions between $0°$ and $360°$ are

$$z = 46.18°, 133.82°$$

Recognising that adding $360°$ to an angle does not change its sine, we see that the solutions between $360°$ and $720°$ are

$$z = 46.18° + 360°, 133.82° + 360°$$
$$= 406.18°, 493.82°$$

Similarly the solutions between $720°$ and $1080°$ are

$$z = 406.18° + 360°, 493.82° + 360°$$
$$= 766.18°, 853.82°$$

Hence

$$z = 46.18°, 133.82°, 406.18°, 493.82°, 766.18°, 853.82°$$

Recall that $z = 3\theta$, that is $\theta = \frac{z}{3}$, and so

$$\theta = 15.39°, 44.61°, 135.39°, 164.61°, 255.39°, 284.61°$$

Example 6.10
Solve

$$\cos \frac{\theta}{2} = -0.4215 \qquad 0° \leq \theta \leq 360°$$

Solution

We define $z = $ ⬛ $\dfrac{\theta}{2}$

As θ moves from $0°$ to $360°$, then z moves from $0°$ to $180°$. Hence the problem may be recast as one of finding z given

$$\cos z = -0.4215 \qquad 0° \leq z \leq 180°$$

In Example 6.8 we solved

$$\cos \theta = -0.4215 \qquad 0° \leq \theta \leq 360°$$

and found $\theta = 114.93°, 245.07°$. Since z is restricted to $0°$ to $180°$, then the only solution is $z = 114.93°$. Recall that $z = \frac{\theta}{2}$, that is $\theta = 2z$, and so

$$\theta = 2 \times 114.93° = 229.86°$$

Example 6.11
Solve

$$\tan 2\theta = -1.4213 \qquad 0° \leq \theta \leq 360°$$

Solution
We define $z = \boxed{}$ $\qquad 2\theta$

As θ ranges through $0°$ to $360°$ then z ranges through $\boxed{}$ $0°$ to $720°$
Hence the problem becomes one of solving

$$\tan z = -1.4213 \qquad 0° \leq z \leq 720°$$

Using a scientific calculator we see

$z = \boxed{}$ $\qquad \tan^{-1}(-1.4213) = -54.87°$

This is outside the range of interest.
 Adding $180°$ to an angle does not change the value of its tangent.
 Hence the required solutions are

$z = \boxed{}$ $\qquad 125.13°, 305.13°, 485.13°, 665.13°$

Recall that $z = 2\theta$, that is $\theta = \frac{z}{2}$, and so

$\theta = \boxed{}$ $\qquad 62.57°, 152.57°, 242.57°, 332.57°$

Exercise

1 Solve the following:
   (a) $\sin 3\theta = 0.7614, 0° \leq \theta \leq 360°$

   (b) $\tan \dfrac{\theta}{2} = 1.0137, 0° \leq \theta \leq 360°$

   (c) $\cos 2\theta = -0.8314, 0° \leq \theta \leq 360°$

   (d) $\sin \dfrac{2\theta}{3} = -0.5000, 0° \leq \theta \leq 360°$

   (e) $\cos \dfrac{\theta}{3} = 0.4162, 0° \leq \theta \leq 360°$

   (f) $\tan \dfrac{4\theta}{3} = -1, 0° \leq \theta \leq 360°$

Solution to exercise

1 (a) $16.53°, 43.47°, 136.53°, 163.47°, 256.53°,$
        $283.47°$
    (b) $90.78°$

    (c) $73.12°, 106.88°, 253.12°, 286.88°$
    (d) $315°$ (e) $196.22°$ (f) $101.25°, 236.25°$

End of block exercises

Solve the following trigonometrical equations:

1 $\sin\theta = 0.7506, 0° \leq \theta \leq 360°$

2 $\sin x = 0.2913, 0 \leq x \leq 2\pi$

3 $3\sin\theta = 2, 0° \leq \theta \leq 360°$

4 $\sin t = -0.3490, 0 \leq t \leq 2\pi$

5 $2\sin x = -1.4600, 0 \leq x \leq 2\pi$

6 $3\sin\theta + 1.5 = 0, 0 \leq \theta \leq 2\pi$

7 $\cos t = 0.4500, 0 \leq t \leq 2\pi$

8 $\cos t = 0.8020, 0 \leq t \leq 2\pi$

9 $3\cos\theta - 1.4216 = 0, 0° \leq \theta \leq 360°$

10 $5\cos x + 3 = 0, 0 \leq x \leq 2\pi$

11 $\tan\theta = 0.4906, 0° \leq \theta \leq 360°$

12 $\tan t = 2.1630, 0 \leq t \leq 4\pi$

13 $\tan z = -0.1421, 0 \leq z \leq 4\pi$

14 $4\tan\phi + 2.5 = 0, 0° \leq \phi \leq 540°$

15 $\sin 4\theta = 0.6000, 0° \leq \theta \leq 180°$

16 $\sin 3\theta = -0.2556, 0° \leq \theta \leq 180°$

17 $\cos\dfrac{t}{3} = -0.4000, 0 \leq t \leq 3\pi$

18 $2\cos 2\theta = -1.3146, 0° \leq \theta \leq 360°$

19 $\tan\dfrac{2\theta}{3} = -1, -180° \leq \theta \leq 180°$

20 $\tan(\theta - 20°) = 0.5614, 0° \leq \theta \leq 360°$

21 $\sin(2\theta + 10°) = 0.7516, 0° \leq \theta \leq 360°$

22 $\cos\left(\dfrac{\theta}{2} - 20°\right) = -0.5516, 0° \leq \theta \leq 360°$

23 $\tan\left(\dfrac{\theta + 30°}{4}\right) = -1.6319, -270° \leq \theta \leq 270°$

Solutions to exercises

1 48.64°, 131.36°

2 0.2956, 2.8460

3 41.81°, 138.19°

4 3.4981, 5.9267

5 3.9599, 5.4649

6 3.6652, 5.7596

7 1.1040, 5.1792

8 2.5014, 3.7818

9 61.71°, 298.29°

10 2.2143, 4.0689

11 26.13°, 206.13°

12 1.1377, 4.2793, 7.4209, 10.5625

13 3.0004, 6.1420, 9.2836, 12.4252

14 147.99°, 327.99°, 507.99°

15 9.22°, 35.78°, 99.22°, 125.78°

16 64.94°, 115.06°

17 5.9469°

18 65.55°, 114.45°, 245.55°, 294.45°

19 -67.5°

20 49.31°, 229.31°

21 19.36°, 60.64°, 199.36°, 240.64°

22 286.95°

23 -264.00°

Engineering waves

Often voltages and currents vary with time, and can be modelled by sine and cosine functions. Important parameters such as amplitude, frequency, period and phase are used in the description of these waves. These terms are described in the following sections.

Two or more waves may be added together, producing a new single wave. The method of doing this is explained and illustrated.

The functions $y = \sin\theta$ and $y = \cos\theta$ were described in Block 4. There we saw that the graphs of these functions look like waves. The angle θ may be measured in degrees or radians.

Voltages and currents encountered in electric circuits usually vary with time, t. Hence we consider sine and cosine waves in which the independent variable is t. For example, consider $y = \sin t$. As t increases from 0 seconds to 2π seconds, one complete cycle is produced. This is illustrated in Figure 7.1.

Figure 7.1
As time t varies from 0 to 2π seconds one complete cycle is produced.

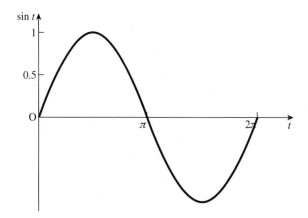

7.3 Amplitude of a wave

We introduced the amplitude in Block 4 and simply recap here. Consider the function $y = A \sin t$ for $A > 0$. Then A is the amplitude of the wave. This is the highest value attained by the wave. Similarly the amplitude of $y = A \cos t$ is A.

Key point   The **amplitude** of both $y = A \sin t$ and $y = A \cos t$ is A.

Example 7.1
State the amplitude of each of the following functions:
(a) $y = 3 \sin t$
(b) $y = \frac{2}{3} \cos t$
(c) $y = -\sin t$

Solution
(a) amplitude $= 3$
(b) amplitude $= \frac{2}{3}$
(c) Noting that the amplitude of a wave is the largest value attained, the amplitude of $-\sin t$ is 1.

7.4 Angular frequency of a wave

Consider the wave $y = A \sin \omega t$. We call ω the **angular frequency** of the wave. The units of ω are radians per second. Noting that t is measured in seconds, then ωt has units of radians. For example, $y = \sin 4t$ has an angular frequency of 4 radians per second. Note that $y = \sin t$ has an angular frequency of 1 radian per second. In like manner the angular frequency of $y = A \cos \omega t$ is ω radians per second.

Key point   The **angular frequency** of $y = A \sin \omega t$ and $y = A \cos \omega t$ is ω radians per second.

Note that the amplitude, A, has no effect upon the angular frequency of a function.

Example 7.2
State the angular frequency of each of the following waves:
(a) $y = 5 \sin 3t$
(b) $y = 7 \cos \dfrac{t}{2}$
(c) $y = \cos \dfrac{2t}{3}$

Solution
(a) Comparing $5 \sin 3t$ with $A \sin \omega t$, we see that $\omega = 3$: that is, the angular frequency is 3 radians per second.

(b) angular frequency = [_____] $\frac{1}{2}$ radian per second

(c) angular frequency = [_____] $\frac{2}{3}$ radians per second

---

Exercises

1   State (i) the amplitude and (ii) the angular frequency of the following waves:
(a) $y = 2 \sin 5t$ (b) $y = 3 \cos 6t$

(c) $y = \sin \dfrac{t}{2}$ (d) $y = \cos \dfrac{4t}{3}$ (e) $y = \dfrac{3}{2} \sin \dfrac{2t}{3}$

(f) $y = -4 \sin \pi t$

Solutions to exercises

1   (a) 2, 5 (b) 3, 6 (c) $1, \frac{1}{2}$ (d) $1, \frac{4}{3}$ (e) $\frac{3}{2}, \frac{2}{3}$ (f) 4, π

7.5   Period of a wave

The time taken to complete one full cycle is called the **period** of the wave. It is closely connected to the angular frequency of the wave.

Consider $y = A \sin \omega t$. When $t = 0$ seconds, then $\omega t = 0$ radians. When $t = \dfrac{2\pi}{\omega}$ seconds, then $\omega t = \omega \left(\dfrac{2\pi}{\omega} \right) = 2\pi$ radians. Hence as t increases by $\dfrac{2\pi}{\omega}$ seconds, the angle, ωt, increases by 2π radians. Recall from Block 4 that a sine function completes one full cycle as the angle increases by 2π radians. Hence $y = A \sin \omega t$ completes a full cycle as t increases by $\dfrac{2\pi}{\omega}$ seconds: that is, the period of y is $\dfrac{2\pi}{\omega}$ seconds. Similarly the period of $y = A \cos \omega t$ is also $\dfrac{2\pi}{\omega}$ seconds. The period is denoted by T.

Key point

The **period** of both $y = A \sin \omega t$ and $y = A \cos \omega t$ is given by $T = \dfrac{2\pi}{\omega}$.

Example 7.3
State the period of each of the following functions:
(a) $y = 3 \sin 6t$
(b) $y = 5.6 \cos \pi t$
(c) $y = 50 \sin 100 \pi t$

Solution

(a) Here $\omega = 6$ and so

$$T = \frac{2\pi}{6}$$

$$= \frac{\pi}{3}$$

The period is $\frac{\pi}{3}$ seconds.

(b) Here $\omega = \pi$ and so

$$T = \boxed{}$$

$$\frac{2\pi}{\pi} = 2$$

The period is 2 seconds.

(c) Here $\omega = \boxed{}$ and so

100π

$$T = \boxed{}$$

$$\frac{2\pi}{100\pi} = 0.02$$

The period is 0.02 seconds.

Example 7.4 Electrical Engineering – Oscilloscope trace

Figure 7.2 shows an oscilloscope trace of a sine wave. State the equation of the wave.

Figure 7.2
Oscilloscope trace.

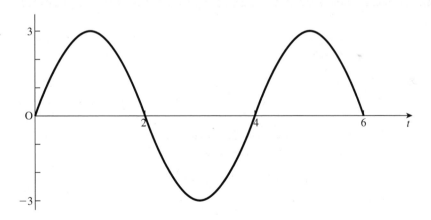

Solution

The wave has an equation of the form

$$y = A \sin \omega t$$

The maximum value of the wave is 3: that is, the amplitude, A, is 3. A full cycle is completed in 4 seconds: that is, $T = 4$. So

$$T = \frac{2\pi}{\omega}$$

$$= 4$$

from which

$$\omega = \frac{\pi}{2}$$

The equation of the wave is $y = 3 \sin \frac{\pi t}{2}$.

Example 7.5 Electrical Engineering – Oscilloscope trace

Figure 7.3 shows an oscilloscope trace of a cosine wave. State the equation of the wave.

Figure 7.3
Oscilloscope trace of a cosine wave.

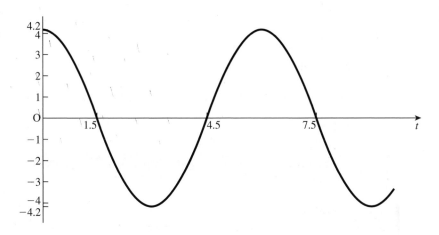

Solution
The wave has an equation of the form

$$y = A \cos \omega t$$

The highest value of the wave is ⬚   4.2

Hence the value of the amplitude is 4.2.

A cycle is completed in ⬚ seconds, and so   6

$$T = \frac{2\pi}{\omega}$$
$$= 6$$

Hence

$$\omega = \boxed{} \qquad\qquad \frac{2\pi}{6} = \frac{\pi}{3}$$

The equation of the wave is therefore ⬚ $4.2 \cos \frac{\pi t}{3}$

7.6 The frequency of a wave

The **frequency**, f, of a wave is the number of cycles completed in 1 second. It is measured in hertz (Hz). One hertz is one cycle per second.

Consider again the function $y = A \sin \omega t$. The period is $\dfrac{2\pi}{\omega}$ seconds: that is, one cycle is completed in $\dfrac{2\pi}{\omega}$ seconds. Hence ω cycles are completed in 2π seconds and $\dfrac{\omega}{2\pi}$ cycles are completed in 1 second. Thus we have

Key point

$$\text{frequency} \; f = \frac{\omega}{2\pi}$$

We note that

$$\text{period } T = \frac{2\pi}{\omega}, \quad \text{frequency } f = \frac{\omega}{2\pi}$$

and so

Key point

$$T = \frac{1}{f}$$

Example 7.6
State the period and frequency of the following waves:
(a) $y = 2 \sin 4t$
(b) $y = 3 \cos 2t$
(c) $y = \sin \pi t$

Solution
(a) Comparing $2 \sin 4t$ with $A \sin \omega t$ we see that $\omega = 4$. Hence the frequency, f, is found from

$$f = \frac{\omega}{2\pi}$$
$$= \frac{4}{2\pi}$$
$$= \frac{2}{\pi}$$
$$= 0.6366 \text{ Hz}$$

So 0.6366 cycles are completed every second. The period, T, is found from

$$T = \frac{1}{f}$$
$$= \frac{\pi}{2}$$
$$= 1.5708$$

It takes 1.5708 seconds to complete one cycle.

(b) Here $\omega = 2$ and so

$$f = \boxed{} \qquad \frac{\omega}{2\pi} = \frac{2}{2\pi} = \frac{1}{\pi} = 0.3183 \text{ Hz}$$

Thus, 0.3183 of a cycle is completed every second.
The period, T, is found using

$$T = \boxed{} \qquad \frac{1}{f} = \pi = 3.1416$$

It takes π seconds to complete one cycle.

(c) Here $\omega = \pi$ and so

$$f = \boxed{} \qquad \frac{\omega}{2\pi} = \frac{\pi}{2\pi} = \frac{1}{2} \text{ Hz}$$

and

$$T = \boxed{} \qquad \frac{1}{f} = 2$$

Thus, 0.5 of a cycle is completed each second. It takes 2 seconds to complete one full cycle.

Exercises

1 State (i) the period and (ii) the frequency of the following waves:
(a) $y = 6 \sin 4t$ (b) $y = \cos 3t$
(c) $y = -4 \sin t$ (d) $y = \frac{1}{2} \cos 100\pi t$
(e) $y = \pi \sin 1.5t$

Solutions to exercises

1 (a) $\dfrac{\pi}{2}, \dfrac{2}{\pi}$ (b) $\dfrac{2\pi}{3}, \dfrac{3}{2\pi}$ (c) $2\pi, \dfrac{1}{2\pi}$ (d) 0.02, 50 (e) $\dfrac{4\pi}{3}, \dfrac{3}{4\pi}$

7.7 Phase and time displacement of a wave

We now introduce waves of the form $y = A \sin(\omega t + \alpha)$ and $y = A \cos(\omega t + \alpha)$. Introducing α has the effect of moving the wave to either the left or the right. Figure 7.4 shows graphs of $y = \sin 2t$ and $y = \sin(2t + 1)$.

From Figure 7.4 we note that the peak of $y = \sin(2t + 1)$ occurs 0.5 seconds before the peak of $y = \sin 2t$. We say $\sin(2t + 1)$ **leads** $\sin 2t$ by 0.5 seconds. The quantity, 0.5 seconds, is known as the **time displacement** of $y = \sin(2t + 1)$.

Figure 7.4
The waves
$y = \sin 2t$ and
$y = \sin(2t + 1)$.

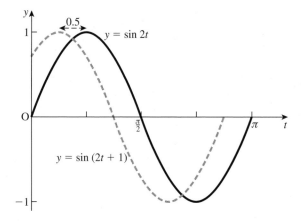

Figure 7.5
The waves
$y = \sin 2t$ and
$y = \sin(2t - 1)$.

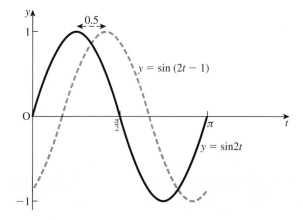

Figure 7.5 shows graphs of $y = \sin 2t$ and $y = \sin(2t - 1)$.

The peak of $y = \sin(2t - 1)$ is reached 0.5 seconds after the peak of $y = \sin 2t$. We say $\sin(2t - 1)$ **lags** $\sin 2t$ by 0.5 seconds. The time displacement of $\sin(2t - 1)$ is -0.5 seconds.

These examples lead us to the following general treatment.

Consider the wave $y = A \sin(\omega t + \alpha)$. The angle α is called the **phase angle**, or simply the **phase**. We note that y may be written as

$$y = A \sin\left[\omega\left(t + \frac{\alpha}{\omega}\right)\right]$$

We call $\dfrac{\alpha}{\omega}$ the **time displacement** of the wave.

Key point

> The **phase** of $y = A \sin(\omega t + \alpha)$ is α radians.
>
> The **time displacement** of $y = A \sin(\omega t + \alpha)$ is $\dfrac{\alpha}{\omega}$ seconds.

In similar manner the phase and time displacement of $y = A \cos(\omega t + \alpha)$ are α radians and $\dfrac{\alpha}{\omega}$ seconds.

Example 7.7

State the phase and time displacement of each of the following waves:

(a) $y = \sin(2t + 1)$

(b) $y = \sin(2t - 1)$

(c) $y = 3\cos\left(\dfrac{t}{2} + 3\right)$

Solution

(a) Here $\omega = 2$, $\alpha = 1$ and so the phase is 1 radian and the time displacement is

$$\frac{\alpha}{\omega} = \frac{1}{2} = 0.5 \text{ seconds}$$

This wave is illustrated in Figure 7.4.

(b) Here $\omega = 2$, $\alpha = -1$ and so the phase is -1 radian and the time displacement is

$$\frac{\alpha}{\omega} = \frac{-1}{2} = -0.5 \text{ seconds}$$

This wave is illustrated in Figure 7.5.

(c) Here $\omega = \boxed{}$ and $\alpha = \boxed{}$ 0.5, 3

and so the phase is 3 radians and the time displacement is $\dfrac{3}{(0.5)} = 6$ seconds.

Exercises

1 State (i) the phase and (ii) the time displacement of the following waves:

(a) $y = 4\sin(3t + 12)$

(b) $y = \sin(4t - 6)$

(c) $y = 2\cos(t + 0.5)$

(d) $y = \sin(0.5t - 2)$

(e) $y = 3\cos\left(\dfrac{t + 4}{3}\right)$

Solutions to exercises

1   (a) 12, 4   (b) -6, -1.5 (c) 0.5, 0.5   (d) -2, -4 (e) $\frac{4}{3}$, 4

7.8   Adding waves of the same frequency

Waves of the same angular frequency may be added together to form a new wave. The new wave has the same angular frequency as the original waves. The trigonometrical identities (see Block 5) are used, especially the formulae for $\sin(A \pm B)$ and $\cos(A \pm B)$. The following examples illustrate the technique.

Example 7.8

Express $2 \sin 3t + 6 \cos 3t$ in the form $A \cos(\omega t + \alpha)$, $\alpha \geq 0$.

Solution

The angular frequency of both $2 \sin 3t$ and $6 \cos 3t$ is 3 radians per second. Hence, on adding the waves, the angular frequency of the sum is also 3 radians per second and so $\omega = 3$. Recall the trigonometrical identity from Block 5:

$$\cos(A + B) = \cos A \cos B - \sin A \sin B$$

and so

$$
\begin{aligned}
2 \sin 3t + 6 \cos 3t &= A \cos(\omega t + \alpha) \\
&= A \cos(3t + \alpha) \\
&= A(\cos 3t \cos \alpha - \sin 3t \sin \alpha) \\
&= (A \cos \alpha) \cos 3t - (A \sin \alpha) \sin 3t \\
&= -(A \sin \alpha) \sin 3t + (A \cos \alpha) \cos 3t
\end{aligned}
$$

We now compare the coefficients of $\sin 3t$ and $\cos 3t$ on the left-hand and right-hand sides.

Comparing the $\sin 3t$ coefficients we have

$$2 = -A \sin \alpha \tag{1}$$

Comparing the $\cos 3t$ coefficients we have

$$6 = A \cos \alpha \tag{2}$$

Equations (1) and (2) must be solved for A and α. To find A, α is eliminated. This is accomplished by squaring equations (1) and (2) and then adding the results.

Squaring equation (1) gives

$$4 = A^2 \sin^2 \alpha \tag{3}$$

Squaring equation (2) gives

$$36 = A^2 \cos^2 \alpha \tag{4}$$

Adding equations (3) and (4) gives

$$
\begin{aligned}
40 &= A^2 \sin^2 \alpha + A^2 \cos^2 \alpha \\
&= A^2(\sin^2 \alpha + \cos^2 \alpha) \\
&= A^2 \text{ using } \sin^2 \alpha + \cos^2 \alpha = 1
\end{aligned}
$$

Hence $A = \sqrt{40}$.

To find α, A is eliminated from equations (1) and (2). This is achieved by dividing equation (1) by equation (2) and using the identity $\dfrac{\sin \alpha}{\cos \alpha} = \tan \alpha$.

Dividing equation (1) by equation (2) gives

$$\frac{2}{6} = -\frac{A \sin \alpha}{A \cos \alpha} = -\tan \alpha$$

Hence

$$\tan \alpha = -\frac{1}{3}$$

Since $\tan \alpha < 0$, then α lies in either the second quadrant or the fourth quadrant. Noting from equation (1) that $\sin \alpha < 0$, then α must be in the fourth quadrant.

Solving $\tan \alpha = -\dfrac{1}{3}$ with α in the fourth quadrant gives $\alpha = 5.9614$. So finally

$$2 \sin 3t + 6 \cos 3t = \sqrt{40} \cos(3t + 5.9614)$$

The resulting wave has an amplitude of $\sqrt{40}$, an angular frequency of 3 radians per second and a phase of 5.9614 radians.

Example 7.9

Express $2 \sin 5t - 5 \cos 5t$ in the form $A \sin(\omega t - \alpha)$, $\alpha \geq 0$.

Solution

The angular frequency of both $2 \sin 5t$ and $5 \cos 5t$ is 5, and so the resultant wave has an angular frequency of 5, that is $\omega = 5$.

Recall the trigonometrical identity for $\sin(A - B)$:

$$\sin(A - B) = \sin A \cos B - \sin B \cos A$$

Hence

$$
\begin{aligned}
2 \sin 5t - 5 \cos 5t &= A \sin(\omega t - \alpha) \\
&= A \sin(5t - \alpha) \\
&= A(\sin 5t \cos \alpha - \sin \alpha \cos 5t) \\
&= (A \cos \alpha) \sin 5t - (A \sin \alpha) \cos 5t
\end{aligned}
$$

Comparing the $\sin 5t$ terms on both sides gives

$$2 = A \cos \alpha \qquad (5)$$

Comparing the $\cos 5t$ terms on both sides gives

$$5 = A \sin \alpha \qquad (6)$$

To eliminate α from equations (5) and (6), the equations are squared and then added. Squaring the equations gives

$$4 = A^2 \cos^2 \alpha$$
$$25 = A^2 \sin^2 \alpha$$

and then adding gives

$$
\begin{aligned}
29 &= A^2 \cos^2 \alpha + A^2 \sin^2 \alpha \\
&= A^2(\cos^2 \alpha + \sin^2 \alpha) \\
&= A^2
\end{aligned}
$$

Hence $A = \sqrt{29}$.

To eliminate A, equation (6) is divided by equation (5) to give

$$
\begin{aligned}
\frac{5}{2} &= \frac{A \sin \alpha}{A \cos \alpha} \\
&= \tan \alpha \\
\tan \alpha &= 2.5
\end{aligned}
$$

We note that $\tan \alpha > 0$ and from equation (5) that $\cos \alpha > 0$ and so α must be in the first quadrant.

$$\alpha = \tan^{-1} 2.5 = 1.1903$$

So

$$2 \sin 5t - 5 \cos 5t = \sqrt{29} \sin(5t - 1.1903)$$

Example 7.10

Express $\cos 2t - 2 \sin 2t$ in the form $A \sin(\omega t + \alpha)$, $\alpha \geq 0$.

Solution

The angular frequency of both $\cos 2t$ and $2 \sin 2t$ is

          2

Hence the angular frequency of the resulting wave is also 2, that is $\omega = 2$.
   Recall from Block 5 the trigonometrical identity

$$\sin(A + B) = \boxed{} \qquad \sin A \cos B + \sin B \cos A$$

Hence

$$\cos 2t - 2 \sin 2t = A \sin(\omega t + \alpha)$$

$$= A \sin(2t + \alpha)$$

$$= \boxed{} \qquad A(\sin 2t \cos \alpha + \sin \alpha \cos 2t)$$

$$= \boxed{} \sin 2t + \boxed{} \cos 2t \qquad A \cos \alpha, A \sin \alpha$$

So we have

$$\cos 2t - 2 \sin 2t = A \cos \alpha \sin 2t + A \sin \alpha \cos 2t$$

Comparing the $\sin 2t$ terms on both sides of the equation gives

$$-2 = \boxed{} \qquad\qquad\qquad (7) \quad A \cos \alpha$$

Comparing the $\cos 2t$ terms on both sides of the equation gives

$$1 = \boxed{} \qquad\qquad\qquad (8) \quad A \sin \alpha$$

Squaring equations (7) and (8) and then adding gives

$$A^2 = \boxed{} \qquad\qquad\qquad\qquad\qquad 5$$

and so $A = \sqrt{5}$.
   To determine α, equation (8) is divided by equation (7). This gives

$$\tan \alpha = \boxed{} \qquad\qquad\qquad\qquad -0.5$$

Recognising that α is in the $\boxed{}$ quadrant we see that $\boxed{}$ second

$$\alpha = \boxed{} \qquad \tan^{-1}(-0.5) = 2.6779$$

Hence

$$\cos 2t - 2 \sin 2t = \boxed{} \qquad \sqrt{5} \sin(2t + 2.6779)$$

Exercises

1   Express $6 \sin 3t - 7 \cos 3t$ in the form
    $A \cos(\omega t + \alpha)$, $\alpha \geq 0$.

2   Express $2 \cos t + 6 \sin t$ in the form
    $A \sin(\omega t - \alpha)$, $\alpha \geq 0$. State the maximum
    value of $2 \cos t + 6 \sin t$.

3   Express $5 \sin 2t + \cos 2t$ in the form
    $A \cos(\omega t - \alpha)$, $\alpha \geq 0$.

4   Show that the maximum value of

    $a \sin \omega t + b \cos \omega t$ is $\sqrt{a^2 + b^2}$.

Solutions to exercises

1   $\sqrt{85} \cos(3t + 3.8502)$

2   $\sqrt{40} \sin(t - 5.9614), \sqrt{40}$

3   $\sqrt{26} \cos(2t - 1.3734)$

End of block exercises

State the amplitude of the functions in questions 1–5.

1   $y = 3 \sin 2t$

2   $y = 2 \cos 3t$

3   $y = \dfrac{4}{3} \sin\left(\dfrac{t}{2}\right)$

4   $y = \cos\left(\dfrac{2t}{3}\right)$

5   $y = \dfrac{5 \cos 3t}{3}$

For questions 6–10 state the angular frequency of
the functions given in questions 1–5.

For questions 11–15 state the period of the functions
in questions 1–5.

For questions 16–20 state the frequency of the
functions in questions 1–5.

State the phase of the functions in questions 21–25.

21   $y = 3 \sin(t + 2)$

22   $y = 2.3 \cos(4t - 2)$

23   $y = \cos\left(\dfrac{3t - 1}{2}\right)$

24   $y = \sin(0.5t + 3)$

25   $y = \sin\left(\dfrac{2t}{3} - \dfrac{\pi}{2}\right)$

For questions 26–30 state the time displacement of
the functions in questions 21–25.

31   Express $3 \sin 5t + 6 \cos 5t$ in the form
     $A \sin(\omega t + \alpha)$, $\alpha \geq 0$.

32   Express $2 \cos 3t - \sin 3t$ in the form
     $A \sin(\omega t + \phi)$, $\phi \geq 0$.

33   Express $\sin 4t + 3 \cos 4t$ in the form
     $A \cos(\omega t + \phi)$, $\phi \geq 0$.

34   Express $\cos t - 7 \sin t$ in the form
     $A \cos(\omega t + \alpha)$, $\alpha \geq 0$.

35   Express $4 \sin 2t + 5 \cos 2t$ in the form
     $A \sin(\omega t - \phi)$, $\phi \geq 0$.

36   Express $5 \sin 3t - 3 \cos 3t$ in the form
     $A \sin(\omega t - \alpha)$, $\alpha \geq 0$.

37   Express $\sin t - 3 \cos t$ in the form
     $A \cos(\omega t - \phi)$, $\phi \geq 0$.

38   Express $-2 \sin 2t + 4 \cos 2t$ in the form
     $A \cos(\omega t - \alpha)$, $\alpha \geq 0$.

Solutions to exercises

1	3	14	3π	26	2	
2	2	15	$\dfrac{2\pi}{3}$	27	$-\dfrac{1}{2}$	
3	$\dfrac{4}{3}$	16	$\dfrac{1}{\pi}$	28	$-\dfrac{1}{3}$	
4	1	17	$\dfrac{3}{2\pi}$	29	6	
5	$\dfrac{5}{3}$	18	$\dfrac{1}{4\pi}$	30	$-\dfrac{3\pi}{4}$	
6	2	19	$\dfrac{1}{3\pi}$	31	$\sqrt{45}\,\sin(5t + 1.1071)$	
7	3	20	$\dfrac{3}{2\pi}$	32	$\sqrt{5}\,\sin(3t + 2.0344)$	
8	$\dfrac{1}{2}$	21	2	33	$\sqrt{10}\,\cos(4t + 5.9614)$	
9	$\dfrac{2}{3}$	22	-2	34	$\sqrt{50}\,\cos(t + 1.4289)$	
10	3	23	$-\dfrac{1}{2}$	35	$\sqrt{41}\,\sin(2t - 5.3871)$	
11	π	24	3	36	$\sqrt{34}\,\sin(3t - 0.5404)$	
12	$\dfrac{2\pi}{3}$	25	$-\dfrac{\pi}{2}$	37	$\sqrt{10}\,\cos(t - 2.8198)$	
13	4π			38	$\sqrt{20}\,\cos(2t - 5.8195)$	

End of chapter exercises

1. Convert the following angles to radians, giving your answer to 4 d.p.:
   (a) $40°$ (b) $100°$ (c) $527°$ (d) $-200°$

2. Convert the following angles in radians to degrees:
   (a) $\dfrac{\pi}{2}$ (b) $\dfrac{\pi}{3}$ (c) $\dfrac{4\pi}{3}$ (d) 1.25π (e) 1.25
   (f) 9.6314 (g) 3

3. Evaluate
   (a) $\operatorname{cosec} 37°$ (b) $\cot 1.3$ (c) $\sec 40°$

4. An arc of a circle, radius 5 cm, subtends an angle of $\frac{3\pi}{4}$ radians at the centre. Calculate the length of the arc.

5. A sector of a circle, radius 9 cm, has an area of 100 cm^2. Calculate the angle subtended at the centre by the sector.

6. Express $6 \sin 2t - 3 \cos 2t$ in the form $A \cos(\omega t - \alpha)$, $\alpha \geq 0$.

7. If $\sin \phi < 0$ and $\cos \phi > 0$, state the quadrant in which ϕ lies.

8. If $\tan \phi < 0$ and $\sin \phi > 0$, state the quadrant in which ϕ lies.

9. Express $\frac{1}{2}\cos t + \sin t$ in the form $A \sin(\omega t - \alpha)$, $\alpha \geq 0$.

10. A voltage source, $v(t)$, varies with time, t, according to

$$v(t) = 50 \sin(\pi t + 10)$$

State (a) the angular frequency, (b) the phase, (c) the amplitude, (d) the period, (e) the time displacement, (f) the frequency of the voltage.

11 Solve

$$3 \cos \theta = 1.2 \quad 0 \le \theta \le 2\pi$$

12 Solve

$$\sin 2\theta = -0.4010 \quad 0 \le \theta \le 2\pi$$

13 Simplify $\sin \theta \cos \theta \tan \theta + \cos^2 \theta$.

14 Express $5 \cos 3t + 2 \sin 3t$ in the form $A \cos(\omega t + \alpha)$, $\alpha \ge 0$.

15 Show that
   (a) $\tan^2 \theta + 1 = \sec^2 \theta$
   (b) $1 + \cot^2 \theta = \text{cosec}^2 \theta$

16 (a) Sketch $y = \cos(x - 20°)$,
       $0° \le x \le 360°$.
   (b) On the same axes, sketch $y = \sin x$.
   (c) Use your graphs to obtain approximate
       solutions of

$$\sin x = \cos(x - 20°)$$

17 A current, $i(t)$, varies with time, t, and is given by

$$i(t) = 30 \cos(t - 0.4) \quad t \ge 0$$

   (a) Find the time when the current is first zero.
   (b) Find the time when the current reaches its
       first peak.

18 Solve

$$\sin \theta \cos 41° + \sin 41° \cos \theta = 0.6100$$
$$0° \le \theta \le 360°$$

19 If $0 \le \theta \le 2\pi$ and $\cos 2\theta < 0$, state the
   range of possible values for θ.

20 Simplify

$$(\sin \theta + \cos \theta)^2 - \sin 2\theta$$

21 Solve

$$\tan\left(\frac{2x}{3}\right) = 0.7 \quad 0 \le x \le 2\pi$$

22 Solve

$$\cos\left(\frac{\theta - 30°}{3}\right) = -0.6010 \quad 0 \le \theta \le 720°$$

23 A voltage, $v(t)$, has the form

$$2 \sin t + \cos t \quad t \ge 0$$

   (a) Calculate the maximum value of v.
   (b) Calculate the first time that this maximum
       value occurs.

Solutions to exercises

1 (a) 0.6981 (b) 1.7453 (c) 9.1979
   (d) −3.4907

2 (a) 90° (b) 60° (c) 240° (d) 225° (e) 71.62°
   (f) 551.84° (g) 171.89°

3 (a) 1.6616 (b) 0.2776 (c) 1.3054

4 11.78 cm

5 2.4691 radians

6 $\sqrt{45} \cos(2t - 2.0344)$

7 fourth quadrant

8 second quadrant

9 1.1180 $\sin(t - 5.8195)$

10 (a) π (b) 10 (c) 50 (d) 2 (e) $\dfrac{10}{\pi}$ (f) 0.5

11 1.1593, 5.1239

12 1.7771, 2.9353, 4.9187, 6.0769

13 1

14 $\sqrt{29} \cos(3t + 5.9027)$

16 (c) 55°, 235°

17 (a) 1.9708 (b) 0.4

18 101.41°, 356.59°

19 $\dfrac{\pi}{4} < \theta < \dfrac{3\pi}{4}, \dfrac{5\pi}{4} < \theta < \dfrac{7\pi}{4}$

20 1

21 0.9161, 5.6285

22 410.82°

23 (a) $\sqrt{5}$ (b) 1.1071

Further trigonometry

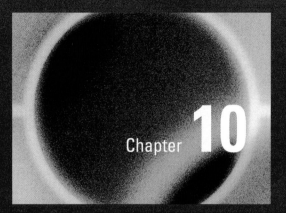

The chapter opens with a statement of **Pythagoras's theorem**. This famous theorem relates the lengths of the sides of a right-angled triangle. A triangle is solved when all its angles and the lengths of all its sides have been found. Methods are covered for solving right-angled triangles.

Blocks 2 and 3 deal with the **sine** and **cosine rules**. These are used to solve any triangle. Conditions under which the rules can be applied are clearly stated.

The application of the **solution of triangles** in surveying is covered in Block 4. Finally trigonometry is applied to the resolution and addition of **forces**.

Chapter 10 contents

Pythagoras's theorem and the solution of right-angled triangles

Introduction

We introduce some common terms and notation used for right-angled triangles. Consider a right-angled $\triangle ABC$ as shown in Figure 1.1.

Figure 1.1
A right-angled
$\triangle ABC$.

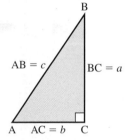

There is a right angle at C; the hypotenuse is AB. The side AC is opposite B. A common notation is to refer to the length of AC as b. Similarly, BC is opposite A and so the length of BC is written as a. Finally, the hypotenuse AB is opposite C and so is labelled c. In summary we have

$$a = BC, \quad b = AC, \quad c = AB$$

The convention of referring to sides of a triangle in such a way is not restricted to right-angled triangles. It is used for any triangle.

1.2 Pythagoras's theorem

Pythagoras's theorem can only be applied to right-angled triangles. Consider the right-angled $\triangle ABC$ in Figure 1.1. Then Pythagoras's theorem states

Key point

$$c^2 = a^2 + b^2$$

Thus the square of the hypotenuse equals the sum of the squares of the other two sides.

Example 1.1

In $\triangle ABC$ in Figure 1.1, AC = 9 cm and BC = 14 cm. Calculate AB.

Solution

We have $b = AC = 9$ and $a = BC = 14$. We need to find AB, that is c. By Pythagoras's theorem we know

$$c^2 = a^2 + b^2$$
$$= 14^2 + 9^2$$
$$= 277$$
$$c = \sqrt{277} = 16.64$$

The hypotenuse, AB, is 16.64 cm.

Example 1.2

Consider $\triangle ABC$ in Figure 1.1. Given AB = 19 cm and AC = 12.5 cm calculate BC.

Solution

We have $c = AB = 19$ and $b = $ 12.5
We seek BC, that is a. Using Pythagoras's theorem

$$c^2 = a^2 + b^2$$
$$19^2 = a^2 + \qquad\qquad\qquad\qquad\qquad\qquad 12.5^2$$
$$a^2 = \qquad\qquad\qquad\qquad\qquad 19^2 - 12.5^2 = 204.75$$
$$a = \qquad\qquad\qquad\qquad\qquad \sqrt{204.75} = 14.31$$

The length of BC is 14.31 cm.

Exercises

1   $\triangle ABC$ has a right angle at B. Given AB = 7 cm, BC = 12 cm, calculate the length of AC.

2   $\triangle PQR$ has a right angle at P. If PR = 3.2 m, QR = 4.9 m, calculate the length of PQ.

3   $\triangle CDE$ has a right angle at E. Given CD = 55 mm and DE = 37 mm, calculate the length of CE.

Solutions to exercises

1   13.89 cm

2   3.71 m

3   40.69 mm

When asked to solve a triangle we need to calculate all the unknown angles and all the unknown sides. Sometimes we are asked to calculate only specified angle(s) and/or side(s). The following examples illustrate the method. You should be aware that the sum of the angles in any triangle is 180°.

Example 1.3

As shown in Figure 1.2, $\triangle ABC$ has a right angle at C, $A = 53°$ and BC = 9 cm. Calculate (a) AC and (b) AB.

Figure 1.2

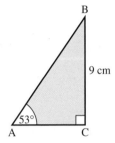

Solution

(a) $\tan A = \dfrac{BC}{AC}$

$\tan 53° = \dfrac{9}{AC}$

$AC = \dfrac{9}{\tan 53°}$

$= 6.7820$

(b) $\sin A = \dfrac{BC}{AB}$

$\sin 53° = \dfrac{9}{AB}$

$AB = \dfrac{9}{\sin 53°}$

$= 11.2692$

Example 1.4

$\triangle XYZ$ has a right angle at Y, $X = 26°$ and XZ = 12 cm.
(a) Sketch $\triangle XYZ$.
(b) Calculate XY.
(c) Calculate YZ.

Solution

(a) $\triangle XYZ$ is illustrated in Figure 1.3.

Figure 1.3

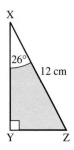

(b) $\cos X = \dfrac{XY}{XZ}$

$\cos 26° = \dfrac{XY}{12}$

$XY = $ ▨▨▨▨▨▨▨▨▨ $12 \cos 26° = 10.7855$

(c) $\sin X = \dfrac{YZ}{XZ}$

$\sin 26° = $ ▨▨▨▨ $\dfrac{YZ}{12}$

$YZ = $ ▨▨▨▨▨▨▨▨ $12 \sin 26° = 5.2605$

Example 1.5
△KLM has a right angle at L, $M = 25°$ and ML = 14 cm. It is illustrated in Figure 1.4. Calculate (a) KL and (b) KM.

Figure 1.4

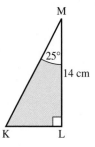

Solution

(a) $\tan M = \dfrac{KL}{ML}$

$\tan 25° = \dfrac{KL}{14}$

$KL = $ ▨▨▨▨▨▨▨▨ 6.5283

(b) $\cos M =$ [____] $\dfrac{ML}{KM} = \dfrac{14}{KM}$

KM = [____] $\dfrac{14}{\cos 25°} = 15.4473$

Example 1.6

$\triangle PQR$ has a right angle at P, $Q = 39°$ and QR $= 18$ cm.
(a) Sketch $\triangle PQR$.
(b) Calculate PQ.
(c) Calculate PR.

Solution

(a) $\triangle PQR$ is illustrated in Figure 1.5.

Figure 1.5

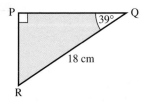

(b) $\cos 39° =$ [____] $\dfrac{PQ}{QR} = \dfrac{PQ}{18}$

PQ = [____] $18 \cos 39° = 13.9886$

(c) $\sin 39° =$ [____] $\dfrac{PR}{QR} = \dfrac{PR}{18}$

PR = [____] $18 \sin 39° = 11.3278$

Example 1.7

$\triangle ABC$ has $C = 90°$, AC $= 14$ cm and BC $= 9$ cm.
(a) Sketch $\triangle ABC$.
(b) Solve $\triangle ABC$.

Solution

(a) $\triangle ABC$ is illustrated in Figure 1.6.

Figure 1.6

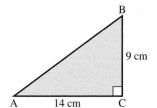

(b) We need to find A, B and AB. First we find A.

$$\tan A = \frac{BC}{AC}$$
$$= \frac{9}{14}$$
$$A = \tan^{-1}\left(\frac{9}{14}\right)$$
$$= 32.74°$$

B may now be found. The sum of the angles in any triangle is 180° and so

$$A + B + C = 180°$$
$$B = 180° - A - C$$
$$= 180° - 32.74° - 90°$$
$$B = 57.26°$$

Finally using Pythagoras's theorem we have

$$(AB)^2 = (BC)^2 + (AC)^2$$
$$= 9^2 + 14^2$$
$$= 277$$
$$AB = \sqrt{277}$$
$$= 16.64$$

Example 1.8
XYZ is a right-angled triangle as shown in Figure 1.7, with XY = 23 cm and YZ = 14 cm. Solve △XYZ.

Figure 1.7

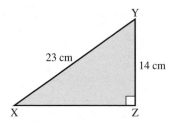

Solution
We need to find X, Y and XZ. We find X first.

$$\sin X = \qquad\qquad\qquad\qquad\qquad \frac{14}{23}$$

$$X = \qquad\qquad\qquad\qquad \sin^{-1}\left(\frac{14}{23}\right) = 37.50°$$

Y may now be found.

$$X + Y + Z = \qquad\qquad\qquad\qquad 180°$$

$$Y = \qquad\qquad\qquad\qquad 52.50°$$

Using Pythagoras's theorem we can find XZ.

$$(XY)^2 = (XZ)^2 + (YZ)^2$$

$(XZ)^2 = $ ⬚ $(XY)^2 - (YZ)^2 = 333$

$XZ = $ ⬚ $\sqrt{333} = 18.25$

Example 1.9 Materials Engineering – Measuring the hardness of a material

We have already noted in Example 7.7 in Chapter 5 that measuring the hardness of a material is important in materials engineering. This is particularly the case in life-critical systems which experience significant stresses, such as aircraft parts where material failure can be, and has been, catastrophic. The Vickers hardness test involves applying a known force through a square-based pyramid-shaped indenter (Figure 1.8(a), where point Q is the apex of the pyramid) into a metal surface. The surface area, A, of the resulting indentation is then calculated and the Vickers hardness (HV) is given by the formula

$$HV = 0.102 \times \frac{F}{A}$$

where F is the applied force (in newtons) and the surface area, A, is measured in mm^2.

To calculate the surface area of the indentation a microscope is used to measure the lengths of the two diagonals, d_1 and d_2 (Figure 1.8(b)). Ideally $d_1 = d_2$, but material imperfections mean that this is not necessarily the case. Their average is calculated as $D = \dfrac{d_1 + d_2}{2}$. We then assume that the diagonal length is accurately represented by the value of D.

(a) Show that the surface area of the indentation is given by

$$A = \frac{D^2}{2\sin(\theta/2)}$$

where θ is the angle between opposite faces of the pyramid indenter as shown in Figure 1.8 (c, d).

(b) A hardness test is performed by using a square-based diamond pyramid indenter in which the angle between the opposite faces of the pyramid is 136° and the applied force is 1000 N. The two diagonal distances are measured as $d_1 = 0.88$ mm, $d_2 = 0.90$ mm. Calculate the surface area, A, of the indentation and deduce the Vickers hardness, HV.

Figure 1.8

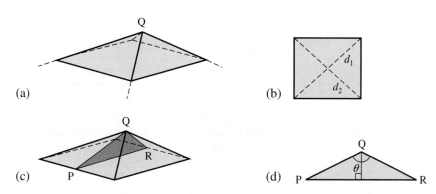

(a)   (b)   (c)   (d)

Solution

(a) Note that the surface area of the indentation is made up of four triangular sides. We first calculate the length of the base of the triangular sides using the known diagonal length D and Pythagoras's theorem. Figure 1.9 shows the base of the pyramid and the diagonal length D. We have labelled the unknown length x.

Figure 1.9

Then, using Pythagoras's theorem,

$$D^2 = x^2 + x^2 = 2x^2$$

$$x^2 = \frac{D^2}{2}$$

from which

$$x = \frac{D}{\sqrt{2}}$$

This is the length of the side of the square base. Figure 1.10 shows one of the triangular sides of the pyramid with the base length now shown.

Figure 1.10

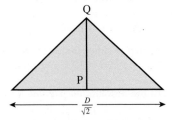

To calculate its area we need to know the length PQ; this can be determined from Figure 1.8(d) using the right-angled triangle shown:

$$\sin \frac{\theta}{2} = \frac{\frac{1}{2} PR}{PQ}$$

and so

$$PQ = \frac{\frac{1}{2} D / \sqrt{2}}{\sin \frac{\theta}{2}} = \frac{D}{2\sqrt{2} \sin \frac{\theta}{2}}$$

Finally we can calculate the area of the triangular side of the pyramid as

$$\text{area} = \frac{1}{2} \times \text{base} \times \text{height}$$

$$= \frac{1}{2} \times \frac{D}{\sqrt{2}} \times \frac{D}{2\sqrt{2} \sin \frac{\theta}{2}}$$

$$= \frac{D^2}{8 \sin \frac{\theta}{2}}$$

Given there are four triangular sides, the surface area of the indentation is then

$$A = \frac{D^2}{2 \sin \frac{\theta}{2}}$$

(b) The average diagonal distance is

$$D = \frac{d_1 + d_2}{2} = \frac{0.88 + 0.90}{2} = 0.89$$

Then

$$A = \frac{D^2}{2 \sin (\theta/2)} = \frac{0.89^2}{2 \sin (136/2)} = 0.43 \text{ mm}^2$$

Thus

$$HV = 0.102 \times \frac{F}{A} = 0.102 \times \frac{1000}{0.43} = 237 \text{ N/mm}^2$$

A hardness number of this magnitude would be typical for a mild steel.

Exercises

Questions 1–3 refer to $\triangle ABC$ in Figure 1.1.

1 Find A given BC = 13 cm and AB = 19 cm.

2 Find B given BC = 14 cm and AB = 21 cm.

3 Find A given AC = 10 cm and BC = 12 cm.

4 $\triangle PQR$ has a right angle at R, $P = 62°$ and PR = 11 cm. Calculate (a) PQ, (b) QR, (c) Q.

5 $\triangle ABC$ has a right angle at A, $B = 25°$ and AC = 17.2 cm. Solve $\triangle ABC$.

Solutions to exercises

1 43.17°

2 48.19°

3 50.19°

4 (a) 23.43 cm (b) 20.69 cm (c) 28°

5 AB = 36.89 cm BC = 40.70 cm C = 65°

End of block exercises

Questions 1–3 refer to $\triangle ABC$ in Figure 1.1.

1 Find AB given AC = 3 cm and BC = 6 cm.

2 Find AC given AB = 12.3 cm and BC = 9.6 cm.

3 Find BC given AB = 190 cm and AC = 142 cm.

4 $\triangle ABC$ has a right angle at C, AB = 21 cm and AC = 17 cm. Calculate the length of BC.

5   △XYZ has a right angle at Z, $Y = 47°$ and
    XY = 17 cm. Calculate (a) XZ, (b) YZ, (c) X.

6   △CDE has a right angle at D, CD = 17 cm
    and DE = 19 cm. Solve △CDE.

7   △KLM has a right angle at M, KL = 14 cm
    and KM = 10 cm. Calculate (a) K, (b) L,
    (c) LM.

8   △RST has a right angle at T, $S = 36°$ and
    ST = 15 cm. Solve △RST.

9   △UVW has a right angle at V, UV = 15 cm
    and VW = 6 cm. Solve △UVW.

10 △CDE has a right angle at E, CE = 19 cm
    and DE = 14 cm. Calculate (a) CD, (b) C,
    (c) D.

11 △KLM has a right angle at K, $L = 51°$ and
    ML = 10 cm. Calculate (a) KL, (b) KM,
    (c) M.

12 △XYZ has a right angle at Y, $X = 42°$ and
    YZ = 13 cm. Calculate (a) XY, (b) XZ, (c) Z.

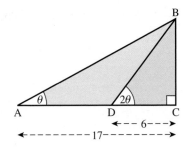

Figure 1.11

13 Figure 1.11 illustrates △ABC and △BDC,
    where C is a right angle, AC = 17 cm and
    DC = 6 cm. Given $\angle BDC = 2\theta$ and
    $\angle BAD = \theta$, find θ.

14 △DEF has a right angle at F, $E = 38°$ and
    DF = 14 cm. Solve △DEF.

15 △XYZ has a right angle at Y, XY = 40 cm
    and YZ = 57 cm. Solve △XYZ.

16 △ABC has a right angle at C, AB = 30 cm
    and AC = 2BC. Solve △ABC.

Solutions to exercises

1   6.71 cm

2   7.69 cm

3   126.24 cm

4   12.33 cm

5   (a) 12.43 cm (b) 11.59 cm (c) 43°

6   $E = 41.82°, C = 48.18°$, CE = 25.50 cm

7   (a) 44.42° (b) 45.58° (c) 9.80

8   $R = 54°$, RT = 10.90 cm, RS = 18.54 cm

9   $W = 68.20°, U = 21.80°$, UW = 16.16 cm

10 (a) 23.60 cm (b) 36.38° (c) 53.62°

11 (a) 6.29 cm (b) 7.77 cm (c) 39°

12 (a) 14.44 cm (b) 19.43 cm (c) 48°

13 28.47°

14 EF = 17.92 cm, DE = 22.74 cm, $D = 52°$

15 $X = 54.94°, Z = 35.06°$, XZ = 69.63 cm

16 $A = 26.57°, B = 63.43°$, BC = 13.42 cm,
    AC = 26.83 cm

Solving triangles using the sine rule

2.1 Introduction

Recall that by solving a triangle we mean calculating all the unknown angles and sides. In Block 1 we solved right-angled triangles. In this block and the next we solve triangles where there is no right angle. This is done by using either the sine rule or the cosine rule. This block focuses on the sine rule.

2.2 The sine rule

Consider any $\triangle ABC$ as shown in Figure 2.1.

Figure 2.1
$a = BC, b = AC,$
$c = AB.$

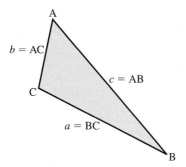

We have used the convention that the side opposite A is labelled a, and so on. The sine rule states:

Key point

$$\frac{a}{\sin A} = \frac{b}{\sin B} = \frac{c}{\sin C}$$

The rule can be used to solve a triangle when we are given either (a) one side and two angles or (b) two sides and one angle that is not included by the given sides. The two cases are illustrated by the following examples.

Example 2.1

Solve $\triangle ABC$ given $A = 46°$, $B = 32°$ and BC = 13 cm.

Solution

We are given two angles and a side and so the triangle can be solved using the sine rule. We have $A = 46°$, $B = 32°$ and $a = $ BC = 13 cm, and need to find C, AB and AC.

First we calculate C.

$$A + B + C = 180°$$
$$C = 180° - A - B$$
$$= 180° - 46° - 32°$$
$$= 102°$$

We now apply the sine rule.

$$\frac{a}{\sin A} = \frac{b}{\sin B} = \frac{c}{\sin C}$$

$$\frac{13}{\sin 46°} = \frac{b}{\sin 32°} = \frac{c}{\sin 102°}$$

so

$$b = \frac{13 \sin 32°}{\sin 46°} \quad \text{and} \quad c = \frac{13 \sin 102°}{\sin 46°}$$

that is

$$b = 9.58 \quad \text{and} \quad c = 17.68$$

Hence $C = 102°$, $b = $ AC = 9.58 cm, $c = $ AB = 17.68 cm.

___

It is worth noting that the largest angle is always opposite the longest side, and the smallest angle is opposite the shortest side.

Example 2.2

In $\triangle ABC$, $A = 21°$, AC = 17 cm and BC = 14 cm. Solve $\triangle ABC$.

Solution

Figure 2.2 illustrates the situation.

Figure 2.2
Two sides and a non-included angle are known.

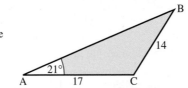

We know $A = 21°$, $a = BC = 14$ and $b = AC = 17$. Since two sides and a non-included angle are given the sine rule can be applied. We need to calculate B, C and AB. The sine rule states

$$\frac{a}{\sin A} = \frac{b}{\sin B} = \frac{c}{\sin C}$$

so

$$\frac{14}{\sin 21°} = \frac{17}{\sin B} = \frac{c}{\sin C}$$

Hence

$\sin B =$
$$\frac{17 \sin 21°}{14} = 0.4352$$

So

$B = \sin^{-1}(0.4352)$

$=$
$\qquad$ 25.80° and 154.20°

Note that there are two possible solutions for B. Both are acceptable.

Case 1: $B = 25.80°$

We calculate C.

$$C = 180° - A - B$$
$$= 133.2°$$

Applying the sine rule gives

$$\frac{a}{\sin A} = \frac{b}{\sin B} = \frac{c}{\sin C}$$

$$\frac{14}{\sin 21°} = \frac{17}{\sin 25.8°} = \frac{c}{\sin 133.2°}$$

So

$c =$
$$\frac{17 \sin 133.2°}{\sin 25.8°} = 28.47$$

Solution 1 is $B = 25.8°$, $C = 133.2°$, $c = AB = 28.47$ cm.

Case 2: $B = 154.20°$

Here $C = 180° - A - B = 4.80°$.

Applying the sine rule gives

$$\frac{a}{\sin A} = \frac{b}{\sin B} = \frac{c}{\sin C}$$

and so

$$\frac{14}{\sin 21°} = \frac{17}{\sin 154.20°} = \frac{c}{\sin 4.8°}$$

Hence

$$c = \boxed{} \qquad\qquad \frac{17 \sin 4.8°}{\sin 154.20°} = 3.27$$

Solution 2 is $B = 154.20°$, $C = 4.80°$, $c = AB = 3.27$ cm.

Example 2.3

In $\triangle ABC$, $C = 42°$, AB = 15 cm and AC = 11 cm. Solve $\triangle ABC$.

Solution

We are given two sides and a non-included angle and so the sine rule can be applied. We have $C = 42°$, $b = AC = 11$ and $c = AB = 15$. We need to calculate A, B and BC. Using the sine rule,

$$\frac{a}{\sin A} = \frac{b}{\sin B} = \frac{c}{\sin C}$$

and so

$$\frac{a}{\sin A} = \frac{11}{\sin B} = \frac{15}{\sin 42°}$$

Hence

$$\sin B = \frac{11 \sin 42°}{15}$$
$$= 0.4907$$
$$B = \sin^{-1}(0.4907)$$
$$= 29.39° \text{ and } 150.61°$$

The solution $B = 150.61°$ is rejected because we are given $C = 42°$ and the sum of the angles in a triangle must equal 180°. We now calculate A.

$$A = 180° - B - C$$
$$= 108.61°$$

Finally

$$a = \frac{15 \sin A}{\sin 42°}$$
$$= \frac{15 \sin 108.61°}{\sin 42°}$$
$$= 21.25$$

Thus we have $A = 108.61°$, $B = 29.39°$, $a = BC = 21.25$ cm.

Example 2.4

Solve $\triangle XYZ$ given $X = 63°$, $Z = 42°$ and YZ = 11.3 cm.

Solution

We are given two angles and a side, so the sine rule can be applied. We have $X = 63°$, $Z = 42°$ and $x = YZ = 11.3$. We need to calculate Y, $XZ = y$ and $XY = z$. First we calculate Y.

$$Y = \boxed{} \qquad\qquad\qquad\qquad\qquad 75°$$

The sine rule states

$$\frac{x}{\sin X} = \frac{y}{\sin Y} = \frac{z}{\sin Z}$$

and so

$$\frac{11.3}{\sin 63°} = \frac{y}{\sin 75°} = \frac{z}{\sin 42°}$$

Hence

$$y = \qquad\qquad\qquad\qquad\qquad \frac{11.3 \sin 75°}{\sin 63°} = 12.25$$

$$z = \qquad\qquad\qquad\qquad\qquad \frac{11.3 \sin 42°}{\sin 63°} = 8.49$$

Example 2.5
Solve $\triangle ABC$ given $C = 40°$, $b = 23$ cm and $c = 19$ cm.

Solution
We are given two sides and a non-included angle and so the sine rule can be applied. We know $C = 40°$, AB $= c = 19$ and AC $= b = 23$. We need to calculate A, B and a. Using the sine rule

$$\frac{a}{\sin A} = \frac{b}{\sin B} = \frac{c}{\sin C}$$

we have

$$\frac{a}{\sin A} = \frac{23}{\sin B} = \frac{19}{\sin 40°}$$

Hence

$$\sin B = \qquad\qquad\qquad\qquad\qquad \frac{23 \sin 40°}{19} = 0.7781$$

$$B = \qquad\qquad\qquad\qquad\qquad 51.09° \text{ or } 128.91°$$

Both solutions are acceptable.

Case 1: $B = 51.09°$

Then $A = \qquad\qquad\qquad\qquad\qquad 180° - 40° - 51.09° = 88.91°$

and from the sine rule

$$a = \qquad\qquad\qquad\qquad\qquad = \frac{19 \sin 88.91°}{\sin 40°} = 29.55$$

Solution 1 is: $A = 88.91°$, $B = 51.09°$, BC $= 29.55$ cm.

Case 2: $B = 128.91°$

Then

$A =$ ⬚ ⬚ $180° - 40° - 128.91° = 11.09°$

and

$a =$ ⬚ ⬚ $\dfrac{19 \sin 11.09°}{\sin 40°} = 5.69$

Solution 2 is $A = 11.09°$, $B = 128.91°$, BC $= 5.69$ cm.

End of block exercises

For questions 1–10 solve $\triangle ABC$ given

1   $A = 36°$, $B = 79°$, AC $= 11.63$ cm

2   AB $= 15$ cm, AC $= 23$ cm, $B = 57°$

3   AB $= 10$ cm, BC $= 6$ cm, $A = 32°$

4   $B = 21°$, $C = 46°$, AB $= 9$ cm

5   $B = 18°$, $C = 110°$, BC $= 12.3$ cm

6   AC $= 29$ cm, BC $= 19$ cm, $B = 49°$

7   $A = 37°$, $B = 47°$, AB $= 17$ cm

8   BC $= 14$ cm, AB $= 20$ cm, $C = 50°$

9   AB $= 27$ cm, AC $= 36$ cm, $B = 17°$

10   $B = 42°$, $C = 93°$, BC $= 13$ cm

Solutions to exercises

1   $C = 65°$, BC $= 6.96$ cm, AB $= 10.74$ cm

2   $C = 33.16°$, $A = 89.84°$, BC $= 27.42$ cm

3   solution 1: $C = 62.03°$, $B = 85.97°$, AC $= 11.29$ cm; solution 2: $C = 117.97°$, $B = 30.03°$, AC $= 5.67$ cm

4   $A = 113°$, BC $= 11.52$ cm, AC $= 4.48$ cm

5   $A = 52°$, AC $= 4.82$ cm, AB $= 14.67$ cm

6   $A = 29.63°$, $C = 101.37°$, AB $= 37.67$ cm

7   $C = 96°$, AC $= 12.50$ cm, BC $= 10.29$ cm

8   $A = 32.43°$, $B = 97.57°$, AC $= 25.88$ cm

9   $A = 150.33°$, $C = 12.67°$, BC $= 60.95$ cm

10   $A = 45°$, AC $= 12.30$ cm, AB $= 18.36$ cm

Solving triangles using the cosine rule

3.1 Introduction

The cosine rule, like the sine rule, can be applied to any triangle. It is used to solve a triangle when we are given either (a) three sides or (b) two sides and the included angle.

3.2 The cosine rule

Key point

For any $\triangle ABC$ the cosine rule states

$$a^2 = b^2 + c^2 - 2bc \cos A$$

$$b^2 = a^2 + c^2 - 2ac \cos B$$

$$c^2 = a^2 + b^2 - 2ab \cos C$$

Example 3.1

In $\triangle ABC$, AB $= 17.3$ cm, BC $= 23.9$ cm and $B = 71°$. Solve $\triangle ABC$.

Solution

Figure 3.1 illustrates the given information.

Figure 3.1
Given two sides and the included angle, the cosine rule can be applied.

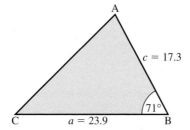

We are given two sides and the included angle and so the cosine rule can be used. We have $a = $ BC $= 23.9$, $c = $ AB $= 17.3$ and $B = 71°$. We need to find A, C and AC: that is, b. We find b first using the cosine rule.

$$b^2 = a^2 + c^2 - 2ac \cos B$$
$$= 23.9^2 + 17.3^2 - 2(23.9)(17.3) \cos 71°$$
$$= 601.27$$
$$b = 24.52$$

We now find A. We can use either the sine rule or the cosine rule to find A. Using the sine rule,

$$\frac{a}{\sin A} = \frac{b}{\sin B}$$

that is

$$\frac{23.9}{\sin A} = \frac{24.52}{\sin 71°}$$

$$\sin A = \frac{23.9 \sin 71°}{24.52}$$

$$= 0.9216$$

$$A = 67.16° \quad \text{or} \quad 112.84°$$

The solution $A = 112.84°$ is rejected because we already have $B = 71°$ and the sum of the angles of a triangle must be $180°$. Finally

$$C = 180° - A - B$$

$$= 41.84°$$

Hence $A = 67.16°, C = 41.84°, AC = 24.52$ cm.

Example 3.2

Solve $\triangle XYZ$ given $XY = 18.4$ cm, $YZ = 19.6$ cm and $XZ = 29.3$ cm.

Solution

Figure 3.2 illustrates the situation.

Figure 3.2
Three sides of the triangle are known and so the cosine rule can be applied.

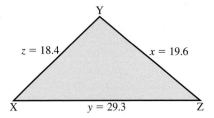

We are given three sides and so the cosine rule can be applied. We have $x = YZ = 19.6$, $y = XZ = 29.3$ and $z = XY = 18.4$. Using the cosine rule we find X.

$$x^2 = y^2 + z^2 - 2yz \cos X$$

$$19.6^2 = 29.3^2 + 18.4^2 - 2(29.3)(18.4) \cos X$$

$$\cos X = \boxed{} \quad \frac{29.3^2 + 18.4^2 - 19.6^2}{2(29.3)(18.4)} = 0.7539$$

$$X = \boxed{} \qquad\qquad\qquad\qquad\qquad 41.07°$$

Now we find Y.

$$y^2 = x^2 + z^2 - 2xz \cos Y$$

$$29.3^2 = 19.6^2 + 18.4^2 - 2(19.6)(18.4) \cos Y$$

$$\cos Y =$$

$$\frac{19.6^2 + 18.4^2 - 29.3^2}{2(19.6)(18.4)} = -0.1882$$

$$Y = \boxed{} \qquad\qquad 100.85°$$

Finally

$$Z = 180° - X - Y$$
$$= 38.08°$$

Hence $X = 41.07°$, $Y = 100.85°$, $Z = 38.08°$.

Example 3.3
In $\triangle RST$, RS = 85 cm, ST = 104 cm and $S = 105°$. Figure 3.3 illustrates the triangle. Solve $\triangle RST$.

Figure 3.3
When given two sides and the included angle, the cosine rule can be applied.

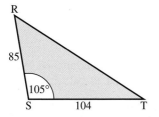

Solution
We are given two sides and the included angle and so the cosine rule can be applied. We have $t = $ RS $= 85$, $r = $ ST $= 104$ and $S = 105°$. We need to calculate R, T and $s = $ RT. First s is found using the cosine rule.

$$s^2 = r^2 + t^2 - \boxed{} \qquad\qquad 2rt \cos S$$

$$= 104^2 + 85^2 - \boxed{} \qquad\qquad 2(104)(85) \cos 105°$$

$$= 22616.92$$

$$s = 150.39$$

Using the sine rule, R is found. We have

$$\frac{r}{\sin R} = \frac{s}{\sin S}$$

from which

$$\frac{104}{\sin R} = \frac{150.39}{\sin 105°}$$

$$\sin R = \boxed{} \qquad \frac{104 \sin 105°}{150.39} = 0.6680$$

$$R = 41.91°$$

Finally $T = 180° - R - S = 33.09°$.
Hence RT $= 150.39$, $R = 41.91°$, $T = 33.09°$.

Example 3.4
In $\triangle$XYZ, XY $= 40$ cm, YZ $= 57$ cm and XZ $= 81$ cm. The triangle is illustrated in Figure 3.4. Solve $\triangle$XYZ.

Figure 3.4
When all three sides of a triangle are known, the cosine rule can be applied.

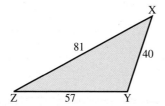

Solution
We are given three sides and so the cosine rule can be applied. We have $x =$ YZ $= 57$, $y =$ XZ $= 81$ and $z =$ XY $= 40$. We need to calculate X, Y and Z. First X is calculated. From the cosine rule we have

$$x^2 = \boxed{} \qquad y^2 + z^2 - 2yz \cos X$$

Substituting in the known values of x, y and z, and rearranging for $\cos X$, gives

$$\cos X = \boxed{} \qquad \frac{81^2 + 40^2 - 57^2}{2(81)(40)} = 0.7580$$

$$X = 40.71°$$

Now Y is found.

$$y^2 = \boxed{} \qquad x^2 + z^2 - 2xz \cos Y$$

Rearranging for $\cos Y$ yields

$$\cos Y = \qquad \frac{57^2 + 40^2 - 81^2}{2(57)(40)} = -0.3754$$

$$Y = \qquad Y = 112.05°$$

Finally $Z = 180° - X - Y = 27.24°$.
    Hence $X = 40.71°$, $Y = 112.05°$, $Z = 27.24°$.

End of block exercises

In questions 1–10 solve $\triangle ABC$ given

1   AB = 42 cm, BC = 37 cm, AC = 26 cm

2   AB = 29 cm, BC = 41 cm, $B = 100°$

3   AC = 105 cm, AB = 76 cm, $A = 29°$

4   BC = 36 cm, AC = 92 cm, $C = 51°$

5   AB = 69 cm, BC = 52 cm, AC = 49 cm

6   AB = 32 cm, BC = 30 cm, AC = 41 cm

7   BC = 17 cm, AC = 27 cm, $C = 45°$

8   AB = 36 cm, BC = 36 cm, $B = 60°$

9   AB = 21 cm, AC = 30 cm, $A = 42°$

10   AB = 80 cm, BC = 70 cm, AC = 62 cm

Solutions to exercises

1   $A = 60.63°$, $B = 37.76°$, $C = 81.61°$

2   AC = 54.18 cm, $A = 48.19°$, $C = 31.81°$

3   $a = $ BC = 53.31 cm, $B = 107.28°$, $C = 43.72°$

4   AB = 74.78 cm, $B = 107.03°$, $A = 21.97°$

5   $A = 48.76°$, $B = 45.11°$, $C = 86.13°$

6   $A = 46.54°$, $B = 82.73°$, $C = 50.73°$

7   $A = 38.75°$, $B = 96.25°$, $c = $ AB = 19.21 cm

8   AC = 36 cm, $A = 60°$, $C = 60°$

9   $a = $ BC = 20.12 cm, $B = 93.69°$, $C = 44.31°$

10   $A = 57.40°$, $B = 48.26°$, $C = 74.33°$

Surveying

This block considers the application of mathematics to surveying. The three areas covered are (a) units to measure angles, (b) angles of elevation and depression, and (c) bearings. Typical calculations in each area are included.

4.2 Units of angles

We have already introduced the units degree and radian in Chapter 9. Surveyors use the degree. For greater accuracy the degree is subdivided into **minutes** denoted ′ and **seconds** denoted ″.

Key point

$$1° = 60 \text{ minutes} = 60'$$
$$1' = 60 \text{ seconds} = 60''$$

The angle 36 degrees 21 minutes and 4 seconds is written as $36°21'04''$. Many calculations express angles in decimal format, for example $36.3511°$. It is useful to be able to convert from decimal format to degrees/minutes/seconds and vice versa. To retain the accuracy required for surveying, angles in decimal format should be given to 4 decimal places.

Example 4.1
Convert $36°21'04''$ to decimal format.

Solution
We note there are 60 minutes in 1 degree and 60 seconds in each minute. Hence there are $60 \times 60 = 3600$ seconds in 1 degree.

To write $36°21'04''$ in decimal format we consider the fractional part, that is $21'04''$, and express this in seconds.

$$21' = 21 \times 60$$
$$= 1260''$$

and so

$$21'04'' = 1264''$$

We now express 1264″ as a decimal fraction.

$$1264'' = \frac{1264°}{3600} = 0.3511°$$

Hence 21′04″ = 0.3511° and so 36°21′04″ = 36.3511°.

Example 4.2
Convert 42.9614° to degree/minute/second format.

Solution
The fractional part 0.9614° needs to be converted to minutes and seconds. Each degree comprises 60 minutes and so to convert from degrees to minutes we multiply by a factor of 60.

$$0.9614° = 0.9614 \times 60'$$
$$= 57.684'$$

We now look at the fractional part, 0.684, and convert this to seconds. Each minute comprises 60 seconds and so to convert from minutes to seconds we multiply by 60.

$$0.684' = 0.684 \times 60''$$
$$= 41.04''$$

Hence 42.9614° may be written as 42°57′41″.

Exercises

1   Express the following angles in degree/minute/second format.
(a) 11.1731° (b) 14.0017° (c) 36.9213°

2   Express the following angles in decimal format.
(a) 12°17′46″ (b) 32°32′56″ (c) 1°1′40″

Solutions to exercises

1   (a) 11°10′23″ (b) 14°0′6″ (c) 36°55′17″

2   (a) 12.2961° (b) 32.5489° (c) 1.0278°

4.3   Angles of elevation and depression

Consider a tower AB and an observer at O, as shown in Figure 4.1.

The angle e is called the **angle of elevation** of the tower from the point O. It is the angle, measured from the horizontal, through which an observer must turn his or her eyes to look at the top of the tower.

Figure 4.1
The angle of
elevation is *e* and
the angle of
depression is *d*.

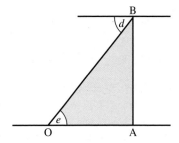

Now consider an observer at the top of the tower, B. The angle *d* is called the
angle of depression. This is the angle, again measured from the horizontal, through
which an observer turns his or her eyes to look at a point O. Note that *e* and *d* have
equal magnitude.

Example 4.3 Civil Engineering – Height of a tower
An observer is 50 m from the base of a vertical tower. The angle of elevation is
11°3′17″. Calculate the height of the tower.

Solution
Figure 4.1 illustrates the situation, with OA = 50 m and *e* = 11°3′17″. First *e* is
expressed in decimal format as 11.0547°. Now consider the right-angled △OAB.

$$\tan e = \tan 11.0547°$$

$$= \qquad\qquad\qquad \frac{AB}{OA} = \frac{AB}{50}$$

$$AB = \qquad\qquad\qquad 50 \tan 11.0547° = 9.77$$

The tower is 9.77 m high.

Example 4.4 Civil Engineering – Angle of depression
An observer is on top of a vertical tower, AB, 23.7 m high. The angle of depression
of a point O is 15°42′18″. Calculate the distance of O from the foot of the tower.

Solution
Figure 4.1 illustrates the problem, with AB = 23.7 m and *d* = 15°42′18″.
    First *d* is expressed in decimal format as

$$\qquad\qquad\qquad\qquad\qquad\qquad\qquad\qquad 15.705°$$

Noting that *e* = *d* we consider △OBA.

$$\tan e = \tan 15.705°$$

$$= \qquad\qquad\qquad \frac{AB}{OA} = \frac{23.7}{OA}$$

$$\text{OA} = \boxed{} \qquad \frac{23.7}{\tan 15.705°} = 84.29$$

The point O is 84.29 m from the foot of the tower.

Example 4.5 Civil Engineering – Finding the height of a tower using three angles of elevation

A, B and C are three points lying in a straight line. The distance AB is known to be x, and the distance BC is known to be y. It is required to find the height, h, of a vertical tower, DE. From A, B and C the angles of elevation to the top of the tower are α, β and γ respectively. Figure 4.2 illustrates the position.

Determine the height, h, in terms of the known quantities, x, y, α, β and γ.

Figure 4.2
The height, h, can be expressed in terms of known distances, x and y, and known angles α, β and γ.

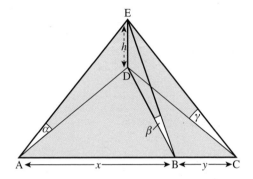

Solution

Consider $\triangle$ADE. In this triangle, $\angle$ADE $= 90°$ and so

$$\tan \alpha = \frac{\text{DE}}{\text{AD}}$$

$$= \frac{h}{\text{AD}}$$

$$\text{AD} = \frac{h}{\tan \alpha}$$

$$= h \cot \alpha$$

Similarly by considering $\triangle$BDE and $\triangle$CDE we see BD $= h \cot \beta$ and CD $= h \cot \gamma$. We now consider $\triangle$ADB and let $\angle$DAB $= \phi$. Figure 4.3 illustrates this triangle.

Figure 4.3
The cosine rule is applied to $\triangle$ADB.

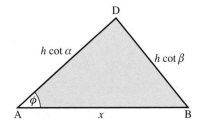

Applying the cosine rule to △ADB gives

$$(h \cot \beta)^2 = (h \cot \alpha)^2 + x^2 - 2xh \cot \alpha(\cos \phi)$$

from which

$$\cos \phi = \frac{(h \cot \alpha)^2 + x^2 - (h \cot \beta)^2}{2xh \cot \alpha} \tag{1}$$

Next we consider △ADC, as illustrated in Figure 4.4.

Figure 4.4
The cosine rule is
applied to △ADC.

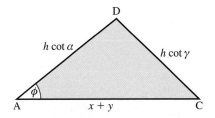

Applying the cosine rule to △ADC gives

$$(h \cot \gamma)^2 = (h \cot \alpha)^2 + (x + y)^2 - 2(x + y)h \cot \alpha (\cos \phi)$$

from which

$$\cos \phi = \frac{(h \cot \alpha)^2 + (x + y)^2 - (h \cot \gamma)^2}{2(x + y)h \cot \alpha} \tag{2}$$

We have two expressions for $\cos \phi$ given by (1) and (2). Equating these yields

$$\frac{(h \cot \alpha)^2 + x^2 - (h \cot \beta)^2}{2xh \cot \alpha} = \frac{(h \cot \alpha)^2 + (x + y)^2 - (h \cot \gamma)^2}{2(x + y)h \cot \alpha}$$

This equation is rearranged to make h the subject.

$$\frac{h^2 \cot^2 \alpha + x^2 - h^2 \cot^2 \beta}{x} = \frac{h^2 \cot^2 \alpha + x^2 + 2xy + y^2 - h^2 \cot^2 \gamma}{x + y}$$

$$(x + y)(h^2 \cot^2 \alpha + x^2 - h^2 \cot^2 \beta) = x (h^2 \cot^2 \alpha + x^2 + 2xy + y^2 - h^2 \cot^2 \gamma)$$

$$-xh^2 \cot^2 \beta + yh^2 \cot^2 \alpha - yh^2 \cot^2 \beta = x^2 y + xy^2 - xh^2 \cot^2 \gamma$$

$$h^2 [x (\cot^2 \gamma - \cot^2 \beta) + y (\cot^2 \alpha - \cot^2 \beta)] = xy(x + y)$$

$$h^2 = \frac{xy (x + y)}{x (\cot^2 \gamma - \cot^2 \beta) + y (\cot^2 \alpha - \cot^2 \beta)}$$

$$h = \sqrt{\frac{xy (x + y)}{x (\cot^2 \gamma - \cot^2 \beta) + y (\cot^2 \alpha - \cot^2 \beta)}}$$

Example 4.6

A, B and C are points lying on a straight line. The distance from A to B is 40 m, the distance from B to C is 70 m. The angles of elevation to the top of a tower from A, B and C are 21°, 27° and 24° respectively. Calculate the height of the tower.

Solution

We use the notation and result of Example 4.5, that is AB $= x = 40$, BC $= y = 70$, $\alpha = 21°$, $\beta = 27°$, $\gamma = 24°$.

If h is the height of the tower then

$$h = \sqrt{\frac{40(70)(40 + 70)}{40\,(\cot^2 24° - \cot^2 27°) + 70(\cot^2 21° - \cot^2 27°)}}$$

$$= 34.88$$

The tower is 34.88 m high.

Exercises

1   From a point 26.3 m from the foot of a tower the angle of elevation to the top of the tower is 29.27°. Calculate the height of the tower.

2   A tower is 36 m high. Calculate the angle of elevation to the top of the tower from a point 50 m from the base of the tower.

3   The angle of depression to the point O from the top of a tower 17 m high is 12°17′. Calculate the distance of O from the foot of the tower.

Solutions to exercises

1   14.74 m

2   35.75°

3   78.08 m

4.4   Bearings

A bearing is the angle that a line makes with some reference direction, usually north. Bearings are measured clockwise from north. (Note the different conventions: in mathematics clockwise angles are considered negative; in surveying clockwise angles are considered positive.) In Figure 4.5, OA has a bearing of 58°31′, OB has a bearing of 100°17′ and OC has a bearing of 301°49′.

Example 4.7

A ship leaves harbour and sails for 19.1 km on a bearing of 47°17′. It then changes direction and travels for 12 km on a bearing of 100°25′. Calculate the distance from the harbour to the ship.

Figure 4.5
Bearings are
measured
clockwise from
north.

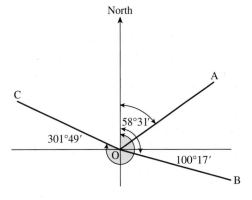

Solution

Figure 4.6 illustrates the ship's movement, starting from the harbour at O.

Figure 4.6
The ship moves
from the harbour at
O to A and then B.

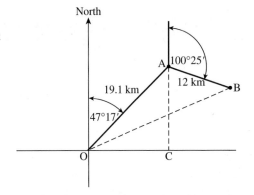

The ship travels 19.1 km on a bearing of $47°17'$. This is represented by OA. It then travels 12 km on a bearing of $100°25'$: this is represented by AB. The final position of the ship is B. We seek the distance OB.

Clearly OA = 19.1 and AB = 12.0. Consider $\triangle OAC$.

$$\angle AOC = 90° - 47°17'$$
$$= 42°43'$$
$$\angle OAC = 90° - \angle AOC$$
$$= 90° - 42°43'$$
$$= 47°17'$$

Now

$$\angle BAC = 180° - 100°25'$$
$$= 79°35'$$

Hence

$$\angle OAB = \angle OAC + \angle BAC$$
$$= 47°17' + 79°35'$$
$$= 126°52'$$
$$= 126.8667°$$

We now apply the cosine rule to △OAB.

$$OB^2 = OA^2 + AB^2 - 2(OA)(AB) \cos 126.8667°$$
$$= (19.1)^2 + 12^2 - 2(19.1)(12) \cos 126.8667°$$
$$= 783.83$$
$$OB = 28.00$$

The ship is 28 km from the harbour.

Example 4.8

A ship travels 17.1 km on a bearing of 120°35′. It then travels 24 km on a bearing of 275°20′.
(a) Find the distance of the ship from its starting position.
(b) Find the bearing the ship must take so that it can travel in a straight line back to its starting position.

Solution

Figure 4.7 illustrates the movement of the ship, starting at O. The final position of the ship is B.

Figure 4.7
The ship travels on a bearing of 120°35′ to A, and then on a bearing of 275°20′ to B.

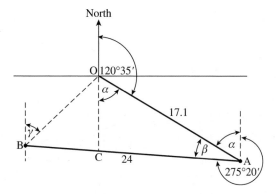

(a) We need to find the distance OB. The angles α and β are determined, and then the cosine rule is applied to △OAB.

$$\alpha = \quad\quad\quad\quad\quad\quad\quad\quad\quad\quad\quad\quad 180° - 120°35′ = 59°25′$$

$$\beta = 360° - 275°20′ - \alpha$$
$$= 360° - 275°20′ - 59°25′$$
$$= 25°15′$$

The cosine rule is applied to △OAB.

$$(OB)^2 = \quad\quad\quad\quad\quad\quad\quad\quad\quad\quad 126.0332$$
$$OB = 11.23$$

The ship is 11.23 km from its starting position.

(b) The bearing that the ship must take to return to its starting point is γ. We note that

$$\angle BOC = \gamma \text{ and } \angle BOA = \boxed{} \qquad\qquad \alpha + \gamma$$

The sine rule is applied to $\triangle OAB$.

$$\frac{24}{\sin BOA} = \frac{11.23}{\sin BAO}$$

$$\frac{24}{\sin(\alpha + \gamma)} = \frac{11.23}{\sin 25°15'}$$

$$\sin(\alpha + \gamma) = \boxed{} \qquad\qquad \frac{24 \sin 25°15'}{11.23} = 0.91163$$

$$\alpha + \gamma = \boxed{} \qquad\qquad 65°44 = 114°16'$$

from which

$$\gamma = 6°19'$$

$$= 54°51'$$

In any triangle the longest side is always opposite the largest angle. Referring to Figure 4.7, in which the longest side is AB = 24km, it follows that alpha+ gamma must equal 114°16' and hence gamma = 54°51' is accepted as the required bearing.

Exercises

1   A ship travels for 10 km on a bearing of 30°. It then follows a bearing of 60° for 20 km. Calculate the distance of the ship from the starting position.

2   An aeroplane flies 150 miles on a bearing of 105° and then 107 miles on a bearing of 217°. Find the bearing that the aeroplane must take to fly directly back to the starting position.

3   A ship travels 50 km from O on a bearing of 290° to get to position A. From A it heads directly to B. Position B is 90 km from O on a bearing of 190°.
   (a) Calculate the distance AB.
   (b) Calculate the bearing the ship must follow from A to arrive directly at B.

Solutions to exercises

1   29.09 km

2   327.07°

3   (a) 110.29 km (b) 163.48°

End of block exercises

1 Convert the following angles to degree/minute/second format:
(a) 39.4613° (b) 101.0913° (c) 1.0036°

2 Convert the following angles to decimal format:
(a) 21°31'46" (b) 19°49'17" (c) 1°1'1"

3 The angle of elevation to the top of an aerial is 21°31'2" when measured from a point 41.6 m from the base of the aerial. Calculate the height of the aerial.

4 The angle of elevation to the top of a tower is 23°0'17" when measured from a point 51.5 m from the base of the tower. Calculate the height of the tower.

5 A vertical tower AB is situated on a ramp, as shown in Figure 4.8.

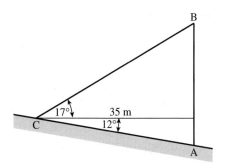

Figure 4.8

From C, the angle of elevation to the top of the tower is 17°; the angle of depression to the foot of the tower is 12°. Given that C is 35 m from the tower, calculate the height of the tower.

6 A ship travels for 9.3 km on a bearing of 36°19'. It then changes its bearing to 100°00' and travels for 12.2 km.

(a) Find the distance of the ship from its starting point.
(b) Calculate the bearing the ship must take to go directly back to its starting point.

7 The angle of elevation to the top of a tower is 19°30' ± 15' when measured from a point 27.00 ± 0.5 m from the base of the tower. Calculate the maximum and minimum possible heights of the tower.

8 A tower AB leans, being 3° from the vertical position. From C the angle of elevation to the top of the tower is 42°. C is 17 m from the base of the tower. Figure 4.9 illustrates the problem.

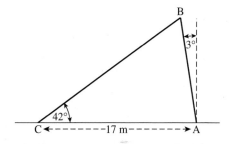

Figure 4.9

Calculate
(a) the length of the tower AB
(b) the height of B above the horizontal AC.

9 A ship starts at O and travels for 10 km on a bearing of 42°10' to arrive at A. From O point B has a bearing of 90° (i.e. due east of O) and is distant 21 km from O. Calculate the bearing the ship must take from A so as to travel directly to B.

Solutions to exercises

1   (a) 39°27′41″ (b) 101°5′29″ (c) 1°0′13″

2   (a) 21.5294° (b) 19.8214° (c) 1.0169°

3   16.40 m

4   21.87 m

5   18.14 m

6   (a) 18.33 km (b) 252°57′12″

7   Minimum height is 9.25 m, maximum height is 9.87 m.

8   (a) 14.64 m (b) 14.62 m

9   117°25′11″

Resolution and resultant of forces

Many engineering problems involve calculations involving forces. To specify a force both its magnitude and its direction must be known. The unit of force is the **newton**, N. For example, a force may be described as 10 N acting at 40° to the horizontal. This force is illustrated in Figure 5.1.

Figure 5.1
A force has magnitude and direction.

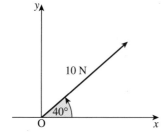

The length of the line represents the magnitude of the force and the direction of the line shows the direction of the force.

5.2 Resolution of a force

A single force may be replaced by two forces acting at right angles to each other. Together these forces are equivalent to the original single force. The process of replacing a single force by two perpendicular forces is called **resolution**. The single force is said to have been **resolved** into two forces at right angles to one another. The following example shows how to resolve a force.

Example 5.1
Resolve the force shown in Figure 5.1 into a horizontal and vertical force.

Solution
Figure 5.2 illustrates the resolution of the 10 N force.

Consider the magnitude of the horizontal component. △OAB contains the relevant information, with OB representing the horizontal force. Then

Figure 5.2
A force may be resolved into two perpendicular forces.

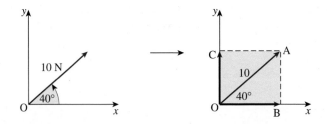

$$\cos 40° = \frac{OB}{OA}$$

$$= \frac{OB}{10}$$

$$OB = 10 \cos 40°$$

$$= 7.66$$

The horizontal force is 7.66 N.

Similarly the vertical force may be calculated using △OCA. OC represents the vertical force. Then

$$\angle COA = 90° - 40°$$

$$= 50°$$

$$\cos 50° = \frac{OC}{OA}$$

$$= \frac{OC}{10}$$

$$OC = 10 \cos 50°$$

$$= 6.43$$

The vertical force is 6.43 N. It is useful to note that the vertical force can also be found by calculating AB in △OAB. In this triangle

$$\sin 40° = \frac{AB}{OA}$$

$$= \frac{AB}{10}$$

$$AB = 10 \sin 40°$$

$$= 6.43$$

The vertical force is 6.43 N.

Hence the 10 N force in Figure 5.1 may be resolved into a 7.66 N force acting horizontally and a 6.43 N force acting vertically.

Example 5.2

A force of 17 N acts at 35° to the negative x axis as shown in Figure 5.3. Resolve the force into two forces, one in the x direction and one in the y direction.

Figure 5.3
A force of 17 N acts at 35° to the negative x axis.

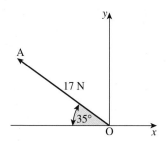

Solution

The horizontal force is represented by OB as shown in Figure 5.4.

Figure 5.4
The horizontal force is represented by OB; the vertical force is represented by BA.

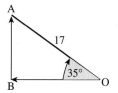

From △OBA

$$\cos 35° = \frac{OB}{OA}$$

$$= \frac{OB}{17}$$

$$OB = \boxed{} \qquad\qquad 17 \cos 35° = 13.93$$

The vertical force is represented by BA. From △OBA

$$\sin 35° = \boxed{} \qquad\qquad \frac{AB}{OA} = \frac{AB}{17}$$

$$AB = \boxed{} \qquad\qquad 17 \sin 35° = 9.75$$

The 17 N force is equivalent to 13.93 N acting in the negative x direction and 9.75 N acting in the y direction.

Example 5.2 illustrates the general result:

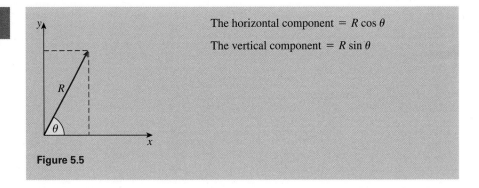

Key point

The horizontal component $= R \cos \theta$

The vertical component $= R \sin \theta$

Figure 5.5

Exercises

1   A 23 N force acts at 71° to the horizontal axis. Resolve the force into forces acting horizontally and vertically.

2   A 40 N force acts at 54° to the horizontal. Resolve the force into horizontal and vertical forces.

3   A 12 N force acts at 43° to the vertical. Resolve the force into horizontal and vertical forces.

Solutions to exercises

1   7.49 N horizontally, 21.75 N vertically

2   23.51 N horizontally, 32.36 N vertically

3   8.18 N horizontally, 8.78 N vertically

5.3   Resolving on an inclined plane

Figure 5.6 illustrates a plane inclined at angle θ to the horizontal. Resting on the plane at O is a body of mass m kg. The body experiences a vertical force mg N due to its weight, as illustrated. Here, g is a constant called the acceleration due to gravity. It is often required to resolve this force into forces that are parallel and perpendicular to the plane.

Figure 5.7 shows the force resolved into forces OA down the plane, and OB perpendicular to the plane. Note that $\angle$AOC has been labelled α and $\alpha = 90° - \theta$. Side OC is the hypotenuse of the right-angled triangle OAC. Furthermore,

$$\sin \alpha = \cos \theta \quad \text{and} \quad \cos \alpha = \sin \theta$$

Figure 5.6

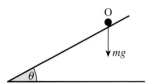

Figure 5.7

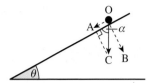

In $\triangle AOC$

$$\cos \alpha = \frac{OA}{OC}$$

so

$$OA = OC \cos \alpha$$
$$= OC \sin \theta$$

Also

$$\sin \alpha = \frac{AC}{OC}$$

so

$$AC = OC \sin \alpha$$
$$= OC \cos \theta$$

Note also that

$$OB = AC = OC \cos \theta$$

So the component parallel to the plane is

$$OA = OC \sin \theta$$
$$= mg \sin \theta$$

The component perpendicular to the plane is

$$OB = OC \cos \theta$$
$$= mg \cos \theta$$

Note the limiting case. If $\theta = 0°$ then the plane is horizontal and intuitively there is no component parallel to the plane. Evaluating the component parallel to the plane, $mg \sin \theta$ with $\theta = 0°$ agrees with the intuitive result.

Key point

For a vertical force mg N and a plane inclined at $\theta°$ to the horizontal:
- Component parallel to plane $= mg \sin \theta$
- Component perpendicular to plane $= mg \cos \theta$

Example 5.3
A body of mass 6 kg rests on a plane inclined at 15° to the horizontal. Find the component of the weight (a) perpendicular to the plane, (b) parallel to the plane.

Solution
Here $m = 6$ and $\theta = 15°$.
(a) The force perpendicular to the plane is

$$mg \cos \theta = 6g \cos 15°$$
$$= 5.7956g \text{ N}$$

(b) The force parallel to the plane is

$$mg \sin \theta = 6g \sin 15°$$
$$= 1.5529g \text{ N}$$

Exercises

1 A body of mass 10 kg rests on a plane inclined at 20° to the horizontal. Calculate the component of the weight
   (a) perpendicular to the plane
   (b) parallel to the plane.

2 A body of mass 1 kg rests on an inclined plane. The force perpendicular to the plane is $0.9g$ N. Calculate the angle of the inclined plane.

Solutions to exercises

1 (a) $9.3969g$ N   (b) $3.420g$ N

2 $25.84°$

5.4   Resultant of a set of forces

In Section 5.2 we saw how a single force can be resolved: that is, expressed as two forces acting at right angles to one another. We now examine a related problem, that of replacing two or more forces acting at a point by a single equivalent force. A single force that is equivalent to two or more forces is called the **resultant**.

Example 5.4 Mechanical Engineering – Forces in a line
A force of 6 N acts in the positive x direction and a force of 2 N acts in the negative x direction. Calculate the resultant.

Solution
The resultant force is $6 - 2 = 4$ N acting in the x direction. Figure 5.8 illustrates this.

Figure 5.8
The resultant is
4 N acting in the
x direction.

Example 5.5 Mechanical Engineering – Forces in a plane
A force of 7 N acts in the x direction and a force of 10 N acts in the negative y direction. Find the resultant force.

Solution
The forces are illustrated in Figure 5.9. The 7 N force is represented by OA; the 10 N force is represented by OC. Let the resultant force be R. The magnitude and direction of R must be found. $\triangle OAB$ contains the relevant information. Note that AB = 10. Let $\angle AOB = \alpha$. Using $\triangle OAB$

$$\tan \alpha = \qquad\qquad \qquad\qquad \frac{AB}{OA} = \frac{10}{7}$$

$$\alpha = \qquad\qquad \qquad\qquad \tan^{-1}\left(\tfrac{10}{7}\right) = 55.0°$$

Figure 5.9
R is the resultant
of the 10 N force
and the 7 N force.

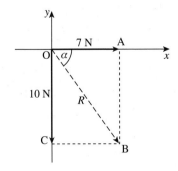

Using Pythagoras's theorem,

$$OB^2 = OA^2 + AB^2$$

$$R^2 = 7^2 + 10^2$$

$$= 149$$

$$R = \sqrt{149}$$

$$= 12.21$$

The resultant is 12.21 N acting at 55° below the x axis.

Example 5.6 Mechanical Engineering – Several forces acting at a point
Figure 5.10 illustrates three forces acting at the origin O.

Figure 5.10
Three forces acting
at the origin.

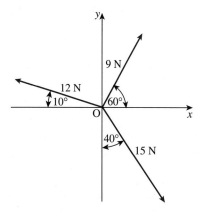

By resolving each force find the resultant.

Solution
Each force is resolved into forces in the *x* and *y* directions.

The 9 N force is resolved into ▨ N acting in the *x* direction
and ▨ N acting in the *y* direction.   9 cos 60°, 9 sin 60°

The 12 N force is resolved into ▨ N in the negative *x*
direction and ▨ N in the *y* direction.   12 cos 10°, 12 sin 10°

The 15 N force is resolved into ▨ N in the *x* direction and
▨ N in the negative *y* direction.   15 sin 40°, 15 cos 40°

The net total force acting in the *x* direction is then

$$9 \cos 60° - 12 \cos 10° + 15 \sin 40° = 2.3241 \text{ N}$$

The net total force acting in the *y* direction is

$$9 \sin 60° + 12 \sin 10° - 15 \cos 40° = -1.6127 \text{ N}$$

The resultant, *R*, may now be found. Figure 5.11 illustrates the situation.
From Figure 5.11

$$\tan \theta = \boxed{}$$ $\dfrac{1.6127}{2.3241}$

$$\theta = \boxed{}$$ 34.76°

The magnitude of the resultant is found using Pythagoras's theorem.

$$R^2 = \boxed{}$$ $(2.3241)^2 + (1.6127)^2$

Figure 5.11
The resultant, R, is found using the net forces in the x and y directions.

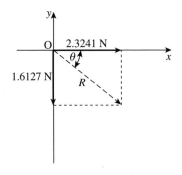

$$R = \boxed{} \qquad\qquad\qquad 2.83$$

The resultant is 2.83 N acting at 34.76° below the x axis.

Example 5.7 Mechanical Engineering – Several forces acting at a point

Figure 5.12 shows three forces acting at the origin O. Calculate the resultant.

Figure 5.12
The resultant of several forces may be found.

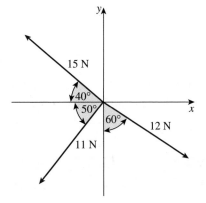

Solution

Each force is resolved into forces along the x and y axes.

The 12 N force is resolved into [   ] N in the x direction and

[   ] N in the negative y direction.   12 sin 60°, 12 cos 60°

The 11 N force is resolved into [   ] N in the negative x direction

and [   ] N in the negative y direction.   11 cos 50°, 11 sin 50°

The 15 N force is resolved into [   ] N in the negative x

direction and [   ] N in the y direction.   15 cos 40°, 15 sin 40°

The net total force in the x direction is [   ]   −8.1690

The net total force in the y direction is [   ]   −4.7847

Figure 5.13 illustrates these forces.

Figure 5.13
The resultant force
is R.

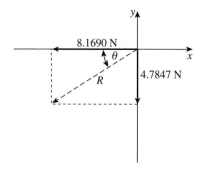

Using Pythagoras's theorem

$$R^2 = \qquad (8.1690)^2 + (4.7847)^2$$

$$R = \qquad 9.47$$

and

$$\tan \theta = \qquad \frac{4.7847}{8.1690}$$

$$\theta = \qquad 30.36°$$

The resultant is a force of 9.47 N acting at 30.36° below the negative x axis.

Exercise

1   Find the resultant of the forces shown in
Figure 5.14.

Figure 5.14

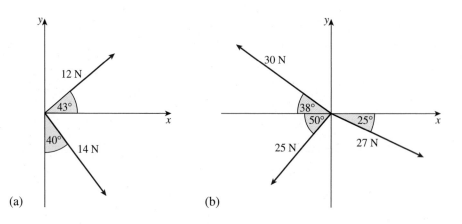

(a)   (b)

Solutions to exercises

1 (a) 17.96 N acting at 8.13° below the positive x axis.

(b) 19.45 N acting at 38.43° below the negative x axis.

End of block exercises

1 Resolve the forces shown in Figure 5.15 into forces along the x and y axes.

2 Find the resultant of the forces shown in Figure 5.16.

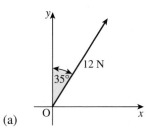

(a)

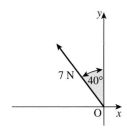

(b)

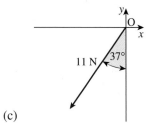

(c)

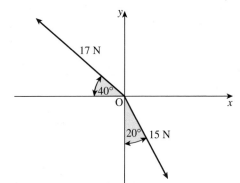

Figure 5.16

3 Find the resultant of the forces shown in Figure 5.17.

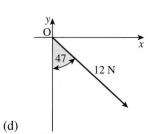

(d)

Figure 5.15

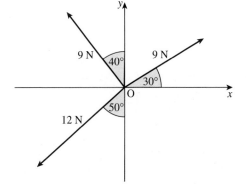

Figure 5.17

4 Find the resultant of the forces shown in Figure 5.18.

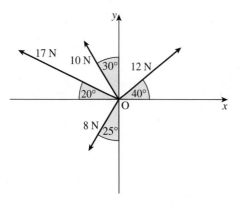

Figure 5.18

6 The forces in Figure 5.20 are in equilibrium. Find R and θ.

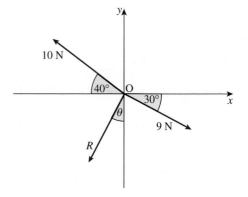

Figure 5.20

5 Forces are said to be in **equilibrium** if their resultant is zero. The forces in Figure 5.19 are in equilibrium. Find R and θ.

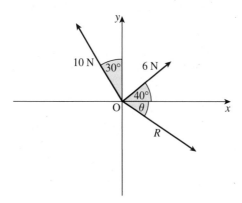

Figure 5.19

Solutions to exercises

1 (a) 6.88 N in the x direction; 9.83 N in the y direction
   (b) 4.50 N in the negative x direction; 5.36 N in the y direction
   (c) 6.62 N in the negative x direction; 8.78 in the negative y direction

   (d) 8.78 N in the x direction; 8.18 N in the negative y direction

2 8.50 N, directed at 21.87° below the negative x axis.

3 $R = 8.07$ N, directed at 27.13° above the negative x axis.	5 $R = 12.52$ N, $\theta = 88.15°$
4 21.29 N, directed at 44.57° above the negative x axis.	6 $R = 1.93$ N, $\theta = 3.97°$

End of chapter exercises

In questions 1–11 ΔABC has a right angle at C.

1   Calculate AB given AC = 9 cm and BC = 15 cm.

2   Calculate sin A given AC = 10 cm and AB = 14 cm.

3   Calculate sin B given AC = 6 cm and BC = 9 cm.

4   Calculate A given AC = 14 cm and BC = 9 cm.

5   Calculate AC given AB = 15 cm and BC = 12 cm.

6   Calculate B given AB = 2(BC).

7   Calculate A given BC is three times AC.

8   Calculate AC given BC = 10 cm and $A = 40°$.

9   Calculate BC given AB = 27 cm and $B = 32°$.

10   Calculate AB given AC = 12 cm and $A = 57°$.

11   Calculate AC given BC = 12 cm and $B = 53°$.

12   A 15 N force acts at 35° to the x axis. Resolve the force into forces in the x and y directions.

13   A 12 N force acts at 40° to the negative y axis. Resolve the force into forces in the x and y directions.

14   Convert the following angles to decimal format: (a) 36°29′42″ (b) 1°2′41″ (c) 10°12′21″

15   Convert the following angles to degree/ minute/second format: (a) 7.3614° (b) 10.0932° (c) 14.9610°

In questions 16–23 solve ΔXYZ given

16   XY = 17 cm, YZ = 23 cm and XZ = 31 cm.

17   YZ = 15 cm, $X = 39°$ and $Y = 75°$.

18   XZ = 85 cm, YZ = 70 cm and $Z = 59°$.

19   $Z = 46°$, YZ = 30 m and XY = 25 m.

20   XY = 71 cm, $Y = 110°$ and $Z = 21°$.

21   XZ = 41 cm, YZ = 29 cm and XY = 52 cm.

22   $Y = 51°$, XY = 25 cm and XZ = 22.5 cm.

23   XY = 100 cm, XZ = 73 cm and $Y = 50°$.

24   Calculate the resultant of the forces in Figure C10.1.

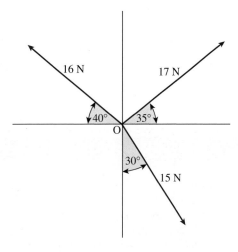

Figure C10.1

25 Calculate the resultant of the forces shown in Figure C10.2.

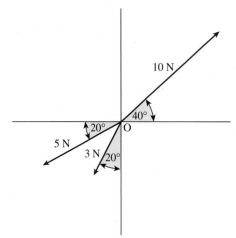

Figure C10.2

26 The angle of elevation to the top, B, of a vertical tower AB is 19°3′ when measured from a point, C, 27.3 m from the base of the tower. Calculate
(a) the height of the tower
(b) the distance BC.

27 Two vertical towers have heights 9 m and 17.2 m and are 42 m apart.
(a) Calculate the angle of elevation from the base of the shorter tower to the top of the taller tower.
(b) Calculate the angle of depression from the top of the taller tower to the top of the shorter tower.

28 A ship travels on a bearing of 40°00′ for 12 km and then changes to a bearing of 270° and travels for 30 km. Calculate

(a) the distance of the ship from its starting point
(b) the bearing the ship must take to return to its starting position.

29 The forces in Figure C10.3 are in equilibrium. Find R and θ.

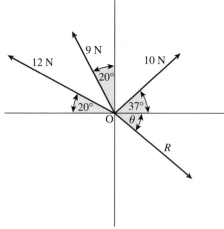

Figure C10.3

30 A tower has a bearing of 37°00′ when measured from a point O, and is 973 m distant from O. A chimney has a bearing of 100°30′ when measured from O and is 1042 m distant from O. Calculate the distance from the tower to the chimney.

31 A point A has a bearing of 45° and is 10 km distant from O. A point B has a bearing of 160° and is 20 km distant from O. A point C is mid-way between A and B. Calculate
(a) the bearing of C from O
(b) the distance of C from O.

Solutions to exercises

1	17.49 cm	4	32.74°
2	0.6999	5	9 cm
3	0.5547	6	60°

7   $71.57°$

8   11.92 cm

9   22.90 cm

10   22.03 cm

11   15.92 cm

12   12.29 N horizontally, 8.60 N vertically

13   x component is 7.71 N; y component is -9.19 N

14   (a) $36.495°$ (b) $1.0447°$ (c) $10.2058°$

15   (a) $7°21'41''$ (b) $10°5'35''$ (c) $14°57'40''$

16   $X = 46.84°$, $Y = 100.54°$, $Z = 32.63°$

17   $Z = 66°$, XZ = 23.02 cm, XY = 21.77 cm

18   XY = 77.43 cm, $X = 50.79°$, $Y = 70.21°$

19   solution 1: $X = 59.6787°$, $Y = 74.3213°$, XZ = 33.46 m; solution 2: $X = 120.3213°$, $Y = 13.6787°$, XZ = 8.22 m

20   $X = 49°$, XZ = 186.17 cm, YZ = 149.52 cm

21   $X = 33.7835°$, $Y = 51.8271°$, $Z = 94.3894°$

22   solution 1: $X = 69.29°$, $Z = 59.71°$, YZ = 27.08 cm; solution 2: $X = 8.71°$, $Z = 120.29°$, YZ = 4.38 cm

23   no triangle possible

24   11.56 N acting at $37.54°$ above positive x axis.

25   $R = 2.87$ N, acting at $47.59°$ to the positive x axis.

26   (a) 9.43 m   (b) 28.88 m

27   (a) $22.27°$ (b) $11.05°$

28   (a) 24.11 km   (b) $112.41°$

29   $R = 19.64$, $\theta = 71.08°$

30   1062 m

31   (a) $130.1198°$ (b) 9.0961 km

Complex numbers

Chapter **11**

Complex numbers are a generalisation of the real number system with which you have been working all your life. Whenever any real non-zero number, whether positive or negative, is squared the result is always positive. Relaxing this condition and allowing numbers that when squared result in negative numbers is the driving force behind the theory of complex numbers. These numbers seem strange when first introduced because we deal with imaginary quantities. However, by persevering a much wider variety of problems can be solved. For example, equations that have no real solutions can have complex solutions as we shall see in this chapter.

Chapter 11 contents

Arithmetic of complex numbers

In this block we explain how the set of real numbers is extended to enable us to find solutions of a greater range of equations. This leads to a study of **complex numbers**, which are useful in a variety of applications, especially alternating current circuit analysis. Complex numbers may seem rather strange at first because you will be dealing with imaginary rather than real quantities. Nevertheless they are so useful in applications that an understanding of them is essential.

1.2 Finding the square root of a negative number

If a real number is squared the answer cannot be negative. For example, squaring both 7 and -7, the result is positive. That is,

$$7^2 = 49 \text{ and } (-7)^2 = 49$$

It is impossible to obtain a negative result by squaring a real number.

Suppose we introduce a new sort of number, called j, with the property that $j^2 = -1$. The number j cannot be a real number because its square is negative. We say it is **imaginary**. Having defined j^2 as -1 it follows that $j = \sqrt{-1}$ and we can use this to write down the square root of any negative number.

Key point

j is an imaginary number such that $j^2 = -1$

Example 1.1
Write down expressions for the square roots of (a) 9, (b) -9.

Solution
(a) $\sqrt{9} = \pm 3$
(b) Noting that $-9 = 9 \times -1$ we can write

$$\sqrt{-9} = \sqrt{9 \times -1}$$
$$= \sqrt{9} \times \sqrt{-1}$$
$$= \pm 3 \times \sqrt{-1}$$

Then using the fact that $\sqrt{-1} = j$ we have

$$\sqrt{-9} = \pm 3j$$

Example 1.2

Write down (a) $\sqrt{100}$, (b) $\sqrt{-100}$, (c) $\sqrt{-\omega^2}$.

Solution

(a) $\sqrt{100} =$ [blank] ± 10

(b) Write $-100 = 100 \times -1$ so that

$$\sqrt{-100} = \sqrt{100 \times (-1)}$$

$$= \text{[blank]}$$ $\pm 10j$

(c) $\sqrt{-\omega^2} = \sqrt{-1 \times \omega^2}$

$$= \text{[blank]}$$ $\pm j\omega$

Example 1.3

Using the fact that $j^2 = -1$ simplify (a) j^3, (b) j^4.

Solution

(a) $j^3 = j^2 \times j$. But $j^2 = -1$ and so $j^3 = -1 \times j = -j$.

(b) $j^4 = j^2 \times j^2 = (-1) \times (-1) = 1$.

Using the imaginary number j it is possible to solve all quadratic equations.

Example 1.4

Use the formula for solving a quadratic equation to solve $2x^2 + x + 1 = 0$.

Solution

We use the formula $x = \dfrac{-b \pm \sqrt{b^2 - 4ac}}{2a}$. With $a = 2, b = 1$ and $c = 1$ we find

$$x = \frac{-1 \pm \sqrt{1^2 - (4)(2)(1)}}{2(2)}$$

$$= \frac{-1 \pm \sqrt{-7}}{4}$$

$$= \frac{-1 \pm \sqrt{7}j}{4}$$

$$= -\frac{1}{4} \pm \frac{\sqrt{7}}{4}j$$

Example 1.5

Use the formula for solving a quadratic equation to solve $x^2 - 10x + 29 = 0$.

Solution

Using the formula $x = \dfrac{-b \pm \sqrt{b^2 - 4ac}}{2a}$ we find

$$x = \boxed{} \qquad\qquad \boxed{\dfrac{10 \pm \sqrt{-16}}{2}}$$

Now using j we can find the square root of -16 and write down two solutions of the equation. Finally simplifying the answers gives

$$x = \boxed{} \quad \text{and} \quad \boxed{} \qquad\qquad \boxed{5 + 2j} \text{ and } \boxed{5 - 2j}$$

Exercises

1   Write down an expression for
    (a) $\sqrt{4}$, (b) $\sqrt{-4}$, (c) $\sqrt{81}$, (d) $\sqrt{-81}$.

2   With the help of a calculator find
    (a) $\sqrt{7}$, (b) $\sqrt{-7}$, (c) $\sqrt{5.32}$, (d) $\sqrt{-5.32}$.

3   Simplify (a) $-j^2$, (b) $(-j)^2$, (c) $(-j)^3$, (d) $-j^3$.

4   Solve the following quadratic equations:
    (a) $x^2 + 1 = 0$ (b) $x^2 + x + 2 = 0$
    (c) $x^2 + 5x + 3 = 0$ (d) $2x^2 + 5x + 3 = 0$
    (e) $2x^2 + x + 3 = 0$

5   Solve the quadratic equation
    $5x^2 - 11x + 13 = 0$.

Solutions to exercises

1   (a) ± 2 (b) $\pm 2j$ (c) ± 9 (d) $\pm 9j$

2   (a) ± 2.646 (b) $\pm 2.646j$ (c) ± 2.307
    (d) $\pm 2.307j$

3   (a) 1 (b) -1 (c) j (d) j

4   (a) $\pm j$ (b) $-\dfrac{1}{2} \pm \dfrac{\sqrt{7}}{2}j$ (c) $-\dfrac{5}{2} \pm \dfrac{\sqrt{13}}{2}$

    (d) $-1, -\dfrac{3}{2}$ (e) $-\dfrac{1}{4} \pm \dfrac{\sqrt{23}}{4}j$

5   $\dfrac{11}{10} \pm \dfrac{\sqrt{139}}{10}j$

1.3   Complex numbers

Real and imaginary parts

In Example 1.5 we found that the solutions of the equation $x^2 - 10x + 29 = 0$ are $5 \pm 2j$. The solutions are known as **complex numbers**. A complex number such as $5 + 2j$ is made up of two parts, a **real part** 5 and an **imaginary part** 2. The imaginary part is the multiple of j.

It is common practice to use the letter z to stand for a complex number and write $z = a + bj$ where a is the real part and b is the imaginary part.

Key point

> If z is a complex number then we write
>
> $$z = a + b\mathrm{j}$$
>
> where a is the real part and b is the imaginary part.

Example 1.6

State the real and imaginary parts of $-11 - 19\mathrm{j}$.

Solution

The real part is ░░░░░ -11

The imaginary part is ░░░░░ -19

Equal complex numbers

Two complex numbers are equal only when their real parts are equal and their imaginary parts are equal. So if $a + b\mathrm{j}$ is equal to $3 - 2\mathrm{j}$ it follows that a must be 3 and b must be -2.

Complex conjugate

In Example 1.5 we solved the quadratic equation $x^2 - 10x + 29 = 0$ and saw that the second solution, $5 - 2\mathrm{j}$, is almost the same as the first, $5 + 2\mathrm{j}$; only the sign of the imaginary part has changed. The number $5 - 2\mathrm{j}$ is said to be the **complex conjugate** of $5 + 2\mathrm{j}$.

In general, to find the complex conjugate of a complex number the sign of the imaginary part is changed from $+$ to $-$, or vice versa. We denote the complex conjugate of z by $\bar{z}$.

Key point

> If $z = a + b\mathrm{j}$, its **complex conjugate**, denoted by $\bar{z}$, is
>
> $$\bar{z} = a - b\mathrm{j}$$

When solving quadratic equations with real coefficients any complex roots occur in complex conjugate pairs just as we saw in Example 1.5.

Example 1.7

A complex number z_1 is given by $z_1 = a_1 + b_1\mathrm{j}$.
(a) State the real and imaginary parts of z_1.
(b) Write down the complex conjugate of z_1.

Solution

(a) The real part of z_1 is a_1. The imaginary part is b_1.
(b) The complex conjugate of z_1 is found by changing the sign of the imaginary part. Thus $\overline{z_1} = a_1 - b_1\mathrm{j}$.

Example 1.8
Write down the complex conjugates of each of the following complex numbers:
(a) $z_1 = 11 + 2j$, (b) $z_2 = 7 - 3j$, (c) $z_3 = 8$, (d) $z_4 = -9j$.

Solution
In each case the sign of the imaginary part is changed. In part (c) there is no imaginary part to alter.

(a) $\overline{z_1} = $ [_____] $\hspace{3cm}$ $11 - 2j$

(b) $\overline{z_2} = $ [_____] $\hspace{3cm}$ $7 + 3j$

(c) $\overline{z_3} = $ [___] $\hspace{3cm}$ 8

(d) $\overline{z_4} = $ [___] $\hspace{3cm}$ $9j$

Exercises

1   If $z = a + bj$ explain what is meant by $\overline{z}$.

2   State the real and imaginary parts of
(a) $3 + 7j$ (b) $a_1 + b_1 j$ (c) $-0.35j$
(d) $\cos\theta + j\sin\theta$ (e) $\cos\omega t + j\sin\omega t$

3   Write down the complex conjugate of
(a) $3 + 13j$ (b) $3 - 13j$ (c) $17j$ (d) 18.5
(e) $3.36 + 2.24j$

4   State the real and imaginary parts of
$jV\sin(\omega t + \phi)$.

Solutions to exercises

1   $\overline{z}$ is the complex conjugate of z. That is,
$\overline{z} = a - bj$.

2   (a) real part 3, imaginary part 7
(b) a_1, b_1 (c) $0, -0.35$ (d) $\cos\theta, \sin\theta$
(e) $\cos\omega t, \sin\omega t$

3   (a) $3 - 13j$ (b) $3 + 13j$ (c) $-17j$
(d) 18.5 (e) $3.36 - 2.24j$

4   Real part 0, imaginary part $V\sin(\omega t + \phi)$.

1.4   Addition and subtraction of complex numbers

Given two complex numbers we can find their sum and difference in an obvious way.

Key point

Addition and subtraction of complex numbers

If $z_1 = a_1 + b_1 j$ and $z_2 = a_2 + b_2 j$ then

$$z_1 + z_2 = (a_1 + a_2) + (b_1 + b_2)j$$

$$z_1 - z_2 = (a_1 - a_2) + (b_1 - b_2)j$$

Note that to add the complex numbers we simply add the real parts together and add the imaginary parts together.

Example 1.9

If $z_1 = 3 + 11j$ and $z_2 = 18 - 21j$ find (a) $z_1 + z_2$, (b) $z_2 - z_1$.

Solution

(a) $z_1 + z_2 = (3 + 11j) + (18 - 21j)$
$= 21 - 10j$
(b) $z_2 - z_1 = (18 - 21j) - (3 + 11j)$
$= 15 - 32j$

Example 1.10

If $z_1 = 3 + 4j$ and $z_2 = 8 + 11j$, find (a) $z_1 + z_2$, (b) $z_1 - z_2$.

Solution

(a) To add the complex numbers just add the real parts and then add the imaginary parts:

$$z_1 + z_2 = $$

$$11 + 15j$$

(b) Subtract the real parts, and then subtract the imaginary parts:

$$z_1 - z_2 = (3 + 4j) - (8 + 11j)$$

$$= $$

$$-5 - 7j$$

Exercises

1 If $z_1 = 1 + j$ and $z_2 = 3 + 2j$ find
(a) $z_1 + z_2$, (b) $z_1 - z_2$, (c) $z_2 - z_1$.

2 If $z = 13 - 11j$ and $w = -2 + 5j$ find
(a) $z + w$, (b) $z - w$, (c) $w - z$.

3 If $z = \alpha + j\beta$ and $w = \gamma + j\delta$ write down
$z + w$.

Solutions to exercises

1 (a) $4 + 3j$ (b) $-2 - j$ (c) $2 + j$

2 (a) $11 - 6j$ (b) $15 - 16j$ (c) $-15 + 16j$

3 $\alpha + \gamma + j(\beta + \delta)$

1.5 Multiplication of complex numbers

To multiply two complex numbers we use the normal rules of algebra and also the fact that $j^2 = -1$. If z_1 and z_2 are the two complex numbers their product is written $z_1 z_2$.

Example 1.11
If $z_1 = 5 - 2j$ and $z_2 = 2 + 4j$ find $z_1 z_2$.

Solution

$$z_1 z_2 = (5 - 2j)(2 + 4j)$$
$$= 10 - 4j + 20j - 8j^2$$

Replacing j^2 by -1 we obtain

$$z_1 z_2 = 10 + 16j - 8(-1)$$
$$= 18 + 16j$$

In general we have the following result:

Key point

Multiplication of two complex numbers

If $z_1 = a_1 + b_1 j$ and $z_2 = a_2 + b_2 j$ then

$$z_1 z_2 = (a_1 + b_1 j)(a_2 + b_2 j)$$
$$= a_1 a_2 + a_1 b_2 j + b_1 a_2 j + b_1 b_2 j^2$$
$$= (a_1 a_2 - b_1 b_2) + j(a_1 b_2 + a_2 b_1)$$

Example 1.12
Find $z_1 z_2$ if $z_1 = 3 - 2j$ and $z_2 = 5 + 3j$.

Solution
$$z_1 z_2 = (3 - 2j)(5 + 3j)$$

$$=$$
$\qquad\qquad\qquad\qquad\qquad\qquad$ $15 + 9j - 10j - 6j^2$

Simplify your result to obtain

$(3 - 2j)(5 + 3j) =$ $\qquad\qquad\qquad\qquad\qquad\qquad$ $21 - j$

Example 1.13
Find $z \bar{z}$ if $z = a + bj$.

Solution
Recall that $\bar{z}$ is the complex conjugate of z.

$$z \bar{z} = (a + bj)(a - bj)$$
$$= a^2 + baj - abj - b^2 j^2$$

$$= a^2 - b^2(-1)$$
$$= a^2 + b^2$$

Note that the result is purely real: multiplying a complex number by its conjugate results in a non-negative real answer. For example, taking $a = 4$ and $b = 3$, we see that $(4 + 3j)(4 - 3j) = 4^2 + 3^2 = 25$.

The result of the previous example is particularly important.

Key point

If $z = a + bj$ then

$$z\bar{z} = a^2 + b^2$$

Exercises

1   If $z_1 = 1 + j$ and $z_2 = 3 + 2j$ find
    (a) $z_1 z_2$, (b) $\bar{z_1}$, (c) $\bar{z_2}$, (d) $z_1 \bar{z_1}$, (e) $z_2 \bar{z_2}$.

2   Express in the form $x + jy$
    (a) $j(3 + 2j)$ (b) $(4 - 2j)j$ (c) $5j(1 - j)$
    (d) $(1 + j)(9 + j)$

3   Find a quadratic equation whose roots are
    $s = -7 + 0.5j$ and $s = -7 - 0.5j$.

4   Find a cubic equation with roots
    $z = 2, z = 1 + j$ and $z = 4 + 2j$. (Note that this cubic equation will have complex coefficients.)

5   Find the real and imaginary parts of
    (a) $7(2 + j)$, (b) $-13(j - 1)$.

6   Find (a) $(4 - 3j)(4 + 3j)$,
    (b) $(5 - 2j)(5 + 2j)$.

Solutions to exercises

1   (a) $1 + 5j$ (b) $1 - j$ (c) $3 - 2j$ (d) 2 (e) 13

2   (a) $-2 + 3j$ (b) $2 + 4j$ (c) $5 + 5j$
    (d) $8 + 10j$

3   For example, $4s^2 + 56s + 197 = 0$

4   For example,
    $$z^3 - (7 + 3j)z^2$$
    $$+ (12 + 12j)z - 4 - 12j = 0$$

5   (a) real 14, imaginary 7 (b) $13, -13$

6   (a) 25 (b) 29

1.6   Division of complex numbers

Division of complex numbers requires a special technique that uses the complex conjugate. Consider the following example.

Example 1.14

Find $\dfrac{z_1}{z_2}$ when $z_1 = 3 + 2j$ and $z_2 = 4 - 3j$.

Solution

We require

$$\frac{z_1}{z_2} = \frac{3 + 2j}{4 - 3j}$$

The special technique is to multiply both numerator and denominator by the complex conjugate of the denominator. Overall, this is equivalent to multiplying by 1 and so the fraction remains unaltered. However, it has the effect of making the denominator purely real as we saw in Example 1.13.

$$\frac{3 + 2j}{4 - 3j} = \frac{3 + 2j}{4 - 3j} \times \frac{4 + 3j}{4 + 3j}$$

$$= \frac{(3 + 2j)(4 + 3j)}{(4 - 3j)(4 + 3j)}$$

$$= \frac{12 + 8j + 9j + 6j^2}{16 + 12j - 12j - 9j^2}$$

$$= \frac{6 + 17j}{25}$$

$$= \frac{6}{25} + \frac{17}{25}j$$

In general we have the following result:

Key point

Division of two complex numbers

$$\frac{z_1}{z_2} = \frac{a_1 + b_1j}{a_2 + b_2j}$$

$$= \frac{a_1 + b_1j}{a_2 + b_2j} \times \frac{a_2 - b_2j}{a_2 - b_2j}$$

$$= \frac{a_1a_2 + b_1b_2 + (a_2b_1 - a_1b_2)j}{a_2^2 + b_2^2}$$

Example 1.15

If $z_1 = 2 + 9j$ and $z_2 = 5 - 2j$ find $\dfrac{z_1}{z_2}$.

Solution

Both numerator and denominator are multiplied by the complex conjugate of the denominator.

So

$$\frac{2 + 9j}{5 - 2j} = \frac{2 + 9j}{5 - 2j} \times \frac{5 + 2j}{5 + 2j}$$

$$= \boxed{} \qquad \frac{10 + 45j + 4j + 18j^2}{25 + 4}$$

Simplify the result to produce

$$\frac{2 + 9j}{5 - 2j} = \boxed{}$$

$$-\frac{8}{29} + \frac{49}{29}j$$

Exercises

1 If $z_1 = 1 + j$ and $z_2 = 3 + 2j$ find

(a) $\dfrac{z_1}{z_2}$ (b) $\dfrac{z_2}{z_1}$ (c) $\dfrac{z_1}{\overline{z_1}}$ (d) $\dfrac{z_2}{\overline{z_2}}$

2 Find (a) $\dfrac{7 - 6j}{2j}$, (b) $\dfrac{3 + 9j}{1 - 2j}$.

3 Express the following in the form $x + jy$:

(a) $\dfrac{1}{j}$ (b) $\dfrac{2}{j}$ (c) $\dfrac{3}{-j}$ (d) $\dfrac{1}{1 + j}$ (e) $\dfrac{2}{1 - j}$

4 If $z = \dfrac{3 + 2j}{2 - j}$ express both (a) $\dfrac{1}{z}$ and

(b) $z + \dfrac{1}{z}$ in the form $x + jy$.

5 Express $\dfrac{1}{\cos \theta - j \sin \theta}$ in the form $a + bj$.

Solutions to exercises

1 (a) $\dfrac{5}{13} + \dfrac{j}{13}$ (b) $\dfrac{5}{2} - \dfrac{j}{2}$ (c) j (d) $\dfrac{5}{13} + \dfrac{12}{13}j$

2 (a) $-3 - \dfrac{7}{2}j$ (b) $-3 + 3j$

3 (a) $-j$ (b) $-2j$ (c) $3j$ (d) $\dfrac{1}{2} - \dfrac{j}{2}$ (e) $1 + 1j$

4 (a) $\dfrac{4}{13} - \dfrac{7}{13}j$ (b) $\dfrac{72}{65} + \dfrac{56}{65}j$

5 $\cos \theta + j \sin \theta$

Computer and calculator exercises

Computer algebra packages allow the user to input and manipulate complex numbers. Investigate how the complex number $a + bj$ is input to the package to which you have access.

1 Use a package to express each of the following numbers in the form $a + bj$:

(a) $(2 - j)^6$ (b) $\dfrac{5}{(3 - 2j)^4}$

2 Solve the equations
(a) $x^3 + 7x^2 + 9x + 63 = 0$
(b) $2x^4 - 3x^3 + 11 = 0$

3 A common requirement in control theory is to find the poles of a rational function, $G(s)$. The poles are the values of s that make the denominator zero. Find the poles of

$$G(s) = \frac{0.25}{s^3 - 2s^2 + 0.1s + 0.006}$$

End of block exercises

1 Write down an expression for (a) $\sqrt{1}$, (b) $\sqrt{-1}$, (c) $\sqrt{49}$, (d) $\sqrt{-49}$.

2 Find $\sqrt{-9.24}$.

3 Solve the equations
(a) $x^2 + x + 1 = 0$, (b) $7s^2 + s + 1 = 0$.

4 Write down the complex conjugate of
(a) $15 + 3j$ (b) $-3j + 2$
(c) $\frac{1}{2}j + 3$ (d) $\frac{1}{2} + \frac{\sqrt{3}}{2}j$

5 Find
(a) $\frac{3 + 2j}{5 + 4j}$, (b) $\frac{5 + 4j}{3 + 2j}$.

6 Find
(a) $z_1 z_2$, (b) $\frac{z_1}{z_2}$, (c) $\frac{z_2}{z_1}$, if $z_1 = 3 + 5j$
and $z_2 = 3 - 5j$.

7 Express the following in the form $x + jy$:
(a) $\frac{5}{-j}$ (b) $\frac{2 + j}{j}$ (c) $\frac{3}{4 - j}$
(d) $\frac{j}{1 + j}$ (e) $\frac{2j}{1 - j}$

8 Write down the complex conjugate of
(a) $\alpha + j\beta$ (b) $\cos\theta + j\sin\theta$
(c) $\cos\theta - j\sin\theta$ (d) $\cos\omega t - j\sin\omega t$

9 Express in the form $x + jy$
(a) $(3 + 2j)j(4 - 3j)$
(b) $(5 + 3j)(2 + j)(2 - j)$
(c) $\frac{(2 - 3j)(6 + 7j)}{j(1 - j)}$

10 Find a quadratic equation whose roots are
$x = -3 + j$ and $x = -3 - j$.

11 Find a cubic equation with roots $x = 1$,
$x = 1 - j$ and $x = 1 + j$.

Solutions to exercises

1 (a) ± 1 (b) $\pm j$ (c) ± 7 (d) $\pm 7j$

2 $\pm 3.04j$

3 (a) $-\frac{1}{2} \pm \frac{\sqrt{3}j}{2}$ (b) $-\frac{1}{14} \pm \frac{3\sqrt{3}}{14}j$

4 (a) $15 - 3j$ (b) $3j + 2$
(c) $-\frac{1}{2}j + 3$ (d) $\frac{1}{2} - \frac{\sqrt{3}}{2}j$

5 (a) $\frac{23}{41} - \frac{2}{41}j$ (b) $\frac{23}{13} + \frac{2}{13}j$

6 (a) 34 (b) $-\frac{8}{17} + \frac{15}{17}j$ (c) $-\frac{8}{17} - \frac{15}{17}j$

7 (a) $5j$ (b) $1 - 2j$ (c) $\frac{12}{17} + \frac{3}{17}j$ (d) $\frac{1}{2} + \frac{1}{2}j$
(e) $-1 + j$

8 (a) $\alpha - j\beta$ (b) $\cos\theta - j\sin\theta$
(c) $\cos\theta + j\sin\theta$ (d) $\cos\omega t + j\sin\omega t$

9 (a) $1 + 18j$ (b) $25 + 15j$ (c) $\frac{29}{2} - \frac{37}{2}j$

10 $x^2 + 6x + 10 = 0$, for example

11 $x^3 - 3x^2 + 4x - 2 = 0$, for example

The Argand diagram and polar form of a complex number

2.1 Introduction

Engineers often find a graphical interpretation of complex numbers useful. This graphical interpretation is known as an Argand diagram. From the Argand diagram two other useful quantities, the modulus and the argument, can be defined. These provide an alternative way of describing complex numbers, known as the polar form.

2.2 The Argand diagram

The complex number $z = a + bj$ is plotted as a point with coordinates (a, b) as shown in Figure 2.1. Because the real part of z is plotted on the horizontal axis we often refer to this as the **real axis**. The imaginary part of z is plotted on the vertical axis and so we refer to this as the **imaginary axis**. Such a diagram is called an **Argand diagram**. Engineers often refer to this diagram as the **complex plane**.

Figure 2.1
An Argand diagram in which the point with coordinates (a, b) represents the complex number $z = a + bj$.

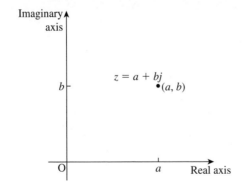

Example 2.1
Plot the complex numbers $z_1 = 2 + 3j$, $z_2 = -3 + 2j$, $z_3 = -3 - 2j$, $z_4 = 2 - 5j$, $z_5 = 6$, $z_6 = j$ on an Argand diagram.

Solution
Figure 2.2 shows the Argand diagram. Note that purely real numbers lie on the real axis. Purely imaginary numbers lie on the imaginary axis. Note also that complex conjugate pairs such as $-3 \pm 2j$ lie symmetrically on opposite sides of the real axis.

Figure 2.2
Argand diagram
for Example 2.1.

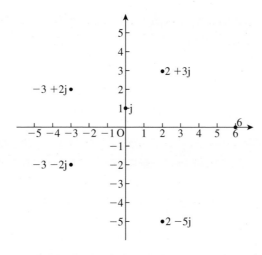

Exercises

1 Plot the following numbers on an Argand diagram:
(a) 3 + 3j (b) 2 − 4j (c) −0.5 (d) 6j
(e) −j (f) −5 − 11j

2 Simplify the following complex numbers and show them on an Argand diagram:
(a) j^2 (b) $-j^2$ (c) $(-j)^2$ (d) j^3 (e) j^4

Solutions to exercises

1 See Figure 2.3.

2 See Figure 2.4.

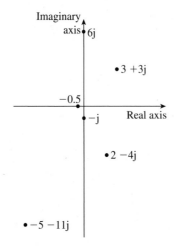

Figure 2.3

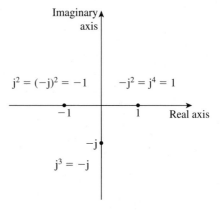

Figure 2.4

2.3 The modulus and argument of a complex number

Modulus

Consider the Argand diagram in Figure 2.5, which shows the complex number $z = a + bj$. The distance of the point (a, b) from the origin is known as the **modulus**, or **magnitude**, of the complex number and is given the symbol r. Alternatively r is written as $|z|$. The plural of modulus is **moduli**. The modulus can be found using Pythagoras's theorem, that is $|z| = r = \sqrt{a^2 + b^2}$. The modulus is never negative.

Figure 2.5
The modulus of
$a + bj$ is the
distance of (a, b)
from the origin O.

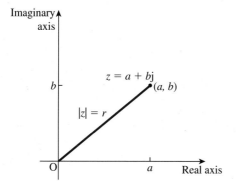

Key point

The **modulus** of the complex number $z = a + bj$ is

$$r = |z| = \sqrt{a^2 + b^2}$$

Graphically, this is the distance of (a, b) from the origin.

Example 2.2

Plot the following complex numbers on an Argand diagram and find their moduli:
(a) $z_1 = 3 + 4j$ (b) $z_2 = -2 + j$ (c) $z_3 = 3j$

Solution

The complex numbers are shown in Figure 2.6.

In each case we can use Pythagoras's theorem to find the modulus.

(a) $|z_1| = \sqrt{3^2 + 4^2} = \sqrt{25} = 5$.

(b) $|z_2| = \sqrt{(-2)^2 + 1^2} = \sqrt{5}$ or 2.236. It is usually sufficient to leave your answer in the surd form as $\sqrt{5}$.

(c) $|z_3| = \sqrt{0^2 + 3^2} = 3$.

Figure 2.6

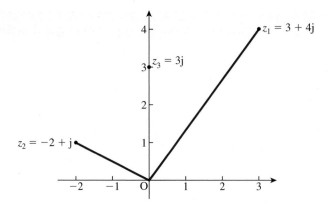

Example 2.3

Find the modulus of each of the following complex numbers:

(a) $z = 5 + 12j$ (b) $z = -6 - 8j$

Solution

(a) $|z| = |5 + 12j| =$ ⬚ $\quad\quad\quad\quad \sqrt{5^2 + 12^2} = 13$

(b) $|z| = |-6 - 8j| =$ ⬚ $\quad\quad\quad \sqrt{(-6)^2 + (-8)^2} = 10$

Argument

Consider Figure 2.7. Given a complex number $z = a + bj$ we can calculate the angle between the positive x axis and a line joining (a, b) to the origin. This angle is called the **argument** of the complex number. It is abbreviated to arg(z) and often given the symbol θ.

Figure 2.7
The argument of a complex number.

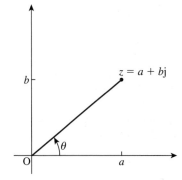

We usually measure this angle so that $-\pi < \theta \leq \pi$, in which case θ is called the **principal value** of the argument. Angles measured anticlockwise from the positive x axis are conventionally positive whereas angles measured clockwise are negative. Several complex numbers with positive and negative arguments are shown in Figure 2.8.

Figure 2.8
Some complex numbers with positive and negative arguments.

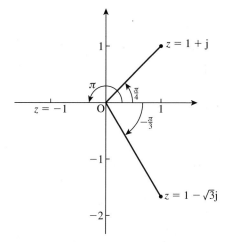

Because the angles in Figure 2.8 can be described in a variety of ways, the argument of a complex number is not unique. For example, the argument of $z = 1 + j$ has been labelled $\frac{\pi}{4}$. This is the principal value of the argument. However, it is important to note that we could have also stated the argument as $\frac{\pi}{4} + 2\pi = \frac{9\pi}{4}$. Repeatedly adding on 2π will produce other valid arguments. This will become important in Block 5 when we use complex numbers to solve equations.

Given $z = a + bj$, then trigonometry can be used to determine θ. Specifically, referring to Figure 2.7,

$$\tan \theta = \frac{b}{a}$$

so that

$$\theta = \tan^{-1}\left(\frac{b}{a}\right)$$

but when using a calculator to find an inverse tangent care must be taken that the solution obtained is in the correct quadrant. Drawing an Argand diagram will always help to identify the correct quadrant. At this stage you may find it helpful to refer back to Chapter 9 on trigonometry, and ensure that you are aware of the sign of the trigonometrical ratios in the four quadrants.

Key point

The argument of the complex number $z = a + bj$ is the angle between the positive x axis and the line joining (a, b) to the origin.

Polar form

The position of a complex number is uniquely determined by giving its modulus and argument. This description is known as the **polar form**. Engineers often write the polar form of a complex number as $z = r\angle\theta$.

The **polar form** of a complex number can be written as

$$z = r\angle\theta$$

where r is the modulus and θ is the argument.

Example 2.4

Find the arguments of the complex numbers in Example 2.2 and pictured in Figure 2.6.

Solution

(a) $z_1 = 3 + 4j$ is in the first quadrant. Its argument is given by

$$\theta = \tan^{-1}\frac{4}{3}$$

Using a calculator we find $\theta = 0.927$ radians.

(b) $z_2 = -2 + j$ is in the second quadrant. To find its argument we seek an angle, θ, in the second quadrant such that $\tan\theta = \frac{1}{-2}$. To calculate this correctly it may help to refer to Figure 2.9 in which α is an acute angle with $\tan\alpha = \frac{1}{2}$. From a calculator $\alpha = 0.464$, and so $\theta = \pi - 0.464 = 2.678$ radians.

Figure 2.9
To calculate θ it is simpler to first calculate α.

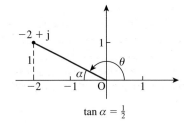

$$\tan\alpha = \tfrac{1}{2}$$

(c) $z_3 = 3j$ is purely imaginary. Its argument is $\frac{\pi}{2}$ radians.

Example 2.5

Sketch an Argand diagram showing the following complex numbers and in each case calculate the argument.

(a) $z = 5 + 12j$ (b) $z = -\sqrt{3} + j$ (c) $z = 2 - 3j$

Solution

(a) Sketch $z = 5 + 12j$ and identify the quadrant in which it lies:

 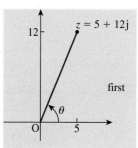

Then arg $z = \theta =$  $\tan^{-1}\left(\dfrac{12}{5}\right) = 1.176$ radians

(b) Sketch $z = -\sqrt{3} + j$ and identify the quadrant in which it lies:

 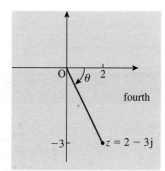

Then arg $z = \theta =$

$$\tan^{-1}\left(-\dfrac{1}{\sqrt{3}}\right) = \dfrac{5\pi}{6} = 2.618 \text{ radians}$$

If you have difficulty calculating the correct angle it is often simpler to calculate first the related acute angle. So, in the diagram above, α is acute and $\tan \alpha = \dfrac{1}{\sqrt{3}}$ from which $\alpha = \dfrac{\pi}{6} = 0.524$ radians. It follows that $\theta = \pi - \alpha = \dfrac{5\pi}{6} = 2.618$ radians.

(c) Sketch $z = 2 - 3j$ and identify the quadrant in which it lies:

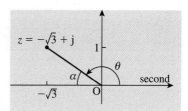

Then arg $(z) = \theta =$

$$\tan^{-1}\left(-\frac{3}{2}\right) = -0.983 \text{ radians}$$

Ensure you have located the angle in the correct quadrant. If you have difficulty calculating the correct angle use the method described in part (b).

Most scientific calculators are able to convert complex numbers given in cartesian form into polar form and vice versa. You should check your calculator to see whether you can carry out these conversions.

Exercises

1 The following complex numbers were plotted on an Argand diagram in question 1 of the previous exercises:
(a) $3 + 3j$ (b) $2 - 4j$ (c) -0.5 (d) $6j$ (e) $-j$
(f) $-5 - 11j$
Express each one in polar form.

2 Show the following complex numbers on an Argand diagram:
(a) $5\angle 30°$ (b) $4\angle 45°$ (c) $2\angle -135°$
(d) $4\angle 270°$ (e) $17\angle 110°$ (f) $1\angle\left(\frac{\pi}{3}\right)$
(g) $4\angle\left(\frac{\pi}{4}\right)$ (h) $2\angle -2.3$

3 (a) If $z = 5$, plot z and jz on an Argand diagram. Deduce that multiplying z by j causes an anticlockwise rotation through $90°$ about the origin.

(b) If $z = 2 + 3j$, plot z and jz. Show that the same behaviour as that in part (a) occurs. Is it true in general that multiplying a complex number by j will rotate it anticlockwise through $90°$?

4 (a) Solve the equation $z^2 - 4z + 13 = 0$ and plot your solutions on an Argand diagram.
(b) Express the solutions in polar form.

5 If $z = a + bj$ show that $|z| = |\bar{z}|$ and that $|z|^2 = z\bar{z}$.

Solutions to exercises

1 (a) $\sqrt{18}\angle\frac{\pi}{4} = 3\sqrt{2}\angle\frac{\pi}{4}$ (b) $2\sqrt{5}\angle -1.107$
(c) $0.5\angle\pi$ (d) $6\angle\left(\frac{\pi}{2}\right)$ (e) $1\angle\left(-\frac{\pi}{2}\right)$
(f) $\sqrt{146}\angle -1.997$

2   See Figure 2.10.

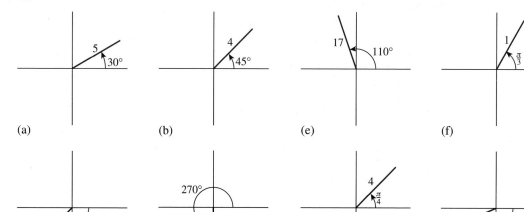

(a)   (b)   (e)   (f)

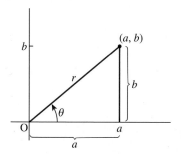

(c)   (d)   (g)   (h)

Figure 2.10

3   (b) Yes, multiplying a number by j has the effect of rotating it anticlockwise by 90°.

4   (a) $2 \pm 3j$ (b) $\sqrt{13}\angle 0.98$, $\sqrt{13}\angle -0.98$

2.4   The form $r(\cos\theta + j\sin\theta)$

Consider Figure 2.11, which shows the complex number $z = a + bj = r\angle\theta$.

Figure 2.11

Using trigonometry we can write

$$\cos\theta = \frac{a}{r} \quad \text{and} \quad \sin\theta = \frac{b}{r}$$

so that, by rearrangement,

$$a = r\cos\theta \quad \text{and} \quad b = r\sin\theta$$

We can use these results to find the real and imaginary parts of a complex number given in polar form:

Key point

If $z = r\angle\theta$, the real and imaginary parts of z are, respectively,

$$a = r\cos\theta \quad \text{and} \quad b = r\sin\theta$$

Using these results we can then write $z = a + bj$ as

$$z = a + bj = r\cos\theta + jr\sin\theta$$
$$= r(\cos\theta + j\sin\theta)$$

This is an alternative way of expressing the complex number with modulus r and argument θ.

Key point

$$z = a + bj$$
$$= r\angle\theta$$
$$= r(\cos\theta + j\sin\theta)$$

Recall that the complex conjugate of $z = a + bj$ is the complex number $\bar{z} = a - bj$, as shown in Figure 2.12.

Figure 2.12
The complex conjugate of $a + bj$.

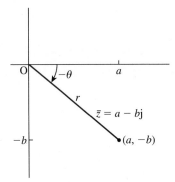

Note that $\bar{z}$ has the same real part as z but the sign of the imaginary part has changed. It follows that we can write $\bar{z} = a - bj = r(\cos\theta - j\sin\theta)$.

Key point

$$\bar{z} = a - bj$$
$$= r(\cos\theta - j\sin\theta)$$

Example 2.6
State the modulus and argument of $z = 17(\cos 3.2 + j\sin 3.2)$.

Solution
Comparing the given complex number with the standard form $r(\cos\theta + j\sin\theta)$ we see that $r = 17$ and $\theta = 3.2$ radians.

Example 2.7

(a) Find the modulus and argument of the complex number $z = 5j$.

(b) Express 5j in the form $r(\cos\theta + j\sin\theta)$.

Solution

(a) On an Argand diagram the complex number 5j lies on the positive vertical axis a distance 5 from the origin. Thus 5j is a complex number with modulus 5 and argument $\frac{\pi}{2}$.

(b) $z = 5j$

$$= 5\left(\cos\frac{\pi}{2} + j\sin\frac{\pi}{2}\right)$$

Example 2.8

(a) Show the complex number $z = 4\angle\left(\frac{\pi}{3}\right)$ on an Argand diagram.

(b) State its modulus and argument.

(c) Find the real and imaginary parts of the complex number and hence express it in the form $a + bj$.

Solution

(a) The complex number $z = 4\angle\left(\frac{\pi}{3}\right)$ is shown in Figure 2.13.

Figure 2.13
The complex number $z = 4\angle\left(\frac{\pi}{3}\right)$.

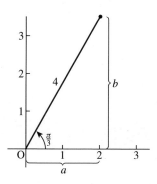

(b) Its modulus is 4 and its argument is $\frac{\pi}{3}$.

(c) To find the real and imaginary parts we can use

$$a = r\cos\theta, \quad b = r\sin\theta$$

so that

$$a = 4\cos\frac{\pi}{3}$$
$$= 2$$

and

$$b = 4\sin\frac{\pi}{3}$$
$$= 3.464$$

Hence the complex number can be written $2 + 3.464j$.

Example 2.9

Find the real and imaginary parts of $z = 3\angle(\frac{\pi}{15})$.

Solution

Writing $z = a + b\mathrm{j}$, the real part is found from $a = r\cos\theta$ and the imaginary part from $b = r\sin\theta$. Thus

$$a = \boxed{} \qquad\qquad 3\cos\left(\frac{\pi}{15}\right) = 2.934$$

$$b = \boxed{} \qquad\qquad 3\sin\left(\frac{\pi}{15}\right) = 0.624$$

Exercises

1   State the modulus and argument of z when

   (a) $z = 13\left(\cos\dfrac{\pi}{3} + \mathrm{j}\sin\dfrac{\pi}{3}\right),$

   (b) $z = 15\left(\cos\dfrac{\pi}{3} - \mathrm{j}\sin\dfrac{\pi}{3}\right).$

2   State the modulus and argument of
   (a) $z = -13$, (b) $z = -4$, (c) $z = -2\mathrm{j}$,
   (d) $z = 3$.

3   Express the following complex numbers in the form $a + b\mathrm{j}$:
   (a) $4\angle 40°$ (b) $5\angle 1.9302$ (c) $4\angle 130°$
   (d) $3\angle 0.7513$

Solutions to exercises

1   (a) $13, \dfrac{\pi}{3}$ (b) $15, -\dfrac{\pi}{3}$

2   (a) modulus 13, argument $-\pi$ (b) $4, -\pi$
   (c) $2, -\dfrac{\pi}{2}$ (d) $3, 0$

3   (a) $3.0642 + 2.5712\mathrm{j}$ (b) $-1.7586 + 4.6805\mathrm{j}$
   (c) $-2.5712 + 3.0642\mathrm{j}$ (d) $2.1924 + 2.0478\mathrm{j}$

2.5   Multiplication and division of complex numbers in polar form

When two complex numbers are given in polar form it is particularly simple to multiply and divide them. This is an advantage of using the polar form.

Key point

If $z_1 = r_1\angle\theta_1$ and $z_2 = r_2\angle\theta_2$ then

$$z_1 z_2 = r_1 r_2 \angle(\theta_1 + \theta_2), \qquad \frac{z_1}{z_2} = \frac{r_1}{r_2}\angle(\theta_1 - \theta_2)$$

Note that to multiply the two numbers we multiply their moduli and add their arguments. To divide, we divide their moduli and subtract their arguments.

Example 2.10

If $z_1 = 5\angle\left(\frac{\pi}{6}\right)$ and $z_2 = 4\angle\left(-\frac{\pi}{4}\right)$ find (a) z_1z_2, (b) $\dfrac{z_1}{z_2}$.

Solution

(a) To multiply the two complex numbers we multiply their moduli and add their arguments. Therefore

$$z_1z_2 = 20\angle\left[\frac{\pi}{6} + \left(-\frac{\pi}{4}\right)\right]$$

$$= 20\angle\left(-\frac{\pi}{12}\right)$$

(b) To divide the two complex numbers we divide their moduli and subtract their arguments. Therefore

$$\frac{z_1}{z_2} = \frac{5}{4}\angle\left[\frac{\pi}{6} - \left(-\frac{\pi}{4}\right)\right]$$

$$= \frac{5}{4}\angle\frac{5\pi}{12}$$

Example 2.11

For z_1 and z_2 in Example 2.10 find $\dfrac{z_2}{z_1}$.

Solution

$$\frac{z_2}{z_1} = \hspace{4cm} \frac{4}{5}\angle\left(-\frac{\pi}{4} - \frac{\pi}{6}\right) = \frac{4}{5}\angle\left(-\frac{5\pi}{12}\right)$$

Let us consider the important and special case of multiplication and division by the imaginary number j.

In polar form

$$j = 1\angle\frac{\pi}{2}$$

If $z = r\angle\theta$, then

$$jz = \left(1\angle\frac{\pi}{2}\right)(r\angle\theta)$$

$$= r\angle\left(\theta + \frac{\pi}{2}\right)$$

Thus multiplying by j has left the modulus of z unchanged but increased the argument by $\frac{\pi}{2}$. This represents an anticlockwise rotation by $\frac{\pi}{2}$ on the Argand diagram.

If $z = r\angle\theta$, then

$$\frac{z}{j} = \frac{r\angle\theta}{1\angle\frac{\pi}{2}}$$

$$= r\angle\left(\theta - \frac{\pi}{2}\right)$$

Thus dividing by j has left the modulus unchanged but reduced the argument by $\frac{\pi}{2}$. This represents a clockwise rotation by $\frac{\pi}{2}$ on the Argand diagram.

Key point	Multiplying a complex number by j causes an anticlockwise rotation by $\frac{\pi}{2}$ on the Argand diagram. Dividing a complex number by j causes a clockwise rotation by $\frac{\pi}{2}$ on the Argand diagram.

These results are very important in electrical engineering applications of complex numbers, in particular phasors, which are described in Block 6.

As a further special case of the multiplication rule consider what happens when both z_1 and z_2 are equal. Suppose $z = r \angle \theta$. Then it follows that $z^2 = r^2 \angle 2\theta$. For example, if

$$z = 5 \angle \frac{\pi}{3}, \text{ then } z^2 = 25 \angle \left(\frac{2\pi}{3} \right)$$

Even more generally, we can state:

Key point	If $z = r \angle \theta$ then $z^n = r^n \angle n\,\theta$.

This is a form of De Moivre's theorem, which is studied in detail in Block 4.

Exercises

1   If $z_1 = 5 \angle 20°$, $z_2 = 3 \angle 35°$ find (a) $z_1 z_2$, (b) $\frac{z_1}{z_2}$, (c) $\frac{z_2}{z_1}$.

2   If $z = 3 - j$ find z^4 expressing your answer in polar form.

3   If $z = 1 + j$ find z^{10}.

4   If $z = -1 - j\sqrt{3}$ find z^5.

5   Consider the complex number $z = r \angle \theta$. By multiplying this by $1 \angle \alpha$ show that the result is a rotation of z through an angle α about the origin.

6   Simplify

$$\frac{\left(4 \angle \left(\frac{2\pi}{3} \right) \right)^2 \left(3 \angle \left(\frac{\pi}{6} \right) \right)^2}{5 \angle \frac{\pi}{4}}$$

Solutions to exercises

1   (a) $15 \angle 55°$ (b) $\frac{5}{3} \angle -15°$ (c) $\frac{3}{5} \angle 15°$

2   $100 \angle -73.72°$

3   $2^5 \angle \frac{5\pi}{2} = 2^5 \angle \frac{\pi}{2}$

4   $2^5 \angle \frac{2\pi}{3} = -16 + 16\sqrt{3}j$

6   $\frac{144}{5} \angle \frac{17\pi}{12}$

End of block exercises

1 Plot each of the following complex numbers on an Argand diagram:
(a) $3 - 3j$ (b) $-4 - 7j$ (c) -5 (d) $4j$

2 If $z = 3 + 8j$ find $|z|$ and arg (z).

3 If $z = -3 - 6j$ find $|z|$ and arg (z).

4 Suppose $z_1 = 3 + 2j$ and $z_2 = 4 + 5j$.
(a) Find $|z_1||z_2|$ and $|z_1 z_2|$. Deduce that these are equal.

(b) Find $\left|\dfrac{z_1}{z_2}\right|$ and $\dfrac{|z_1|}{|z_2|}$. Deduce that these are equal.

5 Find the modulus of the complex number $\cos \theta + j \sin \theta$.

6 Find the real and imaginary parts of each of the following complex numbers:

(a) $2\angle 70°$ (b) $4\angle \pi$ (c) $8\angle\left(-\dfrac{\pi}{2}\right)$
(d) $2\angle 2.306$

7 Simplify

$$\frac{\left(3\angle\left(\dfrac{\pi}{2}\right)\right)^4\left(5\angle\left(\dfrac{\pi}{6}\right)\right)^2}{\left(2\angle\dfrac{\pi}{4}\right)^3}$$

8 Find (a) arg 5, (b) arg -5, (c) arg$(1 + j)$.

Solutions to exercises

2 $\sqrt{73}, 69.4°$

3 $\sqrt{45}, -116.6°$

4 (a) $\sqrt{13}\sqrt{41} = \sqrt{533}$ (b) $\dfrac{\sqrt{13}}{\sqrt{41}}$

5 1

6 (a) 0.684, 1.879 (b) $-4, 0$ (c) $0, -8$
(d) $-1.341, 1.483$

7 $\dfrac{2025}{8}\angle-\dfrac{5\pi}{12}$

8 (a) 0 (b) π (c) $\dfrac{\pi}{4}$

The exponential form of a complex number

3.1 Introduction

In addition to the cartesian and polar forms of a complex number there is a third form in which a complex number may be written – the **exponential form**. In this block we explain the exponential form.

3.2 Some useful power series expansions and Euler's relations

To derive the exponential form we shall need to refer to the power series expansions of $\cos x$, $\sin x$ and e^x. Provided x is measured in radians, the functions $\cos x$ and $\sin x$ can be expressed in the following forms, known as **power series**:

$$\cos x = 1 - \frac{x^2}{2!} + \frac{x^4}{4!} - \cdots$$

$$\sin x = x - \frac{x^3}{3!} + \frac{x^5}{5!} - \cdots$$

Example 3.1
Write down the power series expansions for $\cos \theta$ and $\sin \theta$.

Solution
Simply replace x with θ in the given expansions:

$$\cos \theta = \qquad\qquad\qquad \boxed{1 - \frac{\theta^2}{2!} + \frac{\theta^4}{4!} - \cdots}$$

$$\sin \theta = \qquad\qquad\qquad \boxed{\theta - \frac{\theta^3}{3!} + \frac{\theta^5}{5!} - \cdots}$$

We shall also make use of the power series expansion for e^x. The exponential function has a power series expansion given by

$$e^x = 1 + x + \frac{x^2}{2!} + \frac{x^3}{3!} + \frac{x^4}{4!} + \frac{x^5}{5!} + \cdots$$

Suppose we allow x to be an imaginary number in the expansion of e^x and write $x = j\theta$. We find

$$e^{j\theta} = 1 + j\theta + \frac{j^2\theta^2}{2!} + \frac{j^3\theta^3}{3!} + \frac{j^4\theta^4}{4!} + \frac{j^5\theta^5}{5!} + \cdots$$

$$= 1 + j\theta - \frac{\theta^2}{2!} - j\frac{\theta^3}{3!} + \frac{\theta^4}{4!} + j\frac{\theta^5}{5!} - \cdots$$

since $j^2 = -1, j^3 = -j$, etc. If we now rearrange this to separate the real and imaginary parts we can write

$$e^{j\theta} = 1 - \frac{\theta^2}{2!} + \frac{\theta^4}{4!} - \cdots + j\left(\theta - \frac{\theta^3}{3!} + \frac{\theta^5}{5!} - \cdots\right)$$

The two power series on the right we recognise, from Example 3.1, as $\cos\theta$ and $\sin\theta$, so that we have the following important result, known as **Euler's relation**:

$$e^{j\theta} = \cos\theta + j\sin\theta$$

It is straightforward, and left as an exercise, to show that

$$e^{-j\theta} = \cos\theta - j\sin\theta$$

Key point

Euler's relations

$$e^{j\theta} = \cos\theta + j\sin\theta, \quad e^{-j\theta} = \cos\theta - j\sin\theta$$

Example 3.2

Use Euler's relations to express $\cos\theta$ and $\sin\theta$ in terms of $e^{j\theta}$ and $e^{-j\theta}$.

Solution

Writing down Euler's relations we have

$$e^{j\theta} = \cos\theta + j\sin\theta$$
$$e^{-j\theta} = \cos\theta - j\sin\theta$$

First, adding, and then subtracting, these two identities give

$$e^{j\theta} + e^{-j\theta} = \qquad\qquad\qquad\qquad 2\cos\theta$$

$$e^{j\theta} - e^{-j\theta} = \qquad\qquad\qquad\qquad 2j\sin\theta$$

Rearrange your results to obtain

$$\cos\theta = \qquad\qquad\qquad\qquad \frac{e^{j\theta} + e^{-j\theta}}{2}$$

$$\sin\theta = \qquad\qquad\qquad\qquad \frac{e^{j\theta} - e^{-j\theta}}{2j}$$

The results obtained in Example 3.2 are important:

Key point

$$\cos \theta = \frac{e^{j\theta} + e^{-j\theta}}{2}, \quad \sin \theta = \frac{e^{j\theta} - e^{-j\theta}}{2j}$$

Exercises

1   Express each of the following in terms of trigonometrical functions:
(a) $e^{j\alpha}$ (b) $e^{j\omega t}$ (c) $e^{-j\omega t}$
where α, ω and t are real.

2   Express $\cos \omega t$ in terms of exponential functions.

3   Express $\sin \omega t$ in terms of exponential functions.

4   Show that $3 \left(\dfrac{e^{j\omega} - e^{-j\omega}}{j\omega} \right) = \dfrac{6 \sin \omega}{\omega}$.

Solutions to exercises

1   (a) $\cos \alpha + j \sin \alpha$ (b) $\cos \omega t + j \sin \omega t$
(c) $\cos \omega t - j \sin \omega t$

2   $\dfrac{e^{j\omega t} + e^{-j\omega t}}{2}$

3   $\dfrac{e^{j\omega t} - e^{-j\omega t}}{2j}$

3.3   The exponential form of a complex number

Using the polar form, recall that a complex number with modulus r and argument θ may be written as

$$z = r (\cos \theta + j \sin \theta)$$

It follows immediately from Euler's relations that we can also write this complex number in **exponential form** as $z = re^{j\theta}$. Further, since the complex conjugate of $\cos \theta + j \sin \theta$ is $\cos \theta - j \sin \theta$, the conjugate may be written $\bar{z} = re^{-j\theta}$.

Key point

Exponential form of a complex number
$$z = r (\cos \theta + j \sin \theta) = re^{j\theta}$$
$$\bar{z} = r (\cos \theta - j \sin \theta) = re^{-j\theta}$$

When using this form you should ensure that all angles are measured in radians and not degrees.

Example 3.3

State the modulus and argument of the following complex numbers: (a) $z = 4e^{j\pi/6}$, (b) $z = 0.01e^{0.02j}$, (c) $3e^{-j\pi/2}$.

Solution

(a) The modulus and argument of $4e^{j\pi/6}$ are 4 and $\frac{\pi}{6}$ respectively.

(b) The modulus and argument of $0.01e^{0.02j}$ are 0.01 and 0.02 respectively.

(c) The modulus and argument of $3e^{-j\pi/2}$ are 3 and $-\frac{\pi}{2}$ respectively.

Example 3.4

Express the number $z = 3 + 3j$ in exponential form.

Solution

To express a number in exponential form we must first find its modulus and argument. The modulus of $3 + 3j$ is $\sqrt{3^2 + 3^2} = \sqrt{18}$ or $3\sqrt{2}$. The complex number lies in the first quadrant of the Argand diagram and so its argument θ is given by $\theta = \tan^{-1}\left(\frac{3}{3}\right) = \frac{\pi}{4}$. Thus

$$z = 3 + 3j$$
$$= \sqrt{18}\, e^{j\pi/4}$$

Example 3.5

Find the real and imaginary parts of $z = e^{j\omega t}$ where ω and t are real numbers.

Solution

Using the first of Euler's relations we can write

$$e^{j\omega t} = \qquad\qquad\qquad\qquad \cos \omega t + j \sin \omega t$$

Write down the real and imaginary parts:

$$\cos \omega t,\ \sin \omega t$$

Example 3.6 Control Engineering – Stability of a system

In control theory it is often useful to work with complex numbers in the form

$$z = e^{st}$$

where s is the complex number $\sigma + j\omega$, and where t, ω and σ are real. The quantity ω is a frequency.

(a) Find the real and imaginary parts of z.

(b) Find $|z|$.

Solution

(a) Writing $s = \sigma + j\omega$ we have

$$z = e^{st}$$
$$= e^{(\sigma + j\omega)t}$$
$$= e^{\sigma t + j\omega t}$$

Using the first law of indices we can write this as

$$z = e^{\sigma t}e^{j\omega t}$$
$$= e^{\sigma t}(\cos \omega t + j \sin \omega t)$$

Now because σ and t are real, so too is $e^{\sigma t}$. Thus the real part of z is $e^{\sigma t}\cos \omega t$ and the imaginary part of z is $e^{\sigma t}\sin \omega t$.

(b) Recall that the modulus of $re^{j\theta}$ is r. Hence the modulus of $z = e^{\sigma t}e^{j\omega t}$ is $e^{\sigma t}$, that is

$$|z| = e^{\sigma t}$$

If $\sigma > 0$ then $e^{\sigma t}$ increases as time, t, increases and the system is said to be **unstable**. If $\sigma < 0$, then $e^{\sigma t}$ decreases with t and the system is said to be **stable**.

Exercises

1. State the modulus and argument of each of the following complex numbers:
   (a) $4e^{0.2j}$ (b) $3e^{2\pi j/3}$ (c) $e^{2\pi j}$ (d) $0.35e^{-0.2j}$
   (e) $0.35e^{-0.2}$

2. Express each of the following in the form $re^{j\theta}$:
   (a) $3\angle\left(\dfrac{\pi}{3}\right)$ (b) $\sqrt{2}\angle\left(\dfrac{\pi}{4}\right)$

   (c) $3\angle\left(\dfrac{-\pi}{4}\right)$ (d) $5\angle 0$ (e) $17\angle\left(\dfrac{\pi}{2}\right)$

3. Express each of the following in the form $a + bj$:
   (a) $13e^{j\pi/3}$ (b) $13e^{-j\pi/3}$ (c) $4e^{2\pi j}$

4. Show that $|e^{j\theta}| = 1$.

5. Show that $|e^{-j\theta}| = 1$.

6. Write down the complex conjugate of $e^{j\theta}$.

7. Express $z = -1 + 2j$ in the form $re^{j\theta}$.

8. Show that the real and imaginary parts of $e^{(\alpha + \beta j)t}$ are respectively $e^{\alpha t}\cos \beta t$ and $e^{\alpha t}\sin \beta t$.

9. Express each of the following in exponential form:
   (a) -5 (b) $(1 + j)(1 - 2j)$ (c) $(1 + 8j)^{11}$

   (d) $\dfrac{(1 + 2j)^3}{1 - j}$

Solutions to exercises

1. (a) $4, 0.2$ radians (b) $3, \frac{2\pi}{3}$ radians
   (c) $1, 2\pi$ radians (d) $0.35, -0.2$ radians
   (e) $0.35e^{-0.2}, 0$ radians

2. (a) $3e^{j\pi/3}$ (b) $\sqrt{2}e^{j\pi/4}$ (c) $3e^{-j\pi/4}$ (d) $5e^0 = 5$
   (e) $17e^{j\pi/2}$

3. (a) $6.5 + 11.3j$ (b) $6.5 - 11.3j$ (c) 4

6. $e^{-j\theta}$

7. $\sqrt{5}e^{2.03j}$

9. (a) $5e^{\pi j}$ (b) $\sqrt{10}e^{-0.32j}$ (c) $65^{11/2}e^{-2.94j}$
   (d) $7.91e^{-2.18j}$

End of block exercises

1. Show the following complex numbers on an Argand diagram:
   (a) $5e^{j\pi/3}$ (b) $7e^{2\pi j}$ (c) $4e^{\pi j}$ (d) $e^{j\pi/4}$

2. Find the real and imaginary parts of the following:
   (a) $4e^{j\pi/3}$ (b) $5e^{-j2\pi/3}$ (c) $e^{j\pi/2}$ (d) $e^{-j\pi/2}$

3. Find the real and imaginary parts of
   (a) e^{j-1} (b) $4e^{\alpha + j\beta}$ (c) $e^{-j\pi/6}$ (d) $e^{0.1}$ (e) $je^{0.1}$

4. Show that $\tan \theta = -j\left(\dfrac{e^{j\theta} - e^{-j\theta}}{e^{j\theta} + e^{-j\theta}}\right)$.

5   If $z_1 = 4e^{j\pi/4}$ and $z_2 = 5e^{-j\pi/6}$ find

(a) $z_1 z_2$ (b) $\dfrac{z_1}{z_2}$ (c) $z_1 \overline{z_1}$ (d) $z_1 + z_2$

6   Given $z = 6e^{j0.23}$ express z^3 in exponential and polar forms.

7   If a and b are real, and
$$(a + jb)e^{j\pi/3} + \sqrt{2}e^{-j\pi/4} = 5 + j$$
find a and b.

Solutions to exercises

2   (a) $2, \dfrac{4\sqrt{3}}{2} = 3.464$ (b) $-\dfrac{5}{2}, -\dfrac{5\sqrt{3}}{2}$

(c) $0, 1$ (d) $0, -1$

3   (a) $e^{-1}\cos 1, e^{-1}\sin 1$ (b) $4e^{\alpha}\cos\beta, 4e^{\alpha}\sin\beta$

(c) $\dfrac{\sqrt{3}}{2}, \dfrac{-1}{2}$ (d) $e^{0.1}, 0,$ (e) $0, e^{0.1}$

5   (a) $20e^{j\pi/12}$ (b) $\dfrac{4}{5}e^{j5\pi/12}$ (c) 16

(d) $7.158 + 0.328j$

6   $6^3 e^{0.69j}, 6^3(\cos 0.69 + j\sin 0.69)$

7   $a = \sqrt{3} + 2, b = 1 - 2\sqrt{3}$

De Moivre's theorem

4.1 Introduction

De Moivre's theorem is one of the most important results in the theory of complex numbers. It can be used to find powers of complex numbers, and to prove trigonometrical identities. In this block we introduce and illustrate the use of the theorem. In Block 5 we show how the theorem can be used to solve equations. To understand this block it is essential that you are aware that a complex number can be written in polar form as $z = r(\cos \theta + j \sin \theta)$ or more compactly as $z = r\angle\theta$.

4.2 De Moivre's theorem

We state and use the theorem here. The proof is left as an exercise (see question 4 in the exercises). The theorem involves raising the complex number $\cos \theta + j \sin \theta$ to the power n.

Key point

> **De Moivre's theorem**
>
> $$(\cos \theta + j \sin \theta)^n = \cos n\theta + j \sin n\theta$$

Note that by using the theorem an expression involving powers of trigonometrical functions can be converted into one involving trigonometrical functions of multiple angles. This is useful in proving trigonometrical identities, as we shall see.

The theorem is valid when n is an integer and enables us to write, for example,

$$(\cos \theta + j \sin \theta)^2 \text{ as } \cos 2\theta + j \sin 2\theta$$

It is also valid if n is a fraction and in such a case can be used to find roots of complex numbers. Examples of this can be found in Block 5.

Example 4.1

Use De Moivre's theorem to write $(\cos \theta + j \sin \theta)^3$ in an alternative form.

Solution

Use the theorem with $n = 3$:

$$(\cos \theta + j \sin \theta)^3 = \qquad\qquad\qquad\qquad \cos 3\theta + j \sin 3\theta$$

Example 4.2
Use De Moivre's theorem to write $(\cos\theta + \text{j}\sin\theta)^5$ in an alternative form.

Solution
Use the theorem with $n = 5$:

$(\cos\theta + \text{j}\sin\theta)^5 = $ ⬚ $\cos 5\theta + \text{j}\sin 5\theta$

We can use De Moivre's theorem to obtain powers of a complex number very easily. Consider the following example.

Example 4.3
If $z = r(\cos\theta + \text{j}\sin\theta)$ write down expressions for (a) z^2 and (b) z^3, and use De Moivre's theorem to write your results in an alternative form.

Solution

(a) $z^2 = $ ⬚ $r^2(\cos\theta + \text{j}\sin\theta)^2$

which, using De Moivre's theorem, can be written as

⬚ $r^2(\cos 2\theta + \text{j}\sin 2\theta)$

(b) $z^3 = $ ⬚ $r^3(\cos\theta + \text{j}\sin\theta)^3$

which, using De Moivre's theorem, can be written as

⬚ $r^3(\cos 3\theta + \text{j}\sin 3\theta)$

If we use the abbreviated form $r\angle\theta$ the previous example can be written much more concisely:

Key point

If $z = r\angle\theta$ then

$$z^n = r^n\angle n\theta$$

for example

$$z^2 = r^2\angle 2\theta, \quad z^3 = r^3\angle 3\theta, \quad \text{and so on}$$

Example 4.4
(a) If $z = 3\angle\dfrac{\pi}{4}$ write down z^5.

(b) Express your answer in the form $r(\cos\theta + \text{j}\sin\theta)$ and its cartesian form.

Solution

(a) If $z = 3\angle\frac{\pi}{4}$ then $z^5 =$

$$3^5 \angle\left(5 \times \frac{\pi}{4}\right) = 243\angle\frac{5\pi}{4}$$

(b) $243\angle\frac{5\pi}{4} = 243\left(\cos\frac{5\pi}{4} + j\sin\frac{5\pi}{4}\right) = -171.8 - 171.8j$

Example 4.5

If $z = 7\left(\cos\frac{\pi}{4} + j\sin\frac{\pi}{4}\right)$ find z^2.

Solution

Using De Moivre's theorem with $n = 2$ we can write

$$z^2 = 7^2\left(\cos\frac{\pi}{4} + j\sin\frac{\pi}{4}\right)^2$$

$$= 49\left(\cos\frac{2\pi}{4} + j\sin\frac{2\pi}{4}\right)$$

$$= 49\left(\cos\frac{\pi}{2} + j\sin\frac{\pi}{2}\right)$$

$$= 49j$$

since $\cos\frac{\pi}{2} = 0$ and $\sin\frac{\pi}{2} = 1$.

Example 4.6

(a) Express $z = 2 + 5j$ in polar form.

(b) Hence find $(2 + 5j)^{12}$, leaving your answer in polar form.

Solution

(a) The modulus of $2 + 5j$ is $\sqrt{2^2 + 5^2} = \sqrt{29}$. The number lies in the first quadrant of an Argand diagram and so its argument θ is given by $\theta = \tan^{-1}\frac{5}{2} = 1.190$. Thus $z = 2 + 5j = \sqrt{29}\angle 1.190$.

(b) Using De Moivre's theorem we have

$$(\sqrt{29}\angle 1.190)^{12} = (\sqrt{29})^{12}\angle 12 \times 1.190$$
$$= 29^6\angle 14.283$$

Example 4.7

If $z = \cos\theta + j\sin\theta$ use De Moivre's theorem with $n = -1$ to obtain an expression for $\dfrac{1}{z}$.

Solution

With $n = -1$ we have

$$
\begin{aligned}
z^{-1} &= (\cos\theta + j\sin\theta)^{-1} \\
&= \cos(-\theta) + j\sin(-\theta) \\
&= \cos\theta - j\sin\theta
\end{aligned}
$$

The last line follows since $\cos\theta$ is an even function and $\sin\theta$ is an odd function.

Thus $\dfrac{1}{z} = \cos\theta - j\sin\theta$.

Exercises

1 If $z = 4\angle\dfrac{\pi}{6}$ find z^6 in polar form.

2 Express $z = 2 + 2j$ in polar form and hence find z^8, leaving your answer in polar form. Deduce that $(2 + 2j)^8 = 4096$.

3 Express $z = -3 + 2j$ in polar form and hence find z^6, converting your answer into cartesian form.

4 Consider the complex number $z = \cos\theta + j\sin\theta$.
  (a) Write down the exponential form of this number.

(b) By raising the exponential form to the power n, and using one of the laws of indices, deduce De Moivre's theorem.

5 Use De Moivre's theorem to show that if $z = \cos\theta + j\sin\theta$ then
  (a) $z^n = \cos n\theta + j\sin n\theta$
  (b) $z^{-n} = \cos n\theta - j\sin n\theta$
  Deduce that

$$z^n + \frac{1}{z^n} = 2\cos n\theta$$

and

$$z^n - \frac{1}{z^n} = 2j\sin n\theta$$

Solutions to exercises

1 $4^6\angle\pi$

2 $\sqrt{8}\angle\dfrac{\pi}{4}, 8^4\angle 2\pi$

3 $\sqrt{13}\angle 2.554, 13^3\angle 15.322, -2035 + 827j$

4 (a) $z = e^{j\theta}$ (b) $z^n = (e^{j\theta})^n = e^{jn\theta}$

One of the applications of the theorem is to prove various trigonometrical identities. Consider the following example.

Example 4.8

Use De Moivre's theorem to prove that

(a) $\sin 2\theta = 2 \sin \theta \cos \theta$

(b) $\cos 2\theta = \cos^2 \theta - \sin^2 \theta$

Solution

(a) Using De Moivre's theorem with $n = 2$ we obtain

$$(\cos \theta + j \sin \theta)^2 = \cos 2\theta + j \sin 2\theta$$

Expanding the left-hand side we can write

$$\cos^2 \theta + 2j \sin \theta \cos \theta - \sin^2 \theta = \cos 2\theta + j \sin 2\theta$$

Equating the imaginary parts on both sides gives the required expression for $\sin 2\theta$:

$$\sin 2\theta = 2 \sin \theta \cos \theta$$

(b) Equate the real parts in part (a) to obtain an identity for $\cos 2\theta$:

$$\cos 2\theta = \cos^2 \theta - \sin^2 \theta$$

Example 4.9

Express $\sin 3\theta$ as powers of $\sin \theta$.

Solution

Consider De Moivre's theorem with $n = 3$:

$$(\cos \theta + j \sin \theta)^3 = \cos 3\theta + j \sin 3\theta$$

Expand the expression on the left-hand side:

$$\cos^3 \theta + 3j \cos^2 \theta \sin \theta - 3 \cos \theta \sin^2 \theta - j \sin^3 \theta$$

Compare this result with the right-hand side and equate the imaginary parts:

$$\sin 3\theta = 3 \cos^2 \theta \sin \theta - \sin^3 \theta$$

Finally, noting that $\cos^2 \theta = 1 - \sin^2 \theta$, you should verify that

$$\sin 3\theta = 3 \sin \theta - 4 \sin^3 \theta$$

In the two previous examples you saw how to convert trigonometrical functions of multiple angles into powers of trigonometrical functions. In the previous set of exercises it was shown (question 5) that if $z = \cos\theta + \mathrm{j}\sin\theta$ then

$$z^n + \frac{1}{z^n} = 2\cos n\theta \quad \text{and} \quad z^n - \frac{1}{z^n} = 2\mathrm{j}\sin n\theta$$

These results allow us to obtain a further range of trigonometrical identities. In particular they allow us to convert powers of trigonometrical functions into expressions involving trigonometrical functions of multiple angles.

Key point

If $z = \cos\theta + \mathrm{j}\sin\theta$ then

$$z^n + \frac{1}{z^n} = 2\cos n\theta \quad \text{and} \quad z^n - \frac{1}{z^n} = 2\mathrm{j}\sin n\theta$$

Example 4.10

If $z = \cos\theta + \mathrm{j}\sin\theta$ use the previous key point to write down expressions for

(a) $z + \dfrac{1}{z}$ (b) $z^2 + \dfrac{1}{z^2}$ (c) $z - \dfrac{1}{z}$ (d) $z^2 - \dfrac{1}{z^2}$

Solution

(a) $z + \dfrac{1}{z} = $ ⬚ $2\cos\theta$

(b) $z^2 + \dfrac{1}{z^2} = $ ⬚ $2\cos 2\theta$

(c) $z - \dfrac{1}{z} = $ ⬚ $2\mathrm{j}\sin\theta$

(d) $z^2 - \dfrac{1}{z^2} = $ ⬚ $2\mathrm{j}\sin 2\theta$

Example 4.11

Show that $\cos^2\theta = \frac{1}{2}\cos 2\theta + \frac{1}{2}$.

Solution

Starting with $2\cos\theta = z + \dfrac{1}{z}$ and squaring both sides we obtain

$$4\cos^2\theta = \left(z + \frac{1}{z}\right)^2$$

$$= z^2 + 2 + \frac{1}{z^2}$$

$$= \left(z^2 + \frac{1}{z^2}\right) + 2 \text{ by rearranging}$$

But $z^2 + \dfrac{1}{z^2} = 2\cos 2\theta$ and so

$$4\cos^2\theta = 2\cos 2\theta + 2$$

from which

$$\cos^2\theta = \tfrac{1}{2}\cos 2\theta + \tfrac{1}{2}$$

as required.

Example 4.12

By squaring the result $2j\sin\theta = z - \dfrac{1}{z}$ prove the identity $\sin^2\theta = \tfrac{1}{2} - \tfrac{1}{2}\cos 2\theta$.

Solution

Consider $2j\sin\theta = z - \dfrac{1}{z}$.

Squaring the left-hand side gives $-4\sin^2\theta$. Now square the right-hand side.

$$\left(z - \dfrac{1}{z}\right)^2 = \boxed{} \qquad\qquad z^2 - 2 + \dfrac{1}{z^2}$$

Rearrange this and express it in terms of $\cos 2\theta$ using the previous key point.

$$z^2 - 2 + \dfrac{1}{z^2} = \boxed{} \qquad\qquad \left(z^2 + \dfrac{1}{z^2}\right) - 2 = 2\cos 2\theta - 2$$

Finally we have

$$-4\sin^2\theta = 2\cos 2\theta - 2$$

from which $\sin^2\theta = \tfrac{1}{2} - \tfrac{1}{2}\cos 2\theta$ as required.

Exercises

1   Show that $\cos 4\theta = 8\cos^4\theta - 8\cos^2\theta + 1$.

2   Show that
$$\sin 4\theta = 4\cos^3\theta\sin\theta - 4\cos\theta\sin^3\theta$$

3   Use your answers from questions 1 and 2 to deduce that

$$\tan 4\theta = \dfrac{4\tan\theta - 4\tan^3\theta}{1 - 6\tan^2\theta + \tan^4\theta}$$

4   Show that $\cos^3\theta = \tfrac{1}{4}(\cos 3\theta + 3\cos\theta)$.

End of block exercises

1   Use De Moivre's theorem to show that
$\cos 5\theta = 16\cos^5\theta - 20\cos^3\theta + 5\cos\theta$
and obtain a similar expression for $\sin 5\theta$ in
powers of $\sin\theta$.

2   Show that
$$\sin^4\theta = \tfrac{1}{8}(\cos 4\theta - 4\cos 2\theta + 3)$$

Solutions to exercises

1   $\sin 5\theta = 16\sin^5\theta - 20\sin^3\theta + 5\sin\theta$

Solving equations and finding roots of complex numbers

By using complex numbers, solutions can be obtained to a wide range of equations. There is no single appropriate method for tackling these and so experience and practice are essential. Sometimes it is convenient to work in cartesian form, but on other occasions solutions are easier to obtain in polar form. You should be aware that a polynomial equation of degree n possesses n roots, but some of these can be repeated, or equal, roots. So, when solving a quadratic equation we shall be looking for two roots, when solving a cubic equation we shall be looking for three roots, and so on.

5.2 Solving equations

Example 5.1
Solve the equation $z^3 + 5z = 0$.

Solution
This is a polynomial equation of degree 3. We seek three solutions. The equation can be factorised:

$$z^3 + 5z = z(z^2 + 5) = 0$$

so that one solution is $z = 0$, and others can be obtained by solving $z^2 + 5 = 0$.
   From $z^2 + 5 = 0$ we have

$$z^2 = -5$$
$$z = \pm\sqrt{-5}$$
$$= \pm\sqrt{5}j$$

Thus the solutions of the given equation are $z = 0$ and $z = \pm\sqrt{5}j$.

Example 5.2
Solve the equation $z^2 + 2z + 2 = 0$.

Solution
This is simply a quadratic equation. Apply the formula

$$z = \frac{-b \pm \sqrt{b^2 - 4ac}}{2a}$$

$$= \frac{-2 \pm \sqrt{4 - 4(1)(2)}}{2}$$

$$= \frac{-2 \pm \sqrt{-4}}{2}$$

$$= z = -1 \pm j$$

Often it is more convenient to work in polar form, and so before attempting to solve any more equations it will be useful to review how the polar form of a complex number is obtained. Consider the number 4, illustrated in Figure 5.1(a). Its modulus is 4 and its argument is 0. However, it is important to note that if we increase the argument by 360°, or 2π, we remain at the same location on the Argand diagram as shown in Figure 5.1(b). Indeed we can increase the argument by further multiples of 360° and still represent the same complex number, as shown in Figure 5.1(c).

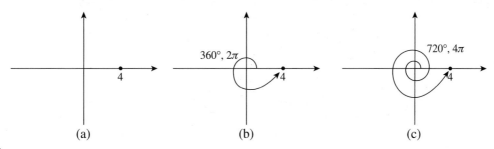

(a) (b) (c)

Figure 5.1
The argument of a complex number takes many values.

Thus in polar form we can write 4 as $4\angle 0$, $4\angle 2\pi$, $4\angle 4\pi$, and so on. We can generalise this as $4\angle 2k\pi$ where $k = 0, 1, 2, \ldots$. In the examples that follow you should bear this in mind when converting complex numbers to polar form.

Example 5.3
Express the number $z = 2 + j$ in polar form, and include the general form for the argument.

Solution
The complex number $z = 2 + j$ is shown in Figure 5.2. Its modulus is $|z| = \sqrt{2^2 + 1^2} = \sqrt{5}$. Its argument is $\tan^{-1}\frac{1}{2} = 0.464$. However, adding on multiples of 2π results in the same complex number, so we can write $\arg(z) = 0.464 + 2k\pi$. Finally the polar form is $z = \sqrt{5}\angle(0.464 + 2k\pi)$.

Figure 5.2

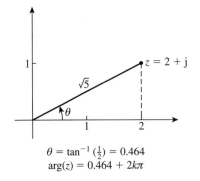

$$\theta = \tan^{-1}\left(\tfrac{1}{2}\right) = 0.464$$
$$\arg(z) = 0.464 + 2k\pi$$

Example 5.4
Solve the equation $z^3 - 4 = 0$.

Solution
This is a polynomial equation of degree 3. We look for three solutions. Note that this equation can be written as $z^3 = 4$. We shall show that by working in polar form it is straightforward to obtain the solutions. You should also note that the equation is equivalent to $z = 4^{1/3}$ so we are finding the cube roots of 4.

As z is a complex number, we can write it in polar form as $z = r\angle\theta$. When we have succeeded in finding r and θ, we shall know z and will have solved the problem.

Consider, first, the left-hand side of

$$z^3 = 4$$

If $z = r\angle\theta$ then write down an expression for z^3:

$$z^3 = \boxed{} \qquad\qquad\qquad r^3\angle 3\theta$$

Now express the right-hand side in polar form including the general term for the argument.

$$4 = \boxed{} \qquad\qquad 4\angle 2k\pi, k = 0, 1, 2, \ldots$$

Therefore, working entirely in polar form, the equation becomes

$$r^3\angle 3\theta = 4\angle 2k\pi$$

Equating moduli on both sides gives $r^3 = 4$ so that $r = 4^{1/3}$, the real cube root of 4 since r is real.

Equating arguments,

$$3\theta = 2k\pi \text{ so that } \theta = \frac{2k\pi}{3}, \quad k = 0, 1, 2, \ldots$$

Taking $k = 0, 1, 2$ in turn gives the three solutions

$$z = 4^{1/3}\angle 0, \quad z = 4^{1/3}\angle\frac{2\pi}{3}, \quad z = 4^{1/3}\angle\frac{4\pi}{3}$$

Figure 5.3
The solutions of
$z^3 - 4 = 0$.

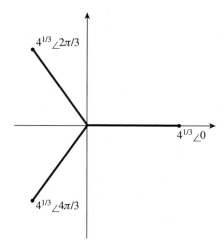

These solutions are illustrated in Figure 5.3.

Note that if k is increased beyond 2, we find the solutions already obtained.

If necessary these solutions can be expressed in cartesian form. Do this for yourself now:

$4^{1/3} \angle 0 =$ ⬚ $4^{1/3}$ or $\sqrt[3]{4}$

$4^{1/3} \angle \dfrac{2\pi}{3} =$ ⬚ $4^{1/3}\left(-\dfrac{1}{2} + \dfrac{\sqrt{3}}{2}j\right)$

$4^{1/3} \angle \dfrac{4\pi}{3} =$ ⬚ $4^{1/3}\left(-\dfrac{1}{2} - \dfrac{\sqrt{3}}{2}j\right)$

Example 5.5
Solve the equation $z^3 + 2 = 0$.

Solution
Note that this equation can be written $z^3 = -2$. Again we work in polar form.

Consider, first, the left-hand side of

$$z^3 = -2$$

If $z = r\angle\theta$ then $z^3 = r^3 \angle 3\theta$.

Now express the right-hand side in polar form. The complex number $z = -2$ is shown in Figure 5.4 to help you do this.

Figure 5.4
The complex
number $z = -2$.

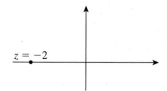

$$-2 = 2\angle(\pi + 2k\pi) \text{ where } k = 0, 1, 2, \ldots$$

Therefore, working entirely in polar form the equation becomes

$$r^3\angle 3\theta = 2\angle(\pi + 2k\pi)$$

Equate moduli on both sides to find r.

$r =$ $2^{1/3}$, the real cube root of 2

Equate arguments, to find θ:

$\theta =$ $\dfrac{\pi}{3} + \dfrac{2k\pi}{3}, k = 0, 1, 2, \ldots$

Taking $k = 0, 1, 2$ in turn gives the three solutions

                           $z = 2^{1/3}\angle\dfrac{\pi}{3}, 2^{1/3}\angle\pi, 2^{1/3}\angle\dfrac{5\pi}{3}$

These solutions are illustrated in Figure 5.5.

Figure 5.5

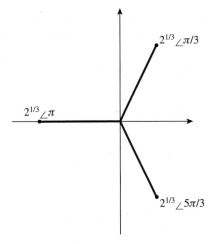

Note that if k is increased beyond 2, we find the solutions already obtained.
If necessary these solutions can be expressed in cartesian form. Do this for yourself now:

$2^{1/3}\angle\dfrac{\pi}{3} =$ $2^{1/3}\left(\dfrac{1}{2} + \dfrac{\sqrt{3}}{2}j\right)$

$$2^{1/3} \angle \pi =$$ 　　　　　　　　$-2^{1/3}$ or $-\sqrt[3]{2}$

$$2^{1/3} \angle \frac{5\pi}{3} =$$ 　　　　　　　$2^{1/3}\left(\frac{1}{2} - \frac{\sqrt{3}}{2}j\right)$

Example 5.6
Solve the equation $z^6 - 2z^3 - 8 = 0$.

Solution
By careful inspection note that this equation can be factorised as

$$z^6 - 2z^3 - 8 = (z^3 - 4)(z^3 + 2)$$
$$= 0$$

(If you cannot spot this directly, try making a substitution, $w = z^3$, and obtain a quadratic equation in w.) Thus $z^3 - 4 = 0$ or $z^3 + 2 = 0$. We have already solved each of these separately in Examples 5.4 and 5.5. Thus the six solutions are those obtained in these examples.

Example 5.7
Solve the equation $z^2 = 2 + j$.

Solution
In general the unknown z will be a complex number and so $z = r\angle\theta$. Then $z^2 = r^2\angle 2\theta$. Now consider the right-hand side of the given equation, $2 + j$. This has been expressed in polar form in Example 5.3. Therefore

$$r^2\angle 2\theta = 5^{1/2}\angle(0.464 + 2k\pi)$$

Comparing moduli on both sides we see that

$$r^2 = 5^{1/2} \quad \text{so that} \quad r = 5^{1/4}$$

Also, comparing arguments,

$$2\theta = 0.464 + 2k\pi$$

from which

$$\theta = \frac{0.464 + 2k\pi}{2}$$
$$= 0.232 + k\pi, k = 0, 1, 2,\ldots$$

The two solutions we seek are obtained by letting $k = 0$ and then $k = 1$. Increasing k further will reproduce solutions already found.

When $k = 0$, $\theta = 0.232$.
When $k = 1$, $\theta = 0.232 + \pi$.

Hence the two solutions are

$$5^{1/4}(\cos 0.232 + j \sin 0.232) \quad \text{and} \quad 5^{1/4}[\cos(0.232 + \pi) + j \sin(0.232 + \pi)]$$

These solutions are shown in the Argand diagram in Figure 5.6. If necessary they can be converted to cartesian form as $1.46 + 0.34j$ and $-1.46 - 0.34j$.

Figure 5.6
The two solutions
of $z^2 = 2 + j$.

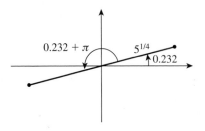

The techniques used in the previous examples can be used to find roots of complex numbers. Suppose we wanted to find the cube roots of j, that is $j^{1/3}$. Writing $z = j^{1/3}$ and cubing both sides we obtain

$$z^3 = j$$

So finding the cube roots of j is equivalent to solving the equation $z^3 = j$. This can be done using the technique of Examples 5.4, 5.5 or 5.7.

Exercises

1 Solve the equation $-z^2 + 3z - 4 = 0$.

2 Solve each of the following equations leaving your answers in polar form:
(a) $z^2 = 1 + j$ (b) $z^2 = 1 - j$
(c) $z^3 = -2 + 3j$

3 Solve the following equations converting your solutions to cartesian form:
(a) $z^3 = -1$ (b) $z^3 = 1$ (c) $z^3 - 6j = 0$

Solutions to exercises

1 $z = \dfrac{3}{2} \pm \dfrac{\sqrt{7}}{2}j$

2 (a) $2^{1/4}\angle\dfrac{\pi}{8}, 2^{1/4}\angle\dfrac{9\pi}{8}$

(b) $2^{1/4}\angle-\dfrac{\pi}{8}, 2^{1/4}\angle\dfrac{7\pi}{8}$

(c) $13^{1/6}\angle 0.720, 13^{1/6}\angle 2.814, 13^{1/6}\angle 4.908$

3 (a) $z = -1, z = \dfrac{1}{2} + \dfrac{\sqrt{3}}{2}j, z = \dfrac{1}{2} - \dfrac{\sqrt{3}}{2}j$

(b) $z = 1, z = -\dfrac{1}{2} \pm j\dfrac{\sqrt{3}}{2}$

(c) $-6^{1/3}j, 6^{1/3}\left(\dfrac{\sqrt{3}}{2} + j\dfrac{1}{2}\right),$

$6^{1/3}\left(-\dfrac{\sqrt{3}}{2} + j\dfrac{1}{2}\right)$

End of block exercises

1   Find the three cube roots of -8.

2   Find the three cube roots of 8.

3   Find the cube roots of $z = \dfrac{(-3 + j)^4}{(2 - j)^2}$.

4   Find all complex numbers which satisfy $z^5 + 4z = 0$ and express your answers in cartesian form.

5   Solve the equation $z^4 + 25 = 0$ showing each solution on an Argand diagram.

6   Find the cube roots of $\sqrt{3} + j$.

7   Find the cube roots of $\sqrt{3} - j$.

Solutions to exercises

1   $-2, 1 \pm \sqrt{3}j$

2   $2, -1 \pm \sqrt{3}j$

3   $z = 2.695 - 0.325j, z = -1.066 + 2.496j,$
    $z = -1.629 - 2.172j$

4   $z = 0, 1 \pm j, -1 \pm j$

5   $z = \dfrac{\sqrt{10}}{2} \pm \dfrac{\sqrt{10}}{2}j, z = -\dfrac{\sqrt{10}}{2} \pm \dfrac{\sqrt{10}}{2}j$

6   $z = 1.241 + 0.219j, z = -0.431 - 1.184j,$
    $z = -0.810 + 0.965j$

7   $z = 1.241 - 0.219j, z = -0.431 + 1.184j,$
    $z = -0.810 - 0.965j$

Phasors

6.1 Introduction

One of the most important applications of complex numbers in engineering is in the study of alternating current (a.c.) circuits. The phase relationships between different components in a circuit can be expressed using complex numbers. A phasor is a quantity that contains information about the peak voltage or peak current in a component together with its phase.

6.2 Phase relationships in *LCR* circuits

The voltage in an a.c. circuit can be expressed in the form

$$v(t) = V \sin(\omega t + \phi)$$

where V is the peak voltage, ω is the angular frequency, ϕ is the phase relative to some reference voltage and t is time. Recall that $\omega = 2\pi f$ where f is the frequency. Similarly a current can be expressed as $i(t) = I \sin(\omega t + \phi)$ where I is the peak current.

Resistor

Consider a circuit containing a resistor of resistance R. Let $v_R(t)$ be the voltage across the resistor. The current, $i(t)$, through the resistor is in phase with this voltage. This means that both the voltage and current are zero at the same time and peak at the same time. This fact is illustrated in the graphs of $i(t)$ and $v_R(t)$ in Figure 6.1.

Figure 6.1
In a circuit containing a resistor, R, the current and voltage are in phase.

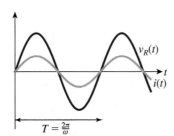

$$T = \frac{2\pi}{\omega}$$

The current is related to the voltage by Ohm's law, which states that $v_R = Ri$. For example, if $i(t) = I \sin \omega t$, then $v_R = IR \sin \omega t$.

Inductor

Consider a circuit containing an inductor having inductance L. Experiment shows that the voltage across the inductor, v_L, leads the current by a phase of $\frac{\pi}{2}$. If $i(t) = I \sin \omega t$, then it can be shown that

$$v_L = \omega L I \sin\left(\omega t + \frac{\pi}{2} \right)$$

This means that the voltage peaks earlier than the current, as illustrated in Figure 6.2.

Figure 6.2
In a circuit containing an inductor, the voltage leads the current by a phase of $\frac{\pi}{2}$.

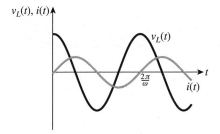

Capacitor

In a circuit containing a capacitor having capacitance C, experiment shows that the voltage across the capacitor, v_C, lags behind the current by a phase of $\frac{\pi}{2}$. It can be shown that if $i(t) = I \sin \omega t$, then

$$v_C = \frac{I}{\omega C} \sin\left(\omega t - \frac{\pi}{2} \right)$$

This means that the voltage peaks after the current, as illustrated in Figure 6.3.

Figure 6.3
In a circuit containing a capacitor, the voltage lags behind the current by a phase of $\frac{\pi}{2}$.

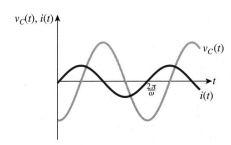

6.3 Phasors

Consider the waveform in Figure 6.4(a), which might represent a voltage, $V_1 \sin \omega t$, in an a.c. circuit. The time it takes to complete one cycle, that is the period, is $\dfrac{2\pi}{\omega}$.

Figure 6.4
Points on a
waveform and
equivalent points
on a rotating arm.

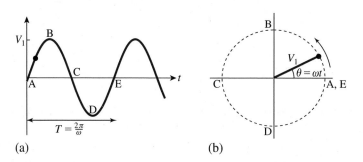

(a)   (b)

An alternative representation of the same physical situation is shown in Figure 6.4(b). It shows an arm rotating anticlockwise. The length of the arm is V_1 and it rotates at ω radians per second.

Moving at an angular velocity of ω radians per second means that it completes

$$1 \text{ radian in } \frac{1}{\omega} \text{ seconds}$$
$$2\pi \text{ radians in } \frac{2\pi}{\omega} \text{ seconds}$$

So it completes a full revolution in $T = \dfrac{2\pi}{\omega}$ seconds. The time it takes to complete a revolution is the same as the time it takes for one cycle of the waveform to be completed in Figure 6.4(a). There is an obvious correspondence between points on the sine wave and positions of the arm. Points A, B, C, D and E illustrate this. For any value of t on the waveform, the angle of the arm is ωt. Such a rotating arm is called a **phasor**.

If we have a second waveform, such as $V_2 \sin(\omega t + \frac{\pi}{2})$, which leads the first by a phase of $\frac{\pi}{2}$, then its rotating arm will lead that of $V_1 \sin \omega t$ by $\frac{\pi}{2}$, as shown in Figure 6.5.

Figure 6.5
Phasor 2 leads
phasor 1 by $\frac{\pi}{2}$.

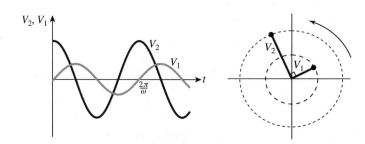

Similarly, if we have a third waveform, such as $V_3 \sin(\omega t - \frac{\pi}{2})$, which lags behind the first by a phase of $\frac{\pi}{2}$, then its rotating arm will lag that of $V_1 \sin \omega t$ by $\frac{\pi}{2}$ as shown in Figure 6.6.

Figure 6.6
Phasor 3 lags
phasor 1 by $\frac{\pi}{2}$.

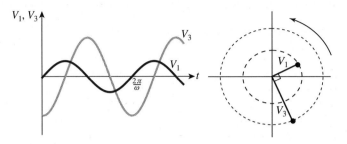

Because these phasors have a length and an angle measured from the positive direction of the x axis they can be represented by complex numbers. We think of the phasor diagram as an Argand diagram. We write a typical phasor in polar form as $\widetilde{V} = V \angle \phi$, where the $\sim$ is used to distinguish the complex quantity, with magnitude V and phase ϕ, from the peak value V. Alternatively we can write $\widetilde{V}$ in cartesian form by stating its real and imaginary parts.

In Block 2 we showed that when a complex number is multiplied by j this has the effect of rotating that number anticlockwise by $\frac{\pi}{2}$ on an Argand diagram.

So, if we want to represent a phase lead of $\frac{\pi}{2}$ we multiply the corresponding phasor by j.

Similarly, dividing a complex number by j has the effect of rotating that number clockwise by $\frac{\pi}{2}$ on an Argand diagram. So, if we want to represent a phase lag of $\frac{\pi}{2}$ we divide the corresponding phasor by j.

Note that

$$\frac{1}{j} = \left(\frac{1}{j}\right) \cdot \left(\frac{-j}{-j}\right)$$

$$= \frac{-j}{1}$$

$$= -j$$

So dividing by j is equivalent to multiplying by $-j$.

If the current through a circuit is represented by $\widetilde{I}$ the voltage across a resistor will be given, from Ohm's law, by $\widetilde{V}_R = R\widetilde{I}$. The voltage across an inductor will be $\widetilde{V}_L = j\omega L \widetilde{I}$ and that across a capacitor will be $\widetilde{V}_C = -\dfrac{j}{\omega C} \widetilde{I}$. Notice the way that multiplication by j achieves the required phase differences for the inductor and capacitor.

In general we write

$$\widetilde{V} = \widetilde{I}Z \text{ where } Z = \begin{cases} R & \text{for a resistor} \\ j\omega L & \text{for an inductor} \\ -\dfrac{j}{\omega C} & \text{for a capacitor} \end{cases}$$

The quantity Z is called the **complex impedance**.

Complex impedances

$$\widetilde{V} = \widetilde{I} Z \text{ where } Z = \begin{cases} R & \text{for a resistor} \\ j\omega L & \text{for an inductor} \\ -\dfrac{j}{\omega C} & \text{for a capacitor} \end{cases}$$

More generally, the complex impedance is written in cartesian form as $Z = R + jX$ where R is the resistance and X is called the **reactance**. The argument of Z gives the phase of the voltage relative to the current. The complex impedance is very important because it can be used in the same way that resistance, R, is used in Ohm's law, which is clear from inspecting the two forms $V = IR$ (Ohm's law) and $\widetilde{V} = \widetilde{I}Z$.

When k circuit elements are placed in series their total impedance, Z, is found by adding the individual impedances, Z_k, that is

$$Z = \sum Z_k$$

When the elements are in parallel

$$\frac{1}{Z} = \sum \frac{1}{Z_k}$$

Example 6.1 Electrical Engineering – A series *LCR* circuit

For the series *LCR* circuit shown in Figure 6.7 the total impedance is the sum of the individual impedances.

$$Z = R + j\omega L - \frac{j}{\omega C}$$

The voltage phasor is

$$\widetilde{V} = \widetilde{I} Z$$
$$= \widetilde{I} R + j\omega L \widetilde{I} - \frac{j}{\omega C} \widetilde{I}$$

Figure 6.7
A series *LCR*
circuit.

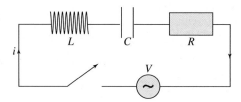

Example 6.2 Electrical Engineering – A series *RC* circuit

The circuit shown in Figure 6.8 consists of an alternating voltage source, a resistor R and a capacitor C in series. Suppose a voltage of 8 mV is applied with a frequency of 10^6 Hz to the circuit in which R is 12 Ω and $C = 100$ pF.
(a) Find the angular frequency ω.
(b) State the supply voltage in phasor form, taking its phase to be zero. This will be the reference phasor.

Figure 6.8
A series *RC*
circuit.

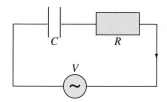

(c) Find the complex impedance for this circuit.
(d) Find the current phasor $\widetilde{I}$ for the circuit.
(e) Find the voltage across each component.

Solution

(a) The angular frequency $\omega = 2\pi f = 2\pi \times 10^6 = 6.2832 \times 10^6$.
(b) The voltage phasor is $\widetilde{V} = 8 \times 10^{-3} \angle 0$.
(c) The complex impedance is the sum of the individual impedances:

$$Z = R - \frac{j}{\omega C}$$

$$= 12 - \frac{j}{6.2832 \times 10^6 \times 10^{-10}}$$

$$= \qquad\qquad\qquad\qquad\qquad\qquad 12 - 1591.55j$$

(d) Since $\widetilde{V} = \widetilde{I} Z$ it follows that $\widetilde{I} = \dfrac{\widetilde{V}}{Z} = \dfrac{8 \times 10^{-3}}{12 - 1591.55j}$.

Simplify this:

$$\widetilde{I} = 3.7897 \times 10^{-8} + 5.0263 \times 10^{-6}j$$

(e) Since the current is the same through each component, we have

$$\widetilde{V}_R = \widetilde{I} R$$
$$= 12\widetilde{I}$$
$$= 4.5477 \times 10^{-7} + 6.0316 \times 10^{-5}j$$

and

$$\widetilde{V}_C = -\frac{j}{\omega C}\widetilde{I}$$
$$= 8.000 \times 10^{-3} - 6.0316 \times 10^{-5}j$$

In polar form $\widetilde{V}_R = 6.0317 \times 10^{-5} \angle (1.5633)$ and $\widetilde{V}_C = 0.008 \angle (-0.0075)$.

We see from the polar form that the voltage across the resistor leads the voltage across the capacitor by $1.5633 + 0.0075 \approx \frac{\pi}{2}$ radians, as expected. It is easy to check from the cartesian forms that $\widetilde{V}_R + \widetilde{V}_C = \widetilde{V}$.

End of block exercises

1 The impedance of a component is
$Z = 12 - 12j$. State (a) the resistance, (b) the reactance, (c) the phase of the voltage relative to the current.

2 The **admittance** is the reciprocal of the impedance. Find the admittance of the component in question 1.

3 The impedance of an *LCR* circuit is

$$Z = R + j\left(\omega L - \frac{1}{\omega C}\right)$$

(a) Find $|Z|$.

(b) From the result of part (a) deduce that the impedance has minimum magnitude when

$$\omega = \sqrt{\frac{1}{(LC)}}$$

(c) Deduce that this minimum value is R.

4 A capacitor and resistor are placed in parallel. Show that the complex impedance of this combination is given by

$$\frac{1}{Z} = \frac{1}{R} + j\omega C$$

Find an expression for Z.

Solutions to exercises

1 (a) 12 (b) -12 (c) $-\dfrac{\pi}{4}$

2 $0.0417 + 0.0417j$

3 (a) $\sqrt{R^2 + \left(\omega L - \dfrac{1}{\omega C}\right)^2}$

4 $Z = \dfrac{R}{1 + j\omega CR}$

End of chapter exercises

1 Find the modulus and argument of (a) $-j$, (b) -3, (c) $1 + j$, (d) $\cos\theta + j\sin\theta$.

2 Represent the following numbers on an Argand diagram:
(a) 13 (b) j^2 (c) $-4 + 2j$ (d) $\pi + \pi j$

3 If $z_1 = 5 + 3j$ and $z_2 = 3 + 2j$ find $z_1 z_2$ and $\dfrac{z_1}{z_2}$.

4 Find the modulus and argument of $-1 + j$.

5 Express $2 + 3j$ in exponential form. Write down its complex conjugate.

6 Sketch an Argand diagram and upon it mark the poles of the rational function

$$G(s) = \frac{3(s - 3)}{(s + 1)(s + 1 + 3j)(s + 1 - 3j)}$$

7 Use De Moivre's theorem to simplify
(a) $(\cos 3\theta + j\sin 3\theta)(\cos 4\theta + j\sin 4\theta)$
(b) $\dfrac{\cos 8\theta + j\sin 8\theta}{\cos 2\theta - j\sin 2\theta}$

8 Express $(\cos\theta + j\sin\theta)^9$ in the form $\cos n\theta + j\sin n\theta$.

9 Use De Moivre's theorem to show that $\cos 3\theta = 4\cos^3\theta - 3\cos\theta$.

10 Solve the equation $z^2 = 4j$ expressing your solutions in cartesian form and using surds.

11 Find $\sqrt[3]{2 + 2j}$.

12 Find the fifth roots of j.

13 Solve the equation $z^4 + 1 = j\sqrt{3}$.

14 Show that

$$32 \sin^6\theta = 10 - 15 \cos 2\theta + 6 \cos 4\theta - \cos 6\theta$$

15 Let $z_1 = r_1(\cos\theta_1 + j\sin\theta_1)$ and let $z_2 = r_2(\cos\theta_2 + j\sin\theta_2)$.
Use trigonometrical identities to prove that

$$z_1z_2 = r_1r_2[\cos(\theta_1 + \theta_2) + j\sin(\theta_1 + \theta_2)]$$

and

$$\frac{z_1}{z_2} = \frac{r_1}{r_2}[\cos(\theta_1 - \theta_2) + j\sin(\theta_1 - \theta_2)]$$

16 The hyperbolic functions $\cosh x$ and $\sinh x$ are defined as

$$\cosh x = \frac{e^x + e^{-x}}{2}, \quad \sinh x = \frac{e^x - e^{-x}}{2}$$

Use these definitions and Euler's relations to prove (a) $\cosh jx = \cos x$, (b) $\sinh jx = j\sin x$, (c) $\cos jx = \cosh x$, (d) $\sin jx = j\sinh x$.

17 Writing $z = re^{j\theta}$ and using the laws of logarithms show that $\ln z = \ln r + j\theta$. Hence interpret $\ln(2 + 2j)$.

18 In each case, draw an Argand diagram and shade the region that satisfies
(a) $\text{Re}(z) > 0$ (b) $\text{Re}(z) \leq -2$
(c) $\text{Im}(z) < 3$ (d) $\text{Im}(z) \geq -3$

where $\text{Re}(z)$ stands for the real part of z, and $\text{Im}(z)$ stands for imaginary part of z.

19 Find the real and imaginary parts of

$$\frac{1}{1 + j\omega}$$

20 Find the modulus and argument of $\dfrac{1}{1 + j\omega}$.

21 Find $\dfrac{1}{1 + j\omega} + \dfrac{1}{2 + j\omega}$.

22 **Electrical Engineering.** The impedance, Z, of a component is given by $Z = 7 + 5j$. The admittance, Y, is given by

$$Y = \frac{1}{Z}$$

Find Y in cartesian form.

23 **Electrical Engineering.** A resistor of resistance, R, and an inductor of inductance, L, are connected in series. If $R = 9$ and $L = 2j$ find
(a) the total impedance, Z
(b) the admittance.

24 **Electrical Engineering.** A resistor of resistance, R, and an inductor of inductance, L, are connected in parallel. If $R = 9$ and $L = 2j$ find
(a) the total impedance, Z
(b) the admittance.

Solutions to exercises

1 (a) $1, -\dfrac{\pi}{2}$ (b) $3, \pi$ (c) $\sqrt{2}, \dfrac{\pi}{4}$ (d) $1, \theta$

3 $9 + 19j, \dfrac{21}{13} - \dfrac{1}{13}j$

4 $\sqrt{2}, \dfrac{3\pi}{4}$

5 $\sqrt{13}e^{0.983j}$. The conjugate is $\sqrt{13}e^{-0.983j}$.

7 (a) $\cos 7\theta + j \sin 7\theta$ (b) $\cos 10\theta + j \sin 10\theta$

8 $\cos 9\theta + j \sin 9\theta$

10 $z = \pm\sqrt{2}(1 + j)$

11 $8^{1/6} \angle \left(\dfrac{\pi}{12} + \dfrac{2\,k\pi}{3} \right), k = 0, 1, 2$

12 $1 \angle \left(\dfrac{\pi}{10} + \dfrac{2\,k\pi}{5} \right), k = 0, 1, 2, 3, 4$

13 $2^{1/4} \angle \left(\dfrac{\pi}{6} + \dfrac{k\pi}{2} \right), k = 0, 1, 2, 3$

17 $\ln(\sqrt{8}) + j\dfrac{\pi}{4}$

19 real part $= \dfrac{1}{1 + \omega^2}$

imaginary part $-\dfrac{\omega}{1 + \omega^2}$

21 modulus $\dfrac{1}{\sqrt{1 + \omega^2}}$, argument $\tan^{-1}(-\omega)$

21 $\dfrac{3 + 2j\omega}{2 + 3j\omega - \omega^2}$

22 $\dfrac{7 - 5j}{74}$

23 (a) $9 + 2j$ (b) $\dfrac{9 - 2j}{85}$

24 (a) $\dfrac{36 + 162j}{85}$ (b) $\dfrac{1}{9} - \dfrac{j}{2}$

Matrices and determinants

A **matrix** is a rectangular array of numbers or expressions. The plural of matrix is matrices. This chapter deals with the addition, subtraction and multiplication of matrices. Unlike numbers, the processes of addition, subtraction and multiplication can be applied to matrices only under specific conditions. Division of matrices is not defined. However, the inverse of a matrix may be found under certain conditions. The inverse matrix and its calculation are explained in Block 4.

Determinants are introduced in Block 3. A **determinant** is a number that is calculated from the elements of a matrix and is used in finding the inverse matrix.

The chapter closes with the application of matrices to **computer graphics**. Multiplication by a matrix can be interpreted as a transformation of a figure.

Chapter 12 contents

Introduction to matrices

1.1 Introduction

A **matrix** is a rectangular array of numbers or expressions usually enclosed in brackets. For example,

$$\begin{pmatrix} 3 & 1 & 9 \\ 0 & -6 & 2 \end{pmatrix}, \quad \begin{pmatrix} \alpha & \beta \\ \gamma & \delta \end{pmatrix}, \quad \begin{pmatrix} 3 - \lambda & 4 \\ 7 & 6 - \lambda \end{pmatrix}$$

are all matrices. Note that the plural of matrix is matrices.

We often denote a matrix by a capital letter, for example

$$A = \begin{pmatrix} 4 & 1 \\ 3 & 6 \\ -1 & 0 \end{pmatrix}, \quad B = \begin{pmatrix} 1 & 0 & -1 & 3 \\ 6 & 1 & 4 & 2 \end{pmatrix}$$

The **size** of a matrix is given by the number of rows and the number of columns. The matrix A has three rows and two columns and so is described as a 3×2 matrix. We say that A is a 'three by two matrix'. Matrix B has two rows and four columns and so is a 2×4 matrix. Notice that the number of rows is always stated first. An $n \times m$ matrix has n rows and m columns.

The individual numbers or expressions in a matrix are called the **elements** and are usually denoted by a small letter. For example, for A, the element in row 1, column 2 is denoted by a_{12}, the element in row 3, column 1 is denoted by a_{31}, and so on. So

$$a_{11} = 4, \quad a_{12} = 1, \quad a_{21} = 3, \quad a_{22} = 6, \quad a_{31} = -1, \quad a_{32} = 0$$

Example 1.1
Given

$$C = \begin{pmatrix} 3 & 0 & 4 & x \\ \lambda & 1 & z & -1 \end{pmatrix}$$

state (a) c_{13}, (b) c_{21}, (c) c_{22}.

Solution
(a) c_{13} is the element in row 1, column 3, and so $c_{13} = 4$.
(b) c_{21} is the element in row 2, column 1, and so $c_{21} = \lambda$.
(c) c_{22} is the element in row 2, column 2, and so $c_{22} = 1$.

Example 1.2 Matrices and digital image processing

A digital image is an image that is represented by an array of numbers. Consider, as a very simple example, the mathematical representation of the symbol π. Very crudely we can picture the symbol as in Figure 1.1.

Figure 1.1
A simple digital representation of π.

We could represent this image of π by a 3×5 matrix such as

$$B = \begin{pmatrix} 1 & 1 & 1 & 1 & 1 \\ 0 & 1 & 0 & 1 & 0 \\ 0 & 1 & 0 & 1 & 0 \end{pmatrix}$$

Note that we have divided the image into a number of areas, or fields, in this case 15 of them. We have then assigned a number to each field: 1 if the field is to be shaded, and 0 otherwise.

Images such as black and white photographs can be 'digitised' in the same way. The image is divided into $m \times n$ fields, where in practice m and n are large integers. Each field is assigned a number that represents the intensity of the image in that area. Often a 64-point grey scale is used so that a range of grey levels can be represented ranging from white at one extreme to black at the other. A digital image is then represented by an $m \times n$ matrix, B say, whose elements are numbers in the range 1–64. In image processing the matrix B is often referred to as $b[m, n]$.

Exercises

1 The matrix, H, is defined by

$$H = \begin{pmatrix} 2 & 1 \\ \alpha & 0 \\ -3 & \beta \end{pmatrix}$$

(a) State the size of H.
(b) State h_{11}, h_{21}, h_{32}.

2 State the number of elements in an $n \times m$ matrix.

Solution to exercises

1 (a) 3×2 (b) $2, \alpha, \beta$

2 nm

1.2 Some special matrices

There are a number of matrices that you will use frequently and which have special names. These are now described.

A **square matrix** has the same number of columns as rows. Both D and E are square matrices where

$$D = \begin{pmatrix} 3 & 1 \\ 4 & 2 \end{pmatrix}, \quad E = \begin{pmatrix} 1 & -1 & 3 \\ 0 & \lambda & 3 \\ -1 & y & 7 \end{pmatrix}$$

The **main diagonal** of a square matrix is the diagonal running from 'top left' to 'bottom right'. The main diagonal of D contains the elements 3 and 2. In general, if A is an $n \times n$ matrix, the main diagonal contains the elements $a_{11}, a_{22}, a_{33}, \ldots, a_{nn}$.

An **identity matrix** is a square matrix with ones on the main diagonal and zeros everywhere else. Identity matrices are usually denoted by I, with a subscript to denote its size. The 2×2 identity matrix is I_2 where

$$I_2 = \begin{pmatrix} 1 & 0 \\ 0 & 1 \end{pmatrix}$$

the 3×3 identity matrix is I_3 where

$$I_3 = \begin{pmatrix} 1 & 0 & 0 \\ 0 & 1 & 0 \\ 0 & 0 & 1 \end{pmatrix}$$

and so on.

The **transpose** of a matrix is obtained by writing rows as columns. The transpose of A is denoted A^{T}. For example, if

$$A = \begin{pmatrix} 6 & 1 & -3 & 0 \\ 9 & 2 & 3 & 7 \end{pmatrix}$$

then the transpose of A, A^{T}, is given by

$$A^{\mathrm{T}} = \begin{pmatrix} 6 & 9 \\ 1 & 2 \\ -3 & 3 \\ 0 & 7 \end{pmatrix}$$

Example 1.3

Given

$$B = \begin{pmatrix} 1 & -2 & 4 \\ 3 & 1 & 0 \\ 7 & 5 & 9 \end{pmatrix}$$

state the transpose of B.

Solution

The transpose is

$$\begin{pmatrix} 1 & 3 & 7 \\ -2 & 1 & 5 \\ 4 & 0 & 9 \end{pmatrix}$$

Exercises

1   State the transpose of C where

$$C = \begin{pmatrix} 9 & 3 \\ 1 & -2 \\ \alpha & 4 \end{pmatrix}$$

2   State the transpose of I_3.

Solutions to exercises

1    $C^{\mathrm{T}} = \begin{pmatrix} 9 & 1 & \alpha \\ 3 & -2 & 4 \end{pmatrix}$

2   I_3

1.3   Addition and subtraction of matrices

Matrices of the same size may be added to and subtracted from one another. To do this, the corresponding elements are added or subtracted.

Example 1.4

Given

$$A = \begin{pmatrix} 3 & 1 & -1 \\ 0 & -2 & 4 \end{pmatrix}, \quad B = \begin{pmatrix} 2 & 6 & 1 \\ 10 & -3 & 6 \end{pmatrix}, \quad C = \begin{pmatrix} -1 & 0 \\ 3 & 2 \end{pmatrix}$$

find, if possible,

(a) $A + B$ (b) $B - C$ (c) $B - A$ (d) $A + C$

Solution

(a)

$$A + B = \begin{pmatrix} 3 & 1 & -1 \\ 0 & -2 & 4 \end{pmatrix} + \begin{pmatrix} 2 & 6 & 1 \\ 10 & -3 & 6 \end{pmatrix}$$

$$= \begin{pmatrix} 3+2 & 1+6 & -1+1 \\ 0+10 & -2-3 & 4+6 \end{pmatrix}$$

$$= \begin{pmatrix} 5 & 7 & 0 \\ 10 & -5 & 10 \end{pmatrix}$$

(b) B is a 2×3 matrix; C is a 2×2 matrix. Since B and C have different sizes then $B - C$ cannot be calculated.

(c)
$$B - A = \begin{pmatrix} 2 & 6 & 1 \\ 10 & -3 & 6 \end{pmatrix} - \begin{pmatrix} 3 & 1 & -1 \\ 0 & -2 & 4 \end{pmatrix}$$

$$= \begin{pmatrix} 2 - 3 & 6 - 1 & 1 - (-1) \\ 10 - 0 & -3 - (-2) & 6 - 4 \end{pmatrix}$$

$$= \begin{pmatrix} -1 & 5 & 2 \\ 10 & -1 & 2 \end{pmatrix}$$

(d) A and C have different sizes and so $A + C$ is not defined.

Example 1.5

If $C = \begin{pmatrix} 7 & 2 \\ -2 & 1 \\ 4 & 3 \end{pmatrix}$ and $D = \begin{pmatrix} -1 & 0 \\ 3 & 4 \\ 9 & 1 \end{pmatrix}$

calculate
(a) $C + D$ (b) $C^{\mathrm{T}} + D^{\mathrm{T}}$ (c) $(C + D)^{\mathrm{T}}$

Solution

(a) $C + D = \begin{pmatrix} 7 & 2 \\ -2 & 1 \\ 4 & 3 \end{pmatrix} + \begin{pmatrix} -1 & 0 \\ 3 & 4 \\ 9 & 1 \end{pmatrix}$

$$= \begin{pmatrix} 6 & 2 \\ 1 & 5 \\ 13 & 4 \end{pmatrix}$$

(b) $C^{\mathrm{T}} + D^{\mathrm{T}} = \begin{pmatrix} 7 & -2 & 4 \\ 2 & 1 & 3 \end{pmatrix} + \begin{pmatrix} -1 & 3 & 9 \\ 0 & 4 & 1 \end{pmatrix}$

$$= \begin{pmatrix} 6 & 1 & 13 \\ 2 & 5 & 4 \end{pmatrix}$$

(c) $C + D$ has been found in (a).

$$(C + D)^{\mathrm{T}} = \begin{pmatrix} 6 & 1 & 13 \\ 2 & 5 & 4 \end{pmatrix}$$

Note that $(C + D)^{\mathrm{T}} = C^{\mathrm{T}} + D^{\mathrm{T}}$.

Example 1.5 illustrates a general rule for matrices.

Key point

The transpose of the sum of two matrices is the same as the sum of the individual matrices transposed:

$$(A + B)^{\mathrm{T}} = A^{\mathrm{T}} + B^{\mathrm{T}}$$

Exercises

1 Given

$$R = \begin{pmatrix} 4 & 1 \\ 6 & 2 \end{pmatrix}, \quad S = \begin{pmatrix} -3 & 2 \\ 7 & -4 \end{pmatrix}$$

find
(a) $R + S$ (b) $S - R$ (c) $R + R^{\mathrm{T}}$
(d) $(R - S)^{\mathrm{T}}$ (e) $(S^{\mathrm{T}})^{\mathrm{T}}$

Solutions to exercises

1 (a) $\begin{pmatrix} 1 & 3 \\ 13 & -2 \end{pmatrix}$ (b) $\begin{pmatrix} -7 & 1 \\ 1 & -6 \end{pmatrix}$ (e) $\begin{pmatrix} -3 & 2 \\ 7 & -4 \end{pmatrix}$

(c) $\begin{pmatrix} 8 & 7 \\ 7 & 4 \end{pmatrix}$ (d) $\begin{pmatrix} 7 & -1 \\ -1 & 6 \end{pmatrix}$

1.4 Multiplication of a matrix by a number

Any matrix can be multiplied by a number. To do this, each element of the matrix is multiplied by the number.

Example 1.6
If

$$A = \begin{pmatrix} 3 & 1 & -1 \\ 0 & -2 & 4 \end{pmatrix}$$

find
(a) $2A$ (b) $\dfrac{1}{2}A$ (c) $-A$

Solution

(a)
$$2A = 2\begin{pmatrix} 3 & 1 & -1 \\ 0 & -2 & 4 \end{pmatrix}$$

$$= \begin{pmatrix} 2 \times 3 & 2 \times 1 & 2 \times (-1) \\ 2 \times 0 & 2 \times (-2) & 2 \times 4 \end{pmatrix}$$

$$= \begin{pmatrix} 6 & 2 & -2 \\ 0 & -4 & 8 \end{pmatrix}$$

(b)
$$\frac{1}{2}A = \frac{1}{2}\begin{pmatrix} 3 & 1 & -1 \\ 0 & -2 & 4 \end{pmatrix}$$

$$= \begin{pmatrix} 1.5 & 0.5 & -0.5 \\ 0 & -1 & 2 \end{pmatrix}$$

(c) $\qquad -A = (-1)\,A$

$$= (-1)\begin{pmatrix} 3 & 1 & -1 \\ 0 & -2 & 4 \end{pmatrix}$$

$$= \begin{pmatrix} -3 & -1 & 1 \\ 0 & 2 & -4 \end{pmatrix}$$

Example 1.7
If

$$A = \begin{pmatrix} 3 & 1 & -1 \\ 0 & -2 & 4 \end{pmatrix} \quad \text{and} \quad B = \begin{pmatrix} 2 & 6 & 1 \\ 10 & -3 & 6 \end{pmatrix}$$

find

(a) $2A + 3B$ (b) $A - 2B$ (c) $\dfrac{1}{2}A + \dfrac{1}{2}B$

Solution

(a) $2A + 3B = \begin{pmatrix} 6 & 2 & -2 \\ 0 & -4 & 8 \end{pmatrix} + \begin{pmatrix} 6 & 18 & 3 \\ 30 & -9 & 18 \end{pmatrix}$

$$= \qquad\qquad\qquad \begin{pmatrix} 12 & 20 & 1 \\ 30 & -13 & 26 \end{pmatrix}$$

(b) $A - 2B = \begin{pmatrix} 3 & 1 & -1 \\ 0 & -2 & 4 \end{pmatrix} - \qquad \begin{pmatrix} 4 & 12 & 2 \\ 20 & -6 & 12 \end{pmatrix}$

$$= \qquad\qquad\qquad \begin{pmatrix} -1 & -11 & -3 \\ -20 & 4 & -8 \end{pmatrix}$$

(c) $\dfrac{1}{2}A + \dfrac{1}{2}B = \dfrac{1}{2}(A + B) = \qquad \begin{pmatrix} 2.5 & 3.5 & 0 \\ 5 & -2.5 & 5 \end{pmatrix}$

Example 1.8
Given

$$A = \begin{pmatrix} 3 & 1 & 4 \\ 2 & 9 & 6 \end{pmatrix}$$

and k is a constant find
(a) A^{T} (b) kA^{T} (c) $(kA)^{\mathrm{T}}$

Solution

(a) $A^\mathrm{T} =$ $\begin{pmatrix} 3 & 2 \\ 1 & 9 \\ 4 & 6 \end{pmatrix}$

(b) $kA^\mathrm{T} = k \begin{pmatrix} 3 & 2 \\ 1 & 9 \\ 4 & 6 \end{pmatrix}$

$= \begin{pmatrix} 3k & 2k \\ k & 9k \\ 4k & 6k \end{pmatrix}$

(c) $kA =$ $\begin{pmatrix} 3k & k & 4k \\ 2k & 9k & 6k \end{pmatrix}$

Hence

$$(kA)^\mathrm{T} = \begin{pmatrix} 3k & 2k \\ k & 9k \\ 4k & 6k \end{pmatrix}$$

Example 1.8 illustrates a general rule:

Key point	$(kA)^\mathrm{T} = kA^\mathrm{T}$

Exercises

1   Given
$$A = \begin{pmatrix} 2 & 1 \\ 3 & -2 \end{pmatrix}, \quad B = \begin{pmatrix} 4 & -1 \\ -2 & 6 \end{pmatrix}$$
find
(a) $3A$ (b) $2B$ (c) $4A + 3B$ (d) $B - 2A$
(e) $2A^\mathrm{T}$ (f) $(2A)^\mathrm{T}$

2   Given
$$A = \begin{pmatrix} 6 & 1 \\ 3 & 7 \end{pmatrix}$$
state
(a) $A - 2I_2$ (b) $A - \lambda I_2$ where λ is a constant.

3   Given
$$A = \begin{pmatrix} 1 & 2 & 3 \\ 4 & 5 & -1 \end{pmatrix}, \quad B = \begin{pmatrix} -2 & 3 \\ 0 & 4 \\ 7 & -1 \end{pmatrix}$$
calculate
(a) $A + B^\mathrm{T}$ (b) $B - 2A^\mathrm{T}$

Solutions to exercises

1 (a) $\begin{pmatrix} 6 & 3 \\ 9 & -6 \end{pmatrix}$ (b) $\begin{pmatrix} 8 & -2 \\ -4 & 12 \end{pmatrix}$

(c) $\begin{pmatrix} 20 & 1 \\ 6 & 10 \end{pmatrix}$ (d) $\begin{pmatrix} 0 & -3 \\ -8 & 10 \end{pmatrix}$

(e) $\begin{pmatrix} 4 & 6 \\ 2 & -4 \end{pmatrix}$ (f) $\begin{pmatrix} 4 & 6 \\ 2 & -4 \end{pmatrix}$

2 (a) $\begin{pmatrix} 4 & 1 \\ 3 & 5 \end{pmatrix}$ (b) $\begin{pmatrix} 6-\lambda & 1 \\ 3 & 7-\lambda \end{pmatrix}$

3 (a) $\begin{pmatrix} -1 & 2 & 10 \\ 7 & 9 & -2 \end{pmatrix}$ (b) $\begin{pmatrix} -4 & -5 \\ -4 & -6 \\ 1 & 1 \end{pmatrix}$

End of block exercises

Questions 1–13 refer to matrices A, B and C where

$$A = \begin{pmatrix} 4 & 1 \\ 9 & -1 \\ 2 & 0 \end{pmatrix}, \quad B = \begin{pmatrix} 1 & 2 & 3 \\ 4 & 5 & -1 \end{pmatrix}, \quad C = \begin{pmatrix} -1 & 3 \\ 0 & 4 \\ 1 & 6 \end{pmatrix}$$

1 State the size of A.

2 State the size of B.

3 State a_{32}, b_{13}, c_{22}.

4 Calculate $5A$.

5 Calculate $-2C$.

6 Calculate $\dfrac{1}{2}C$.

7 State A^{T}.

8 State C^{T}.

9 Calculate $A + 2C$.

10 Calculate $3C - 4A$.

11 Calculate $A + 2B^{\mathrm{T}}$.

12 Calculate $B - 3C^{\mathrm{T}}$.

13 Calculate $\dfrac{2C + 5B^{\mathrm{T}}}{3}$.

14 If D is an $n \times m$ matrix, state the size of (a) D^{T}, (b) $(D^{\mathrm{T}})^{\mathrm{T}}$.

15 If

$$T = \begin{pmatrix} 3 & 1 & 4 \\ -1 & 2 & 6 \\ 7 & 3 & 0 \end{pmatrix}$$

state
(a) $T - 2I$
(b) $T - \lambda I$ where λ is a constant.

Solutions to exercises

1 3×2

2 2×3

3 $0, 3, 4$

4 $\begin{pmatrix} 20 & 5 \\ 45 & -5 \\ 10 & 0 \end{pmatrix}$

5 $\begin{pmatrix} 2 & -6 \\ 0 & -8 \\ -2 & -12 \end{pmatrix}$

6 $\begin{pmatrix} -\frac{1}{2} & \frac{3}{2} \\ 0 & 2 \\ \frac{1}{2} & 3 \end{pmatrix}$

7 $\begin{pmatrix} 4 & 9 & 2 \\ 1 & -1 & 0 \end{pmatrix}$

8 $\begin{pmatrix} -1 & 0 & 1 \\ 3 & 4 & 6 \end{pmatrix}$

9 $\begin{pmatrix} 2 & 7 \\ 9 & 7 \\ 4 & 12 \end{pmatrix}$

10 $\begin{pmatrix} -19 & 5 \\ -36 & 16 \\ -5 & 18 \end{pmatrix}$

11 $\begin{pmatrix} 6 & 9 \\ 13 & 9 \\ 8 & -2 \end{pmatrix}$

12 $\begin{pmatrix} 4 & 2 & 0 \\ -5 & -7 & -19 \end{pmatrix}$

13 $\begin{pmatrix} 1 & \frac{26}{3} \\ \frac{10}{3} & 11 \\ \frac{17}{3} & \frac{7}{3} \end{pmatrix}$

14 (a) $m \times n$ (b) $n \times m$

15 (a) $\begin{pmatrix} 1 & 1 & 4 \\ -1 & 0 & 6 \\ 7 & 3 & -2 \end{pmatrix}$

(b) $7\begin{pmatrix} 3 - \lambda & 1 & 4 \\ -1 & 2 - \lambda & 6 \\ 7 & 3 & -\lambda \end{pmatrix}$

Multiplication of matrices

2.1   Introduction

In Block 1 we saw how a matrix can be multiplied by a number. In this section we look at how one matrix can be multiplied by another matrix. We also specify the conditions under which such a multiplication can take place.

2.2   Conditions needed for two matrices to be multiplied together

Let A and B be two matrices. Then under certain conditions the product AB can be found. The way in which such a product is calculated is explained in the next section; for this section we focus on the conditions needed for the product to exist.

To decide whether the product AB exists we look at the size of each matrix. Suppose A is an $n \times m$ matrix and B is a $p \times q$ matrix. For the product AB to exist it must be that $m = p$: that is, the number of columns in A must be the same as the number of rows in B.

Key point

For the product AB to exist, the number of columns of A must equal the number of rows of B.

When $m = p$ the product AB can be calculated. The result is another matrix whose size is $n \times q$.

Key point

$$\underbrace{A}_{n \times m} \quad \underbrace{B}_{p \times q} \quad = \quad \underbrace{C}_{n \times q}$$

$$\underbrace{}_{m = p}$$

If the number of columns in A differs from the number of rows in B then the product AB cannot be found. We say AB does not exist.

The product AB is distinct from the product BA. For example, if A is a 3×2 matrix and B is a 2×4 matrix then the number of columns in A equals the number of rows in B, that is 2, and so AB can be found. It is a 3×4 matrix:

$$
\underbrace{A}_{3 \times 2} \underbrace{B}_{2 \times 4} = \underbrace{C}_{3 \times 4}
$$

However, let us now consider BA. The number of columns in B is four; the number of rows of A is three and so the product BA cannot be found. Hence the order of the product is crucial, and we say that matrix multiplication is, in general, **not commutative**. In the product AB we say A **premultiplies** B. Alternatively we can say that B **postmultiplies** A.

Products involving three or more matrices are possible. If P, Q and R are three matrices then to calculate PQR we first calculate PQ, and then postmultiply the result by R, that is PQR is treated as $(PQ)R$. Alternatively we can calculate QR and premultiply the result by P, that is we may treat PQR as $P(QR)$. The final result in both cases is the same. This means that matrix multiplication is **associative**.

Finally we note that the product AA is written as A^2, AAA is written as A^3, and so on.

Exercises

Questions 1–3 refer to matrices P, Q and R where P is a 3×2 matrix, Q is a 3×3 matrix and R is a 2×3 matrix.

1 State the size of the following products if they can be found. If they cannot be found then state this.
(a) PQ (b) PR (c) QR (d) RQ (e) P^2
(f) Q^2 (g) RP

2 State the size of each of the following:
(a) P^T (b) Q^T (c) R^T (d) $R^T P^T$ (e) $P^T Q^T$

3 State the size of the following products. If a product cannot be found then state this.
(a) PQR (b) PRQ (c) QPR (d) RQP

4 If A is a matrix, state conditions on A for A^2 to exist.

Solutions to exercises

1 (a) cannot be found (b) 3×3 (c) cannot be found (d) 2×3 (e) cannot be found (f) 3×3 (g) 2×2

2 (a) 2×3 (b) 3×3 (c) 3×2 (d) 3×3 (e) 2×3

3 (a) cannot be found (b) 3×3 (c) 3×3 (d) 2×2

4 A must be a square matrix for A^2 to exist

2.3 How to multiply two matrices

Section 2.2 stated the conditions needed for a product of matrices to exist. In this section we look at how the product is actually calculated. The method is illustrated in Example 2.1.

Example 2.1
Given

$$A = \begin{pmatrix} 3 & 1 & 4 \\ 5 & 2 & -2 \end{pmatrix} \text{ and } B = \begin{pmatrix} 1 & -1 \\ 2 & 0 \\ -6 & 4 \end{pmatrix}$$

calculate the product AB.

Solution
We note that A is a 2×3 matrix, B is a 3×2 matrix and so the product AB exists. The result is a 2×2 matrix, say C.

$$AB = C = \begin{pmatrix} c_{11} & c_{12} \\ c_{21} & c_{22} \end{pmatrix}$$

To find c_{11}, that is the element in row 1, column 1, we use row 1 from A and column 1 from B. The elements in this row and column are multiplied and added thus:

$$\begin{aligned} c_{11} &= 3(1) + 1(2) + 4(-6) \\ &= 3 + 2 - 24 \\ &= -19 \end{aligned}$$

The element c_{12} is in row 1, column 2 of C. It is calculated from row 1 of A and column 2 of B.

$$\begin{aligned} c_{12} &= 3(-1) + 1(0) + 4(4) \\ &= 13 \end{aligned}$$

The element c_{21} is in row 2, column 1 of C. It is calculated from row 2 of A and column 1 of B.

$$\begin{aligned} c_{21} &= 5(1) + 2(2) - 2(-6) \\ &= 21 \end{aligned}$$

The element c_{22} is in row 2, column 2 of C. It is calculated from row 2 of A and column 2 of B.

$$\begin{aligned} c_{22} &= 5(-1) + 2(0) - 2(4) \\ &= -13 \end{aligned}$$

Hence

$$AB = \begin{pmatrix} 3 & 1 & 4 \\ 5 & 2 & -2 \end{pmatrix} \begin{pmatrix} 1 & -1 \\ 2 & 0 \\ -6 & 4 \end{pmatrix}$$

$$= \begin{pmatrix} -19 & 13 \\ 21 & -13 \end{pmatrix}$$

Example 2.1 illustrates the following important point regarding matrix multiplication.

Key point

If $AB = C$ then the element c_{ij} is found from row i of A and column j of B.

Check the working to the following example.

Example 2.2

Given

$$A = \begin{pmatrix} 9 & -1 \\ 2 & 3 \end{pmatrix} \quad \text{and} \quad B = \begin{pmatrix} 4 & 3 & 2 \\ 1 & 2 & -1 \end{pmatrix}$$

calculate AB.

Solution

Note that A has two columns, B has two rows and so the product AB can be found. The result is a 2×3 matrix, C.

$$AB = \begin{pmatrix} 9 & -1 \\ 2 & 3 \end{pmatrix} \begin{pmatrix} 4 & 3 & 2 \\ 1 & 2 & -1 \end{pmatrix}$$

$$= C = \begin{pmatrix} c_{11} & c_{12} & c_{13} \\ c_{21} & c_{22} & c_{23} \end{pmatrix}$$

Now

$$c_{11} = 9(4) - 1(1)$$
$$= 35$$

$$c_{12} = \qquad\qquad 9(3) - 1(2) = 25$$

$$c_{13} = 9(2) - 1(-1)$$
$$= 19$$

$$c_{21} = \qquad\qquad 2(4) + 3(1) = 11$$

$$c_{22} = 2(3) + 3(2)$$
$$= 12$$

$$c_{23} = \qquad\qquad 2(2) + 3(-1) = 1$$

Hence

$$\begin{pmatrix} 9 & -1 \\ 2 & 3 \end{pmatrix} \begin{pmatrix} 4 & 3 & 2 \\ 1 & 2 & -1 \end{pmatrix} = \begin{pmatrix} 35 & 25 & 19 \\ 11 & 12 & 1 \end{pmatrix}$$

Recall the identity matrix, I, given by

$$I_2 = \begin{pmatrix} 1 & 0 \\ 0 & 1 \end{pmatrix}, \quad I_3 = \begin{pmatrix} 1 & 0 & 0 \\ 0 & 1 & 0 \\ 0 & 0 & 1 \end{pmatrix}$$

and so on. If no subscript is given then the size of the identity matrix is taken as that for which any stated product exists.

Example 2.3

Given

$$A = \begin{pmatrix} 3 & -2 \\ 10 & 7 \end{pmatrix}$$

calculate
(a) AI (b) IA

Solution

Since A has two columns the required identity matrix is I_2:

(a) $AI = \begin{pmatrix} 3 & -2 \\ 10 & 7 \end{pmatrix} \begin{pmatrix} 1 & 0 \\ 0 & 1 \end{pmatrix}$

$= \qquad\qquad\qquad \begin{pmatrix} 3 & -2 \\ 10 & 7 \end{pmatrix}$

(b) $IA = \begin{pmatrix} 1 & 0 \\ 0 & 1 \end{pmatrix} \begin{pmatrix} 3 & -2 \\ 10 & 7 \end{pmatrix}$

$= \qquad\qquad\qquad \begin{pmatrix} 3 & -2 \\ 10 & 7 \end{pmatrix}$

Example 2.3 illustrates a general point. For any square matrix A:

Key point

When a square matrix is post- or premultiplied by an identity matrix of the appropriate size the matrix is unchanged.

$$AI = IA = A$$

Example 2.4

Given

$$\begin{pmatrix} 4 & -1 \\ 6 & 5 \end{pmatrix} \begin{pmatrix} x \\ y \end{pmatrix} = \begin{pmatrix} -13 \\ 13 \end{pmatrix}$$

find x and y.

Solution

Multiplying out the matrices on the left-hand side of the equation yields

$$\begin{pmatrix} 4 & -1 \\ 6 & 5 \end{pmatrix} \begin{pmatrix} x \\ y \end{pmatrix} = \qquad\qquad \begin{pmatrix} 4x - y \\ 6x + 5y \end{pmatrix}$$

Hence

$$\begin{aligned} 4x - y &= -13 \\ 6x + 5y &= 13 \end{aligned}$$

Solving these equations simultaneously for x and y yields $x = -2, y = 5$.

Example 2.5

Given

$$A = \begin{pmatrix} 1 & 2 \\ -1 & 3 \\ 2 & 1 \end{pmatrix}$$

find

(a) A^T

(b) AA^T

(c) A^TA

Solution

(a) Recall that A^T denotes the transpose of A.

$$A^T = \qquad\qquad\qquad\qquad\qquad \begin{pmatrix} 1 & -1 & 2 \\ 2 & 3 & 1 \end{pmatrix}$$

(b) $AA^T = \begin{pmatrix} 1 & 2 \\ -1 & 3 \\ 2 & 1 \end{pmatrix} \begin{pmatrix} 1 & -1 & 2 \\ 2 & 3 & 1 \end{pmatrix}$

$$= \qquad\qquad\qquad\qquad\qquad \begin{pmatrix} 5 & 5 & 4 \\ 5 & 10 & 1 \\ 4 & 1 & 5 \end{pmatrix}$$

(c) $A^TA = \begin{pmatrix} 1 & -1 & 2 \\ 2 & 3 & 1 \end{pmatrix} \begin{pmatrix} 1 & 2 \\ -1 & 3 \\ 2 & 1 \end{pmatrix}$

$$= \qquad\qquad\qquad\qquad\qquad \begin{pmatrix} 6 & 1 \\ 1 & 14 \end{pmatrix}$$

Exercises

Questions 1–8 refer to matrices A, B and C where

$$A = \begin{pmatrix} 3 & 1 \\ -1 & 2 \end{pmatrix}, \quad B = \begin{pmatrix} 1 & 4 & 0 \\ 2 & 7 & -1 \end{pmatrix}, \quad C = \begin{pmatrix} -2 & 1 \\ 4 & -1 \\ 0 & 3 \end{pmatrix}$$

Calculate the following products.

1   AB

2   A^2

3   BC

4   CA

5   AA^T

6   BCA

7   ABC

8   $B^T AC^T$

Solutions to exercises

1   $\begin{pmatrix} 5 & 19 & -1 \\ 3 & 10 & -2 \end{pmatrix}$

2   $\begin{pmatrix} 8 & 5 \\ -5 & 3 \end{pmatrix}$

3   $\begin{pmatrix} 14 & -3 \\ 24 & -8 \end{pmatrix}$

4   $\begin{pmatrix} -7 & 0 \\ 13 & 2 \\ -3 & 6 \end{pmatrix}$

5   $\begin{pmatrix} 10 & -1 \\ -1 & 5 \end{pmatrix}$

6   $\begin{pmatrix} 45 & 8 \\ 80 & 8 \end{pmatrix}$

7   $\begin{pmatrix} 66 & -17 \\ 34 & -13 \end{pmatrix}$

8   $\begin{pmatrix} 3 & -1 & 15 \\ 8 & 2 & 54 \\ -4 & 6 & -6 \end{pmatrix}$

End of block exercises

Questions 1–6 refer to matrices A, B, C and D where

$$A = (2 \quad -1 \quad 7), \quad B = \begin{pmatrix} 9 \\ 2 \\ 0 \end{pmatrix}, \quad C = \begin{pmatrix} 3 & 6 \\ -3 & 1 \end{pmatrix}, \quad D = \begin{pmatrix} 1 & 2 & 3 \\ 4 & 5 & 6 \end{pmatrix}$$

1   State the size of each of the following products. If a product does not exist then state this.
(a) AB (b) BA (c) CD (d) DC (e) DB
(f) BD (g) A^2 (h) C^2

2   State the size of each of the following products. If a product does not exist then state this.
(a) AA^T (b) $D^T B$ (c) $D^T C$ (d) AD^T
(e) $(AD^T)^T$

3   State the size of each of the following products. If a product does not exist then state this.
(a) CDB (b) $C^2 D$ (c) DBA (d) $D^T CD$
(e) $AD^T C$

4   Calculate all the products in question 1 that exist.

5   Calculate all the products in question 2 that exist.

6   Calculate all the products in question 3 that exist.

Questions 7–9 refer to matrices A, B, C, D and E. These are given by

$$A = \begin{pmatrix} 4 & -1 \\ 1 & 0 \\ -2 & 2 \end{pmatrix}, \quad B = (2 \quad 1), \quad C = (3 \quad 1 \quad -1), \quad D = \begin{pmatrix} 5 & -2 \\ 1 & 0 \end{pmatrix}, \quad E = \begin{pmatrix} 3 \\ -1 \end{pmatrix}$$

7   State the size of each of the following products if they exist. If a product does not exist then state this.
(a) DE (b) CA (c) DB (d) CA (e) AE

8   State the size of each of the following products if they exist. If a product does not exist then state this.
(a) $A^T A$ (b) AA^T (c) CAD (d) BDE (e) CAB

9   Calculate
(a) AD (b) $E^T D$ (c) CAE

10   If
$$\begin{pmatrix} 3 & 1 \\ 4 & 2 \end{pmatrix} \begin{pmatrix} x \\ y \end{pmatrix} = \begin{pmatrix} 9 \\ 14 \end{pmatrix}$$
calculate x and y.

11   Given
$$\begin{pmatrix} 5 & 2 \\ -3 & 4 \end{pmatrix} \begin{pmatrix} x \\ y \end{pmatrix} = \begin{pmatrix} 19 \\ -27 \end{pmatrix}$$
find x and y.

12   P is an $n \times m$ matrix. Show that the products PP^T and $P^T P$ can always be found. State the size of each product.

13   Given
$$A = \begin{pmatrix} 4 & -2 \\ 1 & 3 \end{pmatrix}, \quad B = \begin{pmatrix} 1 & 3 & 4 \\ -1 & 2 & 0 \end{pmatrix}$$
calculate (a) AB (b) $(AB)^T$ (c) $B^T A^T$. What do you observe from (b) and (c)?

14   This question asks you to show that $(AB)^T = B^T A^T$. A is an $n \times m$ matrix and B is an $m \times p$ matrix given by

$$A = \begin{pmatrix} a_{11} & a_{12} & \cdots & a_{1m} \\ a_{21} & a_{22} & \cdots & a_{2m} \\ \vdots & \vdots & \ddots & \vdots \\ a_{n1} & a_{n2} & \cdots & a_{nm} \end{pmatrix}$$

$$B = \begin{pmatrix} b_{11} & b_{12} & \cdots & b_{1p} \\ b_{21} & b_{22} & \cdots & b_{2p} \\ \vdots & \vdots & \ddots & \vdots \\ b_{m1} & b_{m2} & \cdots & b_{mp} \end{pmatrix}$$

(a) State an expression for the (i, j)th element of AB, that is the element in the ith row and jth column.

(b) State an expression for the (i, j)th element of $(AB)^{\mathrm{T}}$.

(c) State the ith row of B^{T}.

(d) State the jth column of A^{T}.

(e) State an expression for the (i, j)th element of $B^{\mathrm{T}}A^{\mathrm{T}}$.

(f) Hence show that $(AB)^{\mathrm{T}} = B^{\mathrm{T}}A^{\mathrm{T}}$.

Solutions to exercises

1 (a) 1×1 (b) 3×3 (c) 2×3 (d) does not exist (e) 2×1 (f) does not exist (g) does not exist (h) 2×2

2 (a) 1×1 (b) does not exist (c) 3×2 (d) 1×2 (e) 2×1

3 (a) 2×1 (b) 2×3 (c) 2×3 (d) 3×3 (e) 1×2

4 (a) (16) (b) $\begin{pmatrix} 18 & -9 & 63 \\ 4 & -2 & 14 \\ 0 & 0 & 0 \end{pmatrix}$

(c) $\begin{pmatrix} 27 & 36 & 45 \\ 1 & -1 & -3 \end{pmatrix}$ (e) $\begin{pmatrix} 13 \\ 46 \end{pmatrix}$

(h) $\begin{pmatrix} -9 & 24 \\ -12 & -17 \end{pmatrix}$

5 (a) (54) (c) $\begin{pmatrix} -9 & 10 \\ -9 & 17 \\ -9 & 24 \end{pmatrix}$ (d) $(21 \quad 45)$

(e) $\begin{pmatrix} 21 \\ 45 \end{pmatrix}$

6 (a) $\begin{pmatrix} 315 \\ 7 \end{pmatrix}$ (b) $\begin{pmatrix} 87 & 102 & 117 \\ -80 & -109 & -138 \end{pmatrix}$

(c) $\begin{pmatrix} 26 & -13 & 91 \\ 92 & -46 & 322 \end{pmatrix}$

(d) $\begin{pmatrix} 31 & 32 & 33 \\ 59 & 67 & 75 \\ 87 & 102 & 117 \end{pmatrix}$ (e) $(-72 \quad 171)$

7 (a) 2×1 (b) 1×2 (c) does not exist (d) 1×2 (e) 3×1

8 (a) 2×2 (b) 3×3 (c) 1×2 (d) 1×1 (e) does not exist

9 (a) $\begin{pmatrix} 19 & -8 \\ 5 & -2 \\ -8 & 4 \end{pmatrix}$ (b) $(14 - 6)$ (c) (50)

10 $x = 2, y = 3$

11 $x = 5, y = -3$

12 PP^{T} is an $n \times n$ matrix; $P^{\mathrm{T}}P$ is an $m \times m$ matrix.

13 (a) $\begin{pmatrix} 6 & 8 & 16 \\ -2 & 9 & 4 \end{pmatrix}$ (b) $\begin{pmatrix} 6 & -2 \\ 8 & 9 \\ 16 & 4 \end{pmatrix}$

(c) $\begin{pmatrix} 6 & -2 \\ 8 & 9 \\ 16 & 4 \end{pmatrix}$

Conclude that $(AB)^{\mathrm{T}} = B^{\mathrm{T}}A^{\mathrm{T}}$.

14 (a) $\sum_{k=1}^{m} a_{ik}b_{kj}$ (b) $\sum_{k=1}^{m} a_{jk}b_{ki}$

(c) $(b_{1i} \, b_{2i} \, \ldots \, b_{mi})$ (d) $(a_{j1} \, a_{j2} \, \ldots \, a_{jm})$

(e) $\sum_{k=1}^{m} a_{jk}b_{ki}$

(f) From (b) and (e), the (i, j)th elements of $(AB)^{\mathrm{T}}$ and $B^{\mathrm{T}}A^{\mathrm{T}}$ are identical. Since this is true for all values of i and j then the two matrices are equal.

Determinants

3.1 Introduction

We have seen how to add and subtract matrices in Block 1. Block 2 dealt with the multiplication of one matrix by another. Division of one matrix by another is not defined. However, for some matrices it is possible to calculate an inverse matrix. In some ways, use of the inverse matrix takes the place of division. To calculate an inverse matrix requires knowledge of determinants, and it is this topic that is addressed first. All square matrices possess a determinant. We begin with the simplest square matrices: 2×2 matrices.

3.2 Determinant of a 2×2 matrix

Consider the 2×2 matrix A where

$$A = \begin{pmatrix} a & b \\ c & d \end{pmatrix}$$

Then the **determinant** of A is $ad - bc$. The determinant of A is denoted by

$$\det(A), \quad |A| \quad \text{or} \quad \begin{vmatrix} a & b \\ c & d \end{vmatrix}$$

When you have evaluated the determinant of a matrix you will see that the result is no longer a matrix, but a single number or expression.

Key point

If

$$A = \begin{pmatrix} a & b \\ c & d \end{pmatrix}$$

then $\det(A) = ad - bc$. Det(A) may also be written as

$$|A| \quad \text{or} \quad \begin{vmatrix} a & b \\ c & d \end{vmatrix}$$

Example 3.1

Given

$$A = \begin{pmatrix} 3 & 6 \\ -1 & 1 \end{pmatrix}$$

calculate the determinant of (a) A, (b) A^{T}.

Solution

(a)
$$|A| = 3(1) - 6(-1)$$
$$= 9$$

Note that the value of the determinant is the single number, 9.

(b)
$$A^{\text{T}} = \begin{pmatrix} 3 & -1 \\ 6 & 1 \end{pmatrix}$$

Hence

$$|A^{\text{T}}| = \qquad\qquad\qquad\qquad\qquad 3(1) - (-1)6 = 9$$

Example 3.2

If A is a 2 × 2 matrix show that $\det(A) = \det(A^{\text{T}})$.

Solution

Let

$$A = \begin{pmatrix} a & b \\ c & d \end{pmatrix}$$

Then $|A| = ad - bc$.

Now

$$A^{\text{T}} = \qquad\qquad\qquad\qquad \begin{pmatrix} a & c \\ b & d \end{pmatrix}$$

and so

$$|A^{\text{T}}| = \qquad\qquad\qquad\qquad ad - bc$$

Hence $|A| = |A^{\text{T}}|$: that is, the determinant of a matrix and the determinant of its transpose are the same.

Example 3.3

Calculate the determinant of

$$\begin{pmatrix} 4 & 2 \\ 6 & 3 \end{pmatrix}$$

Solution

$$\begin{vmatrix} 4 & 2 \\ 6 & 3 \end{vmatrix} = \qquad\qquad 4(3) - 2(6) = 0$$

A matrix that has a zero determinant is called **singular.**

Example 3.4 Control Engineering – Stability of a system

Many engineering systems can be modelled by a set of simultaneous equations. From these equations, a **state matrix**, A, can be formed which encapsulates crucial information about the system.

The stability of a system is determined by the **poles** of the system. The system poles, s, are determined from the equation

$$|A - sI| = 0$$

where I is the identity matrix.

Given the state matrix is

$$A = \begin{pmatrix} 3 & -2 \\ -1 & 5 \end{pmatrix}$$

determine the system poles.

Solution

The system poles are found from

$$|A - sI| = 0$$

Given

$$A = \begin{pmatrix} 3 & -2 \\ -1 & 5 \end{pmatrix} \quad \text{and} \quad I = \begin{pmatrix} 1 & 0 \\ 0 & 1 \end{pmatrix}$$

then we have

$$A - sI = \begin{pmatrix} 3 & -2 \\ -1 & 5 \end{pmatrix} - s\begin{pmatrix} 1 & 0 \\ 0 & 1 \end{pmatrix} = \begin{pmatrix} 3 - s & -2 \\ -1 & 5 - s \end{pmatrix}$$

So

$$\begin{aligned} |A - sI| &= \begin{vmatrix} 3 - s & -2 \\ -1 & 5 - s \end{vmatrix} \\ &= (3 - s)(5 - s) - 2 \\ &= s^2 - 8s + 13 \end{aligned}$$

The system poles are solutions of $s^2 - 8s + 13 = 0$. Using the quadratic formula, values of s are obtained:

$$s = 2.268, \quad 5.732$$

A system where $s < 0$, or the $\text{Re}(s) < 0$ in the case of complex poles, is said to be **stable**. If $s > 0$, or $\text{Re}(s) > 0$, then the system is **unstable**. Hence the system in Example 3.4 is unstable.

Exercises

1 Calculate the determinant of each of the following matrices.

(a) $\begin{pmatrix} 8 & 3 \\ -3 & 5 \end{pmatrix}$ (b) $\begin{pmatrix} 6 & 10 \\ -3 & -5 \end{pmatrix}$

(c) $\begin{pmatrix} x & 3 \\ y & 4 \end{pmatrix}$ (d) $\begin{pmatrix} 3 - \lambda & 2 \\ 8 & 1 - \lambda \end{pmatrix}$

(e) $\begin{pmatrix} -\lambda & 2 \\ 1 & 1 - \lambda \end{pmatrix}$

2 Calculate the determinant of each of the following matrices.

(a) $\begin{pmatrix} 3 & 6 \\ 2 & 9 \end{pmatrix}$ (b) $\begin{pmatrix} -2 & 1 \\ 1 & 4 \end{pmatrix}$

(c) $\begin{pmatrix} -1 & -1 \\ 4 & -3 \end{pmatrix}$ (d) $\begin{pmatrix} 12 & 20 \\ 6 & 10 \end{pmatrix}$

(e) $\begin{pmatrix} 1 & 0 \\ 3 & \lambda \end{pmatrix}$ (f) $\begin{pmatrix} 3 - \lambda & 4 \\ 6 & 2 - \lambda \end{pmatrix}$

3 Find conditions on a, b and c for

$$\begin{pmatrix} a & b \\ 0 & c \end{pmatrix}$$

to be singular.

4 Under what conditions is the matrix

$$F = \begin{pmatrix} x & 3 \\ 2x & y \end{pmatrix}$$

singular?

5 Find values of λ for

$$\begin{pmatrix} 7 - \lambda & 2 \\ 12 & 5 - \lambda \end{pmatrix}$$

to be singular.

6 Find

$$\begin{vmatrix} \cos \theta & \sin \theta \\ -\sin \theta & \cos \theta \end{vmatrix}$$

7 Find $|I|$ where I is a 2 × 2 identity matrix.

8 **Control Engineering.** Determine the system poles of the following state matrices. In each case state whether the system is stable or unstable.

(a) $\begin{pmatrix} -4 & 5 \\ -5 & 3 \end{pmatrix}$ (b) $\begin{pmatrix} -1 & -2 \\ 5 & 4 \end{pmatrix}$

Solutions to exercises

1 (a) 49 (b) 0 (c) $4x - 3y$ (d) $\lambda^2 - 4\lambda - 13$
(e) $\lambda^2 - \lambda - 2$

2 (a) 15 (b) -9 (c) 7 (d) 0 (e) λ
(f) $\lambda^2 - 5\lambda - 18$

3 Either a or c is 0.

4 F is singular if $x = 0$ or $y = 6$.

5 1, 11

6 $\cos^2 \theta + \sin^2 \theta = 1$, using a trigonometrical identity to simplify.

7 1

8 (a) $s = \dfrac{-1 \pm \sqrt{-51}}{2}$. System is stable.

(b) $s = \dfrac{3 \pm \sqrt{-15}}{2}$. System is unstable.

3.3 Minors and cofactors of a 3 × 3 matrix

Before we can calculate the determinant of a 3 × 3 matrix it is necessary to introduce the minor of an element and the cofactor of an element. Consider a 3 × 3 matrix A where

$$A = \begin{pmatrix} a_{11} & a_{12} & a_{13} \\ a_{21} & a_{22} & a_{23} \\ a_{31} & a_{32} & a_{33} \end{pmatrix}$$

If we choose an element of A, say a_{ij}, and cross out its row and column, that is the ith row and the jth column, we are left with a 2 × 2 matrix. The determinant of this 2 × 2 matrix is called the **minor** of a_{ij}.

Example 3.5

Given

$$A = \begin{pmatrix} 3 & -1 & 6 \\ 9 & -5 & 2 \\ 0 & 4 & 7 \end{pmatrix}$$

calculate the minor of (a) 3, (b) 9, (c) 2.

Solution

(a) The element 3 occurs in row 1, column 1, that is $a_{11} = 3$. Deleting row 1 and column 1 leaves

$$\begin{pmatrix} -5 & 2 \\ 4 & 7 \end{pmatrix}$$

The determinant of this 2 × 2 matrix is $(-5)7 - 2(4) = -43$. The minor of 3 is -43.

(b) The element 9 occurs in row 2, column 1. Deleting these leaves

$$\begin{pmatrix} -1 & 6 \\ 4 & 7 \end{pmatrix}$$

The determinant of this 2 × 2 matrix is

$$(-1)7 - 6(4) = -31$$

The minor of 9 is -31.

(c) The element 2 occurs in row 2, column 3. Deleting these leaves

$$\begin{pmatrix} 3 & -1 \\ 0 & 4 \end{pmatrix}$$

whose determinant is 12. The minor of 2 is 12.

Closely related to the minor of an element is the **cofactor** of an element. The cofactor of a_{ij} is given as

$$\text{cofactor of } a_{ij} = (-1)^{i+j} \times \text{minor of } a_{ij}$$

The term $(-1)^{i+j}$ is either 1 or -1 depending upon the position of the element. The following grid of $+$ and $-$ is an easy way to visualise this. The $+$ and $-$ signs are known as the **place signs** of the elements. If the place sign is $+$ then the cofactor and minor are identical; if the place sign is $-$ then the cofactor is $-$ (minor).

$$\begin{pmatrix} + & - & + \\ - & + & - \\ + & - & + \end{pmatrix}$$

Example 3.6
The matrix A is given as

$$A = \begin{pmatrix} 3 & -1 & 6 \\ 9 & -5 & 2 \\ 0 & 4 & 7 \end{pmatrix}$$

Calculate the cofactor of (a) 3, (b) 9, (c) 2.

Solution
(a) The minor of 3 is -43. Since the place sign is $+$ then the cofactor of 3 is also -43.
(b) The minor of 9 is -31. The place sign is $-$ and so the cofactor is 31.
(c) The minor of 2 is 12. The place sign is $-$ and so the cofactor is -12.

Exercises

1 The matrix D is given by

$$D = \begin{pmatrix} 7 & -8 & 3 \\ 9 & 2 & -2 \\ 1 & -3 & 0 \end{pmatrix}$$

(a) Calculate the minor of (i) 7, (ii) -8, (iii) 3, (iv) 9.
(b) Calculate the cofactor of (i) 2, (ii) -2, (iii) 1, (iv) -3, (v) 0.

2 The matrix G is defined by

$$G = \begin{pmatrix} 2 - \lambda & 3 & 6 \\ 9 & -\lambda & 4 \\ -1 & 0 & 1 - \lambda \end{pmatrix}$$

(a) Calculate the minor of (i) $2 - \lambda$, (ii) 3, (iii) 6.
(b) Calculate the cofactor of (i) 9, (ii) $-\lambda$ (iii) 4.

Solutions to exercises

1 (a) (i) -6 (ii) 2 (iii) -29 (iv) 9
(b) (i) -3 (ii) 13 (iii) 10 (iv) 41 (v) 86

2 (a) (i) $\lambda^2 - \lambda$ (ii) $13 - 9\lambda$ (iii) $-\lambda$
(b) (i) $3\lambda - 3$ (ii) $\lambda^2 - 3\lambda + 8$ (iii) -3

3.4 Determinant of a 3 × 3 matrix

Consider a general 3×3 matrix, A.

$$A = \begin{pmatrix} a_{11} & a_{12} & a_{13} \\ a_{21} & a_{22} & a_{23} \\ a_{31} & a_{32} & a_{33} \end{pmatrix}$$

The determinant of A is found using

$$|A| = a_{11} \times \text{(its cofactor)} + a_{12} \times \text{(its cofactor)} + a_{13} \times \text{(its cofactor)}$$

Calculation of the determinant using this expression is known as **expanding along the first row**.

Key point

$$|A| = a_{11} \times \text{(its cofactor)} + a_{12} \times \text{(its cofactor)} + a_{13} \times \text{(its cofactor)}$$

Example 3.7
Calculate $|A|$ where

$$A = \begin{pmatrix} 3 & -1 & 6 \\ 9 & -5 & 2 \\ 0 & 4 & 7 \end{pmatrix}$$

Solution
We have $a_{11} = 3$, $a_{12} = -1$ and $a_{13} = 6$. The cofactors of these elements are required. The cofactor of 3 is -43 as found in Example 3.6.

The minor of -1 is

$$\begin{vmatrix} 9 & 2 \\ 0 & 7 \end{vmatrix} = 63$$

The place sign of -1 is

$-$

and so the cofactor is

-63

The minor of 6 is

$$\begin{vmatrix} 9 & -5 \\ 0 & 4 \end{vmatrix} = 36$$

The place sign is $+$ and so the cofactor is 36. Hence

$$|A| = 3(-43) + (-1)(-63) + 6(36)$$
$$= 150$$

The determinant of a matrix can also be found by expanding along any row or column. Thus, for example, we could calculate $|A|$ by expanding along the second row thus:

$$|A| = a_{21} \times \text{(its cofactor)} + a_{22} \times \text{(its cofactor)} + a_{23} \times \text{(its cofactor)}$$

or expanding along the third column thus:

$$|A| = a_{13} \times \text{(its cofactor)} + a_{23} \times \text{(its cofactor)} + a_{33} \times \text{(its cofactor)}$$

Example 3.8

The matrix A is defined in Example 3.5 as

$$A = \begin{pmatrix} 3 & -1 & 6 \\ 9 & -5 & 2 \\ 0 & 4 & 7 \end{pmatrix}$$

Evaluate $\det(A)$ by
(a) expanding along the third row
(b) expanding along the second column.

Solution

(a) By expanding along the third row we have

$$|A| = a_{31} \times \text{(its cofactor)} + a_{32} \times \text{(its cofactor)} + a_{33} \times \text{(its cofactor)}$$

Now $a_{31} = 0$, $a_{32} = 4$, $a_{33} = 7$.
The minor of a_{31} is

$$\begin{vmatrix} -1 & 6 \\ -5 & 2 \end{vmatrix} = 28.$$

The cofactor is also 28.

The minor of a_{32} is

$$\begin{vmatrix} 3 & 6 \\ 9 & 2 \end{vmatrix} = -48$$

The cofactor is 48
The minor of a_{33} is

$$\begin{vmatrix} 3 & -1 \\ 9 & -5 \end{vmatrix} = -6$$

The cofactor is -6. Hence

$$|A| = 0(28) + 4(48) + 7(-6)$$

$$= 150$$

(b) We now expand along the second column.
The elements are

$$a_{12} = -1, a_{22} = -5, a_{32} = 4$$

The minor of a_{12} is $\begin{vmatrix} 9 & 2 \\ 0 & 7 \end{vmatrix} = 63$

The cofactor is -63.
    The minor of a_{22} is

$$\begin{vmatrix} 3 & 6 \\ 0 & 7 \end{vmatrix} = 21$$

The cofactor is 21.

The minor of a_{32} is $\begin{vmatrix} 3 & 6 \\ 9 & 2 \end{vmatrix} = -48$

The cofactor is   48
Hence

$$|A| = (-1)(-63) + (-5)(21) + 4(48)$$
$$= 150$$

Exercises

1   The matrix H is defined by

$$H = \begin{pmatrix} 6 & -2 & 4 \\ 3 & -1 & 7 \\ 9 & 8 & -3 \end{pmatrix}$$

Evaluate $|H|$ by expanding
(a) along the first row
(b) along the third row
(c) along the second column.

2   The matrix K is defined by

$$K = \begin{pmatrix} 17 & 9 & -4 \\ 0 & 11 & 3 \\ 0 & 0 & -6 \end{pmatrix}$$

Calculate $|K|$.

3   Calculate the determinant of each of the following matrices:

(a) $\begin{pmatrix} -1 & 6 & 1 \\ -3 & 9 & 2 \\ 6 & 3 & 0 \end{pmatrix}$

(b) $\begin{pmatrix} 5 & 3 & 6 \\ 10 & 11 & 1 \\ -6 & 8 & -7 \end{pmatrix}$

(c) $\begin{pmatrix} 6 & 6 & 3 \\ -1 & 9 & 7 \\ 11 & 21 & 13 \end{pmatrix}$

(d) $\begin{pmatrix} 3 & -6 & 9 \\ 4 & 6 & -6 \\ 5 & 18 & -21 \end{pmatrix}$

(e) $\begin{pmatrix} 1 & 3 & 6 \\ 6 & 3 & 1 \\ 3 & 1 & 6 \end{pmatrix}$

4   Find $\det(B)$ by expanding along the second column, where

$$B = \begin{pmatrix} 1 & 0 & 3 \\ 22 & -1 & 17 \\ -3 & 0 & 19 \end{pmatrix}$$

Solutions to exercises

1 -330

2 -1122

3 (a) 15 (b) 643 (c) 0 (d) 0 (e) -100

4 -28

3.5 Determinant of a 4 × 4 matrix

The determinant of a 4 × 4 matrix is found in a similar way to that for a 3 × 3 matrix. We can expand along any row or column. The minor of an element is the determinant of the 3 × 3 matrix left after deleting the row and column containing the element. The cofactor is then found by multiplying by either 1 or -1 depending upon the place sign.

The place signs for a 4 × 4 matrix are as follows:

$$\begin{pmatrix} + & - & + & - \\ - & + & - & + \\ + & - & + & - \\ - & + & - & + \end{pmatrix}$$

Expanding along the first row of a 4 × 4 matrix, A, for example, would give

$$|A| = a_{11} \times \text{(its cofactor)} + a_{12} \times \text{(its cofactor)}$$
$$+ a_{13} \times \text{(its cofactor)} + a_{14} \times \text{(its cofactor)}$$

Example 3.9
Find $|A|$ by expanding along the first row where

$$A = \begin{pmatrix} 3 & -2 & 4 & 1 \\ 6 & 2 & 0 & 3 \\ 9 & 1 & 1 & -1 \\ 4 & 2 & -2 & 5 \end{pmatrix}$$

Solution
We have $a_{11} = 3$, $a_{12} = -2$, $a_{13} = 4$, $a_{14} = 1$. The cofactors of these elements are found by deleting the appropriate rows and columns.

The minor of a_{11} is

$$\begin{vmatrix} 2 & 0 & 3 \\ 1 & 1 & -1 \\ 2 & -2 & 5 \end{vmatrix} = -6$$

The cofactor is -6.

The minor of a_{12} is

$$\begin{vmatrix} 6 & 0 & 3 \\ 9 & 1 & -1 \\ 4 & -2 & 5 \end{vmatrix} = -48$$

The cofactor is 48.
   The minor of a_{13} is

$$\begin{vmatrix} 6 & 2 & 3 \\ 9 & 1 & -1 \\ 4 & 2 & 5 \end{vmatrix} = -14$$

The cofactor is -14.
   The minor of a_{14} is

$$\begin{vmatrix} 6 & 2 & 0 \\ 9 & 1 & 1 \\ 4 & 2 & -2 \end{vmatrix} = 20$$

The cofactor is -20.
   Hence

$$|A| = 3(-6) + (-2)48 + 4(-14) + 1(-20)$$
$$= -190$$

The sensible choice of row or column can save work.

Example 3.10
Calculate $|T|$ where

$$T = \begin{pmatrix} 13 & 2 & 1 & -9 \\ 4 & 7 & 0 & 6 \\ -1 & 11 & -2 & 3 \\ 4 & -2 & 0 & 6 \end{pmatrix}$$

Solution
Expansion along the third column is efficient as it contains two zeros. The cofactors of these zero elements need not be calculated. We have $t_{13} = 1$ and $t_{33} = -2$.

$$T = t_{13} \times \text{(its cofactor)} + t_{33} \times \text{(its cofactor)}$$

$$= \begin{vmatrix} 4 & 7 & 6 \\ -1 & 11 & 3 \\ 4 & -2 & 6 \end{vmatrix} + (-2) \begin{vmatrix} 13 & 2 & -9 \\ 4 & 7 & 6 \\ 4 & -2 & 6 \end{vmatrix}$$

$$=$$

$$162 - 2(1026) = -1890$$

Exercises

1 Calculate the determinant of each of the following matrices by expanding along an appropriate row or column.

(a) $\begin{pmatrix} 6 & 3 & -2 & 1 \\ 4 & 0 & -2 & 0 \\ 1 & 1 & 7 & 4 \\ 6 & 3 & 0 & 2 \end{pmatrix}$

(b) $\begin{pmatrix} 3 & 17 & 7 & 2 \\ -1 & 6 & 0 & -3 \\ 9 & 6 & 0 & 5 \\ 5 & -3 & 0 & 4 \end{pmatrix}$

(c) $\begin{pmatrix} 2 & -1 & 0 & 6 \\ 6 & -5 & 2 & 1 \\ 0 & 3 & 2 & 0 \\ 1 & 2 & 4 & 5 \end{pmatrix}$

(d) $\begin{pmatrix} 9 & 3 & 5 & -2 \\ 0 & -1 & 6 & 7 \\ 0 & 4 & 9 & 11 \\ -2 & -6 & 3 & 5 \end{pmatrix}$

2 Calculate $|A|$ where

$$A = \begin{pmatrix} \alpha & 17 & 6 & 3 \\ 0 & \beta & 14 & 7 \\ 0 & 0 & \gamma & 13 \\ 0 & 0 & 0 & \delta \end{pmatrix}$$

Solutions to exercises

1 (a) 2 (b) 462 (c) −278 (d) −54

2 $\alpha\beta\gamma\delta$

3.6 Properties of determinants

There are a number of properties of determinants that can be used to simplify their evaluation. Some of these properties are given now.

Property 1

Key point

Suppose A is an $n \times n$ matrix and k a scalar. Suppose the matrix B is obtained by multiplying a *single* row or column of A by k. Then

$$\det(B) = k \det(A)$$

If the matrix A is multiplied by k, that is every element in the matrix is multiplied by k, then we obtain

$$\det(kA) = k^n \det(A)$$

Example 3.11
Suppose

$$A = \begin{pmatrix} 5 & 7 \\ -2 & 3 \end{pmatrix}$$

Obtain B from A by multiplying the second column of A by 4.
(a) Write down B.
(b) Find det(B).
(c) Find det(A).
(d) Verify that det(B) = 4 det(A) as in Property 1.

Solution

(a) $B =$ 　　　　　　　　 $\begin{pmatrix} 5 & 28 \\ -2 & 12 \end{pmatrix}$

(b) $|B| =$ 　　　 116

(c) $|A| =$ 　　　 29

(d) It follows that det(B) = 4 det(A).

Example 3.12
Suppose

$$A = \begin{pmatrix} 1 & 3 \\ 2 & -7 \end{pmatrix}$$

(a) Write down $5A$.
(b) Find det(A).
(c) Find det($5A$).
(d) Verify Property 1, that is det($5A$) = 5^2 det(A).

Solution

(a) $5A =$ 　　　　　　　　 $\begin{pmatrix} 5 & 15 \\ 10 & -35 \end{pmatrix}$

(b) det(A) = 　　　 -13

(c) det($5A$) = 　　　　　　　 $(5)(-35) - (15)(10) = -325$

(d) Then, since $-325 = 25 \times -13$, it follows that det($5A$) = $5^2 \times$ det(A).

Example 3.13
Find det(B) given

$$B = \begin{pmatrix} 6 & -2 & 12 \\ 27 & -15 & 6 \\ 0 & 4 & 7 \end{pmatrix}$$

and given that the determinant of

$$A = \begin{pmatrix} 3 & -1 & 6 \\ 9 & -5 & 2 \\ 0 & 4 & 7 \end{pmatrix}$$

is 150.

Solution
By careful inspection of B note that it is obtained from A by multiplying the first row by 2 and the second row by 3. Hence

$$\det(B) = 2 \times 3 \times \det(A) = 6 \times 150 = 900$$

Property 2

Key point

If B is obtained from A by interchanging two rows or two columns then

$$\det(B) = -\det(A)$$

In other words, interchanging two rows or two columns changes the sign of the determinant. Carrying out two interchanges introduces two sign changes and hence leaves the determinant unchanged.

Example 3.14
Given

$$A = \begin{pmatrix} 3 & -1 & 6 \\ 9 & -5 & 2 \\ 0 & 4 & 7 \end{pmatrix}$$

and $|A| = 150$, find the determinant of each of the following matrices.

(a) $\begin{pmatrix} 0 & 4 & 7 \\ 9 & -5 & 2 \\ 3 & -1 & 6 \end{pmatrix}$ (b) $\begin{pmatrix} 9 & -5 & 2 \\ 0 & 4 & 7 \\ 3 & -1 & 6 \end{pmatrix}$ (c) $\begin{pmatrix} -5 & 9 & 2 \\ -1 & 3 & 6 \\ 4 & 0 & 7 \end{pmatrix}$

Solution
(a) By interchanging the first row and second row of $|A|$ we obtain

$$\begin{vmatrix} 0 & 4 & 7 \\ 9 & -5 & 2 \\ 3 & -1 & 6 \end{vmatrix} = -150$$

(b) Interchanging the first row and second row of the determinant in (a) we obtain

$$\begin{vmatrix} 9 & -5 & 2 \\ 0 & 4 & 7 \\ 3 & -1 & 6 \end{vmatrix} = 150$$

(c) We begin with $|A|$. Interchanging the first column and second column we obtain

$$\begin{vmatrix} -1 & 3 & 6 \\ -5 & 9 & 2 \\ 4 & 0 & 7 \end{vmatrix} = -150$$

We now interchange the first row and second row to produce

$$\begin{vmatrix} -5 & 9 & 2 \\ -1 & 3 & 6 \\ 4 & 0 & 7 \end{vmatrix} = 150$$

Property 3

Adding or subtracting a multiple of one row (or column) to another row (or column) leaves the determinant unchanged.

Example 3.15
Evaluate

$$\begin{vmatrix} 3 & -1 & 6 \\ 9 & -5 & 2 \\ 0 & 4 & 7 \end{vmatrix}$$

Solution
We use Property 3 repeatedly to simplify the determinant.

$$\begin{vmatrix} 3 & -1 & 6 \\ 9 & -5 & 2 \\ 0 & 4 & 7 \end{vmatrix} = \begin{vmatrix} 3 & -1 & 6 \\ 0 & -2 & -16 \\ 0 & 4 & 7 \end{vmatrix} \text{ (subtracting 3} \times \text{first row from second row)}$$

$$= \begin{vmatrix} 3 & 0 & 6 \\ 0 & -2 & -16 \\ 0 & 4 & 7 \end{vmatrix} \left(\text{adding } \frac{1}{3} \text{ of first column to second column} \right)$$

$$= \begin{vmatrix} 3 & 0 & 0 \\ 0 & -2 & -16 \\ 0 & 4 & 7 \end{vmatrix} \text{ (subtracting 2} \times \text{first column from third column)}$$

$$= \begin{vmatrix} 3 & 0 & 0 \\ 0 & -2 & -16 \\ 0 & 0 & -25 \end{vmatrix} \text{ (adding 2} \times \text{second row to third row)}$$

$$= 150 \text{ (by expanding along first row)}$$

Property 4

Key point

If A and B are two square matrices such that AB exists then
$$\det(AB) = \det(A)\det(B)$$

Example 3.16
Given

$$A = \begin{pmatrix} 3 & 2 \\ 1 & 4 \end{pmatrix}, \quad B = \begin{pmatrix} -1 & 3 \\ 0 & 2 \end{pmatrix}$$

verify $\det(AB) = \det(A)\det(B)$.

Solution

$$|A| = 12 - 2 = 10, \quad |B| = -2, \quad |A||B| = -20$$

Now

$$AB = \begin{pmatrix} -3 & 13 \\ -1 & 11 \end{pmatrix}, \quad |AB| = -33 + 13 = -20$$

Hence $\det(AB) = \det(A)\det(B)$.

Property 5

Key point

If two rows or two columns of a matrix are equal, the determinant of the matrix is zero.

Example 3.17
Find the determinant of the matrix

$$A = \begin{pmatrix} 4 & -5 & 2 \\ 4 & -5 & 2 \\ 1 & 2 & 1 \end{pmatrix}$$

Solution
Note that the first two rows are the same. It is easy to check that $|A| = 0$.

Exercises

1 Use the properties of determinants to simplify the following determinants and then evaluate them.

(a) $\begin{vmatrix} 5 & 20 & 9 \\ 3 & 31 & 15 \\ 1 & 24 & 12 \end{vmatrix}$ (b) $\begin{vmatrix} 28 & 12 & 35 \\ 24 & 10 & 31 \\ 15 & 6 & 20 \end{vmatrix}$

(c) $\begin{vmatrix} 113 & 92 & 58 \\ 175 & 144 & 90 \\ 64 & 53 & 33 \end{vmatrix}$

(d) $\begin{vmatrix} 176 & 152 & 101 \\ 229 & 195 & 133 \\ 109 & 95 & 62 \end{vmatrix}$

(e) $\begin{vmatrix} -19 & -15 & -12 \\ -21 & -17 & -14 \\ -23 & -19 & -15 \end{vmatrix}$

(b) $\begin{vmatrix} 19 & 25 & 16 & -30 \\ 7 & 9 & 6 & -8 \\ -11 & -15 & -10 & 22 \\ 21 & 25 & 12 & -15 \end{vmatrix}$

2 Determine the value of:

(a) $\begin{vmatrix} 29 & 39 & -3 & 11 \\ 5 & 3 & 3 & 4 \\ 24 & 36 & -6 & 7 \\ 30 & 59 & -2 & 1 \end{vmatrix}$

3 Verify $\det(GH) = \det(G)\,\det(H)$ where

$$G = \begin{pmatrix} -3 & 1 \\ 2 & 5 \end{pmatrix} \text{ and } H = \begin{pmatrix} 2 & 4 \\ -1 & 3 \end{pmatrix}$$

Solutions to exercises

1 (a) 9 (b) 2 (c) 8 (d) 28 (e) 8

2 (a) 0 (b) 4

Computer and calculator exercises

Many computer algebra packages will calculate the determinant of a matrix. If you have access to such a package try these exercises.

1 Calculate the determinant of

(a) $\begin{pmatrix} 7 & 4 & 11 & 6 \\ 3 & 1 & 7 & -2 \\ 4 & 4 & 5 & 9 \\ 3 & 2 & 1 & 8 \end{pmatrix}$

(b) $\begin{pmatrix} 1 & 2 & 5 & 8 & 3 \\ 3 & 8 & 1 & -2 & 6 \\ 4 & 7 & 5 & -1 & 0 \\ -11 & 11 & 3 & 6 & 7 \\ -1 & 6 & 9 & 4 & 3 \end{pmatrix}$

Solutions to exercises

1 (a) 0 (b) 28118

End of block exercises

1 Find the determinant of each of the following matrices:

(a) $\begin{pmatrix} 6 & 11 \\ 2 & 4 \end{pmatrix}$ (b) $\begin{pmatrix} -1 & 6 \\ 3 & 9 \end{pmatrix}$

(c) $\begin{pmatrix} -1 & 0 \\ 2 & 3 \end{pmatrix}$ (d) $\begin{pmatrix} 3 & 4 \\ -1 & -2 \end{pmatrix}$

(e) $\begin{pmatrix} -7 & -1 \\ -2 & -3 \end{pmatrix}$

2   Find the determinant of each of the following matrices:

(a) $\begin{pmatrix} 3 - \lambda & 4 \\ 2 & 1 - \lambda \end{pmatrix}$ (b) $\begin{pmatrix} -\lambda & 3 \\ 1 & 2 - \lambda \end{pmatrix}$

(c) $\begin{pmatrix} 7 - \lambda & 2 \\ 3 & 2 - \lambda \end{pmatrix}$

(d) $\begin{pmatrix} -1 - \lambda & 1 \\ 6 & 4 - \lambda \end{pmatrix}$

(e) $\begin{pmatrix} 4 - \lambda & 2 \\ -1 & 7 - \lambda \end{pmatrix}$

3   Calculate the values of λ for which each matrix in question 2 is singular.

4   Given

$$A = \begin{pmatrix} 9 & 6 & -1 \\ 3 & 5 & 7 \\ -3 & 2 & -5 \end{pmatrix}$$

calculate (a) the minor of 9, (b) the cofactor of 6, (c) the minor of 3, (d) the cofactor of −5, (e) det(A).

5   Calculate the determinant of each of the following matrices:

(a) $\begin{pmatrix} 3 & 1 & -6 \\ 4 & 2 & 0 \\ 5 & 6 & -1 \end{pmatrix}$ (b) $\begin{pmatrix} 9 & 1 & 1 \\ 4 & 2 & 3 \\ 1 & 6 & 9 \end{pmatrix}$

(c) $\begin{pmatrix} 5 & 1 & 6 \\ 9 & 3 & 2 \\ 1 & -1 & 10 \end{pmatrix}$ (d) $\begin{pmatrix} 1 & 1 & 1 \\ 2 & 2 & 3 \\ 5 & 6 & 8 \end{pmatrix}$

(e) $\begin{pmatrix} 17 & -19 & 31 \\ 18 & 21 & 30 \\ 14 & 20 & 32 \end{pmatrix}$

6   Calculate the determinant of

(a) $\begin{pmatrix} 5 & 2 & 6 & 3 \\ 3 & 9 & 12 & 1 \\ -3 & 1 & 4 & 1 \\ 4 & 1 & 5 & 3 \end{pmatrix}$

(b) $\begin{pmatrix} 6 & 7 & 10 & 1 \\ 11 & 5 & 14 & 2 \\ 12 & 9 & 21 & 3 \\ 7 & 2 & 11 & 2 \end{pmatrix}$

7   If

$$A = \begin{pmatrix} 3 & 8 \\ 2 & 6 \end{pmatrix} \text{ and } B = 3A$$

(a) find $|A|$, (b) find $|B|$,

(c) verify $\det(B) = 3^2 \det(A)$.

8   Given

$$A = \begin{pmatrix} -4 & 2 \\ 1 & 3 \end{pmatrix}, \quad B = \begin{pmatrix} 6 & -1 \\ 5 & 3 \end{pmatrix}$$

verify (a) $\det(A^2) = [\det(A)]^2$,
(b) $\det(AB) = \det(A)\det(B)$.

9   Calculate the determinant of each of the following matrices:

(a) $\begin{pmatrix} 11 & 6 \\ -3 & -1 \end{pmatrix}$ (b) $\begin{pmatrix} 300 & 170 \\ 150 & 85 \end{pmatrix}$

(c) $\begin{pmatrix} 2x & 2y \\ -x & y \end{pmatrix}$

10   Calculate the following determinants:

(a) $\begin{vmatrix} 1 & 2 & 3 \\ 1 & 3 & 2 \\ -2 & 3 & 1 \end{vmatrix}$ (b) $\begin{vmatrix} 17 & 1 & -23 \\ \lambda & 0 & 6 \\ 0 & 0 & \lambda - 1 \end{vmatrix}$

11   Given

$$H = \begin{pmatrix} 3 & -2 & 1 \\ 6 & -1 & 5 \\ -3 & 0 & 9 \end{pmatrix}$$

calculate (a) the minor of −2,
(b) the cofactor of 0.

12   Calculate the values of λ for which the matrix A is singular given

$$A = \begin{pmatrix} 5 - \lambda & -1 \\ 6 & -2 - \lambda \end{pmatrix}$$

13 Calculate $|A|$ where

$$A = \begin{pmatrix} 6 & 19 & -36 & 0 \\ 0 & 4 & 1 & 0 \\ 0 & 2 & 1 & 0 \\ 17 & -21 & 32 & 2 \end{pmatrix}$$

14 Without explicitly evaluating the determinant find $|A|$ if

$$A = \begin{pmatrix} -1 & 3 & -1 \\ 7 & 8 & 7 \\ -15 & 1 & -15 \end{pmatrix}$$

15 **Control Engineering.** A system has a state matrix, A, given by

$$A = \begin{pmatrix} -5 & 2 \\ -11 & -2 \end{pmatrix}$$

Find the system poles and hence determine whether the system is stable or unstable.

Solutions to exercises

1 (a) 2 (b) -27 (c) -3 (d) -2 (e) 19

2 (a) $\lambda^2 - 4\lambda - 5$ (b) $\lambda^2 - 2\lambda - 3$
(c) $\lambda^2 - 9\lambda + 8$ (d) $\lambda^2 - 3\lambda - 10$
(e) $\lambda^2 - 11\lambda + 30$

3 (a) $-1, 5$ (b) $-1, 3$ (c) $1, 8$ (d) $-2, 5$
(e) $5, 6$

4 (a) -39 (b) -6 (c) -28 (d) 27 (e) -408

5 (a) -86 (b) -11 (c) 0 (d) -1 (e) 6234

6 (a) 10 (b) -9

7 (a) 2 (b) 18

9 (a) 7 (b) 0 (c) $2xy + 2xy = 4xy$

10 (a) 14 (b) $\lambda(1 - \lambda)$

11 (a) 69 (b) -9

12 $-1, 4$

13 24

14 0 since two columns are identical.

15 Poles are $s = \dfrac{-7 \pm \sqrt{-79}}{2}$. System is stable.

The inverse of a matrix

4.1 Introduction

This block examines the idea of an inverse matrix. Division of one matrix by another is not defined: use of an inverse matrix is the nearest equivalent of 'matrix division'.

4.2 Definition

Let A be an $n \times n$ matrix. Recall that I_n is the $n \times n$ identity matrix. Suppose that B is an $n \times n$ matrix such that

$$AB = BA = I_n$$

We say that B is the **inverse** of A. We often write A^{-1} to denote the inverse of A. It is important to realise that in this context A^{-1} does not mean $\dfrac{1}{A}$. Division by a matrix is not defined. In what follows we shall write the identity matrix as I, rather than I_n. Its size should be obvious from the context.

> **Key point**
>
> Multiplying a square matrix A by its inverse A^{-1}, if this exists, yields the identity matrix I.
>
> $$A A^{-1} = A^{-1} A = I$$

Example 4.1

Given $A = \begin{pmatrix} 3 & 4 \\ 2 & 3 \end{pmatrix}$

show that

$$B = \begin{pmatrix} 3 & -4 \\ -2 & 3 \end{pmatrix}$$

is the inverse of A.

Solution

We calculate AB and BA.

$$AB = \begin{pmatrix} 3 & 4 \\ 2 & 3 \end{pmatrix} \begin{pmatrix} 3 & -4 \\ -2 & 3 \end{pmatrix}$$

$$= \begin{pmatrix} 1 & 0 \\ 0 & 1 \end{pmatrix}$$

$$= I$$

$$BA = \begin{pmatrix} 3 & -4 \\ -2 & 3 \end{pmatrix} \begin{pmatrix} 3 & 4 \\ 2 & 3 \end{pmatrix}$$

$$= \begin{pmatrix} 1 & 0 \\ 0 & 1 \end{pmatrix}$$

$$= I$$

Hence $AB = BA = I$ and so B is the inverse of A. Note also that A is the inverse of B.

The term 'inverse' can only be applied to square matrices. It is meaningless to talk about the inverse of a 2×3 or 3×7 matrix, for example. However, even though a matrix is square this is no guarantee that its inverse matrix exists. If the determinant of A is 0, that is A is singular, then A does not have an inverse. If the determinant is non-zero, then the inverse can be found.

Key point

If $|A| = 0$, A does not have an inverse.

If $|A| \neq 0$, A does have an inverse.

4.3 Finding the inverse of a 2×2 matrix

Suppose A is a 2×2 matrix and $|A| \neq 0$. Then the inverse of A can be found. If

$$A = \begin{pmatrix} a & b \\ c & d \end{pmatrix}$$

then recall $|A| = ad - bc$. Now A^{-1} is given by

$$A^{-1} = \frac{1}{|A|} \begin{pmatrix} d & -b \\ -c & a \end{pmatrix}$$

Key point

If

$$A = \begin{pmatrix} a & b \\ c & d \end{pmatrix}$$

then

$$A^{-1} = \frac{1}{|A|} \begin{pmatrix} d & -b \\ -c & a \end{pmatrix}$$

Example 4.2

Find A^{-1}, if it exists, when A is given by

(a) $\begin{pmatrix} 3 & 4 \\ 2 & 6 \end{pmatrix}$ (b) $\begin{pmatrix} -1 & 4 \\ 1 & 0 \end{pmatrix}$ (c) $\begin{pmatrix} 1 & 0 \\ 0 & 1 \end{pmatrix}$ (d) $\begin{pmatrix} 3 & -6 \\ -6 & 12 \end{pmatrix}$

Solution

(a) Here $a = 3$, $b = 4$, $c = 2$, $d = 6$ and $|A| = 10$. Then

$$A^{-1} = \frac{1}{10}\begin{pmatrix} 6 & -4 \\ -2 & 3 \end{pmatrix}$$

$$= \begin{pmatrix} 0.6 & -0.4 \\ -0.2 & 0.3 \end{pmatrix}$$

It is useful to check that $AA^{-1} = A^{-1}A = I$.

(b) Here $a = \quad$, $b = \quad$, $c = \quad$, $d = \quad$ $-1, 4, 1, 0$

and

$|A| = \quad$ -4

Hence we have

$A^{-1} = \quad$ $\dfrac{1}{-4}\begin{pmatrix} 0 & -4 \\ -1 & -1 \end{pmatrix}$

$= \quad$ $\begin{pmatrix} 0 & 1 \\ 0.25 & 0.25 \end{pmatrix}$

(c) Here $a = 1$, $b = 0$, $c = 0$, $d = 1$ and $|A| = 1$. We have

$$A^{-1} = \begin{pmatrix} 1 & 0 \\ 0 & 1 \end{pmatrix}$$

Hence the identity matrix is its own inverse.

(d) Here $|A| = 0$. Hence the matrix is singular and does not have an inverse.

Example 4.3

Find the inverse of

$$B = \begin{pmatrix} 3 & -2 \\ 4 & -2 \end{pmatrix}$$

Solution

Here the values of a, b, c and d are given by $3, -2, 4, -2$
The determinant of B is found.

$|B| = \quad$ 2

The inverse, B^{-1}, may now be stated.

$$B^{-1} = \boxed{} \qquad \begin{pmatrix} -1 & 1 \\ -2 & 1.5 \end{pmatrix}$$

Exercises

1 State the condition under which a square matrix does not have an inverse.

2 Find the inverse of the following matrices. If an inverse does not exist, state this.

(a) $\begin{pmatrix} 6 & 1 \\ 4 & 3 \end{pmatrix}$ (b) $\begin{pmatrix} -4 & -1 \\ 3 & 5 \end{pmatrix}$

(c) $\begin{pmatrix} x & -y \\ y & x \end{pmatrix}$ (d) $\begin{pmatrix} 4 & 8 \\ 5 & 10 \end{pmatrix}$

Solutions to exercises

1 If the determinant is zero, then the inverse does not exist.

2 (a) $\dfrac{1}{14} \begin{pmatrix} 3 & -1 \\ -4 & 6 \end{pmatrix}$ (b) $\dfrac{1}{17} \begin{pmatrix} -5 & -1 \\ 3 & 4 \end{pmatrix}$

(c) $\dfrac{1}{x^2 + y^2} \begin{pmatrix} x & y \\ -y & x \end{pmatrix}$ (d) no inverse

4.4   Finding the inverse of a matrix – the general case

We wish to find the inverse of A, an $n \times n$ matrix. We have already noted that, if $|A| = 0$, then the inverse of A does not exist. Suppose $|A| \neq 0$. To find A^{-1} we proceed thus:

1 Find the transpose of A, by interchanging the rows and columns of A.
2 Replace each element of A^{T} by its cofactor. The resulting matrix is known as the **adjoint** of A, written adj(A).
3 The inverse of A is given by

Key point

$$A^{-1} = \frac{\mathrm{adj}(A)}{|A|}$$

Example 4.4

Find the inverse of

$$A = \begin{pmatrix} 3 & 1 & 0 \\ 5 & 2 & -1 \\ 1 & 4 & -2 \end{pmatrix}$$

Solution

We find the transpose, A^T, by interchanging rows and columns.

$$A^T = \begin{pmatrix} 3 & 5 & 1 \\ 1 & 2 & 4 \\ 0 & -1 & -2 \end{pmatrix}$$

Each element of A^T is replaced by its cofactor to yield the adjoint of A.

$$\text{adj}(A) = \begin{pmatrix} 0 & 2 & -1 \\ 9 & -6 & 3 \\ 18 & -11 & 1 \end{pmatrix}$$

Calculating the determinant of A gives $|A| = 9$ and so

$$A^{-1} = \frac{\text{adj}(A)}{|A|}$$

$$= \frac{1}{9} \begin{pmatrix} 0 & 2 & -1 \\ 9 & -6 & 3 \\ 18 & -11 & 1 \end{pmatrix}$$

Example 4.5

Find the inverse of

$$H = \begin{pmatrix} 4 & -2 & 1 \\ 9 & 6 & 0 \\ 3 & -4 & 5 \end{pmatrix}$$

Solution

The transpose, H^T, is found.

$$H^T = \begin{pmatrix} 4 & 9 & 3 \\ -2 & 6 & -4 \\ 1 & 0 & 5 \end{pmatrix}$$

The adjoint of H is obtained by replacing each element of H^T by its cofactor.

$$\text{adj}(H) = \begin{pmatrix} 30 & 6 & -6 \\ -45 & 17 & 9 \\ -54 & 10 & 42 \end{pmatrix}$$

The determinant of H is found.

$$|H| = \boxed{}$$

156

Finally

$$H^{-1} = \frac{\text{adj}(H)}{|H|}$$

$$= \frac{1}{156}\begin{pmatrix} 30 & 6 & -6 \\ -45 & 17 & 9 \\ -54 & 10 & 42 \end{pmatrix}$$

Exercises

1 Calculate, where possible, the inverse of

(a) $\begin{pmatrix} 3 & -1 & 7 \\ 2 & 0 & 1 \\ 5 & -2 & 6 \end{pmatrix}$ (b) $\begin{pmatrix} 1 & 2 & -4 \\ -1 & 3 & 6 \\ 1 & 12 & 0 \end{pmatrix}$ (c) $\begin{pmatrix} -2 & 0 & 3 \\ 4 & 4 & -1 \\ 3 & 1 & 6 \end{pmatrix}$

Solutions to exercises

1 (a) $\frac{1}{15}\begin{pmatrix} -2 & 8 & 1 \\ 7 & 17 & -11 \\ 4 & -1 & -2 \end{pmatrix}$ (b) no inverse (c) $\frac{1}{74}\begin{pmatrix} -25 & -3 & 12 \\ 27 & 21 & -10 \\ 8 & -2 & 8 \end{pmatrix}$

Computer and calculator exercises

1 Use a package to find the inverse of each of the following matrices:

(a) $\begin{pmatrix} 4 & 3 & -2 & 2 \\ 3 & -1 & 0 & 4 \\ 6 & 5 & -3 & 0 \\ 1 & 1 & 2 & 1 \end{pmatrix}$

(b) $\begin{pmatrix} -3 & 10 & 9 & 4 \\ 6 & 0 & 1 & 0 \\ 3 & 2 & -1 & 1 \\ 5 & 2 & 3 & 1 \end{pmatrix}$

(c) $\begin{pmatrix} 1 & 2 & 4 & 6 \\ -1 & -1 & 5 & 0 \\ 9 & 7 & 2 & 3 \\ 14 & 8 & 6 & 1 \end{pmatrix}$

(d) $\begin{pmatrix} 6 & 1 & 0 & 3 & 2 \\ 1 & 4 & -1 & 2 & -3 \\ 0 & 1 & -6 & 6 & 5 \\ 0 & 3 & 2 & 1 & -1 \\ 5 & 1 & -5 & 3 & 4 \end{pmatrix}$

(e) $\begin{pmatrix} -6 & 2 & 0 & 3 & 9 \\ 5 & 1 & 6 & -1 & 4 \\ 7 & 3 & 4 & -2 & 3 \\ 6 & 5 & 5 & 5 & 1 \\ 3 & -2 & -1 & 3 & 4 \end{pmatrix}$

Solutions to exercises

1   (a) $\dfrac{1}{91}\begin{pmatrix} -55 & 27 & 38 & 2 \\ 51 & -30 & -22 & 18 \\ -25 & 4 & 9 & 34 \\ 54 & -5 & -34 & 3 \end{pmatrix}$

(b) $\dfrac{1}{44}\begin{pmatrix} 0 & 8 & 2 & -2 \\ 22 & 86 & 5 & -93 \\ 0 & -4 & -12 & 12 \\ -44 & -200 & 16 & 204 \end{pmatrix}$

(c) $\dfrac{1}{455}\begin{pmatrix} 101 & -192 & -262 & 180 \\ -181 & 272 & 447 & -255 \\ -16 & 107 & 37 & -15 \\ 130 & -130 & -130 & 65 \end{pmatrix}$

(d) $\dfrac{1}{1722}\begin{pmatrix} 188 & 69 & -146 & -141 & 105 \\ -370 & -90 & -134 & 708 & 462 \\ 226 & -201 & 26 & 261 & -231 \\ 456 & 204 & 342 & -342 & -588 \\ -202 & -468 & -8 & 582 & 336 \end{pmatrix}$

(e) $\dfrac{1}{13400}\begin{pmatrix} -640 & -760 & 1348 & 68 & 1172 \\ 850 & -2550 & 2460 & 810 & -1410 \\ -210 & 3310 & -2468 & 462 & -1102 \\ -50 & 150 & -1800 & 1450 & 950 \\ 890 & 10 & 952 & -618 & 778 \end{pmatrix}$

End of block exercises

1   Check that
$$\begin{pmatrix} 2 & 6 \\ -1 & 4 \end{pmatrix}$$
is the inverse of
$$\begin{pmatrix} \dfrac{2}{7} & -\dfrac{3}{7} \\ \dfrac{1}{14} & \dfrac{1}{7} \end{pmatrix}$$

2   Explain why the matrix
$$\begin{pmatrix} x^2 & x \\ xy & y \end{pmatrix}$$
has no inverse.

3   Calculate, if possible, the inverse of each of the following matrices:

(a) $\begin{pmatrix} 4 & 2 \\ 8 & 5 \end{pmatrix}$ (b) $\begin{pmatrix} 6 & 10 \\ 1 & -2 \end{pmatrix}$

(c) $\begin{pmatrix} 1 & 2 \\ 2 & 1 \end{pmatrix}$ (d) $\begin{pmatrix} 5 & -5 \\ 3 & -6 \end{pmatrix}$

(e) $\begin{pmatrix} 10 & -20 \\ -2 & 4 \end{pmatrix}$ (f) $\begin{pmatrix} 0.4 & 2 \\ 0.9 & 3 \end{pmatrix}$

(g) $\begin{pmatrix} -0.5 & -1.5 \\ -2 & -1 \end{pmatrix}$

4 Find the inverse of

$$\begin{pmatrix} x & x \\ x & 2x \end{pmatrix}$$

given $x \neq 0$.

5 A is a square matrix. State a condition that ensures A does not have an inverse.

6 Find, if possible, the inverse of each of the following 2×2 matrices:

(a) $\begin{pmatrix} 3 & -6 \\ 5 & 0 \end{pmatrix}$ (b) $\begin{pmatrix} 5 & 2 \\ 6 & 10 \end{pmatrix}$

(c) $\begin{pmatrix} 1 & 1 \\ -2 & -2 \end{pmatrix}$ (d) $\begin{pmatrix} 9 & 3 \\ -2 & 1 \end{pmatrix}$

(e) $\begin{pmatrix} -3 & 1 \\ 2 & -4 \end{pmatrix}$

7 Calculate the inverse of each of the following matrices. If it is not possible to calculate the inverse then state this clearly.

(a) $\begin{pmatrix} 5 & 6 \\ 2 & 3 \end{pmatrix}$ (b) $\begin{pmatrix} -6 & 0 \\ 3 & -1 \end{pmatrix}$

(c) $\begin{pmatrix} 4 & -2 \\ -3 & 1.5 \end{pmatrix}$

8 Calculate the values of λ for which

$$\begin{pmatrix} 2 - \lambda & 4 \\ 1.5 & 1 - \lambda \end{pmatrix}$$

has no inverse.

9 If

$$\begin{pmatrix} 3 & x \\ 5 & y \end{pmatrix}$$

is the inverse of

$$\begin{pmatrix} 2 & -1 \\ -5 & 3 \end{pmatrix}$$

state the values of x and y.

10 Calculate the inverse of each of the following matrices:

(a) $\begin{pmatrix} 1 & 4 & 3 \\ 5 & 6 & -4 \\ 2 & 0 & 1 \end{pmatrix}$ (b) $\begin{pmatrix} 1 & 2 & 3 \\ -4 & 6 & 1 \\ 0 & 0 & 3 \end{pmatrix}$

(c) $\begin{pmatrix} -3 & -1 & 2 \\ 4 & 7 & 1 \\ 5 & 0 & 3 \end{pmatrix}$

(d) $\begin{pmatrix} 1 & 1 & -3 & 0 \\ 4 & 6 & 1 & 3 \\ 9 & -2 & 0 & 4 \\ 5 & 6 & 3 & 1 \end{pmatrix}$

(e) $\begin{pmatrix} 0 & 1 & 4 & -1 \\ 2 & 7 & 3 & 5 \\ 6 & 3 & -1 & 0 \\ 1 & 2 & 4 & 7 \end{pmatrix}$

11 Find, if possible, the inverse of each of the following matrices:

(a) $\begin{pmatrix} 5 & 2 & -1 \\ 6 & 3 & 1 \\ 5 & 9 & 6 \end{pmatrix}$ (b) $\begin{pmatrix} -4 & 3 & 0 \\ -1 & 1 & 2 \\ 4 & 0 & 1 \end{pmatrix}$

(c) $\begin{pmatrix} 9 & 6 & -9 \\ 7 & 3 & 1 \\ -6 & 4 & 8 \end{pmatrix}$ (d) $\begin{pmatrix} 4 & 2 & -3 \\ 6 & 0 & 5 \\ 2 & 4 & -11 \end{pmatrix}$

(e) $\begin{pmatrix} 4 & -3 & 6 \\ 10 & 9 & 14 \\ 3 & -6 & 0 \end{pmatrix}$

12 Find, if possible, the inverse of each of the following matrices:

(a) $\begin{pmatrix} \lambda & 3 \\ \lambda & 5 \end{pmatrix}$ (b) $\begin{pmatrix} 3 & x \\ x & x^2 \end{pmatrix}$

(c) $\begin{pmatrix} a & bc \\ ab & b^2c \end{pmatrix}$ (d) $\begin{pmatrix} \dfrac{2}{\lambda} & -2 \\ 4 & \lambda \end{pmatrix}$

Solutions to exercises

2 The determinant of the matrix is zero: hence no inverse exists.

3 (a) $\begin{pmatrix} 1.25 & -0.5 \\ -2 & 1 \end{pmatrix}$ (b) $\dfrac{1}{22}\begin{pmatrix} 2 & 10 \\ 1 & -6 \end{pmatrix}$

(c) $\dfrac{1}{3}\begin{pmatrix} -1 & 2 \\ 2 & -1 \end{pmatrix}$ (d) $\dfrac{1}{15}\begin{pmatrix} 6 & -5 \\ 3 & -5 \end{pmatrix}$

(e) no inverse (f) $\dfrac{1}{6}\begin{pmatrix} -30 & 20 \\ 9 & -4 \end{pmatrix}$

(g) $\dfrac{1}{5}\begin{pmatrix} 2 & -3 \\ -4 & 1 \end{pmatrix}$

4 $\dfrac{1}{x}\begin{pmatrix} 2 & -1 \\ -1 & 1 \end{pmatrix}$

5 $|A| = 0$

6 (a) $\dfrac{1}{30}\begin{pmatrix} 0 & 6 \\ -5 & 3 \end{pmatrix}$ (b) $\dfrac{1}{38}\begin{pmatrix} 10 & -2 \\ -6 & 5 \end{pmatrix}$

(c) no inverse (d) $\dfrac{1}{15}\begin{pmatrix} 1 & -3 \\ 2 & 9 \end{pmatrix}$

(e) $-\dfrac{1}{10}\begin{pmatrix} 4 & 1 \\ 2 & 3 \end{pmatrix}$

7 (a) $\dfrac{1}{3}\begin{pmatrix} 3 & -6 \\ -2 & 5 \end{pmatrix}$ (b) $\dfrac{1}{6}\begin{pmatrix} -1 & 0 \\ -3 & -6 \end{pmatrix}$

(c) no inverse

8 $-1, 4$

9 $x = 1, y = 2$

10 (a) $\dfrac{1}{82}\begin{pmatrix} -6 & 4 & 34 \\ 13 & 5 & -19 \\ 12 & -8 & 14 \end{pmatrix}$

(b) $\dfrac{1}{42}\begin{pmatrix} 18 & -6 & -16 \\ 12 & 3 & -13 \\ 0 & 0 & 14 \end{pmatrix}$

(c) $\dfrac{1}{126}\begin{pmatrix} -21 & -3 & 15 \\ 7 & 19 & -11 \\ 35 & 5 & 17 \end{pmatrix}$

(d) $\dfrac{1}{570}\begin{pmatrix} 64 & -90 & 44 & 94 \\ 44 & 45 & -41 & 29 \\ -154 & -15 & 1 & 41 \\ -122 & 225 & 23 & -197 \end{pmatrix}$

(e) $\dfrac{1}{933}\begin{pmatrix} 42 & -105 & 177 & 81 \\ -15 & 193 & -41 & -140 \\ 207 & -51 & 6 & 66 \\ -120 & -11 & -17 & 124 \end{pmatrix}$

11 (a) $\dfrac{1}{56}\begin{pmatrix} -9 & 21 & -5 \\ 31 & -35 & 11 \\ -39 & 35 & -3 \end{pmatrix}$

(b) $\dfrac{1}{23}\begin{pmatrix} 1 & -3 & 6 \\ 9 & -4 & 8 \\ -4 & 12 & -1 \end{pmatrix}$

(c) $\dfrac{1}{606}\begin{pmatrix} -20 & 84 & -33 \\ 62 & -18 & 72 \\ -46 & 72 & 15 \end{pmatrix}$

(d) no inverse

(e) $\dfrac{1}{312}\begin{pmatrix} -84 & 36 & 96 \\ -42 & 18 & -4 \\ 87 & -15 & -66 \end{pmatrix}$

12 (a) $\dfrac{1}{2\lambda}\begin{pmatrix} 5 & -3 \\ -\lambda & \lambda \end{pmatrix}$ (b) $\dfrac{1}{2x^2}\begin{pmatrix} x^2 & -x \\ -x & 3 \end{pmatrix}$

(c) no inverse (d) $\dfrac{1}{10}\begin{pmatrix} \lambda & 2 \\ -4 & \dfrac{2}{\lambda} \end{pmatrix}$

Computer graphics

Many engineering components can now be represented on a computer screen. Modelling components in this way allows calculations of stresses, bending moments, pressure and so forth to be performed before the component is actually manufactured. Hence potential weaknesses and problem areas can be identified at the design stage. Fundamental to representing components is the ability to define points, lines and shapes, for example triangles and rectangles. Points, lines and shapes can be represented by matrices.

Once a component has been drawn it may be necessary to transform it. Transformation includes scaling, translation, shearing, rotation and reflection. Transformations are brought about by the use of matrices.

A point in two dimensions is represented by its coordinates, x and y. These values may be written as a 1×2 matrix, for example $(3 \; -2)$, or as a 2×1 matrix, for example $\begin{pmatrix} 3 \\ -2 \end{pmatrix}$. Such matrices are sometimes referred to as **position matrices** or **position vectors**. In this block we use 2×1 matrices to represent points.

A line can be determined by its two end-points. Hence the matrix

$$\begin{pmatrix} 1 & 3 \\ 2 & 7 \end{pmatrix}$$

represents the line joining the points $\begin{pmatrix} 1 \\ 2 \end{pmatrix}$ and $\begin{pmatrix} 3 \\ 7 \end{pmatrix}$. Similarly a triangle is defined by three points. Figure 5.1 illustrates a triangle.

The triangle can be 'stored' as a 2×3 matrix, T, where

$$T = \begin{pmatrix} 2 & 5 & 3 \\ 3 & 6 & -1 \end{pmatrix}$$

Figure 5.1
A triangle is
defined by three
points.

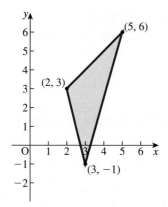

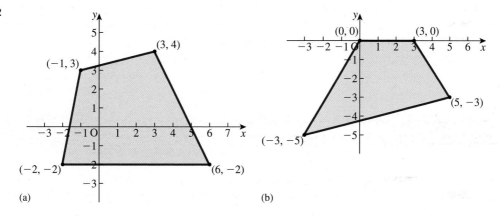

Example 5.1
Write matrices that represent the quadrilaterals shown in Figure 5.2(a), (b).

Figure 5.2

(a)

(b)

Solution
Each quadrilateral is defined by four points.
(a) The quadrilateral can be stored as

$$Q = \begin{pmatrix} -1 & 3 & 6 & -2 \\ 3 & 4 & -2 & -2 \end{pmatrix}$$

Note that other matrices would also be suitable, for example

$$\begin{pmatrix} 3 & 6 & -1 & -2 \\ 4 & -2 & 3 & -2 \end{pmatrix} \quad \text{and} \quad \begin{pmatrix} -2 & 3 & -1 & 6 \\ -2 & 4 & 3 & -2 \end{pmatrix}$$

also describe the quadrilateral.

(b) A possible matrix is $\begin{pmatrix} 0 & 3 & 5 & -3 \\ 0 & 0 & -3 & -5 \end{pmatrix}$

5.3 Matrices as transformations

Figures on a computer screen need to be transformed. Transformation includes scaling, translation, rotation, reflection and shearing. Mathematically, each transformation can be represented by a matrix, called a **transformation matrix**. Since every figure is made up of points we first examine the effect of a transformation matrix on a single point.

Consider the point $X = \begin{pmatrix} x \\ y \end{pmatrix}$ and a transformation matrix, T, given by

$$T = \begin{pmatrix} a & b \\ c & d \end{pmatrix}$$

The matrix T premultiplies the point X. Let the result of this calculation be X^*.

$$TX = \begin{pmatrix} a & b \\ c & d \end{pmatrix} \begin{pmatrix} x \\ y \end{pmatrix}$$

$$= \begin{pmatrix} ax + by \\ cx + dy \end{pmatrix}$$

$$= X^*$$

We can think of the point $\begin{pmatrix} x \\ y \end{pmatrix}$ as having been transformed into the point

$$\begin{pmatrix} ax + by \\ cx + dy \end{pmatrix}$$

by the matrix T.

Example 5.2

The point $\begin{pmatrix} -1 \\ 3 \end{pmatrix}$ is transformed by the matrix $\begin{pmatrix} 1 & 2 \\ -1 & 4 \end{pmatrix}$.

Determine the position of the transformed point.

Solution
We calculate

$$\begin{pmatrix} 1 & 2 \\ -1 & 4 \end{pmatrix} \begin{pmatrix} -1 \\ 3 \end{pmatrix} = \qquad \qquad \begin{pmatrix} 5 \\ 13 \end{pmatrix}$$

The point $\begin{pmatrix} -1 \\ 3 \end{pmatrix}$ is transformed to $\begin{pmatrix} 5 \\ 13 \end{pmatrix}$.

Exercises

1   The point $\begin{pmatrix} 3 \\ -4 \end{pmatrix}$ is transformed by the matrix

$$\begin{pmatrix} 2 & 0 \\ 3 & 1 \end{pmatrix}$$

Find the position of the transformed point.

2   The point $\begin{pmatrix} 5 \\ -1 \end{pmatrix}$ is transformed by the matrix

$$\begin{pmatrix} -1 & 3 \\ 4 & 2 \end{pmatrix}$$

Find the position of the transformed point.

Solutions to exercises

1   $\begin{pmatrix} 6 \\ 5 \end{pmatrix}$

2   $\begin{pmatrix} -8 \\ 18 \end{pmatrix}$

5.4   Transformation of lines

Consider a straight line and a transformation matrix, T, whose elements are constants. If T is applied to every point on the straight line the transformed points form a new straight line; in other words, the transformation of a straight line is another straight line provided the elements of T are constant. Thus to find the transformation of a straight line we need only find the transformation of the end-points of the line and then join these. This is illustrated in Figure 5.3.

Figure 5.3
The transformation of a straight line is another straight line.

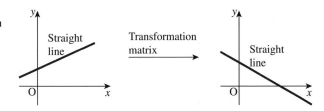

Example 5.3

A straight line joins the points $\begin{pmatrix} 2 \\ -3 \end{pmatrix}$ and $\begin{pmatrix} 4 \\ 2 \end{pmatrix}$. It is transformed by the matrix

$$T = \begin{pmatrix} 1 & 2 \\ -1 & 3 \end{pmatrix}$$

Describe the transformed line.

Solution

The end-points $\begin{pmatrix} 2 \\ -3 \end{pmatrix}$ and $\begin{pmatrix} 4 \\ 2 \end{pmatrix}$ are transformed.

$$\begin{pmatrix} 1 & 2 \\ -1 & 3 \end{pmatrix} \begin{pmatrix} 2 \\ -3 \end{pmatrix} = \qquad \begin{pmatrix} -4 \\ -11 \end{pmatrix}$$

$$\begin{pmatrix} 1 & 2 \\ -1 & 3 \end{pmatrix} \begin{pmatrix} 4 \\ 2 \end{pmatrix} = \qquad \begin{pmatrix} 8 \\ 2 \end{pmatrix}$$

The transformed line joins $\begin{pmatrix} -4 \\ -11 \end{pmatrix}$ and $\begin{pmatrix} 8 \\ 2 \end{pmatrix}$.

5.5 Scaling

Consider a rectangle with vertices at $\begin{pmatrix} 0 \\ 0 \end{pmatrix}$, $\begin{pmatrix} 0 \\ 1 \end{pmatrix}$, $\begin{pmatrix} 1 \\ 1 \end{pmatrix}$ and $\begin{pmatrix} 1 \\ 0 \end{pmatrix}$ and a transformation matrix

$$T = \begin{pmatrix} 4 & 0 \\ 0 & 1 \end{pmatrix}$$

The points $\begin{pmatrix} 0 \\ 0 \end{pmatrix}$, $\begin{pmatrix} 0 \\ 1 \end{pmatrix}$, $\begin{pmatrix} 1 \\ 1 \end{pmatrix}$ and $\begin{pmatrix} 1 \\ 0 \end{pmatrix}$ are transformed by T to $\begin{pmatrix} 0 \\ 0 \end{pmatrix}$, $\begin{pmatrix} 0 \\ 1 \end{pmatrix}$, $\begin{pmatrix} 4 \\ 1 \end{pmatrix}$ and $\begin{pmatrix} 4 \\ 0 \end{pmatrix}$ respectively. This is illustrated in Figure 5.4.

Figure 5.4
An example of
horizontal scaling.

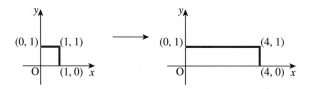

Clearly T has the effect of increasing the horizontal lengths by a factor of 4; the vertical lengths remain unchanged.

In general, scaling is brought about by a diagonal matrix. The first row determines the horizontal scaling; the second row determines the vertical scaling. The matrix

$$\begin{pmatrix} \alpha & 0 \\ 0 & \beta \end{pmatrix}$$

scales the horizontal lengths by a factor of α and the vertical lengths by a factor of β. For horizontal expansion α must be greater than 1; for horizontal compression we require α to be less than 1. Similar comments apply to vertical expansion and compression.

Example 5.4
The zoom facilities of a graphing package expand both horizontal and vertical lengths by a factor of 2.5. Determine the transformation matrix to do this.

Solution
A diagonal matrix is needed for scaling. Since both horizontal and vertical lengths are scaled by a factor of 2.5 the required matrix is

$$\begin{pmatrix} 2.5 & 0 \\ 0 & 2.5 \end{pmatrix}$$

Exercises

1 Write down the transformation matrix that
(a) scales horizontal distances by a factor of 4 and leaves vertical distances unchanged
(b) scales horizontal distances by a factor of 3 and vertical distances by a factor of 7

(c) scales horizontal distances by a factor of 0.7 and vertical distances by a factor of 1.2
(d) leaves horizontal distances unchanged and trebles vertical distances.

Solutions to exercises

1 (a) $\begin{pmatrix} 4 & 0 \\ 0 & 1 \end{pmatrix}$ (b) $\begin{pmatrix} 3 & 0 \\ 0 & 7 \end{pmatrix}$

(c) $\begin{pmatrix} 0.7 & 0 \\ 0 & 1.2 \end{pmatrix}$ (d) $\begin{pmatrix} 1 & 0 \\ 0 & 3 \end{pmatrix}$

5.6   Rotation

Rotation is an important and commonly used transformation. Components can be 'viewed' from different angles using packages with a rotation facility. By convention, anticlockwise movement is considered to be positive. Figure 5.5 illustrates the line from the origin to the point $\begin{pmatrix} x \\ y \end{pmatrix}$ being rotated through an angle θ. We say the point $\begin{pmatrix} x \\ y \end{pmatrix}$ has been rotated about the origin to the new position $\begin{pmatrix} x^* \\ y^* \end{pmatrix}$.

Figure 5.5
The point $\begin{pmatrix} x \\ y \end{pmatrix}$ is rotated anticlock-wise through an angle θ to $\begin{pmatrix} x^* \\ y^* \end{pmatrix}$.

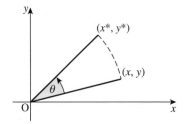

Key point

The transformation matrix, T, that rotates a point anticlockwise about the origin by an angle θ is given by

$$T = \begin{pmatrix} \cos\theta & -\sin\theta \\ \sin\theta & \cos\theta \end{pmatrix}$$

Example 5.5

The point $\begin{pmatrix} 2 \\ 6 \end{pmatrix}$ is rotated anticlockwise about the origin through 40°. Calculate the position of the new point.

Solution

The transformation matrix is

$$T = \begin{pmatrix} \cos 40° & -\sin 40° \\ \sin 40° & \cos 40° \end{pmatrix}$$

and hence the new point is given by

$$\begin{pmatrix} \cos 40° & -\sin 40° \\ \sin 40° & \cos 40° \end{pmatrix}\begin{pmatrix} 2 \\ 6 \end{pmatrix} = \begin{pmatrix} 2\cos 40° - 6\sin 40° \\ 2\sin 40° + 6\cos 40° \end{pmatrix}$$

$$= \begin{pmatrix} -2.3246 \\ 5.8818 \end{pmatrix}$$

Example 5.6
A triangle is defined by the matrix

$$\begin{pmatrix} 1 & 2 & 3 \\ 1 & 5 & 1 \end{pmatrix}$$

It is rotated clockwise about the origin through 120°. Determine the resulting triangle.

Solution
Clockwise angles are considered to be negative and so the transformation matrix, T, is

$$T = \begin{pmatrix} \cos(-120°) & -\sin(-120°) \\ \sin(-120°) & \cos(-120°) \end{pmatrix}$$

$$= \begin{pmatrix} -0.5 & 0.866 \\ -0.866 & -0.5 \end{pmatrix}$$

The transformed triangle is then given by

$$\begin{pmatrix} -0.5 & 0.866 \\ -0.866 & -0.5 \end{pmatrix} \begin{pmatrix} 1 & 2 & 3 \\ 1 & 5 & 1 \end{pmatrix} = \begin{pmatrix} 0.366 & 3.330 & -0.634 \\ -1.366 & -4.232 & -3.098 \end{pmatrix}$$

The transformed triangle is defined by the points $\begin{pmatrix} 0.366 \\ -1.366 \end{pmatrix}$, $\begin{pmatrix} 3.330 \\ -4.232 \end{pmatrix}$ and $\begin{pmatrix} -0.634 \\ -3.098 \end{pmatrix}$.

Example 5.7

The line joining $\begin{pmatrix} -1 \\ 4 \end{pmatrix}$ and $\begin{pmatrix} 3 \\ 1 \end{pmatrix}$ is rotated through 90° anticlockwise about the origin. Determine the new line.

Solution
The transformation matrix, T, is

$$T = \qquad \begin{pmatrix} \cos 90° & -\sin 90° \\ \sin 90° & \cos 90° \end{pmatrix} = \begin{pmatrix} 0 & -1 \\ 1 & 0 \end{pmatrix}$$

The original end-points of the line are $\begin{pmatrix} -1 \\ 4 \end{pmatrix}$ and $\begin{pmatrix} 3 \\ 1 \end{pmatrix}$.

Thus the line is defined by the matrix

$$\begin{pmatrix} -1 & 3 \\ 4 & 1 \end{pmatrix}$$

The transformed end-points are thus

$$\begin{pmatrix} 0 & -1 \\ 1 & 0 \end{pmatrix} \begin{pmatrix} -1 & 3 \\ 4 & 1 \end{pmatrix} = \begin{pmatrix} -4 & -1 \\ -1 & 3 \end{pmatrix}$$

The transformed line joins the points $\begin{pmatrix} -4 \\ -1 \end{pmatrix}$ and $\begin{pmatrix} -1 \\ 3 \end{pmatrix}$.

Exercises

1 State a matrix that rotates points anticlockwise about the origin by 45°.

2 State a matrix that rotates points clockwise about the origin by 200°.

3 The point $\begin{pmatrix} 4 \\ -3 \end{pmatrix}$ is rotated 30° anticlockwise about the origin. State the coordinates of the new point.

4 The point $\begin{pmatrix} -1 \\ 2 \end{pmatrix}$ is rotated 75° clockwise about the origin. State the coordinates of the new point.

Solutions to exercises

1 $\begin{pmatrix} 0.7071 & -0.7071 \\ 0.7071 & 0.7071 \end{pmatrix}$

2 $\begin{pmatrix} -0.9397 & -0.3420 \\ 0.3420 & -0.9397 \end{pmatrix}$

3 $\begin{pmatrix} 4.9641 \\ -0.5981 \end{pmatrix}$

4 $\begin{pmatrix} 1.6730 \\ 1.4836 \end{pmatrix}$

5.7 Reflection

We now consider transformation matrices that effect reflection in the x axis, y axis, the line $y = x$ and the line $y = -x$.

Reflection in the x axis

Figure 5.6 illustrates the line AB reflected in the x axis to produce the line A*B*. Consider the coordinates of A and A*. If the coordinates of A are, say, $\begin{pmatrix} 3 \\ 2 \end{pmatrix}$, then the coordinates of A* will be (3, −2). Clearly, on reflection the x coordinate remains unchanged and the y coordinate changes sign. The transformation matrix to effect this is

$$T = \begin{pmatrix} 1 & 0 \\ 0 & -1 \end{pmatrix}$$

Figure 5.6
The line AB is reflected in the x axis to produce A*B*.

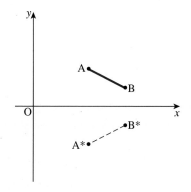

Key point

Reflection in the x axis is brought about by

$$T = \begin{pmatrix} 1 & 0 \\ 0 & -1 \end{pmatrix}$$

Reflection in the y axis

Figure 5.7 illustrates reflection of the line AB in the y axis. Again, considering the coordinates of A and A′, we see that on reflection the x coordinate changes sign and

Figure 5.7
The line AB is reflected in the y axis to produce A′B′.

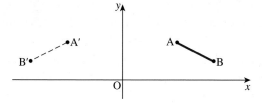

the y coordinate remains unchanged. The transformation matrix is

$$T = \begin{pmatrix} -1 & 0 \\ 0 & 1 \end{pmatrix}$$

Key point

Reflection in the y axis is brought about by

$$T = \begin{pmatrix} -1 & 0 \\ 0 & 1 \end{pmatrix}$$

Reflection in the line $y = x$

Figure 5.8
AB is reflected in
the line $y = x$ to
produce $\overline{A}\,\overline{B}$.

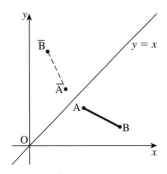

Figure 5.8 illustrates reflection about the line $y = x$. In effect, the x and y coordinates are interchanged. The required transformation matrix is

$$T = \begin{pmatrix} 0 & 1 \\ 1 & 0 \end{pmatrix}$$

Key point

Reflection in $y = x$ is brought about by

$$T = \begin{pmatrix} 0 & 1 \\ 1 & 0 \end{pmatrix}$$

Reflection in the line $y = -x$

Figure 5.9 illustrates reflection in $y = -x$. Here the x and y coordinates are inter-changed and also changed in sign. The transformation matrix is

$$T = \begin{pmatrix} 0 & -1 \\ -1 & 0 \end{pmatrix}$$

Key point

Reflection in $y = -x$ is brought about by

$$T = \begin{pmatrix} 0 & -1 \\ -1 & 0 \end{pmatrix}$$

Figure 5.9
AB is reflected in
$y = -x$ to produce
$\hat{A}\hat{B}$.

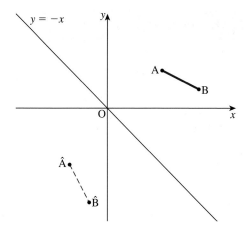

 Example 5.8

The point $A\begin{pmatrix} 3 \\ 2 \end{pmatrix}$ is reflected in the y axis and then the resulting point is reflected in the line $y = -x$. Calculate the coordinates of the final point.

Solution
Reflection in the y axis is effected by the matrix

$$\begin{pmatrix} -1 & 0 \\ 0 & 1 \end{pmatrix}$$

Hence when $A\begin{pmatrix} 3 \\ 2 \end{pmatrix}$ is reflected in the y axis the resulting point is given by

$$\begin{pmatrix} -1 & 0 \\ 0 & 1 \end{pmatrix}\begin{pmatrix} 3 \\ 2 \end{pmatrix} = \begin{pmatrix} -3 \\ 2 \end{pmatrix}$$

Reflection in $y = -x$ is effected by

$$\begin{pmatrix} 0 & -1 \\ -1 & 0 \end{pmatrix}$$

and so the final point is given by

$$\begin{pmatrix} 0 & -1 \\ -1 & 0 \end{pmatrix}\begin{pmatrix} -3 \\ 2 \end{pmatrix} = \begin{pmatrix} -2 \\ 3 \end{pmatrix}$$

The final point is $\begin{pmatrix} -2 \\ 3 \end{pmatrix}$.

Exercises

1 The point $\begin{pmatrix} 3 \\ 5 \end{pmatrix}$ is reflected in the line $y = -x$. State the coordinates of the resulting point.

2 The point $\begin{pmatrix} -1 \\ 4 \end{pmatrix}$ is reflected in the y axis and the resulting point is then reflected in the line $y = x$. State the coordinates of the final point.

3 The point $\begin{pmatrix} a \\ b \end{pmatrix}$ is reflected in the x axis. The resulting point is then reflected in the line $y = -x$. State the coordinates of the final point.

Solutions to exercises

1 $\begin{pmatrix} -5 \\ -3 \end{pmatrix}$

3 $\begin{pmatrix} b \\ -a \end{pmatrix}$

2 $\begin{pmatrix} 4 \\ 1 \end{pmatrix}$

5.8 Shearing

A shear is a distortion. Figure 5.10 illustrates an x direction shear of a square, and Figure 5.11 illustrates a y direction shear.

Figure 5.10
An x direction shear.

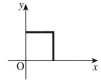
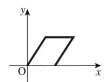

Figure 5.11
A y direction shear.

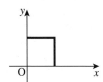
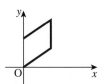

A shear can be represented by a transformation matrix,

$$\begin{pmatrix} a & b \\ c & d \end{pmatrix}$$

The off-diagonal terms, b and c, determine the kind of shear produced. The element b produces a shear in the x direction whereas c produces a shear in the y direction. As mentioned in Section 5.5, the elements a and b are scale factors for the x and y directions.

Example 5.9

A unit square with vertices at $\begin{pmatrix} 0 \\ 0 \end{pmatrix}$, $\begin{pmatrix} 0 \\ 1 \end{pmatrix}$, $\begin{pmatrix} 1 \\ 0 \end{pmatrix}$ and $\begin{pmatrix} 1 \\ 1 \end{pmatrix}$ is sheared by

$$T = \begin{pmatrix} 1 & 0 \\ 3 & 1 \end{pmatrix}$$

(a) Determine the coordinates of the vertices of the figure after shearing.
(b) Sketch both figures.

Solution

(a) The unit square can be represented by

$$\begin{pmatrix} 0 & 0 & 1 & 1 \\ 0 & 1 & 0 & 1 \end{pmatrix}$$

Shearing produces

$$\begin{pmatrix} 1 & 0 \\ 3 & 1 \end{pmatrix} \begin{pmatrix} 0 & 0 & 1 & 1 \\ 0 & 1 & 0 & 1 \end{pmatrix} = \begin{pmatrix} 0 & 0 & 1 & 1 \\ 0 & 1 & 3 & 4 \end{pmatrix}$$

The resulting figure has vertices at $\begin{pmatrix} 0 \\ 0 \end{pmatrix}$, $\begin{pmatrix} 0 \\ 1 \end{pmatrix}$, $\begin{pmatrix} 1 \\ 3 \end{pmatrix}$ and $\begin{pmatrix} 1 \\ 4 \end{pmatrix}$.

(b) See Figure 5.12(a), (b).

Figure 5.12
(a) The original unit square;
(b) the figure after shearing.

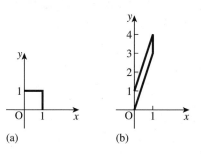

(a)   (b)

Exercises

1   State the effect of the transformation matrix
$$\begin{pmatrix} 1 & 2 \\ 0 & 1 \end{pmatrix}$$
$$\begin{pmatrix} 1 & 0 \\ 4 & 1 \end{pmatrix}$$
Describe the transformed figure.

2   The line joining $\begin{pmatrix} 1 \\ 2 \end{pmatrix}$ and $\begin{pmatrix} 3 \\ 7 \end{pmatrix}$ is transformed by the shearing matrix

Solutions to exercises

1 A shear by a factor of 2 in the x direction.

2 Line joins $\begin{pmatrix} 1 \\ 6 \end{pmatrix}$ and $\begin{pmatrix} 3 \\ 19 \end{pmatrix}$.

5.9 Translation

A translation is a movement in a specific direction by a specific amount. Figure 5.13 illustrates a translation of the line AB to A*B*.

Figure 5.13
AB is translated to A*B*.

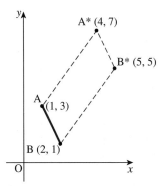

In computer graphics many figures need to be translated, for example a piston moving within a cylinder.

Consider the coordinates of A and B in Figure 5.13. The point $A\begin{pmatrix} 1 \\ 3 \end{pmatrix}$ is translated to the point $A^*\begin{pmatrix} 4 \\ 7 \end{pmatrix}$. The translation has the effect of adding 3 to the x coordinate and adding 4 to the y coordinate. Note that the same change is encountered when B is translated to B*.

Denoting the x translation by t_x and the y translation by t_y we see that if a point $\begin{pmatrix} x \\ y \end{pmatrix}$ is translated to $\begin{pmatrix} x^* \\ y^* \end{pmatrix}$ then

$$x^* = x + t_x$$
$$y^* = y + t_y \tag{1}$$

Note that t_x and/or t_y may be negative.

Mathematically, translations are different from the other transformations. The previous transformations could all be represented by matrix multiplication. Equations (1) suggest that translations are represented by matrix addition.

To enable translation to be represented by matrix multiplication we introduce **homogeneous coordinates**. If the coordinates of a point are $\begin{pmatrix} x \\ y \end{pmatrix}$ then the homogeneous coordinates of that point are

$$\begin{pmatrix} x \\ y \\ 1 \end{pmatrix}$$

The purpose of the extra '1' is to increase the order of the transformation matrix from 2×2 to 3×3. We now introduce a translation matrix, T, where

$$T = \begin{pmatrix} 1 & 0 & t_x \\ 0 & 1 & t_y \\ 0 & 0 & 1 \end{pmatrix}$$

Applying T to

$$X = \begin{pmatrix} x \\ y \\ 1 \end{pmatrix}$$

we have

$$TX = \begin{pmatrix} 1 & 0 & t_x \\ 0 & 1 & t_y \\ 0 & 0 & 1 \end{pmatrix} \begin{pmatrix} x \\ y \\ 1 \end{pmatrix}$$

$$= \begin{pmatrix} x + t_x \\ y + t_y \\ 1 \end{pmatrix}$$

The homogeneous coordinates

$$\begin{pmatrix} x + t_x \\ y + t_y \\ 1 \end{pmatrix}$$

represent the point

$$\begin{pmatrix} x + t_x \\ y + t_y \end{pmatrix}$$

Thus by introducing homogeneous coordinates translation can be represented by matrix multiplication.

Example 5.10
Determine the translation matrix for the translation depicted in Figure 5.13 above.

Solution
Here $t_x = 3$ and $t_y = 4$ and so

$$T = \begin{pmatrix} 1 & 0 & 3 \\ 0 & 1 & 4 \\ 0 & 0 & 1 \end{pmatrix}$$

Example 5.11

The translation matrix

$$T = \begin{pmatrix} 1 & 0 & -1 \\ 0 & 1 & 2 \\ 0 & 0 & 1 \end{pmatrix}$$

is applied to $\begin{pmatrix} 4 \\ -6 \end{pmatrix}$. Calculate the position of the translated point.

Solution

We can proceed in one of two ways. By examining T we see that $t_x = -1$ and $t_y = 2$. Hence

$$\begin{aligned} x^* &= x + t_x \\ &= 4 - 1 \\ &= 3 \\ y^* &= y + t_y \\ &= -6 + 2 \\ &= -4 \end{aligned}$$

The translated point is $\begin{pmatrix} 3 \\ -4 \end{pmatrix}$.

Alternatively we can introduce homogeneous coordinates

$$\begin{pmatrix} 4 \\ -6 \\ 1 \end{pmatrix}$$

Then

$$\begin{pmatrix} 1 & 0 & -1 \\ 0 & 1 & 2 \\ 0 & 0 & 1 \end{pmatrix} \begin{pmatrix} 4 \\ -6 \\ 1 \end{pmatrix} = \qquad \begin{pmatrix} 3 \\ -4 \\ 1 \end{pmatrix}$$

Hence the translated point is $\begin{pmatrix} 3 \\ -4 \end{pmatrix}$.

Exercises

1   A point is translated from $\begin{pmatrix} 1 \\ 6 \end{pmatrix}$ to $\begin{pmatrix} 4 \\ 8 \end{pmatrix}$. State the corresponding translation matrix.

2   A point is translated from $\begin{pmatrix} -3 \\ -1 \end{pmatrix}$ to $\begin{pmatrix} -6 \\ 0 \end{pmatrix}$. State the corresponding translation matrix.

3   State the effect of the translation matrix

$$\begin{pmatrix} 1 & 0 & 4 \\ 0 & 1 & -3 \\ 0 & 0 & 1 \end{pmatrix}$$

4   The line AB is translated 3 units in the x direction with no movement in the y direction. State the corresponding translation matrix.

Solutions to exercises

1 $\begin{pmatrix} 1 & 0 & 3 \\ 0 & 1 & 2 \\ 0 & 0 & 1 \end{pmatrix}$

2 $\begin{pmatrix} 1 & 0 & -3 \\ 0 & 1 & 1 \\ 0 & 0 & 1 \end{pmatrix}$

3 Translation of 4 units in the x direction and 3 units in the negative y direction.

4 $\begin{pmatrix} 1 & 0 & 3 \\ 0 & 1 & 0 \\ 0 & 0 & 1 \end{pmatrix}$

5.10 Composite transformations

We have represented each transformation by a matrix multiplication. Often not one but several transformations are required, and this leads to a product of matrices. Care must be taken in ensuring that the correct order of multiplication is carried out. Recall that if A and B are two matrices then AB and BA are usually not equal.

A sequence of transformations is called a **composite transformation**. Before we can proceed and examine composite transformations we must examine the transformation matrices. The translation matrix is 3×3; all other transformation matrices are 2×2. Multiplication of a 3×3 matrix with a 2×2 matrix is not defined. To overcome this, all the 2×2 matrices are extended so as to become 3×3 matrices and homogeneous coordinates are employed.

Consider any 2×2 transformation matrix

$$\begin{pmatrix} a & b \\ c & d \end{pmatrix}$$

and coordinates $\begin{pmatrix} x \\ y \end{pmatrix}$. The matrix is extended to become

$$T = \begin{pmatrix} a & b & 0 \\ c & d & 0 \\ 0 & 0 & 1 \end{pmatrix}$$

and homogeneous coordinates

$$\begin{pmatrix} x \\ y \\ 1 \end{pmatrix}$$

are used. Note that

$$\begin{pmatrix} a & b \\ c & d \end{pmatrix}\begin{pmatrix} x \\ y \end{pmatrix} = \begin{pmatrix} ax + by \\ cx + dy \end{pmatrix}$$

and

$$\begin{pmatrix} a & b & 0 \\ c & d & 0 \\ 0 & 0 & 1 \end{pmatrix} \begin{pmatrix} x \\ y \\ 1 \end{pmatrix} = \begin{pmatrix} ax + by \\ cx + dy \\ 1 \end{pmatrix}$$

Transforming a point using the extended matrix and homogeneous coordinates has exactly the same effect as using the 2×2 matrix on the coordinates $\begin{pmatrix} x \\ y \end{pmatrix}$. However, all transformation matrices are now 3×3 and so composite transformations, in the form of matrix products, can be calculated.

Example 5.12

The point $\begin{pmatrix} 4 \\ 1 \end{pmatrix}$ is reflected in the line $y = x$ and then rotated anticlockwise about the origin by $70°$.

(a) Calculate the composite transformation matrix.
(b) Calculate the final position of the point.

Solution

(a) Reflection about the line $y = x$ is represented by T_{ref} where

$$T_{\text{ref}} = \begin{pmatrix} 0 & 1 & 0 \\ 1 & 0 & 0 \\ 0 & 0 & 1 \end{pmatrix}$$

Rotation about the origin by $70°$ is represented by T_{rot} where

$$T_{\text{rot}} = \begin{pmatrix} \cos 70° & -\sin 70° & 0 \\ \sin 70° & \cos 70° & 0 \\ 0 & 0 & 1 \end{pmatrix}$$

Hence the composite transformation is given by

$$T_{\text{rot}}\, T_{\text{ref}} = \begin{pmatrix} \cos 70° & -\sin 70° & 0 \\ \sin 70° & \cos 70° & 0 \\ 0 & 0 & 1 \end{pmatrix} \begin{pmatrix} 0 & 1 & 0 \\ 1 & 0 & 0 \\ 0 & 0 & 1 \end{pmatrix}$$

$$= \begin{pmatrix} -\sin 70° & \cos 70° & 0 \\ \cos 70° & \sin 70° & 0 \\ 0 & 0 & 1 \end{pmatrix}$$

(b) The homogeneous coordinates of the point are

$$\begin{pmatrix} 4 \\ 1 \\ 1 \end{pmatrix}$$

The transformed point is then

$$
\begin{pmatrix} -\sin 70° & \cos 70° & 0 \\ \cos 70° & \sin 70° & 0 \\ 0 & 0 & 1 \end{pmatrix} \begin{pmatrix} 4 \\ 1 \\ 1 \end{pmatrix} = \begin{pmatrix} -4\sin 70° + \cos 70° \\ 4\cos 70° + \sin 70° \\ 1 \end{pmatrix}
$$

$$
= \begin{pmatrix} -3.417 \\ 2.308 \\ 1 \end{pmatrix}
$$

The transformed point is $\begin{pmatrix} -3.417 \\ 2.308 \end{pmatrix}$.

Example 5.13

Determine a transformation matrix that rotates a point $30°$ anticlockwise about the point $\begin{pmatrix} 1 \\ 2 \end{pmatrix}$.

Solution

The rotation matrices so far encountered have all represented rotation about the origin. Hence to represent rotation about $\begin{pmatrix} 1 \\ 2 \end{pmatrix}$ we require three transformations (Figure 5.14):

1 translation of $\begin{pmatrix} 1 \\ 2 \end{pmatrix}$ to the origin, T_{trans}

2 rotation of $30°$ anticlockwise, T_{rot}

3 translation from the origin back to $\begin{pmatrix} 1 \\ 2 \end{pmatrix}$, T_{ret}.

Figure 5.14
Rotation about $\begin{pmatrix} 1 \\ 2 \end{pmatrix}$ requires three basic transformations.

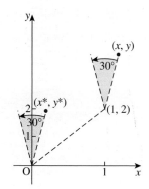

The required transformation matrices are

$$T_{\text{trans}} = \begin{pmatrix} 1 & 0 & -1 \\ 0 & 1 & -2 \\ 0 & 0 & 1 \end{pmatrix}$$

$$T_{\text{rot}} = \begin{pmatrix} 0.8660 & -0.5 & 0 \\ 0.5 & 0.8660 & 0 \\ 0 & 0 & 1 \end{pmatrix}$$

$$T_{\text{ret}} = \begin{pmatrix} 1 & 0 & 1 \\ 0 & 1 & 2 \\ 0 & 0 & 1 \end{pmatrix}$$

The composite transformation is then

$$T_{\text{ret}}\,T_{\text{rot}}\,T_{\text{trans}} = \begin{pmatrix} 1 & 0 & 1 \\ 0 & 1 & 2 \\ 0 & 0 & 1 \end{pmatrix} \begin{pmatrix} 0.8660 & -0.5 & 0 \\ 0.5 & 0.8660 & 0 \\ 0 & 0 & 1 \end{pmatrix} \begin{pmatrix} 1 & 0 & -1 \\ 0 & 1 & -2 \\ 0 & 0 & 1 \end{pmatrix}$$

$$= \begin{pmatrix} 0.8660 & -0.5 & 1.1340 \\ 0.5 & 0.8660 & -0.232 \\ 0 & 0 & 1 \end{pmatrix}$$

Exercises

1 Determine a matrix that shears vertically by a factor of 2 and then rotates anticlockwise about the origin by 30°.

2 Determine a matrix that rotates a figure anti-clockwise through 40° about the point $\begin{pmatrix} 1 \\ 0 \end{pmatrix}$.

Solutions to exercises

1 $\begin{pmatrix} -0.1340 & -0.5 & 0 \\ 2.2321 & 0.8660 & 0 \\ 0 & 0 & 1 \end{pmatrix}$

2 $\begin{pmatrix} 0.7660 & -0.6428 & 0.2340 \\ 0.6428 & 0.7660 & -0.6428 \\ 0 & 0 & 1 \end{pmatrix}$

End of block exercises

1 Write matrices that represent the figures shown in Figure 5.15(a), (b), (c).

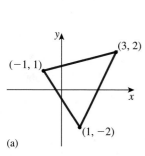

(a)

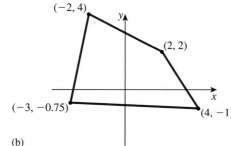

(b)

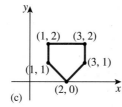

(c)

Figure 5.15

2 Draw the figures represented by each of the following matrices:

(a) $\begin{pmatrix} 2 & -1 \\ 0 & 1 \end{pmatrix}$ (b) $\begin{pmatrix} 1 & 3 & 5 \\ 2 & 4 & 6 \end{pmatrix}$

(c) $\begin{pmatrix} -1 & 2 & -4 & 3 \\ 3 & 4 & 1 & 0 \end{pmatrix}$

3 The line joining $\begin{pmatrix} 1 \\ 7 \end{pmatrix}$ and $\begin{pmatrix} -2 \\ 1 \end{pmatrix}$ is transformed by the matrix

$$\begin{pmatrix} 3 & 2 \\ -1 & 1 \end{pmatrix}$$

Describe the transformed line.

4 The triangle with vertices at $\begin{pmatrix} -1 \\ 1 \end{pmatrix}$, $\begin{pmatrix} 2 \\ 1 \end{pmatrix}$ and $\begin{pmatrix} 3 \\ 4 \end{pmatrix}$ is transformed by the matrix

$$\begin{pmatrix} 2 & 1 \\ 1 & -2 \end{pmatrix}$$

Describe the transformed triangle.

5 The zoom facility of a computer graphics package has the following options:
(a) double horizontal lengths only
(b) double vertical lengths only
(c) double both horizontal and vertical lengths
(d) halve both horizontal and vertical lengths.
Write down the transformation matrix for each option.

6 The point $\begin{pmatrix} -1 \\ 6 \end{pmatrix}$ is rotated anticlockwise about the origin through 70°. Calculate the coordinates of the resulting point.

7 The point $\begin{pmatrix} 3 \\ -2 \end{pmatrix}$ is rotated clockwise about the origin through 40°. By applying a suitable rotation matrix, determine the coordinates of the resulting point.

8 A triangle defined by

$$\begin{pmatrix} 3 & 2 & -1 \\ 1 & 7 & 6 \end{pmatrix}$$

is rotated clockwise about the origin through 90°. Calculate the matrix that defines the transformed triangle.

9 Show that for any transformation matrix, R, that represents a rotation

$$R^T = R^{-1}$$

that is, the transpose of a rotation matrix is equal to its inverse.

10 A rotation matrix, R, represents an anticlockwise rotation about the origin through an angle θ.
(a) State the rotation matrix, S, that represents a clockwise rotation through an angle θ.
(b) Calculate R^{-1}. (Hint: see question 9.)
(c) What do you conclude from (a) and (b)?

11 A line joins $\begin{pmatrix} 1 \\ 4 \end{pmatrix}$ and $\begin{pmatrix} -2 \\ 3 \end{pmatrix}$.
(a) The line is reflected about $y = x$. Describe the resulting line.
(b) The line found in (a) is then reflected in the x axis. Describe the resulting line.

12 Determine a matrix that increases horizontal lengths by a factor of 5, leaving vertical lengths unchanged.

13 The line joining $\begin{pmatrix} -1 \\ 1 \end{pmatrix}$ and $\begin{pmatrix} 4 \\ 2 \end{pmatrix}$ is reflected in the line $y = -x$. Determine the new line.

14 The triangle with vertices at $\begin{pmatrix} 0 \\ 0 \end{pmatrix}$, $\begin{pmatrix} 0 \\ 1 \end{pmatrix}$ and $\begin{pmatrix} 1 \\ 0 \end{pmatrix}$ is sheared by

$$\begin{pmatrix} 1 & 2 \\ 0 & 1 \end{pmatrix}$$

Sketch the resulting triangle.

15 State a matrix that will translate a figure 4 units in the x direction and 5 units in the y direction.

16 A line joins $\begin{pmatrix} 4 \\ -2 \end{pmatrix}$ and $\begin{pmatrix} 5 \\ 9 \end{pmatrix}$. It is translated using the translation matrix

$$\begin{pmatrix} 1 & 0 & 3 \\ 0 & 1 & -1 \\ 0 & 0 & 1 \end{pmatrix}$$

Describe the translated line.

17 Determine a single transformation matrix that will first reflect a figure in the y axis and then translate it 2 units in the x direction and -3 units in the y direction.

18 Determine a single transformation matrix that will first scale a figure by a factor of 2 in the horizontal and 0.5 in the vertical, and then rotate it clockwise about the origin through 90°.

19 Find a matrix that will rotate a figure anti-clockwise through 45° about the point $\begin{pmatrix} -2 \\ 1 \end{pmatrix}$.

Solutions to exercises

1 (a) $\begin{pmatrix} -1 & 3 & 1 \\ 1 & 2 & -2 \end{pmatrix}$

(b) $\begin{pmatrix} -3 & -2 & 2 & 4 \\ -0.75 & 4 & 2 & -1 \end{pmatrix}$

(c) $\begin{pmatrix} 1 & 3 & 3 & 2 & 1 \\ 2 & 2 & 1 & 0 & 1 \end{pmatrix}$

2 See Figure 5.16.
(a)

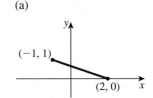

(b)

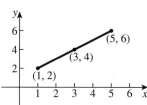

(c)

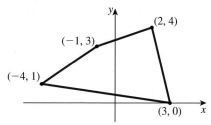

Figure 5.16

3 The transformed line joins $\begin{pmatrix} 17 \\ 6 \end{pmatrix}$ and $\begin{pmatrix} -4 \\ 3 \end{pmatrix}$.

4 The transformed triangle has vertices at

$\begin{pmatrix} -1 \\ -3 \end{pmatrix}$, $\begin{pmatrix} 5 \\ 0 \end{pmatrix}$ and $\begin{pmatrix} 10 \\ -5 \end{pmatrix}$.

5 (a) $\begin{pmatrix} 2 & 0 \\ 0 & 1 \end{pmatrix}$ (b) $\begin{pmatrix} 1 & 0 \\ 0 & 2 \end{pmatrix}$

(c) $\begin{pmatrix} 2 & 0 \\ 0 & 2 \end{pmatrix}$ (d) $\begin{pmatrix} 0.5 & 0 \\ 0 & 0.5 \end{pmatrix}$

6 $\begin{pmatrix} -5.9802 \\ 1.1124 \end{pmatrix}$

7 $\begin{pmatrix} 1.0126 \\ -3.4605 \end{pmatrix}$

8 $\begin{pmatrix} 1 & 7 & 6 \\ -3 & -2 & 1 \end{pmatrix}$

10 (a) $\begin{pmatrix} \cos\theta & \sin\theta \\ -\sin\theta & \cos\theta \end{pmatrix}$

(b) $\begin{pmatrix} \cos\theta & \sin\theta \\ -\sin\theta & \cos\theta \end{pmatrix}$

(c) The inverse of R represents a clockwise rotation through θ.

11 (a) The line joins $\begin{pmatrix} 4 \\ 1 \end{pmatrix}$ and $\begin{pmatrix} 3 \\ -2 \end{pmatrix}$.

(b) The line joins $\begin{pmatrix} 4 \\ -1 \end{pmatrix}$ and $\begin{pmatrix} 3 \\ 2 \end{pmatrix}$.

12 $\begin{pmatrix} 5 & 0 \\ 0 & 1 \end{pmatrix}$

13 The new line joins $\begin{pmatrix} -1 \\ 1 \end{pmatrix}$ and $\begin{pmatrix} -2 \\ -4 \end{pmatrix}$.

15 $\begin{pmatrix} 1 & 0 & 4 \\ 0 & 1 & 5 \\ 0 & 0 & 1 \end{pmatrix}$

16 The line joins $\begin{pmatrix} 7 \\ -3 \end{pmatrix}$ and $\begin{pmatrix} 8 \\ 8 \end{pmatrix}$.

17 $\begin{pmatrix} -1 & 0 & 2 \\ 0 & 1 & -3 \\ 0 & 0 & 1 \end{pmatrix}$

18 $\begin{pmatrix} 0 & 0.5 & 0 \\ -2 & 0 & 0 \\ 0 & 0 & 1 \end{pmatrix}$

19 $\begin{pmatrix} 0.7071 & -0.7071 & 0.1213 \\ 0.7071 & 0.7071 & 1.7071 \\ 0 & 0 & 1 \end{pmatrix}$

End of chapter exercises

1 If

$$A = \begin{pmatrix} 4 & -2 \\ 1 & -3 \end{pmatrix} \quad \text{and} \quad B = \begin{pmatrix} 9 & 3 \\ -2 & 1 \end{pmatrix}$$

find (a) $A + B$, (b) $A - B$, (c) $B - A$, (d) $3A$,

(e) $\dfrac{1}{2} B$, (f) $2A - 4B$.

2 Given

$$A = \begin{pmatrix} 2 & 1 & 3 \\ -1 & 4 & 0 \end{pmatrix} \quad \text{and}$$

$$B = \begin{pmatrix} -1 & 2 & 4 \\ -4 & 1 & 3 \end{pmatrix}$$

calculate (a) $4A$, (b) $-B$, (c) $B - 2A$, (d) $\dfrac{B}{4}$, (e) $A + 3B$, (f) $\alpha A + \beta B$ where α and β are constants.

3 Give an example of (a) a 2×4, (b) a 4×2 matrix.

4 Calculate the number of elements in (a) a 3×4, (b) a 7×2, (c) a $3 \times n$, (d) an $m \times n$ matrix.

5 Calculate the determinant of each of the following matrices:

(a) $\begin{pmatrix} 3 & 1 \\ 4 & 2 \end{pmatrix}$ (b) $\begin{pmatrix} -1 & 3 \\ 2 & 1 \end{pmatrix}$

(c) $\begin{pmatrix} 0 & 1 \\ 1 & 0 \end{pmatrix}$ (d) $\begin{pmatrix} 1 & 0 \\ 0 & 1 \end{pmatrix}$

(e) $\begin{pmatrix} 3 & 2 \\ x & x \end{pmatrix}$ (f) $\begin{pmatrix} \alpha & \beta \\ \beta & \alpha \end{pmatrix}$

6 Find the inverse of each of the following matrices:

(a) $\begin{pmatrix} -1 & 4 \\ 5 & 6 \end{pmatrix}$ (b) $\begin{pmatrix} 6 & -4 \\ 2 & 3 \end{pmatrix}$

(c) $\begin{pmatrix} -3 & 5 \\ 0 & 1 \end{pmatrix}$ (d) $\dfrac{1}{2}\begin{pmatrix} -7 & -2 \\ 4 & 3 \end{pmatrix}$

7 Calculate the determinant of each of the following matrices:

(a) $\begin{pmatrix} 1 & 2 & 1 \\ 0 & 1 & 1 \\ 1 & -1 & 0 \end{pmatrix}$ (b) $\begin{pmatrix} -1 & 3 & 2 \\ 2 & 1 & 0 \\ 1 & 3 & 0 \end{pmatrix}$

(c) $\begin{pmatrix} 3 & -2 & -1 \\ 4 & 1 & 0 \\ 3 & 2 & -1 \end{pmatrix}$ (d) $\begin{pmatrix} 2 & 2 & 1 \\ 1 & 2 & 2 \\ 2 & 1 & 2 \end{pmatrix}$

8 Calculate the inverse of each of the matrices in question 5.

9 Calculate the inverse of each of the matrices in question 7.

10 State conditions under which a square matrix does not possess an inverse.

11 Given

$$A = \begin{pmatrix} 3 & 1 & 4 \\ 0 & 2 & -1 \end{pmatrix} \quad \text{and}$$

$$B = \begin{pmatrix} 9 & -6 \\ 3 & 1 \\ 2 & 0 \end{pmatrix}$$

calculate (a) A^T, (b) B^T, (c) $A + B^T$, (d) $A^T + B$, (e) $2A$, (f) $(2A)^T$, (g) $2(A^T)$.

12 Given

$$A = \begin{pmatrix} 3 \\ 1 \end{pmatrix}, \quad B = \begin{pmatrix} 1 & -3 \\ 2 & 1 \end{pmatrix} \quad \text{and}$$

$$C = (-2 \quad 3)$$

calculate, if possible, (a) AB, (b) BA, (c) AC, (d) CA, (e) BC, (f) CB.

13 Given

$$A = (1 \quad 3 \quad -2), \quad B = \begin{pmatrix} 4 & 2 \\ 1 & 3 \\ -1 & 0 \end{pmatrix}$$

$$C = \begin{pmatrix} 1 & -1 & 0 \\ 4 & 1 & 2 \end{pmatrix}$$

calculate, if possible, (a) $B + C$, (b) $B + C^T$, (c) AB, (d) BA, (e) AC^T, (f) $(AC)^T$, (g) BC, (h) CB.

14 The straight line joining $\begin{pmatrix} -6 \\ 2 \end{pmatrix}$ and $\begin{pmatrix} 1 \\ -4 \end{pmatrix}$ is rotated 50° anticlockwise about the origin. Determine the transformed line.

15 The point $\begin{pmatrix} 2 \\ -3 \end{pmatrix}$ is reflected in the line $y = x$ and then the resulting point is rotated 30° anticlockwise about the origin. Determine the coordinates of the final point.

16 A unit square has vertices at $\begin{pmatrix} 0 \\ 0 \end{pmatrix}$, $\begin{pmatrix} 0 \\ 1 \end{pmatrix}$, $\begin{pmatrix} 1 \\ 0 \end{pmatrix}$ and $\begin{pmatrix} 1 \\ 1 \end{pmatrix}$. It is subject to a shear, and this results in a figure with vertices at $\begin{pmatrix} 0 \\ 0 \end{pmatrix}$, $\begin{pmatrix} 2 \\ 1 \end{pmatrix}$, $\begin{pmatrix} 1 \\ 3 \end{pmatrix}$ and $\begin{pmatrix} 3 \\ 4 \end{pmatrix}$. Determine a transformation that effects this shear.

17 The transformation matrix

$$T = \begin{pmatrix} 1 & 0 & 4 \\ 0 & 1 & 0.5 \\ 0 & 0 & 1 \end{pmatrix}$$

is applied to the line joining $\begin{pmatrix} 0 \\ 1 \end{pmatrix}$ and $\begin{pmatrix} 3 \\ -1 \end{pmatrix}$. Determine the transformed line.

18 The point $\begin{pmatrix} -1 \\ 2 \end{pmatrix}$ is rotated 90° anticlockwise about the origin and then translated 2 units in the *x* direction and 3 units in the *y* direction. Calculate
(a) the composite transformation matrix
(b) the coordinates of the transformed point.

19 The **leading diagonal** of a square matrix, *A*, is the diagonal running from the top left to the bottom right. The sum of the elements on the leading diagonal is called the **trace** of *A*.
(a) Find the trace of

$$A = \begin{pmatrix} 1 & 2 \\ 5 & -3 \end{pmatrix}$$

(b) Find the trace of

$$A = \begin{pmatrix} 4 & 1 & 3 \\ 2 & 7 & 1 \\ 0 & 0 & 12 \end{pmatrix}$$

(c) For an arbitrary 2×2 and 3×3 matrix *A*, show that $\text{trace}(A) = \text{trace}(A^{\text{T}})$.

20 Evaluate

$$\begin{vmatrix} 1 & 4 & 2 & 1 \\ 3 & 0 & 1 & 9 \\ 0 & -1 & 1 & 1 \\ 2 & 1 & 0 & 1 \end{vmatrix}$$

21 (a) A square matrix *A* is said to be **symmetric** if $A = A^{\text{T}}$, and **skew symmetric** if $A = -A^{\text{T}}$. State which, if any, of the following matrices are symmetric and which are skew symmetric.

$$C = \begin{pmatrix} 4 & 7 \\ 0 & -2 \end{pmatrix}$$

$$D = \begin{pmatrix} 0 & -7 & -9 \\ 7 & 0 & 8 \\ 9 & -8 & 0 \end{pmatrix}$$

$$E = \begin{pmatrix} 3 & 4 & 1 \\ 4 & 1 & 0 \\ 1 & 0 & 7 \end{pmatrix}$$

(b) Show that if *A* is a 2×2 or 3×3 skew-symmetric matrix then its diagonal elements must all be zero.

22 (a) A non-singular matrix *A* is said to be **orthogonal** if its transpose is equal to its inverse: that is, $A^{\text{T}} = A^{-1}$. Given that

$$A = \begin{pmatrix} \cos\theta & \sin\theta \\ -\sin\theta & \cos\theta \end{pmatrix}$$

is an orthogonal matrix, write down its inverse, and check your answer by evaluating AA^{-1}.
(b) Show that

$$\begin{pmatrix} \dfrac{3}{5} & -\dfrac{4}{5} \\ \dfrac{4}{5} & \dfrac{3}{5} \end{pmatrix}$$

is an orthogonal matrix.

23 If

$$A = \begin{pmatrix} 0 & 1 \\ -3 & -2 \end{pmatrix}$$

(a) write down the matrix $sI - A$, where *s* is a constant and *I* is the 2×2 identity matrix
(b) find $(sI - A)^{-1}$.

24 **Control engineering.** The state matrix of an engineering system is given by

$$A = \begin{pmatrix} -2 & -1 \\ 16 & 5 \end{pmatrix}$$

Calculate the values of the system poles. Determine whether the system is stable or unstable.

ana.

Solutions to exercises

1 (a) $\begin{pmatrix} 13 & 1 \\ -1 & -2 \end{pmatrix}$ (b) $\begin{pmatrix} -5 & -5 \\ 3 & -4 \end{pmatrix}$

(c) $\begin{pmatrix} 5 & 5 \\ -3 & 4 \end{pmatrix}$ (d) $\begin{pmatrix} 12 & -6 \\ 3 & -9 \end{pmatrix}$

(e) $\begin{pmatrix} 4.5 & 1.5 \\ -1 & 0.5 \end{pmatrix}$ (f) $\begin{pmatrix} -28 & -16 \\ 10 & -10 \end{pmatrix}$

2 (a) $\begin{pmatrix} 8 & 4 & 12 \\ -4 & 16 & 0 \end{pmatrix}$ (b) $\begin{pmatrix} 1 & -2 & -4 \\ 4 & -1 & -3 \end{pmatrix}$

(c) $\begin{pmatrix} -5 & 0 & -2 \\ -2 & -7 & 3 \end{pmatrix}$

(d) $\begin{pmatrix} -0.25 & 0.5 & 1 \\ -1 & 0.25 & 0.75 \end{pmatrix}$

(e) $\begin{pmatrix} -1 & 7 & 15 \\ -13 & 7 & 9 \end{pmatrix}$

(f) $\begin{pmatrix} 2\alpha - \beta & \alpha + 2\beta & 3\alpha + 4\beta \\ -\alpha - 4\beta & 4\alpha + \beta & 3\beta \end{pmatrix}$

3 (a) $\begin{pmatrix} 1 & 2 & 3 & 4 \\ 1 & -1 & 0 & 1 \end{pmatrix}$ (b) $\begin{pmatrix} 1 & 1 \\ 2 & 2 \\ 3 & -1 \\ 4 & 3 \end{pmatrix}$

4 (a) 12 (b) 14 (c) $3n$ (d) mn

5 (a) 2 (b) −7 (c) −1 (d) 1 (e) x (f) $\alpha^2 - \beta^2$

6 (a) $\frac{1}{26}\begin{pmatrix} -6 & 4 \\ 5 & 1 \end{pmatrix}$ (b) $\frac{1}{26}\begin{pmatrix} 3 & 4 \\ -2 & 6 \end{pmatrix}$

(c) $\frac{1}{3}\begin{pmatrix} -1 & 5 \\ 0 & 3 \end{pmatrix}$ (d) $\frac{2}{13}\begin{pmatrix} -3 & -2 \\ 4 & 7 \end{pmatrix}$

7 (a) 2 (b) 10 (c) −16 (d) 5

8 (a) $\begin{pmatrix} 1 & -0.5 \\ -2 & 1.5 \end{pmatrix}$ (b) $\frac{1}{7}\begin{pmatrix} -1 & 3 \\ 2 & 1 \end{pmatrix}$

(c) $\begin{pmatrix} 0 & 1 \\ 1 & 0 \end{pmatrix}$ (d) $\begin{pmatrix} 1 & 0 \\ 0 & 1 \end{pmatrix}$

(e) $\begin{pmatrix} 1 & -\dfrac{2}{x} \\ -1 & \dfrac{3}{x} \end{pmatrix}$ (f) $\dfrac{1}{\alpha^2 - \beta^2}\begin{pmatrix} \alpha & -\beta \\ -\beta & \alpha \end{pmatrix}$

9 (a) $\frac{1}{2}\begin{pmatrix} 1 & -1 & 1 \\ 1 & -1 & -1 \\ -1 & 3 & 1 \end{pmatrix}$

(b) $\frac{1}{10}\begin{pmatrix} 0 & 6 & -2 \\ 0 & -2 & 4 \\ 5 & 6 & -7 \end{pmatrix}$

(c) $\frac{1}{16}\begin{pmatrix} 1 & 4 & -1 \\ -4 & 0 & 4 \\ -5 & 12 & -11 \end{pmatrix}$

(d) $\frac{1}{5}\begin{pmatrix} 2 & -3 & 2 \\ 2 & 2 & -3 \\ -3 & 2 & 2 \end{pmatrix}$

10 If the determinant is zero, then the matrix has no inverse.

11 (a) $\begin{pmatrix} 3 & 0 \\ 1 & 2 \\ 4 & -1 \end{pmatrix}$ (b) $\begin{pmatrix} 9 & 3 & 2 \\ -6 & 1 & 0 \end{pmatrix}$

(c) $\begin{pmatrix} 12 & 4 & 6 \\ -6 & 3 & -1 \end{pmatrix}$ (d) $\begin{pmatrix} 12 & -6 \\ 4 & 3 \\ 6 & -1 \end{pmatrix}$

(e) $\begin{pmatrix} 6 & 2 & 8 \\ 0 & 4 & -2 \end{pmatrix}$ (f) $\begin{pmatrix} 6 & 0 \\ 2 & 4 \\ 8 & -2 \end{pmatrix}$

(g) $\begin{pmatrix} 6 & 0 \\ 2 & 4 \\ 8 & -2 \end{pmatrix}$

12 (a) not possible (b) $\begin{pmatrix} 0 \\ 7 \end{pmatrix}$ (c) $\begin{pmatrix} -6 & 9 \\ -2 & 3 \end{pmatrix}$

(d) -3 (e) not possible (f) $(4 \quad 9)$

13 (a) not possible (b) $\begin{pmatrix} 5 & 6 \\ 0 & 4 \\ -1 & 2 \end{pmatrix}$

(c) $(9 \quad 11)$ (d) not possible (e) $(-2 \quad 3)$

(f) not possible (g) $\begin{pmatrix} 12 & -2 & 4 \\ 13 & 2 & 6 \\ -1 & 1 & 0 \end{pmatrix}$

(h) $\begin{pmatrix} 3 & -1 \\ 15 & 11 \end{pmatrix}$

14 Line joins $\begin{pmatrix} -5.3888 \\ -3.3107 \end{pmatrix}$ to $\begin{pmatrix} 3.7070 \\ -1.8051 \end{pmatrix}$.

15 $\begin{pmatrix} -3.5981 \\ 0.2321 \end{pmatrix}$

16 $\begin{pmatrix} 1 & 2 \\ 3 & 1 \end{pmatrix}$

17 The transformed line joins the points $\begin{pmatrix} 4 \\ 1.5 \end{pmatrix}$ and $\begin{pmatrix} 7 \\ -0.5 \end{pmatrix}$.

18 (a) $\begin{pmatrix} 0 & -1 & 2 \\ 1 & 0 & 3 \\ 0 & 0 & 1 \end{pmatrix}$ (b) $\begin{pmatrix} 0 \\ 2 \end{pmatrix}$

19 (a) -2 (b) 23

20 70

21 D is skew symmetric, E is symmetric and C is neither.

22 $A^{-1} = \begin{pmatrix} \cos\theta & -\sin\theta \\ \sin\theta & \cos\theta \end{pmatrix}$

23 (a) $\begin{pmatrix} s & -1 \\ 3 & s+2 \end{pmatrix}$

(b) $\dfrac{1}{s^2 + 2s + 3} \begin{pmatrix} s+2 & 1 \\ -3 & s \end{pmatrix}$

24 Poles are $s = \dfrac{3 \pm \sqrt{-15}}{2}$. System is unstable.

Using matrices and determinants to solve equations

This chapter looks at some of the ways in which matrices and determinants can be used to solve linear simultaneous equations.

Block 1 introduces **Cramer's rule**, which expresses the solution to simultaneous equations as a ratio of two determinants. The next block shows how simultaneous equations may be written in matrix form. Once in this form an **inverse matrix** method can be used to solve them.

Block 3 explains the method of **Gaussian elimination**. When the simultaneous equations have been written as a matrix, the rows of the matrix are systematically manipulated to yield the solution. The block includes an explanation of the way in which the manipulation is carried out.

Matrices are sometimes used in calculations that model vibrating objects, and in particular those vibrating because of forces in stretched and compressed springs. In such applications engineers are interested in the possible frequencies of vibration. These frequencies are related to what are known as the **eigenvalues** of a matrix. The **eigenvectors** represent the various modes of vibration. Block 4 will explain how the eigenvalues and eigenvectors of a matrix are calculated. As they are usually studied in terms of linear systems, they are included here. In many practical applications, approximate solutions of linear equations are obtained using what are called **numerical methods**. These are introduced in Block 5.

The final block focuses on the application of matrices to the analysis of **electrical networks**. Engineers often need to know the currents in the various branches of a circuit. The idea of branch currents is introduced. A set of simultaneous equations can be formed that models these branch currents, and which can be solved using matrix methods.

Chapter 13 contents

Cramer's rule

1.1 Introduction

In Chapter 12 we introduced determinants. In this block we study Cramer's rule, which uses determinants to solve a set of simultaneous linear equations. Such equations are often referred to as a **system of linear equations**.

1.2 Cramer's rule

Consider a system of two linear simultaneous equations in the two unknowns x and y.

$$a_1 x + b_1 y = k_1$$
$$a_2 x + b_2 y = k_2$$

Here a_1, b_1, k_1, a_2, b_2 and k_2 are known constants and we wish to find the values of x and y.

Cramer's rule is a method of obtaining the solution of equations like these as the ratio of two determinants.

Key point

Cramer's rule states

$$x = \frac{\begin{vmatrix} k_1 & b_1 \\ k_2 & b_2 \end{vmatrix}}{\begin{vmatrix} a_1 & b_1 \\ a_2 & b_2 \end{vmatrix}}, \quad y = \frac{\begin{vmatrix} a_1 & k_1 \\ a_2 & k_2 \end{vmatrix}}{\begin{vmatrix} a_1 & b_1 \\ a_2 & b_2 \end{vmatrix}}$$

Note that the denominator is the same in each case. The denominator is simply the determinant made up of the coefficients of x and y. If the denominator is zero then Cramer's rule cannot be applied. In such a case, either a unique solution to the system does not exist or there is no solution.

Note how the numerator of each fraction is formed. To form the numerator for the x fraction we replace the column of as in

$$\begin{vmatrix} a_1 & b_1 \\ a_2 & b_2 \end{vmatrix}$$

by a column of ks. Similarly the numerator for the y fraction is formed by replacing the column of bs by a column of ks.

Example 1.1 Electrical Engineering – Currents in a circuit

The currents i_1 and i_2 in a simple circuit are connected by the equations

$$i_1 + i_2 = 3$$
$$2i_1 + 3i_2 = 7$$

Calculate i_1 and i_2 using Cramer's rule.

Solution

Using Cramer's rule we have

$$i_1 = \frac{\begin{vmatrix} 3 & 1 \\ 7 & 3 \end{vmatrix}}{\begin{vmatrix} 1 & 1 \\ 2 & 3 \end{vmatrix}} = \frac{2}{1} = 2, \quad i_2 = \frac{\begin{vmatrix} & \\ & \end{vmatrix}}{\begin{vmatrix} 1 & 1 \\ 2 & 3 \end{vmatrix}} = \frac{1}{1} = 1 \qquad \begin{vmatrix} 1 & 3 \\ 2 & 7 \end{vmatrix}$$

The currents in the circuit are $i_1 = 2$, $i_2 = 1$.

Cramer's rule may be extended to systems of linear equations with more than two unknowns. Consider the following linear system with three unknowns, x, y and z.

$$a_1x + b_1y + c_1z = k_1$$
$$a_2x + b_2y + c_2z = k_2$$
$$a_3x + b_3y + c_3z = k_3$$

Then Cramer's rule states

Key point

$$x = \frac{\begin{vmatrix} k_1 & b_1 & c_1 \\ k_2 & b_2 & c_2 \\ k_3 & b_3 & c_3 \end{vmatrix}}{\begin{vmatrix} a_1 & b_1 & c_1 \\ a_2 & b_2 & c_2 \\ a_3 & b_3 & c_3 \end{vmatrix}}, \quad y = \frac{\begin{vmatrix} a_1 & k_1 & c_1 \\ a_2 & k_2 & c_2 \\ a_3 & k_3 & c_3 \end{vmatrix}}{\begin{vmatrix} a_1 & b_1 & c_1 \\ a_2 & b_2 & c_2 \\ a_3 & b_3 & c_3 \end{vmatrix}}, \quad z = \frac{\begin{vmatrix} a_1 & b_1 & k_1 \\ a_2 & b_2 & k_2 \\ a_3 & b_3 & k_3 \end{vmatrix}}{\begin{vmatrix} a_1 & b_1 & c_1 \\ a_2 & b_2 & c_2 \\ a_3 & b_3 & c_3 \end{vmatrix}}$$

Note again that, in all three cases, the denominators are the same. The numerators are formed by replacing a column of as, bs or cs by a column of ks.

Example 1.2 Electrical Engineering – Circuit voltages

The voltages v_1, v_2 and v_3 in an electric circuit are connected by the system

$$v_1 + 2v_2 + v_3 = 2$$
$$2v_1 - v_2 - 2v_3 = 5$$
$$2v_1 + 2v_2 + 3v_3 = 7$$

Use Cramer's rule to determine the three voltages.

Solution

Using Cramer's rule we have:

$$v_1 = \frac{\begin{vmatrix} & & \\ & & \end{vmatrix}}{\begin{vmatrix} 1 & 2 & 1 \\ 2 & -1 & -2 \\ 2 & 2 & 3 \end{vmatrix}} = \frac{}{-13}$$

$$= 3$$

$$\begin{vmatrix} 2 & 2 & 1 \\ 5 & -1 & -2 \\ 7 & 2 & 3 \end{vmatrix}, -39$$

$$v_2 = \frac{}{-13}$$

$$= \frac{13}{-13} = -1$$

$$\begin{vmatrix} 1 & 2 & 1 \\ 2 & 5 & -2 \\ 2 & 7 & 3 \end{vmatrix}$$

$$v_3 = \frac{}{-13}$$

$$= \frac{-13}{-13} = 1$$

$$\begin{vmatrix} 1 & 2 & 2 \\ 2 & -1 & 5 \\ 2 & 2 & 7 \end{vmatrix}$$

The voltages are $v_1 = 3$, $v_2 = -1$ and $v_3 = 1$.

Cramer's rule can be extended to four and more unknowns in an obvious way, but the determinants become too large to make this a sensible approach. Gaussian elimination, described in Block 3, is a practical alternative.

Exercises

1 Solve the following systems of two linear equations using Cramer's rule:

(a) $2x + y = 10$
$3x - 2y = 1$

(b) $-\theta_1 - \theta_2 = 1$
$2\theta_1 + \theta_2 = 1$

(c) $2i_1 + i_2 = 9.3$
$3i_1 - 2i_2 = 5.9$

(d) $3v_1 - v_2 = 6.6$
$2v_1 - 3v_2 = -1.9$

(e) $\alpha - 3\beta = 0.5$
$\beta + 2\alpha = -3.9$

2   Solve the following systems of three linear equations using Cramer's rule:

(a) $2x - y - 2z = 1$
$x + y + 3z = 6$
$-2x + 3z = -1$

(b) $\beta - 2\gamma = -6$
$\alpha + 3\beta + \gamma = 2$
$3\alpha + 2\gamma = 3$

(c) $i_1 + i_2 - i_3 = 1$
$3i_1 - i_2 + 2i_3 = 7$
$-i_1 + 2i_2 - i_3 = -0.5$

(d) $2v_1 - v_2 + v_3 = 5.9$
$v_1 + \frac{1}{2}v_2 - v_3 = 0.7$
$-3v_1 + 2v_3 = -3.1$

(e) $\theta_1 + \theta_2 + 4\theta_3 = 0$
$2\theta_1 - \theta_2 + \theta_3 = 5.4$
$\theta_1 + 2\theta_2 - \theta_3 = 9.2$

Solutions to exercises

1   (a) $x = 3, y = 4$
(b) $\theta_1 = 2, \theta_2 = -3$
(c) $i_1 = 3.5, i_2 = 2.3$
(d) $v_1 = 3.1, v_2 = 2.7$
(e) $\alpha = -1.6, \beta = -0.7$

2   (a) $x = 2, y = 1, z = 1$
(b) $\alpha = -1, \beta = 0, \gamma = 3$
(c) $i_1 = 1.5, i_2 = 1.5, i_3 = 2$
(d) $v_1 = 2.3, v_2 = 0.6, v_3 = 1.9$
(e) $\theta_1 = 4.3, \theta_2 = 1.7, \theta_3 = -1.5$

End of block exercises

1   **Electrical Engineering.** Voltages v_1 and v_2 in a circuit are related by the system of equations

$$3v_1 + 2v_2 = 6$$
$$-4v_1 + 10v_2 = -27$$

Calculate v_1 and v_2 using Cramer's rule.

2   **Electrical Engineering.** The currents i_1, i_2 and i_3 of a circuit are related by

$$2i_1 + i_2 - i_3 = 8$$
$$i_1 - i_2 + i_3 = -5$$
$$3i_1 + 2i_2 = 9$$

Use Cramer's rule to evaluate the three currents.

3   Explain why Cramer's rule cannot be applied to the following system of equations:

$$3x - 2y = 10$$
$$y - 1.5x = -5$$

4   **Mechanical Engineering.** The magnitudes of forces f and g in a mechanical system are connected by the equations

$$3f + 5g - 16.1 = 0$$
$$2f - g - 3.8 = 0$$

Use Cramer's rule to determine the magnitudes of f and g.

Solutions to exercises

1    $v_1 = 3, v_2 = -1.5$

2    $i_1 = 1, i_2 = 3, i_3 = -3$

3   The matrix

$$\begin{pmatrix} 3 & -2 \\ -1.5 & 1 \end{pmatrix}$$

has a zero determinant.

4    $f = 2.7, g = 1.6$

Using the inverse matrix to solve simultaneous equations

2.1 Introduction

Block 1 dealt with Cramer's rule for solving a system of linear equations. The solutions were given in terms of determinants. This block shows how an inverse matrix can be used to solve a system of linear simultaneous equations. You will need to be familiar with the techniques for finding the inverses of 2×2 and 3×3 matrices as described in Chapter 12 Block 4.

2.2 Writing equations in matrix form

Consider the simultaneous equations

$$7x + 2y = 12$$
$$3x + y = 5$$

Recalling how matrices are multiplied we note that the product

$$\begin{pmatrix} 7 & 2 \\ 3 & 1 \end{pmatrix} \begin{pmatrix} x \\ y \end{pmatrix}$$

is equivalent to

$$\begin{pmatrix} 7x + 2y \\ 3x + y \end{pmatrix}$$

Hence we can write the simultaneous equations as

$$\begin{pmatrix} 7 & 2 \\ 3 & 1 \end{pmatrix} \begin{pmatrix} x \\ y \end{pmatrix} = \begin{pmatrix} 12 \\ 5 \end{pmatrix}$$

This is the **matrix form** of the simultaneous equations.

Writing

$$A = \begin{pmatrix} 7 & 2 \\ 3 & 1 \end{pmatrix}, \quad X = \begin{pmatrix} x \\ y \end{pmatrix}, \quad B = \begin{pmatrix} 12 \\ 5 \end{pmatrix}$$

we have the standard form

$$AX = B$$

Note that the elements of A come from the coefficients of x and y and that the elements of B come from the right-hand sides of the equations.

Example 2.1

Write the following system of equations in the standard matrix form $AX = B$.

$$2x + y - z = 4$$
$$x - y + 3z = -2$$
$$2x + 2y - 3z = 9$$

Solution

The elements of A are the coefficients of x, y and z. Hence

$$A = \begin{pmatrix} 2 & 1 & -1 \\ 1 & -1 & 3 \\ 2 & 2 & -3 \end{pmatrix}$$

and

$$X = \begin{pmatrix} x \\ y \\ z \end{pmatrix}, \quad B = \begin{pmatrix} 4 \\ -2 \\ 9 \end{pmatrix}$$

Hence in matrix form the equations are written as

$$\begin{pmatrix} 2 & 1 & -1 \\ 1 & -1 & 3 \\ 2 & 2 & -3 \end{pmatrix} \begin{pmatrix} x \\ y \\ z \end{pmatrix} = \begin{pmatrix} 4 \\ -2 \\ 9 \end{pmatrix}$$

Example 2.2

Express the following system in the form $AX = B$.

$$-x + 3y + 4z = 7$$
$$y - 4x = 7$$
$$y + 3x - z = -9$$

Solution

The equations are rewritten so that the variables x, y and z appear in the same order.

$$-x + 3y + 4z = 7$$
$$-4x + y + 0z = 7$$
$$3x + y - z = -9$$

Then

$$A = \qquad\qquad\qquad\qquad\qquad\qquad\qquad \begin{pmatrix} -1 & 3 & 4 \\ -4 & 1 & 0 \\ 3 & 1 & -1 \end{pmatrix}$$

$$X = \begin{pmatrix} x \\ y \\ z \end{pmatrix}, \quad B = \qquad\qquad\qquad\qquad\qquad \begin{pmatrix} 7 \\ 7 \\ -9 \end{pmatrix}$$

In matrix form the equations are

$$\begin{pmatrix} -1 & 3 & 4 \\ -4 & 1 & 0 \\ 3 & 1 & -1 \end{pmatrix} \begin{pmatrix} x \\ y \\ z \end{pmatrix} = \begin{pmatrix} 7 \\ 7 \\ -9 \end{pmatrix}$$

Exercises

1   Write the following simultaneous equations in matrix form:

(a) $3x - 4y = 9$
    $x + 2y = 6$

(b) $\alpha - \beta = 0.5$
    $3\alpha + 7\beta = 11.6$

(c) $2x - y + 3z = 0$
    $3x + 4y + z = 1$
    $-2x + 3z = -2$

Solutions to exercises

1   (a) $\begin{pmatrix} 3 & -4 \\ 1 & 2 \end{pmatrix} \begin{pmatrix} x \\ y \end{pmatrix} = \begin{pmatrix} 9 \\ 6 \end{pmatrix}$

(b) $\begin{pmatrix} 1 & -1 \\ 3 & 7 \end{pmatrix} \begin{pmatrix} \alpha \\ \beta \end{pmatrix} = \begin{pmatrix} 0.5 \\ 11.6 \end{pmatrix}$

(c) $\begin{pmatrix} 2 & -1 & 3 \\ 3 & 4 & 1 \\ -2 & 0 & 3 \end{pmatrix} \begin{pmatrix} x \\ y \\ z \end{pmatrix} = \begin{pmatrix} 0 \\ 1 \\ -2 \end{pmatrix}$

2.3   Solving equations using the inverse matrix method

Consider a system of equations written in matrix form, that is

$$AX = B$$

where A is a square matrix. Note that A and B are matrices with numerical elements. We wish to find an expression for the unknowns, that is the elements of X. To achieve this we premultiply both sides of the equation by the inverse of A, if it exists, to obtain

$$A^{-1} AX = A^{-1}B \qquad (1)$$

The left-hand side can be simplified by noting that multiplying a matrix by its inverse gives the identity matrix, that is $A^{-1} A = I$. Hence

$$A^{-1} AX = (A^{-1} A) X$$
$$= IX$$

Multiplying a matrix by the identity matrix has no effect and so $IX = X$. So the left-hand side of equation (1) simplifies to

$$A^{-1} AX = IX$$
$$= X$$

and so, from (1),

$$X = A^{-1} B$$

Given $AX = B$, then $X = A^{-1} B$ if A^{-1} exists.

Example 2.3
Solve

$$7x + 2y = 12$$
$$3x + y = 5$$

using the inverse matrix method.

Solution
Writing the equations in matrix form, $AX = B$, gives

$$\begin{pmatrix} 7 & 2 \\ 3 & 1 \end{pmatrix} \begin{pmatrix} x \\ y \end{pmatrix} = \begin{pmatrix} 12 \\ 5 \end{pmatrix}$$

Here

$$A = \begin{pmatrix} 7 & 2 \\ 3 & 1 \end{pmatrix}, \quad B = \begin{pmatrix} 12 \\ 5 \end{pmatrix}, \quad X = \begin{pmatrix} x \\ y \end{pmatrix}$$

We calculate A^{-1}.

$$A^{-1} = \begin{pmatrix} 1 & -2 \\ -3 & 7 \end{pmatrix}$$

Hence

$$X = A^{-1} B$$
$$= \begin{pmatrix} 1 & -2 \\ -3 & 7 \end{pmatrix} \begin{pmatrix} 12 \\ 5 \end{pmatrix}$$
$$= \begin{pmatrix} 2 \\ -1 \end{pmatrix}$$

We have

$$X = \begin{pmatrix} x \\ y \end{pmatrix}$$
$$= \begin{pmatrix} 2 \\ -1 \end{pmatrix}$$

that is $x = 2, y = -1$.

Example 2.4 Electrical Engineering – Circuit currents

The currents i_1 and i_2 in a circuit are connected by the equations

$$2i_1 + i_2 = 8.6$$
$$3i_1 - i_2 = 9.4$$

Use the inverse matrix method to find the values of i_1 and i_2.

Solution
We have $AX = B$ where

$$A = \qquad, \qquad B = \begin{pmatrix} 8.6 \\ 9.4 \end{pmatrix} \quad \text{and} \quad X = \begin{pmatrix} i_1 \\ i_2 \end{pmatrix}$$

$$\begin{pmatrix} 2 & 1 \\ 3 & -1 \end{pmatrix}$$

We calculate A^{-1}.

$$A^{-1} = \qquad$$

$$\begin{pmatrix} \frac{1}{5} & \frac{1}{5} \\ \frac{3}{5} & -\frac{2}{5} \end{pmatrix}$$

So

$$X = A^{-1} B$$

$$= \begin{pmatrix} \frac{1}{5} & \frac{1}{5} \\ \frac{3}{5} & -\frac{2}{5} \end{pmatrix} \begin{pmatrix} 8.6 \\ 9.4 \end{pmatrix}$$

$$= \qquad \qquad \begin{pmatrix} 3.6 \\ 1.4 \end{pmatrix}$$

So $i_1 = 3.6$ and $i_2 = 1.4$.

Example 2.5 Mechanical Engineering – Forces in a pulley system

Forces f_1, f_2 and f_3 in a simple pulley system satisfy the equations

$$-f_1 + 3f_2 + 4f_3 = 7$$
$$f_2 - 4f_1 = 7$$
$$f_2 + 3f_1 - f_3 = -9$$

Solve the system for f_1, f_2 and f_3 .

Solution
The equations are written in matrix form:

$$\begin{pmatrix} -1 & 3 & 4 \\ -4 & 1 & 0 \\ 3 & 1 & -1 \end{pmatrix} \begin{pmatrix} f_1 \\ f_2 \\ f_3 \end{pmatrix} = \begin{pmatrix} 7 \\ 7 \\ -9 \end{pmatrix}$$

Writing

$$A = \begin{pmatrix} -1 & 3 & 4 \\ -4 & 1 & 0 \\ 3 & 1 & -1 \end{pmatrix}$$

we then calculate A^{-1} as

$$A^{-1} = \frac{1}{39} \begin{pmatrix} 1 & -7 & 4 \\ 4 & 11 & 16 \\ 7 & -10 & -11 \end{pmatrix}$$

Hence

$$\begin{pmatrix} f_1 \\ f_2 \\ f_3 \end{pmatrix} = \frac{1}{39} \begin{pmatrix} 1 & -7 & 4 \\ 4 & 11 & 16 \\ 7 & -10 & -11 \end{pmatrix} \begin{pmatrix} 7 \\ 7 \\ -9 \end{pmatrix}$$

$$= \begin{pmatrix} -2 \\ -1 \\ 2 \end{pmatrix}$$

So $f_1 = -2, f_2 = -1, f_3 = 2$.

Exercises

1 Solve the following systems of linear equations using the inverse matrix method:
(a) $3x - y = 4$
   $x + 2y = 13$
(b) $2\alpha - \beta = -11$
   $\frac{1}{2}\alpha + \beta = 1$
(c) $v_1 - 2v_2 = -3.5$
   $3v_1 + 5v_2 = 17$
(d) $2i_1 + i_2 = 3.5$
   $-i_1 + 2i_2 = 3.5$
(e) $p - 2q = -0.7$
   $2p - q = -1.1$

2 State conditions under which the inverse matrix method will fail to find a set of solutions.

3 Given that the inverse of

$$\begin{pmatrix} 2 & -1 & -1 \\ 1 & 2 & -2 \\ 2 & 3 & 1 \end{pmatrix}$$

is

$$\frac{1}{22} \begin{pmatrix} 8 & -2 & 4 \\ -5 & 4 & 3 \\ -1 & -8 & 5 \end{pmatrix}$$

solve the following system of equations:
$$2x - y - z = 7$$
$$x + 2y - 2z = 13$$
$$2x + 3y + z = 9$$

4 Given that the inverse of

$$\begin{pmatrix} 1 & -1 & 2 \\ 4 & 3 & -1 \\ 3 & -1 & 4 \end{pmatrix}$$

is

$$\frac{1}{4} \begin{pmatrix} 11 & 2 & -5 \\ -19 & -2 & 9 \\ -13 & -2 & 7 \end{pmatrix}$$

solve the system of linear equations
$$4x + 3y - z = 3$$
$$3x - y + 4z = 10$$
$$x - y + 2y = 4$$

5 **Electrical Engineering.** Resistances r_1, r_2 and r_3 in a circuit are connected by the equations
$$r_1 - 2r_2 + r_3 = -0.9$$
$$2r_1 - r_2 + r_3 = 0.8$$
$$r_1 - 2r_2 - r_3 = -2.3$$

Use the inverse matrix method to calculate r_1, r_2 and r_3.

Solutions to exercises

1   (a) $x = 3, y = 5$
    (b) $\alpha = -4, \beta = 3$
    (c) $v_1 = 1.5, v_2 = 2.5$
    (d) $i_1 = 0.7, i_2 = 2.1$
    (e) $p = -0.5, q = 0.1$

2   Given $AX = B$, the method will fail if $|A| = 0$;
    that is, if A^{-1} does not exist.

3   $x = 3, y = 2, z = -3$

4   $x = 0, y = 2, z = 3$

5   $r_1 = 0.6, r_2 = 1.1, r_3 = 0.7$

Computer and calculator exercises

1   Using a computer algebra package to calculate
    the relevant inverse matrix and perform the
    matrix multiplication, solve the following
    systems using the inverse matrix method:
    (a) $2x - y + z = -7$
    $$x + 2y + \tfrac{1}{2}z = 19$$
    $$4x + 3z = 0$$

(b) $2\alpha + 4\beta - \gamma = -2$
    $$\alpha + 2\beta + 3\gamma = 20$$
    $$\alpha + \beta = 0.5$$

(c) $i_1 + i_2 + i_3 - i_4 = 2$
    $$i_1 - i_2 + i_3 + i_4 = 6$$
    $$-i_1 + i_2 + i_3 - i_4 = 0$$
    $$-i_1 + i_2 - i_3 + 4i_4 = 14$$

Solutions to exercises

1   (a) $x = 3, y = 9, z = -4$
    (b) $\alpha = -1, \beta = 1.5, \gamma = 6$
    (c) $i_1 = 1, i_2 = 2, i_3 = 3, i_4 = 4$

End of block exercises

1   Use the inverse matrix method to solve
    (a) $4x - 3y = 6$
    $$-2x + y = -4$$

    (b) $\alpha + 2\beta = 2$
    $$3\alpha - 4\beta = -9$$

2   Explain why the inverse matrix method cannot
    be applied to the following system:
    $$3x - y = 7$$
    $$-6x + 2y = -14$$

3   Given that the inverse of
    $$\begin{pmatrix} 1 & -1 & 3 \\ 2 & 1 & 1 \\ -3 & 1 & 4 \end{pmatrix}$$
    is
    $$\frac{1}{29}\begin{pmatrix} 3 & 7 & -4 \\ -11 & 13 & 5 \\ 5 & 2 & 3 \end{pmatrix}$$

solve the following systems:

(a) $x - y + 3z = 3$
$2x + y + z = 7$
$-3x + y + 4z = 0$

(b) $2x + y + z = -1$
$-3x + y + 4z = -7$
$x - y + 3z = -7$

(c) $3x - 3y + 9z = 30$
$-3x + y + 4z = 20$
$4x + 2y + 2z = 6$

Solutions to exercises

1   (a) $x = 3, y = 2$
(b) $\alpha = -1, \beta = 1.5$

2   The matrix

$$\begin{pmatrix} 3 & -1 \\ -6 & 2 \end{pmatrix}$$

has no inverse.

3   (a) $x = 2, y = 2, z = 1$
(b) $x = 0, y = 1, z = -2$
(c) $x = -1, y = 1, z = 4$

Gaussian elimination

3.1 Introduction

Gaussian elimination is a systematic way of simplifying a system of equations so that a solution can easily be found. Rather than work with the actual equations, a matrix, whose elements capture all the properties of the equations, is used.

Manipulation of the matrix is carried out by performing a sequence of elementary row operations. These operations eventually bring the matrix into a form known as echelon form. Once in this form, the solution to the original equations is easily found.

3.2 The augmented matrix

Consider the system of linear equations:

$$2x + 3y = 16$$
$$x - y = -2$$

This system may be represented in the matrix form

$$\begin{pmatrix} 2 & 3 & 16 \\ 1 & -1 & -2 \end{pmatrix}$$

This is an example of an **augmented matrix**. Note that the elements of the augmented matrix comprise the coefficients of x and y and also the right-hand sides of the equations.

Example 3.1
State the augmented matrix for the system

$$3x - y + 2z = 7$$
$$4x + y = 6$$
$$x + 2y - z = 2$$

Solution
The augmented matrix is

$$\begin{pmatrix} 3 & -1 & 2 & 7 \\ 4 & 1 & 0 & 6 \\ 1 & 2 & -1 & 2 \end{pmatrix}$$

Exercises

1   Write the augmented matrix for each of the following systems of equations:

(a) $2x - y = 8$
$x + 2y = 14$

(b) $3x + 2y = 4$
$5x + 2y = 0$

(c) $x - 3y = 10$
$2x + y = -1$

(d) $x - y = 4$
$6x + 3y = 1.5$

(e) $3x + 4y = 50$
$-5x + 2y = -1$

(c) $7x - y - z = 3$
$3x + 2y + z = 4$
$5x + 2y + 3z = 20$

(d) $x - y + z = 2$
$5x + y + 2z = 7$
$4x - 2y - z = -1$

(e) $x - 2y + z = 0$
$-4x + y = 10$
$2y - 7z = 3$

2   Write the augmented matrix for each of the following systems of equations:

(a) $x + y + z = 4$
$3x - 2y + 4z = 25$
$2x + 3y + 2z = 6$

(b) $x - \frac{1}{2}y + 3z = 1$
$3x + 2y + z = -5$
$5x + 2z = -8$

Solutions to exercises

1   (a) $\begin{pmatrix} 2 & -1 & 8 \\ 1 & 2 & 14 \end{pmatrix}$ (b) $\begin{pmatrix} 3 & 2 & 4 \\ 5 & 2 & 0 \end{pmatrix}$

(c) $\begin{pmatrix} 1 & -3 & 10 \\ 2 & 1 & -1 \end{pmatrix}$ (d) $\begin{pmatrix} 1 & -1 & 4 \\ 6 & 3 & 1.5 \end{pmatrix}$

(e) $\begin{pmatrix} 3 & 4 & 50 \\ -5 & 2 & -1 \end{pmatrix}$

(c) $\begin{pmatrix} 7 & -1 & -1 & 3 \\ 3 & 2 & 1 & 4 \\ 5 & 2 & 3 & 20 \end{pmatrix}$

(d) $\begin{pmatrix} 1 & -1 & 1 & 2 \\ 5 & 1 & 2 & 7 \\ 4 & -2 & -1 & -1 \end{pmatrix}$

(e) $\begin{pmatrix} 1 & -2 & 1 & 0 \\ -4 & 1 & 0 & 10 \\ 0 & 2 & -7 & 3 \end{pmatrix}$

2   (a) $\begin{pmatrix} 1 & 1 & 1 & 4 \\ 3 & -2 & 4 & 25 \\ 2 & 3 & 2 & 6 \end{pmatrix}$

(b) $\begin{pmatrix} 1 & -\dfrac{1}{2} & 3 & 1 \\ 3 & 2 & 1 & -5 \\ 5 & 0 & 2 & -8 \end{pmatrix}$

When a system of linear equations is solved, there are three possible outcomes.

Key point

A system of linear equations may have

1 a unique solution
2 an infinite number of solutions
3 no solution.

These three possible outcomes were studied in Block 4 of Chapter 7. In this section we revisit these three possibilities and note the form of the augmented matrix in each case.

Case 1

For example, consider the system

$$x + 3y = 7$$
$$0x + y = 2$$

This system has a unique solution, $x = 1$, $y = 2$. The corresponding augmented matrix is

$$\begin{pmatrix} 1 & 3 & 7 \\ 0 & 1 & 2 \end{pmatrix}$$

Case 2

Consider the system

$$x + 3y = 7$$
$$0x + 0y = 0$$

There is really only one equation here because the second states that $0 = 0$. There is an infinite number of solutions of the first equation. For example, $x = 7$ and $y = 0$; $x = 10$ and $y = -1$. In general, whatever the value of y, the equation is satisfied if $x = 7 - 3y$. We write this as follows: letting $y = t$, say, then $x = 7 - 3t$. This is true for any value of t we choose, so t is called a **free variable**. The corresponding augmented matrix is

$$\begin{pmatrix} 1 & 3 & 7 \\ 0 & 0 & 0 \end{pmatrix}$$

Case 3

Finally consider the system

$$x + 3y = 7$$
$$0x + 0y = 2$$

The second equation seems to state that $0 = 2$. This is nonsense. There are no solutions and the augmented matrix is

$$\begin{pmatrix} 1 & 3 & 7 \\ 0 & 0 & 2 \end{pmatrix}$$

It is worthwhile rereading Block 4 in Chapter 7 before continuing.

Example 3.2
A system has augmented matrix

$$\begin{pmatrix} 1 & -4 & 13 \\ 0 & 1 & -3 \end{pmatrix}$$

State the solution of the system.

Solution
The second row of the augmented matrix is equivalent to the equation

$$0x + 1y = -3$$

that is

$$y = -3$$

The first row of the matrix is equivalent to the equation

$$x - 4y = 13$$

Substituting in $y = -3$ we can obtain x.

$$x - 4(-3) = 13$$
$$x + 12 = 13$$
$$x = 1$$

The system has a unique solution $x = 1, y = -3$.

In Example 3.2 the value of y was found and substituted back into the remaining equation so that x could be found. This technique is known as **back substitution**.

Example 3.3
A system has augmented matrix

$$\begin{pmatrix} 1 & 3 & -6 \\ 0 & 0 & 4 \end{pmatrix}$$

Solve the system.

Solution
The second row of the augmented matrix corresponds to the equation

$$0x + 0y = 4$$

This equation has no solution and so the system has no solution.

Example 3.4
A system has augmented matrix

$$\begin{pmatrix} 1 & 3 & 17 \\ 0 & 0 & 0 \end{pmatrix}$$

Solve the system.

Solution
The second row of the matrix is equivalent to the equation

$$0x + 0y = 0$$

This equation is satisfied for any value of x and y. The first row is equivalent to

$$x + 3y = 17$$

This has an infinite number of solutions: $x = 17 - 3t$, $y = t$.

Example 3.5
A system has augmented matrix

$$\begin{pmatrix} 1 & 2 & -3 & 11 \\ 0 & 1 & -4 & 7 \\ 0 & 0 & 1 & -1 \end{pmatrix}$$

Solve the system.

Solution
From the third row, the equation is

$$0x + 0y + 1z = -1$$

that is

$$z = -1$$

From the second row, the corresponding equation is

$$y - 4z = 7$$

Substituting in $z = -1$ the equation can be solved for y.

$$y - 4(-1) = 7$$
$$y + 4 = 7$$
$$y = 3$$

The equation corresponding to the first row is

$$1x + 2y - 3z = 11$$

By substituting in the values already obtained for y and z, we can determine x.

$$x = \boxed{} \qquad 2$$

The system has a unique solution: $x = 2$, $y = 3$, $z = -1$.

Exercises

1 Solve the systems whose augmented matrices are

(a) $\begin{pmatrix} 1 & 2 & 18 \\ 0 & 1 & 7 \end{pmatrix}$

(b) $\begin{pmatrix} 1 & -2 & 10 \\ 0 & 0 & 0 \end{pmatrix}$

(c) $\begin{pmatrix} 1 & 5 & 6 \\ 0 & 1 & 0 \end{pmatrix}$

(d) $\begin{pmatrix} 1 & 5 & 6 \\ 0 & 0 & 1 \end{pmatrix}$

(e) $\begin{pmatrix} 1 & -2 & -1 & 5 \\ 0 & 1 & 3 & 1 \\ 0 & 0 & 1 & 1 \end{pmatrix}$

(f) $\begin{pmatrix} 1 & 3 & 4 & 2 \\ 0 & 1 & -2 & 6 \\ 0 & 0 & 0 & 0 \end{pmatrix}$

(g) $\begin{pmatrix} 1 & 3 & 6 & 25 \\ 0 & 0 & 1 & 3 \end{pmatrix}$

Solutions to exercises

1 (a) $x = 4, y = 7$ (b) $x = 2t + 10, y = t$
(c) $x = 6, y = 0$ (d) no solution
(e) $x = 2, y = -2, z = 1$

(f) $x = -10t - 16, y = 2t + 6, z = t$
(g) $x = 7 - 3t, y = t, z = 3$

3.4 Row-echelon form of an augmented matrix

We now introduce the **row-echelon form** of an augmented matrix. The matrices given in Examples 3.4 and 3.5 are in row-echelon form. When a matrix is in row-echelon form it is then easy to determine the solution of the system.

Key point

For a matrix to be in row-echelon form:

1 Any rows that consist entirely of zeros are the last rows of the matrix.
2 For a row that is not all zeros, the first non-zero element is a one. We call this a **leading 1**.
3 As you move down the rows of the matrix, the leading 1s move progressively to the right.

Example 3.6

Determine which of the following matrices are in row-echelon form:

(a) $\begin{pmatrix} 1 & 2 & 5 \\ 0 & 1 & 7 \end{pmatrix}$

(b) $\begin{pmatrix} 0 & 1 & 3 & 6 \\ 1 & 4 & -1 & 3 \\ 0 & 0 & 0 & 0 \end{pmatrix}$

(c) $\begin{pmatrix} 0 & 0 & 1 & 6 \\ 0 & 0 & 0 & 0 \end{pmatrix}$

(d) $\begin{pmatrix} 1 & 6 & 9 & 4 \\ 0 & 1 & -3 & 6 \\ 0 & 0 & 1 & 5 \end{pmatrix}$

Solution

(a) This is in row-echelon form. Note that the leading 1 in the second row is further to the right than the leading 1 in the first row.

(b) This is not in row-echelon form since the leading 1 in the first row is further to the right than the leading 1 in the second row.

(c) This is in row-echelon form.

(d) This is in row-echelon form.

Exercises

1   Explain what is meant by row-echelon form.

2   Determine which of the following matrices are in row-echelon form:

(a) $\begin{pmatrix} 1 & 4 & 3 & 9 \\ 0 & 1 & -3 & 6 \end{pmatrix}$

(b) $\begin{pmatrix} 1 & -1 & 0 \\ 0 & 1 & 0 \end{pmatrix}$

(c) $\begin{pmatrix} 1 & 2 & 5 & 6 \\ 1 & 0 & 3 & 6 \\ 0 & 0 & 1 & 4 \end{pmatrix}$

(d) $\begin{pmatrix} 1 & -1 & 3 & 7 \\ 0 & 0 & 0 & 0 \\ 0 & 1 & 5 & 6 \end{pmatrix}$

(e) $\begin{pmatrix} 1 & 0 & 0 & 2 \\ 0 & 0 & 1 & -1 \\ 0 & 1 & 0 & 3 \end{pmatrix}$

Solutions to exercises

2   (a) row-echelon form
    (b) row-echelon form
    (c) not row-echelon form
    (d) not row-echelon form
    (e) not row-echelon form

3.5 Elementary row operations

Given any augmented matrix we would like to be able to transform it into row-echelon form as the solution of the system can then easily be found. The way we effect this transformation is by using **elementary row operations**.

We define three **elementary operations** that we can apply to the equations of the system.

The elementary operations that change a system but leave the solution unaltered are as follows:

1 Change the order of the equations.
2 Multiply or divide an equation by a non-zero constant.
3 Add, or subtract, a multiple of one equation to, or from, another equation.

When applied to a system, elementary row operations may change the equations but the solution to the resulting system remains unchanged. We apply elementary row operations to a simple system to illustrate this.

Consider the system of linear equations

$$2x + 5y = 12 \qquad (1)$$
$$x + y = 3 \qquad (2)$$

The system has solution $x = 1, y = 2$.

If we interchange the order of the equations, the system becomes

$$x + y = 3$$
$$2x + 5y = 12$$

The solution to the system is obviously still $x = 1, y = 2$.

To illustrate the second of the elementary operations consider multiplying equation (1) by, say, 6 and dividing equation (2) by, say, 5. The system is then

$$12x + 30y = 72$$
$$\frac{x}{5} + \frac{y}{5} = \frac{3}{5}$$

You can check that this system still has solution $x = 1, y = 2$.

Finally, to illustrate the third elementary operation, we add 3 times equation (1) to equation (2). The system becomes

$$2x + 5y = 12$$
$$7x + 16y = 39$$

Check that this resulting system has solution $x = 1, y = 2$.

Thus elementary operations may change the system, but they do not change the solutions of a system.

We have been writing a system in the form of an augmented matrix, where each row of the matrix corresponds to an equation of the system. Thus we can apply elementary operations to the rows of an augmented matrix rather than the equations of the system. When elementary operations are applied to the rows of a matrix they are called **elementary row operations**.

Example 3.7

(a) Write down the augmented matrix of the system

$$2x + 5y = 12$$
$$x + y = 3$$

(b) Carry out the following elementary row operations, each time applying the operation to the most recent augmented matrix:
  (i) interchange the rows
  (ii) subtract $2 \times$ the first row from the second row
  (iii) divide the second row by 3.
(c) Hence solve the system.

Solution

(a) The augmented matrix is

$$\begin{pmatrix} 2 & 5 & 12 \\ 1 & 1 & 3 \end{pmatrix}$$

(b) (i) The rows are interchanged. The augmented matrix becomes

$$\begin{pmatrix} 1 & 1 & 3 \\ 2 & 5 & 12 \end{pmatrix}$$

(ii) We now use the augmented matrix in (b) (i). Subtracting $2 \times$ the first row from the second row results in

$$\begin{pmatrix} 1 & 1 & 3 \\ 0 & 3 & 6 \end{pmatrix}$$

(iii) We now use the augmented matrix in (b) (ii). Dividing the second row by 3 gives

$$\begin{pmatrix} 1 & 1 & 3 \\ 0 & 1 & 2 \end{pmatrix}$$

(c) Note that the matrix is now in row-echelon form. From the second row we see that $y = 2$. From the first row we have

$$1x + 1y = 3$$

Substituting $y = 2$ into this equation we obtain $x = 1$. Thus the solution to the original system is $x = 1$, $y = 2$.

The method of solution of Example 3.7 is thus

1 Write down the augmented matrix.
2 Apply elementary row operations to obtain row-echelon form.
3 Solve the system.

This method is known as **Gaussian elimination**.

Example 3.8
Use Gaussian elimination to solve

$$3x - y = 1$$
$$2x + 3y = 19$$

Solution
The augmented matrix is

$$\begin{pmatrix} 3 & -1 & 1 \\ 2 & 3 & 19 \end{pmatrix}$$

We now apply a series of elementary row operations to transform the augmented matrix into row-echelon form.

Subtract the second row from the first row.

$$\begin{pmatrix} 1 & -4 & -18 \\ 2 & 3 & 19 \end{pmatrix}$$

Subtract 2 × the first row from the second row

$$\begin{pmatrix} 1 & -4 & -18 \\ 0 & 11 & 55 \end{pmatrix}$$

Divide the second row by 11.

$$\begin{pmatrix} 1 & -4 & -18 \\ 0 & 1 & 5 \end{pmatrix}$$

The matrix is now in row-echelon form. From the second row we see that $y = 5$. The first row corresponds to the equation

$$1x - 4y = -18$$

Substituting $y = 5$ into this equation allows x to be found.

2

The solution to the system is $x = 2$, $y = 5$.

Example 3.9
Use Gaussian elimination to solve

$$2x + y + 2z = 8$$
$$x - 3y + 3z = -4$$
$$4x + 2y - z = 1$$

Solution
The augmented matrix is

$$\begin{pmatrix} 2 & 1 & 2 & 8 \\ 1 & -3 & 3 & -4 \\ 4 & 2 & -1 & 1 \end{pmatrix}$$

We apply a series of elementary row operations to transform the matrix into row-echelon form.

Interchange the first row and second row.

$$\begin{pmatrix} 1 & -3 & 3 & -4 \\ 2 & 1 & 2 & 8 \\ 4 & 2 & -1 & 1 \end{pmatrix}$$

Subtract $2 \times$ the first row from the second row, subtract $4 \times$ the first row from the third row. This produces

$$\begin{pmatrix} 1 & -3 & 3 & -4 \\ 0 & 7 & -4 & 16 \\ 0 & 14 & -13 & 17 \end{pmatrix}$$

Subtract $2 \times$ the second row from the third row to get

$$\begin{pmatrix} 1 & -3 & 3 & -4 \\ 0 & 7 & -4 & 16 \\ 0 & 0 & -5 & -15 \end{pmatrix}$$

Divide the second row by 7, divide the third row by -5.

$$\begin{pmatrix} 1 & -3 & 3 & -4 \\ 0 & 1 & -\dfrac{4}{7} & \dfrac{16}{7} \\ 0 & 0 & 1 & 3 \end{pmatrix}$$

The matrix is now in row-echelon form. From the third row we see that

$z = \boxed{}$ 3

The second row is equivalent to the equation

$$1y - \frac{4}{7}z = \frac{16}{7}$$

Substituting $z = 3$ and solving for y yields

$y = \boxed{}$ 4

The first row is equivalent to the equation

$\boxed{}$ $1x - 3y + 3z = -4$

Substituting in the values for y and z and solving for x yields

$x = \boxed{}$ -1

The solution to the system is $x = -1$, $y = 4$, $z = 3$.

Exercises

1 Use Gaussian elimination to solve the systems of linear equations given in question 1 of Section 3.2.

2 Use Gaussian elimination to solve the systems of linear equations given in question 2 of Section 3.2.

3 Use Gaussian elimination to solve

$$2x + y = 8$$
$$3x + 1.5y = 12$$

4 Use Gaussian elimination to solve

$$6x - 8y = 10$$
$$-3x + 4y = 4$$

5 Use Gaussian elimination to solve

$$x + 2y + z = 6$$
$$3x - 4y + 2z = 1$$
$$7x - 6y + 5z = 8$$

Solutions to exercises

1 (a) $x = 6, y = 4$ (b) $x = -2, y = 5$
(c) $x = 1, y = -3$ (d) $x = 1.5, y = -2.5$
(e) $x = 4, y = 9.5$

2 (a) $x = 3, y = -2, z = 3$
(b) $x = -2, y = 0, z = 1$
(c) $x = 1, y = -3, z = 7$
(d) $x = \frac{1}{2}, y = \frac{1}{2}, z = 2$
(e) $x = -3, y = -2, z = -1$

3 $x = 4 - \dfrac{t}{2}, y = t$

4 no solution

5 $x = \dfrac{13 - 4t}{5}, y = \dfrac{17 - t}{10}, z = t$

End of block exercises

1 State the augmented matrix of each of the following systems of equations:
(a) $3\alpha - \beta = 4$
$2\beta + 3\gamma = -7$
$\alpha - \beta + \gamma = 11$
(b) $a + b = 3$
$2a - b = 0$
$5a + 3b = 11$

2 Determine which of the following matrices are in echelon form:

(a) $\begin{pmatrix} 1 & 1 & 1 & 0 & 1 \\ 0 & 0 & 0 & 1 & 3 \\ 0 & 0 & 0 & 0 & 0 \end{pmatrix}$

(b) $\begin{pmatrix} 1 & 0 & 1 & 0 & 3 \\ 0 & 1 & 4 & 3 & 6 \\ 0 & 0 & 0 & 0 & 0 \end{pmatrix}$

(c) $\begin{pmatrix} 1 & 3 & 1 & 0 & -4 \\ 0 & 0 & 1 & 0 & 2 \\ 0 & 0 & 0 & 1 & -1 \\ 0 & 0 & 0 & 0 & 0 \end{pmatrix}$

3 Use Gaussian elimination to solve each of the following systems of equations:
(a) $x - 2y = -15$
$y + 2x = 15$

(b) $3x + 4y = 14$
$-2x + y = -24$

4 Use Gaussian elimination to solve

$$2x - y + z = -2$$
$$x + 2y + 3z = -1$$
$$2x + 2y - z = 8$$

5 Use Gaussian elimination to solve
$$2x + y = 10$$
$$x + 0.5y = 5$$

6 Use Gaussian elimination to solve
$$3x + y = -3$$
$$4x + 17y = -4$$

7 Use Gaussian elimination to solve
$$2x - 3y = 12$$
$$6x - 9y = 24$$

8 Use Gaussian elimination to solve
$$2x + y - z = -3$$
$$3x + 2y + z = 6$$
$$x + y + 2z = 8$$

9 Use Gaussian elimination to solve
$$2x + y - z = -3$$
$$3x + 2y + z = 6$$
$$x + y + 2z = 9$$

Solutions to exercises

1 (a) $\begin{pmatrix} 3 & -1 & 0 & 4 \\ 0 & 2 & 3 & -7 \\ 1 & -1 & 1 & 11 \end{pmatrix}$

(b) $\begin{pmatrix} 1 & 1 & 3 \\ 2 & -1 & 0 \\ 5 & 3 & 11 \end{pmatrix}$

2 (a) echelon form (b) echelon form
(c) echelon form

3 (a) $x = 3, y = 9$ (b) $x = 10, y = -4$

4 $x = 1, y = 2, z = -2$

5 $x = 5 - \dfrac{t}{2}, y = t$

6 $x = -1, y = 0$

7 no solution

8 no solution

9 $x = 3t - 12, y = 21 - 5t, z = t$

Eigenvalues and eigenvectors

In this block we consider the meaning and calculation of eigenvalues and eigenvectors. Although such calculations may at first sight appear abstract, eigenvalues have important applications in the area of vibration analysis. Familiarity with the evaluation of determinants and the solution of simultaneous equations by Gaussian elimination is essential.

4.2 Trivial and non-trivial solutions

Before we can discuss eigenvalues some preliminary results about determinants are required. These are now developed.

Consider the simultaneous equations

$$ax + by = 0$$
$$cx + dy = 0$$

where a, b, c and d are constants. Clearly $x = 0$, $y = 0$ is a solution of these equations. We call this the **trivial solution**. Let us explore the relationship that must exist between a, b, c and d so that the system has **non-trivial solutions**. These are solutions other than $x = 0$, $y = 0$. For definiteness we consider two cases with values of a, b, c and d given.

Case 1:

$$5x - 3y = 0$$
$$10x - 2y = 0$$

Case 2:

$$5x - 3y = 0$$
$$10x - 6y = 0$$

Solving Case 1, for example, by Gaussian elimination leads to $x = 0$, $y = 0$ as the only possible solution. Thus the only solution is the trivial solution.

Looking at Case 2 we see that the second equation is simply twice the first equation. This means that the second equation can be deduced from the first equation. In essence, there is only one equation to work with. So we consider the solution of

$$5x - 3y = 0$$

Rearranging we have

$$y = \frac{5}{3}x$$

As long as the y value is $\frac{5}{3}$ of the x value the equation is satisfied. Hence, for example, $x = 1$, $y = \frac{5}{3}$; $x = 2$, $y = \frac{10}{3}$; $x = -5$, $y = -\frac{25}{3}$ are all solutions. In general, the solutions have the form $x = t$, $y = \frac{5}{3}t$ for any value of t. Thus there are an infinite number of non-trivial solutions.

We now return to the system

$$ax + by = 0$$
$$cx + dy = 0$$

As seen, depending upon the values of a, b, c and d the system has either only the trivial solution or an infinity of non-trivial solutions. For there to be an infinity of solutions the second equation must be a multiple of the first. When this is the case then c is a multiple of a and d is the same multiple of b: that is,

$$c = \alpha a, d = \alpha b \text{ for some value of } \alpha$$

In this case, consider the quantity $ad - bc$:

$$ad - bc = a(\alpha b) - b(\alpha a)$$
$$= \alpha ab - \alpha ab$$
$$= 0$$

Hence the condition for non-trivial solutions to exist is that $ad - bc = 0$. Writing the system in matrix form gives

$$\begin{pmatrix} a & b \\ c & d \end{pmatrix} \begin{pmatrix} x \\ y \end{pmatrix} = \begin{pmatrix} 0 \\ 0 \end{pmatrix}$$

or

$$AX = \mathbf{0}$$

where

$$A = \begin{pmatrix} a & b \\ c & d \end{pmatrix}, \quad X = \begin{pmatrix} x \\ y \end{pmatrix} \quad \text{and} \quad \mathbf{0} = \begin{pmatrix} 0 \\ 0 \end{pmatrix}$$

We note that $ad - bc$ is the determinant of A, so non-trivial solutions exist when the determinant of A is zero, that is when A is a singular matrix.

In summary

Key points

Consider the system

$$AX = \mathbf{0}$$

- If $|A| = 0$ the system has non-trivial solutions.
- If $|A| \neq 0$ the system has only the trivial solution.

This key point holds true for a square matrix of any size.

Example 4.1
Decide which system has non-trivial solutions.

(a) $4x - y = 0$
$2x - 3y = 0$

(b) $x + 2y = 0$
$3x + 6y = 0$

Solution

(a) We write the system as

$$\begin{pmatrix} 4 & -1 \\ 2 & -3 \end{pmatrix} \begin{pmatrix} x \\ y \end{pmatrix} = \begin{pmatrix} 0 \\ 0 \end{pmatrix}$$

Let

$$A = \begin{pmatrix} 4 & -1 \\ 2 & -3 \end{pmatrix}$$

Then

$$\det(A) = 4(-3) - (-1)(2) = -10$$

Since the determinant of A is non-zero the system has only the trivial solution.

(b) We write the system as

$$\begin{pmatrix} 1 & 2 \\ 3 & 6 \end{pmatrix} \begin{pmatrix} x \\ y \end{pmatrix} = \begin{pmatrix} 0 \\ 0 \end{pmatrix}$$

Then

$$A = \begin{pmatrix} 1 & 2 \\ 3 & 6 \end{pmatrix}$$

and $|A| = 6 - 6 = 0$. Since the determinant of A is zero the system has non-trivial solutions.

Example 4.2
Determine which system has non-trivial solutions.

(a) $3x - y + z = 0$
$x + 2y + 2z = 0$
$4x + y + 3z = 0$

(b) $3x - y + z = 0$
$x + 2y + 2z = 0$
$5x - y + 3z = 0$

Solution

(a) We have

$$AX = \mathbf{0}$$

where

$$A = \begin{pmatrix} 3 & -1 & 1 \\ 1 & 2 & 2 \\ 4 & 1 & 3 \end{pmatrix} \quad \text{and} \quad X = \begin{pmatrix} x \\ y \\ z \end{pmatrix}$$

Evaluate the determinant of A:

$$|A| = 0$$

Since $|A| = 0$ the system has non-trivial solutions.

(b) Here

$$A = \begin{pmatrix} 3 & -1 & 1 \\ 1 & 2 & 2 \\ 5 & -1 & 3 \end{pmatrix}$$

Evaluate the determinant of A:

$$|A| = 6$$

Since $|A| \neq 0$ the system has only the trivial solution.

Exercises

1 Explain what is meant by the trivial solution of a system of linear equations and what is meant by a non-trivial solution.

2 Determine which of the following systems have non-trivial solutions:
(a) $x - 2y = 0$
$\quad 3x - 6y = 0$
(b) $3x + y = 0$
$\quad 9x + 2y = 0$
(c) $4x - 3y = 0$
$\quad -4x + 3y = 0$
(d) $6x - 2y = 0$
$\quad 2x - \frac{2}{3}y = 0$
(e) $y = 2x$
$\quad x = 3y$

3 Determine which of the following systems have non-trivial solutions:
(a) $x + 2y - z = 0$
$\quad 3x + y + 2z = 0$
$\quad x + y = 0$
(b) $2x - 3y - 2z = 0$
$\quad 3x + y - 3z = 0$
$\quad x - 7y - z = 0$
(c) $x + 2y + 3z = 0$
$\quad 4x - 3y - z = 0$
$\quad 6x + y + 3z = 0$
(d) $x + 3z = 0$
$\quad x - y = 0$
$\quad y + 2z = 0$

Solutions to exercises

2 (a), (c) and (d) have non-trivial solutions.

3 (a) and (b) have non-trivial solutions.

4.3 Eigenvalues

We shall explain the meaning of the term eigenvalue by means of an example.
Consider the system

$$2x + y = \lambda x$$
$$3x + 4y = \lambda y$$

where λ is some unknown constant. Clearly these equations have the trivial solution $x = 0, y = 0$. The equations may be written in matrix form as

$$\begin{pmatrix} 2 & 1 \\ 3 & 4 \end{pmatrix} \begin{pmatrix} x \\ y \end{pmatrix} = \lambda \begin{pmatrix} x \\ y \end{pmatrix}$$

or, using the usual notation,

$$AX = \lambda X$$

We now seek values of λ so that the system has non-trivial solutions. Although it is tempting to write $(A - \lambda)X = \mathbf{0}$ this would be incorrect since $A - \lambda$ is not defined; A is a matrix and λ is a constant. Hence to progress we need to write the right-hand side in a slightly different way. To help us do this we use the 2×2 identity matrix, I.

Now $\lambda \begin{pmatrix} x \\ y \end{pmatrix}$ may be expressed as

$$\lambda \begin{pmatrix} 1 & 0 \\ 0 & 1 \end{pmatrix} \begin{pmatrix} x \\ y \end{pmatrix}$$

since multiplying $\begin{pmatrix} x \\ y \end{pmatrix}$ by the identity matrix leaves it unaltered. So λX may be written as λIX. Hence we have

$$AX = \lambda IX$$

which can be written as

$$AX - \lambda IX = \mathbf{0}$$
$$(A - \lambda I)X = \mathbf{0}$$

Note that the expression $(A - \lambda I)$ is defined since both A and λI are square matrices of the same size.

We have seen in Section 4.2 that for $AX = \mathbf{0}$ to have non-trivial solutions then $|A| = 0$. Hence for

$$(A - \lambda I)X = \mathbf{0}$$

to have non-trivial solutions then

$$|A - \lambda I| = 0$$

Now

$$A - \lambda I = \begin{pmatrix} 2 & 1 \\ 3 & 4 \end{pmatrix} - \lambda \begin{pmatrix} 1 & 0 \\ 0 & 1 \end{pmatrix}$$

$$= \begin{pmatrix} 2 & 1 \\ 3 & 4 \end{pmatrix} - \begin{pmatrix} \lambda & 0 \\ 0 & \lambda \end{pmatrix}$$

$$= \begin{pmatrix} 2 - \lambda & 1 \\ 3 & 4 - \lambda \end{pmatrix}$$

So the condition $|A - \lambda I| = 0$ gives

$$\begin{vmatrix} 2 - \lambda & 1 \\ 3 & 4 - \lambda \end{vmatrix} = 0$$

It follows that

$$(2 - \lambda)(4 - \lambda) - 3 = 0$$
$$\lambda^2 - 6\lambda + 5 = 0$$
$$(\lambda - 1)(\lambda - 5) = 0$$

so that

$$\lambda = 1 \text{ or } 5$$

These are the values of λ that cause the system $AX = \lambda X$ to have non-trivial solutions. They are called **eigenvalues**.

The equation

$$|A - \lambda I| = 0$$

which when written out explicitly is the quadratic equation in λ, is called the **characteristic equation**.

Example 4.3
Find values of λ for which

$$x + 4y = \lambda x$$
$$2x + 3y = \lambda y$$

has non-trivial solutions.

Solution
We write the system as

$$AX = \lambda X$$

where

$$A = \begin{pmatrix} 1 & 4 \\ 2 & 3 \end{pmatrix}, \quad X = \begin{pmatrix} x \\ y \end{pmatrix}$$

To have non-trivial solutions we require

$$|A - \lambda I| = 0$$

Now

$$A - \lambda I = \begin{pmatrix} 1 & 4 \\ 2 & 3 \end{pmatrix} - \lambda \begin{pmatrix} 1 & 0 \\ 0 & 1 \end{pmatrix}$$

$$= \begin{pmatrix} 1 & 4 \\ 2 & 3 \end{pmatrix} - \begin{pmatrix} \lambda & 0 \\ 0 & \lambda \end{pmatrix}$$

$$= \begin{pmatrix} 1 - \lambda & 4 \\ 2 & 3 - \lambda \end{pmatrix}$$

Hence

$$|A - \lambda I| = (1 - \lambda)(3 - \lambda) - 8$$
$$= \lambda^2 - 4\lambda - 5$$

To have non-trivial solutions we require

$$\lambda^2 - 4\lambda - 5 = 0$$
$$(\lambda + 1)(\lambda - 5) = 0$$

which yields

$$\lambda = -1 \quad \text{or} \quad \lambda = 5$$

The given system has non-trivial solutions when $\lambda = -1$ and $\lambda = 5$. These are the eigenvalues.

If A is a 2×2 matrix, the characteristic equation will be a polynomial of degree 2, that is a quadratic equation in λ, leading to two eigenvalues. If A is a 3×3 matrix, the characteristic equation will be a polynomial of degree 3, that is a cubic, leading to three eigenvalues. In general an $n \times n$ matrix gives rise to a characteristic equation of degree n and hence to n eigenvalues.

Key point

The **characteristic equation** of a square matrix A is given by
$$|A - \lambda I| = 0$$

Solutions of this equation are the **eigenvalues** of A. These are the values of λ for which $AX = \lambda X$ has non-trivial solutions.

Example 4.4
Determine the characteristic equation and eigenvalues, λ, in the system

$$\begin{pmatrix} 3 & 1 \\ -1 & 5 \end{pmatrix}\begin{pmatrix} x \\ y \end{pmatrix} = \lambda \begin{pmatrix} x \\ y \end{pmatrix}$$

Solution
In this example the equations have been written in matrix form with

$$A = \begin{pmatrix} 3 & 1 \\ -1 & 5 \end{pmatrix}$$

The characteristic equation is given by

$$|A - \lambda I| = 0$$

$$\left| \begin{pmatrix} 3 & 1 \\ -1 & 5 \end{pmatrix} - \lambda \begin{pmatrix} 1 & 0 \\ 0 & 1 \end{pmatrix} \right| = 0$$

$$\begin{vmatrix} 3 - \lambda & 1 \\ -1 & 5 - \lambda \end{vmatrix} = 0$$

$$(3 - \lambda)(5 - \lambda) + 1 = 0$$

$$\lambda^2 - 8\lambda + 16 = 0$$

The characteristic equation is $\lambda^2 - 8\lambda + 16 = 0$. Solving the characteristic equation gives

$$\lambda^2 - 8\lambda + 16 = 0$$
$$(\lambda - 4)(\lambda - 4) = 0$$
$$\lambda = 4 \text{ (twice)}$$

There is one repeated eigenvalue: $\lambda = 4$.

Example 4.5
Find the eigenvalues λ in the system

$$\begin{pmatrix} 4 & 1 \\ 3 & 2 \end{pmatrix} \begin{pmatrix} x \\ y \end{pmatrix} = \lambda \begin{pmatrix} x \\ y \end{pmatrix}$$

Solution
We form the characteristic equation, $|A - \lambda I| = 0$. Now

$$A - \lambda I = \qquad \qquad \begin{pmatrix} 4 - \lambda & 1 \\ 3 & 2 - \lambda \end{pmatrix}$$

Then

$$|A - \lambda I| = \qquad \qquad \lambda^2 - 6\lambda + 5$$

Solving the characteristic equation $\lambda^2 - 6\lambda + 5 = 0$ gives

$$\lambda = \qquad \qquad 1, 5$$

There are two eigenvalues, $\lambda = 1$ and $\lambda = 5$.

The process of finding the characteristic equation and eigenvalues of a matrix has been illustrated using 2×2 matrices. This same process can be applied to a square matrix of any size.

Example 4.6
Find (a) the characteristic equation and (b) the eigenvalues of A where

$$A = \begin{pmatrix} 1 & 2 & 0 \\ -1 & -1 & 1 \\ 3 & 2 & -2 \end{pmatrix}$$

Solution

(a) We need to calculate $|A - \lambda I|$. Now

$$A - \lambda I = \begin{pmatrix} 1 - \lambda & 2 & 0 \\ -1 & -1 - \lambda & 1 \\ 3 & 2 & -2 - \lambda \end{pmatrix}$$

and

$$\begin{vmatrix} 1 - \lambda & 2 & 0 \\ -1 & -1 - \lambda & 1 \\ 3 & 2 & -2 - \lambda \end{vmatrix} = (1 - \lambda) \begin{vmatrix} -1 - \lambda & 1 \\ 2 & -2 - \lambda \end{vmatrix} - 2 \begin{vmatrix} -1 & 1 \\ 3 & -2 - \lambda \end{vmatrix}$$

$$= (1 - \lambda)[(-1 - \lambda)(-2 - \lambda) - 2] - 2[-1(-2 - \lambda) - 3]$$

Upon simplification this reduces to $-\lambda^3 - 2\lambda^2 + \lambda + 2$. Hence

$$|A - \lambda I| = 0$$

yields

$$-\lambda^3 - 2\lambda^2 + \lambda + 2 = 0$$

which may be written as

$$\lambda^3 + 2\lambda^2 - \lambda - 2 = 0$$

The characteristic equation is $\lambda^3 + 2\lambda^2 - \lambda - 2 = 0$.

(b) We solve the characteristic equation to find the eigenvalues. This cubic can be factorised using the technique described in Chapter 7 Block 3, to give

$$(\lambda + 2)(\lambda + 1)(\lambda - 1)$$

and so the characteristic equation may be written as

$$(\lambda + 2)(\lambda + 1)(\lambda - 1) = 0$$

from which

$$\lambda = -2, -1, 1$$

The eigenvalues are $\lambda = -2, -1, 1$.

Exercises

1. Calculate (i) the characteristic equation and (ii) the eigenvalues of the system $AX = \lambda X$ where A is given by

(a) $\begin{pmatrix} 5 & 6 \\ 2 & 1 \end{pmatrix}$ (b) $\begin{pmatrix} -3 & 4 \\ -4 & 5 \end{pmatrix}$

(c) $\begin{pmatrix} 7 & -2 \\ 1 & 4 \end{pmatrix}$ (d) $\begin{pmatrix} 1 & 3 \\ 4 & -1 \end{pmatrix}$

2. Calculate (i) the characteristic equation and (ii) the eigenvalues of the following 3×3 matrices:

(a) $\begin{pmatrix} 1 & -1 & 2 \\ -3 & -2 & 3 \\ 2 & -1 & 1 \end{pmatrix}$

(b) $\begin{pmatrix} 1 & 0 & -1 \\ 3 & 1 & 4 \\ 0 & 2 & 2 \end{pmatrix}$

(d) $\begin{pmatrix} -2 & 6 & 2 \\ 0 & 3 & 4 \\ 3 & -3 & 5 \end{pmatrix}$

(c) $\begin{pmatrix} 2 & 1 & 2 \\ -1 & 1 & -1 \\ 8 & 3 & 0 \end{pmatrix}$

(e) $\begin{pmatrix} 3 & -2 & 1 \\ 2 & -4 & 3 \\ 16 & -4 & 1 \end{pmatrix}$

Solutions to exercises

1   (a) (i) $\lambda^2 - 6\lambda - 7 = 0$ (ii) $\lambda = -1, 7$
     (b) (i) $\lambda^2 - 2\lambda + 1 = 0$ (ii) $\lambda = 1$ (twice)
     (c) (i) $\lambda^2 - 11\lambda + 30 = 0$ (ii) $\lambda = 5, 6$
     (d) (i) $\lambda^2 - 13 = 0$ (ii) $\lambda = -\sqrt{13}, \sqrt{13}$

     (b) (i) $\lambda^3 - 4\lambda^2 - 3\lambda + 12 = 0$
       (ii) $\lambda = -\sqrt{3}, \sqrt{3}, 4$
     (c) (i) $\lambda^3 - 3\lambda^2 - 10\lambda + 24 = 0$
       (ii) $\lambda = -3, 2, 4$
     (d) (i) $\lambda^3 - 6\lambda^2 + 5\lambda = 0$ (ii) $\lambda = 0, 1, 5$
     (e) (i) $\lambda^3 - 13\lambda + 12 = 0$ (ii) $\lambda = -4, 1, 3$

2   (a) (i) $-\lambda^3 + 7\lambda + 6 = 0$
     (ii) $\lambda = -2, -1, 3$

~ 4.4   Eigenvectors

We have studied the system

$$AX = \lambda X \tag{1}$$

and determined the values of λ for which non-trivial solutions exist. These values of λ are called eigenvalues of the system or, more simply, eigenvalues of A. For each eigenvalue there is a non-trivial solution of the system. This solution is called an **eigenvector**.

Example 4.7
Find the eigenvectors of

$$AX = \lambda X$$

where

$$A = \begin{pmatrix} 4 & 1 \\ 3 & 2 \end{pmatrix} \quad \text{and} \quad X = \begin{pmatrix} x \\ y \end{pmatrix}$$

Solution
We seek solutions of $AX = \lambda X$ that may be written as

$$(A - \lambda I) X = \mathbf{0}$$

The eigenvalues were found in Example 4.5 to be $\lambda = 1, 5$.

First we consider $\lambda = 1$. The system equation becomes

$$(A - \lambda I) X = \mathbf{0}$$
$$(A - I) X = \mathbf{0}$$
$$\left[\begin{pmatrix} 4 & 1 \\ 3 & 2 \end{pmatrix} - \begin{pmatrix} 1 & 0 \\ 0 & 1 \end{pmatrix}\right]\begin{pmatrix} x \\ y \end{pmatrix} = \begin{pmatrix} 0 \\ 0 \end{pmatrix}$$
$$\begin{pmatrix} 3 & 1 \\ 3 & 1 \end{pmatrix}\begin{pmatrix} x \\ y \end{pmatrix} = \begin{pmatrix} 0 \\ 0 \end{pmatrix}$$

Written as individual equations we have

$$3x + y = 0$$
$$3x + y = 0$$

Clearly there is only one equation, which is repeated. As long as $y = -3x$ the equation is satisfied. Thus there are an infinite number of solutions such as $x = 1$, $y = -3$; $x = -5$, $y = 15$, and so on. Generally we write

$$x = t, y = -3t$$

for any number t. Thus the eigenvector corresponding to $\lambda = 1$ is

$$X = \begin{pmatrix} x \\ y \end{pmatrix}$$
$$= \begin{pmatrix} t \\ -3t \end{pmatrix}$$
$$= t\begin{pmatrix} 1 \\ -3 \end{pmatrix}$$

Note that the eigenvector has been determined to within an arbitrary scalar, t. Thus there is an infinity of solutions corresponding to $\lambda = 1$.

We now consider $\lambda = 5$ and seek solutions of the system equation

$$(A - \lambda I) X = \mathbf{0}$$
$$\left[\begin{pmatrix} 4 & 1 \\ 3 & 2 \end{pmatrix} - 5\begin{pmatrix} 1 & 0 \\ 0 & 1 \end{pmatrix}\right]\begin{pmatrix} x \\ y \end{pmatrix} = \begin{pmatrix} 0 \\ 0 \end{pmatrix}$$
$$\begin{pmatrix} -1 & 1 \\ 3 & -3 \end{pmatrix}\begin{pmatrix} x \\ y \end{pmatrix} = \begin{pmatrix} 0 \\ 0 \end{pmatrix}$$

Written as individual equations we have

$$-x + y = 0$$
$$3x - 3y = 0$$

We note that the second equation is simply a multiple of the first so that in essence there is only one equation. Solving $-x + y = 0$ gives $y = x$ for any x. So we write $x = t, y = t$. Hence the eigenvector corresponding to $\lambda = 5$ is

$$X = t\begin{pmatrix} 1 \\ 1 \end{pmatrix}$$

Again the eigenvector has been determined to within an arbitrary scaling constant.

Sometimes the arbitrary scaling constants are not written down; it is understood that they are there. In such a case we say the eigenvectors of the system are

$$X = \begin{pmatrix} 1 \\ -3 \end{pmatrix} \quad \text{and} \quad X = \begin{pmatrix} 1 \\ 1 \end{pmatrix}$$

Example 4.8

Determine the eigenvectors of

$$\begin{pmatrix} 3 & 1 \\ -1 & 5 \end{pmatrix} \begin{pmatrix} x \\ y \end{pmatrix} = \lambda \begin{pmatrix} x \\ y \end{pmatrix}$$

Solution

In Example 4.4 we found that there is only one eigenvalue, $\lambda = 4$. We seek the solution of $(A - \lambda I) X = 0$. Write down the equations that result from letting $\lambda = 4$, firstly in matrix form and then as a single equation.

$$\begin{pmatrix} -1 & 1 \\ -1 & 1 \end{pmatrix} \begin{pmatrix} x \\ y \end{pmatrix} = \begin{pmatrix} 0 \\ 0 \end{pmatrix}$$

$$-x + y = 0$$

Find the infinite number of solutions.

$$x = t, y = t$$

Hence there is one eigenvector:

$$X = t \begin{pmatrix} 1 \\ 1 \end{pmatrix}$$

The concept of eigenvectors is easily extended to matrices of higher order.

Example 4.9

Determine the eigenvectors of

$$\begin{pmatrix} 1 & 2 & 0 \\ -1 & -1 & 1 \\ 3 & 2 & -2 \end{pmatrix} \begin{pmatrix} x \\ y \\ z \end{pmatrix} = \lambda \begin{pmatrix} x \\ y \\ z \end{pmatrix}$$

The eigenvalues were found in Example 4.6.

Solution

From Example 4.6 the eigenvalues are $\lambda = -2, -1, 1$. We consider each eigenvalue in turn.

$\underline{\lambda = -2}$

$$\begin{pmatrix} 1 & 2 & 0 \\ -1 & -1 & 1 \\ 3 & 2 & -2 \end{pmatrix} \begin{pmatrix} x \\ y \\ z \end{pmatrix} = -2 \begin{pmatrix} x \\ y \\ z \end{pmatrix}$$

$$\left[\begin{pmatrix} 1 & 2 & 0 \\ -1 & -1 & 1 \\ 3 & 2 & -2 \end{pmatrix} + 2\begin{pmatrix} 1 & 0 & 0 \\ 0 & 1 & 0 \\ 0 & 0 & 1 \end{pmatrix}\right]\begin{pmatrix} x \\ y \\ z \end{pmatrix} = \begin{pmatrix} 0 \\ 0 \\ 0 \end{pmatrix}$$

$$\begin{pmatrix} 3 & 2 & 0 \\ -1 & 1 & 1 \\ 3 & 2 & 0 \end{pmatrix}\begin{pmatrix} x \\ y \\ z \end{pmatrix} = \begin{pmatrix} 0 \\ 0 \\ 0 \end{pmatrix}$$

We note that the first and last rows are identical. So we have

$$3x + 2y = 0$$
$$-x + y + z = 0$$

Solving these equations gives

$$x = t, \quad y = -\tfrac{3}{2}t, \quad z = \tfrac{5}{2}t$$

Hence the corresponding eigenvector is

$$X = t\begin{pmatrix} 1 \\ -\dfrac{3}{2} \\ \dfrac{5}{2} \end{pmatrix}$$

$\lambda = -1$

We have

$$\left[\begin{pmatrix} 1 & 2 & 0 \\ -1 & -1 & 1 \\ 3 & 2 & -2 \end{pmatrix} + \begin{pmatrix} 1 & 0 & 0 \\ 0 & 1 & 0 \\ 0 & 0 & 1 \end{pmatrix}\right]\begin{pmatrix} x \\ y \\ z \end{pmatrix} = \begin{pmatrix} 0 \\ 0 \\ 0 \end{pmatrix}$$

$$\begin{pmatrix} 2 & 2 & 0 \\ -1 & 0 & 1 \\ 3 & 2 & -1 \end{pmatrix}\begin{pmatrix} x \\ y \\ z \end{pmatrix} = \begin{pmatrix} 0 \\ 0 \\ 0 \end{pmatrix}$$

Thus we have

$$2x + 2y = 0$$
$$-x + z = 0$$
$$3x + 2y - z = 0$$

We note that the third equation can be derived from the first two equations: the first equation minus the second produces the third. If you can't spot this, the equations should be solved by Gaussian elimination. In effect we have only two equations:

$$2x + 2y = 0$$
$$-x + z = 0$$

Solving these gives $x = t, y = -t, z = t$. The eigenvector is

$$X = t\begin{pmatrix} 1 \\ -1 \\ 1 \end{pmatrix}$$

$\underline{\lambda = 1}$

We have

$$\left[\begin{pmatrix} 1 & 2 & 0 \\ -1 & -1 & 1 \\ 3 & 2 & -2 \end{pmatrix} - \begin{pmatrix} 1 & 0 & 0 \\ 0 & 1 & 0 \\ 0 & 0 & 1 \end{pmatrix}\right]\begin{pmatrix} x \\ y \\ z \end{pmatrix} = \begin{pmatrix} 0 \\ 0 \\ 0 \end{pmatrix}$$

$$\begin{pmatrix} 0 & 2 & 0 \\ -1 & -2 & 1 \\ 3 & 2 & -3 \end{pmatrix}\begin{pmatrix} x \\ y \\ z \end{pmatrix} = \begin{pmatrix} 0 \\ 0 \\ 0 \end{pmatrix}$$

Thus we have

$$2y = 0$$
$$-x - 2y + z = 0$$
$$3x + 2y - 3z = 0$$

From the first equation, $y = 0$; putting $y = 0$ into the other equations yields

$$-x + z = 0$$
$$3x - 3z = 0$$

Here the second equation can be derived from the first by multiplying the first by -3. Solving, we have $x = t, z = t$. So the eigenvector is

$$X = t\begin{pmatrix} 1 \\ 0 \\ 1 \end{pmatrix}$$

Exercises

1   Calculate the eigenvectors of the matrices given in question 1 of the previous section.

2   Calculate the eigenvectors of the matrices given in question 2 of the previous section.

Solutions to exercises

1   (a) $t\begin{pmatrix} 1 \\ -1 \end{pmatrix}, t\begin{pmatrix} 1 \\ \frac{1}{3} \end{pmatrix}$ (b) $t\begin{pmatrix} 1 \\ 1 \end{pmatrix}$

(c) $t\begin{pmatrix} 1 \\ 1 \end{pmatrix}, t\begin{pmatrix} 1 \\ \frac{1}{2} \end{pmatrix}$

(d) $t\begin{pmatrix} 1 \\ -\frac{1 + \sqrt{13}}{3} \end{pmatrix}, t\begin{pmatrix} 1 \\ \frac{\sqrt{13} - 1}{3} \end{pmatrix}$

2   (a) $t\begin{pmatrix} 1 \\ 5 \\ 1 \end{pmatrix}, t\begin{pmatrix} 1 \\ 12 \\ 5 \end{pmatrix}, t\begin{pmatrix} 1 \\ 0 \\ 1 \end{pmatrix}$

(b) $t\begin{pmatrix} 1 \\ -5.0981 \\ 2.7321 \end{pmatrix}, t\begin{pmatrix} 1 \\ 0.0981 \\ -0.7321 \end{pmatrix}, t\begin{pmatrix} 1 \\ -3 \\ -3 \end{pmatrix}$

(c) $t\begin{pmatrix} 1 \\ -\frac{1}{3} \\ -\frac{7}{3} \end{pmatrix}, t\begin{pmatrix} 1 \\ -2 \\ 1 \end{pmatrix}, t\begin{pmatrix} 1 \\ -0.8 \\ 1.4 \end{pmatrix}$

(d) $t\begin{pmatrix} 1 \\ \frac{4}{9} \\ -\frac{1}{3} \end{pmatrix}, t\begin{pmatrix} 1 \\ 0.6 \\ -0.3 \end{pmatrix}, t\begin{pmatrix} 1 \\ 1 \\ 0.5 \end{pmatrix}$

(e) $t\begin{pmatrix} 1 \\ \frac{19}{6} \\ -\frac{2}{3} \end{pmatrix}, t\begin{pmatrix} 1 \\ 4 \\ 6 \end{pmatrix}, t\begin{pmatrix} 1 \\ 2 \\ 4 \end{pmatrix}$

End of block exercises

1 Determine which of the following systems have non-trivial solutions:
(a) $2x - y = 0$
$3x - 1.5y = 0$
(b) $6x + 5y = 0$
$5x + 6y = 0$
(c) $-x - 4y = 0$
$2x + 8y = 0$
(d) $7x - 3y = 0$
$1.4x - 0.6y = 0$
(e) $-4x + 5y = 0$
$3x - 4y = 0$

2 Determine which of the following systems have non-trivial solutions:
(a) $3x - 2y + 2z = 0$
$x - y + z = 0$
$2x + 2y - z = 0$
(b) $x + 3y - z = 0$
$4x - y + 2z = 0$
$6x + 5y = 0$
(c) $x + 2y - z = 0$
$x - 3z = 0$
$5x + 6y - 9z = 0$

3 The matrix A is defined by

$$A = \begin{pmatrix} 3 & 2 \\ -3 & -4 \end{pmatrix}$$

(a) Determine the characteristic equation of A.
(b) Determine the eigenvalues of A.
(c) Determine the eigenvectors of A.

(d) Form a new matrix M whose columns are the two eigenvectors of A. M is called a **modal matrix**.
(e) Show that $M^{-1}AM$ is a diagonal matrix, D, with the eigenvalues of A on its leading diagonal. D is called the **spectral matrix** corresponding to the modal matrix M.

4 (a) Show that the matrix

$$A = \begin{pmatrix} 5 & 2 \\ -2 & 1 \end{pmatrix}$$

has only one eigenvalue and determine it.
(b) Calculate the eigenvector of A.

5 The matrix H is given by

$$H = \begin{pmatrix} 4 & -1 & 1 \\ -2 & 4 & 0 \\ -4 & 3 & 1 \end{pmatrix}$$

(a) Find the eigenvalues of H.
(b) Determine the eigenvectors of H.
(c) Form a new matrix M whose columns are the three eigenvectors of H. M is called a **modal matrix**.
(d) Show that $M^{-1}HM$ is a diagonal matrix, D, with the eigenvalues of H on its leading diagonal. D is called the **spectral matrix** corresponding to the modal matrix M.

Solutions to exercises

1   (a), (c) and (d) have non-trivial solutions.

2   (b) and (c) have non-trivial solutions.

3   (a) $\lambda^2 + \lambda - 6 = 0$
      (b) $\lambda = -3, 2$
      (c) $\begin{pmatrix} 1 \\ -3 \end{pmatrix} \begin{pmatrix} 1 \\ -0.5 \end{pmatrix}$

4   (a) $\lambda = 3$
      (b) $\begin{pmatrix} 1 \\ -1 \end{pmatrix}$

5   (a) $\lambda = 2, 3, 4$
      (b) $\begin{pmatrix} 1 \\ 1 \\ -1 \end{pmatrix}, \begin{pmatrix} 1 \\ 2 \\ 1 \end{pmatrix}, \begin{pmatrix} 0 \\ 1 \\ 1 \end{pmatrix}$

Iterative techniques

5.1 Introduction

So far we have met a number of techniques used for solving systems of linear equations – Cramer's rule (Block 1), the inverse matrix method (Block 2) and Gaussian elimination (Block 3) are three such methods.

We now examine two techniques that provide approximate solutions to linear systems. In these, an initial guess of the solution is repeatedly improved, generating a sequence of approximate solutions. With repeated application of the method, under certain conditions, the approximate solutions approach the exact solution. Such methods are known as **iterative methods**. The two iterative methods we look at are Jacobi's method and the Gauss–Seidel method.

5.2 Jacobi's method

We illustrate the method with examples.

Example 5.1
Solve

$$3x - 2y = 9$$
$$x + 4y = -11$$

using Jacobi's iterative method.

Solution
We rewrite the equations so that x and y are the subjects. This yields

$$x = \frac{2}{3}y + 3 \tag{1}$$

$$y = -\frac{x}{4} - \frac{11}{4} \tag{2}$$

We now guess a solution. In practice an educated guess may be possible if realistic estimates of x and y are known. Suppose we guess $x = 0$, $y = -4$. To show this is our initial guess and not the exact solution we write $x_0 = 0$, $y_0 = -4$.

We improve our initial guess by substituting $x_0 = 0$, $y_0 = -4$ into the right-hand side of equations (1) and (2). The values of x and y so obtained are labelled x_1 and y_1.

Hence

$$x_1 = \frac{2}{3}y_0 + 3$$

$$= \frac{2}{3}(-4) + 3$$

$$= 0.3333$$

$$y_1 = -\frac{x_0}{4} - \frac{11}{4}$$

$$= -\frac{0}{4} - \frac{11}{4}$$

$$= -2.7500$$

At this stage we have made one application, or **iteration**, of Jacobi's method. To obtain the next approximate solution, x_2, y_2, we substitute x_1 and y_1 into the right-hand side of equations (1) and (2). This produces

$$x_2 = \frac{2}{3}y_1 + 3$$

$$= \frac{2}{3}(-2.7500) + 3$$

$$= 1.1667$$

$$y_2 = -\frac{x_1}{4} - \frac{11}{4}$$

$$= -\frac{(0.3333)}{4} - \frac{11}{4}$$

$$= -2.8333$$

At this stage two iterations have been performed. We perform another iteration to obtain x_3 and y_3.

$$x_3 = \frac{2}{3}y_2 + 3$$

$$= \frac{2}{3}(-2.8333) + 3$$

$$= 1.1111$$

$$y_3 = -\frac{x_2}{4} - \frac{11}{4}$$

$$= -\frac{(1.1667)}{4} - \frac{11}{4}$$

$$= -3.0417$$

Similarly

$$x_4 = \frac{2}{3}y_3 + 3$$

$$= \frac{2}{3}(-3.0417) + 3$$

$$= 0.9722$$

$$y_4 = -\frac{x_3}{4} - \frac{11}{4}$$
$$= -\frac{(1.1111)}{4} - \frac{11}{4}$$
$$= -3.0278$$

The iterations can be continued. Table 5.1 summarises the results of successive iterations.

Table 5.1

Iteration no. (n)	x_n	y_n
0	0.0000	−4.0000
1	0.3333	−2.7500
2	1.1667	−2.8333
3	1.1111	−3.0417
4	0.9722	−3.0278
5	0.9815	−2.9931
6	1.0046	−2.9954
7	1.0031	−3.0012
8	0.9992	−3.0008
9	0.9995	−2.9998
10	1.0001	−2.9999
11	1.0001	−3.0000
12	1.0000	−3.0000

The value of x_n seems to converge to 1 as n increases; the value of y_n seems to converge to −3 as n increases. It is easy to verify that the exact solution of the given equations is $x = 1$, $y = -3$. So Jacobi's method has produced a sequence of approximate solutions that has converged to the exact solution.

Example 5.2
Find an approximate solution to

$$4x + y = 12$$
$$-2x + 5y = 16$$

using Jacobi's method. Perform five iterations and take $x_0 = 0$, $y_0 = 3$ as your initial guess.

Solution
The equations are rearranged to make x and y the subjects.

$$x = -0.25y + 3$$

$$y = $$
$\qquad\qquad\qquad\qquad\qquad\qquad\qquad\qquad\qquad\qquad\qquad\qquad\qquad\quad 0.4x + 3.2$

The initial guess is $x_0 = 0$, $y_0 = 3$. Then

$$x_1 = -0.25y_0 + 3$$

$$= -0.25(3) + 3$$
$$= 2.25$$

$$y_1 = \boxed{} \qquad 0.4x_0 + 3.2 = 3.2$$

We now find x_2 and y_2.

$$x_2 = \boxed{}$$

$$-0.25y_1 + 3 = -0.25(3.2) + 3 = 2.2$$

$$y_2 = \boxed{}$$

$$0.4x_1 + 3.2 = 0.4(2.25) + 3.2 = 4.1$$

Continuing in this way we find

$$x_3 = \boxed{} , y_3 = \boxed{} \qquad 1.975, 4.08$$

$$x_4 = \boxed{} , y_4 = \boxed{} \qquad 1.98, 3.99$$

$$x_5 = \boxed{} , y_5 = \boxed{} \qquad 2.003, 3.99$$

If the iterations are continued then x_n approaches 2 and y_n approaches 4, the exact solution.

Clearly, this sort of approach is simple to program, and iterative techniques are best implemented on a computer. When writing a program a test is incorporated so that after each iteration a check for convergence is made by comparing successive estimates. The method may not converge, and, even if it does, convergence may be very slow. Other methods with improved rates of convergence are available such as the Gauss–Seidel method, which is covered in Section 5.3.

Exercises

1   Use five iterations of Jacobi's method to find approximate solutions of each system of equations:

(a) $3x - y = 17$
     $2x - 5y = 20$
     Take $x_0 = 3, y_0 = 0$.

(b) $2x + y = 11$
     $x - 3y = -26$
     Take $x_0 = 0, y_0 = 7$.

(c) $-3x + y = 14$
     $2x - 7y = 16$
     Take $x_0 = -4, y_0 = -2$.

(d) $4x + y = 19$
     $-3x + 7y = 40$
     Take $x_0 = 1, y_0 = 5$.

(e) $-3x + 2y = 37$
     $x + 5y = 50$
     Take $x_0 = -3, y_0 = 9$.

Solutions to exercises

1 (a)

Iteration no. (n)	x_n	y_n
0	3.0000	0.0000
1	5.6667	−2.8000
2	4.7333	−1.7333
3	5.0889	−2.1067
4	4.9644	−1.9644
5	5.0119	−2.0142

(b)

Iteration no. (n)	x_n	y_n
0	0.0000	7.0000
1	2.0000	8.6667
2	1.1667	9.3333
3	0.8333	9.0556
4	0.9722	8.9444
5	1.0278	8.9907

(c)

Iteration no. (n)	x_n	y_n
0	−4.0000	−2.0000
1	−5.3333	−3.4286
2	−5.8095	−3.8095
3	−5.9365	−3.9456
4	−5.9819	−3.9819
5	−5.9940	−3.9948

(d)

Iteration no. (n)	x_n	y_n
0	1.0000	5.0000
1	3.5000	6.1429
2	3.2143	7.2143
3	2.9464	7.0918
4	2.9770	6.9770
5	3.0057	6.9902

(e)

Iteration no. (n)	x_n	y_n
0	−3.0000	9.0000
1	−6.3333	10.6000
2	−5.2667	11.2667
3	−4.8222	11.0533
4	−4.9644	10.9644
5	−5.0237	10.9929

5.3 Gauss–Seidel method

The Gauss–Seidel method is very similar to Jacobi's method. The difference comes in the calculation of the y values. When calculating the y values the most recent x value is used. Example 5.3 illustrates this.

Example 5.3
Use the Gauss–Seidel method to find an approximate solution to the system given in Example 5.1. Perform five iterations and take $x_0 = 0$, $y_0 = -4$.

Solution
We arrange the equations so that x and y are the subjects:

$$x = \frac{2}{3}y + 3$$

$$y = -\frac{x}{4} - 2.75$$

We make an initial guess, say $x_0 = 0$, $y_0 = -4$. Now we calculate x_1.

$$x_1 = \frac{2}{3}y_0 + 3$$

$$= \frac{2}{3}(-4) + 3$$

$$= 0.3333$$

Up to this point, the method is identical to Jacobi's method.
When calculating y_1 the most recent value of x, that is x_1, is used.

$$y_1 = -\frac{x_1}{4} - 2.75$$

$$= -\frac{(0.3333)}{4} - 2.75$$

$$= -2.8333$$

This completes the first iteration. We are now ready to calculate x_2 and y_2.

$$x_2 = \frac{2}{3}y_1 + 3$$

$$= \frac{2}{3}(-2.8333) + 3$$

$$= 1.1111$$

When calculating y_2 we use the most recent value of x, that is x_2.

$$y_2 = -\frac{x_2}{4} - 2.75$$

$$= -\frac{(1.1111)}{4} - 2.75$$

$$= -3.0278$$

This completes the second iteration. The process is repeated.

$$x_3 = \frac{2}{3}y_2 + 3$$

$$= \frac{2}{3}(-3.0278) + 3$$

$$= 0.9815$$

$$y_3 = -\frac{x_3}{4} - 2.75$$

$$= -\frac{(0.9815)}{4} - 2.75$$

$$= -2.9954$$

Table 5.2 summarises some further iterations.

Table 5.2

Iteration no. (n)	x_n	y_n
0	0.0000	−4.0000
1	0.3333	−2.8333
2	1.1111	−3.0278
3	0.9815	−2.9954
4	1.0031	−3.0008
5	0.9995	−2.9999
6	1.0001	−3.0000

Note that the Gauss–Seidel method converges more rapidly than Jacobi's method (see Table 5.1). This is because the Gauss–Seidel method uses the most recent value of x in the calculation of the y values.

Example 5.4

Use the Gauss–Seidel method to find an approximate solution of

$$4x + 3y = 0.5$$
$$x - 2y = -9.5$$

Perform five iterations taking $x_0 = 0$, $y_0 = 0$.

Solution

The equations are rearranged to make x and y the subjects.

$$x = \qquad\qquad\qquad\qquad\qquad -0.75y + 0.125$$

$$y = \qquad\qquad\qquad\qquad\qquad 0.5x + 4.75$$

The initial guess is $x_0 = 0$, $y_0 = 0$. So

$$x_1 = -0.75y_0 + 0.125$$
$$= 0.125$$

$$y_1 = \qquad\qquad\qquad\qquad 0.5x_1 + 4.75 = 4.8125$$

and

$$x_2 = \qquad\qquad\qquad\qquad -0.75y_1 + 0.125 = -3.4844$$

$$y_2 = \qquad\qquad\qquad\qquad 0.5x_2 + 4.75 = 3.0078$$

Continuing in this way we find:

$$x_3 = \qquad\qquad , y_3 = \qquad\qquad\qquad -2.1309, 3.6846$$

$$x_4 = \qquad\qquad , y_4 = \qquad\qquad\qquad -2.6384, 3.4308$$

$$x_5 = \qquad\qquad , y_5 = \qquad\qquad\qquad -2.4481, 3.5260$$

Unfortunately, as with all iterative methods, convergence is not guaranteed. However, it can be shown that, if the matrix of coefficients of the equations is **diagonally dominant** (i.e. the modulus of each diagonal element is greater than the sum of the moduli of the other elements in its row), then the Gauss–Seidel method will converge. For further details you should consult a text on numerical methods.

Exercises

1 Find approximate solutions to the linear systems in question 1 of Section 5.2 using five iterations of the Gauss–Seidel method.

Solutions to exercises

1 (a)

Iteration no. (n)	x_n	y_n
0	3.0000	0.0000
1	5.6667	−1.7333
2	5.0889	−1.9644
3	5.0119	−1.9953
4	5.0016	−1.9994
5	5.0002	−1.9999

(b)

Iteration no. (n)	x_n	y_n
0	0.0000	7.0000
1	2.0000	9.3333
2	0.8333	8.9444
3	1.0278	9.0093
4	0.9954	8.9985
5	1.0008	9.0003

(c)

Iteration no. (n)	x_n	y_n
0	−4.0000	−2.0000
1	−5.3333	−3.8095
2	−5.9365	−3.9819
3	−5.9940	−3.9983
4	−5.9994	−3.9998
5	−5.9999	−4.0000

(d)

Iteration no. (n)	x_n	y_n
0	1.0000	5.0000
1	3.5000	7.2143
2	2.9464	6.9770
3	3.0057	7.0025
4	2.9994	6.9997
5	3.0000	7.0000

(e)

Iteration no. (n)	x_n	y_n
0	−3.0000	9.0000
1	−6.3333	11.2667
2	−4.8222	10.9644
3	−5.0237	11.0047
4	−4.9668	10.9994
5	−5.0004	11.0001

End of block exercises

1 Use five iterations of Jacobi's method to find approximate solutions of the following system:

$$-4x + y = 17$$
$$2x + 5y = 19$$

Take $x_0 = -1$ and $y_0 = 3$.

2 Use five iterations of the Gauss–Seidel method to find approximate solutions to the system

$$4x - 3y = 25$$
$$2x + 5y = -7$$

Take $x_0 = 2$, $y_0 = -1$.

3 Consider the system

$$4x + y + z = 6$$
$$2x + 5y - z = 18$$
$$3x - y + 6z = -6$$

(a) Use three iterations of Jacobi's method to find an approximate solution to the given system. Take $x_0 = 0$, $y_0 = 2$, $z_0 = 0$.

(b) Repeat (a) using the Gauss–Seidel method.

Solutions to exercises

1

Iteration no. (n)	x_n	y_n
0	−1.0000	3.0000
1	−3.5000	4.2000
2	−3.2000	5.2000
3	−2.9500	5.0800
4	−2.9800	4.9800
5	−3.0050	4.9920

2

Iteration no. (n)	x_n	y_n
0	2.0000	−1.0000
1	5.5000	−3.6000
2	3.5500	−2.8200
3	4.1350	−3.0540
4	3.9595	−2.9838
5	4.0122	−3.0049

3 (a)

Iteration no. (n)	x_n	y_n	z_n
0	0.0000	2.0000	0.0000
1	1.0000	3.6000	−0.6667
2	0.7667	3.0667	−0.9000
3	0.9583	3.1133	−0.8722

(b)

Iteration no. (n)	x_n	y_n	z_n
0	0.0000	2.0000	0.0000
1	1.0000	3.2000	−0.9667
2	0.9417	3.0300	−0.9658
3	0.9840	3.0133	−0.9898

Electrical networks

6.1 Introduction

In this block we consider electrical networks comprising known resistors and known voltage sources. We wish to determine the current in various parts of the network.

The network can be modelled by a set of simultaneous equations, which may be solved by matrix methods. In order to formulate the simultaneous equations we introduce the idea of a mesh current and a branch current.

6.2 Mesh and branch currents

Figure 6.1 shows an example of an electrical network.

Figure 6.1
An electrical network with mesh currents marked.

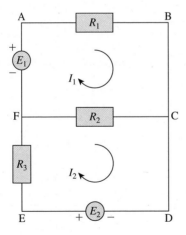

A **mesh** is a loop that does not contain any smaller loops. Figure 6.1 contains two meshes: ABCF and CDEF. Note, for example, that ABCDEF is not a mesh.

We introduce **mesh currents**, I_1 and I_2, as shown in Figure 6.1. Mesh currents are usually denoted as running clockwise although in reality they may run anticlockwise. Upon calculation this would be denoted by the current being negative.

The net current in a particular branch is called the **branch current**. The branch currents are found by combining the mesh currents. For example, the current from C to F is $I_1 - I_2$.

The aim of analysing the network is to determine the mesh currents, I_1 and I_2. Once these are known the branch currents can easily be found.

Extensive use is made of **Kirchhoff's voltage law**, which states that the algebraic sum of voltages around a mesh is zero: that is, the sum of the voltage drops equals the sum of the voltage rises.

Example 6.1 Electrical Engineering – Mesh currents

Figure 6.1 shows a circuit with two meshes, ABCF and CDEF. The values of voltages E_1, E_2 and resistances R_1, R_2 and R_3 are known. The unknowns are the values of the currents I_1 and I_2.

(a) By applying Kirchhoff's voltage law to each mesh formulate two equations in the unknowns I_1 and I_2.
(b) Given $E_1 = 3$, $E_2 = 6$, $R_1 = 1$, $R_2 = 4$ and $R_3 = 2$, find I_1 and I_2.
(c) Find the branch current running from F to C.

Solution

(a) Kirchhoff's voltage law states that the algebraic sum of voltages around a mesh is zero: that is, the sum of the voltage drops equals the sum of the voltage rises. Recall that the voltage drop, V volts, across a resistor of resistance R ohms carrying I amps is given by $V = IR$.

   Consider the mesh ABCF. Starting at A and working clockwise around the mesh we equate voltage drops to voltage rises.

$$R_1 I_1 + (I_1 - I_2)R_2 = E_1$$

This may be written as

$$I_1(R_1 + R_2) - I_2 R_2 = E_1 \tag{1}$$

Consider now the mesh CDEF. Applying Kirchhoff's voltage law we obtain

$$(I_2 - I_1)R_2 + I_2 R_3 = E_2$$

which is then written as

$$-I_1 R_2 + I_2(R_2 + R_3) = E_2 \tag{2}$$

Equations (1) and (2) are written in matrix form as

$$\begin{pmatrix} R_1 + R_2 & -R_2 \\ -R_2 & R_2 + R_3 \end{pmatrix} \begin{pmatrix} I_1 \\ I_2 \end{pmatrix} = \begin{pmatrix} E_1 \\ E_2 \end{pmatrix}$$

(b) Substituting in the values of E_1, E_2, R_1, R_2 and R_3 we get

$$\begin{pmatrix} 5 & -4 \\ -4 & 6 \end{pmatrix} \begin{pmatrix} I_1 \\ I_2 \end{pmatrix} = \begin{pmatrix} 3 \\ 6 \end{pmatrix}$$

These equations can be solved by any of the techniques described in this chapter. You should verify that $I_1 = 3$, $I_2 = 3$.

(c) Consider the branch FC. The net current in FC is $I_2 - I_1 = 0$.

Example 6.2 Electrical Engineering – Mesh currents

Determine the mesh currents I_1, I_2 and I_3 in Figure 6.2 given $E_1 = 5$, $E_2 = 15$, $R_1 = 3$, $R_2 = 5$, $R_3 = 1$, $R_4 = 6$ and $R_5 = 2$.

Figure 6.2
Electrical network
for Example 6.2.

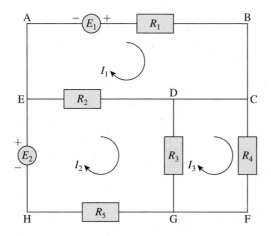

Solution

We use Kirchhoff's voltage law. Recall that the voltage drop, V, across a resistor of resistance R ohms carrying a current I amps is given by $V = IR$.

Consider the mesh ABCDE. For clarity this is shown in Figure 6.3.

Figure 6.3
Mesh ABCDE.

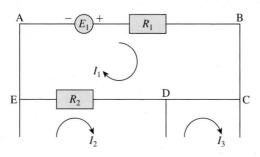

Starting from A and working clockwise around the mesh we apply Kirchhoff's voltage law, equating the voltage rises to the voltage drops. We obtain

$$I_1 R_1 + (I_1 - I_2)R_2 = E_1$$

This is rewritten as

$$I_1(R_1 + R_2) - I_2 R_2 = E_1 \qquad (3)$$

Consider the mesh EDGH, as shown in Figure 6.4.

Figure 6.4
Mesh EDGH.

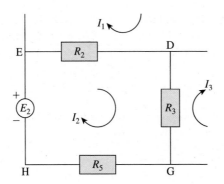

Starting at E and working clockwise we equate the voltage drops and the voltage rises to obtain

$$(I_2 - I_1)R_2 + (I_2 - I_3)R_3 + I_2R_5 = E_2$$

which is then rewritten as

$$-I_1R_2 + I_2(R_2 + R_3 + R_5) - I_3R_3 = E_2 \tag{4}$$

Finally we look at mesh DCFG, as shown in Figure 6.5.

Figure 6.5
Mesh DCFG.

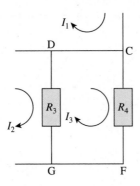

Starting at D and working clockwise we have

$$I_3R_4 + (I_3 - I_2)R_3 = 0$$

which is written as

$$-I_2R_3 + I_3(R_3 + R_4) = 0 \tag{5}$$

The equations (3), (4) and (5) are written in matrix form as

$$\begin{pmatrix} R_1 + R_2 & -R_2 & 0 \\ -R_2 & R_2 + R_3 + R_5 & -R_3 \\ 0 & -R_3 & R_3 + R_4 \end{pmatrix} \begin{pmatrix} I_1 \\ I_2 \\ I_3 \end{pmatrix} = \begin{pmatrix} E_1 \\ E_2 \\ 0 \end{pmatrix}$$

Substituting in the given values we have

$$
\begin{pmatrix}
8 & -5 & 0 \\
-5 & 8 & -1 \\
0 & -1 & 7
\end{pmatrix}
\begin{pmatrix}
I_1 \\
I_2 \\
I_3
\end{pmatrix}
=
\begin{pmatrix}
5 \\
15 \\
0
\end{pmatrix}
$$

The system may be solved using one of the methods explained in this chapter. We find

$$I_1 = 3.0189, \quad I_2 = 3.8302, \quad I_3 = 0.5472$$

Example 6.3 Electrical Engineering – Mesh currents

Figure 6.6 shows an electrical network with the mesh currents I_1, I_2, I_3 and I_4 marked in meshes 1, 2, 3 and 4 respectively. The resistances R_1, R_2, R_3, R_4 and R_5 and the voltage sources E_1, E_2, E_3 and E_4 are known.

Figure 6.6
Electrical network
for Example 6.3.

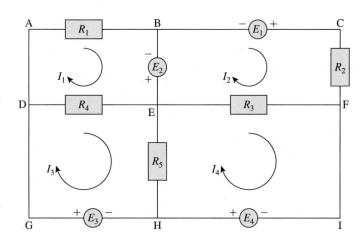

(a) Formulate a matrix equation for the unknowns I_1, I_2, I_3 and I_4.
(b) Given $R_1 = 3$, $R_2 = 4$, $R_3 = 1$, $R_4 = 7$, $R_5 = 4$, $E_1 = 6$, $E_2 = 10$, $E_3 = 5$ and $E_4 = 5$, solve for I_1, I_2, I_3 and I_4.
(c) State the value of the current from E to H.

Solution

(a) We apply Kirchhoff's voltage law to each of the meshes.
   Mesh 1 (ABED)
   We equate voltage drops to voltage rises.

$$(I_1 - I_3)R_4 + I_1 R_1 = E_2$$

which is written as

$$\boxed{} = E_2 \qquad\qquad I_1(R_1 + R_4) - I_3 R_4 \quad (6)$$

Mesh 2 (BCFE)

$$\boxed{} = E_1 \qquad\qquad I_2R_2 + (I_2 - I_4)R_3 + E_2$$

which is written as

$$I_2(R_2 + R_3) - I_4R_3 = E_1 - E_2 \qquad\qquad (7)$$

Mesh 3 (DEHG)

$$(I_3 - I_1)R_4 + (I_3 - I_4)R_5 = E_3$$

which is written as

$$-I_1R_4 + I_3(R_4 + R_5) - I_4R_5 = E_3 \qquad\qquad (8)$$

Mesh 4 (EFIH)

$$\boxed{} = E_4 \qquad\qquad (I_4 - I_3)R_5 + (I_4 - I_2)R_3$$

which is written as

$$-I_2R_3 - I_3R_5 + I_4(R_3 + R_5) = E_4 \qquad\qquad (9)$$

Equations (6), (7), (8) and (9) are written in matrix form as

$$\begin{pmatrix} R_1 + R_4 & 0 & -R_4 & 0 \\ 0 & R_2 + R_3 & 0 & -R_3 \\ -R_4 & 0 & R_4 + R_5 & -R_5 \\ 0 & -R_3 & -R_5 & R_3 + R_5 \end{pmatrix} \begin{pmatrix} I_1 \\ I_2 \\ I_3 \\ I_4 \end{pmatrix} = \begin{pmatrix} E_2 \\ E_1 - E_2 \\ E_3 \\ E_4 \end{pmatrix}$$

(b) Substituting in the given values, the matrix equation becomes

$$\begin{pmatrix} 10 & 0 & -7 & 0 \\ 0 & 5 & 0 & -1 \\ -7 & 0 & 11 & -4 \\ 0 & -1 & -4 & 5 \end{pmatrix} \begin{pmatrix} I_1 \\ I_2 \\ I_3 \\ I_4 \end{pmatrix} = \begin{pmatrix} 10 \\ -4 \\ 5 \\ 5 \end{pmatrix}$$

The system is solved to give

$$I_1 = 4.9217, \quad I_2 = 0.3087, \quad I_3 = 5.6024, \quad I_4 = 5.5437$$

(c) The current running from E to H is

$$\boxed{} \qquad\qquad I_3 - I_4$$

$$I_3 - I_4 = 5.6024 - 5.5437$$
$$= 0.0587$$

Exercises

1 Figure 6.7 shows a circuit with two meshes.

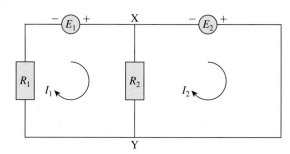

Figure 6.7

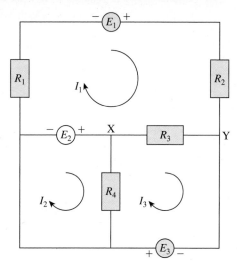

Figure 6.9

(a) Formulate a matrix equation for I_1 and I_2.
(b) Solve for I_1 and I_2 given $E_1 = 3$, $E_2 = 7$, $R_1 = 6$ and $R_2 = 10$.
(c) Calculate the current in the branch XY.

2 Figure 6.8 shows a circuit with three meshes.

(b) Calculate the mesh currents given $E_1 = 5$, $E_2 = 4$, $E_3 = 6$, $R_1 = 5$, $R_2 = 4$, $R_3 = 2$ and $R_4 = 7$.
(c) Calculate the current in the branch XY.

4 (a) Formulate a matrix equation for the mesh currents I_1, I_2, I_3 and I_4 as shown in Figure 6.10.

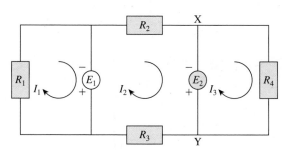

Figure 6.8

(a) Formulate a matrix equation for the mesh currents I_1, I_2 and I_3.
(b) Given $E_1 = 4$, $E_2 = 5$, $R_1 = 3$, $R_2 = 1$, $R_3 = 2$ and $R_4 = 6$, find I_1, I_2 and I_3.
(c) Calculate the current in the branch XY.

3 (a) Formulate a matrix equation for the mesh currents I_1, I_2 and I_3 as shown in Figure 6.9.

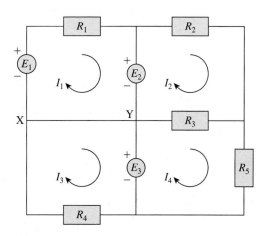

Figure 6.10

(b) Calculate the mesh currents given $E_1 = 5$, $E_2 = 2$, $E_3 = 10$, $R_1 = 6$, $R_2 = 7$, $R_3 = 10$, $R_4 = 1$ and $R_5 = 6$.

(c) Calculate the current in the branch XY.

5 Figure 6.11 shows a circuit with four meshes.

(a) Formulate a matrix equation for the mesh currents I_1, I_2, I_3 and I_4.

(b) Given $E_1 = 5$, $E_2 = 5$, $E_3 = 6$, $E_4 = 12$, $R_1 = 1$, $R_2 = 4$, $R_3 = 2$, $R_4 = 1$, $R_5 = 7$ and $R_6 = 9$, find the mesh currents.

(c) Calculate the current in the branch XY.

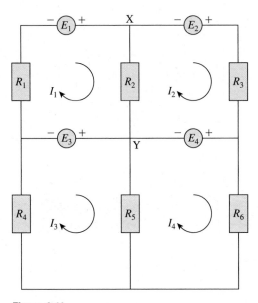

Figure 6.11

Solutions to exercises

1 (a) $\begin{pmatrix} R_1 + R_2 & -R_2 \\ -R_2 & R_2 \end{pmatrix} \begin{pmatrix} I_1 \\ I_2 \end{pmatrix}$

$= \begin{pmatrix} E_1 \\ E_2 \end{pmatrix}$

(b) $I_1 = \frac{5}{3}, I_2 = \frac{71}{30}$

(c) $-\frac{21}{30}$, that is a current of $\frac{21}{30}$ from Y to X.

2 (a) $\begin{pmatrix} R_1 & 0 & 0 \\ 0 & R_2 + R_3 & 0 \\ 0 & 0 & R_4 \end{pmatrix} \begin{pmatrix} I_1 \\ I_2 \\ I_3 \end{pmatrix}$

$= \begin{pmatrix} E_1 \\ E_2 - E_1 \\ -E_2 \end{pmatrix}$

(b) $I_1 = \frac{4}{3}, I_2 = \frac{1}{3}, I_3 = -\frac{5}{6}$

(c) $I_2 - I_3 = \frac{7}{6}$

3 (a)

$\begin{pmatrix} R_1 + R_2 + R_3 & 0 & -R_3 \\ 0 & R_4 & -R_4 \\ -R_3 & -R_4 & R_3 + R_4 \end{pmatrix} \begin{pmatrix} I_1 \\ I_2 \\ I_3 \end{pmatrix}$

$= \begin{pmatrix} E_1 - E_2 \\ E_2 \\ E_3 \end{pmatrix}$

(b) $I_1 = -1.2222, I_2 = 6.7937, I_3 = 6.2222$

(c) $I_3 - I_1 = 5$

4 (a) $$\begin{pmatrix} R_1 & 0 & 0 & 0 \\ 0 & R_2 + R_3 & 0 & -R_3 \\ 0 & 0 & R_4 & 0 \\ 0 & -R_3 & 0 & R_3 + R_5 \end{pmatrix} \begin{pmatrix} I_1 \\ I_2 \\ I_3 \\ I_4 \end{pmatrix}$$

$$= \begin{pmatrix} E_1 - E_2 \\ E_2 \\ -E_3 \\ E_3 \end{pmatrix}$$

(b) $I_1 = 0.5000$, $I_2 = 0.7674$, $I_3 = -10.0000$, $I_4 = 1.1047$

(c) $I_3 - I_1 = -10.5$, that is a current of 10.5 from Y to X.

5 (a) $$\begin{pmatrix} R_1 + R_2 & -R_2 & 0 & 0 \\ -R_2 & R_2 + R_3 & 0 & 0 \\ 0 & 0 & R_4 + R_5 & -R_5 \\ 0 & 0 & -R_5 & R_5 + R_6 \end{pmatrix} \begin{pmatrix} I_1 \\ I_2 \\ I_3 \\ I_4 \end{pmatrix}$$

$$= \begin{pmatrix} E_1 - E_3 \\ E_2 - E_4 \\ E_3 \\ E_4 \end{pmatrix}$$

(b) $I_1 = -2.4286$, $I_2 = -2.7857$, $I_3 = 2.2785$, $I_4 = 1.7468$

(c) $I_1 - I_2 = 0.3571$

End of block exercises

1 Figure 6.12 shows an electric circuit with mesh currents I_1 and I_2 marked.
(a) Determine a matrix equation for I_1 and I_2.
(b) Solve for I_1 and I_2 given $E_1 = 12$, $R_1 = 3$ and $R_2 = 4$.
(c) Calculate the current in branch XY.

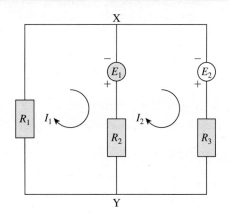

Figure 6.13

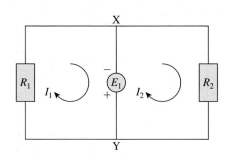

Figure 6.12

2 Figure 6.13 shows an electric circuit with mesh currents I_1 and I_2 marked.
(a) Determine a matrix equation for I_1 and I_2.

(b) Solve for I_1 and I_2 given $E_1 = 12$, $E_2 = 6$, $R_1 = 4$, $R_2 = 3$ and $R_3 = 4$.
(c) Calculate the current in branch XY.

3 Figure 6.14 shows an electric circuit with mesh currents I_1, I_2 and I_3 marked.
(a) Determine a matrix equation for I_1, I_2 and I_3.
(b) Solve for I_1, I_2 and I_3 given $E_1 = 6$, $E_2 = 12$, $R_1 = 2$, $R_2 = 3$, $R_3 = 2$ and $R_4 = 1$.
(c) Calculate the current in branch XY.

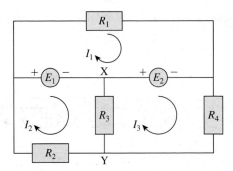

Figure 6.14

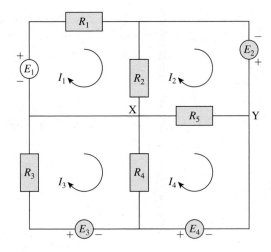

Figure 6.15

4   Figure 6.15 shows an electric circuit with
     mesh currents I_1, I_2, I_3 and I_4 marked.
     (a) Determine a matrix equation for I_1, I_2, I_3
         and I_4.
     (b) Solve for I_1, I_2, I_3 and I_4 given $E_1 = 6$,
         $E_2 = 12, E_3 = 6, E_4 = 10, R_1 = 3$,
         $R_2 = 5, R_3 = 4, R_4 = 5$ and $R_5 = 1$.
     (c) Calculate the current in branch XY.

Solutions to exercises

1   (a) $\begin{pmatrix} R_1 & 0 \\ 0 & R_2 \end{pmatrix} \begin{pmatrix} I_1 \\ I_2 \end{pmatrix} = \begin{pmatrix} E_1 \\ -E_1 \end{pmatrix}$

     (b) $I_1 = 4, I_2 = -3$
     (c) $I_1 - I_2 = 7$

2   (a) $\begin{pmatrix} R_1 + R_2 & -R_2 \\ -R_2 & R_2 + R_3 \end{pmatrix} \begin{pmatrix} I_1 \\ I_2 \end{pmatrix}$

         $= \begin{pmatrix} E_1 \\ E_2 - E_1 \end{pmatrix}$

     (b) $I_1 = 1.65, I_2 = -0.15$
     (c) $I_1 - I_2 = 1.8$

3   (a) $\begin{pmatrix} R_1 & 0 & 0 \\ 0 & R_2 + R_3 & -R_3 \\ 0 & -R_3 & R_3 + R_4 \end{pmatrix} \begin{pmatrix} I_1 \\ I_2 \\ I_3 \end{pmatrix}$

         $= \begin{pmatrix} E_1 + E_2 \\ -E_1 \\ -E_2 \end{pmatrix}$

     (b) $I_1 = 9, I_2 = -\frac{42}{11}, I_3 = -\frac{72}{11}$
     (c) $I_2 - I_2 = \frac{30}{11}$

4   (a) $\begin{pmatrix} R_1 + R_2 & -R_2 & 0 & 0 \\ -R_2 & R_2 + R_5 & 0 & -R_5 \\ 0 & 0 & R_3 + R_4 & -R_4 \\ 0 & -R_5 & -R_4 & R_4 + R_5 \end{pmatrix} \begin{pmatrix} I_1 \\ I_2 \\ I_3 \\ I_4 \end{pmatrix}$

         $= \begin{pmatrix} E_1 \\ E_2 \\ E_3 \\ E_4 \end{pmatrix}$

     (b) $I_1 = 5.5966, I_2 = 7.7546, I_3 = 4.3025$,
         $I_4 = 6.5445$
     (c) $I_4 - I_2 = -1.2101$, that is a current of
         1.2101 from Y to X.

End of chapter exercises

1 Determine which of the following systems have a non-trivial solution:

(a) $2x + 3y = 0$
$4x + 6y = 0$

(b) $9x - y = 0$
$x + 9y = 0$

(c) $x - 3y = 0$
$-3x + 9y = 0$

(d) $\dfrac{x}{2} - \dfrac{2y}{3} = 0$
$3x - 4y = 0$

(e) $-x + 4y = 0$
$2x - 6y = 0$

2 Determine which of the following systems have a non-trivial solution:

(a) $2x - y + 3z = 0$
$x + 2y - z = 0$
$5x + 5z = 0$

(b) $3x + 2y - z = 0$
$2x - 3y + 2z = 0$
$7x - 4y + 3z = 0$

(c) $x + 4y + 3z = 0$
$9x - 10y - z = 0$
$6x - 3y + z = 0$

(d) $3x - y - z = 0$
$x + y + 2z = 0$
$2x - 2y + 3z = 0$

(e) $-x + 2y - 3z = 0$
$4x + z = 0$
$11x + 2y = 0$

3 Figure C13.1 illustrates an electrical network with two mesh currents, I_1 and I_2, shown.

(a) Determine a matrix equation for I_1 and I_2.

(b) Calculate I_1 and I_2 given $E_1 = 5$, $E_2 = 5$, $R_1 = 10$, $R_2 = 6$ and $R_3 = 5$.

(c) Calculate the current in the branch XY.

4 Figure C13.2 illustrates an electrical network with mesh currents I_1, I_2 and I_3 shown.

(a) Determine a matrix equation for I_1, I_2 and I_3.

(b) Calculate I_1, I_2 and I_3 given $E_1 = 5$, $E_2 = 6$, $E_3 = 12$, $R_1 = 15$, $R_2 = 5$ and $R_3 = 10$.

(c) Calculate the current in the branch XY.

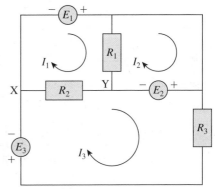

Figure C13.2

5 Figure C13.3 illustrates an electrical network with mesh currents I_1, I_2, I_3 and I_4 shown.

(a) Determine a matrix equation for I_1, I_2, I_3 and I_4.

(b) Calculate I_1, I_2, I_3 and I_4 given $E_1 = 6$, $E_2 = 12$, $R_1 = R_2 = 3$ and $R_3 = R_4 = R_5 = 2$.

(c) Calculate the current in the branch XY.

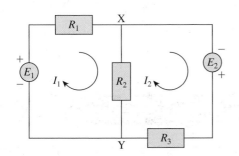

Figure C13.1

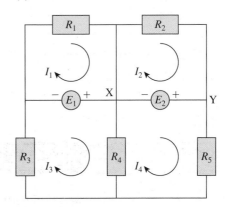

Figure C13.3

6   Solve the following simultaneous equations using Cramer's rule:
   (a) $2x - y = 5$
       $x + y = 1$
   (b) $2x - y = 5$
       $-x + 3y = 0$
   (c) $x + 3y = 1$
       $2x - 5y = -9$

7   Solve the following simultaneous equations using Cramer's rule:
   (a) $i_1 + 2i_2 - i_3 = 7$
       $i_1 + i_2 + 2i_3 = 7$
       $2i_1 - i_3 = 3$
   (b) $3a - b + c = 4$
       $a - 2b + 3c = 2$
       $a + b - 5c = 4$
   (c) $\alpha + 2\beta + \gamma = 1$
       $2\alpha + 3\beta + 2\gamma = 2$
       $a + 4\beta - \gamma = -3$

8   Solve the following simultaneous equations using the inverse matrix method:
   (a) $x + 2y = 11$
       $2x - y = 7$
   (b) $r_1 + 2r_2 = 13.5$
       $3r_1 + r_2 = 23$
   (c) $2\alpha + 3\beta = 5$
       $\alpha + \dfrac{\beta}{2} = 0.5$

9   Solve the following simultaneous equations using the inverse matrix method:
   (a) $x + y - z = -1$
       $2x - y + 2z = 8$
       $2x + 3y - z = -3$
   (b) $2I_1 + I_2 - I_3 = 10$
       $4I_1 + 3I_2 + 2I_3 = 18$
       $3I_1 - 3I_2 + 4I_3 = 0.5$
   (c) $r + s - t = 0.5$
       $-2r - s + 2t = -0.2$
       $r - 4s - 2t = -5$

10 Given that
$$\frac{1}{21}\begin{pmatrix} 4 & 51 & 49 & -28 \\ -9 & -57 & -63 & 42 \\ -6 & -66 & -63 & 42 \\ 4 & 9 & 7 & -7 \end{pmatrix}$$

is the inverse of
$$\begin{pmatrix} 3 & -1 & 4 & 6 \\ 1 & 2 & -1 & 2 \\ 2 & -2 & 3 & -2 \\ 5 & 0 & 4 & 1 \end{pmatrix}$$

solve the following systems of equations:
   (a) $3w - x + 4y + 6z = 19$
       $w + 2x - y + 2z = 11$
       $2w - 2x + 3y - 2z = -11$
       $5w + 4y + z = 2$
   (b) $2\alpha - 2\beta + 3\gamma - 2\delta = 11$
       $\alpha + 2\beta - \gamma + 2\delta = -2$
       $5\alpha + 4\gamma + \delta = 15$
       $3\alpha - \beta + 4\gamma + 6\delta = 1$

11 State which of the following matrices are in echelon form:
   (a) $\begin{pmatrix} 1 & 0 & 1 \\ 0 & 1 & 1 \end{pmatrix}$

   (b) $\begin{pmatrix} 1 & 1 & 1 & 0 \\ 0 & 0 & 0 & 1 \\ 0 & 0 & 0 & 0 \end{pmatrix}$

   (c) $\begin{pmatrix} 1 & 2 & 3 & 1 & 4 \\ 0 & 0 & 1 & -1 & 0 \\ 0 & 0 & 0 & 1 & 2 \end{pmatrix}$

   (d) $\begin{pmatrix} 1 & 0 & 1 & -3 \\ 0 & 0 & 0 & 0 \\ 0 & 1 & 4 & 3 \end{pmatrix}$

   (e) $\begin{pmatrix} 0 & 0 & 1 & 2 & 0 & -4 \\ 0 & 0 & 0 & 0 & 1 & 3 \end{pmatrix}$

12 Use Gaussian elimination to solve the following systems of equations:
   (a) $3i_1 - 2i_2 = 13$
       $2i_1 - 5i_2 = 16$
   (b) $4\alpha - \beta = 4$
       $\dfrac{\alpha}{2} + \dfrac{\beta}{3} = 6$
   (c) $-\dfrac{x}{2} + 3y = 4$
       $2x - 7y = -11$

13 Use Gaussian elimination to solve the following sets of equations:

(a) $2x - y - z = 3$
$5x + 7y + 3z = 10$
$-3x - y + 4z = -17$

(b) $2\alpha - \beta + \gamma = 11$
$\dfrac{\alpha}{2} + \beta + 2\gamma = 6$
$3\beta - 7\gamma = -37$

(c) $i_1 + 2i_2 - 3i_3 = 2$
$-3i_1 + 2i_2 + i_3 = 10$
$2i_1 - 3i_3 = -9$

14 Use Jacobi's method to find the solution of

$$-2x + y = 8.5$$
$$3x + 4y = 1$$

Take $x_0 = -2$, $y_0 = 2$ and perform five iterations.

15 Solve the equations in question 14 using the Gauss–Seidel method; take $x_0 = -2$, $y_0 = 2$ and perform five iterations.

16 Use Jacobi's method to solve

$$4x - y = -9.4$$
$$3x + 5y = 7.9$$

Take $x_0 = -1$, $y_0 = 1.5$ and perform five iterations.

17 Solve the equations of question 16 using the Gauss–Seidel method. Use the same initial values and perform five iterations.

18 The matrix, A, is defined by

$$A = \begin{pmatrix} 3 & 7 \\ -1 & -5 \end{pmatrix}$$

(a) Calculate the characteristic equation of A.
(b) Calculate the eigenvalues of A.
(c) Calculate the eigenvectors of A.

19 (a) Show that the matrix

$$\begin{pmatrix} 5 & -1 \\ 4 & 1 \end{pmatrix}$$

has only one eigenvalue.
(b) Find the eigenvector of the matrix.

20 The matrix A is defined by

$$A = \begin{pmatrix} 3 & 2 & -2 \\ 0 & -4 & 4 \\ 7 & 2 & -2 \end{pmatrix}$$

(a) Calculate the eigenvalues of A.
(b) Calculate the eigenvectors of A.

21 The matrix B is defined by

$$B = \begin{pmatrix} -3 & -1 & 0 \\ 5 & 2 & 1 \\ -5 & 5 & -4 \end{pmatrix}$$

(a) Calculate the eigenvalues of B.
(b) Calculate the eigenvectors of B.

Solutions to exercises

1 (a), (c) and (d) have non-trivial solutions.

2 (a), (b) and (e) have non-trivial solutions.

3 (a) $\begin{pmatrix} R_1 + R_2 & -R_2 \\ -R_2 & R_2 + R_3 \end{pmatrix} \begin{pmatrix} I_1 \\ I_2 \end{pmatrix} = \begin{pmatrix} E_1 \\ E_2 \end{pmatrix}$

(b) $I_1 = 0.6071$, $I_2 = 0.7857$

(c) 0.1786 from Y to X

4 (a) $\begin{pmatrix} R_1 + R_2 & -R_1 & -R_2 \\ -R_1 & R_1 & 0 \\ -R_2 & 0 & R_2 + R_3 \end{pmatrix} \begin{pmatrix} I_1 \\ I_2 \\ I_3 \end{pmatrix}$

$$= \begin{pmatrix} E_1 \\ -E_2 \\ E_2 + E_3 \end{pmatrix}$$

(b) $I_1 = 1.5$, $I_2 = 1.1$, $I_3 = 1.7$
(c) 0.2

5 (a) $$\begin{pmatrix} R_1 & 0 & 0 & 0 \\ 0 & R_2 & 0 & 0 \\ 0 & 0 & R_3 + R_4 & -R_4 \\ 0 & 0 & -R_4 & R_4 + R_5 \end{pmatrix} \begin{pmatrix} I_1 \\ I_2 \\ I_3 \\ I_4 \end{pmatrix}$$

$$= \begin{pmatrix} -E_1 \\ -E_2 \\ E_1 \\ E_2 \end{pmatrix}$$

(b) $I_1 = -2, I_2 = -4, I_3 = 4, I_4 = 5$
(c) 9

6 (a) $x = 2, y = -1$
(b) $x = 3, y = 1$
(c) $x = -2, y = 1$

7 (a) $i_1 = 2, i_2 = 3, i_3 = 1$
(b) $a = 1, b = -2, c = -1$
(c) $\alpha = -1, \beta = 0, \gamma = 2$

8 (a) $x = 5, y = 3$
(b) $r_1 = 6.5, r_2 = 3.5$
(c) $\alpha = -0.5, \beta = 2$

9 (a) $x = 2, y = -2, z = 1$
(b) $I_1 = 3.5, I_2 = 2, I_3 = -1$
(c) $r = 1.2, s = 0.8, t = 1.5$

10 (a) $w = 2, x = -1, y = -3, z = 4$
(b) $\alpha = 1, \beta = 2, \gamma = 3, \delta = -2$

11 (a), (b), (c) and (e) are in echelon form.

12 (a) $i_1 = 3, i_2 = -2$
(b) $\alpha = 4, \beta = 12$
(c) $x = -2, y = 1$

13 (a) $x = 1, y = 2, z = -3$
(b) $\alpha = 2, \beta = -3, \gamma = 4$
(c) $i_1 = 3, i_2 = 7, i_3 = 5$

14

n	x_n	y_n
0	-2	2
1	-3.25	1.75
2	-3.375	2.6875
3	-2.9063	2.7813
4	-2.8594	2.4297
5	-3.0351	2.3496

The exact answer is $x = -3, y = 2.5$.

15

n	x_n	y_n
0	-2	2
1	-3.25	2.6875
2	-2.9063	2.4297
3	-3.0352	2.5264
4	-2.9868	2.4901
5	-3.0049	2.5037

16

n	x_n	y_n
0	-1	1.5
1	-1.975	2.18
2	-1.805	2.765
3	-1.6588	2.663
4	-1.6843	2.5753
5	-1.7062	2.5906

The exact answer is $x = -1.7, y = 2.6$.

17

n	x_n	y_n
0	-1	1.5
1	-1.975	2.765
2	-1.6588	2.5753
3	-1.7062	2.6037
4	-1.6991	2.5994
5	-1.7001	2.6001

18 (a) $\lambda^2 + 2\lambda - 8 = 0$
(b) $-4, 2$

(c) $\begin{pmatrix} -0.7071 \\ 0.7071 \end{pmatrix}, \begin{pmatrix} 0.9899 \\ -0.1414 \end{pmatrix}$

19 (a) $\lambda = 3$

(b) $\begin{pmatrix} 1 \\ 2 \end{pmatrix}$

20 (a) $-4, 0, 1$

(b) $\begin{pmatrix} -0.2747 \\ 0.9615 \\ 0 \end{pmatrix}, \begin{pmatrix} 0 \\ 0.7071 \\ 0.7071 \end{pmatrix}, \begin{pmatrix} 0.1543 \\ 0.6172 \\ 0.7715 \end{pmatrix}$

21 (a) $-4, -3, 2$

(b) $\begin{pmatrix} -0.0902 \\ -0.0902 \\ 0.9919 \end{pmatrix}, \begin{pmatrix} -0.1961 \\ 0 \\ 0.9806 \end{pmatrix}, \begin{pmatrix} 0.1400 \\ -0.7001 \\ -0.7001 \end{pmatrix}$

Vectors

Chapter **14**

In many engineering applications the direction of a physical quantity is of particular interest. **Vectors** are mathematical quantities that contain information about direction as well as magnitude and so are particularly suited to modelling physical situations where direction is important.

Special rules and methods have been developed for handling vectors, and these are described and developed in this chapter.

Chapter 14 contents

Basic concepts of vectors

In engineering, frequent reference is made to physical quantities, such as force, speed and time. For example, we talk of the speed of a car, and the force in a compressed spring. It is useful to separate physical quantities into two types. Quantities of the first type are known as **scalars**. These can be fully described by a single number known as the **magnitude**. An example of a scalar quantity is the mass of an object, so we might state

'the mass of the stone is 3 kg'

It is important to give the units in which the quantity is measured.

Quantities of the second type are those that require the specification of a **direction**, in addition to a magnitude, before they are completely described. These are known as **vectors**. An example of a vector quantity is the force applied to an object to make it move. When the object shown in Figure 1.1 is moved by applying a force to it we can achieve different effects by applying the force in different directions.

Figure 1.1
Applying the force in a different direction will achieve a different effect.

In order to specify the force completely we must state the direction in which the force acts. For example, we might state

'a force of 5 newtons is applied vertically upwards'

Clearly this would achieve a different effect from applying the force horizontally to the right. The direction in which the force acts is crucial.

Special methods have been developed for handling vectors in calculations, giving rise to subjects known as vector algebra, vector geometry and vector calculus. Quantities that are vectors must be manipulated according to certain rules, which are described in this and subsequent blocks.

There are many engineering applications in which vector and scalar quantities play an important role. For example, speed, mass, work, voltage, energy are all scalars, whereas force, acceleration, velocity, electromagnetic field strength are all vectors. Furthermore, when computer software is written to control the position of a robot, the position is described by vectors.

Sometimes confusion can arise because words used in general conversation have specific technical meanings when used in engineering calculations. An example of this ambiguity is the use of the words 'speed' and 'velocity'. In everyday conversation these words have the same meaning and can be used interchangeably. However, in engineering and science they are not the same. **Speed** is a scalar quantity described by giving a single number in appropriate units. So we can make statements such as

'the speed of the car is 40 kilometres per hour'

On the other hand **velocity** is a vector quantity and must be specified by giving a direction as well. So, for example, we can state

'the velocity of the aircraft is 200 metres per second due north'

In engineering calculations, the words 'speed' and 'velocity' should not be used interchangeably. Similar problems arise from use of the words 'mass' and 'weight'. In engineering and science these are different. **Mass** is a scalar that describes the amount of substance in an object. The unit of mass is the kilogram. **Weight** is a vector, the direction of which is vertically downwards. Weight arises through the action of gravity. The unit of weight is the newton.

Displacement and **distance** are related quantities that can cause confusion. Whereas distance is a scalar, displacement is 'directed distance': that is, distance together with a specified direction. So, referring to Figure 1.2, if an object is moved from point A to point B, we can state that the distance moved is 10 metres, but the displacement is 10 metres *in the direction from A to B*.

Figure 1.2
Displacement means directed distance.

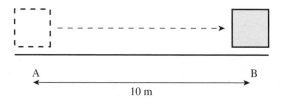

You will meet many other quantities in the course of your studies, and it will be helpful to know which are vectors and which are scalars. Some common quantities and their type are listed in Table 1.1. The common units in which these are measured are also shown.

Table 1.1
Some common
scalar and vector
quantities.

Quantity	Type	SI unit
Distance	Scalar	metre, m
Area	Scalar	metres squared, m^2
Mass	Scalar	kilogram, kg
Temperature	Scalar	kelvin, K
Pressure	Scalar	pascal, Pa
Work	Scalar	joule, J
Energy	Scalar	joule, J
Displacement	Vector	metre, m
Force	Vector	newton, N
Velocity	Vector	metres per second, $m\ s^{-1}$
Acceleration	Vector	metres per second per second, $m\ s^{-2}$

Exercises

1   State, or find out, which of the following are scalars and which are vectors: (a) the volume of a petrol tank, (b) a length measured in metres, (c) a length measured in miles, (d) the angular velocity of a flywheel, (e) the relative velocity of two aircraft, (f) the work done by a force, (g) electrostatic potential, (h) the momentum of an atomic particle.

Solutions to exercises

1   (a), (b), (c), (f), (g) are scalars; (d), (e) and (h) are vectors.

1.2   The mathematical description of vector quantities

Because a vector has a direction as well as a magnitude we can represent a vector by drawing a line. The length of the line represents the magnitude of the vector given some appropriate scale, and the direction of the line represents the direction of the vector. We call such a representation a **directed line segment**. The length of the line is also referred to as the **modulus** of the vector.

For example, Figure 1.3 shows a vector that might represent a velocity of 3 m s^{-1} north-west. Note that the arrow on the vector indicates the direction required.

Figure 1.3
A vector quantity can be represented by drawing a line marked with an arrow.

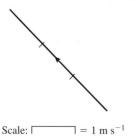

Scale: $\boxed{} = 1 \text{ m s}^{-1}$

More generally, Figure 1.4 shows an arbitrary vector quantity. It is important when writing vectors to distinguish them from scalars, and so various notations are used. In Figure 1.4 we can write the vector as $\overrightarrow{AB}$ where the arrow is used to emphasise that we mean the vector from A to B. In books and other printed work, vectors are often indicated by using a bold typeface such as $\boldsymbol{a}$. It is difficult when handwriting to reproduce the bold face and so it is conventional to underline vector quantities and write $\underline{a}$ instead. So $\overrightarrow{AB}$, $\boldsymbol{a}$ and $\underline{a}$ all represent the same vector in Figure 1.4. In Figure 1.4 the point A is the **tail** of vector $\overrightarrow{AB}$ and point B is referred to as its **head**.

Figure 1.4
Vectors can be written in different ways:
$\overrightarrow{AB} = \boldsymbol{a} = \underline{a}$.

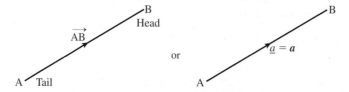

Example 1.1

Consider Figure 1.5, which shows an object being pulled by a force of 5 N at an angle of 60° to the horizontal. Show how this force can be represented by a vector.

Figure 1.5
An object being pulled by a force.

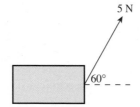

Solution

The force can be represented by drawing a line of length 5 units at an angle of 60° to the horizontal, as shown in Figure 1.6. Note that we have labelled the force $\boldsymbol{F}$. The modulus of the force is 5 newtons. Whenever several forces are involved they can be labelled $\boldsymbol{F}_1$, $\boldsymbol{F}_2$ and so on.

Figure 1.6
The force in
Figure 1.5
represented as a
vector.

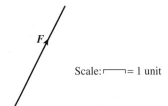

Scale: ⌐────⌐ = 1 unit

When we wish to refer to just the magnitude of a vector we write this using the modulus sign as $|\overrightarrow{AB}|$, or $|\boldsymbol{a}|$, or $|\underline{a}|$. Alternatively we write simply a.

Referring to Figure 1.6, we can write the magnitude of the force $\boldsymbol{F}$ as $|\boldsymbol{F}|$ or simply F (not bold, and without the underline or modulus signs).

Key point

The magnitude, or **modulus**, of a vector $\underline{a} = \overrightarrow{AB}$ can be written as

$$|\boldsymbol{a}|, \quad |\underline{a}|, \quad \text{or} \quad |\overrightarrow{AB}|, \quad \text{or simply} \quad a$$

Equal vectors

In general two vectors are said to be **equal vectors** if they have the same magnitude and direction. So, in Figure 1.7 the vectors $\overrightarrow{CD}$ and $\overrightarrow{AB}$ are equal even though their locations differ.

This is a useful and important property of vectors: a vector can be translated maintaining the same direction and magnitude, without changing the vector itself.

Figure 1.7
Vectors can be
equal even when
their locations
differ.

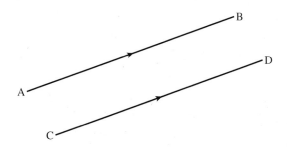

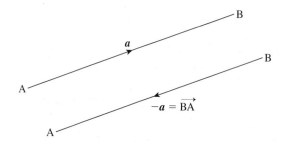

There are exceptions to this property. For example, a vector is sometimes used to represent the position of an object that might be located at a particular point in space. Such a vector cannot be translated freely. Nevertheless most of the vectors we shall meet can be translated, and as such are called **free vectors**.

Negative vectors

The vector $-\boldsymbol{a}$ is a vector in the opposite direction to $\boldsymbol{a}$, but with the same magnitude as $\boldsymbol{a}$, as shown in Figure 1.8. Geometrically, if $\boldsymbol{a} = \overrightarrow{AB}$ then $-\boldsymbol{a} = \overrightarrow{BA}$.

Figure 1.8
The negative vector $-\boldsymbol{a}$ has the opposite direction to $\boldsymbol{a}$.

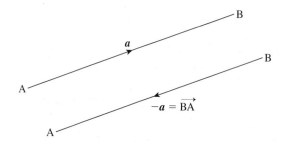

Exercises

1 An object is subject to two forces, one of 3 N vertically downwards, and one of 8 N horizontally to the right. Draw a diagram representing these two forces as vectors.

2 Draw a diagram showing an arbitrary vector $\boldsymbol{F}$. On the diagram show the vector $-\boldsymbol{F}$

3 Vectors $\boldsymbol{p}$ and $\boldsymbol{q}$ are equal vectors. Draw a diagram that might be used to represent $\boldsymbol{p}$ and $\boldsymbol{q}$.

Solutions to exercises

1 See Figure 1.9.

2 See Figure 1.10.

3 See Figure 1.11.

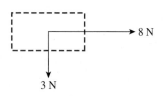

Figure 1.9

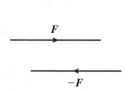

Figure 1.10

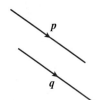

Figure 1.11

Addition of vectors is carried out according to a rule known as the **triangle law**.

Figure 1.12
Two vectors
a and *b*.

Consider Figure 1.12. Suppose we wish to add *b* to *a*. To do this *b* is translated, keeping its direction and length unchanged, until its tail coincides with the head of *a*. Then the sum *a* + *b* is defined by the vector representing the third side of the completed triangle, that is *c* in Figure 1.13. Note from Figure 1.13 that we can write *c* = *a* + *b*.

Figure 1.13
Addition of the
two vectors of
Figure 1.12 using
the triangle law.

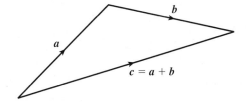

Key point

Vectors are added using the **triangle law**:

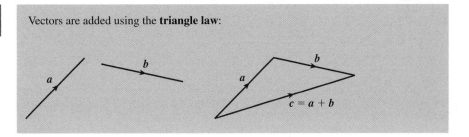

Example 1.2
Redraw the diagram in Figure 1.14 to show the vector sum *a* + *b*.

Figure 1.14

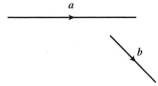

Solution

It is possible, using the triangle law, to prove the following rules, which apply to any three vectors a, b and c:

> $a + b = b + a$ vector addition is commutative.
>
> $a + (b + c) = (a + b) + c$ vector addition is associative.

To see why it is appropriate to add vectors using the triangle law consider the following examples.

Example 1.3 The route taken by an automated vehicle

An unmanned vehicle moves on tracks around a factory floor carrying components from the store at A to workers at C, as shown in Figure 1.15.

Figure 1.15
$\overrightarrow{AB} + \overrightarrow{BC} = \overrightarrow{AC}$.

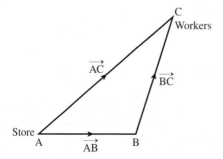

The vehicle may arrive at C either directly or via a second location at B. The movement from A to B can be represented by a displacement vector $\overrightarrow{AB}$. Similarly movement from B to C can be represented by the displacement vector $\overrightarrow{BC}$, and movement from A to C can be represented by $\overrightarrow{AC}$. Since the head of vector $\overrightarrow{AB}$ touches the tail of $\overrightarrow{BC}$ the triangle law can be applied immediately to find the combined effect of the two displacements.

$$\overrightarrow{AB} + \overrightarrow{BC} = \overrightarrow{AC}$$

Example 1.4 Mechanical Engineering – Resultant of two forces acting upon a body

A force F_1 of 2 N acts vertically downwards, and a force F_2 of 3 N acts horizontally to the right, upon the body shown in Figure 1.16. Both forces act through the same point in the body.

Figure 1.16
The combined effect of forces F_1 and F_2 is the resultant R.

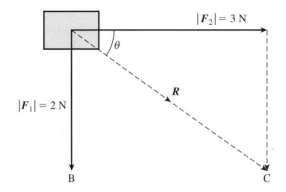

$|F_2| = 3$ N

$|F_1| = 2$ N

θ

R

B

C

We can use vector addition to find the combined effect of the two forces, known as the **resultant force**. Translating F_1 until its tail touches the head of F_2, we complete the triangle as shown. The vector represented by the third side is the resultant, R. We write

$$R = F_2 + F_1$$

and say that R is the **vector sum** of F_2 and F_1. The resultant force acts at an angle of θ below the horizontal where $\tan \theta = \frac{2}{3}$, so that $\theta = 33.7°$, and has magnitude given by Pythagoras's theorem as $\sqrt{13}$ N.

Key point   The **resultant** of two vectors a and b is their **vector sum** $a + b$.

Example 1.5 Mechanical Engineering – Resolving a force into two perpendicular directions

In the previous example we saw that two forces acting upon a body can be replaced by a single force which has the same effect. It is sometimes useful to consider a single force as two forces acting at right angles to each other. Consider the force F in Figure 1.17.

Figure 1.17
Force F has two perpendicular components.

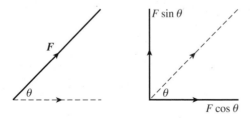

F

θ

$F \sin \theta$

θ

$F \cos \theta$

This force can be replaced by two perpendicular forces as shown. It is a straight-forward use of trigonometry to show that the horizontal force has magnitude $F \cos \theta$. The vertical force has magnitude $F \sin \theta$. We say that F has been **resolved into two perpendicular components**.

For example, Figure 1.18 shows a force of 5 N acting at an angle of 30° to the x axis. It can be resolved into two components, one directed along the x axis with magnitude 5 cos 30° and one perpendicular to this of magnitude 5 sin 30°. Together, these two components have the same effect as the original force.

Figure 1.18

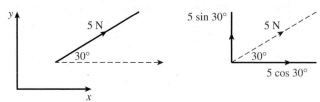

Example 1.6

Consider the force shown in Figure 1.19. Resolve this force into two perpendicular components, one horizontally to the right, and one vertically upwards.

Figure 1.19

Solution

15 cos 40° = 11.49 N horiz., 15 sin 40° = 9.64 N vert.

Subtraction of one vector from another is performed by adding the corresponding negative vector. That is, if we seek $a - b$ we form $a + (-b)$. This is shown geometrically in Figure 1.20.

Figure 1.20
Subtraction of a vector is performed by adding a negative vector.

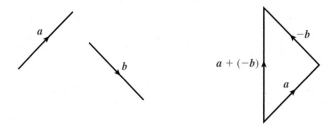

Example 1.7

Figure 1.21 shows vectors $p = \overrightarrow{OP}$ and $q = \overrightarrow{OQ}$. What is the geometrical significance of the vector $q - p$?

Figure 1.21

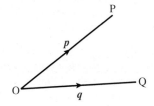

Solution

We know that $\boldsymbol{q} - \boldsymbol{p}$ is interpreted as $\boldsymbol{q} + (-\boldsymbol{p})$.

Start by drawing in the vector $-\boldsymbol{p}$. This is shown in Figure 1.22(a). It has been drawn with its tail touching the head of $\boldsymbol{q}$ so that we can apply the triangle rule for addition. Figure 1.22(b) shows the application of the triangle rule. Note that the resultant $\boldsymbol{q} + (-\boldsymbol{p})$ is the same as the vector $\overrightarrow{PQ}$. This result is very important.

Figure 1.22

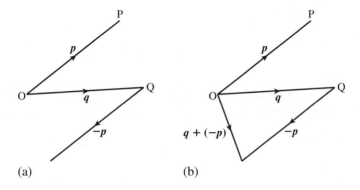

(a)   (b)

Key point

Given two vectors $\boldsymbol{p} = \overrightarrow{OP}$ and $\boldsymbol{q} = \overrightarrow{OQ}$ (Figure 1.23) the vector from P to Q is given by $\boldsymbol{q} - \boldsymbol{p}$.

Figure 1.23

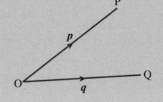

Exercises

1   Vectors $\boldsymbol{p}$ and $\boldsymbol{q}$ represent two perpendicular sides of a square ABCD with $\boldsymbol{p} = \overrightarrow{AB}$ and $\boldsymbol{q} = \overrightarrow{BC}$. Find vector expressions that represent the diagonals of the square $\overrightarrow{AC}$ and $\overrightarrow{BD}$.

2   In the rectangle ABCD, side AB is represented by the vector $\boldsymbol{p}$ and side BC is represented by the vector $\boldsymbol{q}$. State the physical significance of the vectors $\boldsymbol{p} + \boldsymbol{q}$ and $\boldsymbol{p} - \boldsymbol{q}$.

3   An object is positioned at the origin of a set of axes. Two forces act upon it. The first has magnitude 9 N and acts in the direction of the positive y axis. The second has magnitude 4 N and acts in the direction of the negative x axis. Calculate the magnitude and direction of the resultant force.

4   An object moves in the xy plane with a velocity of 15 m s^{-1} in a direction at 48° above the positive x axis. Resolve this velocity into two components, one along the x axis and one along the y axis.

5   Draw a right-angled triangle ABC with the right angle at B. Label $\angle$ACB as θ. Show that
(a) BC = AC cos θ
(b) AB = AC sin θ

Solutions to exercises

1 $p + q, q - p$

2 $p + q$ is the diagonal AC, $p - q$ is the diagonal DB.

3 magnitude $\sqrt{97}$, at an angle $66°$ above the negative x axis

4 10.04 m s^{-1} along the x axis, and 11.15 m s^{-1} along the y axis

---

1.4 Multiplying a vector by a scalar

If k is any positive scalar and a is a vector then ka is a vector in the same direction as a but k times as long. If k is negative, ka is a vector in the opposite direction to a and k times as long. The vector ka is said to be a **scalar multiple** of a. Consider the vectors shown in Figure 1.24.

Figure 1.24
Multiplying a vector by a scalar.

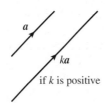

if k is positive

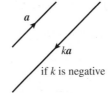

if k is negative

The vector $3a$ is three times as long as a and has the same direction. The vector $\frac{1}{2}r$ is in the same direction as r but is half as long. The vector $-4b$ is in the opposite direction to b and four times as long.

For any scalars k and l, and any vectors a and b, the following rules hold:

Key point

$$k(a + b) = ka + kb$$
$$(k + l)a = ka + la$$
$$k(l)a = (kl)a$$

Unit vectors

A vector that has a modulus of 1 is called a **unit vector**. Unit vectors will play an important role when we come to study cartesian components in Block 2.

If a has modulus 3, say, then a unit vector in the direction of a is $\frac{1}{3}a$, as shown in Figure 1.25.

Figure 1.25

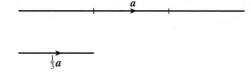

More generally, to obtain a unit vector in the direction of any vector a we divide by its modulus. A unit vector in the direction of a is given the 'hat' symbol $\hat{a}$.

Key point

$$\hat{a} = \frac{a}{|a|}$$

Exercises

1 Draw an arbitrary vector r. On your diagram draw $2r$, $4r$, $-r$, $-3r$ and $\frac{1}{2}r$.

2 In triangle OAB the point P is the midpoint of AB.
   (a) If $\overrightarrow{OA} = a$ and $\overrightarrow{OB} = b$ depict this on a diagram.
   (b) Write down an expression for $\overrightarrow{AB}$ in terms of a and b.
   (c) Write down an expression for $\overrightarrow{AP}$ in terms of a and b.
   (d) Find an expression for $\overrightarrow{OP}$ in terms of a and b.

3 In triangle OAB the point P divides AB in the ratio $m{:}n$. If $\overrightarrow{OA} = a$ and $\overrightarrow{OB} = b$ depict this on a diagram and then find an expression for $\overrightarrow{OP}$ in terms of a and b.

4 Explain what is meant by a unit vector. Given an arbitrary vector n, how is a unit vector having the same direction as n found?

5 If $\hat{e}$ is a unit vector, how would you write a vector in the same direction as $\hat{e}$ but having modulus 8?

Solutions to exercises

2 (b) $b - a$ (c) $\frac{1}{2}(b - a)$
   (d) $a + \frac{1}{2}(b - a) = \frac{1}{2}a + \frac{1}{2}b$

3 $\overrightarrow{OP} = a + \dfrac{m}{m + n}(b - a)$

4 Divide n by its modulus.

5 $8\hat{e}$

End of block exercises

1 Draw three arbitrary vectors a, b and c. By using the triangle law verify that vector addition is associative, that is

$$(a + b) + c = a + (b + c)$$

2 Draw two arbitrary vectors a and b. By using the triangle law verify that vector addition is commutative, that is

$$a + b = b + a$$

3   A force of 13 N acts at an angle of 62°
above the x axis. Resolve this force into two
components: one along the x axis and one
along the y axis.

4   In the triangle ABC, M is the midpoint of BC
and N is the midpoint of AC. Show that
$\overrightarrow{NM} = \frac{1}{2}\overrightarrow{AB}$.

5   A particle is positioned at the origin. Two
forces act on the particle. The first has

magnitude 7 N and acts in the negative x
direction. The second has magnitude 12 N and
acts in the y direction. Calculate the magnitude
and direction of the resultant force.

6   Draw two arbitrary vectors a and b. Verify that

$$2(a + b) = 2a + 2b$$

(This is the first of the rules concerning scalar
multiplication given at the start of Section 1.4.)

Solutions to exercises

3   6.10 N along the x axis, 11.48 N along the
y axis

5   13.9 N at 59.7° to the negative x axis

Cartesian components of vectors

It is useful to be able to describe vectors with reference to various coordinate systems, such as the *xy* plane. So, in this block, we show how this is possible by defining unit vectors in the directions of the *x* and *y* axes. Any other vector in the *xy* plane can be represented as a combination of these **basis vectors**. Such a representation is called a **cartesian form**. You will learn how to calculate the modulus of a vector given in cartesian form. Then, the idea is extended to three-dimensional vectors. This is useful because most engineering problems arise in three-dimensional situations.

The unit vectors *i* and *j*

Figure 2.1 shows a two-dimensional coordinate frame. Any point in the plane of the figure can be defined in terms of its *x* and *y* coordinates.

Figure 2.1
A two-dimensional
coordinate frame.

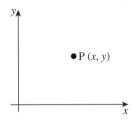

A unit vector pointing in the positive direction of the *x* axis is denoted by *i*. (Note that it is common practice to write this particular unit vector without the 'hat' ^ .) It follows that any vector in the direction of the *x* axis will be a multiple of *i*. Figure 2.2 shows vectors *i*, 2*i*, 5*i* and −3*i*. In general a vector of length *a* in the direction of the *x* axis will be *ai*.

Figure 2.2
All these vectors
are multiples of *i*.

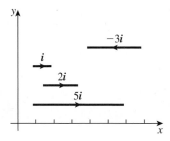

Similarly, a unit vector pointing in the positive direction of the *y* axis is denoted by *j*. Any vector in the direction of the *y* axis will be a multiple of *j*. Figure 2.3 shows *j*, 4*j* and −2*j*. In general a vector of length *b* in the direction of the *y* axis will be *bj*.

Figure 2.3
All these vectors
are multiples of *j*.

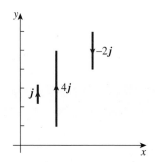

Key point

i represents a unit vector in the direction of the positive *x* axis.
j represents a unit vector in the direction of the positive *y* axis.

Example 2.1
Draw the vectors 5*i* and 4*j*. Use your diagram and the triangle law of addition to add these two vectors together to obtain the sum 5*i* + 4*j*.

Solution
First draw the vectors 5*i* and 4*j*.

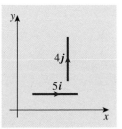

By translating the vectors so that they lie head to tail, find the vector sum 5*i* + 4*j*.

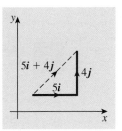

We now generalise the situation in the previous example. Consider Figure 2.4, which shows a vector $\boldsymbol{r} = \overrightarrow{AB}$.

Figure 2.4
$\overrightarrow{AB} = \overrightarrow{AC} + \overrightarrow{CB}$
by the triangle law.

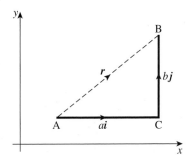

We can regard $\boldsymbol{r}$ as being the resultant of the two vectors $\overrightarrow{AC} = a\boldsymbol{i}$ and $\overrightarrow{CB} = b\boldsymbol{j}$. From the triangle law of vector addition

$$\boldsymbol{r} = \overrightarrow{AB} = \overrightarrow{AC} + \overrightarrow{CB} = a\boldsymbol{i} + b\boldsymbol{j}$$

We conclude that any vector in the xy plane can be expressed in the form $\boldsymbol{r} = a\boldsymbol{i} + b\boldsymbol{j}$. An alternative way of writing this vector is to use **column vector** notation. We write

$$\boldsymbol{r} = \begin{pmatrix} a \\ b \end{pmatrix}$$

A **row vector** would be written as (a, b), but care must be taken not to confuse this form with cartesian coordinates.

Key point

Any vector in the xy plane can be written $\boldsymbol{r} = a\boldsymbol{i} + b\boldsymbol{j}$ or as

$$\boldsymbol{r} = \begin{pmatrix} a \\ b \end{pmatrix}$$

The numbers a and b are called the $\boldsymbol{i}$ and $\boldsymbol{j}$ **components** of $\boldsymbol{r}$.

Addition, subtraction and scalar multiplication of vectors

Example 2.2
(a) Draw an xy plane and show the vectors $a = 2i + 3j$ and $b = 5i + j$.
(b) By translating one of the vectors apply the triangle law to show the sum $a + b$.
(c) Express the resultant in terms of i and j.

Solution
(a) Draw the xy plane and the required vectors. They can be drawn from any point in the plane.

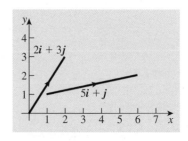

(b) Translate one of the vectors so that they lie head to tail, completing the third side of the triangle to give the resultant $a + b$.
(c) By studying your diagram note that the resultant consists of the two components $7i$ horizontally, and $4j$ vertically. Hence write down an expression for $a + b$.

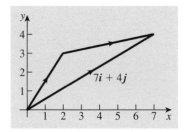

$a + b =$ $7i + 4j$

Note that this result consists of the sum of the respective components of a and b. That is,

$$(2i + 3j) + (5i + j) = 7i + 4j$$

It is important to note from the last example that vectors in cartesian form can be added by simply adding their respective i and j components.

Key point

If $a = a_x i + a_y j$ and $b = b_x i + b_y j$ then

$$a + b = (a_x + b_x)i + (a_y + b_y)j$$

Example 2.3
If $a = 9i + 7j$ and $b = 8i + 3j$ find
(a) $a + b$
(b) $a - b$

Solution
(a) Simply add the respective components:

$a + b =$ \[\qquad\qquad\] $17i + 10j$

(b) Simply subtract the respective components:

$a - b =$ \[\qquad\qquad\] $i + 4j$

Example 2.4
If

$$r = \begin{pmatrix} 7 \\ 11 \end{pmatrix} \quad \text{and} \quad s = \begin{pmatrix} 3 \\ -5 \end{pmatrix}$$

find $r + s$.

Solution
The vectors are added by adding their respective components.

$$\begin{pmatrix} 10 \\ 6 \end{pmatrix}$$

It follows that to multiply any vector a by a scalar k, we multiply each component of a by k.

Example 2.5
If $a = 9i + 3j$ write down (a) $4a$, (b) $-\frac{1}{3}a$.

Solution
(a) $4a =$ \[\qquad\qquad\] $36i + 12j$

(b) $-\frac{1}{3}a =$ \[\qquad\qquad\] $-3i - j$

Key point
If $a = a_x i + a_y j$ then $ka = ka_x i + ka_y j$.

Position vectors

Now consider the special case when r represents the vector from the origin to the point $P(x, y)$ as shown in Figure 2.5. This vector is known as the **position vector** of P.

Key point
The **position vector** of P(x, y) is $r = \overrightarrow{OP} = xi + yj$.

Figure 2.5
$r = xi + yj$ is the position vector of the point P with coordinates (x, y).

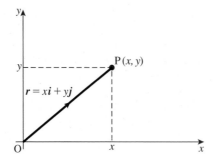

Unlike most vectors, position vectors cannot be freely translated. Because they indicate the position of a point they are fixed vectors in the sense that the tail of a position vector is always located at the origin.

Example 2.6
State the position vectors of the points with coordinates (a) P(2, 4), (b) Q(−1, 5), (c) R(−1, −7), (d) S(8, −4).

Solution
(a) The position vector of P is $2i + 4j$. This could be written

$$\begin{pmatrix} 2 \\ 4 \end{pmatrix}$$

(b) The position vector of Q is $-i + 5j$.
(c) The position vector of R is $-i - 7j$.
(d) The position vector of S is $8i - 4j$.

Example 2.7
Sketch the position vectors $r_1 = 3i + 4j$, $r_2 = -2i + 5j$ and $r_3 = -3i - 2j$.

Solution
The vectors are shown in Figure 2.6. Note that all position vectors start at the origin.

Figure 2.6
The position vectors
$r_1 = 3i + 4j$,
$r_2 = -2i + 5j$ and
$r_3 = -3i - 2j$.

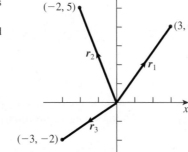

Modulus

The **modulus** of any vector *r* is equal to its length. When $r = xi + yj$ the modulus can be obtained by Pythagoras's theorem. Referring to Figure 2.5, if *r* is the position vector of point P then the modulus is clearly the distance of P from the origin.

Key point

If $r = xi + yj$ then $|r| = \sqrt{x^2 + y^2}$.

Example 2.8
Find the modulus of each of the vectors (a) $r_1 = 3i + 4j$, (b) $r_2 = -2i + 5j$, (c) $r_3 = 9i - 2j$ and (d) $r_4 = -5i - 3j$.

Solution
(a) The modulus of $r_1 = |3i + 4j| = \sqrt{3^2 + 4^2} = \sqrt{25} = 5$.

(b) The modulus of $r_2 = |-2i + 5j| = \sqrt{(-2)^2 + 5^2} = \sqrt{4 + 25} = \sqrt{29}$.

(c) Similarly $|r_3| = \sqrt{9^2 + (-2)^2} = \sqrt{85}$.

(d) $|r_4| = \sqrt{(-5)^2 + (-3)^2} = \sqrt{25 + 9} = \sqrt{34}$.

Example 2.9
Find the modulus of the vector

$$r = \begin{pmatrix} 7 \\ -3 \end{pmatrix}$$

Solution

$$\sqrt{7^2 + (-3)^2} = \sqrt{58}$$

Example 2.10
Point A has coordinates (3, 5). Point B has coordinates (7, 8).
(a) Depict these points on a diagram and state their position vectors.
(b) Find an expression for $\overrightarrow{AB}$.
(c) Find $|\overrightarrow{AB}|$.

Solution
(a) Draw a diagram that shows points A and B.

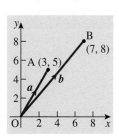

The position vector of A is then

$$\overrightarrow{OA} = a = 3i + 5j$$

The position vector of B is

$$\overrightarrow{OB} = \boldsymbol{b} = 7\boldsymbol{i} + 8\boldsymbol{j}$$

(b) Refer to your figure and use the triangle law to write

$$\overrightarrow{OA} + \overrightarrow{AB} = \overrightarrow{OB}$$

so that

$$\overrightarrow{AB} = \overrightarrow{OB} - \overrightarrow{OA}$$
$$= \boldsymbol{b} - \boldsymbol{a}$$

Hence write down an expression for $\overrightarrow{AB}$ in terms of the unit vectors $\boldsymbol{i}$ and $\boldsymbol{j}$.

$$(7\boldsymbol{i} + 8\boldsymbol{j}) - (3\boldsymbol{i} + 5\boldsymbol{j}) = 4\boldsymbol{i} + 3\boldsymbol{j}$$

(c) The length of $\overrightarrow{AB} = |4\boldsymbol{i} + 3\boldsymbol{j}| =$

$$\sqrt{4^2 + 3^2} = \sqrt{25} = 5$$

Exercises

1 Explain the distinction between a position vector and a more general or free vector.

2 What is meant by the symbols $\boldsymbol{i}$ and $\boldsymbol{j}$?

3 State the position vectors of the points with coordinates (a) P(4, 7), (b) Q(−3, 5), (c) R(0, 3), (d) S(−1, 0).

4 State the coordinates of the point P if its position vector is given as (a) $3\boldsymbol{i} - 7\boldsymbol{j}$, (b) $-4\boldsymbol{i}$, (c) $-0.5\boldsymbol{i} + 13\boldsymbol{j}$, (d) $a\boldsymbol{i} + b\boldsymbol{j}$.

5 Find the modulus of each of the following vectors:
(a) $\boldsymbol{r} = 7\boldsymbol{i} + 3\boldsymbol{j}$ (b) $\boldsymbol{r} = 17\boldsymbol{i}$ (c) $\boldsymbol{r} = 2\boldsymbol{i} - 3\boldsymbol{j}$
(d) $\boldsymbol{r} = -3\boldsymbol{j}$ (e) $\boldsymbol{r} = a\boldsymbol{i} + b\boldsymbol{j}$ (f) $\boldsymbol{r} = a\boldsymbol{i} - b\boldsymbol{j}$

6 Point P has coordinates (7, 8). Point Q has coordinates (−2, 4).
(a) Draw a sketch showing P and Q.
(b) State the position vectors of P and Q.
(c) Find an expression for $\overrightarrow{PQ}$.
(d) Find $|\overrightarrow{PQ}|$.

Solutions to exercises

1 Free vectors can be translated keeping their direction and length the same. Position vectors must always start at the origin.

2 $\boldsymbol{i}$ is a unit vector in the direction of the positive x axis. $\boldsymbol{j}$ is a unit vector in the direction of the positive y axis.

3 (a) $4\boldsymbol{i} + 7\boldsymbol{j}$ (b) $-3\boldsymbol{i} + 5\boldsymbol{j}$ (c) $3\boldsymbol{j}$ (d) $-\boldsymbol{i}$

4 (a) (3, −7) (b) (−4, 0) (c) (−0.5, 13)
(d) (a, b)

5 (a) $\sqrt{58}$ (b) 17 (c) $\sqrt{13}$ (d) 3
(e) $\sqrt{a^2 + b^2}$ (f) $\sqrt{a^2 + b^2}$

6 (b) $\boldsymbol{p} = 7\boldsymbol{i} + 8\boldsymbol{j}, \boldsymbol{q} = -2\boldsymbol{i} + 4\boldsymbol{j}$
(c) $\overrightarrow{PQ} = -9\boldsymbol{i} - 4\boldsymbol{j}$ (d) $|\overrightarrow{PQ}| = \sqrt{97}$

Consider the point P(4, 5) and its position vector $4\boldsymbol{i} + 5\boldsymbol{j}$ as shown in Figure 2.7.

Figure 2.7
The direction cosines of $\overrightarrow{OP}$ are $\cos\alpha$ and $\cos\beta$.

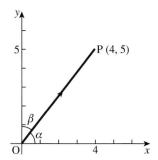

The **direction ratio** of the vector $\overrightarrow{OP}$ is defined to be 4:5. We can interpret this as stating that to move in the direction of the line we must move 5 units in the y direction for every 4 units in the x direction.

The **direction cosines** of the vector $\overrightarrow{OP}$ are the cosines of the angles between the vector and each of the positive axes. Specifically, referring to Figure 2.7 these are

$$\text{direction cosines:} \quad \cos\alpha \quad \text{and} \quad \cos\beta$$

Noting that the length of $\overrightarrow{OP}$ is $\sqrt{4^2 + 5^2} = \sqrt{41}$ we can write

$$\cos\alpha = \frac{4}{\sqrt{41}}, \quad \cos\beta = \frac{5}{\sqrt{41}}$$

It is conventional to label the direction cosines as l and m so that

$$l = \frac{4}{\sqrt{41}}, \quad m = \frac{5}{\sqrt{41}}$$

More generally we have the following result:

Key point

For any vector $\boldsymbol{r} = a\boldsymbol{i} + b\boldsymbol{j}$, its direction ratio is $a{:}b$. Its direction cosines are

$$l = \frac{a}{\sqrt{a^2 + b^2}}, \quad m = \frac{b}{\sqrt{a^2 + b^2}}$$

Example 2.11
Refer back to Example 2.10, in which point A has coordinates (3, 5), and point B has coordinates (7, 8).

(a) Find the direction ratio of the vector $\overrightarrow{AB}$.

(b) Find its direction cosines, l and m.

(c) Show that $l^2 + m^2 = 1$.

Solution

From Example 2.10 we know that $\overrightarrow{AB} = 4i + 3j$.

(a) The direction ratio of $\overrightarrow{AB}$ is therefore 4:3.

(b) The direction cosines are

$$l = \frac{4}{\sqrt{4^2 + 3^2}} = \frac{4}{5}, \quad m = \frac{3}{\sqrt{4^2 + 3^2}} = \frac{3}{5}$$

(c)
$$l^2 + m^2 = \left(\frac{4}{5}\right)^2 + \left(\frac{3}{5}\right)^2$$

$$= \frac{16}{25} + \frac{9}{25}$$

$$= \frac{25}{25}$$

$$= 1$$

The final result in the previous example is true in general:

Key point

If l and m are the direction cosines of a line lying in the xy plane, then

$$l^2 + m^2 = 1$$

Exercises

1 For the vectors in question 6 of the previous section, find
(a) the direction ratio of the vector $\overrightarrow{PQ}$
(b) the direction cosines of $\overrightarrow{PQ}$.

(c) Explain why the direction cosines are both negative.

Solutions to exercises

1 (a) $-9:-4$ (b) $-\dfrac{9}{\sqrt{97}}, -\dfrac{4}{\sqrt{97}}$

(c) The angle between the vector and the positive direction of both axes is greater

than 90° and hence each cosine is negative. (Draw the vector $\overrightarrow{PQ} = -9i - 4j$ to see this.)

2.4   Three-dimensional coordinate frames

The real world is three-dimensional, and in order to solve many engineering problems it is necessary to develop expertise in the mathematics of three-dimensional space. An important application of vectors is their use to locate points in three dimensions. When two distinct points are known we can draw a line between them. Three distinct points that do not lie on the same line form a plane. Vectors can be used to describe points, lines and planes in three dimensions. These mathematical foundations underpin much of the technology associated with computer graphics and the control of robots. In this section we shall introduce the vector methods that underlie these applications.

Figure 2.8 shows a three-dimensional coordinate frame. Note that the third dimension requires the addition of a third axis, the z axis. Although these three axes are drawn in the plane of the paper you should remember that we are now thinking of three-dimensional situations. Just as in two dimensions the x and y axes are perpendicular, in three dimensions the x, y and z axes are perpendicular to each other. We say they are **mutually perpendicular**. There is no reason why we could not have chosen the z axis in the opposite sense to that shown in Figure 2.8. However, it is conventional to choose the directions shown in Figure 2.8. Any point in the three dimensions can be defined in terms of its x, y and z coordinates. Consider the point P with coordinates (x, y, z) as shown. The vector from the origin to the point P is known as the **position vector** of P, $\overrightarrow{OP}$ or $\boldsymbol{r}$. To arrive at P from O we can think of moving x units in the x direction, y units in the y direction and z units in the z direction.

Figure 2.8
The position vector
of the point with
coordinates
(x, y, z) is
$\boldsymbol{r} = x\boldsymbol{i} + y\boldsymbol{j} + z\boldsymbol{k}$.

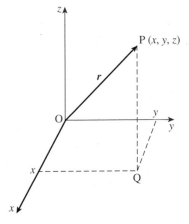

Extending the concept of unit vectors in the direction of each of the axes, a unit vector pointing in the positive direction of the z axis is denoted by $\boldsymbol{k}$. Noting that $\overrightarrow{OQ} = x\boldsymbol{i} + y\boldsymbol{j}$ and that $\overrightarrow{QP} = z\boldsymbol{k}$ we can state

$$\boldsymbol{r} = \overrightarrow{OP}$$

$$= \overrightarrow{OQ} + \overrightarrow{QP}$$

$$= x\boldsymbol{i} + y\boldsymbol{j} + z\boldsymbol{k}$$

We conclude that the position vector of the point with coordinates (x, y, z) is $r = xi + yj + zk$.

In column vector notation we would write

$$r = \begin{pmatrix} x \\ y \\ z \end{pmatrix}$$

Key point

The position vector of $P(x, y, z)$ is $r = \overrightarrow{OP} = xi + yj + zk$ or, alternatively,

$$r = \begin{pmatrix} x \\ y \\ z \end{pmatrix}$$

Example 2.12

State the position vector of the point with coordinates $(9, -8, 6)$.

Solution

The position vector is $9i - 8j + 6k$

Addition and subtraction

Addition and subtraction are carried out in an obvious way.

Example 2.13

If

$$r = \begin{pmatrix} 1 \\ 4 \\ -2 \end{pmatrix} \quad \text{and} \quad s = \begin{pmatrix} 4 \\ -3 \\ 2 \end{pmatrix}$$

find (a) $r + s$, (b) $r - s$.

Solution

(a) $\begin{pmatrix} 5 \\ 1 \\ 0 \end{pmatrix}$

(b) $\begin{pmatrix} -3 \\ 7 \\ -4 \end{pmatrix}$

Modulus

The modulus of the vector $\overrightarrow{OP}$ is equal to the distance OP, which can be obtained by Pythagoras's theorem.

Key point	If $r = xi + yj + zk$ then $\|r\| = \sqrt{x^2 + y^2 + z^2}$.

Example 2.14

Find the modulus of the vector $r = 4i + 2j + 3k$.

Solution

$\|r\| = $

$\sqrt{4^2 + 2^2 + 3^2} = \sqrt{16 + 4 + 9} = \sqrt{29}$

Example 2.15

Points A, B and C have coordinates $(-1, 1, 4)$, $(8, 0, 2)$ and $(5, -2, 11)$ respectively.

(a) Find the position vectors of A, B and C.

(b) Find $\overrightarrow{AB}$ and $\overrightarrow{BC}$.

(c) Find $|\overrightarrow{AB}|$ and $|\overrightarrow{BC}|$.

Solution

(a) Denoting the position vectors of A, B and C by a, b and c respectively, we find

$$a = -i + j + 4k, \quad b = 8i + 2k, \quad c = 5i - 2j + 11k$$

(b) $\overrightarrow{AB} = b - a = 9i - j - 2k, \quad \overrightarrow{BC} = c - b = -3i - 2j + 9k.$

(c) $|\overrightarrow{AB}| = \sqrt{9^2 + (-1)^2 + (-2)^2} = \sqrt{86},$

$|\overrightarrow{BC}| = \sqrt{(-3)^2 + (-2)^2 + 9^2} = \sqrt{94}$

Exercises

1 State the position vector of the point with coordinates $(4, -4, 3)$.

2 Find the modulus of each of the following vectors:
(a) $7i + 2j + 3k$ (b) $7i - 2j + 3k$
(c) $2j + 8k$ (d) $-i - 2j + 3k$
(e) $ai + bj + ck$

3 Points P, Q and R have coordinates $(9, 1, 0)$, $(8, -3, 5)$ and $(5, 5, 7)$ respectively.
(a) Find the position vectors of P, Q and R.
(b) Find $\overrightarrow{PQ}$ and $\overrightarrow{QR}$.
(c) Find $|\overrightarrow{PQ}|$ and $|\overrightarrow{QR}|$.

4 If $a = 5i - 2j + 3k$ and $b = 2i + j - 4k$ write down $b - a$.

Solutions to exercises

1 $4i - 4j + 3k$

2   (a) $\sqrt{62}$ (b) $\sqrt{62}$ (c) $\sqrt{68}$ (d) $\sqrt{14}$
    (e) $\sqrt{a^2 + b^2 + c^2}$

3   (a) $p = 9i + j, q = 8i - 3j + 5k,$
    $r = 5i + 5j + 7k$

(b) $\overrightarrow{PQ} = -i - 4j + 5k,$
    $\overrightarrow{QR} = -3i + 8j + 2k$

(c) $|\overrightarrow{PQ}| = \sqrt{42}, |\overrightarrow{QR}| = \sqrt{77}$

4 $-3i + 3j - 7k$

2.5   Direction ratios and direction cosines in three dimensions

The concepts of direction ratio and direction cosines extend naturally to three dimensions. Consider Figure 2.9.

Consider point P with position vector

$$r = ai + bj + ck$$

Figure 2.9
The direction cosines of $\overrightarrow{OP}$ are $\cos \alpha$, $\cos \beta$ and $\cos \gamma$.

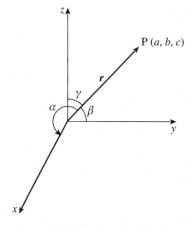

Its direction ratio is

$$a{:}b{:}c$$

This means that to move in the direction of the vector we must must move b units in the y direction and c units in the z direction for every a units in the x direction.

The direction cosines are the cosines of the angles between the vector and each of the axes. It is conventional to label them as l, m and n, and they are given by

$$l = \cos \alpha = \frac{a}{\sqrt{a^2 + b^2 + c^2}}, \quad m = \cos \beta = \frac{b}{\sqrt{a^2 + b^2 + c^2}},$$

$$n = \cos \gamma = \frac{c}{\sqrt{a^2 + b^2 + c^2}}$$

In general we have the following result:

Key point

> For any vector $r = ai + bj + ck$, its direction ratio is $a{:}b{:}c$. Its direction cosines are
>
> $$l = \frac{a}{\sqrt{a^2 + b^2 + c^2}}, \quad m = \frac{b}{\sqrt{a^2 + b^2 + c^2}}, \quad n = \frac{c}{\sqrt{a^2 + b^2 + c^2}}$$
>
> and
>
> $$l^2 + m^2 + n^2 = 1$$

Example 2.16
(a) Calculate the direction ratio of $r = 3i - 2j + k$.
(b) Calculate its direction cosines.

Solution
(a) The direction ratio is $3{:}{-}2{:}1$.
(b) The direction cosines are

$$l = \frac{3}{\sqrt{3^2 + (-2)^2 + 1^2}} = \frac{3}{\sqrt{14}}$$

$$m = \frac{-2}{\sqrt{3^2 + (-2)^2 + 1^2}} = -\frac{2}{\sqrt{14}}$$

$$n = \frac{1}{\sqrt{3^2 + (-2)^2 + 1^2}} = \frac{1}{\sqrt{14}}$$

Exercises

1 Points A and B have position vectors
$a = -3i + 2j + 7k$ and $b = 3i + 4j - 5k$
respectively. Find
(a) $\overrightarrow{AB}$
(b) $|\overrightarrow{AB}|$
(c) the direction ratios of $\overrightarrow{AB}$
(d) the direction cosines of $\overrightarrow{AB}$.
(e) Show that the sum of the squares of the
direction cosines equals 1.

2 Find the direction ratios, the direction cosines
and the angles that the vector $\overrightarrow{OP}$ makes with
each of the axes when P is the point with
coordinates $(2, 4, 3)$.

3 A line is inclined at $60°$ to the x axis and $45°$
to the y axis. Find its inclination to the z axis.

Solutions to exercises

1 (a) $6i + 2j - 12k$ (b) $\sqrt{184}$ (c) $6{:}2{:}{-}12$

   (d) $\dfrac{6}{\sqrt{184}}, \dfrac{2}{\sqrt{184}}, \dfrac{-12}{\sqrt{184}}$

2 $2{:}4{:}3;\ \dfrac{2}{\sqrt{29}}, \dfrac{4}{\sqrt{29}}, \dfrac{3}{\sqrt{29}};\ 68.2°, 42.0°, 56.1°$

3 $60°$ or $120°$

2.6 *N*-dimensional vectors

The vectors we have described so far have been either two- or three-dimensional. However, there are some situations when it is desirable to generalise what we have done to higher dimensions. No geometrical interpretation is then available. Nevertheless we can still refer to the components of a vector and the modulus of a vector.
   Two examples of four-dimensional column vectors are

$$a = \begin{pmatrix} 3 \\ 1 \\ 2 \\ 4 \end{pmatrix} \quad \text{and} \quad b = \begin{pmatrix} 1 \\ 0 \\ 3 \\ 1 \end{pmatrix}$$

The modulus of a, which is generally referred to as its **norm**, is then defined as

$$|a| = \sqrt{3^2 + 1^2 + 2^2 + 4^2}$$

$$= \sqrt{30}$$

An n-dimensional vector will have n components.

Example 2.17 Electronic Engineering – Vector norms and digital signal processing

Signals are quantities used to transmit information. If a signal changes, then this corresponds to information being conveyed. For example, the sequence of symbols

$$00000000000000000000$$

contains no information, whereas

$$00000000010000000000$$

contains information relating to a change occurring in the middle of the sequence. Consider a signal such as that depicted in Figure 2.10, which corresponds to the sequence of symbols 0, 0, 3, 5, 0.

Figure 2.10

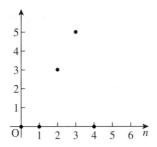

This sequence can also be represented by a vector: for example, we could write it as the **row vector**

$$f = (0, 0, 3, 5, 0)$$

The **norm** of the signal is the modulus of the vector f, defined, by generalising the two- and three-dimensional case, as

$$|f| = \sqrt{0^2 + 0^2 + 3^2 + 5^2 + 0^2}$$
$$= \sqrt{34}$$

More generally a vector, f, with N components can represent a signal of length N, that is

$$f = (f_0, f_1, \ldots, f_{N-1})$$

and its norm is given by

$$|f| = \sqrt{\sum_{n=0}^{N-1} f_n^2}$$

A related quantity is the **power** in a signal, which is given by

$$\frac{1}{N}|f|^2 = \frac{1}{N}\sum_{n=0}^{N-1} f_n^2$$

Therefore the power in the signal $(0, 0, 3, 5, 0)$ is $\frac{34}{5}$.

One of the main problems in signal processing is that of representing a large amount of information by a smaller amount without significantly distorting the original signal. Very often a signal, f, with a large information content is approximated by f', which is a signal with a smaller information content. Then

$$|f - f'|$$

is a measure of the 'distance' or difference between the two signals and is therefore a measure of the quality of the approximation.

End of block exercises

1 If

$$p = \begin{pmatrix} 3 \\ 1 \\ 2 \end{pmatrix} \quad \text{and} \quad q = \begin{pmatrix} 7 \\ 9 \\ 1 \end{pmatrix}$$

find (a) $3p$, (b) $p - q$, (c) $-2q$, (d) $3p - 2q$.

2 If

$$x_1 = \begin{pmatrix} 1 \\ 2 \\ 3 \end{pmatrix}, \quad x_2 = \begin{pmatrix} 1 \\ 0 \\ 1 \end{pmatrix}, \quad x_3 = \begin{pmatrix} 2 \\ 2 \\ 4 \end{pmatrix}$$

show that $x_1 + x_2 - x_3 = 0$ where 0 is the zero vector.

3 A triangle has vertices A, B and C with coordinates $(1, 3, 2)$, $(-1, 5, 9)$ and $(2, 7, 1)$ respectively. Calculate the vectors that represent the sides of the triangle.

4 Find a unit vector in the direction of the vector $i - j$.

5 Find the norm of the row vector $x = (2, 3, 6)$.

6 If

$$a = \begin{pmatrix} 1 \\ 1 \\ 0 \\ 1 \\ 1 \end{pmatrix} \quad \text{and} \quad b = \begin{pmatrix} 3 \\ 2 \\ 1 \\ 0 \\ 1 \end{pmatrix}$$

find the norm of a, the norm of b, and the norm of $a - b$.

Solutions to exercises

1 (a) $\begin{pmatrix} 9 \\ 3 \\ 6 \end{pmatrix}$ (b) $\begin{pmatrix} -4 \\ -8 \\ 1 \end{pmatrix}$

(c) $\begin{pmatrix} -14 \\ -18 \\ -2 \end{pmatrix}$ (d) $\begin{pmatrix} -5 \\ -15 \\ 4 \end{pmatrix}$

3 $\overrightarrow{AB} = -2i + 2j + 7k$,
$\overrightarrow{BC} = 3i + 2j - 8k$,
$\overrightarrow{AC} = i + 4j - k$

4 $\dfrac{1}{\sqrt{2}}(i - j)$

5 7

6 $2, \sqrt{15}, \sqrt{7}$

The scalar product, or dot product

3.1 Introduction

Multiplication of two vectors is defined in two different ways. The first way is known as the **scalar product**. When the scalar product of two vectors is calculated the result is a scalar and not a vector. The second way is known as the **vector product**. When this is calculated the result is a vector. These products are summarised in Table 3.1. The ways in which these products are defined seem rather strange when first met, but the reason for this lies in their applications. In this block we consider only the scalar product.

Table 3.1
The two types of vector multiplication.

Name	Written as	Result
Scalar product	$a \cdot b$	Scalar
Vector product	$a \times b$	Vector

3.2 Definition of the scalar product

Consider the two vectors a and b shown in Figure 3.1.

Figure 3.1
Two vectors separated by an angle θ.

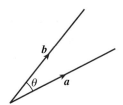

Note that the tails of the two vectors coincide and that the angle between the vectors has been labelled θ. Their scalar product, denoted $a \cdot b$, is defined as

$$a \cdot b = |a|\, |b| \cos \theta$$

It is very important to use the dot in the formula. The dot is the symbol for the scalar product, and is the reason why the scalar product is also known as the **dot product**.

You should never use a $\times$ sign in this context because the $\times$ sign is reserved for the vector product, which is quite different.

 Key point

Scalar product

$$a \cdot b = |a| \, |b| \cos \theta$$

We can remember this formula as: the length of the first vector times the length of the second times the cosine of the angle in between them.

Example 3.1
Vectors a and b are shown in Figure 3.2. The vector a has modulus 6, the vector b has modulus 7, and the angle between them is 60°. Calculate $a \cdot b$.

Figure 3.2
When the tails of a and b coincide the angle between them is 60°.

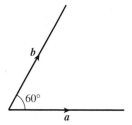

Solution
Identify the angle between the two vectors, and also their moduli. Then apply the formula for the dot product.

$$a \cdot b = |a| \, |b| \cos \theta = (6)(7) \cos 60° = 21$$

The scalar product of a and b is equal to 21. Note that when finding a scalar product the result is always a scalar.

Example 3.2
Find $i \cdot i$, where i is a unit vector in the direction of the positive x axis.

Solution
Because i is a unit vector its modulus is 1. The angle between any two parallel vectors is zero. So the angle between i and itself is zero. Therefore

$$i \cdot i = (1)(1) \cos 0°$$
$$= 1 \text{ since } \cos 0° = 1$$

So the scalar product of i with itself equals 1. It is easy to verify that $j \cdot j = 1$ and $k \cdot k = 1$.

Example 3.3

Find $i \cdot j$, where i and j are unit vectors in the directions of the x and y axes.

Solution

Because i and j are unit vectors they both have a modulus of 1. The angle between the two vectors is 90° as shown in Figure 3.3.

Figure 3.3
The angle between the unit vectors i and j is 90°.

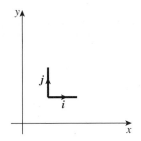

Therefore

$$i \cdot j = (1)(1) \cos 90°$$

$$= 0 \quad \text{since } \cos 90° = 0$$

That is, $i \cdot j = 0$.

More generally, the following results are easily verified:

Key point

$$i \cdot i = j \cdot j = k \cdot k = 1$$
$$i \cdot j = i \cdot k = j \cdot k = 0$$

Exercises

1   Write down the definition of the scalar product of two vectors and illustrate your definition with a diagram.

2   Use the definition of the scalar product to show that, if two vectors are perpendicular, their scalar product is zero.

3   Two vectors have moduli 7 and 13 respectively. The angle between them is 45°. Evaluate their scalar product.

4   Two unit vectors are parallel. What can you deduce about their scalar product?

5   Two vectors have modulus 10 and 12. The angle between them is $\frac{\pi}{3}$. Find their scalar product.

Solutions to exercises

3 64.35

5 60

4 It equals 1.

3.3 Some properties of the scalar product

The scalar product has a number of properties that are necessary for further development.

Commutativity

If $a \cdot b = |a| \, |b| \cos \theta$ then it follows that $b \cdot a = |b| \, |a| \cos \theta$. Since $|b|$, $|a|$ and $\cos \theta$ are numbers, and multiplication of numbers is commutative,

$$|a| \, |b| \cos \theta = |b| \, |a| \cos \theta$$

Clearly then $a \cdot b = b \cdot a$. Thus we can evaluate a scalar product in any order – the operation is commutative.

Distributivity

For three vectors a, b and c

$$a \cdot (b + c) = a \cdot b + a \cdot c$$

This means that the scalar product is distributive over addition.

Example 3.4
Simplify $i \cdot (i + j)$.

Solution
Using the distributivity property,

$$i \cdot (i + j) = i \cdot i + i \cdot j$$

which simplifies to $1 + 0 = 1$.

Multiplication by a scalar

For two vectors a and b, and a scalar k,

$$k(a \cdot b) = (ka) \cdot b$$
$$= a \cdot (kb)$$

Example 3.5
Simplify $i \cdot 7i$.

Solution
Using this rule, with $k = 7$, $i \cdot (7i)$ can be written $7(i \cdot i)$. So

$$i \cdot 7i = 7(i \cdot i)$$
$$= 7(1) = 7$$

Example 3.6
Simplify $(3i + 2j) \cdot (7i + 4j)$.

Solution
Using the distributivity property we can write

$$(3i + 2j) \cdot (7i + 4j) = (3i + 2j) \cdot (7i) + (3i + 2j) \cdot (4j)$$

This can be expanded further to give

$$3i \cdot 7i + 2j \cdot 7i + 3i \cdot 4j + 2j \cdot 4j$$

This simplifies to

$$(3)(7)i \cdot i + (2)(7)j \cdot i + (3)(4)i \cdot j + (2)(4)j \cdot j$$

Finally

$$(3i + 2j) \cdot (7i + 4j) = 21 + 0 + 0 + 8$$
$$= 29$$

Exercise

1   If a and b are perpendicular, simplify

$$(a - 2b) \cdot (3a + 5b)$$

Solution to exercise

1   $3a^2 - 10b^2$

3.4   A formula for finding the scalar product

We can use the previous results to obtain a formula for finding a scalar product when the vectors are given in cartesian form. Suppose $a = a_1 i + a_2 j$ and $b = b_1 i + b_2 j$. Then

$$
\begin{aligned}
a \cdot b &= (a_1 i + a_2 j) \cdot (b_1 i + b_2 j)\\
&= a_1 i \cdot (b_1 i + b_2 j) + a_2 j \cdot (b_1 i + b_2 j)\\
&= a_1 b_1 i \cdot i + a_1 b_2 i \cdot j + a_2 b_1 j \cdot i + a_2 b_2 j \cdot j
\end{aligned}
$$

Now, since $i \cdot i = j \cdot j = 1$, and $i \cdot j = j \cdot i = 0$, we can simplify this to give the following formula:

Key point

If $a = a_1 i + a_2 j$ and $b = b_1 i + b_2 j$ then

$$a \cdot b = a_1 b_1 + a_2 b_2$$

Thus to find the scalar product of two vectors their i components are multiplied together, their j components are multiplied together and the results are added.

Example 3.7
If $a = 7i + 8j$ and $b = 5i - 2j$, find the scalar product $a \cdot b$.

Solution
We use the previous key point and multiply corresponding components together, adding the results.

$$
\begin{aligned}
a \cdot b &= (7i + 8j) \cdot (5i - 2j)\\
&= (7)(5) + (8)(-2)\\
&= 35 - 16\\
&= 19
\end{aligned}
$$

Note that the answer is a scalar.

Example 3.8
If $p = 5i - 3j$ and $q = 2i + j$, find the scalar product $p \cdot q$.

Solution

$$(5)(2) + (-3)(1) = 7$$

This result readily generalises to vectors in three dimensions as follows:

Key point

If $a = a_1 i + a_2 j + a_3 k$ and $b = b_1 i + b_2 j + b_3 k$ then

$$a \cdot b = a_1 b_1 + a_2 b_2 + a_3 b_3$$

Example 3.9
If $a = 5i + 3j - 2k$ and $b = 8i - 9j + 11k$, find $a \cdot b$.

Solution
Corresponding components are multiplied together and the results are added.

$$a \cdot b = $$

$$(5)(8) + (3)(-9) + (-2)(11) = 40 - 27 - 22$$
$$= -9$$

Note again that the result is a scalar: there are no *i*s, *j*s or *k*s in the answer.

Example 3.10
If $p = 4i - 3j + 7k$ and $q = 6i - j + 2k$, find $p \cdot q$.

Solution
Corresponding components are multiplied together and the results are added.

41

Example 3.11
If $r = 3i + 2j + 9k$ find $r \cdot r$. Show that this is the same as $|r|^2$.

Solution

94

Exercises

1 If $a = 2i - 5j$ and $b = 3i + 2j$ find $a \cdot b$ and verify that $a \cdot b = b \cdot a$.

2 If $p = i + 8j + 7k$ and $q = 3i - 2j + 5k$, find $p \cdot q$.

3 Show that the vectors $\frac{1}{2}i + j$ and $2i - j$ are perpendicular.

4 The work done by a force F in moving a body through a displacement r is given by $F \cdot r$. A force $F = 3i + 7k$ causes a body to move from the point with coordinates (1, 1, 2) to the point (7, 3, 5).
   (a) Find the displacement, r, of the body.
   (b) Find the work done by the force.

Solutions to exercises

1 -4

2 22

4 (a) $6i + 2j + 3k$ (b) 39 units.

3.5   Using the scalar product to find the angle between two vectors

Using the scalar product we can find the angle between two vectors.

Example 3.12
Find the angle between the vectors $a = 5i + 3j - 2k$ and $b = 8i - 9j + 11k$ of Example 3.9.

Solution

Their scalar product has already been found to be -9. The modulus of a is

$$\sqrt{5^2 + 3^2 + (-2)^2} = \sqrt{38}$$

The modulus of b is

$$\sqrt{8^2 + (-9)^2 + 11^2} = \sqrt{266}$$

Substituting these into the formula for the scalar product we find

$$a \cdot b = |a|\,|b| \cos \theta$$

$$-9 = \sqrt{38}\,\sqrt{266} \cos \theta$$

from which

$$\cos \theta = \frac{-9}{\sqrt{38}\,\sqrt{266}}$$

$$= -0.0895$$

so that

$$\theta = \cos^{-1}(-0.0895)$$

$$= 95.14°$$

In general, the angle between two vectors can be found from the following formula:

Key point

$$\cos \theta = \frac{a \cdot b}{|a|\,|b|}$$

Exercises

1   Find the angle between $p = 3i - j$ and
$q = -4i + 6j$.

2   Find the angle between the vectors $i - j - k$
and $2i + j + 2k$.

Solutions to exercises

1   142.1°

2   101.1°

3.6 Using the scalar product to find the component of a vector in the direction of another vector

Consider Figure 3.4, which shows arbitrary vectors a and n. Let $\hat{n}$ be a **unit vector** in the direction of n.

Figure 3.4
OQ is the component of a in the direction of n.

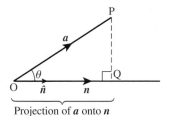

Projection of a onto n

Study the figure carefully and note that a line has been drawn from P to meet n at right angles. The distance OQ is called the **projection** of a onto n, or alternatively 'the component of a in the direction of n'. Simple trigonometry tells us that

$$\cos \theta = \frac{\text{length of projection OQ}}{|a|}$$

so that the length of the projection is $|a| \cos \theta$.

Now by taking the scalar product of a with the unit vector $\hat{n}$ we find

$$a \cdot \hat{n} = |a| |\hat{n}| \cos \theta$$

$$= |a| \cos \theta \quad \text{since } |\hat{n}| = 1$$

We conclude that $a \cdot \hat{n}$ is the length of the projection OQ.

Key point

$a \cdot \hat{n}$ is the length of the projection of a onto n

alternatively

$a \cdot \hat{n}$ is the component of a in the direction of n

Example 3.13
Find the component of the vector $a = 2i + 3j$ in the direction of the vector $n = i + 5j$.

Solution
First we need a unit vector in the direction of n.

$$\hat{n} = \frac{n}{|n|}$$

$$= \frac{1}{\sqrt{26}} (i + 5j)$$

Then the component of *a* in the direction of *n* is

$$a \cdot \hat{n} = (2i + 3j) \cdot \frac{1}{\sqrt{26}}(i + 5j)$$

$$= \frac{1}{\sqrt{26}}(2i + 3j) \cdot (i + 5j)$$

$$= \frac{1}{\sqrt{26}}(2 + 15)$$

$$= \frac{17}{\sqrt{26}}$$

$$= 3.334$$

This is illustrated in Figure 3.5.

Figure 3.5
The projection of *a*
onto *n* is 3.334.

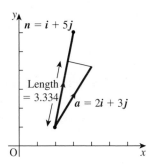

Example 3.14

Figure 3.6 shows a plane containing the point A with position vector *a*. The vector *n* is a perpendicular, or **normal**, to the plane. Find an expression for the perpendicular distance of the plane from the origin.

Figure 3.6
OB is the
perpendicular
distance of the
plane from the
origin.

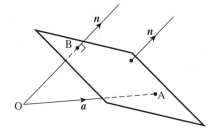

Solution

From the figure note that the perpendicular distance of the plane from the origin, which is OB, is the projection of *a* onto *n*. Thus the perpendicular distance of the plane from the origin is *a* · *n̂*.

Example 3.15 The scalar product and digital signal recognition and identification

A problem arising in digital signal analysis is to attempt to compare a given but unrecognised signal with a collection of known ones. Such recognition and identification problems arise for example when astronomers interpret signals received from outer space and attempt to identify the objects from which they originated, perhaps stars or galaxies.

If the unrecognised signal resembles one of the known ones sufficiently well it may be possible to identify it. To quantify this resemblance we can make use of vector methods.

Suppose that the unrecognised digital signal is represented by the vector x as shown in Figure 3.7 whereas the vectors e_1, e_2, ..., e_N represent known signals. Usually these vectors are **normalised**, that is they are unit vectors. We seek the particular unit vector that most closely resembles x. The 'distance' between the unrecognised vector and any one of the known vectors, e say, is defined as

$$|x - e|$$

Figure 3.7

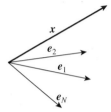

We would like to choose the particular e for which this distance is minimal. First consider the square of this distance $|x - e|^2$. Recall that, for any vector a, then $|a|^2 = a \cdot a$. So,

$$|x - e|^2 = (x - e) \cdot (x - e)$$
$$= x \cdot x + e \cdot e - e \cdot x - x \cdot e$$
$$= |x|^2 + 1 - 2x \cdot e$$

since e is a unit vector and the scalar product is commutative. We see that the square distance, and hence the distance between x and e, is minimal when $x \cdot e$ is maximal. So, the unit vector most closely resembling x is the one having the largest scalar product with x.

For example, suppose an unrecognised signal is represented by

$$x = (4, 3, 6)$$

while the known unit vectors are

$$e_1 = (1, 0, 0), \quad e_2 = \frac{1}{\sqrt{6}}(1, 2, 1) \quad \text{and} \quad e_3 = \frac{1}{3}(2, 2, -1)$$

Evaluating the three scalar products in turn we find

$$x \cdot e_1 = 4, \quad x \cdot e_2 = \frac{16}{\sqrt{6}} = 6.53, \quad x \cdot e_3 = \frac{8}{3} = 2.67$$

We conclude that x most closely resembles e_2.

Exercises

1 (a) What is the component of the vector $2i + 7j$ in the direction of the vector i?

   (b) What is the component of $2i + 7j$ in the direction of the vector j? Interpret your results graphically.

2   Find the component of the vector

$$\begin{pmatrix} 2 \\ 1 \\ 7 \end{pmatrix}$$

in the direction of the vector

$$\begin{pmatrix} 1 \\ 1 \\ 1 \end{pmatrix}$$

3   What is the projection of the vector $2i - j + k$ onto the vector $3i - 2k$?

Solutions to exercises

1   (a) 2 (b) 7. These are just the x and y components.

2   $\dfrac{10}{\sqrt{3}}$

3   $\dfrac{4}{\sqrt{13}}$

End of block exercises

1   If $a = 4i + 6j$ and $b = 3i - 3j$ find $a \cdot b$ and verify that $a \cdot b = b \cdot a$.

2   Find the angle between $p = 2i + j$ and $q = 5i + 11j$.

3   If a and b are perpendicular, simplify $(5a + 2b) \cdot (a + 7b)$.

4   If $a = 5i + 3j + 7k$ and $b = i - 2j - 7k$, find $a \cdot b$.

5   Show that the vectors $7i + 2j + k$ and $i - 4j + k$ are perpendicular.

6   Find the angle between the vectors $2i + j - k$ and $5i + j - 3k$.

7   If

$$r = \begin{pmatrix} x \\ y \\ z \end{pmatrix} \quad \text{and} \quad a = \begin{pmatrix} 3 \\ 2 \\ 5 \end{pmatrix}$$

find $r \cdot a$.

8   Find the component of the vector $7i + 2j - k$ in the direction of the vector $i - j + 2k$.

9   Write down an expression for the component of a in the direction of b.

10 A rhombus is a parallelogram in which all sides have equal length. Use the scalar product to show that the diagonals of a rhombus intersect at 90°.

Solutions to exercises

1   -6

2   $39°$

3   $5a^2 + 14b^2$

4   -50

6   $15°$

7   $3x + 2y + 5z$

8   $\dfrac{3}{\sqrt{6}}$

9   $a \cdot \hat{b}$

The vector product, or cross product

4.1 Introduction

In this block we describe how to find the **vector product** of two vectors. Like the scalar product its definition may seem strange when first met but it is defined in this way because of its many applications. When multiplying vectors using the vector product the result is always a vector. To understand how this vector is formed it is helpful to consider first the right-handed screw rule.

4.2 The right-handed screw rule

Consider the two vectors a and b shown in Figure 4.1.

Figure 4.1
Two vectors separated by an angle θ.

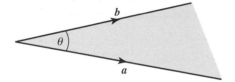

The two vectors lie in a plane; this plane is shaded in Figure 4.1. Figure 4.2 shows the same two vectors and the plane in which they lie together with a unit vector, denoted $\hat{e}$, which is perpendicular to this plane. Imagine turning a right-handed screw in the sense from a towards b as shown. A right-handed screw is one that when turned clockwise enters the material into which it is being screwed. You will see from Figure 4.2 that the screw will advance in the direction of $\hat{e}$.

Figure 4.2
A right-handed screw turned from a towards b will advance in the direction of $\hat{e}$.

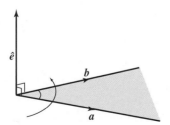

Example 4.1

If a right-handed screw is turned from b towards a, in which direction will it advance? Show this on a diagram similar to Figure 4.2.

Solution

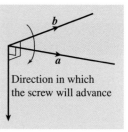

Direction in which the screw will advance

We are now in a position to describe the vector product.

4.3 Definition of the vector product

The result of finding the vector product of two vectors a and b is a vector of modulus $|a|\,|b|\sin\theta$ in the direction of $\hat{e}$, where $\hat{e}$ is a unit vector perpendicular to the plane containing a and b in a sense defined by the right-handed screw rule, as we imagine turning the screwdriver from a towards b, as shown in Figure 4.3.

Figure 4.3
The vector $a \times b$ is perpendicular to the plane containing a and b.

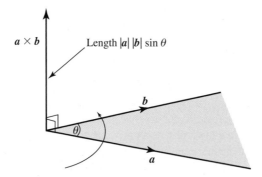

Length $|a|\,|b|\sin\theta$

The symbol we shall use for the vector product is the times sign $\times$, which is why the vector product is also called the cross product. The quantity $a \times b$ is read as a cross b.

Key point

Vector product

$$a \times b = |a|\,|b|\sin\theta\,\hat{e}$$

Note that $|a||b|\sin\theta$ is the modulus of the vector product whereas $\hat{e}$ gives the direction.

Now study Figure 4.4, which is used to illustrate the calculation of $b \times a$. In particular note the direction of $b \times a$ arising through the application of the right-handed screw rule.

Figure 4.4
Calculation of
$b \times a$.

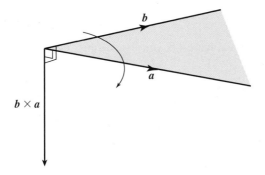

We see that $a \times b$ is not equal to $b \times a$ because their directions are different. In fact $a \times b = -b \times a$.

Example 4.2
If a and b are parallel, show that $a \times b = 0$, the zero vector.

Solution
Use the definition: $a \times b = |a||b|\sin\theta\,\hat{e}$. What is the angle θ given that a and b are parallel?

$$\theta = 0°$$

If a and b are parallel then the angle between them is zero. Consequently $\sin\theta = 0$. It follows that $a \times b = 0$. Note that the result, 0, is the zero vector.

Note in particular the following important results:

Key point

$$i \times i = 0, \quad j \times j = 0, \quad k \times k = 0$$

Example 4.3
Show that $i \times j = k$ and find expressions for $j \times k$ and $k \times i$.

Solution
Note that i and j are perpendicular so that the angle between them is 90°. So the modulus of $i \times j$ is $(1)(1)\sin 90° = 1$. The unit vector perpendicular to i and j in the sense defined by the right-handed screw rule is k as shown in Figure 4.5. Therefore $i \times j = k$ as required.

Figure 4.5
The vector k is perpendicular to both i and j.

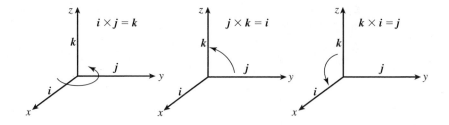

Similarly, by referring to Figure 4.5, you should verify that $j \times k = i$ and $k \times i = j$.

Key point	
	$i \times j = k, \quad j \times k = i, \quad k \times i = j$
	$j \times i = -k, \quad k \times j = -i, \quad i \times k = -j$

Exercises

1   Write down the definition of the vector product of two vectors p and q. Illustrate the definition with a diagram showing clearly the direction of $p \times q$.

2   Explain why, for arbitrary vectors, $p \times q$ is not equal to $q \times p$.

3   Show that if a and b are parallel vectors then their vector product is the zero vector.

4.4   Some properties of the vector product

The vector product has a number of properties that are necessary for further development.

Non-commutativity

We have already seen that $a \times b = -b \times a$.

Distributivity

For three vectors a, b and c

$$a \times (b + c) = a \times b + a \times c$$

This means that the vector product is distributive over addition.

Example 4.4
Simplify $i \times (i + j)$.

Solution
Using the distributivity property,

$$i \times (i + j) = i \times i + i \times j$$

which simplifies to k.

Multiplication by a scalar

For two vectors a and b, and a scalar k

$$k\,(a \times b) = (ka) \times b$$
$$= a \times (kb)$$

Example 4.5
Simplify $i \times 5k$.

Solution
Using this rule,

$$i \times 5k = 5(i \times k)$$
$$= 5(-j)$$
$$= -5j$$

4.5   A formula for finding the vector product

We can use the properties of the vector product to develop a formula for finding the vector product of two vectors given in cartesian form. Suppose $a = a_1 i + a_2 j + a_3 k$ and $b = b_1 i + b_2 j + b_3 k$; then

$$a \times b = (a_1 i + a_2 j + a_3 k) \times (b_1 i + b_2 j + b_3 k)$$
$$= a_1 i \times (b_1 i + b_2 j + b_3 k)$$
$$+ a_2 j \times (b_1 i + b_2 j + b_3 k)$$
$$+ a_3 k \times (b_1 i + b_2 j + b_3 k)$$
$$= a_1 b_1 (i \times i) + a_1 b_2 (i \times j) + a_1 b_3 (i \times k)$$
$$+ a_2 b_1 (j \times i) + a_2 b_2 (j \times j) + a_2 b_3 (j \times k)$$
$$+ a_3 b_1 (k \times i) + a_3 b_2 (k \times j) + a_3 b_3 (k \times k)$$

Using the previous two key points this expression simplifies to

$$a \times b = (a_2 b_3 - a_3 b_2)i - (a_1 b_3 - a_3 b_1)j + (a_1 b_2 - a_2 b_1)k$$

Key point	If $a = a_1i + a_2j + a_3k$ and $b = b_1i + b_2j + b_3k$ then
	$$a \times b = (a_2b_3 - a_3b_2)i - (a_1b_3 - a_3b_1)j + (a_1b_2 - a_2b_1)k$$

Example 4.6
Evaluate the vector product $a \times b$ if $a = 3i - 2j + 5k$ and $b = 7i + 4j - 8k$.

Solution
Identifying $a_1 = 3, a_2 = -2, a_3 = 5, b_1 = 7, b_2 = 4, b_3 = -8$ we find

$$a \times b = [(-2)(-8) - (5)(4)]i - [(3)(-8) - (5)(7)]j$$
$$+ [(3)(4) - (-2)(7)]k$$
$$= -4i + 59j + 26k$$

Example 4.7
Evaluate the vector product $b \times a$ where a and b are the vectors in Example 4.6.

Solution

$4i - 59j - 26k$

Example 4.8
Use the key point to find the vector product of $p = 3i + 5j$ and $q = 2i - j$.

Solution
Note that in this example there are no k components so a_3 and b_3 are both zero. Apply the formula:

$p \times q =$

$-13k$

Exercises

1 Find the vector product of $p = -2i - 3j$ and $q = 4i + 7j$.

2 (a) Find $a \times b$ when $a = 13j$ and $b = -7i + 2j$.
   (b) Verify that $a \times b = -b \times a$.

3 Find $|r_1 \times r_2|$ when $r_1 = 2i + 3j$ and $r_2 = -7i + 2j$.

Solutions to exercises

1 $-2k$

2 (a) $91k$

3 25

4.6 Using determinants to evaluate a vector product

Evaluation of a vector product using the previous formula is very cumbersome. A more convenient and easily remembered method is to use determinants.

The vector product of two vectors $a = a_1 i + a_2 j + a_3 k$ and $b = b_1 i + b_2 j + b_3 k$ can be found by evaluating the determinant:

$$a \times b = \begin{vmatrix} i & j & k \\ a_1 & a_2 & a_3 \\ b_1 & b_2 & b_3 \end{vmatrix}$$

To find the i component of the vector product, imagine crossing out the row and column containing i and finding the determinant of what is left, that is

$$\begin{vmatrix} a_2 & a_3 \\ b_2 & b_3 \end{vmatrix} = a_2 b_3 - a_3 b_2$$

The resulting number is the i component of the vector product. The j component is found by crossing out the row and column containing j and evaluating

$$\begin{vmatrix} a_1 & a_3 \\ b_1 & b_3 \end{vmatrix} = a_1 b_3 - a_3 b_1$$

and then changing the sign of the result. Finally the k component is found by crossing out the row and column containing k and evaluating

$$\begin{vmatrix} a_1 & a_2 \\ b_1 & b_2 \end{vmatrix} = a_1 b_2 - a_2 b_1$$

Key point

If $a = a_1 i + a_2 j + a_3 k$ and $b = b_1 i + b_2 j + b_3 k$ then

$$a \times b = \begin{vmatrix} i & j & k \\ a_1 & a_2 & a_3 \\ b_1 & b_2 & b_3 \end{vmatrix}$$

$$= (a_2 b_3 - a_3 b_2) i - (a_1 b_3 - a_3 b_1) j + (a_1 b_2 - a_2 b_1) k$$

Example 4.9
Find the vector product of $a = 3i - 4j + 2k$ and $b = 9i - 6j + 2k$.

Solution
The two given vectors are represented in the determinant

$$\begin{vmatrix} i & j & k \\ 3 & -4 & 2 \\ 9 & -6 & 2 \end{vmatrix}$$

Evaluating this determinant we obtain

$$a \times b = [-8 - (-12)]i - (6 - 18)j + [-18 - (-36)]k$$

$$= 4i + 12j + 18k$$

Example 4.10
Find the vector product of $a = 7i + j + 2k$ and $b = i + 3j - 2k$.

Solution

$$\begin{vmatrix} i & j & k \\ & & \\ & & \end{vmatrix} = \qquad -8i + 16j + 20k$$

Example 4.11 Mechanical Engineering – The moment of a force
Figure 4.6 shows a force F applied to an object causing the object to rotate around a fixed axis. The force is applied at a point with position vector r. The moment of the force about O is defined to be the vector M_O given by

$$M_O = r \times F$$

Figure 4.6
The moment of a force is found by evaluating a vector product.

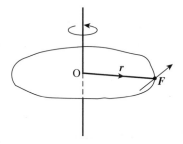

The magnitude of M_O is a measure of the turning effect of the force.
(a) Calculate the moment about O of the force $F = 2i + 3j + k$ newtons applied at the point with position vector $3j$.
(b) Calculate its magnitude.

Solution
(a)

$$M_O = \begin{vmatrix} i & j & k \\ 0 & 3 & 0 \\ 2 & 3 & 1 \end{vmatrix}$$

$$= 3i - 6k$$

(b) The magnitude of this moment is $\sqrt{3^2 + (-6)^2} = \sqrt{45}$ N m.

Example 4.12 The area of a triangle
The area of the triangle shown in Figure 4.7 is given by the formula area $= \frac{1}{2}bc \sin A$. Show that an equivalent formula is

$$\text{area} = \tfrac{1}{2}|\overrightarrow{AB} \times \overrightarrow{AC}|$$

Figure 4.7
The area of a triangle can be found by evaluating a vector product.

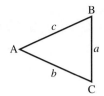

Solution

From the definition of the vector product

$$|\overrightarrow{AB} \times \overrightarrow{AC}| = |\overrightarrow{AB}||\overrightarrow{AC}|\sin A$$

since A is the angle between $\overrightarrow{AB}$ and $\overrightarrow{AC}$. Furthermore $|\overrightarrow{AB}| = c$ and $|\overrightarrow{AC}| = b$. Therefore

$$\text{area} = \tfrac{1}{2}\,bc\,\sin A$$

$$= \tfrac{1}{2}|\overrightarrow{AB} \times \overrightarrow{AC}|$$

as required.

Example 4.13

Find the area of the triangle that has vertices at the following points: A(0, 7, 1), B(1, 3, 2) and C(−2, 0, 3).

Solution

To apply the result of Example 4.12 we need to find the vectors $\overrightarrow{AB}$ and $\overrightarrow{AC}$.

We are given the coordinates of the vertices. Write down the position vectors of these points.

$$\overrightarrow{OA} = 7\boldsymbol{j} + \boldsymbol{k},\ \overrightarrow{OB} = \boldsymbol{i} + 3\boldsymbol{j} + 2\boldsymbol{k} \text{ and } \overrightarrow{OC} = -2\boldsymbol{i} + 3\boldsymbol{k}$$

Hence find vectors $\overrightarrow{AB}$ and $\overrightarrow{AC}$.

$$\overrightarrow{AB} = \boldsymbol{i} - 4\boldsymbol{j} + \boldsymbol{k} \text{ and } \overrightarrow{AC} = -2\boldsymbol{i} - 7\boldsymbol{j} + 2\boldsymbol{k}$$

Now find $\overrightarrow{AB} \times \overrightarrow{AC}$ and apply the formula of Example 4.12.

$$\overrightarrow{AB} \times \overrightarrow{AC} = \qquad\qquad -\boldsymbol{i} - 4\boldsymbol{j} - 15\boldsymbol{k}$$

$$\text{area} = \qquad\qquad \tfrac{11}{2}\sqrt{2}$$

Exercises

1   If $p = 3i + 2j - k$ and $q = 5i - j - k$ find $p \times q$.

2   If $a = 12i + 13j$ and $b = 7i + 3j$ find $a \times b$.

3   If $a = i + 2j + 3k$ and $b = 4i + 3j + 2k$ find $a \times b$. Show that $a \times b \neq b \times a$.

4   Points A, B and C have coordinates $(9, 1, -2)$, $(3, 1, 3)$ and $(1, 0, -1)$ respectively. Find the vector product $\overrightarrow{AB} \times \overrightarrow{AC}$.

5   Find a vector that is perpendicular to both of the vectors $a = i + 2j + 7k$ and $b = i + j - 2k$. Hence find a unit vector that is perpendicular to both a and b.

6   Find a vector that is perpendicular to the plane containing $6i + k$ and $2i + j$.

7   For the vectors $a = 4i + 2j + k$, $b = i - 2j + k$ and $c = 3i - 3j + 4k$, evaluate $a \times (b \times c)$ and $(a \times b) \times c$.

Deduce that, in general, the vector product is not associative.

8   A force of magnitude 3 units acts at the point with coordinates $(1, 2, 3)$. The force is applied in the direction of the vector $3i - j + 4k$. Find the moment of the force about O. What is the moment of the force about the point with coordinates $(1, 2, 3)$?

9   Find the area of the triangle with vertices at the points with coordinates $(1, 2, 3)$, $(4, -3, 2)$ and $(8, 1, 1)$.

10   For the vectors $a = i + 2j + 3k$, $b = 2i - 2j - 5k$ and $c = i - 3j - k$, evaluate
(a) $(a \cdot c)b - (b \cdot c)a$
(b) $(a \times b) \times c$
Deduce that $(a \cdot c)b - (b \cdot c)a = (a \times b) \times c$.

Solutions to exercises

1   $-3i - 2j - 13k$

2   $-55k$

3   $-5i + 10j - 5k$

4   $5i - 34j + 6k$

5   $-11i + 9j - k, \dfrac{1}{\sqrt{203}}(-11i + 9j - k)$

6   $-i + 2j + 6k$

7   $7i - 17j + 6k, -42i - 46j - 3k$

8   $\dfrac{1}{\sqrt{26}}(33i + 15j - 21k), \mathbf{0}$

9   $\frac{1}{2}\sqrt{1106}$

10   $-29i - 10j + k$

End of block exercises

1   Find the vector product of $a = 2i + 4k$ and $b = i - 7k$.

2   If $a = 3i + 2j + k$ and $b = -4i + 6j + 2k$ find $a \times b$. Show that $a \times b \neq b \times a$.

3 Points A, B and C have coordinates $(1, 3, 2)$, $(4, 1, 3)$ and $(0, 1, -1)$ respectively. Find the vector product $\overrightarrow{AB} \times \overrightarrow{AC}$.

4 Find a unit vector that is perpendicular to the plane containing $i + 2k$ and $2i + j + 7k$.

5 For the vectors $a = i - 2j + 3k$, $b = -3i + k$ and $c = 2i + j + 4k$, evaluate $a \times (b \times c)$ and $(a \times b) \times c$.

6 A force of magnitude 4 units acts at the point with coordinates $(4, 3, 2)$. The force is applied in the direction of the vector $i - j$. Find the moment of the force about O.

7 Find the area of the triangle with vertices at the points with coordinates $(4, -3, 2)$, $(5, 0, 5)$ and $(1, 1, 1)$.

Solutions to exercises

1 $18j$

2 $-2i - 10j + 26k$

3 $8i + 8j - 8k$

4 $\dfrac{1}{\sqrt{14}}(-2i - 3j + k)$

5 $-36i + 12k, -34i - 4j + 18k$

6 $4\sqrt{2}i + 4\sqrt{2}j - 14\sqrt{2}k$

7 $\frac{1}{2}\sqrt{458}$

The vector equation of a line and a plane

5.1 Introduction

Vectors are very convenient tools for analysing lines and planes in three dimensions. In this block you will learn how to formulate the vector equation of a line and the vector equation of a plane. Finding the equation of a plane relies heavily on an understanding of the scalar product, which should be revised if necessary.

5.2 The vector equation of a line

You will recall that in two dimensions a straight line has an equation of the form $y = mx + c$, such as $y = 3x - 2$. Values of x and y that satisfy this equation correspond to points that lie on the line. Lines in three dimensions are conveniently described using vectors. If we are given a point through which a line must pass, and if we can find a vector in the direction of the line, then the line can be described uniquely.

Consider the straight line APB shown in Figure 5.1.

Figure 5.1
A straight line passing through points A and B.

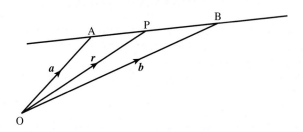

Points A and B are fixed and known points on the line, and have position vectors a and b respectively. Point P is any other arbitrary and variable point on the line, and has position vector r. Note that $\overrightarrow{AB}$ and $\overrightarrow{AP}$ are parallel. It is possible to express the position vector of the arbitrary point in terms of the vectors a and b in what is called the **vector equation of the line**. We shall develop this equation in stages.

Example 5.1
Refer to Figure 5.1. Using the triangle law of vector addition it is possible to write an expression for $\overrightarrow{AB}$ in terms of a and b. Do this now.

Solution

$$a + \overrightarrow{AB} = b \text{ so } \overrightarrow{AB} = b - a$$

Example 5.2

Refer to Figure 5.1. Given $\overrightarrow{AB}$ and $\overrightarrow{AP}$ are parallel, write down a relationship between these vectors.

Solution

$$\overrightarrow{AP} = k\overrightarrow{AB}, \text{ for some scalar } k$$

Using the triangle law of vector addition we can write

$$\overrightarrow{OP} = \overrightarrow{OA} + \overrightarrow{AP}$$

so that

$$r = a + k(b - a)$$

This is the vector equation of the line.

Key point

$$r = a + k(b - a)$$

is the **vector equation of the line** passing through the points with position vectors a and b.

By varying the value of k we can move to any point on the line. For example, when

$$k = 0, \quad r = a, \quad \text{point A}$$
$$k = 1, \quad r = b, \quad \text{point B}$$

If k lies between 0 and 1, the point P lies on the line between A and B. If k is greater than 1, the point P lies on the line beyond B. If k is less than 0, the point P lies on the line beyond A on the opposite side.

Note that the vector $b - a$ is a vector in the direction of the line.

Example 5.3

Write down the vector equation of the line that passes through the points A and B with position vectors $a = 3i + 2j$ and $b = 7i + 5j$ respectively.

Solution

Note that

$$b - a = (7i + 5j) - (3i + 2j)$$
$$= 4i + 3j$$

The equation of the line is then

$$r = a + k(b - a)$$
$$= (3i + 2j) + k(4i + 3j)$$

Alternatively, using column vector notation we could write the equation of the line as

$$r = \begin{pmatrix} 3 \\ 2 \end{pmatrix} + k \begin{pmatrix} 4 \\ 3 \end{pmatrix}$$

This gives the position vector of any point on the line. Note that if $k = 0$ then

$$r = \begin{pmatrix} 3 \\ 2 \end{pmatrix}$$

that is point A. If $k = 1$ then

$$r = \begin{pmatrix} 3 \\ 2 \end{pmatrix} + \begin{pmatrix} 4 \\ 3 \end{pmatrix}$$

$$= \begin{pmatrix} 7 \\ 5 \end{pmatrix}$$

which is point B. Any other value of k will give another point on the line. For example, if $k = 2$ we find

$$r = \begin{pmatrix} 11 \\ 8 \end{pmatrix}$$

which is a point beyond B.

Example 5.4
Write down the vector equation of the line that passes through the points with position vectors $a = 5i - 2j + 3k$ and $b = 2i + j - 4k$.

Solution
Using column vector notation write down $b - a$.

$$\begin{pmatrix} 2 \\ 1 \\ -4 \end{pmatrix} - \begin{pmatrix} 5 \\ -2 \\ 3 \end{pmatrix} = \begin{pmatrix} -3 \\ 3 \\ -7 \end{pmatrix}$$

The equation of the line is then

$$r = a + k(b - a) = \qquad \begin{pmatrix} 5 \\ -2 \\ 3 \end{pmatrix} + k \begin{pmatrix} -3 \\ 3 \\ -7 \end{pmatrix}$$

Example 5.5
On a diagram mark two points M and N. Sketch a straight line through the points M and N, and show their respective position vectors m and n. On your sketch mark an arbitrary point on the line, having position vector r. Write down the vector equation of the line MN.

Solution

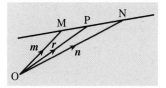

$$r = m + k(n - m)$$

On occasions it is useful to convert the vector form of the equation of a straight line into cartesian form. Suppose we write

$$a = \begin{pmatrix} a_1 \\ a_2 \\ a_3 \end{pmatrix}, \quad b = \begin{pmatrix} b_1 \\ b_2 \\ b_3 \end{pmatrix} \quad \text{and} \quad r = \begin{pmatrix} x \\ y \\ z \end{pmatrix}$$

Then

$$r = a + k(b - a)$$

$$\begin{pmatrix} x \\ y \\ z \end{pmatrix} = \begin{pmatrix} a_1 \\ a_2 \\ a_3 \end{pmatrix} + k\begin{pmatrix} b_1 - a_1 \\ b_2 - a_2 \\ b_3 - a_3 \end{pmatrix}$$

$$= \begin{pmatrix} a_1 + k(b_1 - a_1) \\ a_2 + k(b_2 - a_2) \\ a_3 + k(b_3 - a_3) \end{pmatrix}$$

Equating the individual components we find

$$x = a_1 + k(b_1 - a_1), \quad \text{or equivalently } k = \frac{x - a_1}{b_1 - a_1}$$

$$y = a_2 + k(b_2 - a_2), \quad \text{or equivalently } k = \frac{y - a_2}{b_2 - a_2}$$

$$z = a_3 + k(b_3 - a_3), \quad \text{or equivalently } k = \frac{z - a_3}{b_3 - a_3}$$

Now, since each expression on the right is equal to k we can write

$$\frac{x - a_1}{b_1 - a_1} = \frac{y - a_2}{b_2 - a_2} = \frac{z - a_3}{b_3 - a_3}$$

This is the **cartesian form** of the equation of the straight line that passes through the points with coordinates (a_1, a_2, a_3) and (b_1, b_2, b_3).

Key point

$$\frac{x - a_1}{b_1 - a_1} = \frac{y - a_2}{b_2 - a_2} = \frac{z - a_3}{b_3 - a_3}$$

is the **cartesian form** of the equation of the straight line that passes through the points with coordinates (a_1, a_2, a_3) and (b_1, b_2, b_3).

Example 5.6

(a) Write down the cartesian form of the equation of the straight line that passes through the two points $(9, 3, -2)$ and $(4, 5, -1)$.

(b) State the equivalent vector equation.

Solution

(a)
$$\frac{x - 9}{4 - 9} = \frac{y - 3}{5 - 3} = \frac{z - (-2)}{-1 - (-2)}$$

that is

$$\frac{x - 9}{-5} = \frac{y - 3}{2} = \frac{z + 2}{1}$$

(b) The vector equation is

$$\boldsymbol{r} = \boldsymbol{a} + k(\boldsymbol{b} - \boldsymbol{a})$$

$$= \begin{pmatrix} 9 \\ 3 \\ -2 \end{pmatrix} + k\left[\begin{pmatrix} 4 \\ 5 \\ -1 \end{pmatrix} - \begin{pmatrix} 9 \\ 3 \\ -2 \end{pmatrix} \right] = \begin{pmatrix} 9 \\ 3 \\ -2 \end{pmatrix} + k\begin{pmatrix} -5 \\ 2 \\ 1 \end{pmatrix}$$

Exercises

1 (a) Write down the vector $\overrightarrow{AB}$ joining the points A and B with coordinates $(3, 2, 7)$ and $(-1, 2, 3)$ respectively.
   (b) Find the equation of the straight line through A and B.

2 Write down the vector equation of the line passing through the points with position vectors

$$\boldsymbol{p} = 3\boldsymbol{i} + 7\boldsymbol{j} - 2\boldsymbol{k}$$
$$\boldsymbol{q} = -3\boldsymbol{i} + 2\boldsymbol{j} + 2\boldsymbol{k}$$

Find also the cartesian equation of this line.

3 Find the vector equation of the line passing through $(9, 1, 2)$ and which is parallel to the vector $(1, 1, 1)$.

4 Line I has equation

$$\boldsymbol{r}_1 = \begin{pmatrix} 2 \\ 3 \\ 5 \end{pmatrix} + k\begin{pmatrix} 1 \\ 2 \\ 4 \end{pmatrix}$$

Line II has equation

$$\boldsymbol{r}_2 = \begin{pmatrix} -5 \\ 8 \\ 1 \end{pmatrix} + l\begin{pmatrix} -6 \\ 7 \\ 0 \end{pmatrix}$$

Different values of k give different points on line I. Similarly, different values of l give different points on line II. If the two lines intersect then $\boldsymbol{r}_1 = \boldsymbol{r}_2$ at the point of intersection. If you can find values of k and l which satisfy this condition then the two lines intersect. Show the lines intersect by finding these values and hence find the point of intersection.

Solutions to exercises

1   (a) $-4i - 4k$ (b) $r = \begin{pmatrix} 3 \\ 2 \\ 7 \end{pmatrix} + k \begin{pmatrix} -4 \\ 0 \\ -4 \end{pmatrix}$

3   $r = \begin{pmatrix} 9 \\ 1 \\ 2 \end{pmatrix} + k \begin{pmatrix} 1 \\ 1 \\ 1 \end{pmatrix}$

2   $r = \begin{pmatrix} 3 \\ 7 \\ -2 \end{pmatrix} + k \begin{pmatrix} -6 \\ -5 \\ 4 \end{pmatrix}$;

cartesian form $\dfrac{x - 3}{-6} = \dfrac{y - 7}{-5} = \dfrac{z + 2}{4}$.

4   $k = l = -1$ and the point of intersection is $(1, 1, 1)$.

5.3   The vector equation of a plane

Consider the plane shown in Figure 5.2.

Figure 5.2
This plane passes through A and is perpendicular to n.

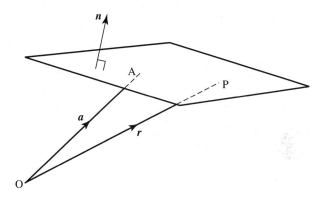

Suppose that A is a known point in the plane and has position vector a. Suppose that P is any other arbitrary point in the plane with position vector r. Clearly the vector $\overrightarrow{AP}$ must lie in the plane. Note that

$$\overrightarrow{AP} = r - a$$

Also shown in Figure 5.2 is a vector that is perpendicular to the plane and denoted by n. Now, because $\overrightarrow{AP}$ and n are perpendicular their scalar product must equal zero, that is

$$\overrightarrow{AP} \cdot n = 0$$
$$(r - a) \cdot n = 0$$

or alternatively, by removing the brackets and rearranging,

$$r \cdot n = a \cdot n$$

This is the equation of the plane passing through the point with position vector a and perpendicular to n.

Key point

A plane passing through the point with position vector a and being perpendicular to the vector n is given by

$$r \cdot n = a \cdot n$$

The previous result is valid whether or not n is a unit vector. However, if $\hat{n}$ is a unit vector then $a \cdot \hat{n}$ represents the perpendicular distance from the origin to the plane, which we usually denote by d. This was shown in Block 3. Hence we can write

$$r \cdot \hat{n} = d$$

This is the **vector equation of a plane** with unit normal $\hat{n}$ and which is a perpendicular distance d from O.

Key point

A plane with unit normal $\hat{n}$ and which is a perpendicular distance d from O is given by

$$r \cdot \hat{n} = d$$

The cartesian form of these equations is obtained by letting $r = xi + yj + zk$ and evaluating the scalar product.

Example 5.7
(a) Find the vector equation of the plane that passes through the point with position vector $3i + 2j + 5k$ and which is perpendicular to $i + k$.
(b) Find the cartesian equation of this plane.

Solution
In this example, $a = 3i + 2j + 5k$ and the normal is $n = i + k$.
(a) Using the previous results we can write down the equation

$$r \cdot n = a \cdot n$$
$$r \cdot (i + k) = (3i + 2j + 5k) \cdot (i + k)$$
$$= 8$$

(b) Writing r as $xi + yj + zk$ we have the cartesian form

$$(xi + yj + zk) \cdot (i + k) = 8$$

so that

$$x + z = 8$$

Exercises

1   Find the equation of a plane that is normal to $8i + 9j + k$ and which is a distance 1 from the origin.

2   Find the equation of a plane that passes through $(8, 1, 0)$ and which is normal to the vector $i + 2j - 3k$.

3   A plane passes through the points A, B and C with coordinates $(1, 0, 1)$, $(-2, 5, 0)$ and $(3, 1, 1)$ respectively.

(a)   Write down $\overrightarrow{AB}$ and $\overrightarrow{AC}$.

(b)   Find the vector product $\overrightarrow{AB} \times \overrightarrow{AC}$.

(c)   Comment upon the direction of $\overrightarrow{AB} \times \overrightarrow{AC}$ in relation to the plane through A, B and C.

(d)   Find the vector and cartesian forms of the equation of the plane containing A, B and C.

Solutions to exercises

1    $r \cdot \dfrac{1}{\sqrt{146}} \begin{pmatrix} 8 \\ 9 \\ 1 \end{pmatrix} = 1$, or

$8x + 9y + z = \sqrt{146}$

2    $r \cdot \begin{pmatrix} 1 \\ 2 \\ -3 \end{pmatrix} = \begin{pmatrix} 8 \\ 1 \\ 0 \end{pmatrix} \cdot \begin{pmatrix} 1 \\ 2 \\ -3 \end{pmatrix}$, that is

$r \cdot \begin{pmatrix} 1 \\ 2 \\ -3 \end{pmatrix} = 10$

3   (a) $-3i + 5j - k$, $2i + j$ (b) $i - 2j - 13k$

(c) it is normal to the plane

(d) $r \cdot \begin{pmatrix} 1 \\ -2 \\ -13 \end{pmatrix} = -12$,

$x - 2y - 13z = -12$

End of block exercises

1   (a) Write down the vector $\overrightarrow{AB}$ joining the points A and B with coordinates $(-2, 1, 3)$ and $(5, 2, 3)$ respectively.

(b) Find the equation of the straight line through A and B.

2   Write down the vector equation of the line passing through the points with position vectors

$$p = 5i + 3j - 2k$$
$$q = 3i + 7j - k$$

Find the cartesian equation of this line.

3   Find the vector equation of the line passing through $(2, 1, 2)$ and which is parallel to the vector $(1, 2, 3)$.

4   Find the vector equation of the line passing through $(2, 1, 2)$ and which is parallel to the

unit vector $\dfrac{1}{\sqrt{14}} (1, 2, 3)$.

5   Find the equation of a plane that passes through $(2, -3, 1)$ and which is normal to the vector $3i - 3j - k$.

6   Find the equation of a plane that passes through the origin and which is normal to the vector $5i + j - k$.

7   What is the distance of the plane

$$r \cdot \begin{pmatrix} 3 \\ 2 \\ 1 \end{pmatrix} = 5$$

from the origin?

8   (a) A plane Π is given by

$$r \cdot \begin{pmatrix} 1 \\ 2 \\ 1 \end{pmatrix} = 6$$

Find the distance of this plane from the origin.

(b) A second plane passes through the point $(3, 2, 1)$ and is parallel to the plane Π. Find its distance from the origin.

(c) Deduce the distance from the point $(3, 2, 1)$ to the plane Π.

9   Write the plane $3x - 3y - z = 14$ in vector form.

Solutions to exercises

1 (a) $7i + j$ (b) $r = \begin{pmatrix} -2 \\ 1 \\ 3 \end{pmatrix} + k\begin{pmatrix} 7 \\ 1 \\ 0 \end{pmatrix}$

2 $r = \begin{pmatrix} 5 \\ 3 \\ -2 \end{pmatrix} + k\begin{pmatrix} -2 \\ 4 \\ 1 \end{pmatrix}$;

$\dfrac{x - 5}{-2} = \dfrac{y - 3}{4} = \dfrac{z + 2}{1}$

3 $r = \begin{pmatrix} 2 \\ 1 \\ 2 \end{pmatrix} + k\begin{pmatrix} 1 \\ 2 \\ 3 \end{pmatrix}$

4 same as 3

5 $r \cdot \begin{pmatrix} 3 \\ -3 \\ -1 \end{pmatrix} = 14$

6 $r \cdot \begin{pmatrix} 5 \\ 1 \\ -1 \end{pmatrix} = 0$

7 $\dfrac{5}{\sqrt{14}}$

8 (a) $\dfrac{6}{\sqrt{6}}$ (b) $\dfrac{8}{\sqrt{6}}$ (c) $\dfrac{2}{\sqrt{6}}$

9 $r \cdot \begin{pmatrix} 3 \\ -3 \\ -1 \end{pmatrix} = 14$

End of chapter exercises

1 Show on a diagram three arbitrary vectors p, q and r. Using the triangle law of addition verify the associativity rule

$$p + (q + r) = (p + q) + r$$

2 A force of 23 newtons acts at an angle of 20° below the horizontal. Resolve this force into two components, one vertical and one horizontal.

3 On a diagram show the arbitrary vectors p and q. Then show the following:
(a) $p + q$ (b) $p - q$ (c) $q - p$
(d) $4q$ (e) $-2q$

4 State the position vectors of the points with coordinates $(9, 1, -1)$ and $(-4, 0, 4)$.

5 If A has coordinates $(4, 3, 0)$ and B has coordinates $(-2, 1, 9)$ find $\vec{AB}$ and $|\vec{AB}|$.

6 Find the modulus of the vector

$$p = 2i - j + 5k$$

7 If A has coordinates $(-4, 2, 1)$ and B has coordinates $(2, 0, 2)$ find the direction ratio of the vector $\vec{AB}$. Find its direction cosines l, m and n and verify that $l^2 + m^2 + n^2 = 1$.

8 Find the angle between the vectors $12i - j$ and $2i + j + k$.

9 Find the projection of the vector $6i + j + 5k$ onto the vector $i - j + 2k$.

10 Points A, B and C have position vectors $(9, 1, 1)$, $(8, 1, 1)$ and $(9, 0, 2)$. Find
(a) the equation of the plane containing A, B and C
(b) the area of the triangle ABC.

11 Find the equation of the plane with normal j that is a distance 2 from the origin.

12 If $a = 3i + j - 2k$ and $b = 13i - j - k$ find
(a) $a + b$ (b) $b - 3a$ (c) $|b|$ (d) $\hat{a}$
(e) $|b - a|$

13 Given that $p = 2i + 2j$ and $q = 7k$ find $p \cdot q$ and interpret this result geometrically.

14 Find the work done by a force of magnitude 10 newtons acting in the direction of the vector $3i + j + 8k$ if it moves a particle from the point $(1, 1, 1)$ to the point $(3, 1, 2)$.

15 Given three vectors a, b and c, their triple scalar product is defined to be $(a \times b) \cdot c$. It can be shown that the modulus of this is the volume of the parallelepiped formed by the three vectors. Find the volume of the parallelepiped formed by the three vectors $a = 3i + j - 2k$, $b = i + 2j - 2k$ and $c = 2i + 5j + k$.

16 If the triple scalar product $(a \times b) \cdot c$ is equal to zero, then
  (i) $a = 0$, or $b = 0$, or $c = 0$ or
  (ii) two of the vectors are parallel, or
  (iii) the three vectors lie in the same plane (they are said to be **coplanar**).

Show that the vectors
$2i - j + k, 3i - 4j + 5k, i + 2j - 3k$
are coplanar.

17 Given three vectors a, b and c, their triple vector product is defined to be $(a \times b) \times c$. For the vectors $a = 4i + 2j + k$, $b = 2i - j + 7k$ and $c = 2i - 2j + 3k$ verify that

$$(a \times b) \times c = (a \cdot c)b - (b \cdot c)a$$

18 Find the volume of the parallelepiped whose edges are represented by the vectors $12i + j + k, 2i$ and $-2j + k$.

19 If

$$a = \begin{pmatrix} 3 \\ 2 \\ 1 \end{pmatrix}, \quad b = \begin{pmatrix} 4 \\ 3 \\ 4 \end{pmatrix} \quad \text{and} \quad c = \begin{pmatrix} 8 \\ 1 \\ 0 \end{pmatrix}$$

show that $a \cdot b \times c = 32$.

Solutions to exercises

2 21.6 N horiz., 7.9 N vert. downwards

4 $9i + j - k, -4i + 4k$

5 $-6i - 2j + 9k, 11$

6 $\sqrt{30}$

7 $6: -2:1, \dfrac{6}{\sqrt{41}}, -\dfrac{2}{\sqrt{41}}, \dfrac{1}{\sqrt{41}}$

8 $38.8°$

9 $\dfrac{15}{\sqrt{6}}$

10 (a) $r \cdot \begin{pmatrix} 0 \\ 1 \\ 1 \end{pmatrix} = 2$ (b) $\dfrac{\sqrt{2}}{2}$

11 $y = 2$

12 (a) $16i - 3k$ (b) $4i - 4j + 5k$ (c) $3\sqrt{19}$
   (d) $\dfrac{1}{\sqrt{14}}(3i + j - 2k)$ (e) $\sqrt{105}$

13 0, vectors are perpendicular

14 $\dfrac{140}{\sqrt{74}} = 16.3$

15 29

18 6

Differentiation

Chapter **15**

Differentiation is one of the most important processes in engineering mathematics. It is the study of the way in which functions change. The function may represent pressure, stress, volume or some other physical variable. For example, the pressure of a vessel may depend upon temperature. As the temperature of the vessel increases, then so does the pressure. Engineers often need to know the rate at which such a variable changes.

Block 1 explains how to calculate the rate of change of a function, $y(x)$, across a range of values of the input variable, x. The rate of change of $y(x)$ at a single point is then developed. This requires the introduction of the idea of taking limits, which is also important in Chapter 18 on the applications of integration.

In practice, most people use a standard table to differentiate functions, and how this is done is explained in Block 2. The chapter closes with a study of repeated differentiation.

Chapter 15 contents

Interpretation of a derivative

1.1 Introduction

Engineers are often interested in the rate at which some variable is changing. For example, an engineer needs to know the rate at which the pressure in a vessel is changing, the rate at which the voltage across a capacitor is changing, or the rate at which the temperature is changing in a chemical reaction. Rapid rates of change of a variable may indicate that a system is not operating normally and is approaching critical values. Alarms may be triggered.

Rates of change may be positive, negative or zero. A positive rate of change means that the variable is increasing; a negative rate of change means that the variable is decreasing. A zero rate of change means that the variable is not changing.

Consider Figure 1.1, which illustrates a variable, $y(x)$.

Figure 1.1
The function $y(x)$ changes at different rates for different values of x.

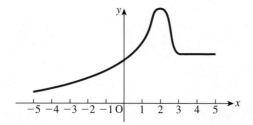

Between $x = -5$ and $x = -3$, y is increasing slowly. Across this interval the rate of change of y is small and positive. Between $x = -3$ and $x = 1$, y is increasing more rapidly; the rate of change of y is positive and fairly large. Between $x = 1$ and $x = 2$, y is increasing very rapidly and so the rate of change is positive and large. From $x = 2$ to $x = 3$, y decreases rapidly; the rate of change is large and negative. From $x = 3$ to $x = 5$, y is constant and so the rate of change on this interval is zero.

The technique for calculating rate of change is called **differentiation**. Often it is not sufficient to describe a rate of change as, for example, 'positive and large' or 'negative and quite small'. A precise value is needed. Use of differentiation provides a precise value or expression for the rate of change of a function.

1.2   Average rate of change across an interval

We see from Figure 1.1 that a function can have different rates of change at different points on its graph. We begin by defining and then calculating the **average rate of change** of a function across an interval. Figure 1.2 shows a function, $y(x)$, and values $x_1, x_2, y(x_1)$ and $y(x_2)$.

Consider x increasing from x_1 to x_2. The change in x is $x_2 - x_1$. As x increases from x_1 to x_2, then y increases from $y(x_1)$ to $y(x_2)$. The change in y is $y(x_2) - y(x_1)$. Then the average rate of change of y across the interval is

$$\frac{\text{change in } y}{\text{change in } x} = \frac{y(x_2) - y(x_1)}{x_2 - x_1}$$

$$= \frac{BC}{AC}$$

From Figure 1.2 we see that $\frac{BC}{AC} = \tan \theta$, which is also the gradient of the straight line or **chord** AB. Hence we see that the average rate of change across an interval is identical to the gradient of the chord across that interval.

Figure 1.2
Average rate of change $= \dfrac{y(x_2) - y(x_1)}{x_2 - x_1}$.

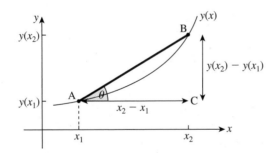

Key point

$$\text{average rate of change of } y = \frac{\text{change in } y}{\text{change in } x}$$

$$= \text{gradient of chord}$$

Example 1.1
Calculate the average rate of change of $y = x^2$ across the interval
(a) $x = 1$ to $x = 4$
(b) $x = -2$ to $x = 0$

Solution
(a) Change in $x = 4 - 1 = 3$.
When $x = 1$, $y = 1^2 = 1$. When $x = 4$, $y = 4^2 = 16$. Hence the change in y is $16 - 1 = 15$. So

$$\text{average rate of change across interval } [1, 4] = \frac{15}{3}$$

$$= 5$$

This means that across the interval [1, 4], on average the y value increases by 5 for every 1 unit increase in x.

(b) Change in $x = 0 - (-2) = 2$. We have $y(-2) = 4$ and $y(0) = 0$ so the change in y is $0 - 4 = -4$. Hence

$$\text{average rate of change} = \frac{-4}{2}$$
$$= -2$$

On average, across the interval $x = -2$ to $x = 0$, y decreases by 2 units for every 1 unit increase in x.

Example 1.2 Electrical Engineering – Voltage across a capacitor

The voltage, $v(t)$, across a capacitor varies with time, t, according to

$$v(t) = 3 + 2e^{-t}$$

Find the average rate of change of voltage as time varies
(a) from $t = 0$ to $t = 2$
(b) from $t = 1$ to $t = 3$

Solution

(a) Change in $t = 2 - 0 = 2$.
　　When $t = 0$, $v = 3 + 2e^0 = 5$.

　　When $t = 2$, $v = $ 　　　　　　　　　　　　$3 + 2e^{-2} = 3.2707$

　　So

　　average rate of change of $v(t) = $

$$\frac{3.2707 - 5}{2} = -0.8647$$

(b) Change in $t = $ 　　　　　　　　　　$3 - 1 = 2$

　　$v(1) = 3 + 2e^{-1}$
　　　　$= 3.7358$

　　$v(3) = $ 　　　　　　　　　　$3 + 2e^{-3} = 3.0996$

　　So

　　average rate of change of $v(t)$ across $[1, 3] = $

$$\frac{3.0996 - 3.7358}{2} = -0.3181$$

Across the interval from $t = 1$ to $t = 3$, the voltage is decreasing but at a slower rate than across the interval from $t = 0$ to $t = 2$.

Exercises

1 Calculate the average rate of change of $y = x^2 + 2x$ from $x = 1$ to $x = 4$.

2 Calculate the average rate of change of $h(t) = 2t^2 - 2t + 1$ from $t = 0$ to $t = 2$.

3 Calculate the average rate of change of $i(t) = 50 \sin t$ from $t = 0$ to $t = \pi$.

4 Calculate the average rate of change of
$$r(x) = \frac{1}{x + 1} \text{ from } x = -3 \text{ to } x = -2.$$

5 Calculate the average rate of change of $z(t) = 4 + 2t^2$ across (a) $t = 1$ to $t = 3$, (b) $t = -1$ to $t = 0$.

6 The temperature, T, of a vessel varies with time, t, according to
$$T(t) = 320 + \frac{65}{t^2}$$
Calculate the average rate of change of T from $t = 2$ to $t = 4$.

Solutions to exercises

1 7

2 2

3 0

4 −0.5

5 (a) 8 (b) −2

6 −6.094

1.3 Rate of change at a point

As mentioned earlier, we often need to know the rate of change of a function at a point, and not simply an average rate of change across an interval.

Refer again to Figure 1.2. Suppose we wish to find the rate of change of y at the point A. The average rate of change across the interval from $x = x_1$ to $x = x_2$ is given by the gradient of the chord AB. This provides an approximation to the rate of change at A.

Suppose the chord AB is extended on both sides, as shown in Figure 1.3. As B is moved closer to A, the gradient of the chord provides a better approximation to the rate of change at A.

Ultimately B is made coincident with A and then the chord AB becomes a tangent to the curve at A. The gradient of this tangent gives the rate of change of y at A:

Key point   rate of change at a point = gradient of tangent to the curve at that point

Calculating the rate of change of a function at a point by measuring the gradient of a tangent is usually not an accurate method. Consequently we develop an exact, algebraic way of finding rates of change.

Figure 1.3
The extended
chord ultimately
becomes the
tangent at A.

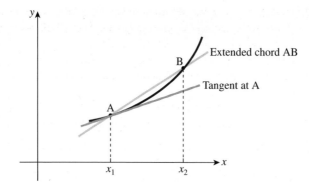

Extended chord AB

Tangent at A

Consider the function $y(x)$ as shown in Figure 1.4.

Figure 1.4
As B approaches
A, $\delta x \rightarrow 0$.

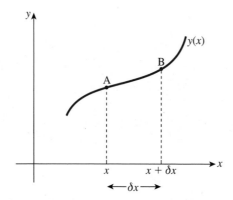

Let A be a point on the curve with coordinates $(x, y(x))$. B is a point on the curve near to A. The x coordinate of B is $x + \delta x$. The term δx is pronounced 'delta x'. It represents a small change in the x direction. The y coordinate of B is $y(x + \delta x)$. We calculate the gradient of the chord AB:

$$\text{gradient of AB} = \frac{\text{change in } y}{\text{change in } x}$$

$$= \frac{y(x + \delta x) - y(x)}{x + \delta x - x}$$

$$= \frac{y(x + \delta x) - y(x)}{\delta x}$$

The change in y, that is $y(x + \delta x) - y(x)$, is also written as δy. So

$$\text{gradient of AB} = \frac{y(x + \delta x) - y(x)}{\delta x}$$

$$= \frac{\delta y}{\delta x}$$

The gradient of AB gives the average rate of change of $y(x)$ across the small interval from x to $x + \delta x$. To calculate the rate of change of $y(x)$ at A we require the gradient of the tangent at A.

Consider A as a fixed point and let B move along the curve towards A. At each position of B we can calculate the gradient of the chord AB. As B gets closer to A, the chord AB approximates more closely to the tangent at A. Also, as B approaches A, the distance δx decreases. To find the gradient of the tangent at A we calculate the gradient of the chord AB and let δx get smaller and smaller. We say δx tends to zero and write this as $\delta x \rightarrow 0$.

As B approaches A, the x difference between A and B gets smaller, that is $\delta x \rightarrow 0$, and likewise the y difference, δy, also gets smaller, so $\delta y \rightarrow 0$. However, the gradient of AB, given by the ratio $\dfrac{\delta y}{\delta x}$, approaches a definite value, called a limit. So we seek the limit of $\dfrac{\delta y}{\delta x}$ as $\delta x \rightarrow 0$. We write this as

$$\lim_{\delta x \to 0} \frac{\delta y}{\delta x}$$

Note that 'limit' has been shortened to 'lim'.

In summary we have

Key point

> rate of change of y = gradient of tangent
>
> $$= \lim_{\delta x \to 0} \frac{\delta y}{\delta x}$$

Let us see this applied to an example.

Example 1.3
Find the rate of change of $y(x) = x^2$.

Solution
Suppose A is the fixed point with coordinates (x, x^2) as shown in Figure 1.5. B is a point on the curve near to A with coordinates $(x + \delta x, (x + \delta x)^2)$. We calculate the gradient of the chord AB.

Figure 1.5
The gradient of the tangent at A is approximated by the gradient of the chord AB.

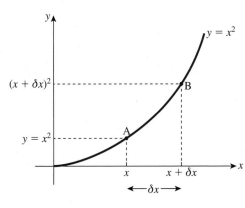

change in $x = \delta x$

change in $y = \delta y$

$\qquad = (x + \delta x)^2 - x^2$

$$= x^2 + 2x\delta x + (\delta x)^2 - x^2$$
$$= 2x(\delta x) + (\delta x)^2$$

$$\text{gradient of chord AB} = \frac{\delta y}{\delta x}$$

$$= \frac{2x(\delta x) + (\delta x)^2}{\delta x}$$

$$= 2x + \delta x$$

This is the average rate of change of $y(x)$ across the small interval from x to $x + \delta x$. To obtain the gradient of the tangent at A, we let $\delta x \to 0$.

$$\text{gradient of tangent at A} = \lim_{\delta x \to 0} (2x + \delta x)$$
$$= 2x$$

Hence the rate of change of x^2 is $2x$.

For example, if $x = 3$, then A is the point $(3, 9)$ and the rate of change of y at this point is 6. Similarly if $x = -1$, A is the point $(-1, 1)$ and the rate of change of $y = -2$.

Exercises

1 Find the rate of change of $y(x) = x^2 + 1$.
   Calculate the rate of change of y when x is
   (a) 6 (b) 3 (c) -2 (d) 0

2 Find the rate of change of $y(x) = x^2 + 2x$.
   Calculate the rate of change of y when x is
   (a) 6 (b) -5 (c) 0

Solutions to exercises

1 $2x$, (a) 12 (b) 6 (c) -4 (d) 0

2 $2x + 2$, (a) 14 (b) -8 (c) 2

1.4 Terminology and notation

The process of finding the rate of change of a given function is called **differentiation**. The function is said to be **differentiated**. If y is a function of the independent variable x, we say that y is differentiated with respect to (w.r.t.) x. The rate of change of a function is also known as the **derivative** of the function.

There is a notation for writing down the derivative of a function. If the function is $y(x)$, we denote the derivative of y by

$$\frac{dy}{dx}$$

pronounced 'dee y by dee x'. Hence

Key point

$$\lim_{\delta x \to 0} \frac{\delta y}{\delta x} = \frac{dy}{dx}$$

Another notation for the derivative is simply y', pronounced y dash. Similarly if the function is $z(t)$ we write the derivative as $\dfrac{dz}{dt}$ or z'. When the independent variable is t, the derivative may also be denoted using the dot notation. Thus, for example, $\dfrac{dz}{dt}$ may be written as $\dot{z}$, pronounced 'z dot'. Sometimes, instead of writing y, a function is written in full: for example, to show the derivative of $\sin 5x$ we write

$$\frac{d(\sin 5x)}{dx}$$

Exercises

1   If x is a function of the independent variable t, write down two ways in which the derivative can be written.

2   If f is a function of x, write down two ways in which the derivative can be written.

Solutions to exercises

1   $\dfrac{dx}{dt}$ or $\dot{x}$

2   $\dfrac{df}{dx}$ or f'

End of block exercises

1   Calculate the average rate of change of
    $y = x^3 - 1$ from
    (a) $x = 1$ to $x = 3$
    (b) $x = 0$ to $x = 2$
    (c) $x = -2$ to $x = 2$

2   The pressure, P atmospheres, in a vessel varies with temperature, T (degrees Celsius), according to

$$P(T) = 120 - 20e^{-T/20}$$

Calculate the average rate of change of pressure as T varies from 10 °C to 100 °C.

3   The current, $i(t)$, in a circuit decays exponentially with time, t, according to the equation

$$i(t) = 5 + 2e^{-t}$$

Calculate the average rate of change of current as t varies from 0 to 3.

4　Explain the meaning of the expression $\dfrac{dy}{dx}$.

5　(a) Calculate the rate of change of
$y(x) = 5 - x^2$.
(b) Calculate the rate of change of y when
$x = -4$.

6　(a) Calculate $\dfrac{dR}{dx}$ when $R(x) = 2x^2$.
(b) Calculate $\dfrac{dR}{dx}$ when $x = 0.5$.

Solutions to exercises

1　(a) 13　(b) 4　(c) 4

2　0.13

3　−0.63

5　(a) $-2x$　(b) 8

6　(a) $4x$　(b) 2

Using a table of derivatives

2.1 Introduction

Block 1 gave a brief introduction to the meaning of a derivative. A derivative is the rate of change of a function. Geometrically we saw that this is given by the gradient of a tangent. If we consider a typical function, as illustrated in Figure 2.1, it is clear that the gradient of a tangent depends upon where the tangent is drawn. For example, tangent A, drawn where x has a value x_1, has a different gradient from tangent B, drawn where x has a value x_2. In other words, the gradient of the tangent is itself a function of x. This was seen in Example 1.3 of Block 1, where the gradient was found to be $2x$.

Figure 2.1
The gradient of a tangent varies along the curve.

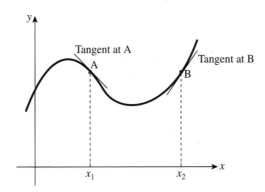

Rather than calculate the derivative of a function as explained in Block 1, it is common practice to use a table of derivatives. This block shows how to use such a table.

2.2 Table of derivatives

Table 2.1 lists some of the common functions used in engineering and their corresponding derivatives.

Example 2.1

Use Table 2.1 to find $\dfrac{dy}{dx}$ when y is given by

(a) $3x$ (b) 3 (c) $3x^2$ (d) $4x^7$

Table 2.1
Common functions and their derivatives

Function	Derivative
constant	0
x	1
kx	k
x^n	nx^{n-1}
kx^n	knx^{n-1}
e^x	e^x
e^{kx}	ke^{kx}
$\ln x$	$\dfrac{1}{x}$
$\ln kx$	$\dfrac{1}{x}$
$\sin x$	$\cos x$
$\sin kx$	$k \cos kx$
$\sin(kx + \alpha)$	$k \cos(kx + \alpha)$
$\cos x$	$- \sin x$
$\cos kx$	$-k \sin kx$
$\cos(kx + \alpha)$	$-k \sin(kx + \alpha)$
$\tan x$	$\sec^2 x$
$\tan kx$	$k \sec^2 kx$
$\tan(kx + \alpha)$	$k \sec^2(kx + \alpha)$
$\sinh x$	$\cosh x$
$\sinh kx$	$k \cosh kx$
$\cosh x$	$\sinh x$
$\cosh kx$	$k \sinh kx$
$\tanh x$	$\operatorname{sech}^2 x$
$\tanh kx$	$k \operatorname{sech}^2 kx$
$\sin^{-1} x$	$\dfrac{1}{\sqrt{1 - x^2}}$
$\sin^{-1} kx$	$\dfrac{k}{\sqrt{1 - k^2 x^2}}$
$\cos^{-1} x$	$\dfrac{-1}{\sqrt{1 - x^2}}$
$\cos^{-1} kx$	$\dfrac{-k}{\sqrt{1 - k^2 x^2}}$
$\tan^{-1} x$	$\dfrac{1}{1 + x^2}$
$\tan^{-1} kx$	$\dfrac{k}{1 + k^2 x^2}$

k, n and α are constants, and all angles are in radians

Solution

(a) We note that $3x$ is of the form kx where $k = 3$. Using Table 2.1 we then have

$$\frac{dy}{dx} = 3.$$

(b) Noting that 3 is a constant we see that $\frac{dy}{dx} = 0$.

(c) We see that $3x^2$ is of the form kx^n, with $k = 3$ and $n = 2$. The derivative, knx^{n-1}, is then $6x^1$, or more simply, $6x$. So if $y = 3x^2$, then $\frac{dy}{dx} = 6x$.

(d) We see that $4x^7$ is of the form kx^n, with $k = 4$ and $n = 7$. Hence the derivative, $\frac{dy}{dx}$, is given by $28x^6$.

Example 2.2

Find $\frac{dy}{dx}$ when y is (a) $\sqrt{x}$, (b) $\frac{3}{x^2}$, (c) $\frac{2}{x}$.

Solution

(a) We write $\sqrt{x}$ as $x^{\frac{1}{2}}$, and use the result for x^n with $n = \frac{1}{2}$. So

$$\frac{dy}{dx} = nx^{n-1}$$
$$= \frac{1}{2}x^{\frac{1}{2}-1}$$
$$= \frac{1}{2}x^{-\frac{1}{2}}$$

This may be written as $\dfrac{1}{2\sqrt{x}}$.

(b) We write $\dfrac{3}{x^2}$ as $3x^{-2}$. Using the result of kx^n we see that

$$\frac{dy}{dx} = knx^{n-1}$$

$$= \rule{4cm}{0.6cm} \qquad\qquad 3(-2)x^{-2-1} = -6x^{-3}$$

(c) We write $\dfrac{2}{x}$ as $2x^{-1}$. Then we see that

$$\frac{dy}{dx} = \rule{4cm}{0.6cm} \qquad\qquad 2(-1)x^{-1-1} = -2x^{-2}$$

Example 2.3

Use Table 2.1 to find $\frac{dz}{dt}$ given z is (a) e^t, (b) e^{3t}, (c) e^{-5t}.

Solution

Although Table 2.1 is written using x as the independent variable, it can be used for any variable.

(a) From Table 2.1, if $y = e^x$, then $\frac{dy}{dx} = e^x$. Hence if $z = e^t$ then $\frac{dz}{dt} = e^t$.

(b) From Table 2.1 we see that when $y = e^{3x}$ then

$$\frac{dy}{dx} = $$

$3e^{3x}$

Hence $\dfrac{dz}{dt} = $

$3e^{3t}$

(c) Using the result for e^{kx} in Table 2.1 we see that when $z = e^{-5t}$,

$$\frac{dz}{dt} = $$

$-5e^{-5t}$

Example 2.4

Find the derivative, $\dfrac{dy}{dx}$, when y is (a) $\sin 3x$, (b) $\cos \dfrac{x}{2}$, (c) $\tan 2x$.

Solution

(a) Using the result for $\sin kx$, and taking $k = 3$, we see that

$$\frac{dy}{dx} = 3 \cos 3x$$

(b) From the result for $\cos kx$, and taking $k = \frac{1}{2}$, we see that

$$\frac{dy}{dx} = $$

$-\dfrac{1}{2} \sin \dfrac{x}{2}$

(c) From the result for $\tan kx$, we see that

$$\frac{dy}{dx} = $$

$2 \sec^2 2x$

Exercises

1 Find the derivative of the following functions:
(a) $9x$ (b) $4x$ (c) $6x^3$ (d) $-3x^2$ (e) $\ln 3t$

2 Find $\dfrac{dz}{dt}$ when z is given by

(a) $\dfrac{4}{t^3}$ (b) $\sqrt{t^3}$ (c) $5t^{-2}$ (d) $-\dfrac{3}{2}t^{3/2}$

3 Find the derivative of each of the following functions:
(a) $\sin 5x$ (b) $\cos 4t$ (c) $\tan 3r$ (d) e^{2v} (e) $\dfrac{1}{e^{3t}}$

4 Find the derivative of the following:
(a) $\cos \dfrac{2x}{3}$ (b) $\sin(-2x)$ (c) $\tan \pi x$ (d) $e^{x/2}$

Solutions to exercises

1 (a) 9 (b) 4 (c) $18x^2$ (d) $-6x$ (e) $\dfrac{1}{t}$

2 (a) $-12t^{-4}$ (b) $\frac{3}{2}t^{1/2}$ (c) $-10t^{-3}$ (d) $-\frac{9}{4}t^{1/2}$

3 (a) $5 \cos 5x$ (b) $-4 \sin 4t$ (c) $3 \sec^2 3r$
(d) $2e^{2v}$ (e) $-3e^{-3t}$

4 (a) $-\dfrac{2}{3} \sin \dfrac{2x}{3}$ (b) $-2 \cos(-2x)$ (c) $\pi \sec^2 \pi x$
(d) $\frac{1}{2}e^{x/2}$

2.3 Extending the table of derivatives

We introduce two simple rules that enable us to extend the range of functions that we can differentiate.

Key point

The derivative of $f(x) \pm g(x)$ is

$$\frac{df}{dx} \pm \frac{dg}{dx}$$

This rule says that to find the derivative of the sum (difference) of two functions, we simply calculate the sum (difference) of the derivatives of each function.

Key point

The derivative of $kf(x)$ is

$$k\frac{df}{dx}$$

This rule tells us that if a function is multiplied by a constant, k, then the derivative is likewise multiplied by the same constant, k.

Example 2.5

Find the derivative of each of the following functions:
(a) $y = 6 \sin 2x$ (b) $y = 6 \sin 2x + 3x^2$ (c) $y = 6 \sin 2x + 3x^2 - 5e^{3x}$

Solution

(a) From Table 2.1, the derivative of $\sin 2x$ is $2 \cos 2x$. Hence the derivative of $6 \sin 2x$ is $6(2 \cos 2x)$, that is $12 \cos 2x$.

$$y = 6 \sin 2x, \quad \frac{dy}{dx} = 6(2 \cos 2x)$$
$$= 12 \cos 2x$$

(b) The function comprises two parts: $6 \sin 2x$ and $3x^2$. We have already differentiated $6 \sin 2x$ in part (a), so we consider the derivative of $3x^2$. The derivative of x^2 is $2x$ and so the derivative of $3x^2$ is $3(2x)$, that is $6x$. These derivatives are now summed.

$$y = 6 \sin 2x + 3x^2, \quad \frac{dy}{dx} = 12 \cos 2x + 6x$$

(c) We differentiate each part of the function in turn.

$$y = 6 \sin 2x + 3x^2 - 5e^{3x}$$
$$\frac{dy}{dx} = 6(2 \cos 2x) + 3(2x) - 5(3e^{3x})$$
$$= 12 \cos 2x + 6x - 15e^{3x}$$

Example 2.6

Find $\dfrac{dy}{dx}$ where y is defined by

(a) $\dfrac{x^6}{2} - 3e^{-2x}$ (b) $4 \cos \dfrac{x}{2} + 9 - 9x^3$

Solution

(a) The derivative of x^6 is $6x^5$. Hence the derivative of $\dfrac{x^6}{2}$ is ▯ $\dfrac{6x^5}{2} = 3x^5$

The derivative of e^{-2x} is ▯ $-2e^{-2x}$

Hence the derivative of $3e^{-2x}$ is

▯ $3(-2)e^{-2x} = -6e^{-2x}$

So given

$$y = \frac{x^6}{2} - 3e^{-2x}$$

then

$\dfrac{dy}{dx} = $ ▯ $3x^5 + 6e^{-2x}$

(b) The derivative of $\cos \dfrac{x}{2}$ is

▯ $-\dfrac{1}{2} \sin \dfrac{x}{2}$

The derivative of 9 is zero. The derivative of $9x^3$ is

▯ $27x^2$

So given

$$y = 4 \cos \frac{x}{2} + 9 - 9x^3$$

then

$\dfrac{dy}{dx} = $ ▯ $-2 \sin \dfrac{x}{2} - 27x^2$

Exercises

1 Find $\dfrac{dy}{dx}$ when y is given by

(a) $4x^6 + 8x^3$ (b) $-3x^4 + 2x^{1.5}$

(c) $\dfrac{9}{x^2} + \dfrac{14}{x} - 3x$ (d) $\dfrac{3 + 2x}{4}$ (e) $(2 + 3x)^2$

(b) $h(v) = 3 \cos 2v - 6 \sin \dfrac{v}{2}$

(c) $m(n) = 4e^{2n} + \dfrac{2}{e^{2n}} + \dfrac{n^2}{2}$

(d) $H(t) = \dfrac{e^{3t}}{2} + 2 \tan 2t$

(e) $S(r) = (r^2 + 1)^2 - 4e^{-2r}$

2 Find the derivative of each of the following functions:

(a) $z(t) = 5 \sin t + \sin 5t$

3   Differentiate the following functions:
(a) $A(t) = (3 + e^t)^2$

(b) $B(s) = \pi e^{2s} + \dfrac{1}{s} + 2 \sin \pi s$

(c) $V(r) = \left(1 + \dfrac{1}{r}\right)^2 + (r + 1)^2$

(d) $M(\theta) = 6 \sin 2\theta - 2 \cos \dfrac{\theta}{4} + 2\theta^2$

(e) $H(t) = 4 \tan 3t + 3 \sin 2t - 2 \cos 4t$

Solutions to exercises

1   (a) $24x^5 + 24x^2$ (b) $-12x^3 + 3x^{0.5}$
(c) $-\dfrac{18}{x^3} - \dfrac{14}{x^2} - 3$ (d) $\frac{1}{2}$ (e) $12 + 18x$

2   (a) $5 \cos t + 5 \cos 5t$ (b) $-6 \sin 2v - 3 \cos \dfrac{v}{2}$

(c) $8e^{2n} - 4e^{-2n} + n$ (d) $\dfrac{3e^{3t}}{2} + 4 \sec^2 2t$

(e) $4r^3 + 4r + 8e^{-2r}$

3   (a) $6e^t + 2e^{2t}$ (b) $2\pi e^{2s} - \dfrac{1}{s^2} + 2\pi \cos \pi s$

(c) $-\dfrac{2}{r^2} - \dfrac{2}{r^3} + 2r + 2$

(d) $12 \cos 2\theta + \dfrac{1}{2} \sin \dfrac{\theta}{4} + 4\theta$

(e) $12 \sec^2 3t + 6 \cos 2t + 8 \sin 4t$

2.4 Evaluating a derivative

Engineers may need to find the rate of change of a function at a particular point: that is, find the derivative of a function at a specific point. We do this by finding the derivative of the function, and then evaluating the derivative at the given value of x. When evaluating, all angles are in radians. Consider a function, $y(x)$. We use the notation $\dfrac{dy}{dx}(0.7)$ or $y'(0.7)$ to denote the derivative of y evaluated at $x = 0.7$.

Example 2.7
Find the value of the derivative of $y = 3x^2$ where $x = 4$. Interpret your result.

Solution

We have $y = 3x^2$ and so $\dfrac{dy}{dx} = 6x$. We now evaluate the derivative.

When $x = 4$, $\dfrac{dy}{dx} = 6(4) = 24$, that is

$$\frac{dy}{dx}(4) = 24$$

The derivative is positive when $x = 4$ and so y is increasing at this point. Thus when $x = 4$, y is increasing at a rate of 24 vertical units per horizontal unit.

Example 2.8 Electrical Engineering

Find the rate of change of current, $i(t)$, given by

$$i(t) = 3e^{-t} + 2 \quad t \geq 0$$

when $t = 0.7$ seconds.

Solution

The rate of change of a function is the same as the derivative of the function, that is $\dfrac{di}{dt}$.

$$\frac{di}{dt} = \boxed{} \qquad\qquad -3e^{-t}$$

When $t = 0.7$

$$\frac{di}{dt} = \boxed{} \qquad\qquad -3e^{-0.7} = -1.4898$$

The derivative is negative and so we know that $i(t)$ is decreasing when $t = 0.7$. Thus, when $t = 0.7$, the current is decreasing at a rate of 1.49 A s^{-1}.

Exercises

1 Calculate the derivative of $y = 3x^2 + e^x$ when $x = 0.5$.

2 Calculate the rate of change of $i(t) = 4 \sin 2t + 3t$ when (a) $t = \frac{\pi}{3}$ (b) $t = 0.6$.

3 Evaluate the rate of change of $H(t) = 5 \sin t - 3 \cos 2t$ when (a) $t = 0$ (b) $t = 1.3$.

Solutions to exercises

1 4.6487

2 (a) -1 (b) 5.8989

3 (a) 5 (b) 4.4305

End of block exercises

1 Find $\dfrac{dy}{dx}$ when y is given by

(a) $7x^5 + 6x^{-2} + \sin 2x$
(b) $3 \cos 4x - 6 \sin 5x$
(c) $e^{3x} + e^{-3x} + 2e^x + 1$

(d) $4 \tan \dfrac{x}{2} + \dfrac{1}{\sqrt{x}}$

(e) $3\sqrt{x} + \dfrac{9}{x} + \dfrac{1}{2} \sin 6x + \ln x$

2 Find the rate of change of the following functions:
(a) $e^t + e^{-t}$ (b) $2 \sin 3t + \ln 2t$ (c) $-3 \cos x$
(d) $\sqrt{r} + 2r^2$ (e) $2e^{-0.5v} + v^3$

3 Find $\dot{x}$ when x is given by
(a) $2t^4 - 3t + 1 + 2 \ln t$
(b) $\sin \pi t - 2 \cos \pi t$
(c) $3 \tan 2t - e^t$ (d) $e^{2t} - e^{-2t} + t$
(e) $t^{3/2} - t^{2/3}$

4 Find the rate of change of each function when $t = 1.2$:
(a) $3t^2 - 2t^3$ (b) $\dfrac{e^{4t}}{2} + 3e^{-t}$

(c) $6 \sin \dfrac{t}{2} + 3 \cos \dfrac{t}{2}$
(d) $2 \tan t - \tan 2t$
(e) $\dfrac{4}{t} + 4 \ln t$

5 Find the derivative of the following:
(a) $e^{2t}(e^t + e^{-t})$ (b) $\dfrac{2t^2 + 1}{t}$
(c) $(2t + 1)(2t - 1)$
(d) $\sin^2 2x + \cos^2 2x + \sin 2x + \cos 2x$
(e) $\dfrac{1}{e^x}$ (f) $\dfrac{\sin x}{\cos x}$

Solutions to exercises

1 (a) $35x^4 - 12x^{-3} + 2 \cos 2x$
(b) $-12 \sin 4x - 30 \cos 5x$
(c) $3e^{3x} - 3e^{-3x} + 2e^x$
(d) $2 \sec^2 \dfrac{x}{2} - \dfrac{1}{2}x^{-3/2}$
(e) $\dfrac{3}{2}x^{-1/2} - \dfrac{9}{x^2} + 3 \cos 6x + \dfrac{1}{x}$

2 (a) $e^t - e^{-t}$ (b) $6 \cos 3t + \dfrac{1}{t}$ (c) $3 \sin x$
(d) $\dfrac{1}{2}r^{-1/2} + 4r$ (e) $-e^{-0.5v} + 3v^2$

3 (a) $8t^3 - 3 + \dfrac{2}{t}$ (b) $\pi \cos \pi t + 2\pi \sin \pi t$
(c) $6 \sec^2 2t - e^t$ (d) $2e^{2t} + 2e^{-2t} + 1$
(e) $\dfrac{3}{2}t^{1/2} - \dfrac{2}{3}t^{-1/3}$

4 (a) -1.44 (b) 242.12 (c) 1.6290 (d) 11.5538
(e) 0.5556

5 (a) $3e^{3t} + e^t$ (b) $2 - \dfrac{1}{t^2}$ (c) $8t$
(d) $2 \cos 2x - 2 \sin 2x$ (e) $-e^{-x}$ (f) $\sec^2 x$

Higher derivatives

3.1 Introduction

Block 2 showed how to calculate the derivative of a function using a table of derivatives. By differentiating the function, $y(x)$, we obtain the derivative, $\dfrac{dy}{dx}$.

The function, $\dfrac{dy}{dx}$, is more correctly called the **first derivative** of y. By differentiating the first derivative, we obtain the **second derivative**; by differentiating the second derivative we obtain the **third derivative** and so on. The second and subsequent derivatives are known as higher derivatives.

Example 3.1

Calculate the first, second and third derivatives of $y = e^{2x} + x^4$.

Solution

The first derivative is $\dfrac{dy}{dx}$.

$$\frac{dy}{dx} = 2e^{2x} + 4x^3$$

To obtain the second derivative we differentiate the first derivative:

$$\text{second derivative} = 4e^{2x} + 12x^2$$

The third derivative is found by differentiating the second derivative:

$$\text{third derivative} = 8e^{2x} + 24x$$

3.2 Notation

Just as there is a notation for the first derivative so there is a similar notation for higher derivatives.

Consider the function, $y(x)$. We know that the first derivative is denoted by $\dfrac{dy}{dx}$ or y'.

The second derivative is calculated by differentiating the first derivative, that is

$$\text{second derivative} = \frac{d}{dx}\left(\frac{dy}{dx}\right)$$

So, the second derivative is denoted by $\dfrac{d^2y}{dx^2}$. This is often written more concisely as y''.

The third derivative is denoted by $\dfrac{d^3 y}{dx^3}$ or y''' and so on. So, referring to Example 3.1 we could have written

$$\frac{dy}{dx} = 2e^{2x} + 4x^3$$

$$\frac{d^2 y}{dx^2} = 4e^{2x} + 12x^2$$

$$\frac{d^3 y}{dx^3} = 8e^{2x} + 24x$$

Key point

If $y = y(x)$

$$\text{first derivative} = \frac{dy}{dx}$$

$$\text{second derivative} = \frac{d^2 y}{dx^2}$$

$$\text{third derivative} = \frac{d^3 y}{dx^3}$$

Derivatives with respect to t are often indicated using a dot notation, so $\dfrac{dx}{dt}$ can be written as $\dot{x}$. Similarly, a second derivative with respect to t can be written as $\ddot{x}$, pronounced x double dot.

Example 3.2

Calculate $\dfrac{d^2 y}{dt^2}$ and $\dfrac{d^3 y}{dt^3}$ given $y = \sin t + \cos t$.

Solution

$$\frac{dy}{dt} = \cos t - \sin t$$

$$\frac{d^2 y}{dt^2} = \rule{3cm}{0.4cm} \qquad\qquad -\sin t - \cos t$$

$$\frac{d^3 y}{dt^3} = \rule{3cm}{0.4cm} \qquad\qquad -\cos t + \sin t$$

We could have used the dot notation and written $\dot{y} = \cos t - \sin t$, and $\ddot{y} = -\sin t - \cos t$.

We may need to evaluate higher derivatives at specific points. We use an obvious notation. The second derivative of $y(x)$, evaluated at, say, $x = 2$, is written as $\dfrac{d^2 y}{dx^2}(2)$, or more simply as $y''(2)$. The third derivative evaluated at $x = -1$ is written as $\dfrac{d^3 y}{dx^3}(-1)$ or $y'''(-1)$.

Example 3.3
Given

$$y(x) = 2 \sin x + 3x^2$$

find (a) $y'(1)$, (b) $y''(-1)$, (c) $y'''(0)$.

Solution
We have

$$y = 2 \sin x + 3x^2$$

$$y' = 2 \cos x + 6x$$

$$y'' = \boxed{} \qquad\qquad -2 \sin x + 6$$

$$y''' = -2 \cos x$$

(a) $y'(1) = 2 \cos 1 + 6(1) = 7.0806$.

(b) $y''(-1) = \boxed{} \qquad\qquad -2 \sin(-1) + 6 = 7.6829$

(c) $y'''(0) = -2 \cos 0 = -2$.

Exercises

1 Find $\dfrac{d^2y}{dx^2}$ where $y(x)$ is defined by
(a) $3x^2 - e^{2x}$ (b) $\sin 3x + \cos x$ (c) $\sqrt{x}$
(d) $e^x + e^{-x}$ (e) $1 + x + x^2 + \ln x$

2 Find $\dfrac{d^3y}{dx^3}$ where y is given in question 1.

3 Calculate $y''(1)$ where $y(t)$ is given by
(a) $t(t^2 + 1)$ (b) $\sin(-2t)$ (c) $2e^t + e^{2t}$
(d) $\dfrac{1}{t}$ (e) $\cos \dfrac{t}{2}$

4 Calculate $y'''(-1)$ of the functions given in question 3.

Solutions to exercises

1 (a) $6 - 4e^{2x}$ (b) $-9 \sin 3x - \cos x$
(c) $-\dfrac{1}{4}x^{-3/2}$ (d) $e^x + e^{-x}$ (e) $2 - \dfrac{1}{x^2}$

2 (a) $-8e^{2x}$ (b) $-27 \cos 3x + \sin x$ (c) $\dfrac{3}{8}x^{-5/2}$
(d) $e^x - e^{-x}$ (e) $\dfrac{2}{x^3}$

3 (a) 6 (b) 3.6372 (c) 34.9928 (d) 2
(e) -0.2194

4 (a) 6 (b) -3.3292 (c) 1.8184 (d) -6
(e) -0.0599

End of block exercises

1 Calculate y'' where y is given by
(a) $\cos 2t - \sin 2t$ (b) $e^{2x} - e^x$ (c) $2x^6 - 3x^7$
(d) $-x^3 + 3x^2$ (e) $9 - \dfrac{9}{x}$

2 Find the fourth derivative of the following functions:
(a) e^{3t} (b) e^{kt}, k constant (c) $\sin 2t$ (d) $\sin kt$, k constant (e) $\cos kt$, k constant

3 Show that $y = e^x + 2x$ satisfies the equation
$$y'' - y' - y = -2 - 2x - e^x$$

4 Evaluate $y'''(0)$ where y is given by
(a) $\sin 3t + t^3$ (b) $2 \cos t + \cos 2t$
(c) $e^{-x}(e^x + 1)$ (d) $3 - 3t^4$ (e) $\dfrac{e^{2x} + 1}{e^x}$

5 The function $y(x)$ is defined by
$$y(x) = x^4 - 3x^3 + 3x^2 + 1$$
Calculate the values of x where $y'' = 0$.

Solutions to exercises

1 (a) $-4 \cos 2t + 4 \sin 2t$ (b) $4e^{2x} - e^x$
(c) $60x^4 - 126x^5$ (d) $-6x + 6$ (e) $-18x^{-3}$

2 (a) $81e^{3t}$ (b) $k^4 e^{kt}$ (c) $16 \sin 2t$ (d) $k^4 \sin kt$
(e) $k^4 \cos kt$

4 (a) -21 (b) 0 (c) -1 (d) 0 (e) 0

5 $\frac{1}{2}$, 1

End of chapter exercises

1 Calculate $\dfrac{dy}{dx}$ where y is given by
(a) $3x^4 - 2x + \ln x$ (b) $\sin 5x - 5 \cos x$
(c) $(x + 1)^2$ (d) $e^{3x} + 2e^{-2x} + 1$
(e) $5 + 5x + \dfrac{5}{x} + 5 \ln x$

2 Find the second derivatives of the functions in question 1.

3 Find $y'(1)$ of the functions in question 1.

4 Find $y''(1)$ of the functions in question 1.

5 Find the rate of change of the following functions:
(a) $\dfrac{3t^3 - t^2}{2t}$ (b) $\ln \sqrt{x}$ (c) $(t + 2)(2t - 1)$
(d) $e^{3v}(1 - e^v)$ (e) $\sqrt{x}(\sqrt{x} - 1)$

6 Find the third and fourth derivatives of y given the second derivative of y is
(a) $\dfrac{2}{e^{3x}}$ (b) $\dfrac{1 + x}{x^2}$ (c) $3 \ln x^2$
(d) $\sin x + \sin(-2x)$ (e) $\dfrac{\cos^2 x + \cos x}{\cos x}$

7 Differentiate $(\sin x + \cos x)^2$. (Hint: use the trigonometrical identities in Table 5.1 of Chapter 9 Block 5.)

8 Verify that
$y = A \sin kx + B \cos kx$, where A, B, k constants is a solution of
$$y'' + k^2 y = 0$$

9 The function $y(x)$ is given by $y(x) = 1 - \cos x$. Find the values of x where (a) $y' = 0$, (b) $y'' = 0$.

10 The function $y(x)$ is given by $y(x) = x^3 - 3x$.
Calculate the intervals on which y is
(a) increasing, (b) decreasing.

11 The function $y(x)$ is given by

$$y(x) = 2x^3 - 9x^2 + 1$$

(a) Calculate the values of x for which $y' = 0$.
(b) Calculate the values of x for which $y'' = 0$.

(c) State the interval(s) on which y is increasing.
(d) State the interval(s) on which y is decreasing.
(e) State the interval(s) on which y' is increasing.
(f) State the interval(s) on which y' is decreasing.

Solutions to exercises

1 (a) $12x^3 - 2 + \dfrac{1}{x}$ (b) $5\cos 5x + 5\sin x$

(c) $2x + 2$ (d) $3e^{3x} - 4e^{-2x}$ (e) $5 - \dfrac{5}{x^2} + \dfrac{5}{x}$

2 (a) $36x^2 - \dfrac{1}{x^2}$ (b) $-25\sin 5x + 5\cos x$

(c) 2 (d) $9e^{3x} + 8e^{-2x}$ (e) $\dfrac{10}{x^3} - \dfrac{5}{x^2}$

3 (a) 11 (b) 5.6257 (c) 4 (d) 59.7153 (e) 5

4 (a) 35 (b) 26.6746 (c) 2 (d) 181.8525 (e) 5

5 (a) $3t - 0.5$ (b) $\dfrac{1}{2x}$ (c) $4t + 3$ (d) $3e^{3v} - 4e^{4v}$

(e) $1 - \dfrac{1}{2\sqrt{x}}$

6 (a) $-6e^{-3x}, 18e^{-3x}$

(b) $-2x^{-3} - x^{-1}, 6x^{-4} + x^{-2}$

(c) $\dfrac{6}{x}, \dfrac{-6}{x^2}$ (d) $\cos x - 2\cos(-2x)$,

$-\sin x - 4\sin(-2x)$ (e) $-\sin x, -\cos x$

7 $2\cos 2x$

9 (a) $\pm n\pi, n = 0, 1, 2, 3, \ldots$

(b) $\dfrac{\pi}{2} \pm n\pi, n = 0, 1, 2, 3, \ldots$

10 (a) $(-\infty, -1)$ and $(1, \infty)$ (b) $(-1, 1)$

11 (a) 0, 3 (b) 1.5 (c) $(-\infty, 0)$ and $(3, \infty)$
(d) $(0, 3)$ (e) $(1.5, \infty)$ (f) $(-\infty, 1.5)$

Techniques and applications of differentiation

Chapter **16**

This chapter continues the study of differentiation started in Chapter 15. The table of derivatives in Chapter 15, although useful, is limited. In this chapter several techniques for expanding the range of functions we can differentiate are introduced.

Methods for differentiating **products** and **quotients** of functions are explained in Block 1. Some functions can be differentiated when the function is written in terms of a new variable. This gives rise to the **chain rule** in Block 2.

Although we have focused on differentiating a function of x, say $y(x)$, it is not always possible to express y explicitly in terms of x. For example, it is impossible to rearrange the equation $e^x + e^y = x^2 + y^3$ to obtain y by itself on the left-hand side. However, by using **implicit differentiation** as explained in Block 3 it is still possible to find an expression for the derivative, $\dfrac{dy}{dx}$.

If a function is expressed **parametrically** the derivative is found by differentiating parametrically. This is covered in Block 4. The final technique, **logarithmic differentiation**, is useful for finding the derivative of products of functions involving many factors, and this is covered in Block 5.

Blocks 6 and 7 turn to some applications of differentiation. The calculation of equations of **tangents and normals** is treated in Block 6. Finally, the important topic of **maximum and minimum values** of a function closes the chapter.

Chapter 16 contents

The product rule and the quotient rule

1.1 Introduction

Chapter 15 introduced the concept of differentiation and the use of a table of derivatives. Clearly every possible function cannot be listed in a table. We need a set of rules, used in conjunction with the table of derivatives, to extend the range of functions that we can differentiate. The product rule and the quotient rule are two such rules.

1.2 The product rule

As its name tells us, the product rule helps us to differentiate a product of functions. Consider the function $y(x)$, where $y(x)$ is the product of two functions, $u(x)$ and $v(x)$, that is

$$y(x) = u(x)v(x)$$

For example, if $y(x) = x^2 \sin x$ then $u(x) = x^2$ and $v(x) = \sin x$. The product rule states:

Key point

If

$$y(x) = u(x)v(x)$$

then

$$\frac{dy}{dx} = \frac{du}{dx}v + u\frac{dv}{dx}$$
$$= u'v + uv'$$

Example 1.1

Find $\dfrac{dy}{dx}$ where $y = x^2 \sin x$.

Solution

We have

$$y = x^2 \sin x = uv$$

and so $u = x^2$ and $v = \sin x$. Hence

$$\frac{du}{dx} = 2x, \quad \frac{dv}{dx} = \cos x$$

Applying the product rule we have

$$\frac{dy}{dx} = \frac{du}{dx}v + u\frac{dv}{dx}$$
$$= 2x\,(\sin x) + x^2\,(\cos x)$$
$$= x\,(2\sin x + x\cos x)$$

Example 1.2
Find y' where $y = e^x \cos x$.

Solution
We have

$$y = e^x \cos x = uv$$

So

$$u = \boxed{}, \quad v = \boxed{}$$

$e^x, \cos x$

and hence

$$\frac{du}{dx} = \boxed{}, \quad \frac{dv}{dx} = \boxed{}$$

$e^x, -\sin x$

Applying the product rule yields

$$\frac{dy}{dx} = \frac{du}{dx}v + u\frac{dv}{dx}$$
$$= e^x \cos x + \boxed{}$$

$e^x(-\sin x)$

$$= e^x(\cos x - \sin x)$$

Example 1.3
Find $\dfrac{d^2y}{dx^2}$ where $y = x^2 \ln x$.

Solution
We have

$$y = x^2 \ln x = uv$$

so

$$u = x^2, \quad v = \ln x$$

Then

$$\frac{du}{dx} = 2x, \quad \frac{dv}{dx} = \quad\rule{2cm}{0.4pt}$$

$$\frac{1}{x}$$

Applying the product rule, we have

$$\frac{dy}{dx} = \frac{du}{dx}v + u\frac{dv}{dx}$$

$$= \rule{3cm}{0.4pt}$$

$$2x \ln x + x$$

To obtain $\dfrac{d^2y}{dx^2}$ we differentiate $\dfrac{dy}{dx}$. Hence we need to find $\dfrac{d}{dx}(2x \ln x + x)$. The derivative of x is simply 1 so let us examine $\dfrac{d}{dx}(2x \ln x)$.

To find $\dfrac{d}{dx}(2x \ln x)$ we use the product rule, with $u = 2x, v = \ln x$. Then

$$\frac{du}{dx} = 2, \quad \frac{dv}{dx} = \frac{1}{x}$$

and so

$$\frac{d}{dx}(2x \ln x) = \frac{du}{dx}v + u\frac{dv}{dx}$$

$$= 2 \ln x + 2x\left(\frac{1}{x}\right)$$

$$= 2 \ln x + 2$$

Finally

$$\frac{d^2y}{dx^2} = \frac{d}{dx}(2x \ln x + x)$$

$$= \rule{3cm}{0.4pt}$$

$$2 \ln x + 3$$

Exercises

1 Find $\dfrac{dy}{dx}$ where y is given by

(a) $x \cos x$ (b) xe^x (c) $\sin x \cos 2x$ (d) $x^3 e^{2x}$
(e) $x^4 \sin 2x$

(c) $(e^x + e^{-2x})(3x^2 - 2x)$ (d) $\sqrt{x}e^x$

(e) $\dfrac{t^2 + 1}{e^t}$

2 Calculate y' where y is given by
(a) $(t^2 + 1)\sin 4t$ (b) $(3t + 7)e^{-2t}$

3 Find the second derivative of the functions in question 1.

Solutions to exercises

1   (a) $\cos x - x \sin x$ (b) $e^x(1 + x)$
(c) $\cos x \cos 2x - 2 \sin x \sin 2x$
(d) $x^2 e^{2x}(3 + 2x)$ (e) $4x^3 \sin 2x + 2x^4 \cos 2x$

(d) $e^x\left(\dfrac{x^{-1/2}}{2} + x^{1/2}\right)$

(e) $-e^{-t}(t^2 - 2t + 1)$

2   (a) $2t \sin 4t + 4(t^2 + 1)(\cos 4t)$
(b) $-e^{-2t}(6t + 11)$
(c) $(e^x - 2e^{-2x})(3x^2 - 2x)$
   $+ (e^x + e^{-2x})(6x - 2)$, which may be written
   as $e^x(3x^2 + 4x - 2) - e^{-2x}(6x^2 - 10x + 2)$

3   (a) $-x \cos x - 2 \sin x$ (b) $e^x(x + 2)$
(c) $-5 \sin x \cos 2x - 4 \cos x \sin 2x$
(d) $2xe^{2x}(2x^2 + 6x + 3)$
(e) $16x^3 \cos 2x - 4x^2(x^2 - 3) \sin 2x$

1.3   The quotient rule

The quotient rule shows us how to differentiate a quotient of functions, for example

$$\frac{\sin x}{x}, \quad \frac{t^2 - 1}{t^2 + 1}, \quad \frac{e^z + z}{\cos z}$$

The quotient rule may be stated thus:

Key point

If

$$y(x) = \frac{u(x)}{v(x)}$$

then

$$\frac{dy}{dx} = \frac{v\dfrac{du}{dx} - u\dfrac{dv}{dx}}{v^2}$$

$$= \frac{vu' - uv'}{v^2}$$

Example 1.4

Find y' given $y = \dfrac{\sin x}{x}$.

Solution

We have

$$y = \frac{\sin x}{x} = \frac{u}{v}$$

so

$$u = \sin x, \quad v = x$$

and so

$$\frac{du}{dx} = \cos x, \quad \frac{dv}{dx} = 1$$

Applying the quotient rule gives

$$\frac{dy}{dx} = \frac{vu' - uv'}{v^2}$$
$$= \frac{x \cos x - \sin x \,(1)}{x^2}$$
$$= \frac{x \cos x - \sin x}{x^2}$$

Example 1.5

Find y' given $y = \dfrac{t^3}{t+1}$.

Solution

We have

$$y = \frac{t^3}{t+1} = \frac{u}{v}$$

and so

$$u = t^3, \quad v = t+1, \quad u' = 3t^2, \quad v' = 1$$

Applying the quotient rule gives

$$\frac{dy}{dt} = \frac{vu' - uv'}{v^2}$$

$$= \qquad\qquad\qquad \frac{(t+1)\,3t^2 - t^3\,(1)}{(t+1)^2}$$

which can be simplified to

$$\frac{dy}{dt} = \qquad\qquad\qquad \frac{t^2(2t+3)}{(t+1)^2}$$

Exercises

1 Find $\dfrac{dy}{dx}$ where y is given by

(a) $\dfrac{e^x}{x}$ (b) $\dfrac{x}{e^x + 1}$ (c) $\dfrac{\cos x}{\sin x}$ (d) $\dfrac{1 - x}{1 + x}$

(e) $\dfrac{\ln x}{x^2}$

2 Find y' when y is given by

(a) $\dfrac{t^2 - 1}{t^2 + 1}$ (b) $\dfrac{e^{2t} + t}{e^t - 1}$ (c) $\dfrac{\sin 3t}{\cos t + t}$

(d) $\dfrac{z + \sin z}{z + \cos z}$ (e) $\dfrac{1 + x + x^2}{1 + x^3}$

Solutions to exercises

1 (a) $\dfrac{e^x(x-1)}{x^2}$ (b) $\dfrac{e^x+1-xe^x}{(e^x+1)^2}$

(c) $-\text{cosec}^2 x$ (d) $\dfrac{-2}{(1+x)^2}$ (e) $\dfrac{1-2\ln x}{x^3}$

2 (a) $\dfrac{4t}{(t^2+1)^2}$

(b) $\dfrac{(e^t-1)(2e^{2t}+1)-(e^{2t}+t)e^t}{(e^t-1)^2}$

(c) $\dfrac{3(\cos t+t)\cos 3t-\sin 3t(-\sin t+1)}{(\cos t+t)^2}$

(d) $\dfrac{(z+1)\cos z+(z-1)\sin z+1}{(\cos z+z)^2}$

(e) $-\left(\dfrac{x^4+2x^3+3x^2-2x-1}{(x^3+1)^2}\right)$

End of block exercises

1 Find the derivative of each of the following:

(a) $(x-1)\sin 2x$ (b) $\dfrac{\sin 2x}{x-1}$ (c) $\dfrac{x-1}{\sin 2x}$

(d) $e^{2x}\sin 3x$ (e) $e^{-2x}\sin 3x$

2 Differentiate the following:

(a) $\dfrac{t^3-t^2}{t^2+1}$ (b) $3\sin 2x\cos x$ (c) $\dfrac{3\cos x}{\sin 2x}$

(d) $\dfrac{e^{3r}}{e^{2r}}$ (e) $(r+1)(r+\sin r)$

3 Find $\dfrac{dH}{dt}$ given

$$H=e^{2t}t^2\sin t$$

4 Find $\dfrac{dR}{dt}$ given

$$R=\dfrac{e^{2t}\sin t}{t^2}$$

Solutions to exercises

1 (a) $2(x-1)\cos 2x+\sin 2x$

(b) $\dfrac{2(x-1)\cos 2x-\sin 2x}{(x-1)^2}$

(c) $\dfrac{\sin 2x-2(x-1)\cos 2x}{\sin^2 2x}$

(d) $e^{2x}(3\cos 3x+2\sin 3x)$

(e) $e^{-2x}(3\cos 3x-2\sin 3x)$

2 (a) $\dfrac{t(t^3+3t-2)}{(t^2+1)^2}$

(b) $6\cos x\cos 2x-3\sin x\sin 2x$

(c) $-3\left(\dfrac{\sin 2x\sin x+2\cos x\cos 2x}{\sin^2 2x}\right)$

(d) e^r (e) $(r+1)\cos r+\sin r+2r+1$

3 $e^{2t}[t^2\cos t+2t(t+1)\sin t]$

4 $\dfrac{e^{2t}[2t\sin t+t\cos t-2\sin t]}{t^3}$

The chain rule

2.1 Introduction

In Block 1 we saw how to differentiate products and quotients of functions. This block introduces the chain rule, which allows us to differentiate an additional class of functions.

2.2 The chain rule

Suppose y is a function of z, that is $y = y(z)$, and that z is a function of x, that is $z = z(x)$. So

$$y = y(z) = y(z(x))$$

Hence y may be considered to be a function of x. For example, if $y(z) = 2z^2 + 3z$ and $z = \cos 2x$ then

$$y = 2(\cos 2x)^2 + 3(\cos 2x)$$

Since y can be considered as a function of x, then $\dfrac{dy}{dx}$ may be found. The chain rule helps us to find $\dfrac{dy}{dx}$. The chain rule states

Key point

If $y = y(z)$ and $z = z(x)$, then

$$\frac{dy}{dx} = \frac{dy}{dz} \times \frac{dz}{dx}$$

Example 2.1

Given $y = z^4$ and $z = 3x + 6$, find $\dfrac{dy}{dx}$.

Solution

We have

$$y = z^4 = (3x + 6)^4$$

and we seek $\dfrac{dy}{dx}$. Now

$$y = z^4, \quad z = 3x + 6$$

and so

$$\frac{dy}{dz} = 4z^3, \quad \frac{dz}{dx} = 3$$

Using the chain rule we have

$$\frac{dy}{dx} = \frac{dy}{dz}\frac{dz}{dx}$$
$$= 4z^3(3)$$
$$= 12z^3$$
$$= 12(3x + 6)^3$$

Sometimes care must be taken to recognise the independent and dependent variables. Example 2.2 uses the chain rule, but it may look unfamiliar.

Example 2.2

Given $z(y) = y^3$ and $y(x) = 2x^2 - x$ find $\dfrac{dz}{dx}$.

Solution

We have

$$z(y) = y^3 \quad \text{so} \quad \frac{dz}{dy} = 3y^2$$

and

$$y(x) = 2x^2 - x \quad \text{so} \quad \frac{dy}{dx} = 4x - 1$$

The chain rule has the form

$$\frac{dz}{dx} = \frac{dz}{dy}\frac{dy}{dx}$$
$$= 3y^2(4x - 1)$$
$$= 3(2x^2 - x)^2 (4x - 1)$$

Example 2.3

Given $y = (x^3 + x)^7$, find $\dfrac{dy}{dx}$.

Solution

We let $z = x^3 + x$ and so

$$y = (x^3 + x)^7 = z^7$$

Then $\dfrac{dy}{dz} = 7z^6$ and $\dfrac{dz}{dx} = 3x^2 + 1$. Applying the chain rule gives

$$\frac{dy}{dx} = \frac{dy}{dz}\frac{dz}{dx}$$
$$= (7z^6)(3x^2 + 1)$$
$$= 7(3x^2 + 1)(x^3 + x)^6$$

Example 2.4

Given $y = \sqrt{x^2 + 1}$ find $\dfrac{dy}{dx}$.

Solution

Let $z = x^2 + 1$ and then

$$y = \sqrt{x^2 + 1} \;=\; \boxed{} \qquad\qquad\qquad \sqrt{z}$$

Then

$$\frac{dy}{dz} = \boxed{}, \quad \frac{dz}{dx} = \boxed{} \qquad\qquad \frac{1}{2}\,z^{-1/2},\; 2x$$

and applying the chain rule gives

$$\frac{dy}{dx} = \frac{dy}{dz}\,\frac{dz}{dx}$$

$$= \boxed{} \qquad\qquad\qquad \frac{1}{2}\,z^{-1/2}\,(2x)$$

$$= \frac{x}{\sqrt{z}}$$

$$= \boxed{} \qquad\qquad\qquad \frac{x}{\sqrt{x^2 + 1}}$$

Example 2.5

Find $\dfrac{dy}{dx}$ given $y = \ln(x^2 + x + 1)$.

Solution

Let $z = x^2 + x + 1$ so that

$$y = \ln(x^2 + x + 1) = \boxed{} \qquad\qquad\qquad \ln z$$

Then

$$\frac{dy}{dz} = \boxed{}, \quad \frac{dz}{dx} = \boxed{} \qquad\qquad \frac{1}{z},\; 2x + 1$$

So

$$\frac{dy}{dx} = \frac{dy}{dz}\,\frac{dz}{dx}$$

$$= \boxed{} \qquad\qquad\qquad \frac{2x + 1}{z}$$

$$= \frac{2x + 1}{x^2 + x + 1}$$

We note that, in the final solution, the numerator is the derivative of the denominator. In general:

If $y = \ln f(x)$ then

$$\frac{dy}{dx} = \frac{f'}{f}$$

Example 2.6

Given $y = \ln(1 - x)$ find $\dfrac{dy}{dx}$.

Solution

Here $f(x) = 1 - x$ and $f'(x) = -1$. So

$$\frac{dy}{dx} = \frac{-1}{1 - x}$$

which may be simplified to

$$\frac{dy}{dx} = \frac{1}{x - 1}$$

Example 2.7

Given $y = 5 \ln(2t - 1)$ find $\dfrac{dy}{dt}$.

Solution

Here $f(t) = 2t - 1$ and so $f'(t) = 2$. Hence

$$\frac{dy}{dt} = 5\left(\frac{2}{2t - 1}\right)$$

$$= \frac{10}{2t - 1}$$

Example 2.8

Given $y = \ln(e^x + \sin x)$ find $\dfrac{dy}{dx}$.

Solution

We have

$$y = \ln(e^x + \sin x)$$
$$= \ln f(x) \quad \text{where } f(x) = e^x + \sin x$$

So

$$\frac{dy}{dx} = \qquad\qquad\qquad\qquad \frac{f'}{f} = \frac{e^x + \cos x}{e^x + \sin x}$$

Exercises

1 Differentiate each of the following functions:
(a) $(x^3 + 2)^6$ (b) $\sqrt{\sin x}$ (c) $(e^x + 1)^7$
(d) $(\cos 2x)^5$ (e) $\ln(x + 1)$

2 Find $\dfrac{dy}{dt}$ where y is given by

(a) $e^{(3t^2)}$ (b) $3e^{t^2}$ (c) $e^{\sin 2t}$ (d) $e^{2 \sin t}$
(e) $2e^{\sin t}$

3 Find the rate of change of y at the specified point:
(a) $y = \ln(3t^2 + 5)$, $t = 1$
(b) $y = \sin(t^2)$, $t = 2$
(c) $y = \cos(t^3 + 1)$, $t = 1$
(d) $y = (t^3 - 1)^{2/3}$, $t = 2$
(e) $y = 4e^{\cos t}$, $t = \dfrac{\pi}{2}$

Solutions to exercises

1 (a) $18x^2(x^3 + 2)^5$ (b) $\dfrac{\cos x}{2\sqrt{\sin x}}$

(c) $7e^x(e^x + 1)^6$

(d) $-10 \sin 2x \cos^4 2x$ (e) $\dfrac{1}{x + 1}$

2 (a) $6te^{3t^2}$ (b) $6te^{t^2}$ (c) $2 \cos 2te^{\sin 2t}$
(d) $2 \cos te^{2 \sin t}$ (e) $2 \cos te^{\sin t}$

3 (a) 0.75 (b) −2.6146 (c) −2.7279
(d) 4.1821 (e) −4

End of block exercises

1 Use the chain rule to differentiate each of the following functions:
(a) $y = (6x^3 - x)^4$ (b) $h = (t^4 - 1)^{1/3}$
(c) $v = \sqrt{9 - 2t}$ (d) $i = \sin(y^3)$
(e) $R = \cos(\sqrt{r})$

2 Find the derivative of each function:
(a) $Y(t) = 5e^{\sin 2t}$ (b) $m(p) = 3 \ln(p^4 + 2)$
(c) $H(r) = 5 \sin(\pi r^2 + 1)$

(d) $x(t) = -3 \cos\left(\dfrac{1}{t}\right)$ (e) $Q(s) = \dfrac{1}{\ln s}$

3 Evaluate $\dfrac{dy}{dx}$ at the specified value of x.
(a) $y = \sqrt{x} + \sin x$, $x = 1$
(b) $y = \sin(x^2 + 1)$, $x = 0.5$
(c) $y = e^{\sqrt{x}}$, $x = 1$

(d) $y = 2 \cos\left(\dfrac{1}{x}\right)$, $x = 1$

(e) $y = \dfrac{1}{(3x^2 + 1)^4}$, $x = 0.5$

4 Differentiate the following functions where a, b and n are constants:
(a) $y = (at + b)^n$ (b) $y = e^{at+b}$
(c) $y = \sin(at + b)$
(d) $y = \cos(at + b)$ (e) $y = \ln(at + b)$

Solutions to exercises

1. (a) $4(18x^2 - 1)(6x^3 - x)^3$

    (b) $\dfrac{4t^3 (t^4 - 1)^{-2/3}}{3}$

    (c) $-(9 - 2t)^{-1/2}$ (d) $3y^2 \cos(y^3)$

    (e) $-\dfrac{1}{2} r^{-1/2} \sin\sqrt{r}$

2. (a) $10 \cos 2t e^{\sin 2t}$ (b) $\dfrac{12 p^3}{p^4 + 2}$

    (c) $10\pi r \cos(\pi r^2 + 1)$

3. (d) $-\dfrac{3}{t^2} \sin\left(\dfrac{1}{t}\right)$ (e) $\dfrac{-1}{s (\ln s)^2}$

    (a) 0.5675 (b) 0.3153
    (c) 1.3591 (d) 1.6829
    (e) -0.7311

4. (a) $an(at + b)^{n-1}$ (b) ae^{at+b}

    (c) $a \cos(at + b)$ (d) $-a \sin(at + b)$

    (e) $\dfrac{a}{at + b}$

Implicit differentiation

3.1 Introduction

So far we have met many functions of the form $y = f(x)$, for example $y = x^2 + 3$, $y = \sin 2x$ and $y = e^{3x} - 2x$. Whenever y is equated to an expression involving only x terms we say that y is expressed **explicitly** in terms of x.

Sometimes we have an equation connecting x and y but it is impossible to write it in the form $y = f(x)$. Examples of this include $x^2 - y^3 + \sin x - \cos y = 1$, $\sin(x + y) + e^x + e^{-y} = x^3 + y^3$. In these cases we say that y is expressed **implicitly** in terms of x.

Whether y is expressed explicitly or implicitly in terms of x we can still differentiate to find the derivative $\dfrac{dy}{dx}$. If y is expressed explicitly in terms of x then $\dfrac{dy}{dx}$ will also be expressed explicitly in terms of x. If y is expressed implicitly in terms of x then $\dfrac{dy}{dx}$ will be expressed in terms of x and y.

3.2 Differentiating $f(y)$ with respect to x

If y is expressed implicitly in terms of x and we wish to find $\dfrac{dy}{dx}$, then we frequently need to differentiate a function of y with respect to x: that is, find $\dfrac{d}{dx}(f(y))$. To do this we use the chain rule. The examples illustrate the technique.

Example 3.1

Find $\dfrac{d}{dx}(y^3)$.

Solution

We use the chain rule. Let $z = y^3$ so that we wish to find $\dfrac{dz}{dx}$. Using the chain rule (see Example 2.2 in Block 2)

$$\frac{dz}{dx} = \frac{dz}{dy}\frac{dy}{dx}$$

Now

$$z = y^3 \text{ and so } \frac{dz}{dy} = 3y^2$$

Hence

$$\frac{dz}{dx} = 3y^2\frac{dy}{dx}$$

Hence

$$\frac{d}{dx}(y^3) = 3y^2\frac{dy}{dx}$$

The result of Example 3.1 can easily be extended to

$$\frac{d}{dx}(y^n) = ny^{n-1}\frac{dy}{dx}$$

Example 3.2

Find $\dfrac{d}{dx}(\sin y)$.

Solution

Let $z = \sin y$ so that we wish to find $\dfrac{dz}{dx}$. We know from the chain rule that

$$\frac{dz}{dx} = \frac{dz}{dy}\frac{dy}{dx}$$

We have $z = \sin y$ so

$$\frac{dz}{dy} = \boxed{} \qquad\qquad \cos y$$

and

$$\frac{d}{dx}(\sin y) = \boxed{} \qquad\qquad \cos y\,\frac{dy}{dx}$$

Generalising the result of Example 3.2 we have

$$\frac{d}{dx}(f(y)) = \frac{df}{dy}\frac{dy}{dx}$$

Example 3.3

Find $\dfrac{d}{dt}(\ln y)$.

Solution

We use the previous result with $f(y) = \ln y$. Then

$$\frac{d}{dt}(\ln y) = \frac{df}{dt}$$

$$= \frac{df}{dy}\frac{dy}{dt}$$

$$= \frac{1}{y}\frac{dy}{dt}$$

Exercises

1   Differentiate the following functions of y with respect to x:
(a) y^2 (b) $3y^4$ (c) $2y^2 - 3y + 1$ (d) $\dfrac{1}{y}$
(e) $\sqrt{y}$

2   Differentiate the following functions of y with respect to x:
(a) $\sin 2y$ (b) $3 \cos y - y$ (c) e^{2y}
(d) $2 \cos 3y$ (e) $\sin^2 y$

3   Differentiate the following functions of y with respect to x:
(a) $(y + 3)^4$ (b) $(y^2 + 3)^4$

Solutions to exercises

1   (a) $2y \dfrac{dy}{dx}$ (b) $12y^3 \dfrac{dy}{dx}$ (c) $4y \dfrac{dy}{dx} - 3 \dfrac{dy}{dx}$

(d) $-\dfrac{1}{y^2} \dfrac{dy}{dx}$ (e) $\dfrac{1}{2\sqrt{y}} \dfrac{dy}{dx}$

2   (a) $2 (\cos 2y) \dfrac{dy}{dx}$ (b) $-3 (\sin y) \dfrac{dy}{dx} - \dfrac{dy}{dx}$

(c) $2e^{2y} \dfrac{dy}{dx}$ (d) $-6 (\sin 3y) \dfrac{dy}{dx}$

(e) $2 \sin y \cos y \dfrac{dy}{dx}$

3   (a) $4 (y + 3)^3 \dfrac{dy}{dx}$ (b) $8y (y^2 + 3)^3 \dfrac{dy}{dx}$

3.3   Finding $\dfrac{dy}{dx}$ implicitly

We illustrate the technique.

Example 3.4

Find $\dfrac{dy}{dx}$ given $y^2 + x = x^3 + 3y$.

Solution
We differentiate each term w.r.t. x.

$$\frac{d}{dx}(y^2) = 2y \frac{dy}{dx}, \quad \frac{d}{dx}(x) = 1, \quad \frac{d}{dx}(x^3) = 3x^2, \quad \frac{d}{dx}(3y) = 3 \frac{dy}{dx}$$

So

$$2y \frac{dy}{dx} + 1 = 3x^2 + 3 \frac{dy}{dx}$$

Rearrangement yields

$$\frac{dy}{dx} = \frac{3x^2 - 1}{2y - 3}$$

Note that $\dfrac{dy}{dx}$ is given in terms of x and y.

Example 3.5

Find y' given

$$x^2y - 3xy^3 = y^2 + 7$$

Solution

We differentiate each term w.r.t. x.

To differentiate x^2y we use the product rule.

$$\frac{d}{dx}(x^2y) = 2xy + x^2\frac{dy}{dx}$$

To differentiate $3xy^3$ we again use the product rule.

$$\frac{d}{dx}(3xy^3) = \boxed{} \qquad\qquad 3y^3 + 9xy^2\frac{dy}{dx}$$

Finally we differentiate $y^2 + 7$.

$$\frac{d}{dx}(y^2 + 7) = \boxed{} \qquad\qquad 2y\frac{dy}{dx}$$

Hence

$$2xy + x^2\frac{dy}{dx} - 3y^3 - 9xy^2\frac{dy}{dx} = 2y\frac{dy}{dx}$$

Rearrangement yields

$$\frac{dy}{dx} = \frac{3y^3 - 2xy}{x^2 - 9xy^2 - 2y}$$

Example 3.6

Find $\dfrac{d^2y}{dx^2}$ if $5x + 3y^2 = 7$.

Solution

First, we calculate $\dfrac{dy}{dx}$. Differentiating each term of the equation w.r.t. x we find

$$5 + 6y\frac{dy}{dx} = 0$$

so

$$\frac{dy}{dx} = -\frac{5}{6y}$$

Now $\dfrac{d^2y}{dx^2}$ means $\dfrac{d}{dx}\left(\dfrac{dy}{dx}\right)$ so we must differentiate $-\dfrac{5}{6y}$ with respect to x, that is

$\dfrac{d}{dx}\left(-\dfrac{5}{6y}\right)$. Here we are differentiating a function of y w.r.t. x. Using the result

$$\frac{d}{dx}(f(y)) = \frac{df}{dy}\frac{dy}{dx}$$

we have

$$\frac{d}{dx}\left(-\frac{5}{6y}\right) = \frac{d}{dy}\left(-\frac{5}{6y}\right)\frac{dy}{dx}$$

$$= \frac{5}{6y^2}\frac{dy}{dx}$$

$$= \frac{5}{6y^2}\left(-\frac{5}{6y}\right) = \frac{-25}{36\,y^3}$$

Exercises

1   Find $\dfrac{dy}{dx}$ given
(a) $x^2 + y^2 = 1$ (b) $3x - y^3 = 10$
(c) $x^2 + 2y^2 + 3x - 7y = 9$
(d) $x^3 - y^3 = x + y$ (e) $\sqrt{x} + \sqrt{y} = 10$

2   Find $\dfrac{dx}{dt}$ given

(a) $4 + \sin x = t$ (b) $3\cos 2x - t^2 = 20$

(c) $\tan x = \dfrac{1}{t}$ (d) $4\sin t - \cos 3x = \cos t$

(e) $\tan 2t - x^2 = \sin 2x$

3   Find $\dfrac{dy}{dx}$ given
(a) $2e^x + 3e^y = 10y$ (b) $e^{2x} - e^{3y} = x + 2y$
(c) $e^x e^y = x^2 y^3$ (d) $e^{2x+3y} + 2x^3 - y^5 = 0$
(e) $2e^{y-x} = 3e^x + y^2$

4   Find $\dfrac{d^2y}{dx^2}$ given

$$x^2 + y^2 = 1$$

5   Find y'' given

$$\sin x + \cos y = 1$$

Solutions to exercises

1   (a) $-\dfrac{x}{y}$ (b) $\dfrac{1}{y^2}$ (c) $\dfrac{2x + 3}{7 - 4y}$

(d) $\dfrac{3x^2 - 1}{3y^2 + 1}$ (e) $-\sqrt{\dfrac{y}{x}}$

2   (a) $\dfrac{1}{\cos x}$ (b) $-\dfrac{t}{3\sin 2x}$ (c) $-\dfrac{\cos^2 x}{t^2}$

(d) $\dfrac{-4\cos t - \sin t}{3\sin 3x}$ (e) $\dfrac{\sec^2 2t}{x + \cos 2x}$

3   (a) $\dfrac{2e^x}{10 - 3e^y}$ (b) $\dfrac{2e^{2x} - 1}{3e^{3y} + 2}$

(c) $\dfrac{2xy^3 - e^x e^y}{e^x e^y - 3x^2 y^2}$

(d) $\dfrac{6x^2 + 2e^{2x+3y}}{5y^4 - 3e^{2x+3y}}$ (e) $\dfrac{3e^x + 2e^{y-x}}{2e^{y-x} - 2y}$

4   $-\dfrac{1}{y^3}$

5   $-\dfrac{\sin^2 y \sin x + \cos^2 x \cos y}{\sin^3 y}$

End of block exercises

1 In which of the following equations is y expressed implicitly in terms of x?

(a) $\sin x + \dfrac{x}{y} = 3$ (b) $\sin y + \dfrac{y}{x} = 3$

(c) $x^2 - y = 10$ (d) $x^2y + y^2x = \sqrt{y}$

(e) $\dfrac{x - y}{x + y} = e^x$

2 Find $\dfrac{dy}{dx}$ given

(a) $x^3 - y^4 = 1$

(b) $2x^2 - 3y^2 + 2x - 7y = 0$

(c) $x^3 - 2y^4 = x$

(d) $2x^2 + y^2 + 3x + 2y + 7 = 0$

(e) $x^3 - y^3 + 3x^2 - y = 0$

3 A circle, centre the origin, radius 5, has equation

$$x^2 + y^2 = 25$$

Find the equation of the tangent to the circle that passes through the point $(3, 4)$.

4 Find $\dfrac{dx}{dt}$ given

(a) $2xt + x^2 - xt^2 = 0$ (b) $x^2t^2 - 3x^2t = t^3$
(c) $4(x + t)(x - t) - 3xt = 0$

(d) $\dfrac{x^2}{t^3} - \dfrac{t^2}{x^2} + x + t = 1$

(e) $\dfrac{x + t}{x - 2t} - \dfrac{3x}{t^2} = 0$

5 Find $\dfrac{dy}{dx}$:

(a) $e^x + e^y = x^3 + y^4$ (b) $x \sin y = y \cos x$
(c) $xe^y - 2ye^x = xy$ (d) $(x + 2y)e^{-y} = x^2y$
(e) $\sin 2x \cos 3y - 2 \sin y \cos x = 0$

6 Differentiate the following expressions w.r.t. x:
(a) $(x + y)^n$ (b) $(x^2 + y)^n$ (c) $(x^2 + y^2)^n$

Solutions to exercises

1 (b) and (d)

2 (a) $\dfrac{3x^2}{4y^3}$ (b) $\dfrac{4x + 2}{6y + 7}$ (c) $\dfrac{3x^2 - 1}{8y^3}$

(d) $-\dfrac{4x + 3}{2y + 2}$ (e) $\dfrac{3x^2 + 6x}{3y^2 + 1}$

3 $y = \dfrac{-3x + 25}{4}$

4 (a) $\dfrac{2x(t - 1)}{2t + 2x - t^2}$ (b) $\dfrac{3t^2 + 3x^2 - 2x^2t}{2xt^2 - 6xt}$

(c) $\dfrac{3x + 8t}{8x - 3t}$ (d) $\dfrac{3x^5 + 2xt^5 - x^3t^4}{2x^4t + 2t^6 + x^3t^4}$

(e) $\dfrac{2xt + 3t^2 + 6x}{6x - t^2 - 6t}$

5 (a) $\dfrac{3x^2 - e^x}{e^y - 4y^3}$ (b) $\dfrac{y \sin x + \sin y}{\cos x - x \cos y}$

(c) $\dfrac{y - e^y + 2ye^x}{xe^y - 2e^x - x}$ (d) $\dfrac{2xye^y - 1}{2 - x - 2y - x^2e^y}$

(e) $\dfrac{2 \cos 2x \cos 3y + 2 \sin y \sin x}{3 \sin 2x \sin 3y + 2 \cos y \cos x}$

6 (a) $n(x + y)^{n-1}\left(1 + \dfrac{dy}{dx}\right)$

(b) $n(x^2 + y)^{n-1}\left(2x + \dfrac{dy}{dx}\right)$

(c) $n(x^2 + y^2)^{n-1}\left(2x + 2y\dfrac{dy}{dx}\right)$

Parametric differentiation

In some circumstances both x and y depend upon a third variable, usually denoted by t. This third variable is referred to as a **parameter**. By eliminating t, y can be determined in terms of x. For example, if $y = (3 + t)^2$ and $x = 2t$, then by eliminating t we have

$$y = (3 + t)^2$$

$$= \left(3 + \frac{x}{2}\right)^2$$

Hence we can consider y as a function of x and so the derivative $\dfrac{dy}{dx}$ can be found.

In some cases, elimination of t is difficult or even impossible. As an example consider $y = e^t + \sin t$, $x = t^3 + \ln t$. Here it is impossible to obtain y in terms of x, but the derivative $\dfrac{dy}{dx}$ can still be found using the chain rule.

Key point

$$\frac{dy}{dx} = \frac{dy}{dt}\frac{dt}{dx}$$

$$= \frac{dy/dt}{dx/dt}$$

Finding $\dfrac{dy}{dx}$ by this method is known as **parametric differentiation**.

Example 4.1

Given $y = (3 + t)^2$ and $x = 2t$ find $\dfrac{dy}{dx}$.

Solution

We could eliminate t thus:

$$y = (3 + t)^2$$

$$= \left(3 + \frac{x}{2}\right)^2$$

$$= 9 + 3x + \frac{x^2}{4}$$

and then differentiate to obtain

$$\frac{dy}{dx} = 3 + \frac{x}{2}$$

Alternatively, using parametric differentiation we have

$$\frac{dy}{dt} = 2(3 + t), \quad \frac{dx}{dt} = 2$$

and so

$$\frac{dy}{dx} = \frac{dy/dt}{dx/dt}$$

$$= \frac{2(3 + t)}{2}$$

$$= 3 + t$$

$$= 3 + \frac{x}{2}$$

Example 4.2

Find $\dfrac{dy}{dx}$ given $y = \sin t + t^2$, $x = e^t + t$.

Solution

In this case elimination of t is impossible. To find $\dfrac{dy}{dx}$ we must use parametric differentiation.

$$\frac{dy}{dt} = \cos t + 2t, \quad \frac{dx}{dt} = e^t + 1$$

and so

$$\frac{dy}{dx} = \frac{dy/dt}{dx/dt}$$

$$= \frac{\cos t + 2t}{e^t + 1}$$

No further simplification is possible.

Example 4.3

(a) Find $\dfrac{dy}{dx}$ given $y = 2t - t^3$, $x = \sin t - \cos t$.

(b) Evaluate $\dfrac{dy}{dx}$ $(t = 0)$.

Solution

(a) $\dfrac{dy}{dt} = \boxed{}$, $\dfrac{dx}{dt} = \boxed{}$

$$2 - 3t^2,\ \cos t + \sin t$$

$$\frac{dy}{dx} = \frac{dy/dt}{dx/dt}$$

$$= \frac{2 - 3t^2}{\cos t + \sin t}$$

(b) $\dfrac{dy}{dx}(t = 0) = \dfrac{2 - 3(0)^2}{\cos 0 + \sin 0}$

$$= 2$$

Exercises

1 Use parametric differentiation to find $\dfrac{dy}{dx}$ given

(a) $x = 1 + t$, $y = 2 + 3t + t^2$
(b) $x = \sin t$, $y = \cos t$ (c) $x = t^2$, $y = t^3$
(d) $x = e^t$, $y = e^t + t$
(e) $x = \sqrt{t}$, $y = 1 + \ln t$

2 Find $\dfrac{dy}{dx}$ given

(a) $x = \sin(t^2)$, $y = t^3 + 1$
(b) $x = (1 + 2t)^4$, $y = (3 - 5t)^7$
(c) $x = 2\cos t$, $y = \cos 2t$
(d) $x = e^{\sin t}$, $y = 1 - t^2$
(e) $x = \ln(1 + 3t)$, $y = 1 + 2t$

Solutions to exercises

1 (a) $3 + 2t$ (b) $-\tan t$ (c) $\dfrac{3t}{2}$ (d) $1 + e^{-t}$

(e) $\dfrac{2}{\sqrt{t}}$

2 (a) $\dfrac{3t}{2\cos(t^2)}$ (b) $\dfrac{-35(3 - 5t)^6}{8(1 + 2t)^3}$ (c) $\dfrac{\sin 2t}{\sin t}$

(d) $\dfrac{-2t}{\cos t e^{\sin t}}$ (e) $\dfrac{2(1 + 3t)}{3}$

End of block exercises

1 Find y' given

(a) $x = 3 + 2t - t^2$, $y = 1 + 5t - t^3$
(b) $x = 1 + \dfrac{1}{t}$, $y = t + \dfrac{1}{t}$

(c) $x = 3\sin t$, $y = \sin t + 3t$
(d) $x = e^{2t}$, $y = e^{-t} + \cos t$
(e) $x = \tan 2t$, $y = e^{2t}$

2 Find $\dfrac{dy}{dx}$ given

(a) $x = \sqrt{1 + 2t}, y = (t^2 + 1)^4$

(b) $x = \sin(t^2 + 1), y = \cos(2t - 3)$

(c) $x = r \sin t, y = r \cos (t^2), r$ constant

(d) $x = \dfrac{1}{1 + 2t}, y = \dfrac{t}{1 + t}$

(e) $x = e^t \sin t, y = e^{-t} \cos t$

Solutions to exercises

1 (a) $\dfrac{5 - 3t^2}{2 - 2t}$ (b) $1 - t^2$ (c) $\dfrac{\cos t + 3}{3 \cos t}$

(d) $\dfrac{-(e^{-t} + \sin t)}{2e^{2t}}$ (e) $e^{2t} \cos^2 2t$

2 (a) $8t\sqrt{1 + 2t}(t^2 + 1)^3$ (b) $\dfrac{-\sin(2t - 3)}{t \cos(t^2 + 1)}$

(c) $-\dfrac{2t \sin (t^2)}{\cos t}$ (d) $-\dfrac{(1 + 2t)^2}{2 (1 + t)^2}$ (e) $-e^{-2t}$

Logarithmic differentiation

The technique of **logarithmic differentiation** is useful when we need to differentiate a cumbersome product. The method involves taking the natural logarithm of the function to be differentiated.

We gather together some important results that are commonly used in logarithmic differentiation.

1. $\dfrac{d}{dx}(\ln y) = \dfrac{1}{y}\dfrac{dy}{dx}$

2. $\dfrac{d}{dx}(\ln f(x)) = \dfrac{1}{f}\dfrac{df}{dx}$

3. $\ln(e^x) = x$

4. $\ln(e^{f(x)}) = f(x)$

5. $\ln AB = \ln A + \ln B$

6. $\ln \dfrac{A}{B} = \ln A - \ln B$

7. $\ln A^n = n \ln A$

The method of logarithmic differentiation is illustrated in the following examples.

Example 5.1

Given $y = t^2(1 + t)^3$ find $\dfrac{dy}{dt}$.

Solution

We could use the product rule to find $\dfrac{dy}{dt}$. However, we shall use logarithmic differentiation to illustrate the technique. Taking the natural logarithm of both sides of the given equation and applying the laws of logarithms yields

$$\ln y = \ln (t^2 (1 + t)^3)$$
$$= \ln t^2 + \ln (1 + t)^3$$
$$= 2 \ln t + 3 \ln (1 + t)$$

Now, both sides are differentiated w.r.t. t.

$$\frac{d}{dt}(\ln y) = \frac{d}{dt}(2 \ln t) + \frac{d}{dt}(3 \ln(1 + t))$$

$$\frac{1}{y}\frac{dy}{dt} = 2\left(\frac{1}{t}\right) + 3\left(\frac{1}{1 + t}\right)$$

$$\frac{dy}{dt} = y\left(\frac{2}{t} + \frac{3}{1 + t}\right)$$

$$= t^2(1 + t)^3\left(\frac{2}{t} + \frac{3}{1 + t}\right)$$

Example 5.2

Given that

$$y = (t + 1)^7(2t + 3)^4(2t - 1)^5$$

find $\dfrac{dy}{dt}$.

Solution

Taking the natural logarithm of both sides and applying the laws of logarithms yields

$$\ln y = \ln((t + 1)^7(2t + 3)^4(2t - 1)^5)$$

$$= \ln(t + 1)^7 + \ln(2t + 3)^4 + \ln(2t - 1)^5$$

$$= 7\ln(t + 1) + 4\ln(2t + 3) + 5\ln(2t - 1)$$

We now differentiate both sides w.r.t. t.

$$\frac{d}{dt}(\ln y) = \frac{d}{dt}(7\ln(t + 1)) + \frac{d}{dt}(4\ln(2t + 3)) + \frac{d}{dt}(5\ln(2t - 1))$$

$$\frac{1}{y}\frac{dy}{dt} = \frac{7}{t + 1} + \frac{8}{2t + 3} + \frac{10}{2t - 1}$$

$$\frac{dy}{dt} = y\left(\frac{7}{t + 1} + \frac{8}{2t + 3} + \frac{10}{2t - 1}\right)$$

$$\frac{dy}{dt} = (t + 1)^7(2t + 3)^4(2t - 1)^5\left(\frac{7}{t + 1} + \frac{8}{2t + 3} + \frac{10}{2t - 1}\right)$$

Example 5.3

Find $\dfrac{dy}{dx}$ given $y = e^{2x}x^3(1 - x)^4$.

Solution

Taking the natural logarithm of both sides and applying the laws of logarithms gives

$$\ln y = \ln(e^{2x}x^3(1 - x)^4)$$

$$= \ln e^{2x} + \ln x^3 + \ln(1 - x)^4$$

$$= \boxed{} + \boxed{} + \boxed{} \qquad\qquad 2x, 3\ln x, 4\ln(1 - x)$$

Differentiating w.r.t. x gives

$$\frac{1}{y}\frac{dy}{dx} = \boxed{} \qquad\qquad 2 + \frac{3}{x} - \frac{4}{1 - x}$$

and so

$$\frac{dy}{dx} = y\left(2 + \frac{3}{x} - \frac{4}{1-x}\right)$$

$$\frac{dy}{dx} = e^{2x}x^3(1-x)^4\left(2 + \frac{3}{x} - \frac{4}{1-x}\right)$$

Example 5.4

Find $\dfrac{dz}{dt}$ given $z(t) = \sqrt{1+t^2}\cos^4 t$.

Solution

Taking the natural logarithm of both sides gives

$$\ln z = \ln\left(\sqrt{1+t^2}\cos^4 t\right)$$

$$= \ln\sqrt{1+t^2} + \ln\cos^4 t$$

$$= \boxed{} + \boxed{} \qquad\qquad \frac{1}{2}\ln(1+t^2) + 4\ln\cos t$$

Differentiating both sides w.r.t. t gives

$$\frac{1}{z}\frac{dz}{dt} = \boxed{} \qquad\qquad \frac{t}{1+t^2} - \frac{4\sin t}{\cos t}$$

So

$$\frac{dz}{dt} = z\left(\frac{t}{1+t^2} - \frac{4\sin t}{\cos t}\right)$$

$$= \sqrt{1+t^2}\cos^4 t\left(\frac{t}{1+t^2} - 4\tan t\right)$$

Exercises

1. Find $\dfrac{dy}{dx}$ where y is given by
   (a) $x^5(3x+7)^9$ (b) $x^8(5x-1)^6$
   (c) $(3x+2)^4(9x-5)^7$
   (d) $(1-3x)^4(2-7x)^6$
   (e) $(5+2x)^4(5x+2)^3$

2. Find the derivative of each of the following functions:
   (a) $z(t) = e^{3t}(2t-5)^3(3t+1)^4$
   (b) $h(t) = 3e^{-6t}t^7(t+6)^3$
   (c) $M(p) = -p^4\sin^5 p$
   (d) $P(r) = 6(1+r^2)^6\sqrt{3+r^2}$
   (e) $y(x) = \dfrac{(7-x)^3}{(3x^2+1)^2}$

3. Find the rate of change of
   $$q(t) = 2e^{-t/2}\cos 2t$$
   when $t = 1$.

4. Find the derivative of the following functions using logarithmic differentiation:
   (a) $a(t) = (1+t^2)^3(1-t^2)^4\sin^3 t$
   (b) $b(r) = e^{-r}\sin^5 2r\cos^4 3r$
   (c) $K(p) = 6p\sqrt{\sin p}\,(\cos 2p)^{1/3}$
   (d) $N(y) = \dfrac{(1+\sqrt{y})^4(6+7y)^3}{\sqrt{1+y}}$
   (e) $x(t) = \dfrac{6}{(2+t)^3(1-t)^4\cos^3 t}$

Solutions to exercises

1 (a) $7x^4(3x + 7)^8(6x + 5)$

(b) $2x^7(5x - 1)^5(35x - 4)$

(c) $33(3x + 2)^3(9x - 5)^6(9x + 2)$

(d) $6(1 - 3x)^3(2 - 7x)^5(35x - 11)$

(e) $7(2x + 5)^3(5x + 2)^2(10x + 13)$

2 (a) $3e^{3t}(2t - 5)^2(3t + 1)^3(6t^2 + t - 23)$

(b) $-6t^6 e^{-6t}(t + 6)^2(3t^2 + 13t - 21)$

(c) $-5p^4 \sin^4 p \cos p - 4p^3 \sin^5 p$

(d) $\dfrac{6r\,(r^2 + 1)^5\,(13r^2 + 37)}{\sqrt{r^2 + 3}}$

(e) $\dfrac{3(x - 7)^2\,(x^2 - 28x - 1)}{(3x^2 + 1)^3}$

3 -1.9537

4 (a) $a\left(\dfrac{6t}{1 + t^2} - \dfrac{8t}{1 - t^2} + 3\cot t\right)$

(b) $b(-1 + 10\cot 2r - 12\tan 3r)$

(c) $K\left(\dfrac{1}{p} + \dfrac{1}{2}\cot p - \dfrac{2}{3}\tan 2p\right)$

(d) $N\left(\dfrac{2}{\sqrt{y} + y} + \dfrac{21}{6 + 7y} - \dfrac{1}{2(1 + y)}\right)$

(e) $x\left(\dfrac{-3}{2 + t} + \dfrac{4}{1 - t} + 3\tan t\right)$

End of block exercises

1 Find $\dfrac{dy}{dx}$ where y is given by

(a) $x^3(3x - 2)^4(5x - 1)^6$

(b) $\sqrt{x}(x^2 + 1)^3\,(x^3 + 1)^2$

(c) $\dfrac{e^{2x}\sin^3 x}{x^4}$ (d) $\dfrac{4e^{-3x}\cos 3x}{(2 - x)^3}$

(e) $\dfrac{3}{e^{2x}\sqrt{\sin x}\,(9 + x^2)}$

2 Find the derivative of each of the following functions:

(a) $y(x) = 3e^{(x^2)}x^2 \sin x$

(b) $h(r) = \dfrac{(3r^2 + 2)^3}{\sqrt{4 - r}\,(9 + r^2)}$

(c) $m(p) = \dfrac{(\sqrt{p} + 2)^2\,e^{-p^2}}{(1 + p^2)^2}$

(d) $x(t) = \dfrac{2\,(t + e^{2t})^4\sqrt{t^2 - 2t^3}}{\sin^2 t}$

(e) $c(q) = e^q\sqrt{\dfrac{q^2 - 2}{3q^2 + 1}}$

Solutions to exercises

1 (a) $y\left(\dfrac{3}{x} + \dfrac{12}{3x - 2} + \dfrac{30}{5x - 1}\right)$

(b) $y\left(\dfrac{1}{2x} + \dfrac{6x}{x^2 + 1} + \dfrac{6x^2}{x^3 + 1}\right)$

(c) $y\left(2 + 3\cot x - \dfrac{4}{x}\right)$

(d) $y\left(-3 - 3\tan 3x + \dfrac{3}{2 - x}\right)$

(e) $y\left(-2 - \dfrac{1}{2}\cot x - \dfrac{2x}{9 + x^2}\right)$

2 (a) $y\left(2x + \dfrac{2}{x} + \cot x\right)$

(b) $h\left(\dfrac{18r}{3r^2 + 2} + \dfrac{1}{2(4 - r)} - \dfrac{2r}{9 + r^2}\right)$

(c) $m\left(\dfrac{1}{p + 2\sqrt{p}} - 2p - \dfrac{4p}{1 + p^2}\right)$

(d) $x\left[4\left(\dfrac{1 + 2e^{2t}}{t + e^{2t}}\right) + \left(\dfrac{1 - 3t}{t - 2t^2}\right) - 2\cot t\right]$

(e) $c\left(1 + \dfrac{q}{q^2 - 2} - \dfrac{3q}{3q^2 + 1}\right)$

Tangents and normals

6.1 Introduction

The derivative gives the rate at which the function is changing. An alternative, but equally valid, interpretation of the derivative is that it represents the gradient of a tangent to the curve. We use this fact when calculating the equation of a tangent to a curve.

A normal is a line that is perpendicular to a tangent through the point of contact. Figure 6.1 illustrates a typical tangent and normal to $y(x)$.

Figure 6.1
A tangent and normal to a curve, $y(x)$.

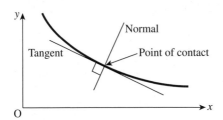

6.2 Calculating the equation of a tangent

Recall that a **tangent** is a straight line that touches a curve in one point only. This point is referred to as the **point of contact**. The derivative of a function can be interpreted as the gradient of a tangent to a curve.

When calculating the equation of a tangent we require its gradient and the coordinates of one point. Usually the coordinates of the point of contact are used.

Example 6.1
Find the equation of the tangent to $y = x^2$ at the point $(2, 4)$.

Solution
Figure 6.2 illustrates the situation. The point of contact is $(2, 4)$. To calculate the gradient of the tangent we find the derivative, $\dfrac{dy}{dx}$.

$$y = x^2 \quad \text{so} \quad \frac{dy}{dx} = 2x$$

Figure 6.2
A tangent to
$y = x^2$ at (2, 4).

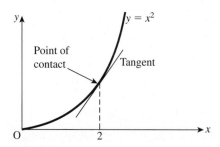

When $x = 2$, $\dfrac{dy}{dx} = 4$. Hence the gradient of the tangent is 4. We are now ready to calculate the equation of the tangent. The tangent is a straight line and so has an equation of the form

$$y = mx + c$$

The gradient, m, is 4 and so

$$y = 4x + c$$

The point of contact, (2, 4), lies on the tangent, that is when $x = 2$, $y = 4$.

$$4 = 4(2) + c$$

$$c = -4$$

Hence the equation of the tangent is $y = 4x - 4$.

Example 6.2
Find the equation of the tangent to $y = 2e^x - x^3$ where $x = 1.5$.

Solution
When $x = 1.5$, $y = $ [____]   5.5884

Thus the point of contact is (1.5, 5.5884).
    The gradient of the tangent is now found.

$$\dfrac{dy}{dx} = \text{[____]}$$ $2e^x - 3x^2$

When $x = 1.5$

$$\dfrac{dy}{dx} = \text{[____]}$$ $2e^{1.5} - 3(1.5)^2 = 2.2134$

and so the gradient of the tangent is 2.2134. The equation of the tangent can now be found. The equation has the form $y = mx + c$ and using the gradient found we have

$$y = 2.2134x + c$$

Applying the point of contact given yields

$5.5884 = $ [____] $2.2134(1.5) + c$

from which

$$c = \boxed{}$$ 2.2683

So the equation of the tangent is $y = 2.2134x + 2.2683$.

Exercises

1   Calculate the equation of the tangent to $y = 3x^2 - x + 9$ where $x = 2$.

2   Calculate the equation of the tangent to $y = \sin x$ where $x = \dfrac{\pi}{4}$.

3   The gradient of a tangent to $y = x^3 + x$ is 1. Calculate the equation of the tangent.

4   (a) Calculate the equation of the tangent to $y = e^{-x}$ where $x = 1.4$.
    (b) Calculate where the tangent cuts the x axis.
    (c) Calculate where the tangent cuts the y axis.

5   Calculate the equation of the tangents to $y = 9 - x^2$ at the points where y crosses the x axis.

Solutions to exercises

1   $y = 11x - 3$

2   $y = \dfrac{1}{\sqrt{2}}\left(x + 1 - \dfrac{\pi}{4}\right) = 0.7071x + 0.1517$

3   $y = x$

4   (a) $y = -0.2466x + 0.5918$
    (b) $(2.4, 0)$
    (c) $(0, 0.5918)$

5   $y = 6x + 18, y = -6x + 18$

6.3   The Newton–Raphson method

The Newton–Raphson method is a technique for generating a sequence of approximations to a solution, or root, of an equation $f(x) = 0$. The method uses differentiation because it relies upon knowledge of the equations of tangents to the graph of $y = f(x)$. It is known as a **numerical method** because it produces a numerical sequence of values that, under certain conditions, get closer and closer to the root. We say that this sequence of values converges. A more detailed discussion of convergence of sequences is given in Chapter 19.

Suppose we are interested in solving an equation that can be written in the form $f(x) = 0$. For example we may wish to solve equations such as

$$x^4 - 3x - 2 = 0 \quad \text{or} \quad e^{-x} - \sin x = 0$$

This problem is equivalent to finding values of x for which the graph of $y = f(x)$ intersects the horizontal axis, as illustrated in Figure 6.3. When x has the particular values x_1, x_2 or x_3 then y has the value of zero. In other words $f(x_1) = 0, f(x_2) = 0$ and $f(x_3) = 0$, and so the roots of $f(x) = 0$ are $x = x_1$, $x = x_2$ and $x = x_3$.

Figure 6.3
Solutions of $f(x) = 0$ can be located by finding values of x where the graph of $y = f(x)$ intersects the x axis.

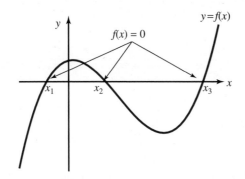

Figure 6.4
The tangent drawn at $P(x_0, f(x_0))$ intersects the x axis at x_1.

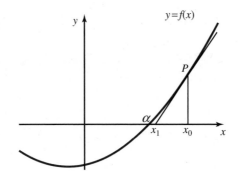

Suppose a root exists at some unknown value $x = \alpha$ as shown in Figure 6.4. Suppose we know or make an estimate of the root, $x = x_0$ say. We can draw a tangent to the curve $y = f(x)$ at the point $P(x_0, f(x_0))$, and use the value of x at the point where the tangent cuts the x axis, x_1 say, as a better estimate of the root.

First we obtain the equation of the tangent to $y = f(x)$ at $P(x_0, f(x_0))$. It will take the form $y = mx + c$. From Section 6.2 we know that its gradient, m, will equal $\dfrac{dy}{dx}$ evaluated at x_0, which, in the context of the Newton–Raphson method, we write as $f'(x_0)$. So the equation becomes $y = f'(x_0)x + c$. We now find the constant c. The tangent passes through the point $(x_0, f(x_0))$, and so $f(x_0) = f'(x_0)x_0 + c$, from which $c = f(x_0) - f'(x_0)x_0$. Since x_0 is a known value we can calculate $f(x_0) - f'(x_0)x_0$, and so c can be evaluated. Finally, the equation of the tangent is

$$y = f'(x_0)x + f(x_0) - f'(x_0)x_0$$

From Figure 6.4 we see that this tangent will cut the horizontal axis when $y = 0$ and $x = x_1$, that is when

$$0 = f'(x_0)x_1 + f(x_0) - f'(x_0)x_0$$

Rearranging this formula to make x_1 the subject gives

$$x_1 = x_0 - \frac{f(x_0)}{f'(x_0)}$$

Recall that x_0 is our first approximation to the root. This formula enables us to calculate a better estimate of the root, denoted by x_1.

Example 6.3
Use the Newton–Raphson formula to find an improved estimate of a solution to the equation $x^4 - 3x - 2 = 0$ given that a first estimate is $x_0 = 1.5$.

Solution
Here $f(x) = x^4 - 3x - 2$ and so, differentiating,

$$f'(x) = \qquad\qquad\qquad 4x^3 - 3$$

Then with a first estimate x_0 the Newton–Raphson formula is

$$x_1 = \qquad\qquad\qquad\qquad x_0 - \frac{x_0^4 - 3x_0 - 2}{4x_0^3 - 3}$$

Substitution of $x_0 = 1.5$ yields our new estimate x_1:

$$x_1 = \qquad\qquad\qquad\qquad$$

$$1.5 - \frac{1.5^4 - 3(1.5) - 2}{4(1.5)^3 - 3} = 1.636905$$

So 1.636905 is an improved estimate of the root.

In practice the formula is used repeatedly with each new estimate used to generate a further improvement. This leads to the following general formula.

Key point

If $x = x_n$ is an approximate root of $f(x) = 0$, then, under certain conditions, an improved estimate is given by

$$x_{n+1} = x_n - \frac{f(x_n)}{f'(x_n)}$$

Unfortunately the sequence of values produced does not always converge. For further details concerning the conditions for convergence and the rate of convergence consult a textbook on numerical methods for engineers.

The Newton–Raphson formula is easy to program in a loop structure. Exit from the loop is usually conditional upon $|x_{n+1} - x_n|$ being smaller than some prescribed very small value. This condition shows that successive approximate roots are very close to each other.

Example 6.4

Apply two more iterations of the Newton–Raphson method to the equation of Example 6.3.

Solution

With $x_1 = 1.636905$ we find

$x_2 =$

$$1.636905 - \frac{1.636905^4 - 3(1.636905) - 2}{4(1.636905)^3 - 3} = 1.618425$$

With $x_2 = 1.618425$ we find

$x_3 =$

$$1.618425 - \frac{1.618425^4 - 3(1.618425) - 2}{4(1.618425)^3 - 3} = 1.618034$$

We see that the numbers generated by this method appear to be getting closer to 1.618. In this example the method has performed well when we note that the true solution is 1.618034 (6 d.p.).

Exercises

1   Use the Newton–Raphson technique to find the value of a root of the following equations correct to two decimal places. An approximate root x_0 is given in each case.

(a) $x^2 - \cos x = 0$, $x_0 = 0.8$
(b) $3x^3 - 5x^2 + 7 = 0$, $x_0 = -0.9$
(c) $x^3 = e^{-x}$, $x_0 = 0.82$
(d) $e^x = \sin 2x$, $x_0 = -1.5$

Solutions to exercises

1   (a) 0.82 (b) −0.95 (c) 0.77 (d) −1.67

 ## Computer and calculator exercise

Use a graphical calculator or computer package to plot a graph of $f(x) = x^3 - 8x^2 + 11x - 2$. Use your graph to estimate roots of the equation

$x^3 - 8x^2 + 11x - 2 = 0$ and use your estimates together with the Newton–Raphson method to find improved estimates.

Solution to exercise

$x = 6.31, 1.48, 0.21$

6.4 Equation of a normal to a curve

A normal is a line that is perpendicular to a tangent and passes through the point of contact. (See Figure 6.1.)

It is useful to state an important result:

Key point

If two lines are perpendicular, the product of their gradients is -1.

Hence, if a tangent has gradient m_1 and the normal has gradient m_2, then $m_1 m_2 = -1$.

Example 6.5
Find the equation of the normal to $y = x^2$ at the point $(2, 4)$.

Solution
We have found from Example 6.1 that the gradient of the tangent at $(2, 4)$ is 4. Hence

$$\text{gradient of normal at } (2, 4) = \frac{1}{-4}$$
$$= -\frac{1}{4}$$

Let the equation of the normal be $y = mx + c$. The gradient, m, has been found to be $-\frac{1}{4}$ and so

$$y = -\frac{x}{4} + c$$

The point of contact, $(2, 4)$, is on the normal so

$$4 = -\frac{2}{4} + c$$
$$c = \frac{9}{2}$$

Hence the equation of the normal is

$$y = -\frac{x}{4} + \frac{9}{2}$$

Example 6.6
Find the equation of the normal to $y = 2x^4 - 3$ at $(1, -1)$.

Solution

We find $\dfrac{dy}{dx}$.

$$\frac{dy}{dx} = \boxed{}$$

$8x^3$

The tangent through $(1, -1)$ has gradient

8

Hence the normal through $(1, -1)$ has gradient

$-\dfrac{1}{8}$

The equation of the normal is

$$y = mx + c$$
$$= -\frac{x}{8} + c$$

The normal passes through $(1, -1)$ and so

$$c = \boxed{}$$

$-\dfrac{7}{8}$

Hence the equation of the normal is

$$y = -\frac{x}{8} - \frac{7}{8}$$

Exercises

1 Calculate the equation of the normal to
   (a) $y = 3x^2 - x$ through $(2, 10)$
   (b) $y = \sqrt{x}$ through $(9, 3)$
   (c) $y = \dfrac{1}{x}$ through $(0.5, 2)$
   (d) $y = 10 - x^4$ through $(1, 9)$
   (e) $y = \dfrac{x}{2} - \dfrac{x^3}{6}$ through $\left(1, \dfrac{1}{3}\right)$

2 Calculate the equation of the normal to
   (a) $y = xe^x$ where $x = 1$
   (b) $y = 2\sin 3x$ where $x = \pi$
   (c) $y = 2\ln x$ where $x = 1$
   (d) $y = (2x + 1)^6$ where $x = 0$
   (e) $y = \dfrac{e^x + 1}{x}$ where $x = 1$

Solutions to exercises

1 (a) $y = -\dfrac{x}{11} + \dfrac{112}{11}$

   (b) $y = -6x + 57$

   (c) $y = \dfrac{x}{4} + \dfrac{15}{8}$

   (d) $y = \dfrac{x + 35}{4}$

   (e) $x = 1$

2 (a) $y = \dfrac{-x + 2e^2 + 1}{2e}$

(b) $y = \dfrac{x - \pi}{6}$

(c) $y = \dfrac{1 - x}{2}$

(d) $y = -\dfrac{x}{12} + 1$

(e) $y = x + e$

End of block exercises

1 Calculate the equation of the tangent to
(a) $y = x^2 + 2$ at $(1, 3)$
(b) $y = 10 - 2x^2$ at $(0, 10)$
(c) $y = 2x^2 - x + 3$ at $(2, 9)$
(d) $y = x^3 + 1$ at $(-1, 0)$
(e) $y = x^3 + x^2 + x - 1$ at $(1, 2)$

2 Calculate the equation of the normal to
(a) $y = x^2 + 2x - 1$ at $(0, -1)$
(b) $y = 6 - x^2$ at $(3, -3)$
(c) $y = 3x^2 - 2x + 11$ at $(1, 12)$
(d) $y = 3x - x^3$ at $(2, -2)$
(e) $y = 2x^3 - 3x + 1$ at $(0, 1)$

3 Calculate the equation of the tangent to
(a) $y = 3 \sin x - \cos x$ where $x = \pi$
(b) $y = 2 \ln x$ where $x = e$
(c) $y = 3e^x + e^{-x}$ where $x = 0$
(d) $y = x^2 + e^{-2x}$ where $x = 0$
(e) $y = \sin 2x + 3 \cos 2x$ where $x = \dfrac{\pi}{4}$

4 Calculate the equation of the normal to
(a) $y = 2 \sin x + x$ where $x = 0$
(b) $y = 3e^{-x} + e^x$ where $x = 1$
(c) $y = 3x + 2 \ln x$ where $x = 1$
(d) $y = 4 \cos x + \sin x$ where $x = \pi$
(e) $y = 3 \ln x + \cos 2x$ where $x = 0.5$

5 Calculate the equation of the tangent to
(a) $y = \sqrt{x + 1}$ where $x = 3$
(b) $y = \sin (2x + \pi)$ where $x = 0$

(c) $y = x^2 e^{-x}$ where $x = 1$
(d) $y = \dfrac{2x + 1}{x + 2}$ where $x = 2$
(e) $y = \dfrac{x}{x^2 + 1}$ where $x = 1$

6 Calculate the equation of the normal to
(a) $y = \cos (x - \pi)$ where $x = \dfrac{\pi}{2}$
(b) $y = x^3 \ln x$ where $x = 1$
(c) $y = \sqrt{x^2 + 1}$ where $x = 2$
(d) $y = \dfrac{e^x}{x}$ where $x = -1$
(e) $y = \sin (x^2)$ where $x = 0.5$

7 Use the Newton–Raphson technique to find the value of a root of the following equations correct to two decimal places. An approximate root x_0 is given in each case.
(a) $x^5 - x^3 + 2 = 0, x_0 = -1.5$
(b) $2 \sin x = \sin 2x, x_0 = 6.2$

8 Apply the Newton–Raphson method to the solution of $x - \tan x = 0$. Show that with an initial estimate of $x_0 = 4$ the sequence of values generated by the method fails to converge, but with an initial estimate $x_0 = 4.6$ the method converges to the root $x = 4.49$ (2 d.p.).

Solutions to exercises

1 (a) $y = 2x + 1$ (b) $y = 10$ (c) $y = 7x - 5$
(d) $y = 3x + 3$ (e) $y = 6x - 4$

2 (a) $y = -\dfrac{x}{2} - 1$ (b) $y = \dfrac{x}{6} - \dfrac{7}{2}$

(c) $y = \dfrac{49 - x}{4}$ (d) $y = \dfrac{x - 20}{9}$

(e) $y = \dfrac{x}{3} + 1$

3 (a) $y = -3x + 3\pi + 1$ (b) $y = \dfrac{2}{e}x$

(c) $y = 2x + 4$ (d) $y = -2x + 1$

(e) $y = -6x + \dfrac{3\pi}{2} + 1$

4 (a) $y = -\dfrac{x}{3}$ (b) $y = -0.6193x + 4.4412$

(c) $y = \dfrac{-x + 16}{5}$ (d) $y = x - 4 - \pi$

(e) $y = -0.2316x - 1.4233$

5 (a) $y = \dfrac{x + 5}{4}$ (b) $y = -2x$

(c) $y = \dfrac{x}{e}$ (d) $y = \dfrac{3x}{16} + \dfrac{7}{8}$

(e) $y = 0.5$

6 (a) $y = -x + \dfrac{\pi}{2}$ (b) $y = -x + 1$

(c) $y = \dfrac{-\sqrt{5}}{2}x + 2\sqrt{5}$

(d) $y = 1.3591x + 0.9913$

(e) $y = -1.0321x + 0.7634$

7 (a) $x = -1.35$ (b) $x = 6.28$

Maximum and minimum values of a function

7.1 Introduction

The maximum and minimum values of a function are often very important. For example, an engineer may need to know the value of the maximum power transferred from a voltage source to a load resistor. Maximum and minimum points are located by examining the derivative of a function, rather than the function itself.

Two tests are described that distinguish between maximum and minimum points. Finally, we explain what is meant by a point of inflexion and how such a point is located.

7.2 Maximum and minimum points

Consider Figures 7.1(a) and (b). The point A is a **local maximum**; the point B is a **local minimum**. Note that A is not the highest point on the graph. However, in the locality of A, A is the highest point. Use of the word 'local' stresses that A is a maximum only in its locality. Similarly B is a minimum in its locality but is not the lowest point on the entire graph.

Figure 7.1
(a), (b) A is a local maximum, B is a local minimum.

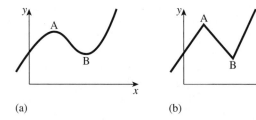

(a)　　　　　　　(b)

When moving away from A along the function, both to the left and to the right, the value of y decreases; when moving away from B along the function, the value of y increases.

It is useful to be able to locate points such as A and B. This is done by referring to the gradient of the function, rather than the function itself.

7.3   Locating maximum and minimum points

Consider Figure 7.2, which shows the curve $y(x)$, together with some tangents. The tangents at A and B are parallel to the x axis, that is $\dfrac{dy}{dx} = 0$ at these points.

Figure 7.2

At A and B,
$\dfrac{dy}{dx} = 0$.

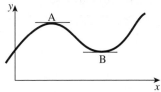

We now consider Figure 7.1(b) again. At A and B it is impossible to draw tangents: that is, they do not exist. Hence at these points $\dfrac{dy}{dx}$ does not exist. In summary we have

Key point

> At maximum and minimum points, $\dfrac{dy}{dx} = 0$ or $\dfrac{dy}{dx}$ does not exist.

So, maximum and minimum points are located by looking for points where $\dfrac{dy}{dx} = 0$ or $\dfrac{dy}{dx}$ does not exist.

7.4   The first-derivative test

When given any function, $y(x)$, we can limit our search for maximum and minimum points to those points where $\dfrac{dy}{dx} = 0$ or $\dfrac{dy}{dx}$ does not exist. We also need to distinguish between a maximum point and a minimum point. To do this we consider y' on either side of the point.

Immediately to the left of a maximum point, such as A in Figure 7.1, $y' > 0$, that is y is increasing. Immediately to the right of a maximum point, $y' < 0$, that is y is decreasing. Thus, in passing from left to right through a maximum point, y' changes from positive to negative.

Now consider a minimum point, such as B in Figure 7.1. Immediately to the left of such a point, $y' < 0$, that is y is decreasing. Immediately to the right, $y' > 0$, that is y is increasing. So, in passing from left to right through a minimum point, y' changes from negative to positive.

This information is contained in the **first-derivative test**.

Key point

> - The first-derivative test distinguishes between maximum and minimum points.
> - To the left of a maximum point, $\dfrac{dy}{dx}$ is positive; to the right $\dfrac{dy}{dx}$ is negative.
> - To the left of a minimum point, $\dfrac{dy}{dx}$ is negative; to the right $\dfrac{dy}{dx}$ is positive.

Example 7.1

Determine the position of any maximum and minimum points of the function $y = x^2 + 1$.

Solution

By differentiation, $\dfrac{dy}{dx} = 2x$. The function $2x$ exists for all values of x. So, we need only look for maximum and minimum points by solving $y' = 0$. So

$$y' = 0$$
$$2x = 0$$

so

$$x = 0$$

At this stage, we know that the only place a maximum or a minimum point can be found is where $x = 0$. We now apply the first-derivative test.

To the left of $x = 0$, x is negative. So $2x$ is negative and hence y' is negative. To the right of $x = 0$, x is positive and so y' is positive. Since y' changes from negative to positive, there must be a minimum point at $x = 0$.

When $x = 0$, $y = 1$ so $(0, 1)$ is a minimum point. Figure 7.3 illustrates this.

Figure 7.3
There is a minimum point at $(0, 1)$.

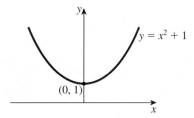

Example 7.2

Determine the position of any maximum and minimum points of the function $y = 2x - x^2$.

Solution

We see that $\dfrac{dy}{dx} = 2 - 2x$, which exists for all x. We look for maximum and minimum points by solving $\dfrac{dy}{dx} = 0$.

$$\frac{dy}{dx} = 0$$
$$2 - 2x = 0$$
$$2 = 2x$$

so

$$x = 1$$

Thus we examine the point where $x = 1$.

We examine the sign of $\dfrac{dy}{dx}$ to the left and to the right of $x = 1$. To determine the sign of $2 - 2x$ to the left and to the right of $x = 1$ we can use one of two techniques. We can sketch a graph of $2 - 2x$ and note the sign on both sides of $x = 1$. Another method is to evaluate $2 - 2x$ just to the left of $x = 1$, say $x = 0.9$, and then evaluate $2 - 2x$ just to the right of $x = 1$, say at $x = 1.1$. When $x = 0.9$, the value of $2 - 2x$ is 0.2; when $x = 1.1$ the value of $2 - 2x$ is -0.2. Since y' changes from positive to negative there must be a maximum at $x = 1$.

When $x = 1$, $y = 1$ and so $(1,1)$ is a maximum point. Figure 7.4 illustrates this.

Figure 7.4
There is a
maximum point
at (1,1).

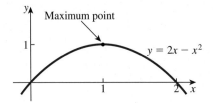

Example 7.3
Determine the position of any maximum and minimum points of the function
$$y = \frac{t^3}{3} - \frac{t^2}{2} - 2t + 3.$$

Solution
We have
$$\frac{dy}{dt} = t^2 - t - 2$$

Clearly, $\dfrac{dy}{dt}$ exists for all values of t. Solving $\dfrac{dy}{dt} = 0$ yields
$$t^2 - t - 2 = 0$$
$$(t - 2)(t + 1) = 0$$
$$t = -1, 2$$

We need to investigate the two points where $t = -1$ and $t = 2$.

$t = -1$
Consider a value just to the left of $t = -1$, say $t = -1.1$. Here

$$\frac{dy}{dt} = (-1.1)^2 - (-1.1) - 2$$
$$= 0.31$$

So, at $t = -1.1$, $\dfrac{dy}{dt}$ is positive.

Just to the right of $t = -1$, say at $t = -0.9$,

$$\frac{dy}{dt} = (-0.9)^2 - (-0.9) - 2$$
$$= -0.29$$

So, $\dfrac{dy}{dt}$ is negative. Since the derivative has changed from positive to negative there must be a maximum point at $t = -1$.

When $t = -1$, $y = \frac{25}{6}$ and so $(-1, \frac{25}{6})$ is a maximum point.

$t = 2$

By considering values just to the left and right of $t = 2$ we see that immediately to the left of $t = 2$, $\dfrac{dy}{dt}$ is negative; immediately to the right, $\dfrac{dy}{dt}$ is positive. Hence at $t = 2$ there is a minimum point.

When $t = 2$, $y = -\frac{1}{3}$ so $(2, -\frac{1}{3})$ is a minimum point. Figure 7.5 illustrates a graph of the function.

Figure 7.5
There is a maximum point at $(-1, \frac{25}{6})$ and a minimum point at $(2, -\frac{1}{3})$.

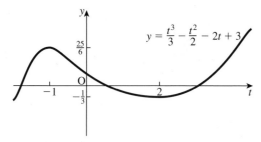

Example 7.4

Determine the position of any maximum and minimum points of the function $y = |t|$.

Solution

Recall that

$$y = |t| = \begin{cases} -t & t < 0 \\ t & t \geq 0 \end{cases}$$

A graph of the modulus function is shown in Figure 8.7 of Chapter 6. Note that there is a corner at $t = 0$. Hence the derivative, $\dfrac{dy}{dt}$, does not exist at $t = 0$ and so this is a possible location of a maximum or minimum point. To the left of $t = 0$, $\dfrac{dy}{dt}$ is negative; to the right of $t = 0$, $\dfrac{dy}{dt}$ is positive and so there is a minimum at $t = 0$.

Example 7.5

Locate the maximum and minimum points of each of the following functions:
(a) $y = x^2 - 6x + 5$
(b) $y = x^3 - 3x$
(c) $y = x^{1/3}$

Solution

(a) Given $y = x^2 - 6x + 5$

$$\frac{dy}{dx} = \boxed{} \qquad\qquad 2x - 6$$

Solving $\dfrac{dy}{dx} = 0$ yields

$$x = \boxed{} \qquad\qquad 3$$

Immediately to the left of $x = 3$,

$\dfrac{dy}{dx}$ is [   ]   negative

Immediately to the right of $x = 3$,

$\dfrac{dy}{dx}$ is [   ]   positive

Hence at $x = 3$, there is a [   ]   minimum

When $x = 3$,

$y =$ [   ] -4

Hence $(3, -4)$ is a minimum point.

(b) Given $y = x^3 - 3x$, here

$y' =$ [   ] $3x^2 - 3$

Solving $y' = 0$ yields

$x =$ [   ] $-1, 1$

Consider $x = -1$.

Immediately to the left of $x = -1$,

y' is [   ]   positive

Immediately to the right of $x = -1$,

y' is [   ]   negative

Hence at $x = -1$ there is a [   ]   maximum

When $x = -1$,

$y =$ [   ]   2

So $(-1, 2)$ is a maximum point.

We now consider $x = 1$.

Immediately to the left of $x = 1$,

y' is [   ]   negative

Immediately to the right of $x = 1$,

y' is [   ]   positive

Hence at $x = 1$ there is a [   ]   minimum

When $x = 1$,

$y =$ [   ] -2

So $(1, -2)$ is a minimum point.

(c) Given $y = x^{1/3}$ then by differentiation we have

$\dfrac{dy}{dx} =$ [   ] $\dfrac{x^{-2/3}}{3} = \dfrac{1}{3x^{2/3}}$

Note that at $x = 0$ the function $\dfrac{1}{3x^{2/3}}$ does not exist so there is no derivative at $x = 0$.

To the left of $x = 0$,

y' is [ ]   positive

To the right of $x = 0$,

y' is [ ]   positive

Since y' does not change sign we conclude there is neither a maximum nor a minimum at $x = 0$.

Example 7.6 Electrical Engineering – Maximum power transfer

Consider the circuit of Figure 7.6 in which a non-ideal voltage source is connected to a variable load resistor with resistance R_L. The source voltage is V and its internal resistance is R_S. Calculate the value of R_L that results in the maximum power being transferred from the voltage source to the load resistor.

Figure 7.6
Maximum power transfer occurs when $R_L = R_S$.

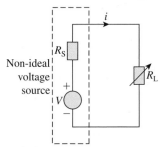

Solution

Let i be the current flowing in the circuit. Using Kirchhoff's voltage law and Ohm's law gives

$$V = i(R_S + R_L)$$

Let P be the power developed in the load resistor. Then

$$P = i^2 R_L$$

$$= \frac{V^2 R_L}{(R_S + R_L)^2}$$

Clearly P depends on the value of R_L. Differentiating we obtain

$$\frac{dP}{dR_L} = V^2 \frac{1\,(R_S + R_L)^2 - R_L\,2\,(R_S + R_L)}{(R_S + R_L)^4}$$

$$= V^2 \frac{R_S - R_L}{(R_S + R_L)^3}$$

Equating $\dfrac{dP}{dR_L}$ to zero gives

$$V^2 \frac{R_S - R_L}{(R_S + R_L)^3} = 0$$

that is

$$R_L = R_S$$

We need to check whether this is a maximum point.

$$\frac{dP}{dR_L} = V^2 \frac{R_S - R_L}{(R_S + R_L)^3}$$

When R_L has a value just to the left of R_S, the derivative is positive. When R_L has a value just to the right of R_S, the derivative is negative. Hence there is a maximum when $R_L = R_S$. Therefore maximum power transfer occurs when the load resistance equals the source resistance.

Exercises

1   Determine the location of all maximum and minimum points of the following functions:
(a) $y = x^2 - 4x$
(b) $y = x^2 - 5x + 4$
(c) $y = 10 + 3x - x^2$
(d) $y = \dfrac{x^3}{3} - \dfrac{x^2}{2} + 1$
(e) $y = x^3 - 27x$

2   Determine the location of all maximum and minimum points of the following functions:
(a) $y = xe^x$
(b) $y = \dfrac{1}{1 + x^2}$
(c) $y = \dfrac{x^5}{5} - \dfrac{x^3}{3}$
(d) $y = \dfrac{x^4}{4} + x + 1$
(e) $y = (1 - \ln x)x$

Solutions to exercises

1   (a) $(2, -4)$ minimum
(b) $\left(\frac{5}{2}, -\frac{9}{4}\right)$ minimum
(c) $\left(\frac{3}{2}, \frac{49}{4}\right)$ maximum
(d) $(0, 1)$ maximum, $\left(1, \frac{5}{6}\right)$ minimum
(e) $(-3, 54)$ maximum, $(3, -54)$ minimum

2   (a) $(-1, -0.3679)$ minimum
(b) $(0, 1)$ maximum
(c) $\left(-1, \frac{2}{15}\right)$ maximum, $\left(1, -\frac{2}{15}\right)$ minimum
(d) $\left(-1, \frac{1}{4}\right)$ minimum
(e) $(1, 1)$ maximum

7.5   The second-derivative test

This is a test to distinguish between maximum and minimum points. Rather than examine the sign of y' on both sides of the points where $y' = 0$ we can instead consider the sign of the second derivative, y'', at these points.

On passing left to right through a maximum point, y' changes from positive to zero to negative. Hence, y' is decreasing and so the derivative of y' is negative, that is $y'' < 0$. Similarly, on passing left to right through a minimum point, y' changes from negative to zero to positive and so y' is increasing. Hence y'' is positive. The second-derivative test summarises this:

- If $y' = 0$ and $y'' < 0$ at a point, then the point is a maximum point.
- If $y' = 0$ and $y'' > 0$ at a point, then the point is a minimum point.
- If $y' = 0$ and $y'' = 0$ the second-derivative test fails and we must return to the first-derivative test.

Example 7.7
Use the second-derivative test to find all maximum and minimum points of

$$y = \frac{x^3}{3} - \frac{x^2}{2} - 6x + 2.$$

Solution
We see that

$$y' = x^2 - x - 6$$
$$= (x + 2)(x - 3)$$

Solving $y' = 0$ yields $x = -2, 3$. Now

$$y'' = 2x - 1$$

The sign of y'' is calculated at both $x = -2$ and $x = 3$.

When $x = -2$, $y'' = -5$. Since $y'' < 0$ then by the second-derivative test there is a maximum point at $x = -2$.

When $x = 3$, $y'' = 5$. Here $y'' > 0$ and so there is a minimum point at $x = 3$.

When $x = -2$, $y = \frac{28}{3}$. When $x = 3$, $y = -\frac{23}{2}$. So $\left(-2, \frac{28}{3}\right)$ is a maximum point; $\left(3, -\frac{23}{2}\right)$ is a minimum point.

Example 7.8
Determine the positions of all maximum and minimum points of $y = x^4$.

Solution
We have $y' = 4x^3$. Solving $y' = 0$ yields $x = 0$. Also we see

$$y'' = 12x^2$$

To apply the second-derivative test we evaluate y'' at $x = 0$. At $x = 0$, $y'' = 0$. Since $y'' = 0$ the second-derivative test fails. We return to the first-derivative test and examine the sign of y' to the left and to the right of $x = 0$.

Immediately to the left of $x = 0$, y' is negative. Immediately to the right of $x = 0$, y' is positive. Hence there is a minimum point at $x = 0$.

When $x = 0$, $y = 0$ and so $(0, 0)$ is a minimum point.

Example 7.9
Determine all maximum and minimum points of

$$y = \frac{x^5}{5} + \frac{x^2}{2} + 1$$

Solution
We have

$y' = $ ⬚ $x^4 + x$

Solving $y' = 0$ yields

$x = $ ⬚ $0, -1$

In order to use the second derivative test we calculate y''.

$y'' = $ ⬚ $4x^3 + 1$

The sign of y'' is calculated at each value of x.
When $x = -1$,

y'' is ⬚   negative

and so there is a ⬚ point at $x = -1$.   maximum
When $x = 0$,

y'' is ⬚   positive

and so there is a ⬚ point at $x = 0$.   minimum
When $x = -1$,

$y = $ ⬚ $\dfrac{13}{10}$

When $x = 0$,

$y = $ ⬚ 1

So $\left(-1, \frac{13}{10}\right)$ is a maximum point; $(0, 1)$ is a minimum point.

Exercises

1   Determine the position of all maximum and minimum points using the second-derivative test.

(a) $y = \dfrac{x^2}{2} - x + 1$

(b) $y = 6 + 2x - \dfrac{3x^2}{2}$

(c) $y = \dfrac{x^3}{3} + \dfrac{3x^2}{2} - 1$

(d) $y = \dfrac{x^4}{4} - \dfrac{x^2}{2}$

Solutions to exercises

1   (a) $\left(1, \frac{1}{2}\right)$, minimum

(b) $\left(\frac{2}{3}, \frac{20}{3}\right)$, maximum

(c) $(0, -1)$, minimum; $\left(-3, \frac{7}{2}\right)$, maximum

(d) $(0, 0)$, maximum; $\left(1, -\frac{1}{4}\right)$, minimum; $\left(-1, -\frac{1}{4}\right)$, minimum

7.6 Points of inflexion

Figure 7.7 shows two curves. In both cases the gradient is increasing as we move along the curve from left to right, that is y' is increasing.

Figure 7.7
The gradient of both curves is increasing; the curves are concave up.

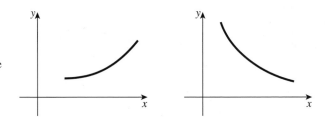

When y' is increasing, then y'' is positive, and the curve is described as **concave up**. Figure 7.8 shows two curves the gradients of which are decreasing as we move along the curve from left to right, that is y' is decreasing. When y' is decreasing, y'' is negative and the curve is said to be **concave down**.

Figure 7.8
The gradient of both curves is decreasing; the curves are concave down.

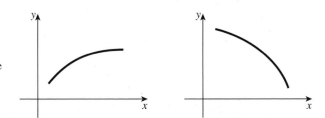

A point at which the concavity changes from concave up to concave down, or vice versa, is called a **point of inflexion**. At such a point either $y'' = 0$ or y'' does not exist. Figure 7.9 illustrates some points of inflexion.

Figure 7.9
(a) There is a point of inflexion at A; (b) there are points of inflexion at A and B; (c) there is a point of inflexion at O.

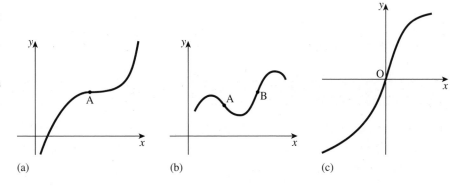

(a)   (b)   (c)

> A curve is **concave up** when $y'' > 0$, and **concave down** when $y'' < 0$. There is a **point of inflexion** where the concavity changes.

Note that not all points where $y'' = 0$ or where y'' does not exist are points of inflexion. However, when searching for points of inflexion, we can limit our search to points where $y'' = 0$ or where y'' does not exist.

Example 7.10
Locate any points of inflexion of $y = x^3$.

Solution
We have

$$y' = 3x^2, \ y'' = 6x$$

A point of inflexion can occur only where either $y'' = 0$ or y'' does not exist. Clearly y'' exists for all values of x but $y'' = 0$ when $x = 0$.

We examine the concavity to the left and right of $x = 0$. To the left of $x = 0$, $6x$ is negative and so $y'' < 0$: that is, the function is concave down. To the right of $x = 0$, $6x$ is positive, and so $y'' > 0$: that is, the function is concave up. Hence the concavity changes at $x = 0$ and so there is a point of inflexion at $x = 0$.

When $x = 0$, $y = 0$ and so $(0, 0)$ is a point of inflexion. Figure 7.10 illustrates $y = x^3$.

Figure 7.10
Since y'' changes sign at $x = 0$, there is a point of inflexion there.

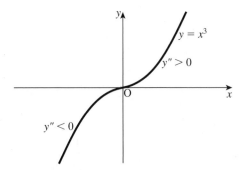

Example 7.11
Determine all maximum points, minimum points and points of inflexion of

$$y = \frac{x^3}{6} - \frac{x^2}{2}$$

Solution
We have

$$y' = \frac{x^2}{2} - x, \qquad y'' = x - 1$$

To locate maximum and minimum points we solve $y' = 0$. This yields

$$\frac{x^2}{2} - x = 0$$

$$x\left(\frac{x}{2} - 1\right) = 0$$

$$x = 0, 2$$

To distinguish between maximum and minimum points we use the second-derivative test at $x = 0$ and at $x = 2$.

$y''(0) < 0$ and so there is a maximum point at $x = 0$.

$y''(2) > 0$ and so there is a minimum point at $x = 2$.

When $x = 0$, then $y = 0$. When $x = 2$, $y = -\frac{2}{3}$. So $(0, 0)$ is a maximum point; $\left(2, -\frac{2}{3}\right)$ is a minimum point.

We now consider points of inflexion. We know that points of inflexion can occur only where $y'' = 0$ or where y'' does not exist. Clearly y'' exists for all values of x but $y'' = 0$ at $x = 1$. To check whether there is a point of inflexion at $x = 1$ we check the concavity to the left and right of $x = 1$.

Immediately to the left of $x = 1$, say at $x = 0.9$, $y'' < 0$. Immediately to the right of $x = 1$, say at 1.1, $y'' > 0$ and so the concavity changes at $x = 1$: that is, there is a point of inflexion at $x = 1$. When $x = 1$, $y = -\frac{1}{3}$, so $\left(1, -\frac{1}{3}\right)$ is a point of inflexion.

In summary we have: $(0, 0)$ maximum point; $\left(2, -\frac{2}{3}\right)$ minimum point; $\left(1, -\frac{1}{3}\right)$ point of inflexion.

Example 7.12

Find all maximum points, minimum points and points of inflexion of

$$y = 1 + 12x + 3x^2 - 2x^3$$

Solution

We calculate y' and y''.

$y' = $ $12 + 6x - 6x^2$

$y'' = $ $6 - 12x$

Solving $y' = 0$ yields

$x = $ $-1, 2$

We apply the second-derivative test to each value of x.

$y''(-1) = $ 18

$y''(2) = $ -18

Hence there is a point at $x = -1$. minimum

There is a point at $x = 2$. maximum

When $x = -1$, $y = -6$. When $x = 2$, $y = 21$.

We seek points of inflexion by considering y''. Solving $y'' = 0$ yields $x = 0.5$. Immediately to the left of $x = 0.5$, say at 0.4,

y'' is positive

Immediately to the right of $x = 0.5$, say at 0.6,

y'' is ⬚ negative

Hence the concavity changes and so there is a point of inflexion at $x = 0.5$.
    When $x = 0.5$,

$y = $ ⬚ 7.5

In summary we have: $(-1, -6)$, minimum point; $(2, 21)$ maximum point; $(0.5, 7.5)$, point of inflexion.

Exercises

1 Determine the position of all points of inflexion of the following functions:

(a) $y = \dfrac{x^3}{6} - x^2$

(b) $y = 1 + 2x^2 - \dfrac{x^3}{6}$

(c) $y = \dfrac{x^4}{24} - \dfrac{x^3}{6} + x + 1$

(d) $y = \dfrac{x^4}{12} - \dfrac{x^2}{2} - x + 1$

(e) $y = x^4$

2 Locate all maximum points, minimum points and points of inflexion of the following functions:

(a) $y = \dfrac{x^3}{3} - 3x^2 + 5x + 1$

(b) $y = -x^5$

(c) $y = x^{1/3}$

Solutions to exercises

1 (a) $\left(2, -\dfrac{8}{3}\right)$ (b) $\left(4, \dfrac{67}{3}\right)$ (c) $(0, 1)$, $\left(2, \dfrac{7}{3}\right)$
(d) $\left(-1, \dfrac{19}{12}\right)$, $\left(1, -\dfrac{5}{12}\right)$
(e) no inflexion points

2 (a) $\left(1, \dfrac{10}{3}\right)$ maximum; $\left(5, -\dfrac{22}{3}\right)$ minimum;
    $(3, -2)$ inflexion
(b) $(0, 0)$ inflexion
(c) $(0, 0)$ inflexion

End of block exercises

1 State when each of the following functions are concave up and when they are concave down:

(a) $y = x^2$

(b) $y = x^3$

(c) $y = \dfrac{x^3}{6} - x^2$

(d) $y = \dfrac{x^2}{2} + \dfrac{x^3}{6}$

2 Find all maximum points, minimum points and points of inflexion of the following functions:

(a) $y = 1 + 4x - \dfrac{3x^2}{2}$

(b) $y = \dfrac{x^3}{3} + x^2 - 3x + 7$

(c) $y = \dfrac{2x^3}{3} - \dfrac{x^2}{2} - x + 6$

(d) $y = 1 + 4x - \dfrac{3x^2}{2} - \dfrac{x^3}{3}$

(e) $y = \dfrac{x^4}{4} - \dfrac{x^2}{2} + 2$

Solutions to exercises

1 (a) always concave up
   (b) concave up on $(0, \infty)$; concave down on $(-\infty, 0)$
   (c) concave down on $(-\infty, 2)$; concave up on $(2, \infty)$
   (d) concave down on $(-\infty, -1)$; concave up on $(-1, \infty)$

2 (a) $\left(\frac{4}{3}, \frac{11}{3}\right)$ maximum
   (b) $\left(1, \frac{16}{3}\right)$ minimum, $(-3, 16)$ maximum, $\left(-1, \frac{32}{3}\right)$ inflexion

(c) $\left(1, \frac{31}{6}\right)$ minimum, $\left(-\frac{1}{2}, \frac{151}{24}\right)$ maximum, $\left(\frac{1}{4}, \frac{275}{48}\right)$ inflexion

(d) $\left(-4, -\frac{53}{3}\right)$ minimum, $\left(1, \frac{19}{6}\right)$ maximum, $\left(-\frac{3}{2}, -\frac{29}{4}\right)$ inflexion

(e) $\left(1, \frac{7}{4}\right)$ minimum, $\left(-1, \frac{7}{4}\right)$ minimum, $(0, 2)$ maximum, $\left(-\frac{1}{\sqrt{3}}, \frac{67}{36}\right)$ inflexion, $\left(\frac{1}{3}, \frac{67}{36}\right)$ inflexion

End of chapter exercises

1 Find $\dfrac{dy}{dx}$ where y is given by
   (a) $x^3 e^{2x}$ (b) $-3 \sin 2x \cos 5x$
   (c) $(x^2 + 1) e^{-x}$
   (d) $\sin x \sin 2x \sin 3x$ (e) $x \tan 3x$

2 Find y' where y is given by
   (a) $\dfrac{x^2 + 1}{2x + 3}$ (b) $\dfrac{\cos 2t}{\sin 3t}$ (c) $\dfrac{r^3}{3e^{2r}}$
   (d) $\dfrac{v + \sin v}{1 + e^v}$ (e) $\dfrac{\sqrt{x}}{x + 1}$

3 Find $\dfrac{dy}{dx}$ when $x = 1$ where y is defined by
   (a) $\dfrac{x \cos x}{\sin x}$ (b) $\dfrac{x^2 + 1}{xe^x}$ (c) $\dfrac{\sin x \cos 2x}{x^2}$
   (d) $\dfrac{1}{x \sin x}$ (e) $\dfrac{4 \tan x}{e^x \sin x}$

4 Find the derivative of the following functions:
   (a) $y = e^{2x} x^3 \sin 3x \cos 2x$
   (b) $y = (x + \sin x)^7$
   (c) $H = \ln(t^2 + 3t - 9)$
   (d) $V(r) = \dfrac{1}{\ln r}$
   (e) $M(b) = \ln b + \ln(b + 1)$

5 Use logarithmic differentiation to differentiate
$$y = x^x$$

6 Find $\dfrac{dy}{dx}$ given
   (a) $x(t) = t^2 + 3$, $y(t) = 2t^2 + t + 1$
   (b) $x(t) = t^2$, $y(t) = t^3 + k$, k constant
   (c) $x(t) = \dfrac{1}{t}$, $y(t) = \sin t$
   (d) $x(t) = 2e^t$, $y(t) = te^t$
   (e) $x(t) = \sqrt{t}$, $y(t) = \sqrt{2t + 1}$

7 Find $\dfrac{dy}{dx}$ given
   (a) $x^2 + x^3 + y^2 - y^3 = 1$
   (b) $2x^2 - y^2 + 3xy - 7x - 10y = 0$
   (c) $xy^2 + \dfrac{y}{x} = e^x$
   (d) $\ln(xy) - \sqrt{x} = \sqrt{y}$
   (e) $x \sin y + y^2 \cos 2x = y$

8 Find $\dfrac{dy}{dx}$ given $\ln(x + y) = k$, k constant.

9 Find y' given $(x^2 - y^3)^6 = e^{xy}$.

10 Find $\dfrac{d^2 y}{dx^2}$ given $y = e^{(x^2)}$.

11 Find the equation of the tangent to $y = 3x^3 - x$ where $x = 1$.

12   Find the equation of the normal to $y = \ln x$ where $x = 2$.

13   Find the equation of the tangent to $y = xe^x$ where $x = 1$.

14   Find the equation of the tangent to $y = \dfrac{1}{x^2}$ where $x = -1$.

15   State the range of values of x for which each of the following functions is (i) concave up (ii) concave down:

(a) $y = \dfrac{x^3}{6} - \dfrac{5x^2}{2} + 3x - 9$

(b) $y = 3 + x + \dfrac{x^2}{2} - \dfrac{x^4}{12}$

(c) $y = e^x - 100x - 100$

(d) $y = (x - 1)^4$

(e) $y = \dfrac{x^2}{2} \ln x - \dfrac{x^2}{4}$

16   Locate all maximum points, minimum points and points of inflexion of

$$y = \dfrac{2x^3}{3} - \dfrac{5x^2}{2} + 1$$

17   Locate all maximum points, minimum points and points of inflexion of

$$y = x^2 e^x$$

Solutions to exercises

1   (a) $x^2 e^{2x}(3 + 2x)$
(b) $15 \sin 2x \sin 5x - 6 \cos 2x \cos 5x$
(c) $-e^{-x}(x - 1)^2$
(d) $\cos x \sin 2x \sin 3x + 2 \sin x \cos 2x \sin 3x + 3 \sin x \sin 2x \cos 3x$
(e) $\tan 3x + 3x \sec^2 3x$

2   (a) $\dfrac{2x^2 + 6x - 2}{(2x + 3)^2}$

(b) $-\dfrac{(2 \sin 2t \sin 3t + 3 \cos 2t \cos 3t)}{\sin^2 3t}$

(c) $\dfrac{r^2}{3}\left(\dfrac{3 - 2r}{e^{2r}}\right)$

(d) $\dfrac{(1 + e^v)(1 + \cos v) - (v + \sin v)e^v}{(1 + e^v)^2}$

(e) $\dfrac{1}{2(x + 1)^2}\left(\dfrac{1 - x}{\sqrt{x}}\right)$

3   (a) -0.7702 (b) -0.7358 (c) -1.0548
(d) -1.9515 (e) 1.5181

4   (a) $x^2 e^{2x}(2x \sin 3x \cos 2x + 3 \sin 3x \cos 2x + 3x \cos 3x \cos 2x - 2x \sin 3x \sin 2x)$

(b) $7(x + \sin x)^6 (1 + \cos x)$

(c) $\dfrac{2t + 3}{t^2 + 3t - 9}$

(d) $-\dfrac{1}{r(\ln r)^2}$ (e) $\dfrac{1}{b} + \dfrac{1}{b + 1}$

5   $x^x (\ln x + 1)$

6   (a) $\dfrac{4t + 1}{2t}$ (b) $\dfrac{3t}{2}$ (c) $-t^2 \cos t$

(d) $\dfrac{1 + t}{2}$ (e) $2\sqrt{\dfrac{t}{2t + 1}}$

7   (a) $\dfrac{2x + 3x^2}{3y^2 - 2y}$ (b) $\dfrac{4x + 3y - 7}{2y - 3x + 10}$

(c) $\dfrac{x^2 e^x + y - x^2 y^2}{x(2x^2 y + 1)}$ (d) $\dfrac{y(\sqrt{x} - 2)}{x(2 - \sqrt{y})}$

(e) $\dfrac{2y^2 \sin 2x - \sin y}{x \cos y + 2y \cos 2x - 1}$

8   -1

9   $\dfrac{12x(x^2 - y^3)^5 - ye^{xy}}{18y^2(x^2 - y^3)^5 + xe^{xy}}$

10 $2e^{(x^2)}(1 + 2x^2)$

11 $y = 8x - 6$

12 $y = -2x + 4.6931$

13 $y = e(2x - 1)$

14 $y = 2x + 3$

15 (a) concave up on $x > 5$; concave down on $x < 5$

(b) concave up on $-1 < x < 1$; concave down on $x > 1$ and $x < -1$
(c) concave up for all values of x
(d) concave up for all values of x except $x = 1$
(e) concave up on $x > e^{-1}$; concave down on $x < e^{-1}$

16 $(0, 1)$ maximum; $\left(\frac{5}{2}, -\frac{101}{24}\right)$ minimum;

$\left(\frac{5}{4}, -\frac{77}{48}\right)$ point of inflexion

17 $(0, 0)$ minimum; $(-2, 0.5413)$ maximum; $(-3.4142, 0.3835)$ and $(-0.5858, 0.1910)$ points of inflexion

Integration

When a function, $f(x)$, is known, we can differentiate it to obtain the derivative, $\dfrac{\mathrm{d}f}{\mathrm{d}x}$. The reverse process is to obtain $f(x)$ from knowledge of its derivative. This process is called **integration.**

However, integration is much more than simply differentiation in reverse. It can be applied to finding areas bounded by curves. Such areas can have various physical interpretations. For example, the area underneath a graph of the velocity of an object against time represents the distance travelled by the object. The area under a graph of current flow into a capacitor against time represents the total charge stored by the capacitor. Clearly, there are many applications where it is required to calculate quantities like these, and this is why a knowledge of integration is important for the engineer.

It is also necessary to regard integration as a process of adding up, or *summation*. Often a physical quantity can be obtained by summing lots of small contributions or elements. For example, the position of the centre of mass of a solid body can be found by adding the contributions from all the small parts of which the body is composed. You will see how this is done using integration in Chapter 18.

In this chapter we lay the foundations, and explain a wide variety of techniques that are necessary to integrate the range of functions commonly met in engineering applications.

Modern software packages enable the easy integration of complicated functions. This is illustrated in Block 4.

Chapter 17 contents

Integration as differentiation in reverse

1.1 Introduction

The topic of **integration** can be approached in several different ways. Perhaps the simplest way of introducing it is to think of it as differentiation in reverse. In some applications we shall know the derivative of a function, but not the function from which it was derived. This is why we need knowledge of integration.

In this block we give a look-up table that you can use to integrate a wide range of functions, and we provide lots of opportunities for you to practise using it. Then rules are given that allow you to integrate a wider range of functions. In particular you will be able to integrate sums, differences and constant multiples of functions.

1.2 Differentiation in reverse

Suppose we differentiate the function $y = x^2$. We obtain $\dfrac{dy}{dx} = 2x$. Integration reverses this process, and we say that the integral of $2x$ is x^2. Pictorially we can regard this as shown in Figure 1.1.

Figure 1.1
Integration can be thought of as differentiation in reverse.

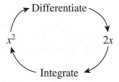

Differentiate

x^2 $2x$

Integrate

The situation is just a little more complicated because there are lots of functions we can differentiate to give $2x$. Here are some of them:

$$x^2 + 4, \quad x^2 - 15, \quad x^2 + 0.5$$

Example 1.1
Write down some more functions that have derivative $2x$.

Solution

                             e.g. $x^2 - 7, x^2 + 0.1$

All these functions have the same derivative, $2x$, because when we differentiate the constant term we obtain zero. Consequently, when we reverse the process, we have no idea what the original constant term might have been. Because of this we include in our answer an unknown constant, c say, called the **constant of integration**. We state that the integral of $2x$ is $x^2 + c$. There is nothing special about the letter c. We might use K for example, but we avoid using letters from the end of the alphabet like x, y and z, which are used for variables.

The symbol for integration is $\int$, known as an **integral sign**. Formally we write

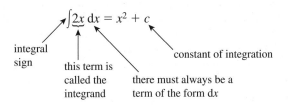

Note that along with the integral sign there is a term of the form $\mathrm{d}x$, which must always be written, and which indicates the name of the variable involved, in this case x. The term $\mathrm{d}x$ must not be interpreted as a product of d and x. Rather it is a notation and as such you should think of $\mathrm{d}x$ as a single quantity.

We say that $2x$ is integrated *with respect to x* to give $x^2 + c$. The function being integrated is called the **integrand**. Technically, integrals of this sort are called **indefinite integrals**, to distinguish them from definite integrals, which are dealt with subsequently. When you find an indefinite integral your answer should always contain a constant of integration.

Exercises

1   (a) Write down the derivatives of each of
     $$x^3, \quad x^3 + 17, \quad x^3 - 21$$
    (b) Deduce that $\int 3x^2 \, \mathrm{d}x = x^3 + c$.

2   Explain what is meant by the term 'integrand'.

3   Explain why, when finding indefinite integrals, a constant of integration is needed.

Solutions to exercises

1   (a) All have derivative $3x^2$.

We could use a table of derivatives to find integrals, but the more common ones are usually found in a 'table of integrals' such as that shown in Table 1.1. You could check many of the entries in this table using your knowledge of differentiation. Try checking some of these for yourself.

When dealing with the trigonometrical functions the variable x must always be measured in radians and not degrees. Note that the fourth entry for integrating a power of x is valid whether n is positive, negative or fractional, but not when $n = -1$. If $n = -1$ use the fifth entry in the table. Various other conditions are detailed in the third column.

Example 1.2
(a) Use the table of integrals to find $\int x^7 \, dx$.
(b) Check the result by differentiating the answer.

Solution
(a) From the table note that

$$\int x^n \, dx = \frac{x^{n+1}}{n+1} + c$$

In words, this states that to integrate a power of x, increase the power by one, and divide the result by the new power. You will find it helpful to remember this rule. With $n = 7$ we find

$$\int x^7 \, dx = \frac{x^8}{8} + c$$

(b) The answer can be differentiated as a check.

$$\frac{d}{dx}\left(\frac{x^8}{8} + c\right) = \frac{1}{8}\frac{d}{dx}x^8$$
$$= \frac{1}{8} \times 8x^7$$
$$= x^7$$

The answer has been verified.

Example 1.3
Find $\int \cos 5x \, dx$.

Solution
From the table note that

$$\int \cos kx \, dx = \frac{\sin kx}{k} + c$$

With $k = 5$ we find

$$\int \cos 5x \, dx = \frac{\sin 5x}{5} + c$$

Check this result for yourself by differentiating it.

Table 1.1
Table of integrals.

Function $f(x)$	Indefinite integral $\int f(x)\,dx$	
constant, k	$kx + c$	
x	$\dfrac{x^2}{2} + c$	
x^2	$\dfrac{x^3}{3} + c$	
x^n	$\dfrac{x^{n+1}}{n+1} + c$	$n \neq -1$
$x^{-1} = \dfrac{1}{x}$	$\ln|x| + c$	
$\sin x$	$-\cos x + c$	
$\cos x$	$\sin x + c$	
$\sin kx$	$\dfrac{-\cos kx}{k} + c$	
$\cos kx$	$\dfrac{\sin kx}{k} + c$	
$\tan kx$	$\dfrac{1}{k} \ln|\sec kx| + c$	
$\sec kx$	$\dfrac{1}{k} \ln|\sec kx + \tan kx| + c$	
e^x	$e^x + c$	
e^{-x}	$-e^{-x} + c$	
e^{kx}	$\dfrac{e^{kx}}{k} + c$	
$\cosh kx$	$\dfrac{1}{k} \sinh kx + c$	
$\sinh kx$	$\dfrac{1}{k} \cosh kx + c$	
$\dfrac{1}{x^2 + a^2}$	$\dfrac{1}{a} \tan^{-1} \dfrac{x}{a} + c$	$a > 0$
$\dfrac{1}{x^2 - a^2}$	$\dfrac{1}{2a} \ln \dfrac{x - a}{x + a} + c$	$|x| > a > 0$
$\dfrac{1}{a^2 - x^2}$	$\dfrac{1}{2a} \ln \dfrac{a + x}{a - x} + c$	$|x| < a$
$\dfrac{1}{\sqrt{x^2 + a^2}}$	$\sinh^{-1}\left(\dfrac{x}{a}\right) + c$	$a > 0$
$\dfrac{1}{\sqrt{x^2 - a^2}}$	$\cosh^{-1}\left(\dfrac{x}{a}\right) + c$	$x \geq a > 0$
$\dfrac{1}{\sqrt{x^2 + k}}$	$\ln(x + \sqrt{x^2 + k}) + c$	
$\dfrac{1}{\sqrt{a^2 - x^2}}$	$\sin^{-1}\left(\dfrac{x}{a}\right) + c$	$-a \leq x \leq a$

In the table the independent variable is always given as x. However, with a little imagination you will be able to use it when other independent variables are involved.

Example 1.4

Find $\int \cos 5t \, dt$.

Solution

We integrated $\cos 5x$ in the previous example. Now the independent variable is t, so simply use the table and read every x as a t. With $k = 5$ we find

$$\int \cos 5t \, dt = \frac{\sin 5t}{5} + c$$

It follows immediately that, for example,

$$\int \cos 5\omega \, d\omega = \frac{\sin 5\omega}{5} + c, \quad \int \cos 5u \, du = \frac{\sin 5u}{5} + c$$

and so on.

Example 1.5

Find $\int \sin \frac{x}{2} \, dx$.

Solution

Note that $\frac{x}{2}$ is equivalent to $\frac{1}{2} x$. Use the table of integrals with $k = \frac{1}{2}$.

$$\int \sin \frac{x}{2} \, dx = \qquad\qquad = -\frac{\cos \dfrac{x}{2}}{\dfrac{1}{2}} + c = -2 \cos \frac{x}{2} + c$$

Example 1.6

Find $\int \sin \frac{2\pi t}{T} \, dt$ where T is a constant.

Solution

With respect to which variable is the integration being carried out?

t

The quantity $\frac{2\pi}{T}$ is a constant. Perform the integration:

$$\int \sin \frac{2\pi t}{T} \, dt =$$

$$= -\frac{\cos \dfrac{2\pi t}{T}}{\dfrac{2\pi}{T}} + c = -\frac{T}{2\pi} \cos \frac{2\pi t}{T} + c$$

Example 1.7

Find $\int 12 \, dx$.

Solution

In this example we are integrating a constant, 12. Using the table we find

$$\int 12 \, dx = 12x + c$$

Note that $\int 12 \, dt$ would be $12t + c$ and so on. This shows the importance of including the term dx or dt in the integral.

Example 1.8

Find $\int dx$.

Solution

This example looks a little different from the earlier ones. But if we think of dx as $1 \, dx$ then we are integrating a constant, 1. Using the table we find

$$\int 1 \, dx = 1x + c \quad \text{or simply} \quad x + c$$

Note that $\int dt$ would be $t + c$.

Example 1.9

Find $\int e^{-3x} \, dx$.

Solution

Use the appropriate entry in the table with $k = -3$:

$$\int e^{-3x} \, dx = \frac{e^{-3x}}{-3} + c = -\frac{1}{3}e^{-3x} + c$$

Example 1.10

Look for an entry in the table of integrals that will enable you to calculate

$$\int \frac{1}{\sqrt{9 - t^2}} \, dt$$

Solution

Write down the appropriate entry.

$$\int \frac{1}{\sqrt{a^2 - x^2}} \, dx = \sin^{-1}\left(\frac{x}{a}\right) + c$$

Select an appropriate value for the constant a and hence find the integral.

$$\text{Take } a = 3, \text{ so that } a^2 = 9. \int \frac{1}{\sqrt{9 - t^2}} \, dt = \sin^{-1}\left(\frac{t}{3}\right) + c$$

Exercises

1. Integrate each of the following functions with respect to x:

   (a) x^9 (b) $x^{1/2}$ (c) x^{-3} (d) $\dfrac{1}{x^4}$ (e) 4 (f) $\sqrt{x}$

   (g) e^{4x} (h) $\cos\dfrac{x}{2}$ (i) $\cos\pi x$

2. Find

   (a) $\int t^2\,dt$ (b) $\int 6\,dt$ (c) $\int \sin 3t\,dt$

   (d) $\int e^{7t}\,dt$ (e) $\int \cos\dfrac{2\pi t}{T}\,dt$ (f) $\int \dfrac{dz}{z}$

3. Find $\int e^t\,dt$.

4. Find

   (a) $\int \dfrac{1}{x^2+25}\,dx$ (b) $\int \dfrac{1}{\sqrt{t^2-4}}\,dt$

5. Find
   (a) $\int \sin\omega\,d\omega$
   (b) $\int \cos 3t\,dt$
   (c) $\int \tan\tfrac{1}{3}x\,dx$
   (d) $\int \tan\dfrac{x}{3}\,dx$

(e) $\int \sinh 0.3x\,dx$

(f) $\int \cos 0.01\pi x\,dx$

(g) $\int \cosh 0.01t\,dt$

(h) $\int x^{7/2}\,dx$

(i) $\displaystyle\int \dfrac{1}{\sqrt[3]{x}}\,dx$

(j) $\int x^{1/4}\,dx$

(k) $\int \sqrt[5]{t}\,dt$

(l) $\int e^{-0.7t}\,dt$

(m) $\displaystyle\int \dfrac{1}{64-x^2}\,dx$

(n) $\displaystyle\int \dfrac{1}{\sqrt{64-x^2}}\,dx$

(o) $\displaystyle\int \dfrac{1}{\sqrt{x^2+8}}\,dx$

6. Find

   (a) $\dfrac{1}{5}\displaystyle\int \cos\dfrac{n\pi t}{5}\,dt$ (b) $\dfrac{1}{5}\displaystyle\int \sin\dfrac{n\pi t}{5}\,dt$

Solutions to exercises

1. (a) $\dfrac{x^{10}}{10}+c$ (b) $\dfrac{2x^{3/2}}{3}+c$ (c) $-\dfrac{1}{2}x^{-2}+c$

   (d) $-\dfrac{1}{3}x^{-3}+c$ (e) $4x+c$ (f) same as (b)

   (g) $\dfrac{e^{4x}}{4}+c$ (h) $2\sin\dfrac{x}{2}+c$ (i) $\dfrac{\sin\pi x}{\pi}+c$

2. (a) $\dfrac{t^3}{3}+c$ (b) $6t+c$ (c) $-\dfrac{\cos 3t}{3}+c$

   (d) $\dfrac{e^{7t}}{7}+c$ (e) $\dfrac{T}{2\pi}\sin\dfrac{2\pi t}{T}+c$ (f) $\ln|z|+c$

3. e^t+c

4. (a) $\dfrac{1}{5}\tan^{-1}\left(\dfrac{x}{5}\right)+c$ (b) $\cosh^{-1}\left(\dfrac{t}{2}\right)+c$

5. (a) $-\cos\omega+c$ (b) $\dfrac{1}{3}\sin 3t+c$

   (c) $3\ln|\sec\dfrac{x}{3}|+c$ (d) $3\ln|\sec\dfrac{x}{3}|+c$

   (e) $\dfrac{10}{3}\cosh 0.3x+c$ (f) $\dfrac{100}{\pi}\sin 0.01\pi x+c$

   (g) $100\sinh 0.01t+c$ (h) $\dfrac{2}{9}x^{9/2}+c$

   (i) $\dfrac{3}{2}x^{2/3}+c$ (j) $\dfrac{4}{5}x^{5/4}+c$ (k) $\dfrac{5}{6}t^{6/5}+c$

   (l) $-1.429e^{-0.7t}+c$ (m) $\dfrac{1}{16}\ln\dfrac{8+x}{8-x}+c$

   (n) $\sin^{-1}\dfrac{x}{8}+c$ (o) $\ln(x+\sqrt{x^2+8})+c$

6. (a) $\dfrac{1}{n\pi}\sin\dfrac{n\pi t}{5}+c$ (b) $-\dfrac{1}{n\pi}\cos\dfrac{n\pi t}{5}+c$

To enable us to find integrals of a wider range of functions than those normally given in a table of integrals we can make use of the following rules.

The integral of *k f*(*x*) where *k* is a constant

A constant factor in an integral can be moved outside the integral sign as follows:

Key point

The integral of a constant multiple of a function

$$\int k f(x)\, \mathrm{d}x = k \int f(x)\, \mathrm{d}x$$

Example 1.11

Find $\int 11x^2\, \mathrm{d}x$.

Solution

$$\int 11x^2\, \mathrm{d}x = 11 \int x^2\, \mathrm{d}x$$
$$= 11\left(\frac{x^3}{3} + c\right)$$
$$= \frac{11x^3}{3} + K$$

where K is a constant.

Example 1.12

Find $\int -5\cos x\, \mathrm{d}x$.

Solution

$$\int -5\cos x\, \mathrm{d}x = -5 \int \cos x\, \mathrm{d}x$$
$$= -5(\sin x + c)$$
$$= -5\sin x + K$$

where K is a constant.

Example 1.13

Find $\int \dfrac{14}{1 + s^2}\, \mathrm{d}s$.

Solution

Use the result in the previous key point to extract the constant factor 14. Then use the table to complete the solution.

$$\int \frac{14}{1 + s^2}\, \mathrm{d}s =$$

$$14 \int \frac{1}{1 + s^2}\, \mathrm{d}s = 14 \tan^{-1} s + c$$

Example 1.14

Find $\int 2\pi m r^3 \, dr$.

Solution

In this example, integration is with respect to which variable?

r

The 2, π and m are all constant factors and can be written in front of the integral sign. Thus

$$\int 2\pi m r^3 \, dr = 2\pi m \int r^3 \, dr$$

$$= 2\pi m \left(\frac{r^4}{4}\right) + c$$

$$= \frac{\pi m r^4}{2} + c$$

The integral of $f(x) \pm g(x)$

When we wish to integrate the sum or difference of two functions, we integrate each term separately as follows:

Key point

The integral of a sum or difference of two functions

$$\int [f(x) \pm g(x)] \, dx = \int f(x) \, dx \pm \int g(x) \, dx$$

Example 1.15

Find $\int (x^3 + \sin x) \, dx$.

Solution

$$\int (x^3 + \sin x) \, dx = \int x^3 \, dx + \int \sin x \, dx$$

$$= \frac{x^4}{4} - \cos x + c$$

Note that only a single constant of integration is needed.

Example 1.16

Find $\int 3t^4 + \sqrt{t} \, dt$.

Solution

You will need to use both of the rules to deal with this integral.

$$\frac{3t^5}{5} + \frac{2t^{3/2}}{3} + c$$

Example 1.17

The hyperbolic sine and cosine functions, $\sinh x$ and $\cosh x$, are defined as follows:

$$\sinh x = \frac{e^x - e^{-x}}{2} \qquad \cosh x = \frac{e^x + e^{-x}}{2}$$

Note that they are simply combinations of the exponential functions e^x and e^{-x}. Find $\int \sinh x \, dx$ and $\int \cosh x \, dx$.

Solution

$$\int \sinh x \, dx = \int \left(\frac{e^x - e^{-x}}{2} \right) dx$$

$$= \qquad\qquad\qquad = \frac{1}{2}(e^x + e^{-x}) + c = \cosh x + c$$

$$\int \cosh x \, dx = \int \left(\frac{e^x + e^{-x}}{2} \right) dx$$

$$= \qquad\qquad\qquad = \sinh x + c$$

We would normally simply use Table 1.1 to integrate hyperbolic functions.

Example 1.18 Structural Engineering – Bending moment in a beam

In Example 8.4 in Chapter 6 we considered a simply supported beam of length L carrying a uniform load w per unit length, and introduced the **bending moment**, M, which is a measure of the internal stress caused when the beam is loaded. M varies with position x along the beam. It can be shown that the bending moment, M, is related to the shear force, V, by the equation $\dfrac{dM}{dx} = V$. In Example 7.8 on p179 we have already seen that $V = \dfrac{w}{2}(L - 2x)$ and hence

$$\frac{dM}{dx} = \frac{w}{2}(L - 2x)$$

from which

$$M(x) = \int \frac{w}{2}(L - 2x) \, dx$$

Evaluate the integral to find an expression for the bending moment.

Solution

Here w is a constant. Performing the integration we find

$$M(x) = \int \frac{w}{2}(L - 2x) \, dx$$

$$= \frac{w}{2} \int (L - 2x) \, dx$$

$$= \frac{w}{2}(Lx - x^2) + c \qquad \text{(where } c \text{ is a constant)}$$

For a simply supported beam it can be shown that $M(0) = 0$ and consequently $c = 0$. Therefore the bending moment at any point is given by

$$M(x) = \frac{w}{2}(Lx - x^2)$$

Further rules for finding more complicated integrals are dealt with in subsequent blocks. However, it is important that you are aware at this stage that the integral of a product of two functions is not the product of two separate integrals:

$$\int f(x) \times g(x)\, dx \neq \int f(x)\, dx \times \int g(x)\, dx$$

You will learn how to integrate products in Block 5 on integration by parts.

Exercises

1. Find $\int 2x - e^x\, dx$.

2. Find $\int 3e^{2x}\, dx$.

3. Find $\int \dfrac{x + \cos 2x}{3}\, dx$.

4. Find $\int 7x^{-2}\, dx$.

5. Find $\int (x + 3)^2\, dx$ (be careful!).

6. Find $\int 3 \tan 2x + 2 \sin 3x\, dx$.

7. Find $\int \dfrac{8}{x^2 + 16}\, dx$.

8. Find $\int 3 \cos n\pi x\, dx$.

9. Find $\int 0.5 \sin n\pi x\, dx$.

10. Find $\int \dfrac{1}{2} \sin \dfrac{2\pi t}{T}\, dt$.

11. Find $\int \dfrac{\sin x + \cos x}{2}\, dx$.

Solutions to exercises

1. $x^2 - e^x + c$

2. $\dfrac{3e^{2x}}{2} + c$

3. $\dfrac{x^2}{6} + \dfrac{\sin 2x}{6} + c$

4. $-\dfrac{7}{x} + c$

5. $\dfrac{x^3}{3} + 3x^2 + 9x + c$

6. $\dfrac{3}{2} \ln|\sec 2x| - \dfrac{2}{3} \cos 3x + c$

7. $2 \tan^{-1}\left(\dfrac{x}{4}\right) + c$

8. $\dfrac{3}{n\pi} \sin n\pi x + c$

9. $-\dfrac{0.5}{n\pi} \cos n\pi x + c$

10. $-\dfrac{T}{4\pi} \cos \dfrac{2\pi t}{T} + c$

11. $\dfrac{-\cos x + \sin x}{2} + c$

End of block exercises

1 Integrate each of the following functions with respect to x:
(a) x^2 (b) x^{17} (c) 6 (d) $\sin 4x$ (e) e^{8x}

2 Integrate each of the following functions with respect to t:
(a) t^3 (b) 7 (c) e^{-3t} (d) e^{-t} (e) $\dfrac{1}{t}$

3 Find the following integrals:
(a) $\int 3t\, dt$ (b) $\int 3t^2\, dt$ (c) $\int 2t^4\, dt$
(d) $\int t + t^2\, dt$

4 Find the following integrals:
(a) $\int x^2 + 4x + 8\, dx$ (b) $\int 2x^2 - 4x + 7\, dx$
(c) $\int 6x^2 - x + 2\, dx$

5 Find $\displaystyle\int \sqrt{t} + \dfrac{1}{t^2}\, dt$.

6 Find $\int e^{-st}\, dt$ where s is a number.

7 Find $\displaystyle\int \dfrac{1}{s^2 + 0.5}\, ds$.

8 Find $\displaystyle\int \dfrac{1}{\sqrt{4 + t^2}}\, dt$.

9 Find $\displaystyle\int \dfrac{1}{\sqrt{25 - x^2}}\, dx$.

10 Find $\displaystyle\int \tan \dfrac{x}{5}\, dx$.

11 Find $\int \sin 100\pi t + \cos 100\pi t\, dt$.

12 Find $\int \cos(m + n)t\, dt$.

13 Find $\int mbx^2\, dx$.

14 Find (a) $\displaystyle\int \dfrac{1}{2}\theta\, d\theta$, (b) $\displaystyle\int 4\cos \dfrac{\theta}{2}\, d\theta$,
(c) $\int \sqrt{2}\, d\theta$.

Solutions to exercises

1 (a) $\dfrac{x^3}{3} + c$ (b) $\dfrac{x^{18}}{18} + c$ (c) $6x + c$
(d) $-\dfrac{\cos 4x}{4} + c$ (e) $\dfrac{e^{8x}}{8} + c$

2 (a) $\dfrac{t^4}{4} + c$ (b) $7t + c$ (c) $-\dfrac{e^{-3t}}{3} + c$
(d) $-e^{-t} + c$ (e) $\ln|t| + c$

3 (a) $\dfrac{3t^2}{2} + c$ (b) $t^3 + c$
(c) $\dfrac{2t^5}{5} + c$ (d) $\dfrac{t^2}{2} + \dfrac{t^3}{3} + c$

4 (a) $\dfrac{x^3}{3} + 2x^2 + 8x + c$
(b) $\dfrac{2x^3}{3} - 2x^2 + 7x + c$
(c) $2x^3 - \dfrac{x^2}{2} + 2x + c$

5 $\dfrac{2t^{3/2}}{3} - \dfrac{1}{t} + c$

6 $-\dfrac{e^{-st}}{s} + c$

7 $\sqrt{2}\tan^{-1}(s\sqrt{2}) + c$

8 $\sinh^{-1}\left(\dfrac{t}{2}\right) + c$

9 $\sin^{-1}\left(\dfrac{x}{5}\right) + c$

10 $5\ln\left|\sec\dfrac{x}{5}\right| + c$

11 $-\dfrac{\cos 100\pi t}{100\pi} + \dfrac{\sin 100\pi t}{100\pi} + c$

12 $\dfrac{\sin(m + n)\, t}{(m + n)} + c$

13 $\dfrac{mbx^3}{3} + c$

14 (a) $\dfrac{\theta^2}{4} + c$ (b) $8\sin\dfrac{\theta}{2} + c$ (c) $\sqrt{2}\theta + c$

Definite integrals

2.1 Introduction

When integration was introduced as the reverse of differentiation in Block 1, the integrals you dealt with were *indefinite integrals*. The result of finding an indefinite integral is usually a function plus a constant of integration. In this block we introduce *definite integrals*, so called because the result will be a definite answer, usually a number, with no constant of integration. Definite integrals have many applications, for example in finding areas bounded by curves, and finding volumes of solids. Applications such as these will be described in later blocks.

2.2 Evaluating definite integrals

Definite integrals can be recognised by numbers written to the upper and lower right of the integral sign. The quantity

$$\int_{x=a}^{x=b} f(x)\,dx$$

is called the definite integral of $f(x)$ from a to b. The numbers a and b are known as the **lower** and **upper limits** of the integral. This integral is commonly written as

$$\int_a^b f(x)\,dx$$

When you evaluate a definite integral the result will usually be a number. To see how to evaluate a definite integral consider the following example.

Example 2.1
Find $\int_1^4 x^2\,dx$.

Solution
First of all the integration of x^2 is performed in the normal way. However, to show we are dealing with a definite integral, the result is usually enclosed in square brackets and the limits of integration are written on the right bracket:

$$\int x^2\,dx = \frac{x^3}{3} + c \quad \text{so that} \quad \int_1^4 x^2\,dx = \left[\frac{x^3}{3} + c\right]_1^4$$

You should always use this convention. Then, the quantity in the square brackets is evaluated, first by letting x equal the value at the upper limit, then by letting x equal the value at the lower limit, and the difference between the resulting values is found:

$$\left[\frac{x^3}{3} + c\right]_1^4 = (\text{evaluate when } x \text{ equals upper limit})$$

$$- (\text{evaluate when } x \text{ equals lower limit})$$

$$= \left(\frac{4^3}{3} + c\right) - \left(\frac{1^3}{3} + c\right)$$

$$= \frac{64}{3} - \frac{1}{3}$$

$$= \frac{63}{3}$$

$$= 21$$

Note that the constants, c, cancel out. This will always happen, and so in future we can ignore them when we are evaluating definite integrals. So, the value of the definite integral $\int_1^4 x^2 \, dx$ is 21.

Example 2.2

Find $\int_0^{\pi/2} \cos x \, dx$.

Solution

Since $\int \cos x \, dx = \sin x + c$ then

$$\int_0^{\pi/2} \cos x \, dx = \left[\sin x\right]_0^{\pi/2}$$

$$= \sin\left(\frac{\pi}{2}\right) - \sin 0$$

$$= 1 - 0$$

$$= 1$$

Always remember that if you use a calculator to evaluate any trigonometrical functions, you must work in radian mode.

Example 2.3

Find $\int_1^2 (x^2 + 1) \, dx$.

Solution

First perform the integration:

$$\left[\frac{x^3}{3} + x\right]_1^2$$

Now insert the limits of integration, the upper limit first, and hence find the value of the integral.

$$\left(\frac{8}{3} + 2\right) - \left(\frac{1}{3} + 1\right) = \frac{10}{3} \text{ or } 3.333 \text{ (3 d.p.)}$$

Example 2.4
This exercise is very similar to the previous one. Note that the limits of integration have been interchanged.
    Find $\int_2^1 (x^2 + 1)\, dx$.

Solution

$$-\frac{10}{3}$$

Note from these two exercises that interchanging the limits of integration changes the sign of the answer.

Key point

If you interchange the limits, you must introduce a minus sign.

$$\int_a^b f(x)\, dx = -\int_b^a f(x)\, dx$$

Example 2.5 The average value of a function
If a function $f(t)$ is defined on the interval $a \le t \le b$ then the **average value** of the function over the interval is defined to be

$$\text{average value} = \frac{\int_a^b f(t)\, dt}{b - a}$$

The average value of a function is found by evaluating a definite integral.
    Find the average value of the function $f(t) = t^2$ across the interval $2 \le t \le 5$.

Solution
Apply the formula for finding the average value:

$$\text{average value} = \qquad\qquad \frac{\int_2^5 t^2\, dt}{5 - 2}$$

Complete the integration to find this average value.

$$\frac{1}{3}\left[\frac{t^3}{3}\right]_2^5 = 13$$

Example 2.6 Civil Engineering – The force on the wall of a tank

Civil engineers are frequently faced with problems involving the pressure and forces on the walls of a tank containing water. Such problems will arise during the design and construction of dams and other underwater structures where it is essential that the structure has sufficient strength to withstand such forces.

Consider the tank of water in Figure 2.1 in the shape of a cuboid with dimensions as shown.

Figure 2.1
Integration can be used to find the force on the wall of a tank or reservoir.

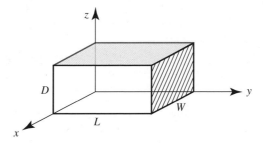

The pressure, p, at a depth h was shown by Pascal in the seventeenth century to be equivalent to $\rho\,g\,h$ where ρ is the density of the water and g is the constant acceleration due to gravity, nominally 9.8 m s^{-2}. With a knowledge of hydrostatics it can be shown that the force exerted on the wall of the tank (hatched in Figure 2.1) is given by the definite integral

$$\int_0^D \rho g W (D - z)\mathrm{d}z$$

(a) Evaluate this integral.
(b) Show that the force on the wall is the same as the product of the average pressure and the area of the wall.

Solution
(a)

$$\int_0^D \rho g W(D - z)\,\mathrm{d}z = \rho g W \int_0^D (D - z)\,\mathrm{d}z \qquad \text{since } \rho,\, g,\, W \text{ are constants}$$

$$= \rho g W \left[Dz - \frac{z^2}{2} \right]_0^D$$

$$= \rho g W \left(D^2 - \frac{D^2}{2} \right)$$

$$= \frac{\rho g W D^2}{2}$$

(b) Given that the pressure at depth h is $\rho g h$, the pressure at the surface is zero and the pressure at the base of the tank is $\rho g D$. Since p varies linearly between these two values, the average pressure is $\frac{1}{2}\rho g D$. The area of the wall is DW. The force on the wall has been shown in part (a) to be $\dfrac{\rho g W D^2}{2}$ which can be written $\frac{1}{2}\rho g D \times DW$ as required.

Exercises

1 Explain why a constant of integration is not needed when evaluating definite integrals.

2 Explain what happens to the value of a definite integral when the upper and lower limits are interchanged.

3 Evaluate

(a) $\int_0^1 x^2 \, dx$ (b) $\int_2^3 \dfrac{1}{x^2} \, dx$

4 Evaluate $\int_1^2 e^x \, dx$.

5 Evaluate $\int_{-1}^1 (1 + t^2) \, dt$.

6 Find $\int_0^{\pi/3} \cos 2x \, dx$.

7 Find $\int_0^\pi \sin x \, dx$.

8 Find $\int_1^3 e^{2t} \, dt$.

9 Evaluate $\int_2^4 x^3 \, dx$. Show that

$\int_2^4 x^3 \, dx = -\int_4^2 x^3 \, dx$.

10 Show that $\int_a^b \sin x \, dx = \int_a^b \sin t \, dt$.

11 Find $\int_0^k 2(kv^3 - v^4) \, dv$.

12 Find $\int_0^{0.001} 100 \, dt$.

13 Find $100 \int_0^{0.01} \sin 100\, n\pi t \, dt$.

14 Find the average value of the function $f(t) = \sin t$ across the interval $0 \leq t \leq \frac{\pi}{2}$.

15 Find $\displaystyle\int_0^1 \dfrac{1}{\sqrt{9 - 4t^2}} \, dt$.

16 Find $\int_0^a 2\pi m r^3 \, dr$.

Solutions to exercises

3 (a) $\frac{1}{3}$ (b) $\frac{1}{6}$

4 $e^2 - e^1 = 4.671$

5 2.667

6 $\dfrac{\sqrt{3}}{4} = 0.4330$

7 2

8 198

9 60

11 $\dfrac{k^5}{10}$

12 0.1

13 $-\dfrac{1}{n\pi}(\cos n\pi - 1)$

14 $\dfrac{2}{\pi}$

15 0.365

16 $\dfrac{m\pi a^4}{2}$

2.3 Some integrals with infinite limits of integration

On occasions, and notably when dealing with Laplace and Fourier transforms, you will come across integrals in which one of the limits is infinite. We avoid a rigorous treatment of such cases here and instead give some commonly occurring examples.

Example 2.7

Find $\int_0^\infty e^{-x}\,dx$.

Solution

The integral is found in the normal way:

$$\int_0^\infty e^{-x}\,dx = \left[-e^{-x}\right]_0^\infty$$

There is no difficulty in evaluating the square bracket at the lower limit. We obtain simply $-e^{-0} = -1$. At the upper limit we must examine the behaviour of $-e^{-x}$ as x gets infinitely large. This is where it is important that you are familiar with the graph of the exponential function. If you refer to the graph you will recognise that as x tends to infinity e^{-x} tends to zero. Consequently the contribution to the integral from the upper limit is zero. So

$$\int_0^\infty e^{-x}\,dx = \left[-e^{-x}\right]_0^\infty$$
$$= (0) - (-e^{-0})$$
$$= 1$$

The value of $\int_0^\infty e^{-x}\,dx$ is 1.

Another way of achieving this result is as follows.

We change the infinite limit to a finite limit, b, say, and then examine the behaviour of the integral as b tends to infinity, written

$$\int_0^\infty e^{-x}\,dx = \lim_{b\to\infty}\int_0^b e^{-x}\,dx$$

So

$$\int_0^b e^{-x}\,dx = \left[-e^{-x}\right]_0^b$$
$$= (-e^{-b}) - (-e^{-0})$$
$$= -e^{-b} + 1$$

Then as b tends to infinity $-e^{-b}$ tends to zero, and the resulting integral has the value 1, as before.

Many integrals having infinite limits cannot be evaluated in a simple way like this, and many cannot be evaluated at all. Fortunately, most of the integrals you will meet will exhibit the sort of behaviour seen in the last example.

Exercises

1 Find $\int_1^\infty e^{-x}\, dx$.

2 Find $\int_0^\infty e^{-2x}\, dx$.

3 Find $\int_2^\infty e^{-3x}\, dx$.

4 Find $\int_{-\infty}^0 e^{7x}\, dx$.

Solutions to exercises

1 e^{-1} or 0.368

2 $\dfrac{1}{2}$

3 $\dfrac{e^{-6}}{3} = 0.0008$ (4 d.p.)

4 $\dfrac{1}{7}$

End of block exercises

1 Find the value of the following integrals:
   (a) $\int_1^2 x^2 + 1\, dx$ (b) $\int_3^4 x^2 + 1\, dx$
   (c) $\int_0^{\pi/4} \sin x\, dx$ (d) $\int_{-1}^1 \cos 2x\, dx$

2 Find the value of the following integrals:
   (a) $\int_0^3 x^2 + 7x - 2\, dx$
   (b) $\int_{-1}^1 2x^2 + x + 3\, dx$

3 Find $\int_1^3 e^t\, dt$.

4 Find $\int_a^b \dfrac{1}{r}\, dr$.

5 Show that $\int_a^b x^3\, dx = \int_a^b t^3\, dt$.

6 Find $\int_0^{0.1} 3 \tan 2x\, dx$.

7 Find $\int_{-1}^0 e^{-2x} + e^{-3x}\, dx$.

8 Find $\int_{\omega_1}^{\omega_2} \sin \omega\, d\omega$.

9 Find $\dfrac{1}{\pi} \int_{-\pi}^{\pi} 1 + t\, dt$.

10 Find $\int_{-b/2}^{b/2} mbx^2\, dx$.

11 Find (a) $\int_0^{2\pi} \dfrac{\theta}{2}\, d\theta$, (b) $\int_0^{\pi/2} 4 \cos \dfrac{\theta}{2}\, d\theta$
    (c) $\int_0^{\pi/2} \sqrt{2}\, d\theta$.

Solutions to exercises

1 (a) $\frac{10}{3}$ (b) $\frac{40}{3}$ (c) 0.293 (d) 0.909

2 (a) 34.5 (b) $\frac{22}{3}$

3 17.367

4 $\ln b - \ln a = \ln\left(\dfrac{b}{a}\right)$

6 0.030

7 9.556

8 $\cos \omega_1 - \cos \omega_2$

9 2

10 $\dfrac{b^4 m}{12}$

11 (a) π^2 (b) $4\sqrt{2}$ (c) $\dfrac{\pi\sqrt{2}}{2}$

BLOCK	3

The area bounded by a curve

3.1 Introduction

One of the important applications of integration is to find the area bounded by a curve. Often such an area can have a physical significance, such as the work done by a motor, or the distance travelled by a vehicle. In this block we explain how such an area is calculated.

3.2 The area bounded by a curve lying above the *x* axis

Consider the graph of the function $y(x)$ shown in Figure 3.1. Suppose we are interested in calculating the area underneath the graph and above the *x* axis, between the points where $x = a$ and $x = b$. When such an area lies entirely above the *x* axis, as is clearly the case in Figure 3.1, this area is given by the definite integral

$$\int_a^b y(x)\,dx$$

Key point

The area under the curve $y(x)$, between $x = a$ and $x = b$, is given by

$$\text{area} = \int_a^b y(x)\,dx$$

when the curve lies entirely above the *x* axis between *a* and *b*.

Figure 3.1
The area bounded by a graph is found by evaluating an integral.

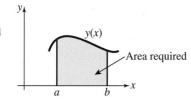

Example 3.1

Calculate the area bounded by the curve $y = \dfrac{1}{x}$ and the x axis, between $x = 1$ and $x = 4$.

Solution

The graph of $y = \dfrac{1}{x}$, for x greater than 0, is shown in Figure 3.2. The area required is shaded. This area lies entirely above the x axis.

Figure 3.2
The required area lies entirely above the x axis.

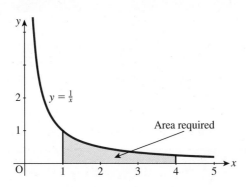

The area is calculated as

$$\text{area} = \int_{1}^{4} \frac{1}{x}\, dx$$
$$= \Big[\ln |x|\Big]_{1}^{4}$$
$$= \ln 4 - \ln 1$$
$$= \ln 4 \quad \text{since } \ln 1 = 0$$
$$= 1.386 \ (3 \text{ d.p.})$$

Example 3.2

Find the area bounded by the curve $y = \sin x$ and the x axis between $x = 0$ and $x = \pi$.

Solution

The required area is shown in Figure 3.3. Note that it lies entirely above the x axis. Calculate the area

$$\int_{0}^{\pi} \sin x\, dx = \Big[-\cos x\Big]_{0}^{\pi} = 2$$

Example 3.3

Find the area under $f(x) = e^{2x}$ from $x = 1$ to $x = 3$ given that the exponential function e^{2x} is always positive.

Solution

Because e^{2x} is positive, the area will lie above the x axis:

$$\text{area} = \qquad\qquad\qquad \int_{1}^{3} e^{2x}\, dx = \left[\frac{e^{2x}}{2}\right]_{1}^{3} = 198$$

Figure 3.3
The area found in
Example 3.2.

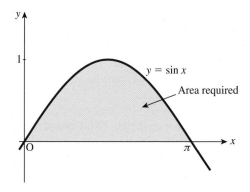

Example 3.4
Figure 3.4 shows the graphs of $y = \sin x$ and $y = \cos x$ for $0 \le x \le \frac{\pi}{2}$. The two graphs intersect at the point where $x = \frac{\pi}{4}$. Find the area that is shaded in Figure 3.4.

Figure 3.4
The required area
is enclosed
between two
curves.

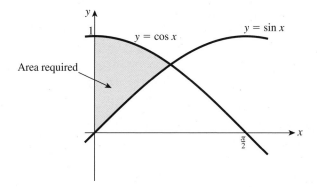

Solution
To find the shaded area we could calculate the area under the graph of $y = \sin x$ for x between 0 and $\frac{\pi}{4}$, and subtract this from the area under the graph of $y = \cos x$ between the same limits. Alternatively the two processes can be combined into one and we can write

$$\text{shaded area} = \int_0^{\pi/4} \cos x - \sin x \, dx$$

$$= [\sin x + \cos x]_0^{\pi/4}$$

$$= \left(\sin \frac{\pi}{4} + \cos \frac{\pi}{4} \right) - (\sin 0 + \cos 0)$$

If you are aware of the standard triangles you will know that $\sin \frac{\pi}{4} = \cos \frac{\pi}{4} = \frac{1}{\sqrt{2}}$, in which case the value of the integral is $\frac{2}{\sqrt{2}} - 1 = 0.414$. Alternatively you can use your calculator to obtain this result directly.

Example 3.5 The area under a velocity–time graph

Figure 3.5 shows a graph of the velocity, v, of an object plotted against time t. It can be shown that the area under such a graph between $t = a$ and $t = b$ represents the distance, s, travelled by the object between these times.

Suppose the velocity (m s^{-1}) of an object is given by the formula $v(t) = 3t^2 + 5$.
(a) State the velocity when $t = 1$ s.
(b) State the velocity when $t = 2$ s.
(c) Find the distance travelled between these times.

Figure 3.5
The area under a
velocity–time
graph represents
distance travelled.

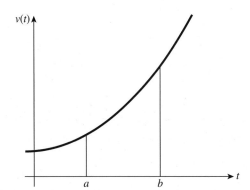

Solution
(a) When $t = 1$, $v =$ [blank] 8 m s^{-1}

(b) When $t = 2$, $v =$ [blank] 17 m s^{-1}

(c) The distance travelled is given by the area under the velocity–time graph.
   Write down the required integral:

[blank] $\displaystyle\int_{1}^{2} (3t^2 + 5)\, dt$

   Finally, evaluate this integral to obtain the distance travelled.

[blank] $[t^3 + 5t]_1^2 = 18 - 6 = 12$ m

We saw in Example 3.5 that the area under a velocity–time graph is interpreted as 'distance travelled'. Example 3.6 illustrates another interpretation of the area under a graph.

When the volume of a gas expands then work is done by the gas. If P denotes pressure, V denotes volume and W denotes work done then

$$W = \int_a^b P \, dV$$

where a is the initial volume and b is the final volume. Note that the integral $\int_a^b P \, dV$ is the area under a graph of P against V, as illustrated in Figure 3.6.

Figure 3.6
Shaded area is given by $\int_a^b P \, dV$ which is the work done by the gas in expanding from volume a to volume b.

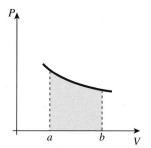

Example 3.6 Thermodynamics – Work done in expansion of a gas

Suppose that the volume, V, and pressure, P, of a gas are related by the equation

$$PV^{1.3} = K$$

where K is a constant. Calculate the work done in expanding from an initial volume of 3 cubic units to a final volume of 10 cubic units.

Solution

$$W = \int_3^{10} P \, dV$$

From $PV^{1.3} = K$ then $P = KV^{-1.3}$ and so

$$W = \int_3^{10} KV^{-1.3} dV$$

$$= K \left[\frac{V^{-0.3}}{-0.3} \right]_3^{10}$$

$$= \frac{-K}{0.3} [10^{-0.3} - 3^{-0.3}]$$

$$= 0.727K$$

Exercises

In each question the required area lies entirely above the horizontal axis, although you should verify this fact for yourself independently.

1 Find the area bounded by the curve $y = x^3$ between $x = 0$ and $x = 2$.

2 Find the area bounded by the curve $y = 3t^2$ between $t = -3$ and $t = 3$.

3 Find the area under $y = \dfrac{1}{x^2}$ between $x = 1$ and $x = 10$.

4 **Mechanical Engineering – Area under a velocity–time graph.** The velocity v (m s^{-1}) of an object varies with time, t, according to
$$v(t) = t^3 + 1$$

(a) Find the distance travelled by the object between $t = 0$ s and $t = 2$ s.
(b) Find the average velocity between $t = 0$ and $t = 2$.

5 **Thermodynamics – Expansion of a gas.** The pressure, P, and volume, V, of a gas are related by
$$PV^2 = 10$$
Calculate the work done when the gas expands from $V = 5$ cubic units to $V = 9$ cubic units.

Solutions to exercises

1 4

2 54

3 0.9

4 (a) 6 m (b) 3 m s^{-1}

5 $\frac{8}{9}$

3.3 The area bounded by a curve, parts of which lie below the x axis

Figure 3.7 shows a graph of $y = -x^2 + 1$. The shaded area is bounded by the x axis and the curve, but lies entirely below the x axis.

Figure 3.7
Areas lying below the x axis need special care.

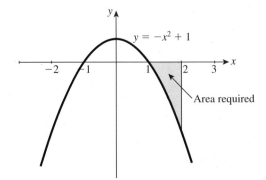

$y = -x^2 + 1$

Area required

Let us evaluate the integral $\int_1^2 -x^2 + 1 \, dx$.

$$\int_1^2 -x^2 + 1 \, dx = \left[-\frac{x^3}{3} + x \right]_1^2$$

$$= \left(-\frac{2^3}{3} + 2 \right) - \left(-\frac{1^3}{3} + 1 \right)$$

$$= -\frac{7}{3} + 1$$

$$= -\frac{4}{3}$$

The evaluation of the area yields a negative quantity. There is no such thing as a negative area. The area is actually $\frac{4}{3}$, and the negative sign is an indication that the area lies below the *x* axis.

If an area contains parts both above and below the horizontal axis, care must be taken when trying to calculate this area. It is necessary to determine which parts of the graph lie above the horizontal axis and which lie below. Separate integrals need to be calculated for each 'piece' of the graph. This idea is illustrated in the next example.

Example 3.7

Find the total area enclosed by the curve $y = x^3 - 5x^2 + 4x$ and the *x* axis between $x = 0$ and $x = 3$.

Solution

We need to determine which parts of the graph, if any, lie above and which lie below the *x* axis. To do this it is helpful to consider where the graph cuts the *x* axis. So we consider the function $x^3 - 5x^2 + 4x$ and look for its zeros.

$$x^3 - 5x^2 + 4x = x(x^2 - 5x + 4)$$
$$= x(x - 1)(x - 4)$$

So the graph cuts the *x* axis when $x = 0$, $x = 1$ and $x = 4$. Also, when *x* is large and positive, *y* is large and positive since the term involving x^3 dominates. When *x* is large and negative, *y* is large and negative for the same reason. With this information we can sketch a graph showing the required area. If you have access to a graphics calculator or computer package this is a trivial matter. The graph is shown in Figure 3.8.

Figure 3.8
This area must be calculated in two parts.

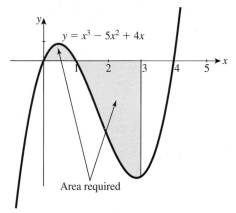

$y = x^3 - 5x^2 + 4x$

Area required

From the graph we see that the required area lies partly above the x axis (when $0 \leq x \leq 1$) and partly below (when $1 \leq x \leq 3$). So we evaluate the integral in two parts.

First,

$$\int_0^1 x^3 - 5x^2 + 4x \, dx = \left[\frac{x^4}{4} - \frac{5x^3}{3} + \frac{4x^2}{2} \right]_0^1$$
$$= \left(\frac{1}{4} - \frac{5}{3} + 2 \right) - (0)$$
$$= \frac{7}{12}$$

This is the part of the required area that lies above the x axis.

Second,

$$\int_1^3 x^3 - 5x^2 + 4x \, dx = \left[\frac{x^4}{4} - \frac{5x^3}{3} + \frac{4x^2}{2} \right]_1^3$$
$$= \left(\frac{81}{4} - \frac{135}{3} + 18 \right) - \left(\frac{1}{4} - \frac{5}{3} + 2 \right)$$
$$= -\frac{81}{12} - \frac{7}{12}$$
$$= -\frac{88}{12}$$
$$= -\frac{22}{3}$$

This represents the part of the required area that lies below the x axis. The actual area is $\frac{22}{3}$.

Combining the results of the two separate calculations we can find the total area bounded by the curve:

$$\text{area} = \frac{7}{12} + \frac{22}{3}$$
$$= \frac{95}{12}$$

Example 3.8

(a) Sketch the graph of $y = \sin 2x$ for $0 \leq x \leq \pi$.

(b) Find the total area bounded by the curve and the x axis between $x = \frac{\pi}{3}$ and $x = \frac{3\pi}{4}$.

Solution

Sketch the graph and indicate the required area noting where the graph crosses the x axis.

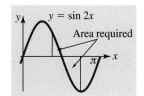

Perform the integration in two parts to obtain the required area.

For the area above the x axis evaluate the appropriate integral.

$$\int_{\pi/3}^{\pi/2} \sin 2x \, dx = \left[-\frac{\cos 2x}{2} \right]_{\pi/3}^{\pi/2} = \frac{1}{4}$$

Now evaluate the integral for finding the area below the x axis.

$$\int_{\pi/2}^{3\pi/4} \sin 2x \, dx = -\frac{1}{2}$$

Finally write down the total area required.

$$\frac{1}{4} + \frac{1}{2} = \frac{3}{4}$$

Exercises

1 Find the total area enclosed between the x axis and the curve $y = x^3$ between $x = -1$ and $x = 1$.

2 Find the area under $y = \cos 2t$ from $t = 0$ to $t = 0.5$.

3 Find the area enclosed by $y = 4 - x^2$ and the x axis from (a) $x = 0$ to $x = 2$, (b) $x = -2$ to $x = 1$, (c) $x = 1$ to $x = 3$.

4 Calculate the area enclosed by the curve $y = x^3$ and the line $y = x$.

5 Find the area bounded by $y = e^x$, the y axis, the x axis and the line $x = 2$.

6 Find the area enclosed between $y = x(x - 1)(x - 2)$ and the x axis.

7 Find the area enclosed by the graph of $y = \dfrac{1}{\sqrt{9 - 4t^2}}$ between $t = 0$ and $t = 1$.

Solutions to exercises

1   0.5

2   0.4207

3   (a) $\frac{16}{3}$ (b) 9 (c) 4

4   0.5

5    $e^2 - 1$

6    $\frac{1}{2}$

7   0.365

End of block exercises

1   Find the area enclosed by $y = 4t - t^2$ and the t axis between $t = 1$ and $t = 2$.

2   Find the area enclosed between the curves $y = x^2 - 2x$ and $y = 4 - x^2$.

3   Find the area enclosed by the curve $y = -x^2 + 6x - 5$ and the x axis.

4   Find the total area enclosed by the graph of the function $y = x^3 - 4x$ and the x axis.

Solutions to exercises

1    $3\frac{2}{3}$

2   9

3   10.667

4   8

Computational approaches to integration

4.1 Introduction

Computer software packages are readily available that can integrate. The results that they produce fall into two categories, **symbolic** and **numerical**.

For many indefinite integrals, $\int f(x)\,dx$, it is possible for the computer to obtain a symbolic answer in terms of common functions such as exponential, trigonometrical or polynomial functions. Furthermore, the computer can evaluate *exactly* many definite integrals, $\int_a^b f(x)\,dx$, often by leaving an answer in the form of a fraction, or in surd form, or using common mathematical constants such as π and e. We shall illustrate such use of computer software.

For some integrals it is impossible to obtain a symbolic answer. However, there are techniques for approximating definite integrals. Such techniques are called **numerical methods**, and two of these, the trapezium rule and Simpson's rule, are explained here. Because such methods require substantial calculation they are very laborious to perform by hand. So, they are best implemented using a computer. We shall be content to illustrate the principles involved on simple examples, and then use software to tackle more substantial problems.

4.2 Use of symbolic algebra packages to find integrals exactly

Computer software designed for tackling mathematical problems invariably has the facility to integrate. This is certainly true of Matlab and Maple. You should work through the following examples using the package to which you have access. You may need to refer to local documentation. Some of the packages may require you to call, or load, an additional piece of software (e.g. a symbolic toolkit). You will note from the examples below that the particular commands and syntax required by the different packages vary, but often only slightly. Most packages provide extensive on-line help and examples to which you should refer for further details as required.

Example 4.1 Finding indefinite integrals exactly

Find $\int x \cos^2 x\,dx$.

Solution

Maple

In Maple the command to perform this integration is

```
int(x*cos(x)^2,x);
```

and Maple outputs

$$x\left(\frac{1}{2}\cos(x)\sin(x) + \frac{1}{2}x\right) + \frac{1}{4}\cos(x)^2 - \frac{1}{4}x^2$$

Note that Maple can supply a symbolic answer, but gives no constant of integration – it is assumed that the user will remember to add this.

Matlab

In Matlab, to perform this calculation it is necessary first to declare that x is a symbol using the command

```
syms x
```

Then the command to integrate $x\cos^2 x$ is

```
int(x*cos(x)^2,x)
```

and Matlab outputs

```
ans=
x*(1/2*cos(x)*sin(x)+1/2*x)+1/4*cos(x)^2-1/4*x^2
```

This is equivalent to the answer produced by Maple although it is not set out in as friendly a form.

Example 4.2 Finding definite integrals exactly

Find $\displaystyle\int_1^2 \frac{1}{t}\,dt$.

Solution

Maple

In Maple the command to perform this integration is

```
int(1/t,t=1..2);
```

and Maple outputs

$$\ln(2)$$

Note that, when possible, the output from the symbolic algebra package is an exact result, that is ln 2, as opposed to a decimal approximation.

Matlab
In Matlab, the symbol t must be declared before integrating:

```
syms t
int(1/t,1,2)
```

and Matlab outputs

```
ans = log(2)
```

Again, this answer is equivalent to that produced by Maple although given in a different form.

Example 4.3

Find $\int_0^3 \sqrt{x}\,dx$.

Solution

Maple
The command to perform this integration is

```
int(sqrt(x), x=0..3);
```

and Maple outputs

$$2\sqrt{3}$$

Once again, note the exact result produced by the package, rather than a decimal approximation. There is the facility in all packages to produce a decimal approximation if this is required.

Matlab
```
syms x
int(sqrt(x),0,3)
```

and the output is

```
ans = 2*3^(1/2)
```

Computer packages can find indefinite integrals when one or both limits is infinity. Consider the following example.

Example 4.4 Finding definite integrals with infinite limits of integration

Find $\int_{-\infty}^{\infty} e^{-x^2} \, dx$.

Solution

Maple
The Maple command to evaluate this integral is

```
int(exp(-x^2),x = -infinity..infinity);
```

and Maple outputs the exact solution

$$\sqrt{\pi}$$

Matlab
```
syms x
int(exp(-x^2),-inf,inf)
```

with output

```
ans = pi^(1/2)
```

Although symbolic algebra packages are extremely powerful, there are nevertheless some integrals that are beyond their scope. In such cases you will either be warned that an explicit solution could not be found, or the expression you input may be returned to you unevaluated. For example, try to use a package to find $\int x^x \, dx$.

Computer and calculator exercises

Computer algebra packages are sophisticated products. Because they are designed for professionals their scope extends way beyond an introductory textbook like this one. Some of the computer output may well be given in terms of functions with which you are not familiar. This is normal, and when necessary you will need to explore these functions further.

Use a symbolic algebra package to find the following integrals exactly:

1 $\int_2^3 \frac{1}{\sqrt{x}} \, dx$

2 $\int x \sin^2 x \, dx$

3 $\int t e^{-st} \, dt$

4 $\int_0^1 \frac{\sin x}{x} \, dx$

5 $\int_{-1}^1 e^{-j\omega t} \, dt$ where j is the complex number with $j^2 = -1$.

Solutions to exercises

1 $2\sqrt{3} - 2\sqrt{2}$

2 $x\left(-\frac{1}{2}\cos(x)\sin(x) + \frac{1}{2}x\right)$
$-\frac{1}{4}\cos(x)^2 - \frac{1}{4}x^2 + c$

3 $\frac{-ste^{-st} - e^{-st}}{s^2} + c$

4 Si(1). This result may be given in terms of a function you will probably not be familiar with, known as the sine integral. See if you can use the computer to evaluate a decimal approximation to this as 0.946 (3 d.p.).

5 $\frac{j(e^{-j\omega} - e^{j\omega})}{\omega}$ or the alternative form $\frac{2\sin\omega}{\omega}$.

4.3 The trapezium rule and Simpson's rule

We have seen the power of computer packages for performing some definite and indefinite integrals. We have also seen that not all functions can be integrated, and so not all integrals can be found exactly. However, techniques do exist for finding approximations to definite integrals. These techniques are called **numerical methods**.

Trapezium rule

We have seen in Block 3 that the definite integral $\int_a^b f(x) \, dx$ can be thought of as the area lying under the graph of $f(x)$ for $a \le x \le b$. This observation leads us to the following method for approximating the integral. We estimate the area lying under the graph and use this as an estimate of the integral. In the trapezium rule, the area under $f(x)$ between $x = a$ and $x = b$ is divided into several vertical strips, and the area of each strip is estimated by assuming that it has the shape of a trapezium. Consider Figure 4.1. Suppose that we divide the area under $y = f(x)$ from $x = a$ to $x = b$ into n strips. We denote the width of each strip by h. Note that since the distance from $x = a$ to $x = b$ is $b - a$, then the width of each strip is $\frac{b - a}{n}$. The lengths of the sides of the strips are denoted by $y_0, y_1, \ldots, y_n$.

Figure 4.1
The area under $f(x)$ is divided into n strips, each of width h. The area of the typical strip shown is
$$\frac{h}{2}(y_i + y_{i+1}).$$

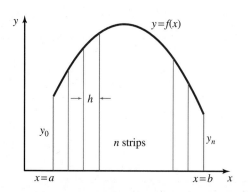
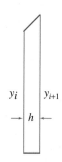

Now the area of the first trapezium of width h, which has sides of heights y_0 and y_1, is $\frac{h}{2}(y_0 + y_1)$. Similarly the area of the second strip is $\frac{h}{2}(y_1 + y_2)$. We continue in this way until the areas of all the strips have been found. The area of the final strip is $\frac{h}{2}(y_{n-1} + y_n)$.

Consequently we can estimate the total area as

$$\text{area} \approx \frac{h}{2}(y_0 + y_1) + \frac{h}{2}(y_1 + y_2) + \frac{h}{2}(y_2 + y_3) + \cdots + \frac{h}{2}(y_{n-1} + y_n)$$

which can be simplified to

$$\text{area} \approx \frac{h}{2}(y_0 + 2y_1 + 2y_2 + \cdots + 2y_{n-1} + y_n)$$

Key point

Trapezium rule: with n strips, and $h = \dfrac{b - a}{n}$

$$\int_a^b f(x)\,dx \approx \frac{h}{2}(y_0 + 2y_1 + 2y_2 + \cdots + 2y_{n-1} + y_n)$$

Example 4.5

Use the trapezium rule with eight strips to estimate $\displaystyle\int_1^2 \frac{1}{x}\,dx$. Work throughout to six decimal places and quote your final result to four decimal places.

Solution

In this example $a = 1$, $b = 2$ and $n = 8$. We are dividing the area under the graph of $y = \dfrac{1}{x}$ from $x = 1$ to $x = 2$ into eight strips, and so the width of each strip will be $h = \dfrac{2 - 1}{8} = 0.125$. We must calculate the lengths of the sides of each of the strips, $y_0, y_1, \ldots, y_8$. These are simply the respective y values calculated from $y = \dfrac{1}{x}$. It is helpful, and conventional, to set these out as in Table 4.1. The table has been laid out in the way shown because all y values other than the first and last will be multiplied by 2.

Table 4.1

x	$y = 1/x$	
1	$y_0 = 1.000000$	
1.125		$y_1 = 0.888889$
1.25		$y_2 = 0.800000$
1.375		$y_3 = 0.727273$
1.5		$y_4 = 0.666667$
1.625		$y_5 = 0.615385$
1.75		$y_6 = 0.571429$
1.875		$y_7 = 0.533333$
2	$y_8 = 0.500000$	
$y_0 + y_n$	1.500000	
$y_1 + y_2 + \cdots + y_{n-1}$		4.802976

Using the data in the table we can apply the trapezium rule formula:

$$\int_1^2 \frac{1}{x}\, dx \approx \frac{0.125}{2}(1.5000000 + 2(4.802976)) = 0.694122$$

So, to four decimal places we estimate the integral to be 0.6941. Compare this with the exact value $\ln 2 = 0.6931$ (4 d.p.).

Simpson's rule

Simpson's rule is a more sophisticated technique, which results in a more accurate estimate of the definite integral. It can be applied only when there is an even number of strips.

We state and use Simpson's rule without deriving it. For specific details refer to a textbook on numerical methods for engineers.

Key point

Simpson's rule: with n (even) strips, and $h = \dfrac{b - a}{n}$

$$\int_a^b f(x)\, dx \approx \frac{h}{3}(y_0 + 4y_1 + 2y_2 + 4y_3 + \cdots + 2y_{n-2} + 4y_{n-1} + y_n)$$

It can be helpful to remember this formula as

$$\int_a^b f(x)\, dx \approx \frac{h}{3}(\text{first} + \text{last} + 4(\text{odd-numbered terms}) + 2(\text{even-numbered terms}))$$

Example 4.6

Use Simpson's rule with eight strips to estimate $\displaystyle\int_1^2 \frac{1}{x}\, dx$.

As in the previous example it is helpful to set out the data as in Table 4.2. It has been laid out in this way because y_1, y_3, etc., will be multiplied by 4, whereas y_2, y_4, etc., will be multiplied by 2.

Table 4.2

x	$y = 1/x$		
1	$y_0 = 1.000000$		
1.125		$y_1 = 0.888889$	
1.25			$y_2 = 0.800000$
1.375		$y_3 = 0.727273$	
1.5			$y_4 = 0.666667$
1.625		$y_5 = 0.615385$	
1.75			$y_6 = 0.571429$
1.875		$y_7 = 0.533333$	
2	$y_8 = 0.500000$		
first + last, $y_0 + y_8$	1.500000		
sum of odd terms, $y_1 + y_3 + \cdots + y_7$		2.764880	
sum of even terms, $y_2 + y_4 + \cdots + y_6$			2.038096

Solution

Using the data in the table we can apply Simpson's rule:

$$\int_1^2 \frac{1}{x}\, dx \approx \frac{0.125}{3}(1.500000 + 4(2.764880) + 2(2.038096)) = 0.693155$$

So, to four decimal places we can estimate the integral to be 0.6932. This compares well with the exact answer ln 2, which is 0.6931 to 4 d.p.

Computer algebra packages have the facility to carry out numerical integration, which means that the user can avoid lengthy and tedious arithmetical calculations.

Example 4.7 Using a computer package for numerical integration

Use a computer package to evaluate $\int_1^4 x^x\, dx$.

Solution

We have seen previously that no symbolic expression exists for $\int x^x\, dx$. However, computer packages can apply methods such as Simpson's rule.

Maple

The command to perform the required numerical evaluation is

```
evalf(int(x^x,x = 1..4));
```

and Maple outputs the result 113.3356 (4 d.p.).

Matlab

We need a command quad, which stands for quadrature, the technical name for numerical integration.

```
Q = quad('x.^x',1,4)
```

and Matlab calculates the value of the integral

$$\text{ans} = 113.3356$$

Exercises

1 Use the trapezium rule with the number of strips specified to approximate the following definite integrals:

(a) $\int_0^1 \tan^2 x \, dx$, 4 strips

(b) $\int_1^{1.6} \dfrac{e^{2x}}{x^2} \, dx$, 3 strips

2 Use Simpson's rule with the number of strips specified to approximate the following definite integrals:

(a) $\int_0^{0.8} \tan^2 x \, dx$, 8 strips

(b) $\int_1^2 \sqrt{1 + x^3} \, dx$, 10 strips

Solutions to exercises

1 (a) 0.611069 (6 d.p.) (b) 4.906229 (6 d.p.)

2 (a) 0.229685 (6 d.p.) (b) 2.129862 (6 d.p.)

Computer and calculator exercises

1 Use software to evaluate

(a) $\int (\ln x)^2 \, dx$ (b) $\int \sin^3 x \, dx$

(c) $\int_0^1 \sin z^3 \, dz$ (d) $\int_0^{0.1} \dfrac{x^3 + x}{\cos x} \, dx$

Solutions to exercises

1 (a) $x(\ln x)^2 - 2x \ln x + 2x + c$

(b) $-\dfrac{1}{3} \sin^2 x \cos x - \dfrac{2}{3} \cos x + c$

(c) 0.23385 (5 d.p.) (d) 0.00504 (5 d.p.)

End of block exercise

1 Use a package to find an exact expression for the area bounded by $y = 5 - x^2$ and lying above the x axis.

Solution to exercise

1 $\dfrac{20\sqrt{5}}{3}$

Integration by parts

5.1 Introduction

The technique known as **integration by parts** is used to integrate a product of two functions, for example

$$\int e^{2x} \sin 3x \, dx \quad \text{and} \quad \int_0^1 x^3 e^{-2x} \, dx$$

Note that in the first example the integrand is the product of the functions e^{2x} and $\sin 3x$, and in the second example the integrand is the product of the functions x^3 and e^{-2x}. Note also that we can change the order of the terms in the product if we wish and write

$$\int \sin 3x \, e^{2x} \, dx \quad \text{and} \quad \int_0^1 e^{-2x} x^3 \, dx$$

It is often possible to find integrals involving products using a method known as **integration by parts** – you can think of this as a *product rule* for integrals. This technique is described in the block.

5.2 The integration by parts formula

The integration by parts formula states:

Key point

For indefinite integrals:

$$\int u\left(\frac{dv}{dx}\right) dx = uv - \int v\left(\frac{du}{dx}\right) dx$$

For definite integrals:

$$\int_a^b u\left(\frac{dv}{dx}\right) dx = [uv]_a^b - \int_a^b v\left(\frac{du}{dx}\right) dx$$

Study the formula carefully and note the following observations:

- The formula replaces one integral, the one on the left, with a different integral, that on the right. The intention is that the latter is simpler to evaluate.

- The integrand on the left-hand side is the product of the terms u and $\dfrac{dv}{dx}$. To apply the formula to a particular example we must let one function in the product equal u. We must be able to differentiate this function to find $\dfrac{du}{dx}$. We let the other function in the product equal $\dfrac{dv}{dx}$. We must be able to integrate this function, to find v.

Consider the following example.

Example 5.1

Find $\int x \sin x \, dx$.

Solution

Compare the required integral with the formula for integration by parts: we see that it makes sense to choose

$$u = x \quad \text{and} \quad \frac{dv}{dx} = \sin x$$

It follows that

$$\frac{du}{dx} = 1 \quad \text{and} \quad v = \int \sin x \, dx = -\cos x$$

(When integrating $\dfrac{dv}{dx}$ to find v there is no need to worry about a constant of integration. When you become confident with the method, you may like to think about why this is the case.) Applying the formula we obtain

$$\int x \sin x \, dx = uv - \int v \left(\frac{du}{dx} \right) dx$$

$$= x \left(-\cos x \right) - \int (-\cos x) \cdot 1 \, dx$$

$$= -x \cos x + \int \cos x \, dx$$

$$= -x \cos x + \sin x + c$$

Example 5.2

Find $\int (5x + 1) \cos 2x \, dx$.

Solution

Let $u = 5x + 1$ and $\dfrac{dv}{dx} = \cos 2x$. Now calculate $\dfrac{du}{dx}$ and v.

$$\frac{du}{dx} = 5 \quad \text{and} \quad v = \int \cos 2x \, dx = \frac{\sin 2x}{2}$$

Substitute these results into the formula for integration by parts:

$$\frac{(5x + 1) \sin 2x}{2} - \int \frac{\sin 2x}{2} \, 5 \, dx$$

Finish the problem by finding the final integral:

$$\frac{(5x + 1)\sin 2x}{2} + \frac{5}{4}\cos 2x + c$$

Example 5.3

Find $\int_0^2 xe^x \, dx$.

Solution

We let $u = x$ and $\dfrac{dv}{dx} = e^x$. Then $\dfrac{du}{dx} = 1$ and $v = e^x$. Using the formula for integration by parts we obtain

$$\int_0^2 xe^x \, dx = [xe^x]_0^2 - \int_0^2 e^x \cdot 1 \, dx$$

$$= 2e^2 - [e^x]_0^2$$

$$= 2e^2 - [e^2 - 1]$$

$$= e^2 + 1 \text{ (or 8.389 to 3 d.p.)}$$

Sometimes it is necessary to apply the formula more than once, as the next example shows.

Example 5.4

Find $\int_0^2 x^2 e^x \, dx$.

Solution
We let

$$u = x^2 \quad \text{and} \quad \frac{dv}{dx} = e^x$$

Then

$$\frac{du}{dx} = 2x \quad \text{and} \quad v = e^x$$

Using the formula for integration by parts we find

$$\int_0^2 x^2 e^x \, dx = [x^2 e^x]_0^2 - \int_0^2 2x e^x \, dx$$

$$= 4e^2 - 2\int_0^2 xe^x \, dx$$

The remaining integral must be integrated by parts also, but we have just done this in the previous example. So

$$\int_0^2 x^2 e^x \, dx = 4e^2 - 2[e^2 + 1]$$

$$= 2e^2 - 2$$

$$= 12.778 \text{ (3 d.p.)}$$

Example 5.5

Find $\int_0^{\pi/4} (4 - 3x) \sin x \, dx$.

Solution

Take $u = 4 - 3x$ and $\dfrac{dv}{dx} = \sin x$. State $\dfrac{du}{dx}$.

$$\frac{du}{dx} = -3$$

State v.

$$v = -\cos x$$

Now apply the formula.

$$\int_0^{\pi/4} (4 - 3x) \sin x \, dx =$$

$$[(4 - 3x)(-\cos x)]_0^{\pi/4} - 3\int_0^{\pi/4} \cos x \, dx$$

Perform the final integral and apply the limits:

$$[(4 - 3x)(-\cos x)]_0^{\pi/4} - 3[\sin x]_0^{\pi/4}$$

Evaluate this result:

$$0.716$$

Example 5.6

Let I stand for the integral

$$\int e^{-x} \sin x \, dx$$

Calculate this integral.

Solution

In this example you will find that it does not matter which term in the product is differentiated and which is integrated. Verify this for yourself later.

Take $u = e^{-x}$ and $\dfrac{dv}{dx} = \sin x$.

State $\dfrac{du}{dx}$.

$$\frac{du}{dx} = -e^{-x}$$

State v.

$$v = -\cos x$$

Apply the integration by parts formula.

$$I = \int e^{-x} \sin x \, dx =$$

$$e^{-x}(-\cos x) - \int (-\cos x)(-e^{-x}) \, dx$$

This simplifies to $I = -e^{-x} \cos x - \int e^{-x} \cos x \, dx$. This integral must now be evaluated by parts. Still with $u = e^{-x}$, and taking $\dfrac{dv}{dx} = \cos x$, perform the integration.

$$I = -e^{-x} \cos x - \left[e^{-x} \sin x + \int e^{-x} \sin x \, dx \right]$$

At this stage it looks as though we have gone round in a circle because the integral remaining is the same as the one we started with. However, writing this as I we find

$$I = -e^{-x} \cos x - e^{-x} \sin x - I$$

so that

$$2I = -e^{-x} \cos x - e^{-x} \sin x$$

from which

$$I = \frac{-e^{-x} \cos x - e^{-x} \sin x}{2}$$

We conclude that

$$I = \frac{-e^{-x} \cos x - e^{-x} \sin x}{2} + c$$

Exercises

In some questions it may be necessary to apply the formula more than once.

1 Find (a) $\int x \sin(2x) \, dx$, (b) $\int t e^{3t} \, dt$,
   (c) $\int x \cos x \, dx$.

2 Find $\int (x + 3) \sin x \, dx$.

3 By writing $\ln x$ as $1 \times \ln x$ find $\int \ln x \, dx$.

4 Find (a) $\int \tan^{-1} x \, dx$, (b) $\int -7x \cos 3x \, dx$,
   (c) $\int 5x^2 e^{3x} \, dx$, (d) $\int (x + 1) e^x \, dx$.

5 Find (a) $\int x \cos kx \, dx$, (b) $\int z^2 \cos kz \, dz$,
   where k is a constant.

6 Find $\int t e^{-st} \, dt$ where s is a constant.

7 Find $\int t^2 e^{-st} \, dt$ where s is a constant.

8 Evaluate the following definite integrals:

(a) $\int_0^1 x \cos 2x \, dx$ (b) $\int_0^{\pi/2} x \sin 2x \, dx$

(c) $\int_{-1}^1 te^{2t} \, dt$

9 Find $\int_1^2 (x + 2) \sin x \, dx$.

10 Find $\int_0^1 (x^2 - 3x + 1)e^x \, dx$.

11 Show that

$$\int e^{ax} \cos bx \, dx$$

$$= \frac{e^{ax} (a \cos bx + b \sin bx)}{a^2 + b^2} + c$$

Solutions to exercises

1 (a) $\dfrac{\sin 2x}{4} - \dfrac{x \cos 2x}{2} + c$

(b) $e^{3t}\left(\dfrac{t}{3} - \dfrac{1}{9}\right) + c$

(c) $\cos x + x \sin x + c$

2 $-(x + 3) \cos x + \sin x + c$

3 $x \ln x - x + c$

4 (a) $x \tan^{-1} x - \dfrac{\ln(x^2 + 1)}{2} + c$

(b) $-\dfrac{7 \cos 3x}{9} - \dfrac{7x \sin 3x}{3} + c$

(c) $\dfrac{5e^{3x}(9x^2 - 6x + 2)}{27} + c$

(d) $xe^x + c$

5 (a) $\dfrac{\cos kx}{k^2} + \dfrac{x \sin kx}{k} + c$

(b) $\dfrac{2z \cos kz}{k^2} + \dfrac{z^2 \sin kz}{k} - \dfrac{2 \sin kz}{k^3} + c$

6 $\dfrac{-e^{-st}(st + 1)}{s^2} + c$

7 $\dfrac{-e^{-st}(s^2t^2 + 2st + 2)}{s^3} + c$

8 (a) 0.1006 (b) $\dfrac{\pi}{4} = 0.7854$ (c) 1.9488

9 3.3533

10 -0.5634

End of block exercises

1 Find the following integrals:

(a) $\int t \sin 4t \, dt$

(b) $\int te^{-5t} \, dt$

(c) $\int 3xe^{3x} \, dx$

(d) $\int (2 + 3x)e^x \, dx$

(e) $\int (7 - x)\sin x \, dx$

2 Evaluate the following definite integrals:

(a) $\int_0^1 2x^2 e^x \, dx$

(b) $\int_{-1}^1 (x + 5)e^{5x} \, dx$

(c) $\int_0^{\pi} e^{-x} \sin x \, dx$

3 Let I_n stand for the integral $\int \sin^n x \, dx$, so for example I_4 means $\int \sin^4 x \, dx$. By writing $\sin^n x$ as $\sin^{n-1} x \sin x$ and using integration by parts show that

$$I_n = -\frac{1}{n} \sin^{n-1} x \cos x + \frac{n-1}{n} I_{n-2}$$

Such a formula is called a **reduction formula**. Used repeatedly it can be used to reduce an integral such as $\int \sin^4 x \, dx$ to one involving $\sin^2 x$ and then to one involving $\sin^0 x$ or simply 1. Use the reduction formula to find $\int \sin^6 x \, dx$.

4 Obtain a reduction formula for $\int \cos^n x \, dx$.

Solutions to exercises

1 (a) $\dfrac{\sin 4t}{16} - \dfrac{t \cos 4t}{4} + c$

(b) $-\dfrac{e^{-5t}(5t + 1)}{25} + c$

(c) $\dfrac{e^{3x}(3x - 1)}{3} + c$

(d) $e^x(3x - 1) + c$

(e) $(x - 7)\cos x - \sin x + c$

2 (a) 1.437 (b) 172.154 (c) 0.522

3 $I_6 = -\frac{1}{6} \sin^5 x \cos x - \frac{5}{24} \sin^3 x \cos x$

$\qquad -\frac{5}{16} \sin x \cos x + \frac{5x}{16} + c$

4 $I_n = \dfrac{1}{n} \cos^{n-1} x \sin x + \dfrac{n-1}{n} I_{n-2}$

Integration by substitution

6.1 Introduction

The technique described in this block involves making a substitution in order to simplify an integral or make it more amenable to integration. There are many different ways in which a substitution can be made. For example, we may let a new variable, u say, equal a more complicated part of the function we are trying to integrate. The choice of which substitution to make often relies upon experience: don't worry if at first you cannot see an appropriate substitution. This skill develops with practice. However, it is not simply a matter of changing the variable – care must be taken with the term dx, as we shall see.

Integrals of the form $\int \dfrac{f'(x)}{f(x)}\, dx$ can be dealt with in a particularly simple way as we show in Section 6.4. In Section 6.5 we provide a table of suggested substitutions to be tried in some more difficult cases.

6.2 Making a substitution

The technique of integrating by substitution is illustrated in the following example.

Example 6.1

Find $\int (3x + 5)^6\, dx$.

Solution

First, look at the function we are trying to integrate: $(3x + 5)^6$. It looks quite a complicated function to integrate. Suppose we introduce a new variable, u, such that $u = 3x + 5$. Doing this means that the function we must integrate becomes u^6. Would you not agree that this looks a much simpler function to integrate than $(3x + 5)^6$? There is a slight complication, however. The new function of u must be integrated with respect to u and not with respect to x. This means that we must take care of the term dx correctly. It is not simply a matter of changing the dx to du. From the substitution

$$u = 3x + 5$$

note, by differentiation, that

$$\frac{du}{dx} = 3$$

It follows that we can write

$$dx = \frac{1}{3} du$$

This is the way in which the dx is changed into du.

The required integral then becomes

$$\int (3x + 5)^6 \, dx = \int u^6 \frac{1}{3} \, du$$

The factor of $\frac{1}{3}$, being a constant, means that we can write

$$\int (3x + 5)^6 \, dx = \frac{1}{3} \int u^6 \, du$$

$$= \frac{1}{3} \frac{u^7}{7} + c$$

$$= \frac{u^7}{21} + c$$

To finish off we must rewrite this answer in terms of the original variable x and replace u by $3x + 5$:

$$\int (3x + 5)^6 \, dx = \frac{(3x + 5)^7}{21} + c$$

Example 6.2
By making the substitution $u = 2x - 7$ find $\int \cos(2x - 7) \, dx$.

Solution
First differentiate the given substitution.

$$\frac{du}{dx} = \boxed{}$$

$$2$$

Hence write down an expression for dx.

$$\boxed{}$$

$$dx = \frac{1}{2} du$$

Now the integral can be written entirely in terms of the variable u:

$$\int \cos(2x - 7) \, dx = \boxed{}$$

$$\frac{1}{2} \int \cos u \, du$$

Complete the integration:

$$\boxed{}$$

$$\frac{1}{2} \sin u + c = \frac{1}{2} \sin(2x - 7) + c$$

Example 6.3

By means of the substitution $u = 1 - x^2$, find the integral $\int x\sqrt{1 - x^2}\, dx$.

Solution

First differentiate $u = 1 - x^2$.

$$\frac{du}{dx} = \boxed{} \qquad\qquad\qquad\qquad -2x$$

This implies that $x\, dx = -\frac{1}{2}\, du$, so $\sqrt{1 - x^2} = \sqrt{u}$ and $x\, dx = -\frac{1}{2}\, du$.

Rewrite the integral in terms of u.

$$\int x\sqrt{1 - x^2}\, dx = \boxed{} \qquad\qquad -\frac{1}{2}\int \sqrt{u}\, du$$

Complete the integration: $\boxed{}$

$$-\frac{1}{2}\frac{u^{3/2}}{3/2} + c = -\frac{1}{3} u^{3/2} + c$$

Finally rewrite the answer in terms of the original variable, x.

$$\boxed{} \qquad\qquad -\frac{(1 - x^2)^{3/2}}{3} + c$$

Sometimes it may be possible to complete the square to obtain a recognisable standard form.

Example 6.4

Find $\displaystyle \int \frac{1}{s^2 + 6s + 18}\, ds$.

Solution

By completing the square in the denominator we can write the integral as

$$\int \frac{1}{(s + 3)^2 + 9}\, ds$$

Referring to the table of integrals, Table 1.1 in Block 1, note that

$$\int \frac{1}{x^2 + a^2}\, dx = \frac{1}{a}\tan^{-1}\left(\frac{x}{a}\right) + c$$

This standard form suggests the substitution $x = s + 3$, $\dfrac{dx}{ds} = 1$ from which $dx = ds$. Write down and complete the resulting integral:

$$\int \frac{1}{x^2 + 3^2}\,dx$$

$$\frac{1}{3}\tan^{-1}\left(\frac{s + 3}{3}\right) + c$$

Exercises

1 Use a substitution to find

(a) $\int (4x + 1)^7\,dx$ (b) $\int t^2 \sin(t^3 + 1)\,dt$
(hint: let $u = t^3 + 1$) (c) $\int \sin(3x - 1)\,dx$
(d) $\int e^{2x-3}\,dx$ (e) $\int x(2x^2 + 7)^4\,dx$
(f) $\int \sin^2 4t \cos 4t\,dt$

2 Find (a) $\int 3e^{-x-1}\,dx$, (b) $\int 4 \sin(3x + 7)\,dx$.

3 By making the substitution $u = 3^x$, and using the laws of logarithms, find $\int 3^x\,dx$.

4 Find (a) $\displaystyle\int \frac{1}{(x + 1)^2}\,dx$, (b) $\displaystyle\int \frac{1}{(2x + 3)^3}\,dx$.

5 Find $\int \cos(1 - x)\,dx$.

6 By factorising the denominator and making a substitution find

$$\int \frac{1}{t^2 - 2t + 1}\,dt$$

Solutions to exercises

1 (a) $\dfrac{(4x + 1)^8}{32} + c$ (b) $-\dfrac{\cos(t^3 + 1)}{3} + c$

(c) $-\dfrac{\cos(3x - 1)}{3} + c$ (d) $\dfrac{e^{2x-3}}{2} + c$

(e) $\dfrac{(2x^2 + 7)^5}{20} + c$ (f) $\dfrac{\sin^3 4t}{12} + c$

2 (a) $-3e^{-x-1} + c$ (b) $-\frac{4}{3}\cos(3x + 7) + c$

3 $\dfrac{1}{\ln 3}\,3^x + c$

4 (a) $-\dfrac{1}{x + 1} + c$ (b) $-\dfrac{1}{4}\cdot\dfrac{1}{(2x + 3)^2} + c$

5 $-\sin(1 - x) + c$

6 $-\dfrac{1}{t - 1} + c$

6.3 Substitution and definite integrals

If you are dealing with definite integrals (ones with limits of integration) you must be particularly careful when you substitute. Consider the following example.

Example 6.5

Find $\int_2^3 t \sin(t^2)\,dt$ by making the substitution $u = t^2$.

Solution

Note that if $u = t^2$ then $\dfrac{du}{dt} = 2t$ so that $dt = \dfrac{du}{2t}$. We find

$$\int_{t=2}^{t=3} t \sin(t^2) \, dt = \int_{t=2}^{t=3} t \sin u \, \frac{du}{2t}$$

$$= \frac{1}{2} \int_{t=2}^{t=3} \sin u \, du$$

An important point to note is that the limits of integration are limits on the variable t, not on u. To emphasise this they have been written explicitly as $t = 2$ and $t = 3$. When we integrate with respect to the variable u, the limits must be written in terms of u too. From the substitution $u = t^2$, note that

$$\text{when } t = 2, u = 4 \text{ and when } t = 3, u = 9$$

so the integral becomes

$$\frac{1}{2} \int_{u=4}^{u=9} \sin u \, du = \frac{1}{2} [-\cos u]_4^9$$

$$= \frac{1}{2}(-\cos 9 + \cos 4)$$

$$= 0.129$$

Example 6.6

By making the substitution $u = t + 1$ find $\int_1^4 t\sqrt{t + 1} \, dt$.

Solution

State the corresponding limits on u:

$2, 5$

Make the substitution and obtain an integral in terms of u:

$$\int_2^5 (u - 1)\sqrt{u} \, du$$

This integration can be performed by removing the brackets and integrating each term separately. Finish the solution:

$$\left[\frac{2}{5} u^{5/2} - \frac{2}{3} u^{3/2}\right]_2^5 = 14.530$$

Exercises

1 Use a substitution to find

(a) $\int_1^2 (2x + 3)^7 dx$ (b) $\int_0^1 3t^2 e^{t^3} dt$

Solutions to exercises

1   (a) 3.3589×10^5 (b) 1.7183

6.4   Integrals of the form $\int \frac{f'(x)}{f(x)} dx$

Example 6.7

Find $\int \frac{(3x^2 + 1)}{x^3 + x + 2} dx$.

Solution

Let us consider what happens when we make the substitution $z = x^3 + x + 2$. Note that

$$\frac{dz}{dx} = 3x^2 + 1$$

so that we can write

$$dz = (3x^2 + 1) dx$$

Then

$$\int \frac{(3x^2 + 1)}{x^3 + x + 2} dx = \int \frac{dz}{z}$$

$$= \ln|z| + c$$

$$= \ln|x^3 + x + 2| + c$$

Note that, in the previous example, the numerator of the integrand is the derivative of the denominator. The result is the logarithm of the denominator. This is a special case of the following rule, which should be remembered:

Key point

$$\int \frac{df/dx}{f(x)} dx = \ln|f(x)| + c$$

Example 6.8

Write down, purely by inspection, the following integrals:

(a) $\int \frac{1}{x + 1} dx$ (b) $\int \frac{2x}{x^2 + 8} dx$ (c) $\int \frac{1}{x - 3} dx$

Solution

(a) [blank] $\ln|x + 1| + c$

(b) [blank] $\ln|x^2 + 8| + c$

(c) [blank] $\ln|x - 3| + c$

Example 6.9

Evaluate $\displaystyle\int_2^4 \frac{3t^2 + 2t}{t^3 + t^2 + 1}\,dt$ by inspection.

Solution

[blank]

$$\Big[\ln|t^3 + t^2 + 1|\Big]_2^4 = \ln 81 - \ln 13 = 1.83$$

Sometimes it is necessary to make slight adjustments to the integrand to obtain a form for which the previous rule is suitable. Consider the next example.

Example 6.10

Find $\displaystyle\int \frac{x^2}{x^3 + 1}\,dx$.

Solution

In this example the derivative of the denominator is $3x^2$ whereas the numerator is just x^2. We adjust the numerator as follows:

$$\int \frac{x^2}{x^3 + 1}\,dx = \frac{1}{3}\int \frac{3x^2}{x^3 + 1}\,dx$$

Note that this sort of procedure is possible only because we can move constant factors through the integral sign. It would be wrong to try to move terms involving x in a similar way. Then

$$\frac{1}{3}\int \frac{3x^2}{x^3 + 1}\,dx = \frac{1}{3}\ln|x^3 + 1| + c$$

Example 6.11

Adjust the numerator of the integral $\displaystyle\int \frac{1}{1 - x}\,dx$ in order to find the integral.

Solution

$$\int \frac{1}{1 - x}\,dx = \text{[blank]} \qquad\qquad -\int \frac{-1}{1 - x}\,dx$$

Finish the calculation:

$$\int \frac{1}{1 - x}\,dx = \text{[blank]} \qquad\qquad -\ln|1 - x| + c$$

---

1 Write down the result of finding the following integrals:

(a) $\int \dfrac{1}{x}\,dx$ (b) $\int \dfrac{2t}{t^2 + 1}\,dt$

(c) $\int \dfrac{1}{2x + 5}\,dx$ (d) $\int \dfrac{2}{3x - 2}\,dx$

Solutions to exercises

1 (a) $\ln|x| + c$ (b) $\ln|t^2 + 1| + c$

(c) $\frac{1}{2}\ln|2x + 5| + c$ (d) $\frac{2}{3}\ln|3x - 2| + c$

6.5 Some harder examples of integration by substitution

The range of possible substitutions is so wide and varied that it is impossible to give examples of every type here. However, Table 6.1 provides some suggestions for evaluating $\int f(x)\,dx$ when $f(x)$ takes particular forms.

Table 6.1

$f(x)$ contains	make the substitution:		
$\sqrt{a^2 - x^2}$	$x = a\sin\theta$		$\dfrac{dx}{d\theta} = a\cos\theta$
	$x = a\tanh u$		$\dfrac{dx}{du} = a\,\text{sech}^2 u$
$\sqrt{a^2 + x^2}$	$x = a\sinh u$		$\dfrac{dx}{du} = a\cosh u$
	$x = a\tan\theta$		$\dfrac{dx}{d\theta} = a\sec^2\theta$
$\sqrt{x^2 - a^2}$	$x = a\cosh u$		$\dfrac{dx}{du} = a\sinh u$
	$x = a\sec\theta$		$\dfrac{dx}{d\theta} = a\sec\theta\tan\theta$
$\sin x,\ \cos x$	$\sin x = \dfrac{2t}{1 + t^2}$, $\cos x = \dfrac{1 - t^2}{1 + t^2}$, where $t = \tan\dfrac{x}{2}$		$\dfrac{dx}{dt} = \dfrac{2}{1 + t^2}$
$\sin^2 x,\ \cos^2 x$	$\sin x = \dfrac{t}{\sqrt{1 + t^2}}$, $\cos x = \dfrac{1}{\sqrt{1 + t^2}}$, where $t = \tan x$		$\dfrac{dx}{dt} = \dfrac{1}{1 + t^2}$

Example 6.12

By making a hyperbolic substitution, find $\int \sqrt{1 + x^2} \, dx$. You will also need to make use of the hyperbolic identities:

$$\cosh^2 u - \sinh^2 u = 1, \quad \cosh^2 u = \frac{1 + \cosh 2u}{2}$$

Solution

Study the form of the integrand and use Table 6.1 to select an appropriate substitution.

$$x = \sinh u$$

It follows that $\dfrac{dx}{du} = \cosh u$.

Make the substitution and use the first of the identities to obtain an integral in terms of u.

$$\int \sqrt{1 + \sinh^2 u} \, \cosh u \, du = \int \cosh^2 u \, du$$

The second of the identities allows us to rewrite the integral in the form

$$\int \frac{1 + \cosh 2u}{2} \, du$$

which can be integrated directly. Complete the calculation.

$$\frac{1}{2} u + \frac{\sinh 2u}{4} + c$$

From $x = \sinh u$ it follows that $u = \sinh^{-1} x$ and so in terms of the original variable, x, we have

$$\frac{1}{2} \sinh^{-1} x + \frac{1}{4} \sinh(2 \sinh^{-1} x) + c$$

This integral arises when calculating the length of a quadratic curve as you will see in Chapter 18 Block 5.

Example 6.13

Find $\displaystyle \int \frac{dx}{1 + \cos x + \sin x}$.

Solution

By noting the form of the integrand and referring to Table 6.1 we let

$$\sin x = \frac{2t}{1 + t^2} \quad \text{and} \quad \cos x = \frac{1 - t^2}{1 + t^2}, \quad \text{where } t = \tan \frac{x}{2}$$

$$\frac{dx}{dt} = \frac{2}{1 + t^2}$$

Make these substitutions and simplify the result to obtain an integral in terms of t.

$$\int \frac{dt}{1 + t}$$

This can be integrated using the technique of Section 6.4.

$$\ln(1 + t) + c$$

Finally, reverting to the original variable x we have

$$\int \frac{dx}{1 + \cos x + \sin x} = \ln\left(1 + \tan \frac{1}{2}x\right) + c$$

Exercises

1 Find $\displaystyle\int \frac{x}{\sqrt{4 - x^2}} \, dx$.

2 Find $\displaystyle\int \sqrt{x^2 - 1} \, dx$.

3 Find $\displaystyle\int_0^{\pi/2} \frac{1}{2 + \cos x} \, dx$.

4 Refer to Figure 6.1 from which $\tan x = t$.

Show that (a) $\sin x = \dfrac{t}{\sqrt{1 + t^2}}$,

Figure 6.1

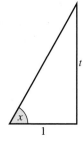

(b) $\cos x = \dfrac{1}{\sqrt{1 + t^2}}$, (c) $\dfrac{dx}{dt} = \dfrac{1}{1 + t^2}$.

5 Refer to Figure 6.2 and use it to obtain the $t = \tan \dfrac{x}{2}$ substitution given in Table 6.1.

Figure 6.2

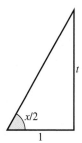

6 Show that

$$\int_0^{\pi/2} \frac{1}{3\cos x + 4\sin x} \, dx = \frac{1}{5} \ln 6.$$

Solutions to exercises

1   $-\sqrt{(4 - x^2)} + c$

2   $\dfrac{\sinh(2 \cosh^{-1} x)}{4} - \dfrac{\cosh^{-1} x}{2} + c$

3   $\dfrac{\pi \sqrt{3}}{9}$

End of block exercises

1   Find $\displaystyle\int \dfrac{1}{4x - 3}\, dx$.

2   Find $\displaystyle\int_3^4 \dfrac{t}{\sqrt{t^2 - 3}}\, dt$.

3   Find $\displaystyle\int_0^{\pi/3} \cos^3 x \sin x\, dx$.

4   Find $\displaystyle\int \dfrac{3}{(x + 2)^2}\, dx$.

5   Find $\displaystyle\int \dfrac{7}{(2 - x)^3}\, dx$.

6   Find $\displaystyle\int (9 - 2t)^7\, dt$.

7   Find $\displaystyle\int \dfrac{1}{\sqrt{3 + y}}\, dy$.

8   Find $\displaystyle\int \dfrac{\ln x}{x}\, dx$ by making the substitution $z = \ln x$.

Solutions to exercises

1   $\frac{1}{4} \ln(4x - 3) + c$

2   1.156

3   0.234

4   $-\dfrac{3}{x + 2} + c$

5   $\dfrac{7}{2(2 - x)^2} + c$

6   $-\dfrac{(9 - 2t)^8}{16} + c$

7   $2(3 + y)^{1/2} + c$

8   $\dfrac{(\ln x)^2}{2} + c$

Integration using partial fractions

7.1 Introduction

Often the technique of partial fractions can be used to write an algebraic fraction as the sum of simpler fractions. On occasions this means that we can then integrate a complicated algebraic fraction. We shall explore this approach in this block. A thorough understanding of the various forms that partial fractions can take is an essential prerequisite.

7.2 Integration using partial fractions

Sometimes expressions which at first sight look impossible to integrate using the techniques already met may in fact be integrated by first expressing them as simpler partial fractions, and then using the techniques described earlier in this chapter. Consider the following example.

Example 7.1

By expressing the integrand as the sum of its partial fractions find

$$\int \frac{2x + 11}{(x + 7)^2} \, dx$$

Solution

The denominator of the integrand contains a repeated linear factor $(x + 7)^2$. The appropriate form of partial fractions is

$$\frac{2x + 11}{(x + 7)^2} = \frac{A}{x + 7} + \frac{B}{(x + 7)^2}$$

Then

$$2x + 11 = A(x + 7) + B$$

from which, by letting $x = -7$, we find $-3 = B$. Equating coefficients of x, for example, gives $A = 2$. The integral becomes

$$\int \frac{2}{x + 7} - \frac{3}{(x + 7)^2} \, dx$$

The first integral is of the form $\int \dfrac{f'(x)}{f(x)} \, dx$ covered in Section 6.4. Thus

$$\int \frac{2}{x + 7} \, dx = 2 \ln|x + 7| + c$$

The second integral can be evaluated by making a substitution $u = x + 7, \dfrac{du}{dx} = 1$, to give

$$\int \frac{3}{(x + 7)^2} \, dx = \int \frac{3}{u^2} \, du$$

$$= -\frac{3}{u} + c$$

$$= -\frac{3}{x + 7} + c$$

Finally, putting both parts together we find

$$\int \frac{2x + 11}{(x + 7)^2} \, dx = 2 \ln|x + 7| + \frac{3}{x + 7} + K$$

Example 7.2

Express

$$\frac{23 - x}{(x - 5)(x + 4)}$$

as the sum of its partial fractions. Hence find

$$\int \frac{23 - x}{(x - 5)(x + 4)} \, dx$$

Solution

First produce the partial fractions (hint: write the fraction in the form $\dfrac{A}{x - 5} + \dfrac{B}{x + 4}$):

$$A = 2, B = -3$$

Then integrate each term separately.

$$2 \ln|x - 5| - 3 \ln|x + 4| + c$$

Exercises

By expressing the following in partial fractions evaluate the given integral. Remember to select the correct form for the partial fractions.

1    $\int \dfrac{1}{x^3 + x} \, dx$

2    $\int \dfrac{13x - 4}{6x^2 - x - 2} \, dx$

3 $\displaystyle\int \frac{1}{(x + 1)(x - 5)}\,dx$

4 $\displaystyle\int \frac{2x}{(x - 1)^2\,(x + 1)}\,dx$

5 $\displaystyle\int \frac{1}{x^2 - 2x - 1}\,dx$

Solutions to exercises

1 $\ln|x| - \frac{1}{2}\ln|x^2 + 1| + c$

2 $\frac{3}{2}\ln|2x + 1| + \frac{2}{3}\ln|3x - 2| + c$

3 $\frac{1}{6}\ln|x - 5| - \frac{1}{6}\ln|x + 1| + c$

4 $-\frac{1}{2}\ln|x + 1| + \frac{1}{2}\ln|x - 1| - \dfrac{1}{x - 1} + c$

5 $\dfrac{1}{2\sqrt{2}}\ln\left(\dfrac{x - 1 - \sqrt{2}}{x - 1 + \sqrt{2}}\right)$

End of block exercises

1 Find $\displaystyle\int \frac{3x - 17}{x^2 - 2x - 3}\,dx.$

2 Find $\displaystyle\int \frac{x + 3}{(x + 1)^2}\,dx.$

3 Find $\displaystyle\int \frac{t}{(t + 3)(t - 2)}\,dt.$

4 Find $\displaystyle\int \frac{3 - x}{(x - 2)^2\,(x + 1)}\,dx.$

5 Find $\displaystyle\int \frac{15x - 6}{(1 - 2x)(2 - x)}\,dx.$

6 Find $\displaystyle\int \frac{2}{(1 + t)(1 + 3t)}\,dt.$

7 Find $\displaystyle\int \frac{8 - x}{(x - 2)^2\,(x + 1)}\,dx.$

Solutions to exercises

1 $5\ln(x + 1) - 2\ln(x - 3) + c$

2 $\ln(x + 1) - \dfrac{2}{x + 1} + c$

3 $\dfrac{3}{5}\ln(t + 3) + \dfrac{2}{5}\ln(t - 2) + c$

4 $\dfrac{4}{9}\ln(x + 1) - \dfrac{4}{9}\ln(x - 2) - \dfrac{1}{3(x - 2)} + c$

5 $8\ln(x - 2) - \dfrac{\ln(2x - 1)}{2} + c$

6 $\ln(3t + 1) - \ln(t + 1) + c$

7 $\ln\dfrac{x + 1}{x - 2} - \dfrac{2}{x - 2} + c$

Integration of trigonometrical functions

8.1 Introduction

Integrals involving trigonometrical functions are commonplace in engineering mathematics. This is especially true when modelling waves, and alternating current circuits. When the root-mean-square (r.m.s.) value of a waveform or signal is to be calculated, you will often find this results in an integral of the form

$$\int \sin^2 t \, dt$$

In Fourier analysis it is necessary to find integrals such as

$$\int \sin mt \cos nt \, dt$$

In this block you will learn how such integrals can be evaluated.

Simple integrals involving trigonometrical functions have already been dealt with in Block 1. See what you can remember.

Example 8.1

Write down the following integrals:

(a) $\int \sin x \, dx$ (b) $\int \cos x \, dx$ (c) $\int \sin 2x \, dx$ (d) $\int \cos 2x \, dx$

Solution

(a) $-\cos x + c$

(b) $\sin x + c$

(c) $-\dfrac{\cos 2x}{2} + c$

(d) $\dfrac{\sin 2x}{2} + c$

The basic rules from which these results can be derived are summarised here:

Key point

$$\int \sin kx \, dx = -\frac{\cos kx}{k} + c, \quad \int \cos kx \, dx = \frac{\sin kx}{k} + c$$

8.2 The integration of more complicated trigonometrical functions

In engineering applications it is often necessary to integrate functions involving powers of the trigonometrical functions such as

$$\int \sin^2 x \, dx \quad \text{or} \quad \int \cos^2 \omega t \, dt$$

Note that these integrals cannot be obtained directly from the formulae in the key point above, which involve multiples of the variable x rather than powers of trigonometrical functions. However, by making use of trigonometrical identities the integrands can be rewritten in an alternative form. It is often not clear which identities are useful, and each case needs to be considered individually. Experience and practice are essential. Work through the following example.

 Example 8.2

Use the trigonometrical identity

$$\sin^2 \theta = \tfrac{1}{2} (1 - \cos 2\theta)$$

to express the integral $\int \sin^2 x \, dx$ in an alternative form.

Solution

$$\int \tfrac{1}{2} (1 - \cos 2x) \, dx$$

Note from the last example that the trigonometrical identity was used to convert a power of $\sin x$ into a function involving $\cos 2x$, which can be integrated directly.

 Example 8.3

Find the integral $\int \sin^2 x \, dx$.

Solution

$$\frac{1}{2} \left(x - \frac{\sin 2x}{2} + c \right) = \frac{1}{2} x - \frac{1}{4} \sin 2x + K \text{ where } K = \frac{c}{2}$$

 Example 8.4

Use the trigonometrical identity $\sin 2x = 2 \sin x \cos x$ to find

$$\int \sin x \cos x \, dx$$

Solution

Use the identity to rewrite the integrand.

$$\int \frac{1}{2} \sin 2x \, dx$$

Complete the integration.

$$\int \frac{1}{2} \sin 2x \, dx =$$

$$-\frac{1}{4} \cos 2x + c$$

Example 8.5

Find $\int 2 \sin 7x \cos 3x \, dx$.

Solution

In this example you will need to make use of the trigonometrical identity

$$2 \sin A \cos B = \sin(A + B) + \sin(A - B)$$

Use this identity to rewrite the integral:

$$\int \sin 10x + \sin 4x \, dx$$

Now complete the calculation:

$$-\frac{\cos 10x}{10} - \frac{\cos 4x}{4} + c$$

Example 8.6

Find $\int \sin^3 x \cos^3 x \, dx$.

Solution

To perform this calculation you will need to use a trigonometrical identity, and also use a substitution (see Block 6).

Make the substitution $u = \sin x$.

Write down $\dfrac{du}{dx}$.

$$\cos x$$

Write $\cos^3 x$ as $\cos^2 x \cos x$ and use the trigonometrical identity $\cos^2 x = 1 - \sin^2 x$.
Try to obtain an integral in terms of u.

$$\int u^3 - u^5 \, du$$

Complete the calculation:

$$\frac{\sin^4 x}{4} - \frac{\sin^6 x}{6} + c$$

If you have worked through the previous examples you will have realised that there is no single approach for tackling such integrals. Sometimes it is appropriate to make a substitution; sometimes a trigonometrical substitution is appropriate. You may need to integrate by parts. When faced with a particular integral to evaluate you may need to have several attempts and try different methods. Carrying out a range of practice exercises will help build your confidence.

Exercises

You will need to refer to a table of trigonometrical identities to answer these questions.

1   Find $\int \cos^2 x \, dx$.

2   Find $\int_0^{\pi/2} \cos^2 t \, dt$.

3   Find $\int (\cos^2 \theta + \sin^2 \theta) \, d\theta$.

4   Use the identity
$\sin(A + B) + \sin(A - B) = 2 \sin A \cos B$ to
find $\int \sin 3x \cos 2x \, dx$.

5   Find $\int (1 + \tan^2 x) \, dx$.

6   The mean square value of a function $f(t)$ over the interval $t = a$ to $t = b$ is defined to be

$$\frac{1}{b - a} \int_a^b (f(t))^2 \, dt$$

Find the mean square value of $f(t) = \sin t$ over the interval $t = 0$ to $t = 2\pi$.

7   Find $\int \sin mt \sin nt \, dt$ where m and n are constants and $m \neq n$.

8   Find $\int \dfrac{\sin^3 x}{\cos x} + \sin x \cos x \, dx$.

Solutions to exercises

1   $\frac{1}{2} x + \frac{1}{4} \sin 2x + c$

2   $\dfrac{\pi}{4}$

3   $\theta + c$

4   $-\dfrac{\cos 5x}{10} - \dfrac{\cos x}{2} + c$

5   $\tan x + c$

6   $\frac{1}{2}$

7   $\dfrac{1}{2} \left(\dfrac{\sin(m - n)t}{m - n} - \dfrac{\sin(m + n)t}{m + n} \right) + c$

8   $\int \tan x \, dx = \ln \sec x + c$

End of block exercises

1   Find $\int \sin^2 2x \, dx$.

2   Find $\int \sin 3x \cos 3x \, dx$.

3   Find $\int 5 \sin t \cos t \, dt$.

4   Find $\int \cos \omega t \sin \omega t \, dt$.

5 Find $\int_{-\pi}^{\pi} \sin^2 t \, dt$.

6 Find $\int_{-\pi}^{\pi} \cos^2 3t \, dt$.

7 Find $\int \cos 3x \cos 5x \, dx$.

8 Find $\int \sin t \sin 2t \, dt$.

9 Find $\int \sin^2 x + \cos^2 x \, dx$.

10 Find $\int \cos \pi \, dx$.

Solutions to exercises

1 $\dfrac{x}{2} - \dfrac{\sin 4x}{8} + c$

2 $-\dfrac{1}{12} \cos 6x + c$

3 $-\dfrac{5}{4} \cos 2t + c$

4 $-\dfrac{\cos 2\omega t}{4\omega} + c$

5 π

6 π

7 $\dfrac{\sin 8x}{16} + \dfrac{\sin 2x}{4} + c$

8 $\dfrac{\sin t}{2} - \dfrac{\sin 3t}{6} + c$

9 $x + c$

10 $(\cos \pi)x + c = -x + c$

End of chapter exercises

1 Find the following integrals using an appropriate technique selected from this chapter:

(a) $\displaystyle\int \dfrac{1}{4 + x^2} \, dx$

(b) $\int \cos^2 x \, dx$

(c) $\displaystyle\int \dfrac{1}{4 - t} \, dt$

(d) $\int_0^{\pi} \cos t \, dt$

(e) $\int_1^{1.2} \tan x \, dx$

(f) $\displaystyle\int \dfrac{2x^2 + x + 2}{x^3 + x} \, dx$

(g) $\int \sin 3t \cos 3t \, dt$

(h) $\displaystyle\int \dfrac{1}{16 - x^2} \, dx$

(i) $\displaystyle\int \dfrac{3}{\sqrt{t^2 + 0.25}} \, dt$

(j) $\int_{-1}^{1} \cosh 3t \, dt$

(k) $\displaystyle\int_1^3 \dfrac{1}{1 + x^2} \, dx$

(l) $\displaystyle\int \dfrac{1}{\sqrt{100 - 4x^2}} \, dx$

(m) $\displaystyle\int x \cos \dfrac{x}{2} \, dx$

(n) $\int 3^x \, dx$

(o) $\int \sec t \, dt$

(p) $\int \sin 5x \cos 2x \, dx$

(q) $\displaystyle\int \dfrac{1}{x(x - 3)} \, dx$

(r) $\int \dfrac{1}{t^2(t-1)}\, dt$

(s) $\int \dfrac{\ln x}{x}\, dx$

(t) $\int t\sqrt{3t-2}\, dt$

(u) $\int \dfrac{1}{5+4\cos x}\, dx$

(v) $\int_0^2 t^2\, e^t\, dt$

(w) $\int \cos x\sqrt{\sin x}\, dx$

(x) $\int \dfrac{x^2}{x^3+1}\, dx$

(y) $\int \dfrac{8t+10}{4t^2+8t+3}\, dt$

(z) $\int \dfrac{7}{(e^t)^7}\, dt$

2 By making the substitution $u^2 = x$ find

$$\int \sin\sqrt{x}\, dx.$$

3 Use the substitution $t = \dfrac{\pi}{4} - u$ to show that

$$\int_0^{\pi/4} \ln(1+\tan t)\, dt = \frac{1}{8}\pi \ln 2.$$

4 Find $\displaystyle\int_0^1 \dfrac{1+2x}{1+x^2}\, dx$.

5 Find $\displaystyle\int \dfrac{dx}{(1-x)\sqrt{x}}$.

6 By means of the substitution $x = \sin^2\theta$ find

$$\int \sqrt{\dfrac{x}{1-x}}\, dx.$$

7 Find $\displaystyle\int \sqrt{\dfrac{x}{x-2}}\, dx$.

8 Find $\displaystyle\int_0^{\pi/4} \sin^2 t\, dt$.

9 Use the substitution $u = e^x - 1$ to find

$$\int \dfrac{e^{2x}}{e^x-1}\, dx.$$

10 Find $\displaystyle\int_0^{0.3} \dfrac{1}{x^2-3x+2}\, dx$.

11 Use integration by parts to find $\int \sec^3 x\, dx$.

12 If $I_n = \int x^n\, e^{2x}\, dx$ show that

$$I_n = \dfrac{x^n e^{2x}}{2} - \dfrac{n}{2} I_{n-1}.$$

13 Find $\displaystyle\int \dfrac{2x+7}{x^2+7x}\, dx$.

14 Find $\displaystyle\int \dfrac{\cos t - \sin t}{\sin t + \cos t}\, dt$.

15 Find $\displaystyle\int \dfrac{4}{3-x}\, dx$.

16 Find $\displaystyle\int \dfrac{1}{s^2-2s+5}\, ds$.

17 Find $\int_0^\infty e^{-2t}\, dt$.

18 Find $\int_0^\infty e^{-x}\cos x\, dx$.

19 Find $\displaystyle\int \dfrac{1}{x^2-4x+3}\, dx$.

20 **Mechanical Engineering – Distance travelled by a projectile.** The velocity, $v(t)$, in m s^{-1}, of a projectile is given by

$$v(t) = 10e^{-t} \qquad t \geq 0$$

(a) Calculate the distance travelled in the first 3 seconds.
(b) Calculate the average speed over the first 3 seconds.

21 **Thermodynamics.** The pressure, P, and volume, V, of a gas are connected by

$$PV^{0.9} = 10^6$$

Calculate the work done by the gas as it expands in volume from 100 to 1000.

Solutions to exercises

1 (a) $\frac{1}{2}\tan^{-1}\left(\dfrac{x}{2}\right) + c$

(b) $\dfrac{1}{2}\left(x + \dfrac{\sin 2x}{2}\right) + c$ (c) $-\ln(4 - t) + c$

(d) 0 (e) 0.4 (f) $\tan^{-1}x + 2\ln x + c$

(g) $-\dfrac{\cos 6t}{12} + c$ (h) $\dfrac{1}{8}\ln\dfrac{4 + x}{4 - x} + c$

(i) $3\sinh^{-1}2t + c$ (j) 6.679 (k) 0.464

(l) $\dfrac{1}{2}\sin^{-1}\dfrac{x}{5} + c$

(m) $2x\sin\left(\dfrac{x}{2}\right) + 4\cos\left(\dfrac{x}{2}\right) + c$

(n) $\dfrac{1}{\ln 3}\,3^x + c$ (o) $\ln|\sec x + \tan x| + c$

(p) $-\dfrac{\cos 7x}{14} - \dfrac{\cos 3x}{6} + c$

(q) $\dfrac{-1}{3}\ln x + \dfrac{1}{3}\ln(x - 3) + c$

(r) $\dfrac{1}{t} - \ln\dfrac{t}{t - 1} + c$ (s) $\dfrac{1}{2}(\ln x)^2 + c$

(t) $\dfrac{2(3t - 2)^{3/2}(9t + 4)}{135} + c$

(u) $\dfrac{2}{3}\tan^{-1}\left(\dfrac{\tan(x/2)}{3}\right) + c$

(v) 12.778 (w) $\dfrac{2}{3}(\sin x)^{3/2} + c$

(x) $\dfrac{1}{3}\ln(x^3 + 1) + c$

(y) $\dfrac{1}{2}\ln(2t + 3) + \dfrac{3}{2}\ln(2t + 1) + c$

(z) $-e^{-7t} + c$

2 $2\sin\sqrt{x} - 2\sqrt{x}\cos\sqrt{x} + c$

4 1.479

5 $-\ln\left(\dfrac{\sqrt{x} - 1}{\sqrt{x} + 1}\right) + c$

6 $(x - 1)\sqrt{\dfrac{x}{1 - x}} + \tan^{-1}\left(\sqrt{\dfrac{x}{1 - x}}\right) + c$

7 $\dfrac{2\sqrt{x - 2}\,(x + 4)}{3} + c$

8 $\dfrac{\pi}{8} - \dfrac{1}{4}$

9 $e^x + \ln(e^x - 1) + c$

10 0.194

11 $\frac{1}{2}\ln|\tan x + \sec x| + \frac{1}{2}\tan x\sec x + c$

13 $\ln(x^2 + 7x) + c$

14 $\ln(\sin t + \cos t) + c$

15 $-4\ln(3 - x) + c$

16 $\frac{1}{2}\tan^{-1}\left(\dfrac{s - 1}{2}\right) + c$

17 0.5

18 0.5

19 $\frac{1}{2}\ln(x - 3) - \frac{1}{2}\ln(x - 1) + c$

20 (a) 9.502 (b) 3.167

21 4.104×10^6

Applications of integration

Chapter **18**

In the previous chapter integration was introduced, and you were shown a wide variety of techniques for obtaining integrals. In this chapter you will meet several important applications of those techniques.

The first block is particularly important for it introduces integration as the **limit of a sum**. This involves obtaining a physical quantity by considering it to be made up of contributions from lots of small parts. The sum of these contributions is calculated using integration. An understanding of Block 1 is essential if you want to be able to apply the techniques to a wider range of problems than those described here.

Chapter 18 contents

Integration as the limit of a sum

1.1 Introduction

In Chapter 17, integration was introduced as the reverse of differentiation. A more rigorous treatment would show that integration is a process of adding or 'summation'. By viewing integration from this perspective it is possible to apply the techniques of integration to finding areas, volumes, centres of gravity and many other important physical quantities.

The content of this block is important because it is here that integration is defined properly. A thorough understanding of the process involved is essential if you need to apply the techniques to practical problems.

1.2 The limit of a sum

Consider the graph of the positive function $y(x)$ shown in Figure 1.1. Suppose we are interested in finding the area under the graph between $x = a$ and $x = b$.

Figure 1.1
The area under a curve.

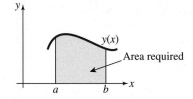

One way in which this area can be approximated is to divide it into a number of rectangles, find the area of each rectangle, and then add all these individual areas up. This is illustrated in Figure 1.2(a), which shows the area divided into n rectangles, and Figure 1.2(b), which shows the dimensions of a typical rectangle that is located at $x = x_k$.

Figure 1.2
(a) The area approximated by n rectangles; (b) a typical rectangle.

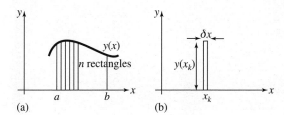

First, refer to Figure 1.2(a) and state the distance from a to b:

$$b - a$$

In Figure 1.2(a) the area has been divided into n rectangles. If n rectangles span the distance from a to b state the width of each rectangle:

$$\frac{b - a}{n}$$

It is conventional to label the width of each rectangle as δx. Suppose we label the x coordinates at the left-hand side of the rectangles as x_1, x_2 up to x_n. A typical rectangle, the kth rectangle, is shown in Figure 1.2(b). Note that its height is $y(x_k)$. Calculate its area:

$$y(x_k) \times \delta x$$

The sum of the areas is then

$$y(x_1)\delta x + y(x_2)\delta x + y(x_3)\delta x + \cdots + y(x_n)\delta x$$

which we write concisely using sigma notation (introduced in Chapter 5 Block 1) as

$$\sum_{k=1}^{n} y(x_k)\delta x$$

This, of course, gives us an estimate of the area under the curve, but it is not exact. To improve the estimate we must take a large number of very thin rectangles. So, what we want to find is the value of this sum when n tends to infinity and δx tends to zero. We write this as

$$\lim_{n \to \infty} \sum_{k=1}^{n} y(x_k)\delta x$$

The lower and upper limits on the sum correspond to the first and last rectangle where $x = a$ and $x = b$ respectively, and so we can write this limit in the equivalent form

$$\lim_{\delta x \to 0} \sum_{x=a}^{x=b} y(x)\delta x \tag{1}$$

If this sum can actually be found, it is called the **definite integral** of $y(x)$, from $x = a$ to $x = b$, and it is written $\int_a^b y(x)\,dx$. You are already familiar with the mechanism for finding definite integrals, which were studied in Chapter 17 Block 2.

Therefore we have the following definition:

Key point

The definite integral

$$\int_a^b y(x)\,dx$$

is defined as

$$\lim_{\delta x \to 0} \sum_{x=a}^{x=b} y(x)\delta x$$

Note that the quantity δx represents the thickness of a small but finite rectangle. When we have taken the limit as δx tends to zero to obtain the integral, we write $\mathrm{d}x$.

This process of dividing an area into very small regions, performing a calculation on each region, and then adding the results by means of an integral is very important. This will become apparent when finding volumes, centres of gravity, moments of inertia, etc., in the following blocks.

Example 1.1

The area under the graph of $y = x^2$ between $x = 0$ and $x = 1$ is to be found using the technique just described. If the required area is approximated by a large number of thin rectangles, the limit of the sum of their areas is given from equation (1) as

$$\lim_{\delta x \to 0} \sum_{x=0}^{x=1} y(x)\delta x$$

Write down the integral that this sum defines and evaluate it to obtain the area under the curve.

Solution

The limit of the sum defines the integral $\int_0^1 y(x)\,\mathrm{d}x$. Here $y = x^2$ and so

$$\int_0^1 x^2\,\mathrm{d}x = \left[\frac{x^3}{3}\right]_0^1$$

$$= \frac{1}{3}$$

To show that taking the limit of a sum actually works we have provided a detailed example for you to work through.

Example 1.2

Use the idea of the limit of a sum to find the area under the graph of $y = x^2$ between $x = 0$ and $x = 1$.

Solution

Refer to the graph in Figure 1.3 to help you answer the questions.

If the interval between $x = 0$ and $x = 1$ is divided into n rectangles what is the width of each rectangle?

Figure 1.3
The area under $y = x^2$ is approximated by a number of thin rectangles.

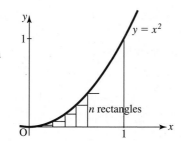

$$\frac{1}{n}$$

Mark this on Figure 1.3.
    What is the x coordinate at the left-hand side of the first rectangle?

$$0$$

What is the x coordinate at the left-hand side of the second rectangle?

$$\frac{1}{n}$$

What is the x coordinate at the left-hand side of the third rectangle?

$$\frac{2}{n}$$

Mark these coordinates on Figure 1.3.
    What is the x coordinate at the left-hand side of the kth rectangle?

$$\frac{k-1}{n}$$

Given that $y = x^2$, what is the y coordinate at the left-hand side of the kth rectangle?

$$\left(\frac{k-1}{n}\right)^2$$

The area of the kth rectangle is its height $\times$ its width. Write down the area of the kth rectangle:

$$\left(\frac{k-1}{n}\right)^2 \times \frac{1}{n} = \frac{(k-1)^2}{n^3}$$

To find the total area of the n rectangles we must add up all these individual areas:

$$\text{total area of rectangles} = \sum_{k=1}^{n} \frac{(k-1)^2}{n^3}$$

This sum can be simplified and then calculated as follows. You will need to make use of the formulae for the sum of the first k integers, and the sum of the squares of the first k integers. Details of these will be given in Chapter 19 Block 2. For completeness we state the relevant formulae now:

$$\sum_{k=1}^{n} 1 = n, \quad \sum_{k=1}^{n} k = \frac{n}{2}(n+1), \quad \sum_{k=1}^{n} k^2 = \frac{n}{6}(n+1)(2n+1)$$

Then, the total area of the rectangles is given by

$$\sum_{k=1}^{n} \frac{(k-1)^2}{n^3} = \frac{1}{n^3} \sum_{k=1}^{n} (k-1)^2$$

$$= \frac{1}{n^3} \sum_{k=1}^{n} k^2 - 2k + 1$$

$$= \frac{1}{n^3} \left(\sum_{k=1}^{n} k^2 - 2 \sum_{k=1}^{n} k + \sum_{k=1}^{n} 1 \right)$$

$$= \frac{1}{n^3} \left(\frac{n}{6} (n+1)(2n+1) - 2\frac{n}{2}(n+1) + n \right)$$

$$= \frac{1}{n^2} \left(\frac{(n+1)(2n+1)}{6} - (n+1) + 1 \right)$$

$$= \frac{1}{n^2} \left(\frac{(n+1)(2n+1)}{6} - n \right)$$

$$= \frac{1}{n^2} \left(\frac{(n+1)(2n+1)}{6} - \frac{6n}{6} \right)$$

$$= \frac{1}{6n^2} (2n^2 - 3n + 1)$$

$$= \frac{1}{3} - \frac{1}{2n} + \frac{1}{6n^2}$$

Note that this is a formula for the total area of the n rectangles. It is an estimate of the area under the graph of $y = x^2$. Now, as n gets larger, the terms $\frac{1}{2n}$ and $\frac{1}{6n^2}$ become small and will eventually tend to zero. If we now let n tend to infinity we obtain the exact answer:

$$\lim_{n \to \infty} \left(\frac{1}{3} - \frac{1}{2n} + \frac{1}{6n^2} \right) = \boxed{} \qquad \frac{1}{3}$$

The required area is $\frac{1}{3}$.

The area has been found as **the limit of a sum**.

In the calculations that follow in subsequent blocks the need to evaluate complicated limits like this is avoided by performing the integration using the techniques of Chapter 17. Nevertheless it will still be necessary to go through the process of dividing a region into small sections, performing a calculation on each section and then adding the results, in order to obtain the integral required.

Exercises

There are deliberately few exercises in this block because in practice integrals are evaluated using the techniques of Chapter 17 and not via taking the limits of sums explicitly. What is important though is an understanding of how the appropriate sum is formed.

1　Find the area under $y = x + 1$ from $x = 0$ to $x = 10$ using the limit of a sum.

2　Find the area under $y = 3x^2$ from $x = 0$ to $x = 2$ using the limit of a sum.

3 Write down, but do not evaluate, the integral defined by the limit as $\delta x \to 0$, or $\delta y \to 0$, of the following sums:

(a) $\sum_{x=0}^{x=1} x^3 \, \delta x$ (b) $\sum_{x=0}^{x=4} 4\pi x^2 \, \delta x$

(c) $\sum_{y=0}^{y=1} y^3 \, \delta y$ (d) $\sum_{x=0}^{x=1} 6mx^2 \, \delta x$

Solutions to exercises

1 60

2 8

3 (a) $\int_0^1 x^3 \, dx$ (b) $4\pi \int_0^4 x^2 \, dx$ (c) $\int_0^1 y^3 \, dy$

(d) $\int_0^1 6mx^2 \, dx$

End of block exercises

1 Use the method of the limit of a sum to find the area bounded by the curve $y = x^3$ and the x axis between $x = a$ and $x = b$. You may assume the result

$$\sum_{k=1}^{n} k^3 = \frac{n^2}{4}(n + 1)^2$$

2 Write down, but do not evaluate, the integral defined by the limit as $\delta x \to 0$, $\delta y \to 0$, or $\delta s \to 0$ of the following sums:

(a) $\sum_{x=1}^{x=5} \pi x^3 \, \delta x$ (b) $\sum_{x=0}^{x=a} \pi x^2 \, \delta x$

(c) $\sum_{y=0}^{y=2} m(2 - y)y\delta y$

(d) $\sum_{x=-b/2}^{x=b/2} mbx^2 \, \delta x$

(e) $\sum_{s=0}^{s=1} s^2 \, \delta s$

Solutions to exercises

1 $\dfrac{(b^4 - a^4)}{4}$

2 (a) $\int_1^5 \pi x^3 \, dx$ (b) $\int_0^a \pi x^2 \, dx$

(c) $\int_0^2 m(2 - y)y \, dy$ (d) $\int_{-b/2}^{b/2} mbx^2 \, dx$

(e) $\int_0^1 s^2 \, ds$

Volumes of revolution

2.1 Introduction

In this block we show how the concept of integration as the limit of a sum can be used to find volumes of solids formed when areas are rotated around the x or y axes.

2.2 Volumes generated by rotating areas about the x axis

Figure 2.1 shows a graph of the function $y = 2x$ for x between 0 and 3.

Figure 2.1
A graph of the function $y = 2x$, for $0 \leq x \leq 3$.

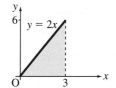

Imagine rotating the shaded area under $y = 2x$ a full revolution around the x axis. The body so formed is a cone as shown in Figure 2.2. Such a three-dimensional body is known as a **solid of revolution** or a **volume of revolution**.

Figure 2.2
When the area under $y = 2x$ is rotated around the axis, a solid is generated.

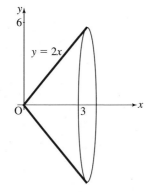

Example 2.1

Find the volume of the cone generated by rotating the area under $y = 2x$, for $0 \leq x \leq 3$, around the x axis, as shown in Figure 2.2.

Solution

In order to find the volume of this solid we assume that it is composed of lots of thin circular discs all aligned perpendicular to the x axis, such as that shown in Figure 2.3. From Figure 2.3 we note that a typical disc has radius y, which in this example equals $2x$, and thickness δx.

Figure 2.3
The cone is divided into a number of thin circular discs.

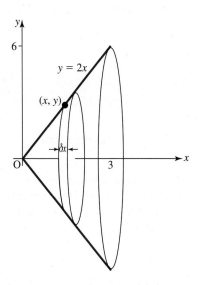

The volume of a circular disc is the circular area multiplied by the thickness. Write down an expression for the volume of this typical disc:

$$\pi(2x)^2 \, \delta x = 4\pi x^2 \, \delta x$$

To find the total volume we must sum the contributions from all discs and find the limit of this sum as the number of discs becomes infinite and δx becomes zero. That is,

$$\lim_{\delta x \to 0} \sum_{x=0}^{x=3} 4\pi x^2 \, \delta x$$

This is the definition of a definite integral. Write down the corresponding integral.

$$\int_0^3 4\pi x^2 \, dx$$

Find the required volume by performing the integration:

$$\left[\frac{4\pi x^3}{3} \right]_0^3 = 36\pi$$

Example 2.2

A graph of the function $y = x^2$ for x between 0 and 4 is shown in Figure 2.4. The area under the graph is rotated around the x axis to produce the solid shown. Find its volume.

Figure 2.4
The solid of revolution is divided into a number of thin circular discs.

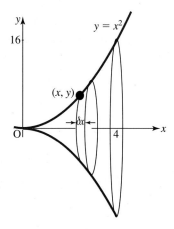

Solution

As in Example 2.1, the volume is considered to be composed of lots of circular discs of radius y, which in this example is equal to x^2, and thickness δx.

The volume of a circular disc is the circular area multiplied by the thickness.

Write down the volume of each disc:

$$\pi(x^2)^2\, \delta x = \pi x^4\, \delta x$$

Write down the expression that results by summing the volumes of all such discs:

$$\sum_{x=0}^{x=4} \pi x^4\, \delta x$$

Write down the integral that results from taking the limit of the sum as $\delta x \to 0$:

$$\int_0^4 \pi x^4\, dx$$

Perform the integration to find the volume of the solid:

$$\frac{4^5\,\pi}{5} = 204.8\pi$$

In general, suppose the area under the graph of $y(x)$ between $x = a$ and $x = b$ is rotated about the x axis, and the solid so formed is considered to be composed of lots of circular discs of thickness δx.

Write down an expression for the radius of a typical disc:

y

Write down an expression for the volume of a typical disc:

$$\pi y^2 \, \delta x$$

The total volume is found by summing these individual volumes and taking the limit as δx becomes zero:

$$\lim_{\delta x \to 0} \sum_{x=a}^{x=b} \pi y^2 \, \delta x$$

Write down the definite integral that this sum defines:

$$\int_a^b \pi y^2 \, dx$$

Key point

If the area under the graph of $y(x)$, between $x = a$ and $x = b$, is rotated about the x axis the volume of the solid formed is

$$\int_a^b \pi y^2 \, dx$$

Exercises

1   When the area under the graph of $y(x)$ between $x = a$ and $x = b$ is rotated around the x axis, show that the volume of the solid formed is $\int_a^b \pi y^2 \, dx$.

2   Find the volume of the solid formed when the area under $y = x^2$ between $x = 1$ and $x = 2$ is rotated about the x axis.

3   The area under the parabola $y^2 = 4x$ for $0 \leq x \leq 1$ is rotated around the x axis. Find the volume of the solid formed.

Solutions to exercises

2   $\dfrac{31\pi}{5}$

3   2π

2.3   Volumes generated by rotating areas about the y axis

We can obtain a different solid of revolution by rotating an area around the y axis instead of around the x axis. In Figure 2.5 the area between the graph of $y(x)$ and the y axis has been rotated about the y axis.

To find the volume of this solid it is divided into a number of circular discs as before, but this time the discs are horizontal. The radius of a typical disc is x and

Figure 2.5
A solid can be generated by rotation around the *y* axis.

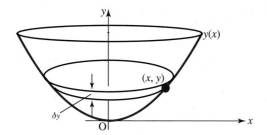

its thickness is δy. The volume of the disc will be $\pi x^2 \delta y$ where δy is the thickness of the disc.

The total volume is found by summing these individual volumes and taking the limit as $\delta y \to 0$. If the lower and upper limits on *y* are *c* and *d*, we find

$$\lim_{\delta y \to 0} \sum_{y=c}^{y=d} \pi x^2 \, \delta y$$

which is the definite integral

$$\int_c^d \pi x^2 \, dy$$

Key point

If the area between the graph of $y(x)$ and the *y* axis between $y = c$ and $y = d$ (Figure 2.6) is rotated about the *y* axis the volume of the solid formed is

$$\int_c^d \pi x^2 \, dy$$

Figure 2.6
Rotating the shaded area about the *y* axis produces a volume of revolution.

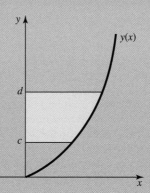

Example 2.3
Find the volume generated when the area between $y = x^2$ and the y axis between $y = 0$ and $y = 1$ is rotated around the y axis.

Solution
Write down the required integral.

$$\int_0^1 \pi x^2 \, dy$$

Because $y = x^2$ this integral can be written entirely in terms of y. Do this now, and then evaluate the integral.

$$\int_0^1 \pi y \, dy = \left[\frac{\pi y^2}{2} \right]_0^1 = \frac{\pi}{2}$$

Exercises

1   The area between $y = x^2$ and the y axis between $y = 1$ and $y = 4$ is rotated about the y axis. Find the volume of the solid formed.

2   The area between $y = 2 - 2x$ and the y axis between $y = -2$ and $y = 2$ is rotated around the y axis. Find the volume of revolution.

Solutions to exercises

1   $\dfrac{15\pi}{2}$

2   $\dfrac{16\pi}{3}$

End of block exercises

1   The area between the parabola $y^2 = 4ax$ and the x axis for $0 < x < a$ is rotated around the x axis. Show that the volume of the solid formed is $2\pi a^3$.

2   The equation of a circle with centre the origin and radius a is $x^2 + y^2 = a^2$. By considering that portion of the circle that lies in the first quadrant, and rotating this around the x axis, show that the volume of a hemisphere of radius a is $\dfrac{2\pi a^3}{3}$.

3   Find the volume obtained when the area between $y = \sin x$ and the x axis for $0 < x < \pi$ is rotated around the x axis.

4   The area under $y = \dfrac{1}{x}$ between $x = 1$ and $x = 3$ is rotated about the x axis. Find the volume of revolution.

5 Consider the area bounded by the curve $y = x^2 - 1$ between $x = 1$ and $x = 2$ and the x axis. If this area is rotated around the y axis show that the volume of the solid generated is $\frac{9\pi}{2}$.

6 The area between $y = 1 - \frac{1}{2}x$ and the x axis for $0 \le x \le 2$ is rotated around the y axis. Find the volume of the solid generated.

Solutions to exercises

3 $\dfrac{\pi^2}{2}$

4 $\dfrac{2\pi}{3}$

6 $\dfrac{4\pi}{3}$

Calculating centres of mass

3.1 Introduction

In this block we show how the idea of integration as the limit of a sum can be used to find the centre of mass of an object such as a thin plate, like a sheet of metal. Such a plate is also known as a *lamina*. An understanding of a 'moment' is necessary, and so this concept is introduced as well.

3.2 The centre of mass of a collection of point masses

Suppose we have a collection of masses located at a number of known points. The **centre of mass** is the point where, for many purposes, all the mass can be assumed to be located.

For example, if two objects each of mass m are placed at the locations $x = 1$ and $x = 2$, as shown in Figure 3.1(a), then the total mass, $2m$, can be assumed to be located at $x = 1.5$ as shown in Figure 3.1(b). This is the point where we could imagine placing a pivot to achieve a perfectly balanced system.

Figure 3.1
(b) The location of the centre of mass of the objects in (a).

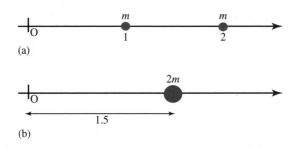

To think of this another way, if a pivot is placed at the origin, as on a see-saw, then the two masses at $x = 1$ and $x = 2$ together have the same **turning effect** or **moment** as a single mass $2m$ located at $x = 1.5$. This is illustrated in Figure 3.2.

Before we can calculate the position of the centre of mass of a collection of masses it is important to define moment more precisely.

Given a mass M located a distance d from O, as shown in Figure 3.3, its moment about O is defined to be

$$\text{moment} = Md$$

In words, the moment of the mass about O is the mass multiplied by its distance from O. The unit of moment will therefore be kg m when the mass is measured in kilograms and the distance is measured in metres.

Figure 3.2
The single object of mass 2m has the same turning effect as the two objects each of mass m.

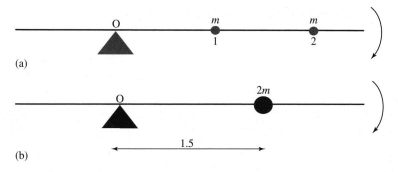

(a)

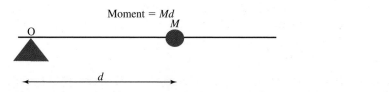

(b)

Figure 3.3
The moment of the mass M about O is Md.

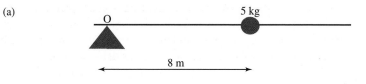

Moment = Md

Example 3.1 Mechanical Engineering – Moments

Calculate the moment of the mass about O in each of the cases shown in Figures 3.4 and 3.5.

Figure 3.4

(a)

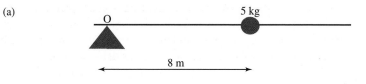

Figure 3.5

(b)

Solution

(a)   40 kg m

(b)   50 kg m

Intuition tells us that a large moment corresponds to a large turning effect. A mass placed 8 metres from the origin has a smaller turning effect than the same mass placed 10 metres from the origin. This is of course why it is easier to rock a see-saw by pushing it at a point further from the pivot.

For a collection of masses the moment of the total mass located at the centre of mass is equal to the sum of the moments of the individual masses. This result enables us to calculate the position of the centre of mass, as we shall see in the following examples. It is conventional to label the x coordinate of the centre of mass as $\bar{x}$, pronounced 'x bar'.

Key point	The moment of the total mass located at the centre of mass is equal to the sum of the moments of the individual masses.

Example 3.2 Mechanical Engineering – Centre of mass

Objects of mass m and $3m$ are placed at the locations shown in Figure 3.6. Find the distance $\bar{x}$ of the centre of mass from the origin O.

Figure 3.6

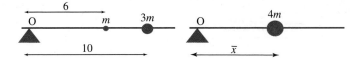

Solution

First calculate the sum of the individual moments:

$$6(m) + 10(3m) = 36m$$

The moment of the total mass about O is

$$(4m)\bar{x}$$

Now, the moment of the total mass is equal to the sum of the moments of the individual masses. Write down the equation satisfied by $\bar{x}$.

$$36m = 4m\bar{x}$$

Solve for $\bar{x}$.

$$\bar{x} = 9$$

So the centre of mass is located a distance 9 units along the x axis.

Example 3.3 Mechanical Engineering – Centre of mass

Obtain an equation for the location of the centre of mass of two objects of masses m_1 and m_2 located at $x = x_1$ and $x = x_2$ respectively, as shown in Figure 3.7.

Figure 3.7

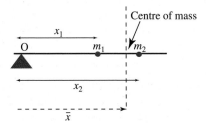

Solution
Referring to Figure 3.7, first write down an expression for the sum of the individual moments:

$$m_1x_1 + m_2x_2$$

Write down an expression for the total mass:

$$m_1 + m_2$$

Write down an expression for the moment of the total mass:

$$(m_1 + m_2)\bar{x}$$

The moment of the total mass is equal to the sum of the moments of the individual masses. Write down the equation satisfied by $\bar{x}$.

$$(m_1 + m_2)\bar{x} = m_1x_1 + m_2x_2$$

Finally write down an expression for $\bar{x}$.

$$\bar{x} = \frac{m_1x_1 + m_2x_2}{m_1 + m_2}$$

The formula obtained in the previous example can be generalised very easily to deal with the situation of n masses, $m_1, m_2, \ldots, m_n$, located at $x_1, x_2, \ldots, x_n$:

Key point

The centre of mass of $m_1, m_2, \ldots, m_n$ located at $x_1, x_2, \ldots, x_n$ is

$$\bar{x} = \frac{\displaystyle\sum_{i=1}^{n} m_i x_i}{\displaystyle\sum_{i=1}^{n} m_i}$$

Example 3.4 Mechanical Engineering – Centre of mass
Calculate the centre of mass of the four masses distributed as shown in Figure 3.8.

Figure 3.8

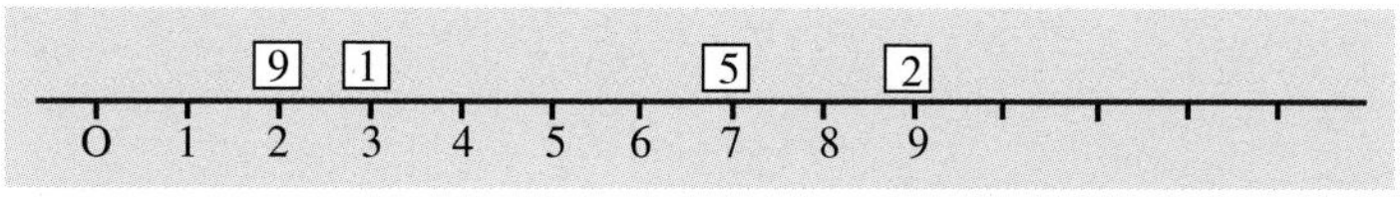

9 1   5   2
O 1 2 3 4 5 6 7 8 9

Solution

$$\bar{x} = \frac{\displaystyle\sum_{i=1}^{4} m_i x_i}{\displaystyle\sum_{i=1}^{4} m_i}$$

$$= \boxed{} \qquad \frac{(9)(2) + (1)(3) + (5)(7) + (2)(9)}{9 + 1 + 5 + 2}$$

$$= \boxed{} \qquad\qquad\qquad\qquad \frac{74}{17}$$

The centre of mass is located a distance $\frac{74}{17} \approx 4.35$ units along the x axis from O.

When n masses are distributed, not just along the x axis but anywhere in the xy plane (Figure 3.9), the coordinates of their centre of mass are given as follows:

Key point

The centre of mass of $m_1, m_2, \ldots, m_n$ located at $(x_1, y_1), (x_2, y_2), \ldots, (x_n, y_n)$ has coordinates

$$\bar{x} = \frac{\displaystyle\sum_{i=1}^{n} m_i x_i}{\displaystyle\sum_{i=1}^{n} m_i} , \bar{y} = \frac{\displaystyle\sum_{i=1}^{n} m_i y_i}{\displaystyle\sum_{i=1}^{n} m_i}$$

Figure 3.9
These masses
are distributed
throughout the
xy plane.

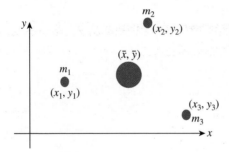

Example 3.5 Mechanical Engineering – Centre of mass

Masses of 5 kg, 3 kg and 9 kg are located at the points with coordinates $(-1, 1)$, $(4, 3)$ and $(8, 7)$ respectively. Find the coordinates of their centre of mass.

Solution
Here $n = 3$.

$$\bar{x} = \frac{\displaystyle\sum_{i=1}^{3} m_i x_i}{\displaystyle\sum_{i=1}^{3} m_i}$$

$$= \frac{5(-1) + 3(4) + 9(8)}{5 + 3 + 9}$$

$$= \frac{79}{17}$$

$$= 4.65$$

$$\bar{y} = \boxed{} \qquad \frac{5(1) + 3(3) + 9(7)}{17} = 4.53$$

Hence the centre of mass is located at the point (4.65, 4.53).

Exercises

1 Explain what is meant by the centre of mass of a collection of point masses.

2 Find the x coordinate of the centre of mass of five identical masses placed at $x = 2$, $x = 5$, $x = 7$, $x = 9$, $x = 12$.

3 Derive the formula for $\bar{y}$ given in the previous key point.

4 **Mechanical Engineering – Centre of mass.** Calculate the position of the centre of mass of 2 kg placed at $x = 1$, 3 kg placed at $x = 4$, 1 kg placed at $x = 6$ and 6 kg placed at $x = -5$.

5 **Mechanical Engineering – Centre of mass.** Masses of 3 kg, 2 kg, 2 kg and 4 kg are located at points with coordinates $(1, 3)$, $(-2, 0)$, $(4, -1)$ and $(-3, 4)$, respectively. Calculate the coordinates of the centre of mass.

Solutions to exercises

2 $\bar{x} = 7$

4 $\bar{x} = -\frac{5}{6}$

5 $\bar{x} = -\frac{5}{11}, \bar{y} = \frac{23}{11}$

3.3 Finding the centre of mass of a plane, uniform lamina

In the previous section you calculated the centre of mass of several individual point masses. Suppose we are interested in the centre of mass of a thin sheet of material, such as a plane sheet of metal. Such an object is called a **lamina**. The mass is not located at individual points. Rather, it is distributed continuously throughout the lamina. In what follows we assume that the mass is distributed uniformly throughout the lamina. In this section you will see how integration as the limit of a sum is used to find the centre of mass.

Figure 3.10 shows a lamina where the centre of mass has been marked at point G with coordinates $(\bar{x}, \bar{y})$. If the total mass of the lamina is M then the moments about

Figure 3.10
The centre of mass of the lamina is located at G$(\bar{x}, \bar{y})$.

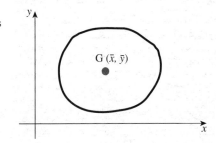

the y and x axes are respectively $M\bar{x}$ and $M\bar{y}$. Our approach to locating the position of G is to divide the lamina into many small pieces, find the mass of each piece and the moment of each piece about the axes. The sum of the moments of the individual pieces about the y axis will be equal to $M\bar{x}$. The sum of the moments of the individual pieces about the x axis will be equal to $M\bar{y}$.

There are no formulae that can be memorised for finding the centre of mass of a lamina because of the wide variety of possible shapes. Instead you should be familiar with the technique for deriving the centre of mass.

An important preliminary concept is 'mass per unit area.'

Mass per unit area

Suppose we have a uniform lamina and suppose we select a piece of the lamina that has area equal to 1 unit. Let m stand for the mass of such a piece. Then m is called the **mass per unit area**. The mass of any other piece can be expressed in terms of m. For example, an area of 2 units must have mass $2m$. An area of 3 units must have mass $3m$, and so on. Any other portion of the lamina that has area A has mass mA.

Key point	If a lamina has mass per unit area m, then the mass of part of the lamina having area A is Am.

We shall illustrate the calculation of centre of mass in the following examples.

Example 3.6 Mechanical Engineering – Centre of mass of a lamina
Consider the plane sheet, or lamina, in Figure 3.11. Find the location of its centre of mass.

Figure 3.11
By symmetry the centre of mass of this lamina lies on the x axis.

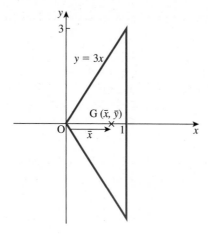

Solution
First inspect the figure and note the symmetry of the lamina. Purely from the symmetry, what must be the y coordinate, $\bar{y}$, of the centre of mass?

$\bar{y} = 0$ since the centre of mass must lie on the x axis

Let m stand for the mass per unit area of the sheet. The total area is 3 units. The total mass is therefore $3m$. Its moment about the y axis is $3m\bar{x}$ where $\bar{x}$ is the x coordinate of the centre of mass.

To find $\bar{x}$ first divide the sheet into a large number of thin vertical slices. In Figure 3.12 a typical slice has been highlighted. Note that the slice has been drawn from the point P on the graph of $y = 3x$. The point P has coordinates (x, y). The thickness of the slice is δx. This notation is consistent with that introduced in Block 1.

Figure 3.12
A typical slice of this sheet has been shaded.

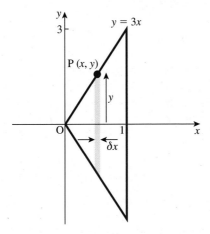

Assuming that the slice is rectangular in shape, write down its area.

$2y\delta x$

Writing m as the mass per unit area, write down the mass of the slice.

$(2y\delta x)m$

The centre of mass of this slice lies on the x axis. So the slice can be assumed to be a point mass, $2ym\delta x$, located a distance x from O.

Write down the moment of the mass of the slice about the y axis:

$(2y\delta x)mx$

By adding up contributions from all such slices in the lamina we obtain the sum of the moments of the individual masses:

$$\sum_{x=0}^{x=1} 2mxy\delta x$$

The limits on the sum are chosen so that all slices are included.

Write down the integral defined by letting $\delta x \rightarrow 0$:

$$\int_{x=0}^{x=1} 2mxy \, dx$$

Note that this represents the sum of all the individual moments taken throughout the entire lamina. Noting that $y = 3x$ express the integral in terms of x and evaluate it:

$$\int_0^1 6mx^2 \, dx = [2mx^3]_0^1 = 2m$$

This must equal the moment of the total mass acting at the centre of mass.
Hence

$$3m\bar{x} = 2m$$

from which

$$\bar{x} = \frac{2}{3}$$

Finally the coordinates of the centre of mass are $\left(\frac{2}{3}, 0\right)$.

Example 3.7 Mechanical Engineering – Centre of mass of a lamina
Find the centre of mass of the plane lamina shown in Figure 3.13.

Figure 3.13
We assume that the shaded slice is rectangular.

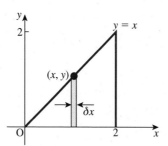

Solution
The coordinates of $\bar{x}$ and $\bar{y}$ must be calculated separately.
To calculate $\bar{x}$
Let m equal the mass per unit area. Write down the total area, the total mass and its moment about the y axis:

$$2, 2m, 2m\bar{x}$$

To calculate $\bar{x}$ the lamina is divided into thin slices; a typical slice is shown in Figure 3.13. Assume that it is rectangular.
   Write down the height of the typical strip shown in Figure 3.13.

$$y$$

Write down the area of the typical strip.

$$y\delta x$$

Write down the mass of the typical strip.

$$(y\delta x)m$$

Write down the moment about the y axis of the typical strip.

$(y\delta x)mx$

The sum of the moments of all strips is

$$\sum_{x=0}^{x=2} mxy\delta x$$

Write down the integral that follows as $\delta x \to 0$.

$$\int_0^2 mxy \, dx$$

In this example, $y = x$. Substitute this for y in the integral, and evaluate it.

$$\int_0^2 mx^2 \, dx = \frac{8}{3} m$$

Equating the sum of individual moments and the total moment gives

$$2m\bar{x} = \frac{8}{3} m$$

from which

$$\bar{x} = \frac{4}{3}$$

To calculate $\bar{y}$
We will illustrate two alternative ways of calculating $\bar{y}$.

Referring to Figure 3.13, the centre of mass of the slice must lie half-way along its length: that is, its y coordinate is $\frac{y}{2}$. Assume that all the mass of the slice, $ym\delta x$, acts at this point. Then its moment about the x axis is $ym\delta x \frac{y}{2}$. Adding contributions from all slices gives the sum

$$\sum_{x=0}^{x=2} \frac{y^2 m}{2} \delta x$$

Write down the integral which is defined as $\delta x \to 0$.

$$\int_0^2 \frac{my^2}{2} \, dx$$

which we can write as

$$m\int_0^2 \frac{y^2}{2} \, dx$$

In this example $y = x$, so the integral becomes

$$m \int_0^2 \frac{x^2}{2} \, dx$$

which equals $\dfrac{4m}{3}$. This is the sum of the individual moments about the x axis and must equal the moment of the total mass about the x axis which is $2m\bar{y}$. Therefore

$$2m\bar{y} = \frac{4m}{3}$$

from which

$$\bar{y} = \frac{2}{3}$$

Finally, the coordinates of the centre of mass are $\left(\dfrac{4}{3}, \dfrac{2}{3} \right)$.

Consider now an alternative way of finding $\bar{y}$.

This time the lamina is divided into a number of horizontal slices; a typical slice is shown in Figure 3.14.

The length of the typical slice shown is $2 - x$.

Figure 3.14
A typical horizontal slice is shaded.

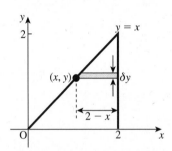

Write down its area, its mass and its moment *about the x axis*

$$(2 - x)\delta y, \; m(2 - x)\delta y, \; m(2 - x)y\delta y$$

Write down the expression for the sum of all such moments and the corresponding integral as $\delta y \to 0$.

$$\sum_{y=0}^{y=2} m(2 - x)y\delta y, \quad \int_0^2 m(2 - x)y \, dy$$

Now, since $y = x$ the integral can be written entirely in terms of y as

$$\int_0^2 m(2 - y)y \, dy$$

Evaluate the integral:

$$\frac{4m}{3}$$

As before the total mass is $2m$, and its moment about the x axis is $2m\bar{y}$. Hence

$$2m\bar{y} = \frac{4m}{3}$$

from which

$$\bar{y} = \frac{2}{3}$$

as before.

Example 3.8 Mechanical Engineering – Centre of mass of a lamina
Find the position of the centre of mass of a uniform semicircular lamina of radius a.

Solution
The lamina is shown in Figure 3.15. The equation of a circle with centre the origin, radius a, is $x^2 + y^2 = a^2$. By symmetry $\bar{x}$ must lie on the y axis. However, it is necessary to calculate $\bar{y}$.

Figure 3.15
A typical
horizontal strip
is shaded.

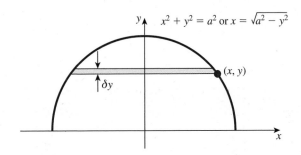

The lamina is divided into a number of horizontal strips, and a typical strip is shown. Assume that each strip is rectangular. Writing the mass per unit area as m, identify the area and the mass of the strip:

$2x\delta y, 2xm\delta y$

Write down the moment of the mass about the x axis:

$2xmy\delta y$

Write down the expression representing the sum of the moments of all strips and the corresponding integral obtained as $\delta y \to 0$.

$$\sum_{y=0}^{y=a} 2xmy\delta y, \int_0^a 2xmy \, dy$$

Now since $x = \sqrt{a^2 - y^2}$ the integral becomes

$$\int_0^a 2my\sqrt{a^2 - y^2} \, dy$$

Evaluate this integral by making the substitution $u^2 = a^2 - y^2$ to obtain the total moment.

$$\frac{2ma^3}{3}$$

The total area is half that of a circle of radius a, that is $\frac{1}{2}\pi a^2$. The total mass is $\frac{1}{2}\pi a^2 m$. Its moment is $\frac{1}{2}\pi a^2 m\bar{y}$.

Hence

$$\frac{1}{2}\pi a^2 m\bar{y} = \frac{2ma^3}{3}$$

from which

$$\bar{y} = \frac{4a}{3\pi}$$

Exercise

1   Find the centre of mass of a lamina bounded by $y^2 = 4x$ and $x = 9$.

Solution to exercise

1   $\left(\dfrac{27}{5}, 0\right)$

End of block exercises

1   Find the position of the centre of mass of a lamina in the shape of a triangle formed by $y = 1 - x$, for $0 < x < 1$, and the x and y axes.

2   Show that the distance of the centre of mass of a uniform solid right circular cone, of height h, from its plane base is $\dfrac{h}{4}$. (Hint: you will need to introduce a mass per unit volume, or density, and divide the cone into a large number of discs.)

3   Show that the centre of mass of a uniform triangular lamina ABC is at a distance $\frac{1}{3}h$ from BC where h is the perpendicular distance from A to BC.

Solutions to exercises

1   $\left(\dfrac{1}{3}, \dfrac{1}{3}\right)$

Moment of inertia

4.1 Introduction

Figure 4.1 shows a lamina, or plane sheet, which is allowed to rotate about an axis perpendicular to the plane of the lamina and through O. The **moment of inertia** about this axis is a measure of how difficult it is to rotate the lamina. It plays the same role when dealing with rotating bodies as the mass of an object plays when dealing with motion in a line. An object with large mass needs a large force to achieve a given acceleration. Similarly, an object with large moment of inertia needs a large turning force to achieve a given angular acceleration. A knowledge of the moments of inertia of laminae, and also of solid bodies, is essential for understanding their rotational dynamics.

Figure 4.1
A lamina rotating about an axis through O.

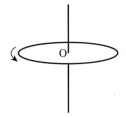

In this block we show how the idea of integration as the limit of a sum can be used to find the moment of inertia of a lamina.

4.2 Calculating the moment of inertia

Suppose a lamina is divided into a large number of small pieces or *elements*. A typical piece is shown in Figure 4.2.

Figure 4.2
The moment of inertia of the small element is $\delta m r^2$.

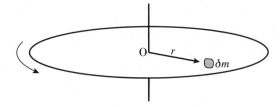

It has mass δm, and is located a distance r from the axis O. The moment of inertia of this small piece is defined to be $\delta m r^2$, that is the mass multiplied by the square of its distance from the axis of rotation. To find the total moment of inertia we sum the individual contributions to give

$$\sum r^2 \delta m$$

where the sum must be taken in such a way that all parts of the lamina are included. As $\delta m \to 0$ we obtain the following integral as the definition of moment of inertia, I:

Key point

moment of inertia $I = \displaystyle\int r^2 \, dm$

where the limits of integration are chosen so that the entire lamina is included.

The unit of moment of inertia is kg m^2.

We shall illustrate how the moment of inertia is actually calculated in practice, in the following examples.

Example 4.1 Mechanical Engineering – Moment of inertia of a square

Calculate the moment of inertia about the y axis of the square lamina of mass M and width b, shown in Figure 4.3. The moment of inertia about the y axis is a measure of the resistance to rotation around this axis.

Figure 4.3
A square lamina rotating about the y axis.

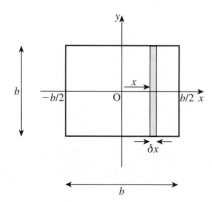

Solution

Let the mass per unit area of the lamina be m. Then, because its total area is b^2, its total mass is $b^2 m$, which must equal the given mass, M.

Imagine that the lamina has been divided into a large number of thin vertical strips. A typical strip is shown in Figure 4.3. The strips are chosen in this way because each point on a particular strip is approximately the same distance from the axis of rotation (the y axis).

Referring to Figure 4.3 write down the width of each strip.

δx

Write down the area of the strip.

$$b\delta x$$

With m as the mass per unit area write down the mass of the strip.

$$mb\delta x$$

The distance of the strip from the y axis is x. Write down its moment of inertia.

$$mbx^2\delta x$$

Adding contributions from all strips gives the expression $\sum mbx^2\delta x$, where the sum must be such that the entire lamina is included. As $\delta x \to 0$ the sum defines an integral. Write down this integral.

$$I = \int_{-b/2}^{b/2} mbx^2 \, dx$$

Note that the limits on the integral have been chosen so that the whole lamina is included. Then

$$I = mb \int_{-b/2}^{b/2} x^2 \, dx$$

$$= \qquad\qquad mb\left[\frac{x^3}{3}\right]_{-b/2}^{b/2}$$

$$= \qquad\qquad \frac{mb^4}{12}$$

Noting that $M = b^2m$ then we can write I as $\dfrac{Mb^2}{12}$.

Example 4.2 Mechanical Engineering – Moment of inertia of a circular disc

Find the moment of inertia of a circular disc of mass M and radius a about an axis passing through its centre and perpendicular to the disc.

Solution

Figure 4.4 shows the disc lying in the plane of the paper. Imagine that the axis of rotation is coming out of the paper through O and is perpendicular to the disc. The disc can be considered to be spinning in the plane of the paper. Because of the circular symmetry the disc is divided into concentric rings of width δr. A typical ring is shown in Figure 4.5. Note that each point on the ring is approximately the same distance from the axis of rotation.

The ring has radius r and inner circumference $2\pi r$. Imagine cutting the ring and opening it up. Its area will be approximately that of a long thin rectangle of length $2\pi r$ and width δr. If m is the mass per unit area write down an expression for the mass of the ring.

$$2\pi rm\delta r$$

Figure 4.4
A circular disc
rotating about an
axis through O.

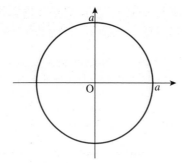

Figure 4.5
The lamina is
divided into many
circular rings.

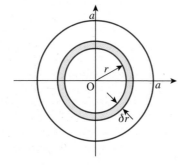

The moment of inertia of the ring about O is its mass multiplied by the square of its distance from the axis of rotation. This is $(2\pi rm\delta r) \times r^2 = 2\pi r^3 m\delta r$.

The contribution from all rings must be summed. This gives rise to the sum

$$\sum_{r=0}^{r=a} 2\pi r^3 m\delta r$$

Note the way that the limits have been chosen so that all rings are included in the sum. As $\delta r \to 0$ the limit of the sum defines the integral

$$\int_0^a 2m\pi r^3 \, dr$$

Evaluate this integral to give the moment of inertia I.

$$I = \left[\frac{2m\pi r^4}{4}\right]_0^a = \frac{m\pi a^4}{2}$$

Write down the radius and area of the whole disc.

$a, \pi a^2$

With m as the mass per unit area, write down the mass of the disc.

$\pi a^2 m$

But this total mass is M. Hence I can be written $\dfrac{Ma^2}{2}$.

Exercises

1. The moment of inertia about a diameter of a sphere of radius 1 m and mass 1 kg is found by evaluating the integral

$$\frac{3}{8} \int_{-1}^{1} (1 - x^2)^2 \, dx$$

Show that the moment of inertia of the sphere is $\frac{2}{5}$ kg m^2.

2. Find the moment of inertia of the lamina in Figure 4.3 about one of its sides.

3. Calculate the moment of inertia of a uniform thin rod of mass M and length l about a perpendicular axis of rotation at its end.

4. Calculate the moment of inertia of the rod in question 3 about an axis through its centre and perpendicular to the rod.

5. The **parallel axis theorem** states that the moment of inertia about any axis is equal to the moment of inertia about a parallel axis through the centre of mass, plus the mass of the body $\times$ the square of the distance between the two axes. Verify this theorem for the rod in questions 3 and 4.

6. The **perpendicular axis theorem** applies to a lamina lying in the xy plane. It states that the moment of inertia of the lamina about the z axis is equal to the sum of the moments of inertia about the x and y axes. Suppose a thin circular disc of mass M and radius a lies in the xy plane and the z axis passes through its centre. The moment of inertia of the disc about this axis is $\frac{1}{2}Ma^2$.
   (a) Use this theorem to find the moment of inertia of the disc about the x and y axes.
   (b) Use the parallel axis theorem to find the moment of inertia of the disc about a tangential axis parallel to the plane of the disc.

Solutions to exercises

2. $\dfrac{Mb^2}{3}$

3. $\dfrac{1}{3}Ml^2$

4. $\dfrac{1}{12}Ml^2$

6. (a) The moments of inertia about the x and y axes must be the same by symmetry, and are equal to $\frac{1}{4}Ma^2$.

   (b) $\dfrac{5Ma^2}{4}$

End of block exercises

1. A lamina of mass M has profile $y = 8 - x^2$ for $-2\sqrt{2} < x < 2\sqrt{2}$. Find the moment of inertia of the lamina about the y axis.

2. A uniform rectangular lamina of mass M is defined by the region $-a < x < a$ and $-b < y < b$. It is allowed to rotate about the y axis. Show that its moment of inertia is $\frac{1}{3}Ma^2$.

3. If the lamina in question 2 is allowed to rotate about the x axis, show that its moment of inertia is $\frac{1}{3}Mb^2$.

4   If the lamina in question 2 rotates about the line $x = -a$ show that the relevant moment of inertia is $\frac{4}{3}Ma^2$.

5   Prove that the moment of inertia of a uniform ring of mass M and radius a, about an axis through its centre and perpendicular to its plane, is Ma^2.

6   Use the parallel axis theorem given in the previous exercises to show that the moment of inertia of a thin circular disc of mass M and radius a, about an axis perpendicular to the plane of the disc and passing through a point on the circumference, is $\frac{3}{2}Ma^2$.

7   When a body rotates about a fixed axis with angular speed ω radians per second, its kinetic energy, KE, is given by KE $= \frac{1}{2}I\omega^2$, where I is the moment of inertia of the body about the fixed axis. Show that the kinetic energy of a rod of length l rotating about an axis through its midpoint is $\frac{1}{24}Ml^2\omega^2$.

8   A lamina has a profile in the first quadrant given by $y^2 = 4ax$, for $0 \le x \le b$.
    (a) Show that the area of the lamina is $\frac{4}{3}a^{1/2}b^{3/2}$.
    (b) Show that its moment of inertia about the y axis is $\frac{3}{7}Mb^2$.
    (c) Show that its moment of inertia about the x axis is $\frac{4}{5}Mab$.

Solution to exercise

1   $\dfrac{8M}{5}$

The length of a curve and the area of a surface of revolution

5.1 Introduction

Integration can be used to find the length of a curve and the area of the surface generated when a curve is rotated around an axis. In this block we state and use the formulae for doing this.

5.2 The length of a curve

Given a curve with equation $y = f(x)$, then the length of the curve between the points where $x = a$ and $x = b$ is given by the formula

Key point

$$\text{length of curve} = \int_a^b \sqrt{1 + \left(\frac{dy}{dx}\right)^2}\, dx$$

The proof of this formula is left as an exercise (question 1).

Because of the form of the integrand, and in particular the square root, integrals of this type are often difficult to calculate and, in practice, approximate rather than exact methods are needed to perform the integration. We shall illustrate the application of the formula by an example which could be calculated in a much simpler way, before looking at some harder examples.

Example 5.1
Find the length of the curve $y = 3x + 2$ between $x = 1$ and $x = 5$.

Solution
In this example, the curve is in fact a straight line, although we shall proceed as though it had been more complicated.

Notice from the formula in the key point that it is necessary to find $\frac{dy}{dx}$. Do this first.

$$\frac{dy}{dx} = 3$$

Applying the formula we find

$$\text{length of curve} = \int_1^5 \sqrt{1 + (3)^2}\,dx$$

$$= \int_1^5 \sqrt{10}\,dx$$

$$= \left[\sqrt{10}\,x\right]_1^5$$

$$= (5 - 1)\sqrt{10}$$

$$= 4\sqrt{10}$$

$$= 12.65$$

Thus the length of the curve $y = 3x + 2$ between the points where $x = 1$ and $x = 5$ is 12.65 units.

Example 5.2 Mechanical Engineering – Length of a curve

Find the length of the catenary $y = \cosh x$ between $x = 0$ and $x = 2$.

Solution

First find $\dfrac{dy}{dx}$.

$$\frac{dy}{dx} = \sinh x$$

Hence write down the required integral:

$$\int_0^2 \sqrt{1 + \sinh^2 x}\,dx$$

This integral can be evaluated by making use of the hyperbolic identity

$$\cosh^2 x - \sinh^2 x = 1$$

Write down the integral that results after applying this identity.

$$\int_0^2 \cosh x\,dx$$

Perform the integration for yourself to find the required length.

$$[\sinh x]_0^2 = 3.63$$

Thus the length of $y = \cosh x$ between $x = 0$ and $x = 2$ is 3.63 units.

The final example is more complicated still and requires the use of a hyperbolic substitution and knowledge of the hyperbolic identities.

Example 5.3

Find the length of the curve $y = x^2$ between $x = 0$ and $x = 3$.

Solution

Given $y = x^2$ then $\dfrac{dy}{dx} = 2x$. Apply the formula to obtain the integral required:

$$\int_0^3 \sqrt{1 + 4x^2}\, dx$$

Make the substitution $x = \dfrac{1}{2} \sinh u, \dfrac{dx}{du} = \dfrac{1}{2} \cosh u$, to obtain an integral in terms of u.

$$\int_0^{\sinh^{-1} 6} \sqrt{1 + \sinh^2 u}\; \frac{1}{2} \cosh u\, du$$

Use the hyperbolic identity $\cosh^2 u - \sinh^2 u = 1$ to rewrite this integral:

$$\frac{1}{2} \int_0^{\sinh^{-1} 6} \cosh^2 u\, du$$

Another hyperbolic identity is $\cosh^2 u = \dfrac{1}{2}(\cosh 2u + 1)$. Apply this identity to rewrite the integrand.

$$\frac{1}{4} \int_0^{\sinh^{-1} 6} \cosh 2u + 1\, du$$

Finally, performing the integration we can now complete the calculation:

$$\frac{1}{4} \int_0^{\sinh^{-1} 6} \cosh 2u + 1\, du = \frac{1}{4}\left[\frac{\sinh 2u}{2} + u \right]_0^{\sinh^{-1} 6}$$

$$= 9.75$$

Thus the length of the curve $y = x^2$ between $x = 0$ and $x = 3$ is 9.75 units.

Exercises

1   Figure 5.1 shows the portion of the curve $y(x)$ between $x = a$ and $x = b$. A small piece of this curve has been selected and can be considered as the hypotenuse of a triangle with base δx and height δy.

    (a) Use Pythagoras's theorem to find the length of the hypotenuse.

    (b) By summing all such contributions between $x = a$ and $x = b$, and letting $\delta x \to 0$, obtain the integral expression for the total length of the curve.

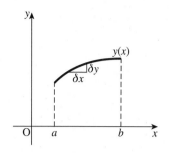

Figure 5.1

2   Find the length of the line $y = 2x + 7$ between $x = 1$ and $x = 3$ using the technique of this section. Verify your result from your knowledge of the straight line.

3   Find the length of $y = x^{3/2}$ between $x = 0$ and $x = 5$.

4   Calculate the length of the curve $y^2 = 4x^3$ between $x = 0$ and $x = 2$.

Solutions to exercises

1   (a) $\sqrt{(\delta x)^2 + (\delta y)^2}$

(b) $\sum_{x=a}^{x=b} \sqrt{(\delta x)^2 + (\delta y)^2}$ which, by taking out a factor of $(\delta x)^2$ from under the square root, gives

$$\sum_{x=a}^{x=b} \sqrt{1 + \left(\frac{\delta y}{\delta x}\right)^2}\, \delta x$$

and then letting $\delta x \to 0$ defines the integral

$$\int_a^b \sqrt{1 + \left(\frac{dy}{dx}\right)^2}\, dx$$

2   $2\sqrt{5} = 4.47$

3   12.41

4   6.06 (first quadrant only)

5.3   The area of a surface of revolution

Given a curve with equation $y = f(x)$, then the area of the surface generated by rotating that part of the curve between the points where $x = a$ and $x = b$ around the x axis is given by the formula

Key point

$$\text{area of surface} = \int_a^b 2\pi y \sqrt{1 + \left(\frac{dy}{dx}\right)^2}\, dx$$

The proof of this formula is left as an exercise (question 1).

Example 5.4
Find the area of the surface generated when the part of the curve $y = x^3$ between $x = 0$ and $x = 4$ is rotated around the x axis.

Solution

The area of surface is given by

$$\text{area} = \int_a^b 2\pi y \sqrt{1 + \left(\frac{dy}{dx}\right)^2} \, dx$$

$$= \int_0^4 2\pi x^3 \sqrt{1 + (3x^2)^2} \, dx$$

$$= \int_0^4 2\pi x^3 \sqrt{1 + 9x^4} \, dx$$

This integral can be found by making a substitution $u = 1 + 9x^4$, $\dfrac{du}{dx} = 36x^3$ so that $x^3 \, dx = \dfrac{du}{36}$. When $x = 0$, $u = 1$ and when $x = 4$, $u = 2305$.

Write down the corresponding integral in terms of u.

$$\frac{\pi}{18} \int_1^{2305} \sqrt{u} \, du$$

Perform the integration.

$$\frac{\pi}{18} \left[\frac{2}{3} u^{3/2} \right]_1^{2305}$$

Apply the limits of integration to find the area.

$$\frac{\pi}{27} [(2305)^{3/2} - 1]$$

Exercises

1   Figure 5.2 shows the portion of the curve $y(x)$ between $x = a$ and $x = b$, which is rotated around the x axis through 360°. A small disc of the solid of revolution has been selected. Its radius is y and so its circumference has length $2\pi y$. The surface area of this disc is then approximately $2\pi y \sqrt{(\delta x)^2 + (\delta y)^2}$. Obtain the formula for the total surface area.

2   The line $y = x$ between $x = 0$ and $x = 1$ is rotated around the x axis.
   (a) Find the area of the surface generated.

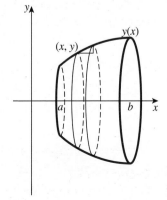

Figure 5.2

(b) Verify this result by finding the curved surface area of the corresponding cone. (The curved surface area of a cone of radius r and slant height l is $\pi r l$.)

3   Find the area of the surface generated when $y = \sqrt{x}$ for $1 \leq x \leq 2$ is rotated completely about the x axis.

Solutions to exercises

2    $\pi\sqrt{2}$

3   8.28

End of block exercises

1   Calculate the length of the curve $y^2 = 8x^3$ between $x = 1$ and $x = 3$.

2   When a curve is defined parametrically by $x = x(t), y = y(t)$, show that
(a) the length of the curve between the points where $t = t_1$ and $t = t_2$ is

$$\int_{t_1}^{t_2} \sqrt{\left(\frac{dx}{dt}\right)^2 + \left(\frac{dy}{dt}\right)^2}\, dt$$

(b) the area of the surface of revolution is

$$\int_{t_1}^{t_2} 2\pi y \sqrt{\left(\frac{dx}{dt}\right)^2 + \left(\frac{dy}{dt}\right)^2}\, dt$$

3   Show that the area of the surface generated when $y = \dfrac{x^3}{9}$ between $x = 0$ and $x = 2$ is rotated around the x axis is $\dfrac{98\pi}{81}$.

4   The curve shown in Figure 5.3 is called a **cycloid**. It is the path traced out by a point on the rim of a wheel that rolls along the ground without slipping. This curve has parametric equations

$$x = \theta - \sin\theta, \quad y = 1 - \cos\theta$$

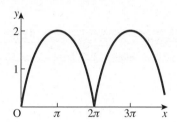

Figure 5.3
A cycloid.

and the graph shows the part of the curve generated when θ lies between 0 and 4π.
(a) Using the result from question 2(a) show that the length of one arch of this curve, that is $0 \leq \theta \leq 2\pi$, is 8. $\Bigg($ Hint: to integrate $\sqrt{1 - \cos\theta}$ use the trigonometric identity $\sin^2 A = \dfrac{1 - \cos 2A}{2}.\Bigg)$

(b) Using the result from question 2(b) show that the area of the surface generated when the arch between $0 \leq \theta \leq 2\pi$ is rotated completely around the x axis is $\dfrac{64\pi}{3}$.

Solution to exercise

1   12.04

The mean value and root-mean-square value of a function

6.1 Introduction

Currents and voltages often vary with time. Engineers may wish to know the average value of such a current or voltage over some particular time interval. The average value of a time-varying function is defined in terms of an integral. An associated quantity is the root-mean-square (r.m.s.) value of a function. The r.m.s. value of a current is used in the calculation of the power dissipated by a resistor.

6.2 Average value of a function

Suppose a function $f(t)$ is defined on the interval $a \leq t \leq b$. The area, A, under the graph of $f(t)$ is given by the integral

$$A = \int_a^b f(t)\,\mathrm{d}t$$

This is illustrated in Figure 6.1.

Figure 6.1
The area under the curve from $t = a$ to $t = b$ and the area of the rectangle are equal.

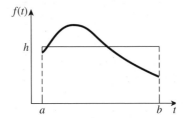

On Figure 6.1 we have also drawn a rectangle with base spanning the interval $a \leq t \leq b$ and which has the same area as that under the curve. Suppose the height of the rectangle is h. Then

$$\text{area of rectangle} = \text{area under curve}$$

$$h(b - a) = \int_a^b f(t)\,\mathrm{d}t$$

$$h = \frac{1}{b - a} \int_a^b f(t)\,\mathrm{d}t$$

The value of h is the **average value** of the function across the interval $a \leq t \leq b$.

Key point

$$\text{average value} = \frac{1}{b-a} \int_a^b f(t)\, dt$$

The average value depends upon the interval chosen. If the values of a or b are changed, then the average value of the function across the interval from a to b will change as well. Note that the average value is also referred to as the **mean value**.

Example 6.1
Find the average value of $f(t) = t^2$ across the interval $1 \leq t \leq 3$.

Solution
Here $a = 1$ and $b = 3$.

$$\text{average value} = \frac{1}{b-a} \int_a^b f(t)\, dt$$

$$= \frac{1}{3-1} \int_1^3 t^2\, dt$$

$$= \frac{1}{2} \left[\frac{t^3}{3} \right]_1^3$$

$$= \frac{1}{2} \left(9 - \frac{1}{3} \right)$$

$$= \frac{1}{2} \left(\frac{26}{3} \right)$$

$$= \frac{13}{3}$$

Example 6.2
Find the average value of $f(t) = t^2$ across the interval $2 \leq t \leq 5$.

Solution
Here $a = 2$ and $b = 5$.

$$\text{average value} = \qquad\qquad\qquad\qquad \frac{1}{5-2} \int_2^5 t^2\, dt$$

$$= \qquad\qquad\qquad\qquad\qquad\qquad\qquad\qquad\qquad 13$$

Example 6.3 Electrical Engineering – A thyristor-firing circuit

Figure 6.2 shows a simple circuit to control the voltage across a load resistor, R_L. This circuit has many uses, one of which is to adjust the level of lighting in a room. The circuit has an alternating current power supply with peak voltage, V_S. The main control element is the thyristor. This device is similar in many ways to a diode. It has a very high resistance when it is reverse biased and a low resistance when it is forward biased. However, unlike a diode, this low resistance depends on the thyristor being 'switched on' by the application of a gate current. The point at which the thyristor is switched on can be varied by varying the resistor, R_G. Figure 6.3 shows a typical waveform of the voltage, v_L, across the load resistor.

Figure 6.2
A thyristor-firing circuit.

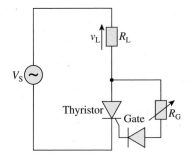

Figure 6.3
Load voltage waveform.

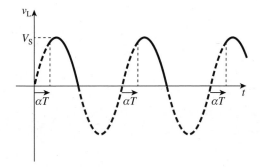

The point at which the thyristor is turned on in each cycle is characterised by the quantity αT, where $0 \le \alpha \le 0.25$ and T is the period of the waveform. This restriction on α reflects the fact that if the thyristor has not turned on when the supply voltage has peaked in the forward direction then it will never turn on.

Calculate the average value of the waveform over a period and comment on the result.

Solution

The average value of load voltage is

$$\frac{1}{T}\int_0^T v_L\,dt = \frac{1}{T}\int_{\alpha T}^{T/2} V_S \sin\left(\frac{2\pi t}{T}\right) dt$$

$$= \frac{V_S}{T}\frac{T}{2\pi}\left[-\cos\left(\frac{2\pi t}{T}\right)\right]_{\alpha T}^{T/2}$$

$$= \frac{V_S}{2\pi}(1 + \cos 2\pi\alpha)$$

If $\alpha = 0$, then the average value is $\dfrac{V_S}{\pi}$, the maximum value for this circuit. If $\alpha = 0.25$, then the average value is $\dfrac{V_S}{2\pi}$, which shows that delaying the turning on of the thyristor reduces the average value of the load voltage.

Exercises

1 Calculate the average value of the given functions across the specified interval:
(a) $f(t) = 1 + t$ across $[0, 2]$
(b) $f(x) = 2x - 1$ across $[-1, 1]$
(c) $f(t) = t^2$ across $[0, 1]$
(d) $f(t) = t^2$ across $[0, 2]$
(e) $f(z) = z^2 + z$ across $[1, 3]$

2 Calculate the average value of the given functions over the specified interval:
(a) $f(x) = x^3$ across $[1, 3]$
(b) $f(x) = \dfrac{1}{x}$ across $[1, 2]$
(c) $f(t) = \sqrt{t}$ across $[0, 2]$
(d) $f(z) = z^3 - 1$ across $[-1, 1]$
(e) $f(t) = \dfrac{1}{t^2}$ across $[-3, -2]$

3 Calculate the average value of the following:
(a) $f(t) = \sin t$ across $\left[0, \frac{\pi}{2}\right]$
(b) $f(t) = \sin t$ across $[0, \pi]$
(c) $f(t) = \sin \omega t$ across $[0, \pi]$
(d) $f(t) = \cos t$ across $\left[0, \frac{\pi}{2}\right]$
(e) $f(t) = \cos t$ across $[0, \pi]$
(f) $f(t) = \cos \omega t$ across $[0, \pi]$
(g) $f(t) = \sin \omega t + \cos \omega t$ across $[0, 1]$

4 Calculate the average value of the following functions:
(a) $f(t) = \sqrt{t + 1}$ across $[0, 3]$
(b) $f(t) = e^t$ across $[-1, 1]$
(c) $f(t) = 1 + e^t$ across $[-1, 1]$

Solutions to exercises

1 (a) 2 (b) -1 (c) $\dfrac{1}{3}$ (d) $\dfrac{4}{3}$ (e) $\dfrac{19}{3}$

2 (a) 10 (b) 0.6931 (c) 0.9428 (d) -1 (e) $\dfrac{1}{6}$

3 (a) $\dfrac{2}{\pi}$ (b) $\dfrac{2}{\pi}$ (c) $\dfrac{1}{\pi\omega}[1 - \cos(\pi\omega)]$ (d) $\dfrac{2}{\pi}$

(e) 0 (f) $\dfrac{\sin(\pi\omega)}{\pi\omega}$ (g) $\dfrac{1 + \sin\omega - \cos\omega}{\omega}$

4 (a) $\dfrac{14}{9}$ (b) 1.1752 (c) 2.1752

6.3 Root-mean-square value of a function

If $f(t)$ is defined on the interval $a \le t \le b$, the **root-mean-square** (r.m.s.) value is given by the following formula:

Key point

$$\text{r.m.s. value} = \sqrt{\dfrac{1}{b - a} \int_a^b [f(t)]^2 \, dt}$$

The r.m.s. value depends upon the interval chosen. If the values of a or b are changed, then the r.m.s. value of the function across the interval from a to b will change as well. Note that when finding an r.m.s. value the function must be squared before it is integrated. The expression

$$\frac{1}{b-a} \int_a^b [f(t)]^2 \, dt$$

is the mean value of $[f(t)]^2$ over the interval from a to b, that is the **mean square value**.

Example 6.4

Find the r.m.s. value of $f(t) = t^2$ across the interval from $t = 1$ to $t = 3$.

Solution

$$
\begin{aligned}
\text{r.m.s. value} &= \sqrt{\frac{1}{b-a} \int_a^b [f(t)]^2 \, dt} \\[2mm]
&= \sqrt{\frac{1}{3-1} \int_1^3 [t^2]^2 \, dt} \\[2mm]
&= \sqrt{\frac{1}{2} \int_1^3 t^4 \, dt} \\[2mm]
&= \sqrt{\frac{1}{2} \left[\frac{t^5}{5} \right]_1^3} \\[2mm]
&= \sqrt{\frac{1}{2} \left(\frac{3^5}{5} - \frac{1}{5} \right)} \\[2mm]
&= \sqrt{\frac{1}{2} \left(\frac{243}{5} - \frac{1}{5} \right)} \\[2mm]
&= \sqrt{\frac{242}{10}} \\[2mm]
&= 4.92
\end{aligned}
$$

Example 6.5 Electrical Engineering – r.m.s. value of a sinusoidal current

The r.m.s. value is often used in the analysis of circuits. A sinusoidal current, $i(t)$, is given by

$$i(t) = \sin t$$

Calculate the r.m.s. value of $i(t)$ across the interval $t = 0$ to $t = 2\pi$.

Solution

Here $a = 0$ and $b = 2\pi$.

$$\text{r.m.s. value} = \sqrt{\frac{1}{2\pi} \int_0^{2\pi} \sin^2 t \, dt}$$

The integral of $\sin^2 t$ is performed by using trigonometrical identities to rewrite it in the alternative form $\frac{1}{2}(1 - \cos 2t)$. This technique was described in Chapter 17 Block 8.

$$
\begin{aligned}
\text{r.m.s. value} &= \sqrt{\frac{1}{2\pi}\int_0^{2\pi}\frac{(1-\cos 2t)}{2}\,dt} \\
&= \sqrt{\frac{1}{4\pi}\int_0^{2\pi}(1-\cos 2t)\,dt} \\
&= \sqrt{\frac{1}{4\pi}\left[t - \frac{\sin 2t}{2}\right]_0^{2\pi}} \\
&= \sqrt{\frac{1}{4\pi}(2\pi)} \\
&= \sqrt{\frac{1}{2}} \\
&= 0.707
\end{aligned}
$$

Thus the r.m.s. value is 0.707.

In the previous example the amplitude of the sine wave was 1, and the r.m.s. value was 0.707. In general, if the amplitude of a sine wave is A, its r.m.s. value is $0.707A$.

Key point

The r.m.s. value of any sinusoidal waveform taken across an interval equal to one period is $0.707 \times$ the amplitude of the waveform.

Exercises

1 Calculate the r.m.s. values of the functions in question 1 in the previous exercises.

2 Calculate the r.m.s. values of the functions in question 2 in the previous exercises.

3 Calculate the r.m.s. values of the functions in question 3 in the previous exercises.

4 Calculate the r.m.s. values of the functions in question 4 in the previous exercises.

Solutions to exercises

1 (a) 2.0817 (b) 1.5275 (c) 0.4472 (d) 1.7889
   (e) 6.9666

2 (a) 12.4957 (b) 0.7071 (c) 1 (d) 1.0690
   (e) 0.1712

3 (a) 0.7071 (b) 0.7071

   (c) $\sqrt{\dfrac{1}{2} - \dfrac{\sin \pi\omega \cos \pi\omega}{2\pi\omega}}$ (d) 0.7071

   (e) 0.7071 (f) $\sqrt{\dfrac{1}{2} + \dfrac{\sin \pi\omega \cos \pi\omega}{2\pi\omega}}$

   (g) $\sqrt{1 + \dfrac{\sin^2 \omega}{\omega}}$

4 (a) 1.5811 (b) 1.3466 (c) 2.2724

End of block exercises

1   Find the average value of the following functions across the specified interval:
(a) $f(t) = 3 - t$ across $[0, 4]$
(b) $f(t) = t^2 - 2$ across $[1, 3]$
(c) $f(t) = t + \dfrac{1}{t}$ across $[1, 4]$
(d) $f(t) = \sqrt{t} + 1$ across $[0, 4]$
(e) $f(t) = t^{2/3}$ across $[0, 1]$

2   Calculate the average value of the following:
(a) $f(t) = 2\sin 2t$ across $\left[0, \dfrac{\pi}{2}\right]$

(b) $f(t) = A\sin 4t$ across $\left[0, \dfrac{\pi}{2}\right]$

(c) $f(t) = \sin t + \cos t$ across $[0, \pi]$

(d) $f(t) = \cos\left(\dfrac{t}{2}\right)$ across $\left[0, \dfrac{\pi}{2}\right]$

(e) $f(t) = \sin t \cos t$ across $[0, \pi]$

3   Calculate the average value of the following functions:
(a) $f(t) = Ae^{kt}$ across $[0, 1]$

(b) $f(t) = \dfrac{1}{e^{3t}}$ across $[0, 2]$

(c) $f(t) = 3 - e^{-t}$ across $[1, 3]$
(d) $f(t) = e^t + e^{-t}$ across $[0, 2]$
(e) $f(t) = t + e^t$ across $[0, 2]$

4   Find the average and r.m.s. values of $A\cos t + B\sin t$ across
(a) $[0, 2\pi]$
(b) $[0, \pi]$

5   Find the r.m.s. values of the functions in question 1.

6   Find the r.m.s. values of the functions in question 2.

7   Find the r.m.s. values of the functions in question 3.

Solutions to exercises

1   (a) 1 (b) $\dfrac{7}{3}$ (c) 2.9621 (d) $\dfrac{7}{3}$ (e) $\dfrac{3}{5}$

2   (a) $\dfrac{4}{\pi}$ (b) 0 (c) $\dfrac{2}{\pi}$ (d) 0.9003 (e) 0

3   (a) $A\left(\dfrac{e^k - 1}{k}\right)$ (b) 0.1663 (c) 2.8410
(d) 3.6269 (e) 4.1945

4   (a) average $= 0$, r.m.s. $= \sqrt{\dfrac{A^2 + B^2}{2}}$
(b) average $= \dfrac{2B}{\pi}$, r.m.s. $= \sqrt{\dfrac{A^2 + B^2}{2}}$

5   (a) 1.5275 (b) 3.2965 (c) 3.0414
(d) 2.3805 (e) 0.6547

6   (a) 1.4142 (b) $\dfrac{A}{\sqrt{2}}$ (c) 1 (d) 0.9046
(e) 0.3536

7   (a) $\sqrt{\dfrac{e^{2k} - 1}{2k}}$ (b) 0.2887 (c) 2.8423
(d) 3.9554 (e) 4.8085

End of chapter exercises

1 Find the mean value of each of the following functions across the interval stated:
(a) $f(t) = \sin t, [0, \pi]$
(b) $f(t) = t^3, [1, 5]$
(c) $f(t) = e^{3t}, [0, 2]$
(d) $f(t) = 20 + 2 \sin \pi t, [0, 0.5]$

2 Calculate the root-mean-square value of each of the following functions across the interval stated:
(a) $f(t) = t, [0, 10]$
(b) $f(t) = \cos 2t, [0, \pi]$
(c) $f(t) = \sin 2t, [0, \pi]$
(d) $f(t) = 10 \sin 50\pi t, [0, 0.01]$

3 The acceleration, a, of an object varies with time, t, according to the formula $a = t^2 + 3t$. Given that velocity v is given by $v = \int a(t) \, dt$, find the velocity after 5 seconds given that the object is at rest at $t = 0$.

4 Find the volume of the solid generated when the curve $y = 3x^2$ for $0 \le x \le 1$ is rotated around the y axis.

5 Find the volume of the solid generated when $y = \dfrac{1}{x}$ for $1 \le x \le 5$ is rotated about the x axis.

6 Consider the cycloid defined by $x = a(\theta - \sin\theta), y = a(1 - \cos\theta)$ where a is a constant. Show that the length of this curve for values of the parameter θ between 0 and 2π is $8a$. (Hint: see Block 5, End of block exercises, question 4.)

7 Find the length of the curve $y = \cosh x$ for $-1 \le x \le 1$.

Solutions to exercises

1 (a) $\dfrac{2}{\pi}$ (b) 39 (c) 67.07 (d) $\dfrac{4}{\pi} + 20 = 21.2732$

2 (a) 5.7735 (b) $\dfrac{1}{\sqrt{2}}$ (c) $\dfrac{1}{\sqrt{2}}$
(d) $\sqrt{50} = 7.07106$

3 79.2

4 $\dfrac{3\pi}{2}$

5 $\dfrac{4\pi}{5}$

7 $2 \sinh 1 = 2.35$

Sequences and series

A **sequence** is a set of numbers written down in a specific order. If a continuous function or waveform is measured, or sampled, at periodic intervals we obtain a sequence of values. Sequences also arise when an attempt is made to find approximate solutions of equations that model physical phenomena. Such approximation is often necessary if a solution is to be obtained using a digital computer.

Sequences are important in engineering mathematics because they can be used to describe discrete time signals. These are signals that have a non-zero value only at certain specific instants of time. For example, in digital computers calculations are carried out at fixed intervals of time governed by an electronic clock.

A **series** is formed when the terms of a sequence are added. Series are important because the solutions of some mathematical problems can be expressed as series. Two notable groups of series are known as Taylor series and Maclaurin series. These are used extensively to provide approximations or estimates of function values.

In this chapter we discuss properties common to all sequences and series and look at some particular and important examples.

Chapter 19 contents

Sequences and series

A **sequence** is a set of numbers written down in a specific order. For example, 1, 3, 5, 7, 9 and $-1, -2, -3, -4, -5$ are both sequences. Each number in the sequence is called a **term** of the sequence. Sometimes we use the symbol . . . to indicate that the sequence continues. For example, 1, 2, 3, . . . , 20 is the sequence of integers from 1 to 20 inclusive. The sequence 1, 3, 5, 7, 9, . . . can be assumed to continue indefinitely.

It is necessary to introduce a notation for handling sequences. Consider again the sequence of odd numbers 1, 3, 5, 7, 9, ... Suppose we let $x[1]$ stand for the first term, $x[2]$ stand for the second term and so on. Then

$$x[1] = 1, \quad x[2] = 3, \quad x[3] = 5 \text{ and so on}$$

Clearly $x[k]$ will stand for the kth term. Note that by inspection we can write down a formula for the kth term. It is $x[k] = 2k - 1$. Substitute some values of k for yourself to check this.

To denote the full sequence concisely we write

$$x[k] = 2k - 1 \text{ where } k = 1, 2, 3, \ldots$$

Some books use the alternative notation $x_1 = 1, x_2 = 3, x_3 = 5$ and $x_k = 2k - 1$.

Example 1.1
Consider the sequence of even numbers 2, 4, 6, 8, ... By inspection find a formula for the kth term of this sequence.

Solution

$$x[k] = 2k, \text{ where } k = 1, 2, 3, \ldots$$

Example 1.2
Write down the first five terms of the sequence given by $x[k] = (-1)^{k+1}\dfrac{1}{k}$ where $k = 1, 2, 3, \ldots$.

Solution

$$x[1] = (-1)^2 \frac{1}{1} = 1$$

$$x[2] = (-1)^3 \frac{1}{2} = -\frac{1}{2}$$

Write down the third, fourth and fifth terms.

$$\frac{1}{3}, -\frac{1}{4}, \frac{1}{5}$$

A **series** is obtained when the terms of a sequence are added. For example, if we add the terms of the sequence 1, 3, 5, 7, 9 we obtain the series

$$1 + 3 + 5 + 7 + 9$$

We use **sigma notation**, first introduced in Chapter 5 Block 1, to write a series concisely. For example, the sum of the first 10 odd numbers can be written

$$\sum_{k=1}^{10} 2k - 1$$

where the lowermost and uppermost values of k are placed below and above the sigma sign respectively. Similarly, the sum of the first six even numbers can be written

$$\sum_{k=1}^{6} 2k$$

Example 1.3

Write out explicitly the terms of the series $\sum_{k=1}^{5} \dfrac{1}{k}$.

Solution

$$1 + \frac{1}{2} + \frac{1}{3} + \frac{1}{4} + \frac{1}{5}$$

In this block we start by discussing the behaviour of sequences and series that have an infinite number of terms. Then two particular types of sequence are described: the arithmetic sequence and the geometric sequence. These are particularly straightforward, and there are simple formulae for finding their sums.

Exercises

1   Explain carefully the distinction between a sequence and a series.

2   Write out fully the following sums:

   (a) $\sum_{k=1}^{6} k$ (b) $\sum_{k=1}^{k=4} k^2$ (c) $\sum_{k=1}^{k=5} k^3$

3   A sequence is defined by

   $$x[k] = \frac{k^2}{2} + k, \quad k = 0, 1, 2, 3, \ldots$$

   State the first five terms.

4   A sequence is given by $5, \frac{5}{8}, \frac{5}{27}, \frac{5}{64}, \ldots$. Write down an expression to denote the full sequence.

5   Write out explicitly the series

   $$\sum_{k=1}^{4} \frac{1}{(2k + 1)(2k + 3)}$$

Solutions to exercises

2 (a) $1 + 2 + 3 + 4 + 5 + 6$
   (b) $1^2 + 2^2 + 3^2 + 4^2$
   (c) $1^3 + 2^3 + 3^3 + 4^3 + 5^3$

3 $0, \frac{3}{2}, 4, \frac{15}{2}, 12$

4 $x[k] = \dfrac{5}{k^3}, k = 1, 2, 3, \ldots$, for example

5 $\frac{1}{(3)(5)} + \frac{1}{(5)(7)} + \frac{1}{(7)(9)} + \frac{1}{(9)(11)}$

1.2   The limit of an infinite sequence

Some sequences stop after a finite number of terms. These are called **finite sequences**. Others continue indefinitely, and these are called **infinite sequences**. Sometimes the terms of an infinite sequence get closer and closer to a fixed value. For example, the terms of the sequence $x[k] = \dfrac{1}{k}$, for $k = 1, 2, 3, \ldots$, are

$$1, \frac{1}{2}, \frac{1}{3}, \frac{1}{4}, \frac{1}{5}, \ldots$$

The terms appear to be getting smaller. What do you think will be the eventual behaviour of the sequence as k gets large?

terms approach the value 0

We say that 'as k tends to infinity, $x[k]$ tends to zero', or 'the **limit** of $x[k]$ as k tends to infinity is zero' and write this as

$$\lim_{k \to \infty} x[k] = 0$$

When a sequence possesses a limit as k tends to infinity, it is said to **converge**.

Example 1.4

Write down the first four terms of the sequence $x[k] = \dfrac{1}{k^2}$, for $k = 1, 2, 3, \ldots$. What is the behaviour of the sequence as k tends to infinity?

Solution

The first four terms are

$$1, \frac{1}{4}, \frac{1}{9}, \frac{1}{16}$$

As k tends to infinity, $\dfrac{1}{k^2}$ tends to zero.

Example 1.5

Explore the behaviour of the sequence $x[k] = \sqrt{2 + \dfrac{1}{k}}, k = 1, 2, 3, \ldots,$ as $k \to \infty.$

Solution

Write down the first few terms of this sequence:

$$\sqrt{3}, \sqrt{2\tfrac{1}{2}}, \sqrt{2\tfrac{1}{3}}, \ldots$$

As k tends to infinity the term $\dfrac{1}{k}$ tends to zero. So $\lim\limits_{k \to \infty} x[k] = \sqrt{2}.$

When a sequence does not possess a limit it is said to **diverge**. The sequence of odd numbers diverges because the terms of the sequence get larger and larger and so the sequence does not possess a limit. The oscillating sequence $-1, 1, -1, 1, -1, 1, \ldots$ is divergent because it does not possess a limit.

Example 1.6

(a) Write out the first four terms of the sequence $x[k] = \sqrt{k}, k = 1, 2, 3, \ldots.$
(b) Find $\lim\limits_{k \to \infty} x[k].$

Solution

(a) $1, \sqrt{2}, \sqrt{3}, \sqrt{4}.$
(b) The terms of this sequence increase indefinitely. The sequence therefore diverges.

Example 1.7

Find the limit of the sequence $x[k] = \dfrac{3k^2 + 2k - 7}{9k^2 - 7k}$ as $k \to \infty.$

Solution

The approach to tackling a problem like this is to rewrite $x[k]$ in a form in which we can sensibly let $k \to \infty.$ Dividing both numerator and denominator by the highest power of k, that is k^2, gives

$$x[k] = \frac{3 + \dfrac{2}{k} - \dfrac{7}{k^2}}{9 - \dfrac{7}{k}}$$

Now because $\dfrac{1}{k}$ and $\dfrac{1}{k^2}$ both tend to zero as k tends to infinity it follows that

$$\lim_{k \to \infty} x[k] = \lim_{k \to \infty} \frac{3 + \dfrac{2}{k} - \dfrac{7}{k^2}}{9 - \dfrac{7}{k}}$$

$$= \frac{3}{9}$$

$$= \frac{1}{3}$$

Exercises

1   Find if possible the limit of each of the
following sequences:

(a) $x[k] = \dfrac{1}{k+1}, k = 1, 2, 3, \ldots$

(b) $x[k] = k^2, k = 1, 2, 3, \ldots$

(c) $x[k] = \dfrac{k-1}{k+1}, k = 1, 2, 3, \ldots$

(d) $x[k] = \left(\dfrac{1}{3}\right)^k, k = 1, 2, 3, \ldots$

(e) $x[k] = \dfrac{2k+3}{4k+2}, k = 1, 2, 3, \ldots$

Solutions to exercises

1   (a) 0 (b) diverges (c) 1 (d) 0 (e) $\dfrac{1}{2}$

1.3   The sum of an infinite series

When the terms of an infinite sequence are added we obtain an **infinite series**. It may seem strange to try to add together an infinite number of terms but under some circumstances their sum is finite and can be found. For example, consider the infinite series formed from the sequence $x[k] = \dfrac{1}{2^k}, k = 0, 1, 2, \ldots$.

$$1 + \frac{1}{2} + \frac{1}{4} + \frac{1}{8} + \cdots$$

We can calculate the sum of n terms, S_n, for various values of n. For example, the sum of just the first term is

$$S_1 = 1$$

The sum of the first two terms is

$$S_2 = 1 + \frac{1}{2}$$
$$= 1.5$$

Similarly $S_3 = 1.75, S_4 = 1.875, \ldots, S_{10} = 1.9980$. The sequence $S_1, S_2, S_3, \ldots$ is called the **sequence of partial sums**. As we calculate S_n for larger and larger values of n, we note that S_n gets nearer and nearer to 2. We write S to stand for the sum of an infinite number of terms, so

$$S = \sum_{k=0}^{\infty} \frac{1}{2^k}$$
$$= 2$$

For any infinite series, say $\sum_{k=1}^{\infty} x[k]$, we can form the sequence of partial sums:

$$S_1 = x[1], S_2 = x[1] + x[2], S_3 = x[1] + x[2] + x[3], \ldots$$

If the sequence S_n converges to a limit S we say that the infinite series has sum S, or that it has converged to S.

In general it is difficult to determine whether or not an infinite series has a finite sum. It is possible to establish tests or **convergence criteria** to help decide whether or not a given series converges or diverges, but for these you must refer to a more advanced book.

Example 1.8

Calculate the first six partial sums of the series

$$\sum_{k=1}^{\infty} (-1)^{k+1} \frac{1}{k}$$

Solution

The first six terms of the series are

$$1 - \frac{1}{2} + \frac{1}{3} - \frac{1}{4} + \frac{1}{5} - \frac{1}{6}$$

Form the first six partial sums for yourself working to 4 d.p.

> 1, 0.5000, 0.8333, 0.5833, 0.7833, 0.6167

If you proceed to calculate many more terms you will see that the sequence of partial sums converges to 0.6931, which is in fact ln 2.

Exercises

1. Calculate the first four partial sums of the series $\sum_{k=0}^{\infty} \frac{1}{k!}$. (Recall $0! = 1$.)

2. Calculate the first four partial sums of the series $\sum_{k=0}^{\infty} (-1)^k \frac{1}{(2k)!}$.

Solutions to exercises

1. 1, 2, 2.5, 2.6667 (in fact this converges to $e = 2.7183$).

2. 1, 0.5, 0.5417, 0.5403 (this converges to $\cos 1 = 0.5403$).

1.4 Arithmetic sequences

An **arithmetic sequence** is a sequence of numbers where each new term after the first is formed by adding a fixed amount called the **common difference** to the previous term in the sequence. For example, the sequence

$$3, 5, 7, 9, 11, \ldots$$

is an arithmetic sequence. Note that having chosen the first term to be 3, each new term is found by adding 2 to the previous term, so the common difference is 2.

The common difference can be negative: for example, the sequence

$$2, -1, -4, -7, \ldots$$

is an arithmetic sequence with first term 2 and common difference -3.

In general we can write an arithmetic sequence as follows:

Key point

Arithmetic sequence

$$a, a + d, a + 2d, a + 3d, \ldots$$

where the first term is a and the common difference is d. Some important results concerning arithmetic sequences now follow:

Key point

The nth term of an arithmetic sequence is given by

$$a + (n - 1)d$$

Key point

The sum of the first n terms of an arithmetic sequence is

$$S_n = \frac{n}{2}[2a + (n - 1)d]$$

The sum of the terms of an arithmetic sequence is known as an **arithmetic series**.

Example 1.9

Find the 17th term of an arithmetic sequence with first term 5 and common difference 2.

Solution

The nth term is $a + (n - 1)d$.

Hence

17th term =

$$5 + (17 - 1)2 = 5 + 32 = 37$$

Example 1.10

Find the sum of the first 23 terms of the arithmetic sequence

$$4, -3, -10, \ldots$$

Solution

First identify a and d.

$$a = 4, d = -7$$

Then, using $S_n = \dfrac{n}{2}[2a + (n - 1)d]$, find the sum of 23 terms:

$$\frac{23}{2}[2(4) + (23 - 1)(-7)] = -1679$$

Exercises

1 Find the 23rd term of an arithmetic sequence with first term 2 and common difference 7.

2 Find the sum of the first five terms of the arithmetic sequence with first term 3 and common difference 5.

3 Write down the 10th and 19th terms of the arithmetic sequence
(a) $8, 11, 14, \ldots$ (b) $8, 5, 2, \ldots$

4 An arithmetic sequence is given by

$$b, \frac{2b}{3}, \frac{b}{3}, 0, \ldots$$

(a) State the sixth term.
(b) State the kth term.
(c) If the 20th term has a value of 15, find b.

Solutions to exercises

1 156

2 65

3 (a) $35, 62$ (b) $-19, -46$

4 (a) $-\dfrac{2b}{3}$ (b) $\dfrac{b(4 - k)}{3}$ (c) $-\dfrac{45}{16}$

1.5 Geometric sequences

A **geometric sequence** is a sequence of numbers where each term after the first is found by multiplying the previous term by a fixed number called the **common ratio**. For example, the sequence

$$1, 3, 9, 27, \ldots$$

is a geometric sequence with first term 1 and common ratio 3. Each term after the first is found by multiplying the preceding term by 3.

The common ratio could be a fraction and it might be negative.

Example 1.11

Write down the first few terms of the geometric sequence with first term 2 and common ratio $-\frac{1}{3}$.

Solution

$$2, -\frac{2}{3}, \frac{2}{9}, -\frac{2}{27}, \ldots$$

In general we can write a geometric sequence as follows:

Geometric sequence

$$a, ar, ar^2, ar^3, \ldots$$

where the first term is a and the common ratio is r.

Some important results concerning geometric sequences now follow:

The nth term of a geometric sequence is given by

$$ar^{(n-1)}$$

The sum of the first n terms of a geometric sequence is

$$S_n = \frac{a(1 - r^n)}{1 - r} \text{ (valid only if } r \neq 1)$$

The sum of the terms of a geometric sequence is known as a **geometric series**.

Example 1.12

Find the seventh term of the geometric sequence

$$2, -6, 18, \ldots$$

Solution

First identify a and the common ratio r.

$$a = 2, r = -3$$

Then, the nth term is ar^{n-1}, so the seventh term is

$$(2)(-3)^6 = 1458$$

If the common ratio in a geometric series is less than 1 in modulus (i.e. $-1 < r < 1$), the sum of an infinite number of terms converges and can be calculated. This is known as the **sum to infinity**, S_∞.

Key point

$$S_\infty = \frac{a}{1 - r} \text{ provided } -1 < r < 1$$

Example 1.13
Find the sum to infinity of the geometric sequence with first term 3 and common ratio $\frac{1}{2}$.

Solution

$$S_\infty = \frac{a}{1 - r}$$

$$= \boxed{} \qquad\qquad 6$$

Exercises

1 Find the seventh term of a geometric sequence with first term 2 and common ratio 3.

2 Find the sum of the first five terms of the geometric sequence with first term 3 and common ratio 2.

3 Find the sum of the infinite geometric series with first term 2 and common ratio $\frac{1}{2}$.

4 A geometric sequence has first term 1. The ninth term exceeds the fifth term by 240. Find possible values for the eighth term.

5 The sum to infinity of a geometric sequence is four times the first term. Find the common ratio.

6 A geometric series has $S_3 = \frac{37}{8}$ and $S_6 = \frac{3367}{512}$. Find the first term and the common ratio.

7 A geometric sequence is given by $1, \frac{1}{2}, \frac{1}{4}, \ldots$. What is its common ratio?

Solutions to exercises

1 1458

2 93

3 4

4 ± 128

5 $\frac{3}{4}$

6 $2, \frac{3}{4}$

7 $\frac{1}{2}$

End of block exercises

1 State whether the following sequences are arithmetic, geometric or neither:
(a) $1, -1, -3, -5, \ldots$ (b) $4, 2, 1, 0.5, \ldots$
(c) $6, 7, 8, 9, \ldots$ (d) $4, 5, 7, 10, \ldots$
(e) $1, 0.1, 0.01, 0.001, \ldots$
(f) $1, -1, 1, -1, 1, \ldots$ (g) $1, 1, 1, 1, \ldots$

2 An arithmetic sequence has first term -3 and common difference 4. State (a) the 10th term, (b) the 300th term.

3 An arithmetic series has first term 4 and common difference $\frac{1}{2}$. Find (a) the sum of the first 20 terms, (b) the sum of the first 100 terms.

4 A geometric sequence has first term -2 and common ratio $-\frac{3}{4}$. State the 20th term.

5 A geometric series has first term equal to 3 and a common ratio of 1.5. Calculate the sum of
(a) the first 10 terms, (b) the first 50 terms,
(c) the 30th to the 49th terms inclusive.

6 Find the limit as $k \to \infty$, if it exists, of each of the following:
(a) $x[k] = k^5$ (b) $x[k] = \sin k$
(c) $x[k] = \dfrac{k^2 + k}{k^2 + k + 1}$
(d) $x[k] = \dfrac{3k^2 + k}{k^2 + k + 1}$

7 Show that the sum of the first n terms of an arithmetic sequence, S_n, can be expressed as
$$S_n = \frac{n}{2}(\text{first term} + \text{last term})$$

Solutions to exercises

1 (a) arithmetic (b) geometric (c) arithmetic
(d) neither (e) geometric (f) geometric
(g) arithmetic and geometric

2 (a) 33 (b) 1193

3 (a) 175 (b) 2875

4 8.4566×10^{-3}

5 (a) 339.99 (b) 3.8257×10^9
(c) 2.5497×10^9

6 (a) diverges (b) diverges (c) 1 (d) 3

Sums of whole numbers, their squares and cubes

2.1 Introduction

There are a number of series that arise in the solution of certain problems and it is useful to know their sums. In this block we give these series.

2.2 The sum of the first n positive integers

Key point

The sum of the first n positive integers is given by

$$\sum_{k=1}^{n} k = 1 + 2 + 3 + \cdots + n$$

$$= \frac{n}{2}(n + 1)$$

In fact this series is an arithmetic series with first term 1 and common difference 1, as discussed in Block 1.

Example 2.1

Find the sum of the first 100 integers, $1 + 2 + 3 + \cdots + 100$.

Solution

Use the formula with $n = 100$:

$$\frac{100}{2}(101) = 5050$$

2.3 The sum of the squares of the first n positive integers

Key point

The sum of the squares of the first n positive integers is given by

$$\sum_{k=1}^{n} k^2 = 1^2 + 2^2 + 3^2 + \cdots + n^2$$

$$= \frac{n}{6}(n + 1)(2n + 1)$$

Example 2.2

Find the sum of the squares of the first 10 integers, $1^2 + 2^2 + 3^2 + \cdots + 10^2$.

Solution

Use the formula with $n = 10$:

$$\frac{10}{6}(11)(21) = 385$$

2.4 The sum of the cubes of the first n positive integers

The sum of the cubes of the first n positive integers is given by

$$\sum_{k=1}^{n} k^3 = 1^3 + 2^3 + 3^3 + \cdots + n^3$$

$$= \left(\frac{n(n + 1)}{2}\right)^2$$

Example 2.3

Find the sum of the cubes of the first 12 integers, $1^3 + 2^3 + 3^3 + \cdots + 12^3$.

Solution

Use the formula with $n = 12$:

$$\left(\frac{12(13)}{2}\right)^2 = 6084$$

Exercises

1 Find the sum of the first 20 positive integers.

2 Find the sum of the squares of the first 20 positive integers.

3 Find the sum of the cubes of the first six positive integers.

4 Find the sum of the first 40 positive integers.

Solutions to exercises

1 210

2 2870

3 441

4 820

End of block exercises

1   Find $\sum_{k=1}^{32} k^2$.

2   Find $\sum_{k=1}^{18} k$.

3   Find the sum of the integers from 18 to 96 inclusive.

4   Find the sum of the integers from -200 to -6 inclusive.

5   Find the sum of the integers from -50 to 24 inclusive.

6   Find $7^2 + 8^2 + \cdots + 21^2$.

Solutions to exercises

1   11440

2   171

3   4503

4   -20085

5   -975

6   3220

Pascal's triangle and the binomial theorem

3.1 Introduction

It is frequently necessary to expand expressions of the form $(a + b)^2$ or $(a + b)^3$. The expansion of an expression of the form $(a + b)^n$ is called a **binomial expansion**. When the power n is small it is possible to multiply out the brackets. If n is large this approach becomes tedious. An alternative and very simple approach is to use **Pascal's triangle**, which is described in the next section.

When n is very large, even Pascal's triangle is tedious, and the **binomial theorem** provides an alternative.

3.2 Pascal's triangle

Consider the triangle of numbers shown below. Every entry is obtained by adding the two entries on either side in the preceding row, always starting and finishing a row with a 1.

Key point

Pascal's triangle

```
                    1
                1       1
            1       2       1
        1       3       3       1
    1       4       6       4       1
  1     5      10      10       5      1
  ⋮     ⋮      ⋮       ⋮       ⋮      ⋮      ⋮
```

Example 3.1
Complete the next two rows in the triangle. Add these rows to the triangle above.

1 6 15 20 15 6 1

1 7 21 35 35 21 7 1

Compare the third and fourth rows of the triangle with the expansions of $(a + b)^2$ and $(a + b)^3$.

$$(a + b)^2 = a^2 + 2ab + b^2, \quad (a + b)^3 = a^3 + 3a^2b + 3ab^2 + b^3$$

Note that Pascal's triangle gives the coefficients in the expansions. Furthermore, the terms in these expansions are composed of decreasing powers of a and increasing powers of b. If we want the expansion of $(a + b)^4$ the row in the triangle beginning 1, 4 will provide the coefficients. We simply insert the appropriate powers of a and b, starting with the highest power of a, namely a^4. Thus

$$(a + b)^4 = a^4 + 4a^3b + 6a^2b^2 + 4ab^3 + b^4$$

Example 3.2
Use Pascal's triangle to expand $(a + b)^5$.

Solution

$$a^5 + 5a^4b + 10a^3b^2 + 10a^2b^3 + 5ab^4 + b^5$$

Example 3.3
Find $(\cos x + \sin x)^3$.

Solution
The answer can be written down immediately:

$$(\cos x + \sin x)^3 = \cos^3 x + 3\cos^2 x \sin x + 3 \cos x \sin^2 x + \sin^3 x$$

Exercises

1   Use Pascal's triangle to expand $(a + b)^7$.

2   Expand $(a + b)^4$. By letting $a = 1$ and $b = 3y$ expand $(1 + 3y)^4$.

Solutions to exercises

1    $a^7 + 7a^6b + 21a^5b^2 + 35a^4b^3 + 35a^3b^4 + 21a^2b^5 + 7ab^6 + b^7$

2    $1 + 12y + 54y^2 + 108y^3 + 81y^4$

3.3   The binomial theorem when n is a positive integer

The binomial theorem is concerned with the expansion of $(a + b)^n$. It is useful to separate the discussion into two cases, the first when n is a positive integer, and the second when n is negative or fractional. Section 3.4 will deal with the second case.

When n is a positive integer, the expansion of $(a + b)^n$ contains a finite number of terms. For example, we have seen in Example 3.2 that

$$(a + b)^5 = a^5 + 5a^4b + 10a^3b^2 + 10a^2b^3 + 5ab^4 + b^5$$

and so the expansion contains six terms. The binomial theorem is useful when we want to expand $(a + b)^n$ for large n. To use the theorem you will need to recall factorial notation first introduced in Chapter 5 Block 1.

$$2! = 2 \times 1, \quad 3! = 3 \times 2 \times 1, \quad 4! = 4 \times 3 \times 2 \times 1, \quad \text{and so on}$$

Key point

The binomial theorem states that when n is a positive integer

$$(a + b)^n = a^n + na^{n-1}b + \frac{n(n - 1)}{2!}a^{n-2}b^2$$
$$+ \frac{n(n - 1)(n - 2)}{3!}a^{n-3}b^3 + \cdots + b^n$$

This is a sum of a finite number of terms, that is the series stops. The last term is b^n.

Example 3.4
(a) Write out the first three terms of the expansion of $(a + b)^{21}$.
(b) What would be the last term?

Solution
(a) Using the theorem with $n = 21$ we can write down the first three terms:

$$(a + b)^{21} = a^{21} + 21a^{20}b + \frac{21(20)}{2!}a^{19}b^2 + \cdots$$

(b) The last term is b^{21}.

The theorem is often quoted for the particular case when $a = 1$ and $b = x$:

Key point

Binomial theorem (alternative form)

$$(1 + x)^n = 1 + nx + \frac{n(n - 1)}{2!}x^2 + \frac{n(n - 1)(n - 2)}{3!}x^3 + \cdots + x^n$$

Example 3.5
Use the theorem to write down the expansion of $(1 + x)^4$.

Solution
Use the expansion with $n = 4$:

$$1 + 4x + 6x^2 + 4x^3 + x^4$$

Exercises

1 Use the binomial theorem to expand

(a) $(1 + x)^3$ (b) $\left(1 + \dfrac{x}{3}\right)^4$ (c) $\left(1 - \dfrac{x}{2}\right)^5$.

2 Write down the first three terms in the expansion of $(x + 3y)^{12}$.

Solutions to exercises

1 (a) $1 + 3x + 3x^2 + x^3$

(b) $1 + \dfrac{4x}{3} + \dfrac{2x^2}{3} + \dfrac{4x^3}{27} + \dfrac{x^4}{81}$

(c) $1 - \dfrac{5x}{2} + \dfrac{5x^2}{2} - \dfrac{5x^3}{4} + \dfrac{5x^4}{16} - \dfrac{x^5}{32}$

2 $x^{12} + 36x^{11}y + 594x^{10}y^2 + \cdots$

3.4 The binomial theorem when *n* is not a positive integer

The theorem can be applied when n is not a positive integer provided that x lies between -1 and 1. However, when n is not a positive integer, the series is *infinite:*

Key point

The binomial theorem states that when n is not a positive integer

$$(1 + x)^n = 1 + nx + \frac{n(n - 1)}{2!}x^2 + \frac{n(n - 1)(n - 2)}{3!}x^3 + \cdots$$

$$\text{only for } -1 < x < 1$$

This is an infinite series.

Example 3.6

Use the binomial theorem to write down the first four terms in the expansion of $(1 + x)^{1/2}$.

Solution

Use the theorem with $n = \dfrac{1}{2}$.

$$1 + \frac{1}{2}x + \frac{\left(\frac{1}{2}\right)\left(-\frac{1}{2}\right)}{2!}x^2 + \frac{\left(\frac{1}{2}\right)\left(-\frac{1}{2}\right)\left(-\frac{3}{2}\right)}{3!}x^3 + \cdots$$

Simplify your result:

$$1 + \frac{x}{2} - \frac{x^2}{8} + \frac{x^3}{16} + \cdots$$

Note that $(1 + x)^{1/2}$ and $1 + \dfrac{x}{2} - \dfrac{x^2}{8} + \dfrac{x^3}{16} + \cdots$ have the same numerical value only when x lies in the range -1 to 1. If x is outside this range the two expressions have different values.

Example 3.7

Expand $\dfrac{1}{1 + x}$ in ascending powers of x up to and including the term in x^3.

Solution

$\dfrac{1}{1 + x}$ can be written $(1 + x)^{-1}$. Use the binomial theorem with $n = -1$:

$$1 - x + x^2 - x^3 + \cdots$$

This expansion is valid provided $-1 < x < 1$.

Example 3.8

Obtain the first four terms in the expansion of $\dfrac{1}{(l - x)^2}$ in powers of $\dfrac{x}{l}$.

Solution

Note that we can rewrite $\dfrac{1}{(l - x)^2}$ in the following way:

$$\frac{1}{(l - x)^2} = \frac{1}{l^2 \left(1 - \dfrac{x}{l}\right)^2}$$

$$= \frac{1}{l^2}\left(1 - \frac{x}{l}\right)^{-2}$$

The binomial theorem can then be used to expand $\left(1 - \dfrac{x}{l}\right)^{-2}$:

$$\left(1 - \frac{x}{l}\right)^{-2} = 1 + (-2)\left(-\frac{x}{l}\right) + \frac{(-2)(-3)}{2!}\left(-\frac{x}{l}\right)^2$$

$$+ \frac{(-2)(-3)(-4)}{3!}\left(-\frac{x}{l}\right)^3 + \cdots$$

$$= 1 + 2\left(\frac{x}{l}\right) + 3\left(\frac{x}{l}\right)^2 + 4\left(\frac{x}{l}\right)^3 + \cdots$$

Hence

$$\frac{1}{(l-x)^2} = \frac{1}{l^2}\left[1 + 2\left(\frac{x}{l}\right) + 3\left(\frac{x}{l}\right)^2 + 4\left(\frac{x}{l}\right)^3 + \cdots\right]$$

Exercises

1 Obtain the first five terms in the expansion of $(1 + x)^{-2}$. State the range of values of x for which the expansion is valid.

2 Obtain a quadratic approximation to $(1 - 2x)^{1/2}$ using the binomial theorem. For what range of values of x is your expansion valid?

Solutions to exercises

1 $1 - 2x + 3x^2 - 4x^3 + 5x^4$, $-1 < x < 1$

2 $1 - x - \frac{1}{2}x^2 + \cdots$, $-\frac{1}{2} < x < \frac{1}{2}$

End of block exercises

1 (a) Use the binomial theorem to expand $(1 + y)^5$.
   (b) Use the result from part (a) to expand $(1 + 3y)^5$.

2 Expand (a) $(1 - z)^3$, (b) $(2 - z)^3$, (c) $(2 + z)^3$, (d) $(1 + 2z)^3$, (e) $(z + 3)^3$.

3 Expand $(\cos\theta + j\sin\theta)^5$ where $j^2 = -1$.

4 Write down the first two terms in the binomial expansion of

$$\left(1 + \frac{x}{5}\right)^{1/4}$$

5 Expand $\left(2 - \frac{1}{x}\right)^5$.

Solutions to exercises

1 (a) $1 + 5y + 10y^2 + 10y^3 + 5y^4 + y^5$
   (b) $1 + 15y + 90y^2 + 270y^3 + 405y^4 + 243y^5$

2 (a) $1 - 3z + 3z^2 - z^3$
   (b) $8 - 12z + 6z^2 - z^3$
   (c) $8 + 12z + 6z^2 + z^3$
   (d) $1 + 6z + 12z^2 + 8z^3$
   (e) $z^3 + 9z^2 + 27z + 27$

3 $\cos^5\theta - 10\cos^3\theta\sin^2\theta + 5\cos\theta\sin^4\theta + j(\sin^5\theta - 10\cos^2\theta\sin^3\theta + 5\cos^4\theta\sin\theta)$

4 $1 + \dfrac{x}{20}$

5 $32 - \dfrac{80}{x} + \dfrac{80}{x^2} - \dfrac{40}{x^3} + \dfrac{10}{x^4} - \dfrac{1}{x^5}$

Taylor series and Maclaurin series

4.1 Introduction

Suppose the value of a function and the value of its derivatives are known at a particular point. From this information it is possible to obtain values of the function around that point, using a series called a **Taylor series**.

When the particular point is the origin, it is possible to obtain function values at neighbouring points using a **Maclaurin series**. Thus the Maclaurin series is a special case of the Taylor series.

These two series are explored in this block.

4.2 Maclaurin series

Suppose we have a function $f(x)$ from which we find the value of the function when $x = 0$, that is $f(0)$. Suppose also that by successive differentiation we find the value of the derivatives of $f(x)$ at $x = 0$, that is $f'(0), f''(0)$, and so on.

The **Maclaurin series** formula enables us to find the value of the function at a point, x, close to the origin:

Key point

> **Maclaurin series**
>
> $$f(x) = f(0) + xf'(0) + \frac{x^2}{2!}f''(0) + \frac{x^3}{3!}f'''(0) + \cdots$$

This is an infinite series, although often we can approximate $f(x)$ by using just a finite number of the terms as we shall show.

Consider the following example.

Example 4.1

Obtain the Maclaurin series for $f(x) = e^x$.

Solution

Here $f(x) = e^x$. Evaluate $f(0)$:

$$f(0) = e^0 = 1$$

Differentiate $f(x)$ and so find $f'(0)$.

$$f'(x) = e^x, f'(0) = 1$$

We can continue in this way to show that

$$f''(0) = 1, f'''(0) = 1, \text{ and so on}$$

Applying the Maclaurin series formula,

$$f(x) = f(0) + xf'(0) + \frac{x^2}{2!}f''(0) + \frac{x^3}{3!}f'''(0) + \cdots$$

$$e^x = 1 + x + \frac{x^2}{2!} + \frac{x^3}{3!} + \cdots$$

This is an infinite series, which is equal to e^x for all values of x. We can write it concisely using sigma notation as

$$e^x = \sum_{n=0}^{\infty} \frac{x^n}{n!}$$

It is also known as the **power series expansion** of e^x about the origin.

We have shown in the previous example that

$$e^x = \sum_{n=0}^{\infty} \frac{x^n}{n!} = 1 + x + \frac{x^2}{2!} + \frac{x^3}{3!} + \cdots$$

By taking successive partial sums of this Maclaurin series we can obtain polynomials that are approximations to e^x, known as **Maclaurin polynomials**. Denoting these polynomials by $p_0(x)$, $p_1(x)$ and so on, we have

$p_0(x) = 1,$ by taking just the first term

$p_1(x) = 1 + x,$ by taking the first two terms

$p_2(x) = 1 + x + \dfrac{x^2}{2!},$ and so on

Each Maclaurin polynomial is a function that approximates the original function, in this case e^x. As more and more terms are included the approximation improves. To see this more clearly study the graphs in Figure 4.1. The first shows $f(x) = e^x$ together with the graph of $p_0 = 1$. The second and third graphs show $f(x) = e^x$ together with $p_1(x) = 1 + x$ and $p_2(x) = 1 + x + \dfrac{x^2}{2!}$. Notice that close to the origin you would expect p_1 and p_2 to give good approximate values of e^x, but the quality of the approximation will deteriorate as you move further from the origin.

Figure 4.1
$f(x) = e^x$ together with a number of Maclaurin polynomials.

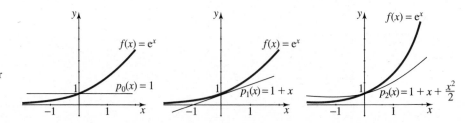

Example 4.2

(a) Find the Maclaurin series for $f(x) = \sin x$, where x is measured in radians.

(b) State a cubic approximation to $\sin x$.

Solution

(a) Obtain the successive derivatives of $f(x)$ and evaluate them when $x = 0$. Remember that, whenever it is necessary to differentiate trigonometrical functions, work must be carried out in radians and not in degrees.

$f(x) = \sin x, \quad f(0) = \quad$ 0

$f'(x) = \quad, f'(0) = \quad$ $\cos x, 1$

$f''(x) = \quad, f''(0) = \quad$ $-\sin x, 0$

$f'''(x) = \quad, f'''(0) = \quad$ $-\cos x, -1$

Continuing in this way and then substituting into the Maclaurin series formula you will find

$$\sin x = x - \frac{x^3}{3!} + \frac{x^5}{5!} - \frac{x^7}{7!} + \cdots$$

This is the power series expansion of $\sin x$ about the origin.

(b) Taking the first two non-zero terms of the Maclaurin series gives the cubic polynomial

$$p_3(x) = x - \frac{x^3}{3!}$$

A useful exercise would be for you to use a computer package to draw graphs of $\sin x$ and the first few Maclaurin polynomials to see the way in which the approximation improves as the degree of the polynomial increases.

The power series expansions obtained in Examples 4.1 and 4.2 are valid for any x. This means that the value of the series and the value of the function used to generate it are equal for every value of x. However, some functions have a power series expansion that is valid only for a limited range of values of x. For example, the Maclaurin expansion of $f(x) = \dfrac{1}{1+x}$ is valid only when $-1 < x < 1$.

Some common power series expansions are shown in Table 4.1.

The small-angle approximations

The Maclaurin series for $\sin x$ and $\cos x$ can be used to approximate the trigonometrical ratios when x is small and measured in radians.

Key point

Small-angle approximations
If x is small and measured in radians:

$$\sin x \approx x \quad \text{and} \quad \cos x \approx 1 - \frac{x^2}{2}$$

Use your calculator to verify these approximations. For example, show that $\sin 0.3 \approx 0.3$.

Table 4.1
Some common
power series
expansions.

$$e^x = 1 + \frac{x}{1!} + \frac{x^2}{2!} + \frac{x^3}{3!} + \cdots \text{ for all } x$$

$$\sin x = x - \frac{x^3}{3!} + \frac{x^5}{5!} - \frac{x^7}{7!} + \cdots \text{ for all } x$$

$$\cos x = 1 - \frac{x^2}{2!} + \frac{x^4}{4!} - \frac{x^6}{6!} + \cdots \text{ for all } x$$

$$\log_e(1 + x) = x - \frac{x^2}{2} + \frac{x^3}{3} - \frac{x^4}{4} + \cdots \text{ for } -1 < x \le 1$$

$$(1 + x)^p = 1 + px + \frac{p(p - 1)}{2!}x^2 + \frac{p(p - 1)(p - 2)}{3!}x^3 + \cdots \text{ for } -1 < x < 1$$

$$\tan x = x + \frac{x^3}{3} + \frac{2x^5}{15} + \frac{17x^7}{315} + \cdots \text{ for } -\frac{\pi}{2} < x < \frac{\pi}{2}$$

$$\sinh x = x + \frac{x^3}{3!} + \frac{x^5}{5!} + \frac{x^7}{7!} + \cdots \text{ for all } x$$

$$\cosh x = 1 + \frac{x^2}{2!} + \frac{x^4}{4!} + \frac{x^6}{6!} + \cdots \text{ for all } x$$

Exercises

1 Derive the Maclaurin series for $f(x) = \cos x$.

2 Use the power series expansion of e^x to show that

$$e^{2x} = 1 + 2x + 2x^2 + \frac{4x^3}{3} + \cdots$$

3 Use the power series expansion of $\cos x$ to show that

$$\cos\frac{x}{2} = 1 - \frac{x^2}{8} + \frac{x^4}{384} - \frac{x^6}{46080} + \cdots$$

4 Determine the Maclaurin series expansion for
$$f(x) = \frac{1}{1 + x}.$$

5 Find the Maclaurin expansion for $\sin^2 x$.
(Hint: use a trigonometrical identity and the series for $\sin x$.)

6 Obtain the Maclaurin series expansion for $f(x) = \cosh x$.

7 (a) Obtain a quadratic Maclaurin polynomial approximation, $p_2(x)$, to $f(x) = \cos 2x$.
(b) Compare the approximate value given by $p_2(1)$ with actual value $f(1)$.

Solutions to exercises

4 $\quad 1 - x + x^2 - x^3 + \cdots = \sum_{n=0}^{\infty} (-1)^n x^n$

5 $\quad x^2 - \frac{1}{3}x^4 + \frac{2}{45}x^6 - \cdots$

6 $\quad 1 + \frac{x^2}{2!} + \frac{x^4}{4!} + \frac{x^6}{6!} + \cdots$

7 (a) $1 + 2x^2$ (b) $f(1) = 3.7622$, $p_2(1) = 3$

4.3 Taylor series

The Taylor series is very similar to the Maclaurin series. The only difference is that the expansion is now taken about any point, $x = a$ say, not necessarily the origin. So if we know the value of a function and its derivatives when $x = a$ we can use the Taylor series formula to find the value of the function at nearby points.

Key point

> **Taylor series**
>
> $$f(x) = f(a) + (x - a)f'(a) + \frac{(x - a)^2}{2!}f''(a) + \frac{(x - a)^3}{3!}f'''(a) + \cdots$$

Example 4.3

Find the Taylor series expansion of $f(x) = \sqrt{x}$ about the point $x = 4$.

Solution

Here, the known point is $x = 4$. We must evaluate $f(x)$ and its derivatives at this point:

$$f(4) = \sqrt{4}$$
$$= 2$$

$$f'(x) = \boxed{}$$

$$\frac{1}{2}x^{-1/2} = \frac{1}{2\sqrt{x}}$$

and so $f'(4) = \dfrac{1}{2\sqrt{4}} = \dfrac{1}{4}$.

Now obtain the second and third derivatives of $f(x)$.

$$f''(x) = -\frac{1}{4(\sqrt{x})^3}, \; f'''(x) = \frac{3}{8\,(\sqrt{x})^5}$$

Evaluate the second and third derivatives at $x = 4$.

$$f''(4) = -\frac{1}{32}, \; f'''(4) = \frac{3}{256}$$

Substitute these values into the Taylor series formula:

$$f(x) = f(4) + (x - 4)f'(4) + \frac{(x - 4)^2}{2!}f''(4) + \frac{(x - 4)^3}{3!}f'''(4) + \cdots$$

$$\sqrt{x} = 2 + (x - 4)\frac{1}{4} + \frac{(x - 4)^2}{2}\left(-\frac{1}{32}\right) + \frac{(x - 4)^3}{6}\left(\frac{3}{256}\right) + \cdots$$

$$= 2 + \frac{1}{4}(x - 4) - \frac{1}{64}(x - 4)^2 + \frac{1}{512}(x - 4)^3 + \cdots$$

This is the Taylor series expansion of $\sqrt{x}$ about $x = 4$.

In the previous example we calculated the Taylor series of $\sqrt{x}$. By taking successive partial sums of this series we can obtain polynomials that are approximations to $\sqrt{x}$, known as **Taylor polynomials**. Denoting these polynomials by $p_0(x)$, $p_1(x)$ and so on, we have

$p_0(x) = 2,$ by taking just the first term

$p_1(x) = 2 + \frac{1}{4}(x - 4),$ by taking the first two terms

$p_2(x) = 2 + \frac{1}{4}(x - 4) - \frac{1}{64}(x - 4)^2,$ and so on

Each Taylor polynomial is a function that approximates the original function, in this case $\sqrt{x}$ when x is close to the expansion point, $x = 4$. As more and more terms are included the approximation improves.

The first graph in Figure 4.2 shows $f(x) = \sqrt{x}$ together with a graph of $p_0(x) = 2$. The second and third graphs show $f(x) = \sqrt{x}$ together with $p_1(x)$ and $p_2(x)$. Notice how the approximation improves as the degree of the polynomial increases.

Figure 4.2
$f(x) = \sqrt{x}$
together with a
number of Taylor
polynomials.

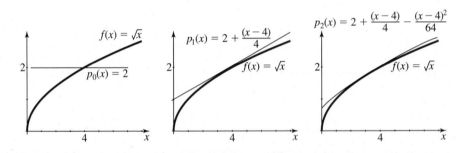

Exercises

1   Show that the first four terms in the Taylor series expansion of $f(x) = \tan x$ about $x = \dfrac{\pi}{4}$ are

$$1 + 2\left(x - \frac{\pi}{4}\right) + 2\left(x - \frac{\pi}{4}\right)^2 + \frac{8}{3}\left(x - \frac{\pi}{4}\right)^3$$

2   Find the Taylor series for $f(x) = x + e^x$ about $x = 1$.

Solutions to exercises

2   $f(x) = (1 + e)x + e\left(\dfrac{(x - 1)^2}{2!} + \dfrac{(x - 1)^3}{3!} + \dfrac{(x - 1)^4}{4!} + \cdots\right)$

Computer and calculator exercises

1. (a) Calculate the second-order Taylor polynomial, $p_2(x)$, generated by $y(x) = x^3$ about $x = 0$.
    (b) Draw $y(x)$ and $p_2(x)$ for $-2 \leq x \leq 2$.

2. (a) Calculate the second-order Taylor polynomial, $p_2(x)$, generated by $y(x) = \sin x$ about $x = 0$.
    (b) Draw $y(x)$ and $p_2(x)$ for $-2 \leq x \leq 2$.

3. (a) Calculate the second-order Taylor polynomial, $p_2(x)$, generated by
    $$y(x) = \sin\left(\frac{1}{x}\right) \text{ about } x = 3$$
    (b) Draw $y(x)$ and $p_2(x)$ for $1 \leq x \leq 5$.

4. (a) Calculate the second-order Taylor polynomial, $p_2(x)$, generated by $y(x) = e^{\cos x}$ about $x = 0$.
    (b) Draw $y(x)$ and $p_2(x)$ for $-2 \leq x \leq 2$.

5. Draw $y(x) = \dfrac{1}{x}$ for $1 \leq x \leq 8$. On the same axes draw the fourth-order Taylor polynomial generated by $y(x)$ about $x = 3$.

6. Draw $y = \tan x$ for $-1 \leq x \leq 1$. Using the same axes, draw the fifth-order Taylor polynomial generated by $y(x)$ about $x = 0$.

7. Draw $y(x) = \ln x$ for $0.5 \leq x \leq 10$. On the same axes, draw the third-, fourth- and fifth-order Taylor polynomials generated by $y(x)$ about $x = 1$.

End of block exercises

1. Use the Maclaurin expansion of e^x to find the value of e correct to four decimal places.

2. Obtain the Maclaurin series for $\ln(1 - 4x)$.

3. Obtain the Maclaurin series for $f(x) = \sinh x$.

4. Obtain a small-angle approximation for $\tan x$ and verify it by example.

5. Use the Maclaurin series for $\dfrac{1}{1-x}$ to deduce that for $\dfrac{1}{1-x^2}$.

Solutions to exercises

1. 2.7183

2. $-4x - 8x^2 - \dfrac{64}{3}x^3 - \cdots$

4. $\tan x = x$

5. $1 + x^2 + x^4 + \cdots$

End of chapter exercises

1. Write out explicitly the first three non-zero terms of the series
    $$\sum_{k=0}^{\infty} k(k + 1)(k + 3)$$

2. Use the binomial theorem to show that
    $$\frac{1}{1 + x^2} = 1 - x^2 + x^4 - x^6 + \cdots$$
    $$\text{for } -1 < x < 1$$

By integrating both sides of this expansion, assuming that this is permissible, show that $\tan^{-1} x$ has power series expansion

$$\tan^{-1} x = x - \frac{1}{3}x^3 + \frac{1}{5}x^5 - \frac{1}{7}x^7 + \cdots$$

$$\text{for } -1 < x < 1$$

3 (a) If $y = \ln \cos x$ show that y satisfies the equation

$$y''' = -2\, y''y'$$

(b) Hence obtain the Maclaurin expansion of $y = \ln \cos x$ as far as the term in x^5.

4 Find $10^3 + 11^3 + \cdots + 20^3$.

5 Expand $\left(3 - \dfrac{2}{x}\right)^5$.

6 Use the Maclaurin series for e^x to show that the sum of the infinite series

$$1 + \frac{1}{1!} + \frac{1}{2!} + \frac{1}{3!} + \cdots$$

is e.

7 Use the Maclaurin series for $\ln(1 + x)$ to show that the sum of the infinite series

$$1 - \frac{1}{2} + \frac{1}{3} - \frac{1}{4} + \cdots$$

is $\ln 2$.

8 Find $\displaystyle\sum_{k=1}^{4} k^2$.

9 Expand $(1 - 2x)^4$.

10 By considering $x + x^2$ as a single term, expand $(1 + x + x^2)^3$.

11 Use the binomial theorem to expand $(9 + x^2)^{1/2}$.

12 By considering the Maclaurin expansions of $\sin kx$ and $\cos kx$ where k is a constant, evaluate, if possible,

(a) $\displaystyle\lim_{x \to 0} \frac{\sin kx}{x}$ (b) $\displaystyle\lim_{x \to 0} \frac{\cos kx - 1}{x}$

(c) $\displaystyle\lim_{x \to 0} \frac{\sin kx}{1 - \cos kx}$

Solutions to exercises

1 (1) (2) (4) + (2) (3) (5) + (3) (4) (6)

3 $-\dfrac{1}{2}x^2 - \dfrac{1}{12}x^4$

4 42075

5 $243 - \dfrac{810}{x} + \dfrac{1080}{x^2} - \dfrac{720}{x^3} + \dfrac{240}{x^4} - \dfrac{32}{x^5}$

8 30

9 $1 - 8x + 24x^2 - 32x^3 + 16x^4$

10 $1 + 3x + 6x^2 + 7x^3 + 6x^4 + 3x^5 + x^6$

11 $3 + \dfrac{x^2}{6} - \dfrac{x^4}{216} + \cdots$, valid for $-3 < x < 3$

12 (a) k (b) 0 (c) not defined

Differential equations

Chapter **20**

The solution of problems concerning the motion of objects, the flow of charged particles, heat transport, etc., often involves discussion of relations of the form

$$\frac{\mathrm{d}x}{\mathrm{d}t} = f(x, t) \quad or \quad \frac{\mathrm{d}q}{\mathrm{d}t} = g(q, t)$$

In the first equation, x might represent distance. For this case $\frac{\mathrm{d}x}{\mathrm{d}t}$ is the rate of change of distance with respect to time, t, that is speed. In the second equation, q might be charge and $\frac{\mathrm{d}q}{\mathrm{d}t}$ the rate of flow of charge, that is current. These are examples of **differential equations**, so called because they are equations involving the derivatives of various quantities. Such equations arise out of situations in which change is occurring. To solve such a differential equation means to find the function $x(t)$ or $q(t)$ when we are given the differential equations. Solutions to these equations may be **analytical** in that we can write down an answer in terms of common elementary functions such as e^t, $\sin t$ and so on. Alternatively, the problem may be so difficult that only **numerical methods** are available, which produce approximate solutions. Modern software packages are available that can produce analytical and numerical solutions. We shall introduce some of these in Block 4.

In engineering, differential equations are most commonly used to model **dynamic systems**. These are systems that change with time. Examples include an electronic circuit with time-dependent currents and voltages, a chemical production line in which pressures, tank levels, flow rates, etc., vary with time, and a semiconductor device in which hole and electron densities change with time.

There are a wide variety of differential equations that occur in engineering applications, and consequently there are a wide variety of solution techniques available. In order to be able to apply an appropriate technique it is essential that you can identify the sort of equation you are dealing with. Block 1 gives details of the terminology associated with differential equations. Knowledge of this will help in the selection of an appropriate method. Later blocks describe these methods in detail.

Chapter 20 contents

Basic concepts of differential equations

An equation is a mathematical expression that contains an unknown quantity that we try to find. A **differential equation** is an equation that contains the derivative of an unknown expression. For example, a very simple differential equation is

$$\frac{dy}{dx} = x^2$$

This equation contains the derivative of y, which represents the unknown expression that we seek. Clearly y is an expression that when differentiated gives x^2.

In order to solve a differential equation it is important to identify certain of its features. This is because there are different techniques applicable to different sorts of differential equations. In this block we introduce terminology associated with differential equations that then allows these features to be described.

Dependent and independent variables

Recall that in a function such as $y = x^2 + 3x$ we say that x is the **independent variable** and y is the **dependent variable**, since the value of y depends upon the choice we have made for x. When solving a differential equation it is essential that you can identify the dependent and independent variables.

In the differential equation $\frac{dx}{dt} = x + t^2$, t is the independent variable and x is the dependent variable. In the equation $\frac{dy}{dx} = y + \sin x$, x is the independent variable and y is the dependent variable. Note that the dependent variable is always the variable being differentiated.

Example 1.1

For each of the following differential equations, state which variable is dependent and which is independent.

(a) $\dfrac{d^2y}{dx^2} + \dfrac{dy}{dx} = x$

(b) $\dfrac{dx}{dt} = (xt)^5$

(c) $RC\dfrac{dv_c}{dt} + v_c = V_S;$ R, C, V_S are constants.

Solution
The dependent and independent variables are respectively:

(a) ▨   dep.: y; ind.: x

(b) ▨   dep.: x; ind.: t

(c) ▨   dep.: v_c; ind.: t

Order

A differential equation may contain derivatives of various orders, for example first
derivatives, second derivatives, etc. The **order** of a differential equation is the order
of its highest derivative. Thus in the equation

$$\frac{d^2y}{dx^2} + \frac{dy}{dx} = y$$

the highest derivative is the second derivative, $\dfrac{d^2y}{dx^2}$, so this equation is a **second-
order differential equation**. In the equation

$$\frac{dv}{dt} + v^2 = 0$$

the highest derivative is the first derivative $\dfrac{dv}{dt}$, so this is a **first-order differential
equation**.

Example 1.2
State the order of the following differential equations:

(a) $\dfrac{d^2y}{dx^2} + \left(\dfrac{dy}{dx}\right)^3 = x^7$

(b) $\dfrac{dx}{dt} = (xt)^5$

Solution
(a) Look for the term involving the highest derivative.

The order of $\dfrac{d^2y}{dx^2} + \left(\dfrac{dy}{dx}\right)^3 = x^7$ is

▨   2

Note that although $\dfrac{dy}{dx}$ is raised to the power 3, it is not the highest derivative.

The highest derivative is $\dfrac{d^2y}{dx^2}$.

(b) The order of $\dfrac{dx}{dt} = (xt)^5$ is

| | 1 |

Linear

A differential equation is said to be **linear** if the dependent variable and its derivatives occur to the first power only and if there are no products involving the dependent variable and/or its derivatives. There should be no non-linear functions of the dependent variable, such as sine, exponential, etc. A differential equation that is not linear is said to be **non-linear**. The linearity of a differential equation is not determined or affected by the presence of non-linear terms involving the independent variable.

The equation

$$\frac{d^2y}{dx^2} - y = e^x$$

is linear, but the equations

$$\left(\frac{dy}{dx}\right)^2 = y \quad \text{and} \quad y\frac{dy}{dx} = x$$

are both non-linear, the former because the derivative of the dependent variable is raised to the power 2, and the latter because of the product $y\dfrac{dy}{dx}$.

The equation

$$\frac{dy}{dx} + y = e^y$$

is non-linear because it contains a non-linear function, e^y, of the dependent variable. However the equation

$$\frac{dy}{dx} + y = e^x$$

is linear.

It is particularly important to know whether an equation is linear or not, because in general non-linear equations can be much harder to solve than linear ones.

The solution of first-order linear equations is described in Block 3.

The solution of second-order linear equations is described in Blocks 5 and 6.

Example 1.3
Decide whether or not the following equations are linear:

(a) $\sin x\dfrac{dy}{dx} + y = x$

(b) $\sin y\dfrac{dy}{dx} + y = x$

(c) $\dfrac{d^2y}{dx^2} = -y^2$

(d) $\dfrac{dx}{dt} + x = t^3$

(e) $\dfrac{dy}{dx} + \sin y = 0$

Solution

(a) ░░░░░   linear

(b) ░░░░░   non-linear

(c) ░░░░░   non-linear

(d) ░░░░░   linear

(e) ░░░░░   non-linear

Note in part (b) that the equation is non-linear because of the product term involving the dependent variable and its derivative $\left(\sin y \dfrac{dy}{dx} \right)$; however, it is also non-linear because of the non-linear function of the dependent variable, $\sin y$.

Constant-coefficient linear equations

A linear differential equation has **constant coefficients** if the coefficients of the dependent variable and its derivatives are constants.

So

$$5\dfrac{d^2y}{dx^2} - 3\dfrac{dy}{dx} + 8y = x^2$$

is a constant-coefficient equation. The coefficient of y is 8. The coefficients of the first and second derivatives of y are -3 and 5 respectively.

However,

$$x\dfrac{d^2y}{dx^2} + 7\dfrac{dy}{dx} + 4y = 0$$

is not a constant-coefficient equation since the coefficient of the second derivative of y is x (i.e. not constant).

Example 1.4 Chemical Engineering – A liquid system

Figure 1.1 shows a tank of liquid. The tank has a constant cross-sectional area A. The liquid can flow out from the tank through a valve near the base. As it does so, the height, or **head**, h, of liquid in the tank will reduce. Let q stand for the rate at which liquid flows out of the tank. Under certain conditions the rate of outflow is proportional to the head, so that $q = kh$ where k is a constant of proportionality. Situations like this one arise frequently in the chemical engineering industry. By considering the rate of change of the volume of liquid in the tank, obtain a differential equation with h as the dependent variable, and time, t, as the independent variable.

Solution

The volume, V, of liquid in the tank at any time is

$$V = Ah$$

Figure 1.1
Modelling a liquid
system can give
rise to a constant-
coefficient
equation.

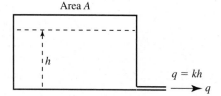

Now, the volume of liquid in the tank changes because liquid is flowing out. Specifically

the rate at which this volume changes = rate of flow in − rate of flow out

This is the **law of conservation of mass**. The rate of change of volume is $\dfrac{dV}{dt}$. There is no flow into the tank and liquid flows out at a rate q. Hence

$$\frac{dV}{dt} = -q$$

But $V = Ah$ and A is constant, so the rate of change of volume is $A\dfrac{dh}{dt}$.
Therefore

$$A\frac{dh}{dt} = -q$$

We are also given that $q = kh$ and so

$$A\frac{dh}{dt} = -kh$$

This is a first-order differential equation with dependent variable h and independent variable t. It is linear and has constant coefficients. The unknown function that we seek is $h(t)$. You will see how this equation can be solved to find the head, h, at any time, t, in Block 2.

Example 1.5 Electrical Engineering – An *LCR* circuit

Figure 1.2 shows an *LCR* circuit. This is a circuit comprising an inductor of inductance L, a capacitor of capacitance C, and a resistor of resistance R placed in series.

Figure 1.2
Modelling an *LCR*
circuit can give
rise to a constant-
coefficient
differential
equation.

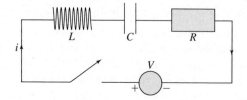

When a constant voltage source, V, is applied, it can be shown that the current, i, through the circuit satisfies the differential equation

$$L\frac{d^2i}{dt^2} + R\frac{di}{dt} + \frac{1}{C}i = 0$$

This equation can be derived using Kirchhoff's voltage law, and the individual laws for each component. Because L, R and C are constants, this is a constant-coefficient equation. Furthermore it is linear and second order. When you have learnt how to solve such equations you will be able to find the current in the circuit.

Exercises

1   Explain the distinction between a dependent variable and an independent variable.

2   In the differential equation

$$\frac{d\theta}{dt} + \theta = \sin t$$

state which is the dependent variable and which is the independent variable.

3   Explain what is meant by the order of a differential equation.

4   State the order of the equations

(a) $y'' + 2y' + 7y = 0$ (b) $\left(\frac{dx}{dt}\right)^3 + x = 0$

5   Explain what is meant by a linear differential equation.

6   Is the equation $\frac{d^2y}{dt^2} + y = \sin t$ linear?

Solutions to exercises

2   θ is the dependent variable and t is the independent variable.

4   (a) second (b) first

6   yes

1.3   The solution of a differential equation

Given a differential equation such as

$$\frac{dy}{dx} = y$$

a **solution** is found when we have obtained an expression for y in terms of x that can be substituted into the equation to make both sides equal. That is, a solution is an expression that **satisfies** the differential equation. Note that a solution is an expression of the dependent variable in terms of the independent variable.

Example 1.6

Show that $y = 5e^{2x}$ is a solution of the equation $\dfrac{d^2y}{dx^2} - \dfrac{dy}{dx} = 2y$.

Solution

We are given an expression for y in terms of x, namely $y = 5e^{2x}$. We need to substitute this into both sides of the given equation. To do this we need the derivatives $\dfrac{dy}{dx}$ and $\dfrac{d^2y}{dx^2}$.

Write expressions for $\dfrac{dy}{dx}$ and $\dfrac{d^2y}{dx^2}$.

$$\frac{dy}{dx} = 10e^{2x}$$

$$\frac{d^2y}{dx^2} = 20e^{2x}$$

Now substitute these expressions into the left-hand side of the given equation $\dfrac{d^2y}{dx^2} - \dfrac{dy}{dx}$:

$$20e^{2x} - 10e^{2x} \text{ which equals } 10e^{2x}$$

Now substitute $y = 5e^{2x}$ into the right-hand side of the given equation, $2y$:

$$2y = 10e^{2x}$$

Note that both the left- and right-hand sides are the same and equal to $10e^{2x}$ when $y = 5e^{2x}$. We conclude that $y = 5e^{2x}$ is a solution of the equation.

Example 1.7

Show that $y = e^x$ is *not* a solution of $\dfrac{dy}{dx} = 2y$.

Solution

Differentiate the given expression and substitute into the left-hand side:

$$\frac{dy}{dx} = e^x$$

Substitute for y on the right-hand side:

$$2y = 2e^x$$

So the left-hand side is e^x and the right-hand side is $2e^x$. But e^x is not the same as $2e^x$ and so the given expression does not satisfy the equation and is therefore not a solution.

There are many different expressions that can satisfy a differential equation: that is, there are many solutions. A solution from which all possible solutions can be found is called the **general solution**.

The general solution of the equation $\dfrac{dy}{dx} = y$ is $y = Ce^x$, where C is any constant. C is called an **arbitrary constant**. By choosing different values of C, different solutions are obtained. Every solution of the equation can be obtained from this general solution.

Conditions

To determine a particular value for the constant C we need to be given more information in the form of a **condition**. For example, if we are told that at $x = 0$, $y = 4$ then from $y = Ce^x$ we have

$$4 = Ce^0$$
$$= C$$

so that $C = 4$. Therefore $y = 4e^x$ is the solution of the differential equation that additionally satisfies the condition $y(0) = 4$. This is called a **particular solution**.

When the solution of a differential equation is sought, and a condition is to be satisfied at the leftmost point of the interval of interest, such a condition is called an **initial condition**. The problem of solving a differential equation subject to an initial condition is often referred to as an **initial value problem**.

Example 1.8

Show that $x(t) = 2e^{3t}$ is a solution of the initial value problem

$$\frac{dx}{dt} - 2x = 2e^{3t}, \quad x(0) = 2$$

Solution

The solution requires us to do two things. First we must check that $x = 2e^{3t}$ satisfies the differential equation, and second we must check that it satisfies the initial condition: that is, x must equal 2 when $t = 0$.

Differentiate $x(t) = 2e^{3t}$.

$$6e^{3t}$$

Substitute for $\dfrac{dx}{dt}$ and x in the left-hand side of the given equation.

$$\frac{dx}{dt} - 2x = \qquad\qquad\qquad\qquad 6e^{3t} - 2(2e^{3t})$$

which simplifies to $2e^{3t}$, the same as the right-hand side. Thus $x = 2e^{3t}$ is a solution of the differential equation.

We now check that the initial condition is satisfied. $x(0) = 2e^0 = 2$ and so the condition is indeed satisfied.

Exercises

1 Verify that $y = 3 \sin 2x$ is a solution of

$$\frac{d^2y}{dx^2} + 4y = 0$$

2 Verify that $3e^x$, Axe^x, $Axe^x + Be^x$, where A, B are constants, all satisfy the differential equation

$$\frac{d^2y}{dx^2} - 2\frac{dy}{dx} + y = 0$$

3 Verify that $x = t^2 + A \ln t + B$ is a solution of

$$t\frac{d^2x}{dt^2} + \frac{dx}{dt} = 4t$$

4 Verify that $y = A\cos x + B \sin x$ satisfies the differential equation

$$\frac{d^2y}{dx^2} + y = 0$$

Verify also that $y = A \cos x$ and $y = B \sin x$ each individually satisfy the equation.

5 The general solution of the equation

$$\frac{d^2x}{dt^2} - 3\frac{dx}{dt} + 2x = 0$$

is given by

$$x = Ae^t + Be^{2t}$$

Find the particular solution that satisfies $x = 3$ and $\dfrac{dx}{dt} = 5$ when $t = 0$.

6 The general solution of

$$\frac{d^2y}{dx^2} - 2\frac{dy}{dx} + y = 0$$

is $y = Axe^x + Be^x$. Find the particular solution satisfying $y(0) = 0$, $\dfrac{dy}{dx}(0) = 1$.

Solutions to exercises

5 $x = e^t + 2e^{2t}$

6 $y = xe^x$

1.4   Solving a differential equation by direct integration

When a differential equation has a particularly simple form it is easy to solve it by integration.

Suppose we have a first-order equation of the form

$$\frac{dy}{dx} = f(x)$$

Examples include $\dfrac{dy}{dx} = 3x^4$ and $\dfrac{dy}{dx} = 4 \sin 2x + 3 \cos 2x$. That is, $\dfrac{dy}{dx}$ is equal to a function of x only. Then y is given by

$$y = \int f(x)\,dx$$

Key point

$$\text{if} \quad \frac{dy}{dx} = f(x) \quad \text{then} \quad y = \int f(x)\, dx$$

Consider the following example.

Example 1.9

Obtain the general solution of the equation

$$\frac{dy}{dx} = \cos x + \sin x$$

Solution

Here, the right-hand side is a function of x only.

$$y = \int \cos x + \sin x\, dx$$

$$= \sin x - \cos x + c$$

This is the general solution. Notice that it contains one arbitrary constant.

The same method can be applied to second-order equations of the form

$$\frac{d^2y}{dx^2} = f(x)$$

Here again, the right-hand side is a function of x only.

Example 1.10

Obtain the general solution of the equation

$$\frac{d^2y}{dx^2} = 5e^{2x}$$

Solution

Integrating once gives

$$\frac{dy}{dx} = \int 5e^{2x}\, dx$$

$$\frac{dy}{dx} = \frac{5e^{2x}}{2} + A$$

where A is an arbitrary constant. This gives $\dfrac{dy}{dx}$. We must integrate again to find y.

$$y = \int \left(\frac{5e^{2x}}{2} + A \right) dx$$

$$= \frac{5e^{2x}}{4} + Ax + B$$

where B is another arbitrary constant.

This is the general solution. Notice that it contains two arbitary constants. This is because the differential equation is second order.

The technique described in this section is applicable only when the right-hand side is a function of the independent variable only. If the dependent variable appears on the right, such as in the equation

$$\frac{dy}{dx} = xy$$

you will need to apply other techniques such as those described in Blocks 2 and 3.

 Example 1.11 Dynamics – The vertical motion of a projectile
When a projectile such as a ball is travelling vertically under the action of gravity its motion is described by the differential equation

$$\frac{d^2y}{dt^2} = -g$$

where g is a constant called the **acceleration due to gravity**. The dependent variable y is the vertical displacement of the projectile, and the independent variable is time t. This equation is a statement of Newton's second law of motion. Solving the equation gives y in terms of t, that is the displacement as a function of time.

(a) Integrate the differential equation twice, to obtain the general solution for y.

(b) Apply the following initial conditions: $y = 0$ at $t = 0$, $\frac{dy}{dt} = v_0$ at $t = 0$, in order to obtain a particular solution.

Solution
(a) Integrate once:

$$\frac{dy}{dt} = -gt + A$$

Integrate again:

$$y = -\frac{gt^2}{2} + At + B$$

This is the general solution.
(b) Apply the condition $y = 0$ when $t = 0$ to obtain a value for B.

$$B = 0$$

So $y = -\dfrac{gt^2}{2} + At$. If this is differentiated, $\dfrac{dy}{dt} = -gt + A$, which is a result already obtained. Applying the condition $\dfrac{dy}{dt} = v_0$ when $t = 0$ gives a value for A:

$$A = v_0$$

So the particular solution is $y = -\dfrac{gt^2}{2} + v_0 t$.

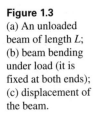

Example 1.12 Mechanical Engineering – Bending of a beam

Figure 1.3(a) illustrates an unloaded beam of length L. The beam is fixed at both ends. When a load (force) is applied the beam bends as illustrated in Figure 1.3(b).

Let the weight of the load being supported per unit length be w. Referring to Figure 1.3(b) we see that, when loaded, the beam bends in the negative y direction. The magnitude of the bending is the crucial quantity and this is illustrated in Figure 1.3(c). We refer to this as the displacement of the beam, denoted $H(x)$. Note that in Figure 1.3(c) the H axis is positive in the downward direction. It is possible to show that

$$K \frac{d^4 H}{dx^4} = w$$

Figure 1.3
(a) An unloaded beam of length L;
(b) beam bending under load (it is fixed at both ends);
(c) displacement of the beam.

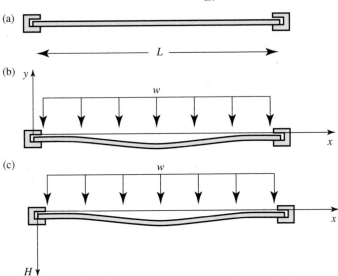

where K is known as the structural rigidity of the beam and is assumed to be constant. It is the product of Young's modulus and the moment of inertia of the beam about its central axis.

(a) If w is constant, find an expression for the displacement of the beam.
(b) The beam is fixed rigidly at both ends. Hence when $x = 0$ and $x = L$ the displacement is zero, that is $H(0) = H(L) = 0$. Additionally at these end-points the gradient of the beam remains zero, so $H'(0) = H'(L) = 0$. Find the displacement at any point x.
(c) Show that the maximum displacement of the beam occurs half-way along its length.

Solution

(a) We are given

$$K \frac{d^4 H}{dx^4} = w$$

Repeated integration produces

$$\frac{d^3 H}{dx^3} = \frac{w}{K} x + A \tag{1}$$

$$\frac{d^2 H}{dx^2} = \frac{w}{K} \frac{x^2}{2} + Ax + B \tag{2}$$

$$\frac{d H}{dx} = \frac{w}{K} \frac{x^3}{6} + A \frac{x^2}{2} + Bx + C \tag{3}$$

$$H(x) = \frac{w}{K} \frac{x^4}{24} + A \frac{x^3}{6} + B \frac{x^2}{2} + Cx + D \tag{4}$$

where A, B, C and D are constants of integration.

(b) We now apply the given conditions to find the unknown constants A, B, C and D.
Applying $H(0) = 0$ to equation (4) yields

$$0 = D$$

Applying $H'(0) = 0$ to equation (3) yields

$$0 = C$$

Applying $H(L) = 0$ to equation (4) with $C = D = 0$ yields

$$0 = \frac{w}{K} \frac{L^4}{24} + A \frac{L^3}{6} + B \frac{L^2}{2} \tag{5}$$

Applying $H'(L) = 0$ to equation (3) with $C = 0$ yields

$$0 = \frac{w}{K} \frac{L^3}{6} + A \frac{L^2}{2} + BL \tag{6}$$

Equations (5) and (6) can be solved to express A and B in terms of the known constants w, K and L. This produces

$$A = -\frac{w L}{2K}, \quad B = \frac{wL^2}{12K}$$

Replacing the expressions for A and B and the values of C and D into equation (4) and simplifying produces

$$H(x) = \frac{wx^2}{24K} (x^2 - 2Lx + L^2) = \frac{wx^2}{24K} (x - L)^2$$

(c) We now seek the point along the beam at which the displacement is a maximum. Recall that turning points of a function are located by equating the first derivative to zero, that is $H' = 0$; the nature of the turning point is then determined using the second-derivative test.

Using equation (3) we have

$$\frac{dH}{dx} = \frac{wx^3}{6K} - \frac{wLx^2}{4K} + \frac{wL^2x}{12K} = \frac{wx}{12K}(2x^2 - 3xL + L^2)$$

which may be expressed as

$$\frac{dH}{dx} = \frac{wx}{12K}(2x - L)(x - L)$$

so $dH/dx = 0$ when $x = 0, L/2, L$.

The conditions $H'(0) = H'(L) = 0$ are given so we investigate that case where $x = L/2$ using the second-derivative test.

Using equation (2) we have

$$\frac{d^2H}{dx^2} = \frac{wx^2}{2K} - \frac{wLx}{2K} + \frac{wL^2}{12K}$$

Substituting $x = L/2$ and simplifying yields

$$\frac{d^2H}{dx^2}(L/2) = -\frac{wL^2}{24K}$$

We note that this is negative and so H is a maximum at $x = L/2$. Hence the maximum displacement occurs at $x = L/2$, that is half-way along the beam. This accords with intuition.

Exercises

1   Obtain the general solutions of the following differential equations:

(a) $\dfrac{dy}{dx} = 3x + 2$ (b) $\dfrac{dy}{dx} = e^{3x}$

(c) $\dfrac{d^2y}{dx^2} = k$, constant   (d) $\dfrac{d^2y}{dx^2} = -\sin x$

2   Find the particular solution of

$$\frac{dx}{dt} = t^2$$

that satisfies the condition $x(3) = 0$.

Solutions to exercises

1 (a) $y = \dfrac{3x^2}{2} + 2x + c$ (b) $y = \dfrac{e^{3x}}{3} + c$

(c) $y = \dfrac{kx^2}{2} + Ax + B$

(d) $y = \sin x + Ax + B$

2 $x = \dfrac{t^3}{3} - 9$

End of block exercises

1 If $y = Ae^{2x}$ is the general solution of $\dfrac{dy}{dx} = 2y$, find the particular solution satisfying $y(0) = 3$. What is the particular solution satisfying $\dfrac{dy}{dx} = 2$ when $x = 0$?

2 Identify the dependent and independent variables of the following differential equations. Give the order of the equations and state which are linear.

(a) $\dfrac{dy}{dx} + 9y = 0$

(b) $\left(\dfrac{dy}{dx}\right)\left(\dfrac{d^2y}{dx^2}\right) + 3\dfrac{dy}{dx} = 0$

(c) $\dfrac{d^3x}{dt^3} + 5\dfrac{dx}{dt} = \sin x$

3 Show that $x(t) = 7\cos 3t - 2\sin 2t$ is a solution of

$$\dfrac{d^2x}{dt^2} + 2x = -49\cos 3t + 4\sin 2t$$

4 The general solution of

$$\dfrac{d^2x}{dt^2} = -\omega^2 x$$

is $x = Ae^{j\omega t} + Be^{-j\omega t}$, where $j^2 = -1$. Verify that this is indeed a solution. What is the particular solution satisfying $x(0) = 0$, $\dfrac{dx}{dt}(0) = 1$? Express the general solution and the particular solution in terms of trigonometrical functions.

Solutions to exercises

1 $y(x) = 3e^{2x}$, $y(x) = e^{2x}$

2 (a) y is the dependent variable; x is the independent variable; first order, linear
(b) y is the dependent variable; x is the independent variable; second order, non-linear

(c) x is the dependent variable; t is the independent variable; third order, non-linear

4 particular solution: $\dfrac{\sin \omega t}{\omega}$; general solution:

$$(A + B)\cos \omega t + (A - B)j \sin \omega t$$

Separation of variables

Separation of variables is a technique commonly used to solve first-order differential equations. It is so called because we try to rearrange the equation to be solved in such a way that all terms involving the dependent variable (y say) appear on one side of the equation, and all terms involving the independent variable (x say) appear on the other. Then it is a matter of integration to complete the solution. It is not possible to rearrange all first-order equations in this way, so this technique is not always appropriate. Further, it is not always possible to perform the integration even if the variables are separable. In this block you will learn how to decide whether the method is appropriate, and how to apply it in such cases.

2.2 Separating the variables

In this section we consider differential equations that can be written in the form

$$\frac{dy}{dx} = f(x)g(y)$$

Note that the right-hand side is a product of a function of x and a function of y. Examples of such equations are

$$\frac{dy}{dx} = x^2y^3, \quad \frac{dy}{dx} = y^2 \sin x \quad \text{and} \quad \frac{dy}{dx} = (\ln x)(\sec y)$$

You may have to think carefully about some equations to recognise this form. For example, the equation

$$\frac{dy}{dx} = \frac{y}{x+1}$$

can be written as

$$\frac{dy}{dx} = \left(\frac{1}{x+1}\right)y$$

and so is of the form $\dfrac{dy}{dx} = f(x)g(y)$.

Not all first-order equations can be written in this form. For example, it is not possible to rewrite the equation

$$\frac{dy}{dx} = x^2 + y^3$$

in the form

$$\frac{dy}{dx} = f(x)g(y)$$

Example 2.1

Which of the following differential equations do you think can be written in the form $\frac{dy}{dx} = f(x)g(y)$?

If possible, rewrite each equation in this form.

(a) $\dfrac{dy}{dx} = \dfrac{x^2}{y^2}$ (b) $\dfrac{dy}{dx} = 4x^2 + 2y^2$ (c) $y\dfrac{dy}{dx} + 3x = 7$

Solution

(a) $\qquad\qquad\qquad\qquad\qquad\qquad\qquad\qquad\qquad\qquad \dfrac{dy}{dx} = x^2\left(\dfrac{1}{y^2}\right)$

(b) $\qquad\qquad\qquad\qquad\qquad\qquad\qquad$ cannot be written in the stated form

(c) $\qquad\qquad\qquad\qquad\qquad\qquad\qquad\qquad \dfrac{dy}{dx} = (7 - 3x) \times \dfrac{1}{y}$

The variables involved need not be x and y. Other equations of this type are

$$\frac{dz}{dt} = te^z, \quad \frac{d\theta}{dt} = -\theta \quad \text{and} \quad \frac{dv}{dr} = \frac{v}{r^2}$$

Given a differential equation in the form $\dfrac{dy}{dx} = f(x)g(y)$ we can divide through by $g(y)$ to obtain

$$\frac{1}{g(y)}\frac{dy}{dx} = f(x)$$

Note that in this form terms involving y and $\dfrac{dy}{dx}$ have been placed on the left, and terms involving just x have been placed on the right.

If we now integrate both sides of this equation with respect to x we obtain

$$\int \frac{1}{g(y)}\frac{dy}{dx}\, dx = \int f(x)\, dx$$

that is

$$\int \frac{1}{g(y)}\, dy = \int f(x)\, dx$$

We have **separated the variables** because the left-hand side contains only the variable y, and the right-hand side contains only the variable x. We can now try to integrate each side separately. If we can actually perform the required integrations we shall obtain a relationship between y and x. Examples of this process are given in the next section.

<table>
<tr><td>Key point</td><td>

Separation of variables
The solution of the equation

$$\frac{dy}{dx} = f(x)g(y)$$

is found from

$$\int \frac{1}{g(y)}\,dy = \int f(x)\,dx$$

</td></tr>
</table>

2.3   Applying the method of separation of variables

The method is illustrated in the following example.

Example 2.2

Solve the first-order equation $\dfrac{dy}{dx} = 3ye^x$ by separation of variables.

Solution
Note that the right-hand side has the form $f(x)g(y)$, where $f(x) = e^x$ and $g(y) = 3y$. Dividing both sides by $g(y)$ we find

$$\frac{1}{3y}\frac{dy}{dx} = e^x$$

The left-hand side now contains terms involving only y and $\dfrac{dy}{dx}$. The right-hand side is a function of x. Integrating both sides with respect to x,

$$\int \frac{1}{3y}\frac{dy}{dx}\,dx = \int e^x\,dx$$

that is

$$\int \frac{1}{3y}\,dy = \int e^x\,dx$$

Performing the two integrations gives

$$\frac{1}{3}\ln y = e^x + c$$

You might think that there should be a constant of integration on the left too. You are quite right, but the two constants can be combined into a single constant and so we need write only one. We now have a relationship between y and x as required. This is the general solution. In this case it could be rearranged to give y explicitly if required.

Example 2.3

Use the method of separation of variables to solve the differential equation

$$\frac{dy}{dx} = \frac{3x^2}{y}$$

Solution

The equation already has the form $\frac{dy}{dx} = f(x)g(y)$ where $f(x) = 3x^2$ and $g(y) = \frac{1}{y}$. Dividing both sides by $g(y)$ we find

$$y\frac{dy}{dx} = 3x^2$$

Integrating both sides with respect to x gives

$$\int y\frac{dy}{dx}\,dx = \int 3x^2\,dx$$

that is

$$\int y\,dy = \int 3x^2\,dx$$

Note that the left-hand side is an integral involving just y; the right-hand side is an integral involving just x. After integrating we find

$$\frac{y^2}{2} = x^3 + c$$

where c is a constant of integration. We now have a relationship between y and x as required. This is the general solution. Often it is sufficient to leave your answer in this form but you may also be required to obtain an explicit relation for y in terms of x. In this particular case

$$y^2 = 2x^3 + 2c$$

so that

$$y = \pm\sqrt{2x^3 + D} \quad \text{where } D = 2c$$

Example 2.4

Use the method of separation of variables to solve the differential equation

$$\frac{dy}{dx} = \frac{\cos x}{\sin 2y}$$

Solution

Separate the variables so that terms involving y and $\frac{dy}{dx}$ appear on the left, and terms involving just x appear on the right:

$$\sin 2y\frac{dy}{dx} = \cos x$$

Then integrate both sides with respect to x to obtain

$$\int \sin 2y \, \frac{dy}{dx} \, dx = \int \cos x \, dx$$

that is

$$\int \sin 2y \, dy = \int \cos x \, dx$$

Now integrate both sides:

$$-\frac{\cos 2y}{2} = \sin x + c$$

Finally, rearrange, if possible, to obtain an expression for y in terms of x:

$$y = \frac{1}{2} \cos^{-1} (D - 2 \sin x) \quad \text{where } D = -2c$$

This is the general solution.

Example 2.5 Mechanical Engineering – Simple harmonic motion

When an object such as a mass vibrating on a spring, or a pendulum oscillating under the action of gravity, moves in **simple harmonic motion** its velocity v satisfies a differential equation of the form

$$v \frac{dv}{dx} = -k^2 x$$

where x is displacement and k is a constant. Solve this equation.

Solution

The variables are already separated because v and $\dfrac{dv}{dx}$ are on the left, and the right-hand side is a function of x only.

Integrate both sides with respect to x.

$$\int v \frac{dv}{dx} \, dx = \int -k^2 x \, dx$$

This can be written

$$\int v \, dv = \int -k^2 x \, dx$$

Complete the integration to find v.

$$\frac{v^2}{2} = -k^2 \frac{x^2}{2} + c$$

This is the general solution. If we wished to we could write v explicitly in terms of x.

Example 2.6 Chemical Engineering – A liquid system

The differential equation describing the head, h, of liquid in a tank (see Example 1.4 in Block 1) is given by

$$A \frac{dh}{dt} = -kh$$

Solve this differential equation given that at $t = 0$ the head of liquid is h_0. A and k are constants.

Solution

Separate the variables h and t:

$$\frac{A}{h} \frac{dh}{dt} = -k$$

Both sides are then integrated.

$$A \int \frac{dh}{h} = -k \int dt$$

Perform the integration to obtain the general solution:

$$A \ln h = -kt + C$$

This is the general solution. By imposing the condition $h = h_0$ at $t = 0$ find the particular solution.

$$C = A \ln h_0$$

Thus

$$A \ln h = -kt + A \ln h_0$$

Rearranging and using the laws of logarithms to combine the two log terms,

$$A \ln h - A \ln h_0 = A \ln \frac{h}{h_0}$$

$$= -kt$$

Rearrange this to obtain h:

$$h = h_0 e^{-kt/A}$$

We interpret this result as saying that the head reduces exponentially from its initial value of h_0 to a final value of zero as we would expect on physical grounds.

Example 2.7 Manufacturing Engineering – Heat transfer during quenching

During the manufacture of steel components it is often necessary to quench them in a large bath of liquid in order to cool them down. This reduces the temperature of the component to the temperature of the liquid. If θ is the temperature of the component in excess of the liquid temperature it can be shown that the variation of θ with time, t, satisfies the first-order differential equation

$$\frac{d\theta}{dt} = -k\theta$$

where k is a constant that depends upon the volume and surface area of the component, its specific heat capacity, and the heat transfer coefficient between the component and the liquid. Solve this equation given the initial condition that at $t = 0$ the temperature excess is θ_0.

Solution

Separate the variables.

$$\frac{1}{\theta}\frac{d\theta}{dt} = -k$$

Integrate both sides with respect to t.

$$\ln\theta + C = -kt$$

Apply the initial condition.

$$\ln\theta_0 + C = 0 \text{ so } C = -\ln\theta_0$$

Hence

$$\ln\theta - \ln\theta_0 = -kt$$

from which

$$\ln\frac{\theta}{\theta_0} = -kt$$

Finally, obtain θ as a function of t.

$$\theta = \theta_0 e^{-kt}$$

This means that the temperature excess of the component decays exponentially from its initial value θ_0 to zero. When the temperature excess is zero, the component has cooled to the temperature of the bath.

Exercises

1   Solve the equation

$$\frac{dy}{dx} = \frac{e^{-x}}{y}$$

2   Solve the equation

$$\frac{dy}{dx} = 3x^2 e^{-y}$$

subject to the condition $y(0) = 1$.

3   Find the general solution of the following equations:

(a) $\dfrac{dy}{dx} = 3$ (b) $\dfrac{dy}{dx} = \dfrac{6\sin x}{y}$

4   Find the general solution of the equation

$$\frac{dx}{dt} = t(x - 2)$$

Find the particular solution that satisfies the condition $x(0) = 5$.

5   Some equations that do not appear to be separable can be made so by means of a suitable substitution. In this exercise you will solve the equation

$$\frac{dy}{dx} = \frac{y^2}{x^2} + \frac{y}{x} + 1$$

by means of the substitution $z = \dfrac{y}{x}$.

(a) If $z = \dfrac{y}{x}$, then $y = zx$. Using the product rule for differentiation show that

$$\frac{dy}{dx} = z + x\frac{dz}{dx}$$

(b) Make the substitution to obtain an equation involving only z and x. Solve this by separation of variables. Hence obtain the solution of the given equation.

6 The equation

$$iR + L\frac{di}{dt} = E$$

where R, L and E are constants arises in electrical circuit theory. This equation can be solved by separation of variables. Find the solution that satisfies the condition $i(0) = 0$.

7 Solve the equation $\dot{x} = 4xt$.

Solutions to exercises

1 $y = \pm\sqrt{D - 2e^{-x}}$

2 $y = \ln(x^3 + e)$

3 (a) $y = 3x + C$ (b) $\dfrac{y^2}{2} = C - 6\cos x$

4 $x = 2 + Ae^{t^2/2}, x = 2 + 3e^{t^2/2}$

5 $z = \tan(\ln Dx)$ so that $y = x\tan(\ln Dx)$

6 $i = \dfrac{E}{R}(1 - e^{-t/\tau})$ where $\tau = L/R$

7 $x = Ae^{2t^2}$

End of block exercises

1 Find the general solution of the following equations:

(a) $\dfrac{dy}{dx} = kx$ (b) $\dfrac{dy}{dx} = -ky$ (c) $\dfrac{dy}{dx} = y^2$

(d) $y\dfrac{dy}{dx} = \sin x$ (e) $y\dfrac{dy}{dx} = x + 2$

(f) $x^2\dfrac{dy}{dx} = 2y^2 + yx$ (g) $\dfrac{dx}{dt} = \dfrac{t^4}{x^5}$

2 Find the general solution of the following:

(a) $\dfrac{dx}{dt} = xt$ (b) $\dfrac{dy}{dx} = \dfrac{x}{y}$

(c) $t\dfrac{dx}{dt} = \tan x$ (d) $\dfrac{dx}{dt} = \dfrac{x^2 - 1}{t}$

Solutions to exercises

1 (a) $\dfrac{kx^2}{2} + c$ (b) Ae^{-kx} (c) $-\dfrac{1}{x + c}$

(d) $y^2 = 2(C - \cos x)$ (e) $y^2 = x^2 + 4x + c$

(f) $\dfrac{x}{A - 2\ln x}$ (g) $\dfrac{x^6}{6} = \dfrac{t^5}{5} + c$

2 (a) $x = Ae^{t^2/2}$ (b) $y^2 = x^2 + C$

(c) $x = \sin^{-1}(kt)$ (d) $x = \dfrac{1 + At^2}{1 - At^2}$

Solving first-order linear equations using an integrating factor

3.1 Introduction

The standard form of a first-order linear differential equation is

$$\frac{dy}{dx} + P(x)y = Q(x) \tag{1}$$

$P(x)$ and $Q(x)$ are given functions of x, which may be constants. All first-order linear equations can be written in this form. Note that the coefficient of $\frac{dy}{dx}$ is 1. An example is

$$\frac{dy}{dx} + x^2y = 3 + x$$

in which $P(x) = x^2$ and $Q(x) = 3 + x$.

Other variables will be used. For example,

$$\frac{dx}{dt} + x \sin t = t^2$$

is a first-order linear equation in standard form with dependent variable x and independent variable t. Here, $P(t) = \sin t$ and $Q(t) = t^2$.

Key point

> All first-order linear differential equations can be written in standard form as
>
> $$\frac{dy}{dx} + P(x)y = Q(x)$$

Equations of this type can always be solved by multiplying through by a function known as the **integrating factor** for the equation. In this block you will learn how to determine the integrating factor and how to obtain the solution of the original equation.

3.2 Writing equations in standard form

In order to use the technique described in this block you must write the equation in the standard form.

Example 3.1

Write $x^2 \frac{dy}{dx} + 3xy = x^3$ in standard form. Identify $P(x)$ and $Q(x)$.

Solution

Note that this is a first-order equation. It is linear, and so the integrating factor method is appropriate. Study the standard form (1) of a first-order linear equation and note that the coefficient of the derivative term is 1. We divide the given equation through by x^2 to achieve this.

$$\frac{dy}{dx} + \frac{3}{x}y = x$$

Compare this with the standard form to identify $P(x)$ and $Q(x)$.

$P(x) = $ $\dfrac{3}{x}$

$Q(x) = $ x

Example 3.2

Compare the equation $\dfrac{dy}{dx} - 2y = x$ with the standard form and identify $P(x)$ and $Q(x)$.

Solution

The coefficient of the derivative term is already 1.

$P(x) = $ -2, in this case, a constant

$Q(x) = $ x

The equation is already written in standard form.

Example 3.3 Electrical Engineering – Current in a circuit

The equation $L\dfrac{di}{dt} + Ri = E\sin t$, where L, R and E are constants, models the current, $i(t)$, in a series circuit. Write the equation in standard form. Identify P and Q.

Solution

Note that this is a first-order equation. It is linear, and so the integrating factor method is appropriate. Here the dependent variable is i and the independent variable is t. Study the standard form (1) and note that the coefficient of the derivative term is 1. Dividing the given equation through by L we can achieve this:

$$\frac{di}{dt} + \frac{R}{L}i = \frac{E\sin t}{L}$$

This is the standard form. Write down $P(t)$ and $Q(t)$.

$P(t) = $ $\dfrac{R}{L}$, which is constant

$Q(t) = $ $\dfrac{E\sin t}{L}$

Exercises

1 Write each of the following equations in standard form:

(a) $x\dfrac{\mathrm{d}y}{\mathrm{d}x} + 2xy = xe^{-2x}$

(b) $\dfrac{\mathrm{d}y}{\mathrm{d}x} + 3y - \sin x = 0$

(c) $iR + L\dfrac{\mathrm{d}i}{\mathrm{d}t} = \sin \omega t$

Solutions to exercises

1 (a) $\dfrac{\mathrm{d}y}{\mathrm{d}x} + 2y = e^{-2x}$ (b) $\dfrac{\mathrm{d}y}{\mathrm{d}x} + 3y = \sin x$

(c) $\dfrac{\mathrm{d}i}{\mathrm{d}t} + \dfrac{R}{L}i = \dfrac{1}{L}\sin \omega t$

3.3 Finding an integrating factor

The idea behind the method is to multiply the equation

$$\frac{\mathrm{d}y}{\mathrm{d}x} + P(x)y = Q(x)$$

through by some function, as yet unknown, called an **integrating factor**. Let this integrating factor be μ. Then

$$\mu\frac{\mathrm{d}y}{\mathrm{d}x} + \mu P(x)y = \mu Q(x) \tag{2}$$

Suppose we choose μ in a special way so that the left-hand side of this equation can be written more simply as

$$\frac{\mathrm{d}}{\mathrm{d}x}(\mu y)$$

This special way of choosing μ is described below. The equation becomes

$$\frac{\mathrm{d}}{\mathrm{d}x}(\mu y) = \mu Q(x)$$

In this form the equation is said to be **exact**. It follows immediately, by integrating both sides, that

$$\mu y = \int \mu Q(x)\,\mathrm{d}x \tag{3}$$

So by knowing $Q(x)$ and μ, and performing the integration on the right, and dividing through by μ, we can obtain y.

Choosing μ

We want to choose μ so that the left-hand side of (2) is equal to $\dfrac{d}{dx}(\mu y)$. That is,

$$\mu\frac{dy}{dx} + \mu P(x)y = \frac{d}{dx}(\mu y)$$

Using the product rule for differentiation we can expand the product on the right to give

$$\mu\frac{dy}{dx} + \mu P(x)y = \mu\frac{dy}{dx} + y\frac{d\mu}{dx}$$

Comparing the coefficients of y on both sides shows that we must choose μ to satisfy

$$\mu P(x) = \frac{d\mu}{dx}$$

This is a first-order equation for μ, which can be solved by separating the variables as follows:

$$\int\frac{d\mu}{\mu} = \int P(x)\,dx$$

from which

$$\ln\mu = \int P(x)\,dx$$

We have ignored the constant of integration. When you are experienced at this technique, try including a constant of integration and you will see that it makes no difference to the solution.

It follows that

$$\mu = e^{\int P(x)\,dx}$$

Thus knowing $P(x)$ we can use this formula to find the integrating factor. Once μ is known y can be determined from the key point.

Key point

> **The integrating factor method**
> If the linear differential equation is in the standard form
>
> $$\frac{dy}{dx} + P(x)y = Q(x)$$
>
> then the integrating factor is
>
> $$\mu = e^{\int P(x)\,dx}$$
>
> and y is given by
>
> $$\mu y = \int \mu Q(x)\,dx$$

Example 3.4

Find the integrating factor for the equations

(a) $x\dfrac{dy}{dx} + 2xy = xe^{-2x}$ (b) $\dfrac{dy}{dx} - (\tan x)y = 1$

Solution

In each case write the equation in standard form and identify $P(x)$. The integrating factor is found from $\mu = e^{\int P(x)\, dx}$.

(a) $P(x) = 2;\ \mu = e^{2x}$

(b) $P(x) = -\tan x;\ \mu = e^{\ln \cos x} = \cos x$

Example 3.5

Use the integrating factors found in Example 3.4 to find the general solutions to the following differential equations:

(a) $x\dfrac{dy}{dx} + 2xy = xe^{-2x}$ (b) $\dfrac{dy}{dx} - (\tan x)y = 1$

Solution

(a) The standard form is $\dfrac{dy}{dx} + 2y = e^{-2x}$ for which the integrating factor is $\mu = e^{2x}$.
Note that $Q(x) = e^{-2x}$.
From the previous key point we have

$$\mu y = \int \mu Q(x)\, dx$$

so that

$$e^{2x}y = \int e^{2x}e^{-2x}\, dx$$

$$= \int 1\, dx$$

$$= x + C$$

Therefore

$$y = \frac{(x + C)}{e^{2x}}$$

$$= (x + C)e^{-2x}$$

This is the general solution.

(b) The equation is in standard form. The integrating factor is $\mu = \cos x$ and $Q(x) = 1$. From the key point we have

$$\mu y = \int \mu Q(x)\, dx$$

so that

$$(\cos x)\, y = \int \cos x \, dx$$

$$= \sin x + C$$

giving

$$y = \frac{\sin x}{\cos x} + \frac{C}{\cos x}$$

$$= \tan x + C \sec x$$

This is the general solution.

Example 3.6

(a) Use the integrating factor method to solve the equation $\dfrac{dy}{dx} + \dfrac{y}{x} = 1$.

(b) Find the particular solution where $y(1) = 3$.

Solution

(a) Refer to the standard form and identify $P(x)$ and $Q(x)$.

$P(x) = $ ▢ $\hspace{6cm} \dfrac{1}{x}$

$Q(x) = $ ▢ $\hspace{6cm} 1$

The integrating factor is found from the formula $\mu = e^{\int P(x)\, dx}$. Find the integrating factor:

▢ $\hspace{5cm} \mu = e^{\int (1/x)\, dx} = e^{\ln x} = x$

Knowing the integrating factor y can be found using the previous key point, then

$$\mu y = \int \mu Q(x)\, dx$$

$$= \int x \times 1 \, dx$$

$$= \frac{x^2}{2} + C$$

where C is the constant of integration. So

$$\mu y = xy = \frac{x^2}{2} + C$$

so that

$$y = \frac{x}{2} + \frac{C}{x}$$

is the general solution.

(b) We now apply the condition $y(1) = 3$ in order to find the constant C and thus obtain the particular solution.

We know that when $x = 1$, $y = 3$, and so

$$3 = \frac{1}{2} + \frac{C}{1}$$

that is $C = \frac{5}{2}$. Thus the particular solution is $y = \frac{x}{2} + \frac{5}{2x}$.

Exercises

1 Solve the equation

$$x^2 \frac{dy}{dx} + xy = 1$$

2 Find the solution of the equation

$$x \frac{dy}{dx} - y = x$$

subject to the condition $y(1) = 2$.

3 Find the general solution of the equation

$$\frac{dy}{dt} + (\tan t)\, y = \cos t$$

4 Solve the equation

$$\frac{dy}{dt} + (\cot t)\, y = \sin t$$

5 The temperature θ (measured in degrees) of a body immersed in an atmosphere of varying temperature satisfies the equation

$$\frac{d\theta}{dt} + 0.1\theta = 5 - 2.5t$$

Find the temperature at time t if $\theta = 60°$ when $t = 0$.

6 In an LR circuit with applied voltage $E = 10(1 - e^{-0.1t})$ the current i satisfies the equation

$$L \frac{di}{dt} + Ri = 10(1 - e^{-0.1t})$$

If the initial current is i_0 find i subsequently.

Solutions to exercises

1 $y = \frac{1}{x} \ln x + \frac{C}{x}$

2 $y = x \ln x + 2x$

3 $y = (t + C) \cos t$

4 $y = \left(\frac{1}{2}t - \frac{1}{4} \sin 2t + C \right) \operatorname{cosec} t$

5 $\theta = 300 - 25t - 240e^{-0.1t}$

6 $i = \frac{10}{R} - \left(\frac{100}{10R - L} \right) e^{-0.1t}$

$$+ \left(i_0 + \frac{10L}{R(10R - L)} \right) e^{-Rt/L}$$

End of block exercises

1 Find the general solution of the following equations:

(a) $\dfrac{dy}{dx} + y = 1$ (b) $\dfrac{dy}{dx} + 2y = 6$

(c) $\dfrac{dx}{dt} + 6x = 4$ (d) $\dfrac{dy}{dx} - 3y = 2$

(e) $\dfrac{dy}{dx} = 6y + 9$ (f) $\dfrac{dx}{dt} = 3x - 8$

2 Use an integrating factor to obtain the general solution of $iR + L\dfrac{di}{dt} = \sin \omega t$ where R, L and ω are constants.

Solutions to exercises

1 (a) $y = 1 + ce^{-x}$ (b) $y = 3 + ce^{-2x}$

(c) $x = \frac{2}{3} + ce^{-6t}$ (d) $y = ce^{3x} - \frac{2}{3}$

(e) $y = ce^{6x} - \frac{3}{2}$ (f) $x = \frac{8}{3} + ce^{3t}$

2 $\dfrac{L[(R/L)\sin \omega t - \omega \cos \omega t]}{R^2 + L^2\omega^2} + ce^{-Rt/L}$

Computational approaches to differential equations

Computer software packages are readily available that can solve differential equations. You may recall that when using software to perform integration the results produced fall into two categories, **symbolic** and **numerical**. This is also true when we use computers to solve differential equations.

For many equations it is possible for the computer to obtain a symbolic answer in terms of common functions, and we shall illustrate this shortly. However, for many equations it is impossible to obtain a symbolic answer and so techniques, known as **numerical methods**, exist for obtaining approximate solutions. One technique for solving first-order equations is Euler's method. In general, numerical techniques are laborious to perform by hand and so are best implemented using a computer. We shall be content to illustrate Euler's method using simple examples, and then rely upon software to tackle more substantial problems.

4.2 Use of symbolic algebra packages to solve first-order differential equations exactly

Many software packages designed for tackling mathematical problems have the facility to solve first-order equations. You should enquire whether the package(s) to which you have access can solve differential equations symbolically. You may need to refer to local documentation. Some of the packages require additional software to be loaded. For example, the standard version of Matlab does not come with a symbolic processor, but requires a Symbolic Math Toolbox.

Example 4.1 Finding an exact, general solution

Use a computer algebra package to solve the differential equation $\dfrac{dy}{dx} - xy = 0$. This is a first-order equation, which can be solved using the method of separation of variables.

20

Solution

Maple
In Maple the command **dsolve** can be used to find the general solution of a differential equation with dependent variable y and independent variable x. y must be input as $y(x)$, and the derivative term is input as diff($y(x)$, x), which stands for the derivative of $y(x)$ with respect to x. So, to solve the equation $\dfrac{dy}{dx} - xy = 0$ the Maple command is

```
dsolve(diff(y(x),x)-x*y(x) = 0);
```

and Maple outputs

$$y(x) = \_C1 e^{\left(\frac{1}{2}x^2\right)}$$

The output $\_C1$ is the arbitrary constant, and so we can interpret this general solution as simply $y = Ae^{\frac{1}{2}x^2}$.

Matlab
Assuming the Symbolic Math Toolbox is available, the function **dsolve** computes symbolic solutions to differential equations. In Matlab the symbol D is used to denote the derivative. The command

```
dsolve('Dy-x*y=0','x')
```

generates the result

```
C1*exp(1/2*x^2)
```

Note the requirement to put the differential equation and the independent variable between apostrophes. Clearly the output is not as user-friendly as that provided by Maple, although another Matlab command, **pretty**, can be applied to generate more familiar output.

The power of software packages lies in their ability to handle symbolic constants in problems. Consider the following example.

Example 4.2 Electrical Engineering – *RL* circuit with sinusoidal input
In question 2, End of block exercises, Block 3 the integrating factor method was used to solve the equation

$$iR + L\frac{di}{dt} = \sin \omega t$$

where R, L and ω are constants, in order to find the current $i(t)$ in an *RL* circuit. Use a computer algebra package to obtain the general solution of this equation.

Solution

> *Maple*
> In Maple the relevant command is
>
> `dsolve(i(t)*R+L*diff(i(t),t) = sin(w*t));`
>
> and Maple produces the following output:
>
> $$i(t) = e^{(-\frac{Rt}{L})} \_C1 + \frac{-L\,w\cos(wt) + R\sin(wt)}{R^2 + w^2 L^2}$$
>
> where $\_C1$ represents the arbitrary constant in the general solution. Compare this solution with the one stated analytically in question 2 on page 970.

> *Matlab*
> In Matlab the symbol i is used to denote $\sqrt{-1}$ and so should not be used here. Instead we use y as the dependent variable. The command
>
> `dsolve('y*R+L*Dy = sin(w*t)','t')`
>
> results in the output
>
> `(-L*w*cos(w*t) + R*sin(w*t) + exp(-1/L*R*t)*C1*R^2 +`
>
> `exp(-1/L*R*t)*C1*w^2*L^2)/(R^2 + w^2*L^2)`
>
> This is equivalent to that produced by Maple, although the format is not so user-friendly.

Example 4.3 Finding an exact, particular solution

Find the particular solution of $\dfrac{dy}{dx} - x^2 y = 0$ that satisfies the initial condition $y(0) = 3$.

Solution

Software packages will have the facility to handle initial conditions.

> *Maple*
> In Maple the appropriate command is
>
> `dsolve({diff(y(x),x) − x^2*y(x) = 0,y(0) = 3});`
>
> and Maple outputs the particular solution
>
> $$y(x) = 3e^{(\frac{1}{3}x^3)}$$

> *Matlab*
> The command
>
> `dsolve('Dy-x^2*y = 0','y(0) = 3','x')`
>
> results in the output
>
> `3*exp(1/3*x^3)`

 Computer and calculator exercises

1 Use a computer algebra package to find the general solutions of the following differential equations:

(a) $\dfrac{dy}{dx} = \dfrac{xy}{x^2 + 1}$

(b) $m\dfrac{dv}{dt} = -v$ where m is a constant

2 Use a computer algebra package to find the exact, particular solution of the following equations:

(a) $\dfrac{d\theta}{dt} = -k\theta, \theta(0) = \theta_0$

(b) $iR + L\dfrac{di}{dt} = E, i(0) = 0$, where R, L and E are constants

Solutions to exercises

1 (a) $y(x) = C\sqrt{x^2 + 1}$
  (b) $v(t) = Ce^{-t/m}$

2 (a) $\theta(t) = \theta_0\, e^{-kt}$ (b) $i(t) = \dfrac{E}{R} - \dfrac{Ee^{(-Rt/L)}}{R}$

4.3 Euler's method

Although symbolic algebra packages are extremely powerful, there are nevertheless some differential equations that they cannot solve exactly. In such cases it may be possible to use the computer to obtain an approximate solution. There are many sophisticated techniques available for doing this, the details of which can be found in textbooks on numerical methods. The simplest technique is Euler's method, which we explain here.

Euler's method is a numerical technique for finding approximate solutions of differential equations having the form

$$\frac{dy}{dx} = f(x, y), \quad y(x_0) = y_0$$

Note that to use this method we need to be given an initial condition $y(x_0) = y_0$. When solving this equation numerically we use the fact that $\dfrac{dy}{dx}$ is the gradient of the tangent to $y(x)$. We know from the given equation that the point (x_0, y_0) lies on the graph of the solution. We also know that the gradient of the solution is $\dfrac{dy}{dx}$, that is $f(x, y)$. So, at the point (x_0, y_0) the graph of the solution has a tangent line with gradient $f(x_0, y_0)$. This is illustrated in Figure 4.1.

Figure 4.1
A tangent line is used to obtain an estimate of the true solution.

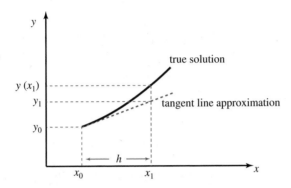

The point (x_0, y_0) lies both on the solution curve and on the tangent line. Consider moving to a nearby point on the tangent line. Let the nearby point be (x_1, y_1), where $x_1 = x_0 + h$. The value of h and hence x_1 is known. Then (x_1, y_1) is on the tangent line, but not on the solution curve. When $x = x_1$ the value of the solution is $y(x_1)$. We do not know the value of $y(x_1)$. We can, however, use y_1 as an approximation to $y(x_1)$.

Referring to Figure 4.1 we note that the gradient of the tangent line is $\dfrac{y_1 - y_0}{h}$, and so

$$\frac{y_1 - y_0}{h} = f(x_0, y_0)$$

It follows by rearrangement that $y_1 = y_0 + hf(x_0, y_0)$. So, knowing the values of x_0, y_0, the function $f(x, y)$ and the value of h we can calculate y_1. Note from Figure 4.1 that $y(x_1)$ is the true, but unknown, solution, whereas y_1 is now the known, but approximate, solution.

We can use the point (x_1, y_1) as a new starting point to find an approximate solution at $x_2 = x_1 + h$, and using an identical argument this is $y_2 = y_1 + hf(x_1, y_1)$.

The process is repeated using the general formula $y_{n+1} = y_n + hf(x_n, y_n)$. The result is a sequence of values, y_n, which approximates the true solution $y(x)$ at the points where $x = x_n$.

Key point

Euler's method

An approximate solution of $\dfrac{dy}{dx} = f(x, y)$, subject to the initial condition

$y(x_0) = y_0$, is given by

$$y_{n+1} = y_n + hf(x_n, y_n), \quad y_0 = y(x_0)$$

where h is the step-size, or **increment**, in x.

Example 4.4

Use three steps of Euler's method to obtain an approximate solution of $\dfrac{dy}{dx} = \dfrac{x}{y}$, subject to the initial condition $y(1) = 2$, at the points where $x = 1.1$, 1.2 and 1.3. Compare the results with the exact solution.

Solution

We start from the given initial condition, $x_0 = 1$, $y_0 = y(x_0) = y(1) = 2$. Then applying Euler's method with $h = 0.1$, and $f(x, y) = \dfrac{x}{y}$, we find

$$y_1 = y_0 + hf(x_0, y_0)$$

$$= y_0 + h\left(\frac{x_0}{y_0}\right)$$

$$= 2 + 0.1\left(\frac{1}{2}\right)$$

$$= 2.05$$

So, at the point $x_1 = 1.1$ the solution is approximated by $y_1 = 2.05$. These values are recorded in Table 4.1.

Applying the formula again:

$$y_2 = y_1 + hf(x_1, y_1)$$

$$= y_1 + h\left(\frac{x_1}{y_1}\right)$$

$$= 2.05 + 0.1\left(\frac{1.1}{2.05}\right)$$

$$= 2.103659 \text{ (6 d.p.)}$$

So, at the point $x_2 = 1.2$ the solution is approximated by $y_2 = 2.103659$.

Applying the formula one more time:

$$y_3 = y_2 + hf(x_2, y_2) = y_2 + h\left(\frac{x_2}{y_2}\right) = \qquad \qquad 2.160702$$

These results and the true solutions, for comparison, are given in Table 4.1. Note that the true solution can be obtained by separation of variables and is $y = \sqrt{x^2 + 3}$.

Table 4.1
For Example 4.4.

x_n	y_n	true: $y(x_n)$ (to 6 d.p.)
$x_0 = 1$	$y_0 = 2$	2
$x_1 = 1.1$	$y_1 = 2.05$	2.051828
$x_2 = 1.2$	$y_2 = 2.103659$	2.107131
$x_3 = 1.3$	$y_3 = 2.160702$	2.165641

Clearly this rather crude approach can be improved by taking a smaller increment in x, but the price we pay is an increase in the number of calculations required.

We have provided details of the simplest numerical method for solving a first-order differential equation. More sophisticated formulae exist, but for a detailed treatment you should refer to a textbook on numerical methods for engineers.

Exercises

1   Use five steps of Euler's method to find an approximate solution of the initial value problem $\dfrac{dy}{dx} = x \sin y$, $y(0) = 2$ using $h = 0.2$, for $0 \le x \le 1$. Work throughout to six decimal places. Hence approximate $y(1)$.

2   Use five steps of Euler's method to find an approximate solution of the initial value problem $\dfrac{dx}{dt} = \dfrac{x + x^2}{t}$, $x(1) = -5$ using $h = 0.01$. Work throughout to six decimal places. Hence approximate $x(1.05)$.

Solutions to exercises

1   $y(1) \approx 2.339282$

2   $x(1.05) \approx -4.169132$

4.4   Use of software to find numerical solutions of differential equations

As we stated in the previous section, symbolic algebra packages use sophisticated techniques to find accurate, approximate solutions of initial value problems. For full details of the underlying algorithms you will need to consult the documentation accompanying your package. We shall be content to illustrate their application in a simple example.

Example 4.5 Finding a numerical solution of a first-order equation

Find a numerical solution of the differential equation of Example 4.3, $\dfrac{dy}{dx} - x^2 y = 0$, subject to the initial condition $y(0) = 3$.

Solution

A numerical solution produces a sequence of values of the dependent variable, at values of the independent variable specified by the user. For example, suppose we wish to know the value of y when $x = 0, 0.2, 0.4, 0.6, 0.8$ and 1.

Maple

```
dsolve({diff(y(x),x)-x^2*y(x)=0,y(0)=3},
type=numeric,output=array([0,0.2,0.4,0.6,0.8,1]));
```

and Maple outputs an array containing the *x* and *y* values:

$$[x \quad y(x)]$$

0	3.
.2	3.00801944725338
.4	3.06469041449842
.6	3.22396357678057
.8	3.55828609702334
1	4.18683969904184

You will find that producing a graph of the output using the software is straightforward.

Matlab

Using Matlab, it is first necessary to write and save a file, known as an m.file, to store the right-hand side of the differential equation $\dfrac{dy}{dx} = f(x, y)$. The second line below represents the right-hand side of the equation $\dfrac{dy}{dx} = x^2 y.$

```
function ode1=example1(x,y)
ode1=x^2*y
```

After saving this m.file the following command will obtain the numerical solution, subject to the initial condition $y(0) = 3$:

```
[x,y]=ode45('example1',[0,0.2,0.4,0.6,0.8,1],3)
```

resulting in

$$y = \begin{bmatrix} 3.0000 \\ 3.0080 \\ 3.0647 \\ 3.2240 \\ 3.5583 \\ 4.1868 \end{bmatrix}$$

Computer and calculator exercises

1 Use a computer algebra package to find an approximate numerical solution of the following initial value problems at the values of the independent variable specified:

(a) $\dfrac{dy}{dx} - \dfrac{1}{\ln y} = 0$, $y(0) = 2$,

at $x = 0, 0.1, 0.2, \ldots, 0.5$.

(b) $\dfrac{dx}{dt} = \sin(x^2)$, $x(0) = 1$, at

$t = 0, 0.25, 0.5, 0.75, 1$.

Solutions to exercises

1                    $[x, y(x)]$ $[t, x(t)]$

(a)
$\begin{bmatrix}
0 & 2. \\
.1 & 2.13759470217855 \\
.2 & 2.26437308827318 \\
.3 & 2.38299104649312 \\
.4 & 2.49515771401761 \\
.5 & 2.60205536137248
\end{bmatrix}$

(b)
$\begin{bmatrix}
0 & 1. \\
.25 & 1.23412636541130 \\
.5 & 1.47072990280928 \\
.75 & 1.63173824177242 \\
1 & 1.71208213688752
\end{bmatrix}$

End of block exercises

1 Using software, obtain a symbolic solution of $L\dfrac{di}{dt} + Ri = 10(1 - e^{-0.1t})$ when $i(0) = i_0$.

2 Using software, obtain a symbolic solution of $iR + L\dfrac{di}{dt} = E$ where R, L and E are constants subject to the initial condition $i(0) = 0$.

3 Use a package to find the general solution of $\dfrac{dT}{d\theta} = \mu(T - K)$ where μ and K are constants.

4 Use Euler's method to find a numerical solution of $\dfrac{dy}{dx} = -2y$, $y(0) = 1$, for $0 \le x \le 0.5$. First take $h = 0.1$, then $h = 0.05$, and compare your answers at $x = 0.5$ with the exact solution obtained by separating the variables. Work throughout to six decimal places.

5 Use Euler's method to find a numerical solution of $\dfrac{dy}{dx} = \dfrac{xy}{x^2 + 2}$ subject to $y(1) = 3$. Take $h = 0.1$ and hence approximate $y(1.5)$. Obtain the true solution using the method of separation of variables. Work throughout to six decimal places.

Solutions to exercises

1. $i(t) = \dfrac{10}{R} - \left(\dfrac{100}{10R - L}\right) e^{-0.1t}$

   $\qquad + \left(i_0 + \dfrac{10L}{R(10R - L)}\right) e^{-Rt/L}$

2. $i(t) = \dfrac{E}{R}(1 - e^{-Rt/L})$

3. $T(\theta) = K + A e^{\mu\theta}$

4. $h = 0.1$: $y(0.5) \approx 0.327680$.
   $h = 0.05$: $y(0.5) \approx 0.348678$. True solution
   $y = e^{-2x}$: $y(0.5) = 0.367879$

5. 3, 3.1, 3.206231, 3.318076, 3.434973,
   $y(1.5) \approx 3.556411$. True solution:
   $y = \sqrt{3(x^2 + 2)^{1/2}}$, $y(1.5) = 3.570714$

Second-order linear constant-coefficient equations I

In this block we start to learn how to solve second-order differential equations of a particular type – those that are linear and which have constant coefficients. Such equations are used widely in the modelling of physical phenomena – for example, in the analysis of vibrating systems, and the analysis of electric circuits.

The solution of these equations is achieved in stages. The first stage is to find what is called a **complementary function**, and is the subject of this block. The second stage is to find a **particular integral**. This is dealt with in the following block. Finally, the sum of the complementary function and the particular integral forms the general solution of a second-order linear differential equation. It is important that you understand and use this terminology. A knowledge of complex numbers, and particularly the exponential form, is essential.

Key point

> **Solving a second-order linear equation**
>
> 1 Find the complementary function, y_{cf}.
> 2 Find a particular integral, y_{pi}.
> 3 The general solution is the sum of the complementary function and the particular integral:
> $$y = y_{cf} + y_{pi}$$

The general form of a second-order linear equation that has constant coefficients is

$$a\frac{d^2y}{dx^2} + b\frac{dy}{dx} + cy = f(x) \tag{1}$$

where a, b, c are constants. An example of such an equation is

$$3\frac{d^2y}{dx^2} - 2\frac{dy}{dx} + 4y = e^{-x}\sin x$$

When $f(x)$ is not identically zero, the equation is said to be **inhomogeneous**. An important relative of this equation is found by replacing the function $f(x)$ with zero to give

$$a\frac{d^2y}{dx^2} + b\frac{dy}{dx} + cy = 0 \tag{2}$$

This is the **homogeneous** form of equation (1). The homogeneous form is found by ignoring the term that is independent of y, or its derivatives.

To find the general solution of equation (1), it is first necessary to solve equation (2). The general solution of (2) is called the **complementary function** and will always contain two arbitrary constants. We shall denote this solution by y_{cf}.

The technique for finding the complementary function is described in this block.

Example 5.1

Which of the following are constant-coefficient equations? Which are homogeneous?

(a) $\dfrac{d^2y}{dx^2} + 4\dfrac{dy}{dx} + 3y = e^{-2x}$

(b) $x\dfrac{d^2y}{dx^2} + 2y = 0$

(c) $\dfrac{d^2x}{dt^2} + 3\dfrac{dx}{dt} + 7x = 0$

(d) $\dfrac{d^2y}{dx^2} + 4\dfrac{dy}{dx} + 4y = 0$

Solution

(a) ⬚   const. coeff., not homogeneous

(b) ⬚   not const. coeff., homogeneous

(c) ⬚   const. coeff. and homogeneous

(d) ⬚   const. coeff. and homogeneous

Example 5.2

What is a complementary function?

Solution

⬚   A complementary function is the general solution of a homogeneous, linear differential equation.

Example 5.3

The complementary function for the equation

$$\frac{d^2y}{dx^2} + y = 0$$

is $y = A\cos x + B\sin x$ for any constants A and B. Verify that this is indeed a solution.

Solution

The given equation is homogeneous, linear and constant coefficient. By differentiating $y = A\cos x + B\sin x$ twice and substituting into the equation it is easy to verify that this is a solution.

Find $\dfrac{dy}{dx}$ and $\dfrac{d^2y}{dx^2}$.

$$y' = -A \sin x + B \cos x, \; y'' = -A \cos x - B \sin x$$

Substitute the expressions for y and y'' into the left-hand side of the differential equation and simplify your answer.

$$-A \cos x - B \sin x + A \cos x + B \sin x = 0$$

So the left-hand side simplifies to zero. The right-hand side of the differential equation is zero. Hence $y = A \cos x + B \sin x$ is a solution.

Note that $y = A \cos x + B \sin x$ contains two arbitrary constants. It is the general solution and therefore the complementary function for this homogeneous equation.

Exercises

1. Write down the homogeneous form of the following equations:

   (a) $\dfrac{d^2y}{dx^2} + 3\dfrac{dy}{dx} - 2y = \sin x$

   (b) $\dfrac{d^2x}{dt^2} + 7x - e^t \cos t = 0$

   (c) $\dfrac{d^2x}{dt^2} = -\omega^2 x$, where ω is a constant.

2. State what is meant by the term 'complementary function'.

Solutions to exercises

1. (a) $\dfrac{d^2y}{dx^2} + 3\dfrac{dy}{dx} - 2y = 0$ (b) $\dfrac{d^2x}{dt^2} + 7x = 0$ (c) is already homogeneous

5.3 Finding the complementary function

To find the complementary function we must make use of the following property.

If $y_1(x)$ and $y_2(x)$ are any two (linearly independent) solutions of a linear, homogeneous second-order differential equation then the general solution $y_{cf}(x)$ is

$$y_{cf}(x) = Ay_1(x) + By_2(x)$$

where A and B are constants.

We see that the second-order linear ordinary differential equation has two arbitrary constants in its general solution. The functions $y_1(x)$ and $y_2(x)$ are **linearly independent** if one is not simply a multiple of the other.

This property gives us a method for constructing the complementary function. If we can find two independent solutions of a homogeneous differential equation, we can form the complementary function by simply adding constant multiples of the two solutions.

Example 5.4

(a) Verify that $y_1 = e^{4x}$ and $y_2 = e^{2x}$ both satisfy the constant-coefficient homogeneous equation

$$\frac{d^2y}{dx^2} - 6\frac{dy}{dx} + 8y = 0 \tag{3}$$

(b) Write down the general solution of this equation and so form the complementary function.

Solution

(a) If $y_1 = e^{4x}$, differentiation yields

$$\frac{dy_1}{dx} = 4e^{4x}$$

and similarly,

$$\frac{d^2y_1}{dx^2} = 16e^{4x}$$

Substitution into the left-hand side of equation (3) gives

$$16e^{4x} - 6(4e^{4x}) + 8e^{4x}$$

which simplifies to zero. The right-hand side is zero. So $y_1 = e^{4x}$ is indeed a solution. Similarly if $y_2 = e^{2x}$, then

$$\frac{dy_2}{dx} = 2e^{2x} \quad \text{and} \quad \frac{d^2y_2}{dx^2} = 4e^{2x}$$

Substitution into the left-hand side of equation (3) gives

$$4e^{2x} - 6(2e^{2x}) + 8e^{2x}$$

which simplifies to zero, so $y_2 = e^{2x}$ is also a solution of equation (3).

(b) So we know two independent solutions of the equation: these are e^{2x} and e^{4x}. They are linearly independent because e^{2x} is not a multiple of e^{4x}. From the property stated above we have

$$y_{cf}(x) = Ae^{4x} + Be^{2x}$$

as the general solution of equation (3). This is the complementary function.

Key point	**Finding a complementary function**

If two independent solutions, y_1 and y_2, of a homogeneous second-order linear equation can be found, the complementary function is found by adding constant multiples of them:

$$y_{cf}(x) = Ay_1(x) + By_2(x)$$

where A and B are arbitrary constants.

Example 5.5

Substitute $y = e^{kx}$, where k is a constant, into the equation

$$\frac{d^2y}{dx^2} - \frac{dy}{dx} - 6y = 0$$

in order to find the values of k that make $y = e^{kx}$ a solution. Hence state the general solution.

Solution

As suggested we substitute $y = e^{kx}$. Differentiating we find

$$\frac{dy}{dx} = ke^{kx} \quad \text{and} \quad \frac{d^2y}{dx^2} = k^2e^{kx}$$

Substitution into the given equation yields

$$k^2e^{kx} - ke^{kx} - 6e^{kx} = 0$$

that is

$$(k^2 - k - 6)e^{kx} = 0$$

Since the function e^{kx} can never be zero it follows that

$$k^2 - k - 6 = 0 \tag{4}$$

that is

$$(k - 3)(k + 2) = 0$$

so that $k = 3$ or $k = -2$. That is to say, if $y = e^{kx}$ is to be a solution of the differential equation k must be either 3 or -2. We therefore have found two solutions:

$$y_1(x) = e^{3x} \quad \text{and} \quad y_2(x) = e^{-2x}$$

These two functions are linearly independent because e^{-2x} is not a multiple of e^{3x}, and therefore the general solution is

$$y_{cf}(x) = Ae^{3x} + Be^{-2x}$$

This is the complementary function for the given differential equation. Equation (4) for determining k is called the **auxiliary equation**.

Example 5.6

By substituting $y = e^{kx}$, find values of k so that y is a solution of

$$\frac{d^2y}{dx^2} - 3\frac{dy}{dx} + 2y = 0$$

Hence write down two solutions and the general solution of this equation. Hence state the complementary function.

Solution

By substituting $y = e^{kx}$ find the auxiliary equation:

$$k^2 - 3k + 2 = 0$$

This can be factorised:

$$(k - 1)(k - 2) = 0$$

So the required values of k are 1 and 2. Write down the two solutions:

$$y = e^x \quad \text{and} \quad y = e^{2x}$$

Finally write down the general solution:

$$y_{cf}(x) = Ae^x + Be^{2x}$$

The complementary function is therefore $y_{cf}(x) = Ae^x + Be^{2x}$.

Example 5.7

Find the auxiliary equation of the differential equation

$$a\frac{d^2y}{dx^2} + b\frac{dy}{dx} + cy = 0$$

where a, b and c are constants.

Solution

We try a solution of the form $y = e^{kx}$ so that

$$\frac{dy}{dx} = ke^{kx} \quad \text{and} \quad \frac{d^2y}{dx^2} = k^2e^{kx}$$

Substitution into the given differential equation yields

$$ak^2e^{kx} + bke^{kx} + ce^{kx} = 0$$

that is

$$(ak^2 + bk + c)e^{kx} = 0$$

Because e^{kx} cannot be zero it follows that

$$ak^2 + bk + c = 0$$

This is the required auxiliary equation.

Key point

The auxiliary equation of

$$a\frac{d^2y}{dx^2} + b\frac{dy}{dx} + cy = 0$$

is $ak^2 + bk + c = 0$. The solutions of the auxiliary equation give values of k that make $y = e^{kx}$ a solution of the differential equation.

Example 5.8

Write down, but do not solve, the auxiliary equations of the following differential equations:

(a) $\dfrac{d^2y}{dx^2} + \dfrac{dy}{dx} + y = 0$ (b) $2\dfrac{d^2y}{dx^2} + 7\dfrac{dy}{dx} - 3y = 0$

(c) $4\dfrac{d^2y}{dx^2} + 7y = 0$ (d) $\dfrac{d^2y}{dx^2} + \dfrac{dy}{dx} = 0$

Solution

(a) $\hspace{5cm}$ $k^2 + k + 1 = 0$

(b) $\hspace{5cm}$ $2k^2 + 7k - 3 = 0$

(c) $\hspace{5cm}$ $4k^2 + 7 = 0$

(d) $\hspace{5cm}$ $k^2 + k = 0$

Solution of the auxiliary equation gives the values of k that we seek. Note that the auxiliary equation is a quadratic equation and so has two roots. Clearly the nature of the roots will depend upon the values of a, b and c. If $b^2 > 4ac$ the roots will be real and distinct. The two values of k thus obtained, k_1 and k_2, will allow us to write down two independent solutions:

$$y_1(x) = e^{k_1x}, \quad y_2(x) = e^{k_2x}$$

and so the general solution of the differential equation will be

$$y(x) = Ae^{k_1x} + Be^{k_2x}$$

Key point

If the auxiliary equation has real, distinct roots k_1 and k_2, the complementary function will be

$$y_{cf}(x) = Ae^{k_1x} + Be^{k_2x}$$

On the other hand, if $b^2 = 4ac$ the two roots of the auxiliary equation will be equal and this method will therefore yield only one independent solution. In this case, special treatment is required. If $b^2 < 4ac$ the two roots of the auxiliary equation will be complex: that is, k_1 and k_2 will be complex numbers. The procedure for dealing with such cases will become apparent in the following examples.

Example 5.9

Find the general solution of

$$\frac{d^2y}{dx^2} + 3\frac{dy}{dx} - 10y = 0$$

Solution

Write down the auxiliary equation:

$$k^2 + 3k - 10 = 0$$

Factorise to find the solutions:

$$(k - 2)(k + 5) = 0, k = 2 \text{ and } k = -5$$

Thus there exist two solutions $y_1 = e^{2x}$ and $y_2 = e^{-5x}$. We can write the general solution as

$$y(x) = Ae^{2x} + Be^{-5x}$$

This is the complementary function.

Example 5.10

Find the general solution of

$$\frac{d^2y}{dx^2} + 4y = 0$$

Solution

Write down the auxiliary equation:

$$k^2 + 4 = 0$$

It follows that

$$k^2 = -4$$

so that

$$k = \pm 2j$$

that is, we have complex roots. The two independent solutions of the equation are thus

$$y_1(x) = e^{2jx}, \quad y_2(x) = e^{-2jx}$$

so that the general solution can be written in the form

$$y(x) = Ae^{2jx} + Be^{-2jx}$$

However, in cases such as this, it is usual to rewrite the solution in the following way. Recall that Euler's relations (Chapter 11 Block 3) give

$$e^{2jx} = \cos 2x + j \sin 2x \quad \text{and} \quad e^{-2jx} = \cos 2x - j \sin 2x$$

so that

$$y(x) = A(\cos 2x + \mathrm{j} \sin 2x) + B(\cos 2x - \mathrm{j} \sin 2x)$$
$$= (A + B) \cos 2x + (A\mathrm{j} - B\mathrm{j}) \sin 2x$$

If we now relabel the constants such that

$$A + B = C \quad \text{and} \quad A\mathrm{j} - B\mathrm{j} = D$$

we can write the general solution in the form

$$y(x) = C \cos 2x + D \sin 2x$$

This is the complementary function.

Example 5.11

Given $a\dfrac{\mathrm{d}^2 y}{\mathrm{d}x^2} + b\dfrac{\mathrm{d}y}{\mathrm{d}x} + cy = 0$, write down the auxiliary equation. If the roots of the auxiliary equation are complex (one root will always be the complex conjugate of the other) and are denoted by

$$k_1 = \alpha + \beta\mathrm{j}, \qquad k_2 = \alpha - \beta\mathrm{j}$$

show that the general solution can be written as

$$y(x) = \mathrm{e}^{\alpha x}(C \cos \beta x + D \sin \beta x)$$

where C and D are arbitrary constants.

Solution
Write down the auxiliary equation:

$$ak^2 + bk + c = 0$$

If $k_1 = \alpha + \beta\mathrm{j}, k_2 = \alpha - \beta\mathrm{j}$ then the general solution is

$$y = A\mathrm{e}^{(\alpha+\beta\mathrm{j})x} + B\mathrm{e}^{(\alpha-\beta\mathrm{j})x}$$

where A and B are arbitrary constants. Using the laws of indices this is rewritten as

$$y = A\mathrm{e}^{\alpha x} \mathrm{e}^{\beta\mathrm{j}x} + B\mathrm{e}^{\alpha x} \mathrm{e}^{-\beta\mathrm{j}x}$$
$$= \mathrm{e}^{\alpha x}(A\mathrm{e}^{\beta\mathrm{j}x} + B\mathrm{e}^{-\beta\mathrm{j}x})$$

Then, using Euler's relations, we obtain

$$y = \mathrm{e}^{\alpha x}(A \cos \beta x + A\mathrm{j} \sin \beta x + B \cos \beta x - B\mathrm{j} \sin \beta x)$$
$$= \mathrm{e}^{\alpha x}[(A + B) \cos \beta x + (A\mathrm{j} - B\mathrm{j}) \sin \beta x]$$

Writing $C = A + B$ and $D = A\mathrm{j} - B\mathrm{j}$, we find

$$y = \mathrm{e}^{\alpha x}(C \cos \beta x + D \sin \beta x)$$

This is the required solution.

Key point

If the auxiliary equation has complex roots, $\alpha + \beta\mathrm{j}$ and $\alpha - \beta\mathrm{j}$, then the complementary function is

$$y_{\mathrm{cf}} = \mathrm{e}^{\alpha x}(C \cos \beta x + D \sin \beta x)$$

Example 5.12
Find the general solution of $\dfrac{d^2y}{dx^2} + 2\dfrac{dy}{dx} + 4y = 0$.

Solution
The auxiliary equation is

$$k^2 + 2k + 4 = 0$$

This equation has complex roots. Find them.

$$k = -1 \pm \sqrt{3}j$$

Using the key point above with $\alpha = -1$ and $\beta = \sqrt{3}$ write down the general solution.

$$y = e^{-x}(C\cos\sqrt{3}\,x + D\sin\sqrt{3}\,x)$$

Example 5.13 Mechanical Engineering – Simple harmonic motion
The behaviour of a wide range of oscillating and vibrating mechanical and electrical systems is described by a differential equation of the form

$$\frac{d^2x}{dt^2} = -\omega^2x$$

where ω is a constant. This is known as the **simple harmonic motion** equation. Find its general solution.

Solution
Writing the equation in the form $\dfrac{d^2x}{dt^2} + \omega^2x = 0$ we recognise it as second order, constant coefficient and homogeneous. Write down the auxiliary equation.

$$k^2 + \omega^2 = 0$$

This has solutions $k = \pm\omega$j. So there are complex roots with real part zero and imaginary parts $\pm\omega$. From the previous key point write down the general solution.

$$x = C\cos\omega t + D\sin\omega t$$

The fact that the solution consists of a sine and a cosine wave is indicative of the fact that the differential equation describes an oscillating or vibrating system.

Key point

The simple harmonic motion equation and its solution are, respectively,

$$\frac{d^2x}{dt^2} = -\omega^2x$$

and

$$x = C\cos\omega t + D\sin\omega t$$

It is often sufficient simply to quote this solution when faced with solving a simple harmonic motion problem.

When the auxiliary equation has equal roots special treatment is required. This is because the auxiliary equation provides only one independent solution, and we seek two to construct the general solution of any second-order differential equation. The second solution is found by multiplying the first by x, as shown in the following example.

Example 5.14

The auxiliary equation of $a\dfrac{d^2y}{dx^2} + b\dfrac{dy}{dx} + cy = 0$ is $ak^2 + bk + c = 0$. Suppose this equation has equal roots $k = k_1$. Verify that $y = xe^{k_1 x}$ is a solution of the differential equation. Hence state the general solution.

Solution

We must check that $y = xe^{k_1 x}$ is a solution. Using the product rule find $\dfrac{dy}{dx}$ and $\dfrac{d^2y}{dx^2}$.

$$\dfrac{dy}{dx} = \qquad\qquad\qquad\qquad e^{k_1 x} + xk_1 e^{k_1 x} = e^{k_1 x}(1 + k_1 x)$$

$$\dfrac{d^2y}{dx^2} = $$

$$(1 + k_1 x)\, k_1 e^{k_1 x} + e^{k_1 x}\, k_1 = e^{k_1 x}(k_1^2 x + 2k_1)$$

Substitution into the left-hand side of the differential equation yields

$$a(k_1^2 x + 2k_1)e^{k_1 x} + b(1 + k_1 x)e^{k_1 x} + cxe^{k_1 x}$$

that is

$$e^{k_1 x}[a(k_1^2 x + 2k_1) + b(1 + k_1 x) + cx]$$

which can be rewritten as

$$e^{k_1 x}[(ak_1^2 + bk_1 + c)x + 2ak_1 + b] \qquad\qquad (5)$$

But $ak_1^2 + bk_1 + c = 0$ since k_1 satisfies the auxiliary equation. Also,

$$k_1 = \frac{-b \pm \sqrt{b^2 - 4ac}}{2a}$$

but since the roots are equal, then $b^2 - 4ac = 0$ and hence $k_1 = -\dfrac{b}{2a}$. So $2ak_1 + b = 0$, and the expression (5) simplifies to zero. We conclude that $y = xe^{k_1 x}$ is a solution of $a\dfrac{d^2y}{dx^2} + b\dfrac{dy}{dx} + cy = 0$ when the roots of the auxiliary equation are equal.

Since k_1 is a root of the auxiliary equation, $y = e^{k_1 x}$ is a solution of the differential equation. The second independent solution is $y = xe^{k_1 x}$. So the general solution is

$$y = Ae^{k_1 x} + Bxe^{k_1 x}$$
$$= (A + Bx)e^{k_1 x}$$

Key point

If the auxiliary equation has two equal roots, k_1, the complementary function is

$$y_{cf} = (A + Bx)e^{k_1 x}$$

Example 5.15

Obtain the general solution of the equation

$$\frac{d^2y}{dx^2} + 8\frac{dy}{dx} + 16y = 0$$

Solution

As before, a trial solution of the form $y = e^{kx}$ yields an auxiliary equation

$$k^2 + 8k + 16 = 0$$

This equation factorises so that

$$(k + 4)(k + 4) = 0$$

and we obtain equal roots, that is $k = -4$ (twice). If we proceed as before, writing $y_1(x) = e^{-4x}$ and $y_2(x) = e^{-4x}$, it is clear that the two solutions are not independent. We need to find a second independent solution. Using the result of the previous example we conclude that, because the roots of the auxiliary equation are equal, the second independent solution is $y_2 = xe^{-4x}$. The general solution is then

$$y(x) = (A + Bx)e^{-4x}$$

This is the complementary function.

Key points

Summary

Second-order linear constant-coefficient homogeneous equation:

$$a\frac{d^2y}{dx^2} + b\frac{dy}{dx} + cy = 0$$

Auxiliary equation:

$$ak^2 + bk + c = 0$$

Roots of auxiliary equation	Complementary function
k_1, k_2, both real and different	$y = Ae^{k_1x} + Be^{k_2x}$
$\alpha \pm j\beta$, complex conjugate pair	$y = e^{\alpha x}(A \cos \beta x + B \sin \beta x)$
$k_1 = k_2 = k$, equal roots	$y = (A + Bx)e^{kx}$

Exercises

1 Obtain the general solutions, that is the complementary functions, of the following homogeneous equations:

(a) $\dfrac{d^2y}{dx^2} - 3\dfrac{dy}{dx} + 2y = 0$

(b) $\dfrac{d^2y}{dx^2} + 7\dfrac{dy}{dx} + 6y = 0$

(c) $\dfrac{d^2x}{dt^2} + 5\dfrac{dx}{dt} + 6x = 0$

(d) $\dfrac{d^2y}{dt^2} + 2\dfrac{dy}{dt} + y = 0$

(e) $\dfrac{d^2y}{dx^2} - 4\dfrac{dy}{dx} + 4y = 0$

(f) $\dfrac{d^2y}{dt^2} + \dfrac{dy}{dt} + 8y = 0$

2 Find the auxiliary equation for the differential equation

$$L\frac{d^2i}{dt^2} + R\frac{di}{dt} + \frac{1}{C}i = 0$$

Hence write down the complementary function.

3 Find the complementary function of the equation

$$\frac{d^2y}{dx^2} + \frac{dy}{dx} + y = 0$$

Solutions to exercises

1 (a) $y = Ae^x + Be^{2x}$ (b) $y = Ae^{-x} + Be^{-6x}$
(c) $x = Ae^{-2t} + Be^{-3t}$ (d) $y = Ae^{-t} + Bte^{-t}$
(e) $y = Ae^{2x} + Bxe^{2x}$
(f) $y = e^{-0.5t}(A \cos 2.78t + B \sin 2.78t)$

2 $Lk^2 + Rk + \frac{1}{C} = 0$;

$i(t) = Ae^{k_1 t} + Be^{k_2 t}$,

$k_1, k_2 = \dfrac{-R \pm \sqrt{\dfrac{R^2 C - 4L}{C}}}{2L}$

3 $e^{-x/2}\left(A \cos\dfrac{\sqrt{3}\,x}{2} + B \sin\dfrac{\sqrt{3}\,x}{2}\right)$

End of block exercises

1 Obtain the general solutions, that is the complementary functions, of the following homogeneous equations:

(a) $\dfrac{d^2y}{dx^2} - 2\dfrac{dy}{dx} + y = 0$

(b) $\dfrac{d^2y}{dt^2} + \dfrac{dy}{dt} + 5y = 0$

(c) $\dfrac{d^2y}{dx^2} + \dfrac{dy}{dx} - 2y = 0$

(d) $\dfrac{d^2y}{dx^2} + 9y = 0$

(e) $\dfrac{d^2y}{dx^2} - 2\dfrac{dy}{dx} = 0$

(f) $\dfrac{d^2x}{dt^2} - 16x = 0$

Solutions to exercises

1 (a) $y = Ae^x + Bxe^x$
(b) $x = e^{-0.5t}(A \cos 2.18t + B \sin 2.18t)$
(c) $y = Ae^{-2x} + Be^x$
(d) $y = A \cos 3x + B \sin 3x$
(e) $y = A + Be^{2x}$
(f) $x = Ae^{4t} + Be^{-4t}$

Second-order linear constant-coefficient equations II

6.1 Introduction

We stated in Block 5 that the general solution of an inhomogeneous equation is the sum of the complementary function and a particular integral. We have seen how to find the complementary function in the case of a constant-coefficient equation. We shall now deal with the problem of finding a particular integral.

The **particular integral** is *any* solution of the inhomogeneous equation. The remarkable thing is that we need only find any solution of the inhomogeneous equation, and from it, together with the complementary function, the general solution can be constructed.

There are a number of advanced techniques available for finding particular integrals but we shall adopt a simpler strategy. Since any solution will do we shall try to find one by a combination of educated guesswork and trial and error.

6.2 A particular integral

Given a second-order inhomogeneous differential equation

$$a\frac{d^2y}{dx^2} + b\frac{dy}{dx} + cy = f(x) \tag{1}$$

a **particular integral** is any function, which we will denote by $y_{pi}(x)$, which satisfies the equation: that is, any function that, when substituted into the left-hand side and simplified, results in the function on the right.

Example 6.1

Show that $y = \sin 2x$ is a particular integral of

$$\frac{d^2y}{dx^2} + 7y = 3\sin 2x$$

Solution

We substitute $y = \sin 2x$ into the given equation to see whether it is a solution. First do the differentiation: if $y = \sin 2x$ then

$$\frac{dy}{dx} = \boxed{} \qquad\qquad 2\cos 2x$$

$$\frac{d^2y}{dx^2} = \boxed{} \qquad\qquad -4\sin 2x$$

Now substitute the expressions for y and $\dfrac{d^2y}{dx^2}$ into the left-hand side of the differential

equation $\dfrac{d^2y}{dx^2} + 7y$, and simplify the result:

$$-4\sin 2x + 7(\sin 2x) = 3\sin 2x$$

which is the same as the right-hand side of the differential equation. So $y = \sin 2x$ is a solution of the inhomogeneous equation and so is a particular integral. We can write $y_{\text{pi}} = \sin 2x$.

Example 6.2

Show that

$$y = -\frac{1}{4}e^{2x}$$

is a particular integral of

$$\frac{d^2y}{dx^2} - \frac{dy}{dx} - 6y = e^{2x} \tag{2}$$

Solution

Starting with $y = -\dfrac{1}{4}e^{2x}$, find $\dfrac{dy}{dx}$ and $\dfrac{d^2y}{dx^2}$:

$$\frac{dy}{dx} = -\frac{1}{2}e^{2x}, \frac{d^2y}{dx^2} = -e^{2x}$$

Substitute the expressions for y, $\dfrac{dy}{dx}$ and $\dfrac{d^2y}{dx^2}$ into the left-hand side of (2) to show that y satisfies the inhomogeneous equation.

$$-e^{2x} - \left(-\frac{1}{2}e^{2x}\right) - 6\left(-\frac{1}{4}e^{2x}\right), \text{ which simplifies to } e^{2x}$$

Therefore $y = -\dfrac{1}{4}e^{2x}$ is a particular integral and we can write $y_{\text{pi}}(x) = -\dfrac{1}{4}e^{2x}$.

Example 6.3

What is a particular integral?

Solution

A particular integral is *any* solution of an inhomogeneous differential equation.

Exercises

1　Show that $y = 4$ is a particular integral of the equation $\dfrac{d^2y}{dx^2} - 5\dfrac{dy}{dx} + 6y = 24$.

2　Explain what is meant by a particular integral.

3　Show that $x = \dfrac{1}{15} e^{2t}$ is a particular integral of the equation $\ddot{x} + 4\dot{x} + 3x = e^{2t}$.

6.3 Finding a particular integral

In the previous section we explained what is meant by a particular integral. Now we look at how one is actually found. In fact our method is rather crude. It involves trial and error and educated guesswork. We try solutions that are of the same general form as the inhomogeneous term, $f(x)$, on the right-hand side. As a guide, use Table 6.1. The trial solutions all contain constants. These can be adjusted to force the trial solution to be an actual solution of the inhomogeneous equation.

Table 6.1
Trial solutions to find the particular integral.

Inhomogeneous term $f(x)$	Trial solution
Constant	Constant
$ax^r + \cdots + bx + c$	$\alpha x^r + \cdots + \beta x + \gamma$
$a \cos kx$	$\alpha \cos kx + \beta \sin kx$
$a \sin kx$	$\alpha \cos kx + \beta \sin kx$
ae^{kx}	αe^{kx}
$a \cosh kx$	$\alpha \cosh kx + \beta \sinh kx$
$a \sinh kx$	$\alpha \cosh kx + \beta \sinh kx$

Example 6.4
In each case you are given the inhomogeneous term $f(x)$. State the appropriate form of solution to try when finding a particular integral.
(a) $f(x) = 7e^{2x}$
(b) $f(x) = 5$
(c) $f(x) = 7x - 3$
(d) $f(x) = \cos 2x$

Solution
Use Table 6.1 to select an appropriate trial solution.

(a)　$y = \alpha e^{2x}$

(b)　$y = C$, constant

(c)　$y = \alpha x + \beta$

(d)　$y = \alpha \cos 2x + \beta \sin 2x$

Example 6.5
In each case you are given the inhomogeneous term $f(x)$. State the appropriate form of solution to try when finding a particular integral.
(a) $f(x) = x$
(b) $f(x) = e^{2x}$
(c) $f(x) = x + e^{2x}$

Solution
(a) $y = \alpha x + \beta$

(b) $y = \gamma e^{2x}$

(c) This time the inhomogeneous term is a combination of an exponential and a polynomial, so this is the form the trial solution should take.

$$y = \gamma e^{2x} + \alpha x + \beta$$

Example 6.6
Find a particular integral of the equation

$$\frac{d^2 y}{dx^2} - \frac{dy}{dx} - 6y = e^{2x} \tag{3}$$

Solution
We shall attempt to find a solution of the inhomogeneous problem by trying a function of the same form as that on the right-hand side. In particular, let us try $y(x) = \alpha e^{2x}$, where α is a constant that we shall now determine. If $y(x) = \alpha e^{2x}$ then

$$\frac{dy}{dx} = 2\alpha e^{2x} \quad \text{and} \quad \frac{d^2 y}{dx^2} = 4\alpha e^{2x}$$

Substitution in equation (3) gives

$$4\alpha e^{2x} - 2\alpha e^{2x} - 6\alpha e^{2x} = e^{2x}$$

that is

$$-4\alpha e^{2x} = e^{2x}$$

so that y will be a solution if α is chosen so that $-4\alpha = 1$, that is $\alpha = -\dfrac{1}{4}$.

Therefore the particular integral is $y_{pi}(x) = -\dfrac{1}{4} e^{2x}$.

Example 6.7
By trying a solution of the form $y = \alpha e^{-x}$ find a particular integral of the equation

$$\frac{d^2 y}{dx^2} + \frac{dy}{dx} - 2y = 3e^{-x}$$

Solution

If $y = \alpha e^{-x}$, write down expressions for $\dfrac{dy}{dx}$ and $\dfrac{d^2y}{dx^2}$.

$$\frac{dy}{dx} = -\alpha e^{-x}, \frac{d^2y}{dx^2} = \alpha e^{-x}$$

Substitute into the given equation to find α, and hence the particular integral.

$$-2\alpha = 3, \alpha = -\frac{3}{2}; y_{pi}(x) = -\frac{3}{2}e^{-x}$$

Example 6.8

Obtain a particular integral of the equation

$$\frac{d^2y}{dx^2} - 6\frac{dy}{dx} + 8y = x$$

Solution

In the last example, we found that a fruitful approach was to assume a solution in the same form as that on the right-hand side. Suppose we assume a solution $y(x) = \alpha x$ and proceed to determine α. This approach will actually fail, but let us see why. If $y(x) = \alpha x$ then $\dfrac{dy}{dx} = \alpha$ and $\dfrac{d^2y}{dx^2} = 0$. Substitution into the differential equation yields

$$0 - 6\alpha + 8\alpha x = x$$

and α ought now to be chosen so that this expression is true for all x. If we equate the coefficients of x we find $8\alpha = 1$ so that $\alpha = \frac{1}{8}$, but with this value of α the constant terms are inconsistent (i.e. $-\frac{6}{8}$ on the left, but zero on the right). Clearly a particular integral of the form αx is not possible. The problem arises because differentiation of the term αx produces constant terms that are unbalanced on the right-hand side. So, we try a solution of the form

$$y(x) = \alpha x + \beta$$

with α and β constants. This is consistent with the recommendation in Table 6.1. Proceeding as before, $\dfrac{dy}{dx} = \alpha$, $\dfrac{d^2y}{dx^2} = 0$. Substitution in the differential equation now gives

$$0 - 6\alpha + 8(\alpha x + \beta) = x$$

Equating coefficients of x we find

$$8\alpha = 1 \tag{4}$$

and equating constant terms we find

$$-6\alpha + 8\beta = 0 \tag{5}$$

From equation (4), $\alpha = \dfrac{1}{8}$ and then from equation (5)

$$-6\left(\frac{1}{8}\right) + 8\beta = 0$$

so that

$$8\beta = \frac{3}{4}$$

that is

$$\beta = \frac{3}{32}$$

The required particular integral is

$$y_{\text{pi}}(x) = \frac{x}{8} + \frac{3}{32}$$

Example 6.9
Find a particular integral for the equation

$$\frac{d^2 y}{dx^2} - 6\frac{dy}{dx} + 8y = 3\cos x$$

Solution
First try to decide on an appropriate form for the trial solution. Refer to Table 6.1 if necessary.

$$y = \alpha \cos x + \beta \sin x$$

We shall try a solution of the form

$$y(x) = \alpha \cos x + \beta \sin x$$

Differentiating, we find

$$\frac{dy}{dx} = -\alpha \sin x + \beta \cos x, \quad \frac{d^2 y}{dx^2} = -\alpha \cos x - \beta \sin x$$

Substitution into the differential equation gives

$$(-\alpha \cos x - \beta \sin x) - 6(-\alpha \sin x + \beta \cos x) + 8(\alpha \cos x + \beta \sin x) = 3 \cos x$$

which simplifies to

$$\qquad \sin x + \qquad \cos x = 3 \cos x$$

$$7\beta + 6\alpha, 7\alpha - 6\beta$$

Equating coefficients of $\sin x$,

$$7\beta + 6\alpha = 0$$

and equating coefficients of $\cos x$, we find

$$7\alpha - 6\beta = 3$$

Solve these simultaneously to find α and β, and hence the particular integral.

$$\alpha = \frac{21}{85}, \beta = -\frac{18}{85}$$

The particular integral is

$$y_{\text{pi}}(x) = \frac{21}{85}\cos x - \frac{18}{85}\sin x$$

Exercises

1   Find a particular integral for the equation

$$\frac{d^2x}{dt^2} - 3\frac{dx}{dt} + 2x = 5e^{3t}$$

2   Find a particular integral for the equation

$$\frac{d^2x}{dt^2} - x = 4e^{-2t}$$

3   Find a particular integral for
$y'' + y' + y = 2 + x + \cos x.$

Solutions to exercises

1   $x_{\text{pi}} = 2.5e^{3t}$

2   $x_{\text{pi}} = \dfrac{4}{3}e^{-2t}$

3   $y_{\text{pi}} = 1 + x + \sin x$

6.4   Finding the general solution of a second-order inhomogeneous equation

The general solution of a second-order linear inhomogeneous equation is the sum of its complementary function and a particular integral. In Block 5 you learnt how to find a complementary function, and in the previous section you learnt how to find a particular integral. We now put these together to find the general solution.

Key point

The general solution of a constant-coefficient ordinary differential equation

$$a\frac{d^2y}{dx^2} + b\frac{dy}{dx} + cy = f(x)$$

is the sum of the particular integral and the complementary function:

$$y = y_{\text{pi}} + y_{\text{cf}}$$

y_{pi} contains no arbitrary constants; y_{cf} contains two arbitrary constants.

Example 6.10
Find the general solution of

$$\frac{d^2y}{dx^2} + 3\frac{dy}{dx} - 10y = 3x^2$$

Solution
The complementary function has already been found in Example 5.9 to be

$$y_{cf} = Ae^{2x} + Be^{-5x}$$

By noting that $3x^2$ is a polynomial of degree 2, the particular integral is found by trying a solution of the form

$$y = ax^2 + bx + c$$

Differentiate this trial solution twice:

$$\frac{dy}{dx} = 2ax + b, \frac{d^2y}{dx^2} = 2a$$

Substitute into the inhomogeneous equation and by equating coefficients of x^2, x and constants, write down the equations satisfied by a, b and c.

$$2a + 3(2ax + b) - 10(ax^2 + bx + c) = 3x^2$$

Equate coefficients of x^2:

$$-10a = 3$$

Equate coefficients of x:

$$6a - 10b = 0$$

Equate constants:

$$2a + 3b - 10c = 0$$

Solve these to find a, b and c.

$$a = -\frac{3}{10}, b = -\frac{9}{50}, c = -\frac{57}{500}$$

Hence write down the particular integral:

$$y_{pi}(x) = -\frac{3}{10}x^2 - \frac{9}{50}x - \frac{57}{500}$$

Thus the general solution is

$$y = y_{pi}(x) + y_{cf}(x)$$

$$= -\frac{3}{10}x^2 - \frac{9}{50}x - \frac{57}{500} + Ae^{2x} + Be^{-5x}$$

Example 6.11

Find the particular solution of $y'' + y' - 12y = 4e^{2x}$ that satisfies $y(0) = 7$ and $y'(0) = 0$.

Solution

The solution must be found in three parts:

1   Solve the homogeneous equation to find the complementary function.
2   Find a particular integral of the inhomogeneous equation, and add this to the complementary function to find the general solution.
3   Apply the initial conditions to find the arbitrary constants.

First of all we find the complementary function. The homogeneous equation is $y'' + y' - 12y = 0$ and so the auxiliary equation is $k^2 + k - 12 = (k - 3)(k + 4) = 0$. Therefore $k = 3$ and $k = -4$ and so the complementary function is

$$y = Ae^{3x} + Be^{-4x}$$

The particular integral is found by trying a solution of the form $y = \alpha e^{2x}$.

If $y = \alpha e^{2x}$, $y' = 2\alpha e^{2x}$ and $y'' = 4\alpha e^{2x}$. Substitute these expressions into the differential equation

$$4\alpha e^{2x} + 2\alpha e^{2x} - 12\alpha e^{2x} = 4e^{2x}$$

so that

$$4\alpha + 2\alpha - 12\alpha = 4$$

from which $-6\alpha = 4$, that is $\alpha = -\dfrac{2}{3}$. The general solution is then

$$y = Ae^{3x} + Be^{-4x} - \frac{2}{3}e^{2x}$$

The particular solution is found by applying the initial conditions.

Apply the condition $y(0) = 7$.

$$7 = A + B - \frac{2}{3}$$

To apply the condition $y'(0) = 0$ we must differentiate the general solution:

$$\frac{dy}{dx} = \qquad\qquad 3Ae^{3x} - 4Be^{-4x} - \frac{4}{3}e^{2x}$$

Now apply the condition $y'(0) = 0$.

$$0 = 3A - 4B - \frac{4}{3}$$

Solve these simultaneous equations.

$$A = \frac{32}{7} \text{ and } B = \frac{65}{21}$$

Finally, the particular solution is

$$y = \frac{32}{7} e^{3x} + \frac{65}{21} e^{-4x} - \frac{2}{3} e^{2x}$$

Example 6.12 Electrical Engineering – An *LC* circuit with sinusoidal input

The differential equation governing the flow of current in the series *LC* circuit shown in Figure 6.1 when subject to an applied voltage $v(t) = V_0 \sin \omega t$ is

$$L\frac{d^2i}{dt^2} + \frac{1}{C} i = \omega V_0 \cos \omega t$$

Obtain its general solution.

Figure 6.1
A series *LC* circuit.

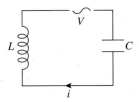

Solution
The homogeneous equation is

$$L\frac{d^2i}{dt^2} + \frac{i}{C} = 0$$

Letting $i = e^{kt}$ we find the auxiliary equation is $Lk^2 + \frac{1}{C} = 0$ so that $k^2 = -\frac{1}{LC}$

and so $k = \pm\frac{j}{\sqrt{LC}}$. Therefore the complementary function is

$$i_{cf} = A \cos\frac{t}{\sqrt{LC}} + B \sin\frac{t}{\sqrt{LC}}$$

where A and B are arbitrary constants. To find a particular integral, try $i = E \cos \omega t + F \sin \omega t$, where E and F are constants. We find

$$\frac{di}{dt} = -\omega E \sin \omega t + \omega F \cos \omega t$$

$$\frac{d^2i}{dt^2} = -\omega^2 E \cos \omega t - \omega^2 F \sin \omega t$$

Substitution into the inhomogeneous equation yields

$$L(-\omega^2 E \cos \omega t - \omega^2 F \sin \omega t) + \frac{1}{C}(E \cos \omega t + F \sin \omega t) = \omega V_0 \cos \omega t$$

which is simplified to

$$\boxed{} \sin \omega t + \boxed{} \cos \omega t = \omega V_0 \cos \omega t$$

$$-\omega^2 LF + \frac{F}{C}, \; -\omega^2 LE + \frac{E}{C}$$

Equating coefficients of $\sin \omega t$ gives

$$-\omega^2 LF + \frac{F}{C} = 0$$

from which $F = 0$.

Equating coefficients of $\cos \omega t$ gives

$$-\omega^2 LE + \frac{E}{C} = \omega V_0$$

from which

$$E = \frac{CV_0 \, \omega}{1 - \omega^2 \, LC}$$

Hence the particular integral is

$$i_{\text{pi}} = \frac{CV_0 \, \omega}{1 - \omega^2 \, LC} \cos \omega t$$

Finally, the general solution is

$$i = i_{\text{cf}} + i_{\text{pi}}$$

$$= A\cos \frac{t}{\sqrt{LC}} + B\sin \frac{t}{\sqrt{LC}} + \frac{CV_0 \omega}{1 - \omega^2 LC} \cos \omega t$$

Exercises

1   Find the general solution of the following equations:

(a) $\dfrac{d^2 x}{dt^2} - 2\dfrac{dx}{dt} - 3x = 6$

(b) $\dfrac{d^2 y}{dx^2} + 5\dfrac{dy}{dx} + 4y = 8$

(c) $\dfrac{d^2 y}{dt^2} + 5\dfrac{dy}{dt} + 6y = 2t$

(d) $\dfrac{d^2 x}{dt^2} + 11\dfrac{dx}{dt} + 30x = 8t$

2   Obtain the general solution of

$$y'' - y' - 2y = 6$$

3   Obtain the general solution of the equation

$$\frac{d^2 y}{dx^2} + 3\frac{dy}{dx} + 2y = 10 \cos 2x$$

Find the particular solution satisfying

$$y(0) = 1, \quad \frac{dy}{dx}(0) = 0$$

Solutions to exercises

1   (a) $x = Ae^{-t} + Be^{3t} - 2$
    (b) $y = Ae^{-x} + Be^{-4x} + 2$
    (c) $y = Ae^{-2t} + Be^{-3t} + \dfrac{t}{3} - \dfrac{5}{18}$
    (d) $x = Ae^{-6t} + Be^{-5t} + 0.267t - 0.0978$

2   $y = Ae^{2x} + Be^{-x} - 3$

3   $y = Ae^{-2x} + Be^{-x} + \frac{3}{2} \sin 2x - \frac{1}{2} \cos 2x;$
    $\frac{3}{2} e^{-2x} + \frac{3}{2} \sin 2x - \frac{1}{2} \cos 2x$

Occasionally you will come across a differential equation

$$a\frac{d^2y}{dx^2} + b\frac{dy}{dx} + cy = f(x)$$

for which the inhomogeneous term, $f(x)$, forms part of the complementary function. One such example is the equation

$$\frac{d^2y}{dx^2} - \frac{dy}{dx} - 6y = e^{3x}$$

It is straightforward to check that the complementary function is $y_{cf} = Ae^{3x} + Be^{-2x}$. Note that the first of these terms has the same form as the inhomogeneous term, e^{3x}, on the right-hand side of the differential equation.

You should verify for yourself that trying a particular integral of the form $y(x) = \alpha e^{3x}$ will not work in a case like this. Can you see why?

Instead, try a particular integral of the form $y(x) = \alpha x e^{3x}$. Verify that

$$\frac{dy}{dx} = \alpha e^{3x}(3x + 1) \quad \text{and} \quad \frac{d^2y}{dx^2} = \alpha e^{3x}(9x + 6)$$

Substitute these expressions into the differential equation to find $\alpha = \frac{1}{5}$. Finally, the particular integral is $y_{pi}(x) = \frac{1}{5}xe^{3x}$ and so the general solution to the differential equation is

$$y = Ae^{3x} + Be^{-2x} + \frac{1}{5}xe^{3x}$$

In general, if $f(x)$ is part of the complementary function we try $y = \alpha x f(x)$ for the particular integral.

End of block exercises

1   Find the general solution of the following equations:

(a) $\dfrac{d^2y}{dx^2} + 2\dfrac{dy}{dx} + 3y = 2\sin 2x$

(b) $\dfrac{d^2y}{dt^2} + \dfrac{dy}{dt} + y = 4\cos 3t$

(c) $\dfrac{d^2y}{dx^2} + 9y = 4e^{8x}$

(d) $\dfrac{d^2x}{dt^2} - 16x = 9e^{6t}$

2   Find a particular integral for the equation

$$\frac{d^2y}{dx^2} + \frac{dy}{dx} + y = 1 + x$$

3   Find the general solution of

(a) $\dfrac{d^2x}{dt^2} - 6\dfrac{dx}{dt} + 5x = 3$

(b) $\dfrac{d^2x}{dt^2} - 2\dfrac{dx}{dt} + x = e^t$

Solutions to exercises

1 (a) $y = e^{-x}(A \sin \sqrt{2}x + B \cos \sqrt{2}x)$

$\quad -\dfrac{8}{17} \cos 2x - \dfrac{2}{17} \sin 2x$

(b) $y = e^{-0.5t}(A \cos 0.866t + B \sin 0.866t)$
$\quad -0.438 \cos 3t + 0.164 \sin 3t$

(c) $y = A \cos 3x + B \sin 3x + 0.0548e^{8x}$

(d) $x = Ae^{4t} + Be^{-4t} + \dfrac{9}{20}e^{6t}$

2 $y_{pi} = x$

3 (a) $x = Ae^t + Be^{5t} + \dfrac{3}{5}$

(b) $x = Ae^t + Bte^t + \dfrac{1}{2}t^2e^t$

End of chapter exercises

1 Give one example each of a first-order linear equation, first-order non-linear equation, second-order linear equation, second-order non-linear equation.

2 By integrating twice find the general solution of $y'' = 12x^2$.

3 Find a first-order equation satisfied by $x = Ae^{-2t}$.

4 Classify the following equations, specifying the order and type (linear or non-linear):

(a) $\dfrac{d^2y}{dx^2} - \dfrac{dy}{dx} = x^2$

(b) $\dfrac{dy}{dt} + \cos y = 0$

5 Find a second-order differential equation that is satisfied by

$$y = A \cosh 2x + B \sinh 2x$$

6 Integrate twice the differential equation

$$\frac{d^2y}{dx^2} = \frac{w}{2}(lx - x^2)$$

where w and l are constants, to find a general solution for y.

7 Integrate the equation $\dfrac{dy}{dx} = 3x^2$ subject to the condition $y(1) = 4$ in order to find the particular solution.

8 Two tanks containing a liquid are placed in series so that the first discharges into the second and the second discharges into a waste outlet. Let $q_1(t)$ and $q_2(t)$ be the flow rates out of the two tanks respectively, and let the height of liquid in each of the tanks be $h_1(t)$ and $h_2(t)$ respectively. The two tanks are identical and each has a constant cross-sectional area A. The outflow from each tank is proportional to the height of liquid in the tank. At $t = 0$ the height of liquid in the first tank is h_0 and the second tank is empty.
(a) Derive and solve the differential equation for $h_1(t)$.
(b) Hence find $q_1(t)$.
(c) Derive and solve the differential equation for $h_2(t)$.
(d) Hence find $q_2(t)$.

9 The charge, q, on a capacitor in an LCR series circuit satisfies the second-order differential equation

$$L\frac{d^2q}{dt^2} + R\frac{dq}{dt} + \frac{1}{C}q = E$$

where L, R, C and E are constants. Show that if $2L = CR^2$ the general solution of this equation is

$$q = $$

$$e^{-t/(CR)}\left(A \cos \frac{1}{CR}t + B \sin \frac{1}{CR}t\right) + CE$$

If $i = \dfrac{dq}{dt} = 0$ and $q = 0$ when $t = 0$ show

that the current in the circuit is

$$i = \frac{2E}{R} e^{-t/(CR)} \sin\frac{1}{CR} t$$

10 Obtain the general solution of the differential equation

$$\frac{dT}{d\theta} - \mu T = -\mu K$$

where μ and K are constants.

11 The equation governing the flow of current i in a series LR circuit with applied constant voltage E is

$$L\frac{di}{dt} + Ri = E$$

(a) Solve this equation subject to the condition $i(0) = 0$.
(b) State the final value of the current.
(c) Find the time taken for the current to reach 95% of its final value.

Solutions to exercises

2   $y = x^4 + Ax + B$

3   $x' = -2x$

4   (a) second-order linear
    (b) first-order non-linear due to term $\cos y$

5   $\dfrac{d^2y}{dx^2} = 4y$

6   $y = \dfrac{w}{2}\left(\dfrac{lx^3}{6} - \dfrac{x^4}{12}\right) + Ax + B$

7   $y = x^3 + 3$

8   (a) $h_1 = h_0 e^{-kt/A}$ (b) $q_1 = kh_0 e^{-kt/A}$
    (c) $h_2 = \dfrac{kh_0 t}{A} e^{-kt/A}$ (d) $q_2 = \dfrac{k^2 h_0 t}{A} e^{-kt/A}$

10 $T = K + Ae^{\mu\theta}$

11 (a) $i = \dfrac{E}{R}\left(1 - e^{-Rt/L}\right)$ (b) $\dfrac{E}{R}$
    (c) about $\dfrac{3L}{R}$

Functions of more than one variable and partial differentiation

In many engineering applications a physical quantity depends upon several other quantities that can vary independently. For example, the temperature at a point in a kiln will vary both with time and with the position of the point in the kiln. Such quantities are described mathematically using functions of more than one variable. In this chapter we explain the notation used to represent such functions. We go on to explain two ways in which these functions can be visualised. The techniques of calculus, previously applied to functions of one variable, are now extended to functions of several variables. Differentiation involving several variables is called **partial differentiation**. Finally we illustrate an application of partial differentiation to finding maximum, minimum and saddle points of a function of two variables.

Chapter 21 contents

Functions of two independent variables, and their graphs

1.1 Introduction

We have explained in Chapter 6 that a function is a mathematical rule that operates upon an input to produce a single output. We saw that the input is referred to as the independent variable because we are free, within reason, to choose its value. The output is called the dependent variable because its value *depends* upon the value of the input. Commonly we use the letter x to represent the input, y the output and f the function, in which case we write $y = f(x)$. Examples include $y = 5x^2$, $y = \dfrac{1}{x}$ and so on. In such cases, there is a single independent variable, x.

We now move on to consider examples in which there are two independent variables. This means that there will be two inputs to a function, each of which can be chosen independently. Once these values have been chosen, the function rule will be used to process them in order to produce a single output called the dependent variable. Notice that whilst there is now more than one input, there is still a *single* output.

We then illustrate how these functions can be visualised in the form of contour plots and three-dimensional graphs. User-friendly computer software can be used for such visualisation, and you should enquire about packages available for your use.

1.2 Functions of two variables

The block diagram in Figure 1.1 depicts the function $f(x, y) = 3x + 4y$. This function has two independent inputs, denoted by x and y. The function rule is 'multiply the input x by 3, multiply the input y by 4, and add the results'. There is a single output $3x + 4y$.

Figure 1.1
A block diagram
of the function
$f(x, y) = 3x + 4y$

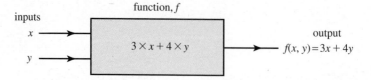

The notation $f(x, y) = 3x + 4y$ indicates that, with inputs x and y, the function, f, produces an output of $3x + 4y$. The inputs to the function are placed in brackets, after the f. The expression $f(x, y)$ is read as 'f is a function of x and y'. As before, the inputs to a function are also known as the **arguments** of the function.

Key point	A function of two variables is a rule that produces a single output when values of two independent variables are chosen.

Example 1.1

A function f is defined by $f(x, y) = 3x + 4y$. Calculate the output when the input values are $x = 5$ and $y = 6$.

Solution

Refer to the block diagram in Figure 1.1. When the input x takes the value 5, and the input y takes the value 6, the output is $3 \times 5 + 4 \times 6 = 15 + 24 = 39$. We could write this as $f(5, 6) = 39$.

Example 1.2

A function f is defined by $f(x, y) = 11x^2 - 7y + 2$. Calculate the output when the inputs are $x = -2$ and $y = 3$.

Solution

In this example we are required to find $f(-2, 3)$. This can be calculated directly from the expression for $f(x, y)$, without reference to a block diagram. Substitution into the function rule produces

$$f(-2, 3) = 11 \times (-2)^2 - 7 \times (3) + 2 = 44 - 21 + 2 = 25$$

Example 1.3

A function is defined by $f(x, y) = \dfrac{x}{y}$. Calculate the output when the inputs are $x = 12$ and $y = 3$.

Solution

$f(12, 3) = $ ⬚　　　　　　　　　　　　　　$\dfrac{12}{3} = 4$

Example 1.4

A function is defined by $f(x, y) = e^{x+y}$.
(a) Calculate the output when the inputs are $x = -1$ and $y = 2$.
(b) Show that this function can be written in the equivalent form $f(x, y) = e^x e^y$.

Solution

(a) $f(-1, 2) = $ ⬚　　　　　　　　　$e^{-1+2} = e^1 = e = 2.718\,(3 \text{ d.p.})$

(b) Using the first law of indices we can write e^{x+y} as $e^x e^y$.

Note that in the previous examples both x and y are independent variables. We can choose a value for y quite independently of the value we have chosen for x. This is quite different from the case of a function of a single variable when, for example, if we write $y = 5x^2$, choosing x automatically determines y.

It is common to introduce another symbol to stand for the output. So, in Example 1.1 we may write $z = f(x, y) = 3x + 4y$ or simply $z = 3x + 4y$. In Example 1.2 we can write simply $z = 11x^2 - 7y + 2$. Here the value of z depends upon the values chosen for x and y, and so z is referred to as the dependent variable.

Functions of two variables are introduced because they arise naturally in engineering applications. Consider the following example, which illustrates this.

Example 1.5

The volume V of the cylinder shown in Figure 1.2 is given by the formula $V = \pi r^2 h$, where r is the radius and h is the height of the cylinder. Suppose we choose a value for the radius, and a value for the height. Note that we can choose these values independently. We can then use the formula to determine the volume of the cylinder. We can regard V as depending upon the two independent variables r and h and write $V = f(r, h) = \pi r^2 h$. Here, the dependent variable is V since the value of the volume depends upon the values chosen for the radius and the height.

Figure 1.2
The volume, V, depends upon the two independent variables r and h, and is given by $V = \pi r^2 h$.

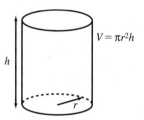

$V = \pi r^2 h$

h

r

1.3 Contour plots of functions of two variables

By now you will be familiar with the way in which a function of one variable is represented graphically. For example, you have seen in Chapter 6 how the graph of the function $y = f(x)$ is drawn by plotting the independent variable x on the horizontal axis, and the dependent variable y on the vertical axis. Given a value of x, the function rule enables us to calculate a value for y, and the point with coordinates (x, y) is then plotted. Joining all such points produces a graph of the function.

When two independent variables are involved, as in $z = f(x, y)$, there are various ways in which we can represent this graphically. One way is to use a **contour plot**, as we shall show in Examples 1.6 and 1.7.

A contour plot is a two-dimensional visualisation of a function of two variables. You may already be familiar with contour plots in the context of maps. Contours are drawn on the map to show the location of points that are the same height above sea level. In the example that follows, contours show the location of points that have the same temperature. This is analogous to the plotting of isotherms on a weather map.

Example 1.6 Mechanical Engineering – Temperature distribution in a metal plate

There are many occasions in the engineering design process when it is necessary to analyse heat flow. For example an engineer may be interested in the effect of different insulation materials on heat losses through the roof of a building. In a nuclear reactor heat is generated in the fuel rods, which are then often cooled by water. It is critical to be able to calculate the rate of heat flow from the rods to the water if a life-threatening situation such as meltdown of the reactor is to be avoided. In applications like these an engineer will meet functions of more than one variable.

As an example consider a two-dimensional metal plate, defined by the region $0 \leq x \leq 1$, $0 \leq y \leq 1$, which contains no internal sources of heat. Suppose the temperature is maintained at $0\,°C$ on three of the plate boundaries. Suppose it varies with x along the fourth boundary at $y = 1$ according to $T = \sin \pi x$, as shown in Figure 1.3.

Figure 1.3
A metal plate with temperatures imposed on its boundaries.

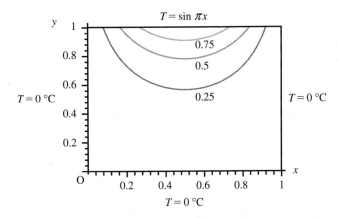

A typical problem is to use this information to determine the temperature at any point in the plate. Knowing the equations for heat conduction, and some advanced techniques in calculus, it is possible to show that the temperature, T, at any point (x, y) is given by the function of two variables

$$T(x, y) = \frac{\sinh \pi y \sin \pi x}{\sinh \pi}$$

The function $\sinh \pi y$ is a hyperbolic function, detailed in Chapter 8 Block 1. It is straightforward, using a computer graphics package, to produce a contour plot from this function that shows curves of constant temperature. In Figure 1.3 we have plotted contours showing curves upon which the temperature is $0.25\,°C$, $0.5\,°C$, and $0.75\,°C$.

Example 1.7 Fluid Dynamics – Fluid flow around a corner

The analysis of the ways in which fluids flow around objects is essential in fields such as aeronautical and marine engineering, and the study of renewable energy systems. A useful quantity is the **streamfunction**, ψ. Under particular conditions ψ is a function of the two spatial variables x and y: that is, $\psi = \psi(x, y)$. Plotting curves upon which ψ is constant gives a useful visual representation because fluid flows along these curves.

For example, the streamfunction $\psi(x, y) = 4(x^2 - y^2)$ gives the streamlines for flow around a corner. Curves along which ψ is constant are illustrated in Figure 1.4.

Figure 1.4
The streamlines of a flow around a corner are described by a function of two variables, $\psi = 4(x^2 - y^2)$.

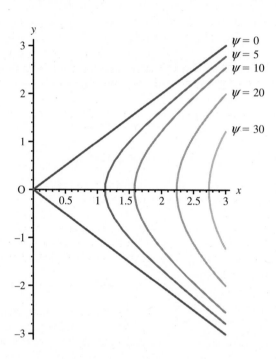

Exercises

1. Given $z = f(x, y) = 7x + 2y$ find the output when $x = 8$ and $y = 2$.

2. If $z = f(x, y) = -11x + y$ find (a) $f(2, 3)$, (b) $f(11, 1)$.

3. If $z = f(x, y) = 3e^x - 2e^y + x^2y^3$ find $z(1, 1)$.

4. If $w = g(x, y) = 7 - xy$ find the value of the dependent variable w when $x = -3$ and $y = -9$.

5. If $z = f(x, y) = \sin(x + y)$ find $f(20°, 30°)$ where the inputs are angles measured in degrees.

6. If $f(x, t) = e^{2xt}$ find $f(0.5, 3)$.

7. Using the temperature distribution function
$$T(x, y) = \frac{\sinh \pi y \sin \pi x}{\sinh \pi}$$ in Example 1.6, evaluate the temperature at the centre of the plate.

Solutions to exercises

1   60

2   (a) -19 (b) -120

3    $e^1 + 1 = 3.718$ (3 d.p.)

4   -20

5   0.766 (3 d.p.)

6   20.086 (3 d.p.)

7   0.199 °C (3 d.p.)

Computer and calculator exercises

1   Matlab code that can be used to generate the contour plot in Figure 1.3 is

```
[x, y] = meshgrid (0:0.1:1,
0:0.1:1);
z = sinh (pi*y).*sin(pi*x) /
sinh (pi);
contour (x, y, z, [0.25, 0.5,
0.75]);
```

If you have access to Matlab, reproduce this figure and then modify the code to produce a contour plot of the function $\psi = 4(x^2 - y^2)$ shown in Figure 1.4. Note that x^2 will be coded as

```
x.^2
```

2   Maple code that can be used to generate Figure 1.4 is

```
with (plots):
contourplot (4*(x^2 - y^2),
x = 0..3, y = -3..3,
contours = [0, 5, 10, 20, 30]);
```

If you have access to Maple, reproduce this figure and modify the code to produce the contour plot in Figure 1.3.

3   The function $\psi(x, y) = y\left(1 - \dfrac{1}{x^2 + y^2}\right)$ represents the streamline pattern of fluid flow past a cylinder. Investigate whether a computer graph-plotting package to which you have access can produce a contour plot of this flow.

1.4   Three-dimensional graphs

As an alternative to a contour plot, a graph can be drawn in three dimensions – a task that is difficult to do in the two-dimensional plane of the paper. We now need two axes for the independent variables and a third axis for the dependent variable. Computer software is readily available for plotting graphs in three dimensions, so rather than attempt this manually we shall be content with understanding the process involved in producing the graph, and illustrating this with several examples.

Three axes are drawn at right angles and these are labelled x, y and z as shown in Figure 1.5. There is more than one way of doing this, but the usual convention is shown here. When labelled in this way the axes are said to form a **right-handed set**. The axes intersect at the origin O.

Figure 1.5
Three
perpendicular axes,
labelled x, y and z.

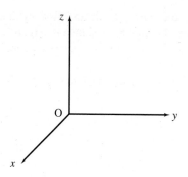

Example 1.8

Given the function of two variables $z = f(x, y) = 3x + 4y$ calculate the value of the function when x and y take the following values. Represent each point graphically.
(a) $x = 3$, $y = 2$ (b) $x = 5$, $y = 0$ (c) $x = 0$, $y = 5$

Solution

In each case we use the function rule $z = f(x, y) = 3x + 4y$ to calculate the corresponding value of z.
(a) $z = 3(3) + 4(2) = 17$ (b) $z = 3(5) + 4(0) = 15$ (c) $z = 3(0) + 4(5) = 20$
It is conventional to write the coordinates of the points in the form (x, y, z) so that the three points in this example are $(3, 2, 17)$, $(5, 0, 15)$ and $(0, 5, 20)$.

Each of the points can then be drawn as shown in Figure 1.6, where they have been labelled A, B and C respectively. Notice that the z coordinate of point A, which is 17, gives the height of A above the point $(3, 2)$ in the xy plane. Similar comments apply to points B and C.

Figure 1.6
Three points in
three-dimensional
space with
coordinates
$A(3, 2, 17)$,
$B(5, 0, 15)$ and
$C(0, 5, 20)$.

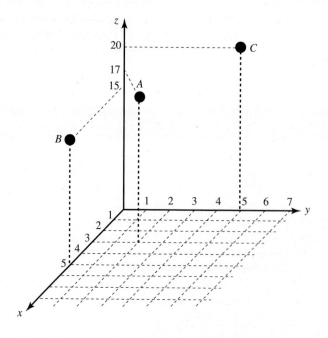

If we were to continue selecting more points and plotting them we would find that all the points lie in a plane. This plane is shown in Figure 1.7. Notice that in each case the z coordinate is the height of the plane above the point (x, y) in the xy plane.

For example, the point C is 20 units above the point $(0, 5)$ in the xy plane.

Figure 1.7
The plane
$z = 3x + 4y$.

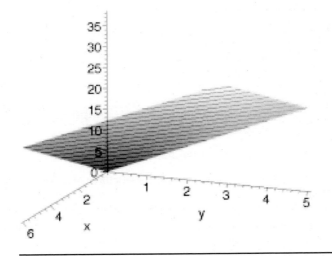

In the previous example we saw that when we drew a graph of the function of two variables $z = 3x + 4y$ we obtained a plane. In more general cases the graph of $z = f(x, y)$ will be a curved surface, and the z coordinate is the height of the surface above the point (x, y). Some more examples of functions of two variables and their graphs are shown in Figures 1.8 and 1.9. It would be a useful exercise for you to try to reproduce these graphs for yourself using computer software.

Figure 1.8
The function
$z = x^2 + y^2$.

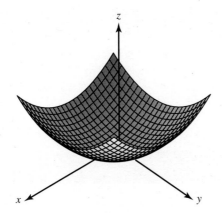

Figure 1.9
The function
$z = \sin(x^2 + y^2)$.

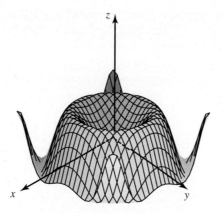

Example 1.9 Mechanical Engineering –Temperature distribution in a metal plate

In Example 1.6 we saw that the function $T(x, y) = \dfrac{\sinh \pi y \sin \pi x}{\sinh \pi}$ represents the temperature distribution in a metal plate when three of its boundaries are maintained at $0\,°C$ and the fourth has a temperature given by $T = \sin \pi x$. A graph of this function is shown in Figure 1.10. Observe the temperature distribution along the line $y = 1$, given by $T = \sin \pi x$.

Figure 1.10

The function
$$T(x, y) = \frac{\sinh \pi y \sin \pi x}{\sinh \pi}.$$

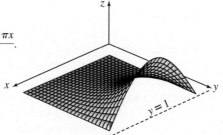

Exercises

1 Given the function $z = 3x - 6y$, find the z coordinate corresponding to each of the following points, which lie in the xy plane:
(a) $(4, 4)$ (b) $(0, 3)$ (c) $(5, 0)$

Plot the points on a three-dimensional graph.

Solutions to exercises

1 (a) -12 (b) -18 (c) 15

Computer and calculator exercises

1   The following Matlab code can be used to generate a three-dimensional plot as shown in Figure 1.10.

```
[x,y] = meshgrid (0:0.1:1,
0:0.1:1);
z = sinh(pi*y).*sin(pi*x)/
sinh(pi);
mesh(z);
```

If you have access to Matlab generate a plot for yourself. Plot a graph of the surface $z = \sin(x^2 + y^2)$ as shown in Figure 1.9.

2   The Maple code

```
plot3d(x^2+y^2,x = -3..3,
y = -3..3);
```

can be used to plot the graph of $z = x^2 + y^2$ as shown in Figure 1.8. Try to reproduce this graph for yourself.

End of block exercises

1   Find the value of the function
$f(x, y) = \sqrt{x^2 + y^2}$ when $x = 2$ and $y = 6$.

2   If $f(x, y) = x^2 + 3xy$ find $f(2, 3)$.

3   If $V = \ln(x^2 + y^2)$ find the value of V when $x = 3$ and $y = 7$.

4   Functions of three (or more) independent variables can be defined in an analogous way to functions of two variables. Suppose $u(x, y, z)$ is a function of the three independent variables x, y and z, and is defined by

$$u = \frac{1}{\sqrt{x^2 + y^2 + z^2}}$$

Find u when $x = 1, y = 2$ and $z = 3$. Can the value of u be determined when $x = y = z = 0$?

Solutions to exercises

1   $\sqrt{40}$

2   22

3   $V = \ln 58 = 4.060$ (3 d.p.)

4   $u(1, 2, 3) = \frac{1}{\sqrt{14}}$. No, because division by zero is not allowed.

Partial differentiation

Before attempting this section it is essential that you have a thorough understanding of differentiation of functions of one variable, and you should revise Chapters 15 and 16 if necessary. In particular, you may need to refer frequently to Table 2.1 in Chapter 15, which gives the derivatives of some common functions.

In Chapter 15 we introduced differentiation of a function of a single variable and showed how the derivative of such a function can be calculated. Recall that this gives useful information about the gradient of the graph of the function at different points. In this section we explain how functions of two variables are differentiated.

Also in Chapter 15 we differentiated y, which was a function of x, to obtain the derivative $\dfrac{\mathrm{d}y}{\mathrm{d}x}$. Note that the dependent variable (in this case y) is differentiated with respect to the independent variable (in this case x).

Consider now z, which depends upon two variables x and y. Recall that z is then the dependent variable, and x and y are the independent variables. Hence z can be differentiated with respect to x to produce a derivative, and it can also be differentiated with respect to y to produce another, different, derivative. So for functions of two variables there are two derivatives: we can no longer talk about *the* derivative of z. This is a fundamental difference between functions of one variable and functions of two variables.

When differentiating functions of two variables we refer to this process as **partial differentiation**, and instead of using a normal letter d as in $\dfrac{\mathrm{d}y}{\mathrm{d}x}$, we use a curly d instead and write ∂. Do not be put off by this notation – you will soon get used to it. When we differentiate z with respect to (w.r.t.) x we denote the resulting derivative by $\dfrac{\partial z}{\partial x}$. An alternative notation in common use is to write this derivative as z_x. When we differentiate z w.r.t. y we denote the derivative produced by $\dfrac{\partial z}{\partial y}$. An alternative notation is to write this derivative as z_y.

Key point

If $z = f(x, y)$ then

the (first) partial derivative of z with respect to x is denoted $\dfrac{\partial z}{\partial x}$ or z_x,

the (first) partial derivative of z with respect to y is denoted $\dfrac{\partial z}{\partial y}$ or z_y.

We now explain how these derivatives are calculated.

2.2 Partial differentiation with respect to x

Suppose we have a function $z = 5x + 11$. You will recall that its derivative is $\dfrac{dz}{dx} = 5$.
Note that the derivative of the constant 11 is zero. Similarly if $z = 6x + 12$, $\dfrac{dz}{dx} = 6$.
Here the derivative of the constant 12 is zero. We are now ready to introduce partial differentiation with respect to x.

Consider the function $z = 5x + y$. When we differentiate with respect to x we treat any occurrence of the variable y *as though it were a constant*. Hence in this case the derivative of y is zero, and we write

$$\text{if } z = 5x + y \quad \text{then} \quad \frac{\partial z}{\partial x} = 5 + 0 = 5$$

Similarly,

$$\text{if } z = 7x^2 - y \quad \text{then} \quad \frac{\partial z}{\partial x} = 14x - 0 = 14x$$

If y is treated as a constant, then so too will be multiples of y, such as $7y$ and $-3y$. Furthermore, any functions of y, such as y^2 and e^y, will also be regarded as constants.

Example 2.1

Calculate $\dfrac{\partial z}{\partial x}$ when $z = 9x + 2$.

Solution

We must differentiate $z = 9x + 2$ with respect to x. This function is particularly simple because y does not appear at all. The derivative of $9x$ is 9, and the derivative of the constant 2 is zero. Hence $\dfrac{\partial z}{\partial x} = 9$.

Example 2.2

Calculate $\dfrac{\partial z}{\partial x}$ when $z = 9x + y$.

Solution

To find $\dfrac{\partial z}{\partial x}$ we treat y as though it is a constant. Imagine it is just a number like the '2' in the previous example. Then $\dfrac{\partial z}{\partial x} = 9$.

Example 2.3

Calculate $\dfrac{\partial z}{\partial x}$ when $z = y - 19x$.

Solution

Treat y as though it is a constant.

$$\frac{\partial z}{\partial x} = $$

-19

Example 2.4

Calculate $\dfrac{\partial z}{\partial x}$ when $z = 3x + 4y + 11$.

Solution

In this case 11 is a constant and we treat y, and hence $4y$, as a constant.

$$\frac{\partial z}{\partial x} = $$

3

Example 2.5

Calculate $\dfrac{\partial z}{\partial x}$ when $z = 4x^2 - 3y$.

Solution

$\dfrac{\partial z}{\partial x} = 8x - 0 = 8x$. The quantity $-3y$ is treated as a constant.

Example 2.6

Calculate $\dfrac{\partial z}{\partial x}$ when $z = x^2 - y^2$.

Solution

y, and hence y^2, is treated as a constant. When $-y^2$ is differentiated with respect to x the result will be zero. Hence if $z = x^2 - y^2$ then $\dfrac{\partial z}{\partial x} = 2x$.

Example 2.7

Calculate $\dfrac{\partial z}{\partial x}$ when $z = 5y^3 - x^4$.

Solution

y, and hence $5y^3$, is treated as a constant.

$$\frac{\partial z}{\partial x} = $$

$-4x^3$

Example 2.8

Calculate $\dfrac{\partial z}{\partial x}$ when $z = 3x - 4y^3$.

Solution

If $z = 3x - 4y^3$, then $\dfrac{\partial z}{\partial x} = 3$.

Recall that

- if $4x$ is differentiated with respect to x the result is 4,
- if $5x$ is differentiated with respect to x the result is 5.

Extending this to the following function of two variables:

- if $z = yx$ is differentiated with respect to x the result is y.

We see that because y is treated as if it is a constant then $\frac{\partial z}{\partial x} = y$.

Example 2.9

Calculate $\frac{\partial z}{\partial x}$ when $z = yx^2$.

Solution

To find $\frac{\partial z}{\partial x}$ imagine that you were trying to differentiate $3x^2$ say. The result would be $3(2x) = 6x$. Hence if $z = yx^2$ then $\frac{\partial z}{\partial x} = y(2x)$ or simply $2xy$.

Example 2.10

Calculate $\frac{\partial z}{\partial x}$ when $z = 3yx^2$.

Solution

Treating y as a constant we see that this function is of the form constant $\times\ x^2$ and so $\frac{\partial z}{\partial x} = (3y)(2x) = 6xy$.

Example 2.11

(a) Calculate $\frac{\partial z}{\partial x}$ when $z = 3x^2 + 4xy + 11$.

(b) Evaluate $\frac{\partial z}{\partial x}$ at the point $(x, y) = (-2, 5)$.

(c) Evaluate $\frac{\partial z}{\partial x}$ at the point $(x, y) = (9, -3)$.

Solution

(a) $\frac{\partial z}{\partial x} =$ ______ $6x + 4y$

(b) When $x = -2$ and $y = 5$, $\frac{\partial z}{\partial x} = 6(-2) + 4(5) = 8$.

(c) When $x = 9$ and $y = -3$

$\frac{\partial z}{\partial x} =$ ______ 42

Example 2.12

Calculate $\frac{\partial z}{\partial x}$ when (a) $z = x^2e^y$, (b) $z = x^2 + e^y$, (c) $z = 3x^2 \cos y$, (d) $z = 3x^2 + \cos y$, (e) $z = 3x^2 + 4x \sin y$.

21

Solution

(a) y and hence e^y are treated as a constant. Then $\dfrac{\partial z}{\partial x} = 2xe^y$.

(b) $\dfrac{\partial z}{\partial x} = 2x$

(c) y and hence $\cos y$ are treated as a constant. Then $\dfrac{\partial z}{\partial x} = 6x \cos y$.

(d) $\dfrac{\partial z}{\partial x} = 6x$

(e) $\dfrac{\partial z}{\partial x} = 6x + 4 \sin y$

Example 2.13

Calculate $\dfrac{\partial z}{\partial x}$ when (a) $z = \sin 3x$, (b) $z = \sin yx$.

Solution

(a) If $z = \sin 3x$ then $\dfrac{\partial z}{\partial x} = 3 \cos 3x$.

(b) If $z = \sin yx$ then $\dfrac{\partial z}{\partial x} = y \cos yx$ because y is treated as a constant. Compare this result with that in part (a).

> **Key point**
>
> The (first) partial derivative with respect to x of a function $z = f(x, y)$ is denoted by $\dfrac{\partial z}{\partial x}$ and is calculated by differentiating the function with respect to x and treating y as though it were a constant.

2.3 Partial differentiation with respect to y

When we differentiate a function $f(x, y)$ with respect to y we treat any occurrence of the variable x *as though it were a constant*. The partial derivative with respect to y of a function $z = f(x, y)$ is denoted by $\dfrac{\partial z}{\partial y}$. Consider the following examples.

Example 2.14

Find $\dfrac{\partial z}{\partial y}$ when $z = 3y^4 + 4x^2 + 8$.

Solution

When calculating $\dfrac{\partial z}{\partial y}$ we treat any occurrence of x as if it were a constant. So the term $4x^2$ is treated as a constant and its partial derivative with respect to y is zero. That is,

$$\text{if} \quad z = 3y^4 + 4x^2 + 8 \quad \text{then} \quad \frac{\partial z}{\partial y} = 12y^3 + 0 + 0 = 12y^3$$

Example 2.15

Find $\dfrac{\partial z}{\partial y}$ when $z = 3x^2 y$.

Solution

Because x is treated as a constant, we are dealing with a function of the form $z = \text{constant} \times y$. The derivative with respect to y will be simply the constant factor. That is,

$$\text{if} \quad z = 3x^2 y \quad \text{then} \quad \frac{\partial z}{\partial y} = 3x^2$$

Example 2.16

(a) Find $\dfrac{\partial z}{\partial y}$ when $z = 4xy^3$.

(b) Evaluate $\dfrac{\partial z}{\partial y}$ when $(x, y) = (6, 11)$.

Solution

(a) Because $4x$ is treated as a constant, we are dealing with a function of the form $z = \text{constant} \times y^3$.

$\dfrac{\partial z}{\partial y} =$ [_____] $4x \times (3y^2) = 12xy^2$

(b) When $x = 6$ and $y = 11$, $\dfrac{\partial z}{\partial y} =$ [_____] $12(6)(11^2) = 8712$

Key point

The (first) partial derivative with respect to y of a function $z = f(x, y)$ is denoted by $\dfrac{\partial z}{\partial y}$ and is calculated by differentiating the function with respect to y and treating x as though it were a constant.

It will be necessary to work with symbols other than z, x and y. Consider the following example.

Example 2.17

Consider the function $w = f(p, t)$. Find $\dfrac{\partial w}{\partial p}$ and $\dfrac{\partial w}{\partial t}$ when $w = 3t^7 + 4pt + p^2$.

Solution

When differentiating with respect to p, any occurrence of t is treated as if it were a constant. So $\dfrac{\partial w}{\partial p} = 4t + 2p$.

When differentiating with respect to t, any occurrence of p is treated as if it were a constant. So $\dfrac{\partial w}{\partial t} = 21t^6 + 4p$.

Exercises

1 In each case, given $z = f(x, y)$, find $\dfrac{\partial z}{\partial x}$ and $\dfrac{\partial z}{\partial y}$.

(a) $z = 5x + 11y$ (b) $z = -7y - 14x$
(c) $z = 8x$ (d) $z = -5y$
(e) $z = 3x + 8y - 2$ (f) $z = 17 - 3x + 2y$
(g) $z = 8$ (h) $z = 8 - 3y$
(i) $z = 2x^2 - 7y$ (j) $z = 9 - 3y^3 + 7x$
(k) $z = 9 - 9(x - y)$ (l) $z = 9(x + y + 3)$

2 In each case, given $z = f(x, y)$ find z_x and z_y.
(a) $z = xy$ (b) $z = 3xy$ (c) $z = -9yx$
(d) $z = x^2y$ (e) $z = 9x^2y$ (f) $z = 8xy^2$

3 If $z = 9x + y^2$ evaluate $\dfrac{\partial z}{\partial x}$ and $\dfrac{\partial z}{\partial y}$ at the point $(4, -2)$.

4 Find $\dfrac{\partial z}{\partial x}$ and $\dfrac{\partial z}{\partial y}$ when

(a) $z = e^{2x}$ (b) $z = e^{5y}$ (c) $z = e^{xy}$
(d) $z = 4e^{2y}$

5 If $y = x \sin t$ find $\dfrac{\partial y}{\partial x}$ and $\dfrac{\partial y}{\partial t}$.

Solutions to exercises

1 (a) $\dfrac{\partial z}{\partial x} = 5, \dfrac{\partial z}{\partial y} = 11$ (b) $-14, -7$

(c) $8, 0$ (d) $0, -5$ (e) $3, 8$ (f) $-3, 2$
(g) $0, 0$ (h) $0, -3$ (i) $4x, -7$
(j) $7, -9y^2$ (k) $-9, 9$ (l) $9, 9$

2 (a) $z_x = y, z_y = x$ (b) $3y, 3x$ (c) $-9y, -9x$
(d) $2xy, x^2$ (e) $18xy, 9x^2$ (f) $8y^2, 16xy$

3 $9, -4$

4 (a) $\dfrac{\partial z}{\partial x} = 2e^{2x}, \dfrac{\partial z}{\partial y} = 0$ (b) $0, 5e^{5y}$

(c) ye^{xy}, xe^{xy} (d) $0, 8e^{2y}$

5 $\dfrac{\partial y}{\partial x} = \sin t, \dfrac{\partial y}{\partial t} = x \cos t$

2.4 Partial derivatives requiring the product, quotient and chain rules

Consider the following more demanding examples, which use the rules developed in Chapter 16.

Example 2.18

Find $\dfrac{\partial z}{\partial x}$ and $\dfrac{\partial z}{\partial y}$ when $z = yxe^{2x}$.

Solution

To find $\dfrac{\partial z}{\partial x}$ we treat y as constant. We are dealing with a function of the form constant $\times xe^{2x}$. Note that this function itself contains a product of the functions x and e^{2x} and so we must use the product rule for differentiation. The derivative of xe^{2x} is $e^{2x}(1) + x(2e^{2x}) = e^{2x}(1 + 2x)$. Hence

$$\text{if} \quad z = yxe^{2x} \quad \text{then} \quad \dfrac{\partial z}{\partial x} = y(e^{2x}(1 + 2x)) = ye^{2x}(1 + 2x)$$

To find $\dfrac{\partial z}{\partial y}$ we treat x as constant. In turn, this means that xe^{2x} is constant too. This time the calculation is much simpler because we are dealing with a function of the form $z = \text{constant} \times y$. So,

$$\text{if} \quad z = yxe^{2x} \quad \text{then} \quad \frac{\partial z}{\partial y} = xe^{2x}$$

Example 2.19

Find $\dfrac{\partial z}{\partial x}$ and $\dfrac{\partial z}{\partial y}$ when $z = \dfrac{ye^x}{x}$.

Solution

To find $\dfrac{\partial z}{\partial x}$ we treat y as constant. We are dealing with a function of the form constant $\times \dfrac{e^x}{x}$. Note that this function itself contains a quotient of the functions e^x and x and so we must use the quotient rule for differentiation. The derivative of $\dfrac{e^x}{x}$ is

$$\frac{e^x x - e^x(1)}{x^2} = \frac{e^x(x - 1)}{x^2}$$

Hence

$$\text{if} \quad z = \frac{ye^x}{x} \quad \text{then} \quad \frac{\partial z}{\partial x} = \frac{ye^x(x - 1)}{x^2}$$

To find $\dfrac{\partial z}{\partial y}$ we treat x as constant. In turn, this means that $\dfrac{e^x}{x}$ is constant too. This calculation is simple because we are dealing with a function of the form $z = \text{constant} \times y$. So,

$$\text{if} \quad z = \frac{ye^x}{x} \quad \text{then} \quad \frac{\partial z}{\partial y} = \frac{e^x}{x}$$

Example 2.20

Find (a) $\dfrac{\partial z}{\partial x}$ and (b) $\dfrac{\partial z}{\partial y}$ given $z = \sin(3x - y^2)$.

Solution

We use the chain rule. Let $V = 3x - y^2$ so that $z = \sin V$

(a) Now

$$\frac{\partial z}{\partial x} = \frac{dz}{dV} \frac{\partial V}{\partial x} = \cos V (3) = 3 \cos(3x - y^2)$$

(b)

$$\frac{\partial z}{\partial y} = \frac{dz}{dV} \frac{\partial V}{\partial y} = \cos V (-2y) = -2y \cos(3x - y^2)$$

Exercises

1 Find $\dfrac{\partial z}{\partial x}$ and $\dfrac{\partial z}{\partial y}$ when

(a) $z = yxe^x$ (b) $z = xye^y$ (c) $z = xe^{xy}$

(d) $z = ye^{xy}$ (e) $z = x^2 \sin(xy)$

(f) $z = y \cos(xy)$ (g) $z = x \ln(xy)$

(h) $z = 3xy^3e^x$

2 Find the first partial derivatives of $u = \dfrac{1}{x^2 + y^2}$.

3 Find the first partial derivatives of $u = \dfrac{x}{x^2 + y^2}$.

Solutions to exercises

1 (a) $ye^x(x + 1)$, xe^x (b) ye^y, $xe^y(y + 1)$

(c) $e^{xy}(xy + 1)$, x^2e^{xy} (d) y^2e^{xy}, $e^{xy}(xy + 1)$

(e) $x^2y \cos(xy) + 2x \sin(xy)$, $x^3 \cos(xy)$

(f) $-y^2 \sin(xy)$, $-yx \sin(xy) + \cos(xy)$

(g) $1 + \ln(xy)$, $\dfrac{x}{y}$ (h) $3e^xy^3(x + 1)$, $9xy^2e^x$

2 $-\dfrac{2x}{(x^2 + y^2)^2}$, $-\dfrac{2y}{(x^2 + y^2)^2}$

3 $\dfrac{y^2 - x^2}{(x^2 + y^2)^2}$, $\dfrac{-2xy}{(x^2 + y^2)^2}$

End of block exercises

1 The derivative of $f(x)$ with respect to x is defined as $\dfrac{df}{dx} = \lim_{\delta x \to 0} \dfrac{f(x + \delta x) - f(x)}{\delta x}$ (see Chapter 15 Block 1). In an analogous manner we define the partial derivatives of $f(x, y)$ by

$$\frac{\partial f}{\partial x} = \lim_{\delta x \to 0} \frac{f(x + \delta x, y) - f(x, y)}{\delta x}$$

$$\frac{\partial f}{\partial y} = \lim_{\delta y \to 0} \frac{f(x, y + \delta y) - f(x, y)}{\delta y}$$

Use these definitions to find the first partial derivatives of (a) $f(x, y) = x^2y$, (b) $f(x, y) = x^2 + 3xy$, from first principles.

2 If $r = \sqrt{x^2 + y^2}$ find $\dfrac{\partial r}{\partial x}$ and $\dfrac{\partial r}{\partial y}$.

3 The derivative of $f(x) = \tan^{-1} kx$ is $\dfrac{k}{1 + k^2x^2}$.

If $\phi = \tan^{-1} \dfrac{y}{x}$ find $\dfrac{\partial \phi}{\partial y}$.

Solutions to exercises

1 (a) $2xy$, x^2 (b) $2x + 3y$, $3x$

2 $\dfrac{x}{(x^2 + y^2)^{1/2}}$, $\dfrac{y}{(x^2 + y^2)^{1/2}}$

3 $\dfrac{1/x}{1 + (1/x)^2y^2} = \dfrac{x}{x^2 + y^2}$

Higher-order derivatives

3.1 Introduction

Just as a function of one variable has a second derivative found by differentiating the first derivative, so too does a function of two variables. If $z = f(x, y)$ the first partial derivatives are $\dfrac{\partial z}{\partial z}$ and $\dfrac{\partial z}{\partial y}$. The second partial derivatives are found by differentiating the first partial derivatives. We can do this in various ways, as we show in the next section.

3.2 Finding higher derivatives

We can differentiate first partial derivatives either with respect to x or with respect to y to obtain various second partial derivatives, as summarised in the next key point.

Key point

$$\text{differentiating } \frac{\partial z}{\partial x} \text{ w.r.t. } x \text{ produces } \frac{\partial}{\partial x}\left(\frac{\partial z}{\partial x}\right) = \frac{\partial^2 z}{\partial x^2}$$

$$\text{differentiating } \frac{\partial z}{\partial x} \text{ w.r.t. } y \text{ produces } \frac{\partial}{\partial y}\left(\frac{\partial z}{\partial x}\right) = \frac{\partial^2 z}{\partial y \partial x}$$

$$\text{differentiating } \frac{\partial z}{\partial y} \text{ w.r.t. } x \text{ produces } \frac{\partial}{\partial x}\left(\frac{\partial z}{\partial y}\right) = \frac{\partial^2 z}{\partial x \partial y}$$

$$\text{differentiating } \frac{\partial z}{\partial y} \text{ w.r.t. } y \text{ produces } \frac{\partial}{\partial y}\left(\frac{\partial z}{\partial y}\right) = \frac{\partial^2 z}{\partial y^2}$$

The **second partial derivatives** of z are

$$\frac{\partial^2 z}{\partial x^2}, \quad \frac{\partial^2 z}{\partial y \partial x}, \quad \frac{\partial^2 z}{\partial x \partial y}, \quad \frac{\partial^2 z}{\partial y^2}$$

These are sometimes written in the alternative, briefer form

$$z_{xx}, \quad z_{xy}, \quad z_{yx}, \quad z_{yy}$$

respectively.

Example 3.1
Given $z = 3xy^3 - 2xy$ find all the second partial derivatives of z.

Solution
First of all the first partial derivatives must be found.

$$\frac{\partial z}{\partial x} = 3y^3 - 2y, \quad \frac{\partial z}{\partial y} = 9xy^2 - 2x$$

Then each of these is differentiated with respect to x:

$$\frac{\partial}{\partial x}\left(\frac{\partial z}{\partial x}\right) = \frac{\partial}{\partial x}(3y^3 - 2y) = 0. \text{ Thus } \frac{\partial^2 z}{\partial x^2} = 0.$$

$$\frac{\partial}{\partial x}\left(\frac{\partial z}{\partial y}\right) = \frac{\partial}{\partial x}(9xy^2 - 2x) = 9y^2 - 2. \text{ Thus } \frac{\partial^2 z}{\partial x \partial y} = 9y^2 - 2.$$

Now, each of the first partial derivatives must be differentiated with respect to y.

$$\frac{\partial}{\partial y}\left(\frac{\partial z}{\partial x}\right) = \frac{\partial}{\partial y}(3y^3 - 2y) = 9y^2 - 2. \text{ Thus } \frac{\partial^2 z}{\partial y \partial x} = 9y^2 - 2.$$

$$\frac{\partial}{\partial y}\left(\frac{\partial z}{\partial y}\right) = \frac{\partial}{\partial y}(9xy^2 - 2x) = 18xy. \text{ Thus } \frac{\partial^2 z}{\partial y^2} = 18xy.$$

Note that $\frac{\partial^2 z}{\partial y \partial x} = \frac{\partial^2 z}{\partial x \partial y}$. It is usually the case that the result is the same either way.

Example 3.2
Find all second partial derivatives of $z = \sin(xy)$.

Solution
First of all the first partial derivatives are found.

$$\frac{\partial z}{\partial x} = y\cos(xy), \quad \frac{\partial z}{\partial y} = x\cos(xy)$$

Then each of these is differentiated with respect to x:

$$\frac{\partial^2 z}{\partial x^2} = \frac{\partial}{\partial x}\left(\frac{\partial z}{\partial x}\right) = -y^2\sin(xy)$$

$$\frac{\partial^2 z}{\partial x \partial y} = \frac{\partial}{\partial x}\left(\frac{\partial z}{\partial y}\right) = -xy\sin(xy) + \cos(xy)$$

Note here the need to use the product rule to differentiate $x\cos(xy)$ with respect to x.
Now, each of the first partial derivatives must be differentiated with respect to y.

$$\frac{\partial^2 z}{\partial y \partial x} = \frac{\partial}{\partial y}\left(\frac{\partial z}{\partial x}\right) = -xy\sin(xy) + \cos(xy)$$

$$\frac{\partial^2 z}{\partial y^2} = \frac{\partial}{\partial y}\left(\frac{\partial z}{\partial y}\right) = -x^2\sin(xy)$$

Example 3.3

Consider the function $\phi(x, y) = 2xy$.

(a) Find $\dfrac{\partial \phi}{\partial x}$, $\dfrac{\partial \phi}{\partial y}$, $\dfrac{\partial^2 \phi}{\partial x^2}$ and $\dfrac{\partial^2 \phi}{\partial y^2}$.

(b) Using the result of part (a) evaluate $\dfrac{\partial^2 \phi}{\partial x^2} + \dfrac{\partial^2 \phi}{\partial y^2}$.

Solution

(a) $\dfrac{\partial \phi}{\partial x} = \rule{1cm}{0.4cm}, \dfrac{\partial \phi}{\partial y} = \rule{1cm}{0.4cm}$ \hfill $2y \quad 2x$

$\dfrac{\partial^2 \phi}{\partial x^2} = \rule{1cm}{0.4cm}, \dfrac{\partial^2 \phi}{\partial y^2} = \rule{1cm}{0.4cm}$ \hfill $0, 0$

(b) $\dfrac{\partial^2 \phi}{\partial x^2} + \dfrac{\partial^2 \phi}{\partial y^2} = \rule{1cm}{0.4cm}$ \hfill 0

Note from the last result that $\phi = 2xy$ satisfies the equation $\dfrac{\partial^2 \phi}{\partial x^2} + \dfrac{\partial^2 \phi}{\partial y^2} = 0$. This equation is a **partial differential equation**. It occurs in the modelling of heat conduction, fluid flow and electrostatics problems, and is known as **Laplace's equation**.

Example 3.4 Temperature of fluid in a pipe

Laplace's equation is often used in the modelling of the temperature of a fluid that is flowing in an insulated pipe. If T is the temperature of the fluid then under some simplifying assumptions T satisfies the equation

$$\frac{\partial^2 T}{\partial x^2} + \frac{\partial^2 T}{\partial y^2} = 0$$

This is known as Laplace's equation.

Show that

$$T = k - \ln(x^2 + y^2)$$

where k is a constant, is a solution.

Solution

$$\frac{\partial T}{\partial x} = \frac{-2x}{x^2 + y^2}, \quad \frac{\partial T}{\partial y} = \frac{-2y}{x^2 + y^2} \quad \text{using the chain rule}$$

$$\frac{\partial^2 T}{\partial x^2} = \frac{-2(x^2 + y^2) + (2x)(2x)}{(x^2 + y^2)^2} = \frac{2x^2 - 2y^2}{(x^2 + y^2)^2} \quad \text{using the quotient rule}$$

Similarly

$$\frac{\partial^2 T}{\partial y^2} = \frac{2y^2 - 2x^2}{(x^2 + y^2)^2}$$

So

$$\frac{\partial^2 T}{\partial x^2} + \frac{\partial^2 T}{\partial y^2} = \frac{2x^2 - 2y^2}{(x^2 + y^2)^2} + \frac{2y^2 - 2x^2}{(x^2 + y^2)^2} = 0$$

Hence $T = k - \ln(x^2 + y^2)$ is a solution of Laplace's equation as required.

Exercises

1 Find all the second partial derivatives in each of the following cases:
(a) $z = xy$ (b) $z = 7xy$
(c) $z = 8x + 9y + 10$
(d) $z = 8y^2x + 11$ (e) $z = -2y^3x^2$
(f) $z = x + y$

2 Find all the second partial derivatives in each of the following cases:
(a) $z = \dfrac{1}{x}$ (b) $z = \dfrac{y}{x}$
(c) $z = \dfrac{x}{y}$ (d) $z = \dfrac{1}{x} + \dfrac{1}{y}$

3 Find all the second partial derivatives in each of the following cases:
(a) $z = x \sin y$ (b) $z = y \cos x$
(c) $z = ye^{2x}$ (d) $z = ye^{-x}$

4 Find all the second partial derivatives in each of the following cases:
(a) $z = 8e^{xy}$ (b) $z = -3e^x \sin y$
(c) $z = 4e^y \cos x$

5 Find all the second partial derivatives in each of the following cases:
(a) $z = \ln x$ (b) $z = \ln y$ (c) $z = \ln xy$
(d) $z = x \ln y$ (e) $z = y \ln x$

Solutions to exercises

1 (a) $\dfrac{\partial^2 z}{\partial x^2} = 0, \dfrac{\partial^2 z}{\partial x \partial y} = \dfrac{\partial^2 z}{\partial y \partial x} = 1, \dfrac{\partial^2 z}{\partial y^2} = 0$
(b) 0, 7, 0 (c) 0, 0, 0 (d) 0, 16y, 16x
(e) $-4y^3, -12y^2x, -12yx^2$ (f) 0, 0, 0

2 (a) $\dfrac{2}{x^3}, 0, 0$ (b) $\dfrac{2y}{x^3}, -\dfrac{1}{x^2}, 0$ (c) $0, -\dfrac{1}{y^2}, \dfrac{2x}{y^3}$
(d) $\dfrac{2}{x^3}, 0, \dfrac{2}{y^3}$

3 (a) $0, \cos y, -x \sin y$ (b) $-y \cos x, -\sin x, 0$
(c) $4ye^{2x}, 2e^{2x}, 0$ (d) $ye^{-x}, -e^{-x}, 0$

4 (a) $8y^2e^{xy}, 8e^{xy}(xy + 1), 8x^2e^{xy}$ (b) $-3e^x \sin y$,
$-3e^x \cos y, 3e^x \sin y$ (c) $-4e^y \cos x$,
$-4e^y \sin x, 4e^y \cos x$

5 (a) $-\dfrac{1}{x^2}, 0, 0$ (b) $0, 0, -\dfrac{1}{y^2}$ (c) $-\dfrac{1}{x^2}, 0, -\dfrac{1}{y^2}$
(d) $0, \dfrac{1}{y}, -\dfrac{x}{y^2}$ (e) $-\dfrac{y}{x^2}, \dfrac{1}{x}, 0$

End of block exercises

1 If $u(x, t) = e^{-2t} \cos x$ find $\dfrac{\partial u}{\partial t}$ and $\dfrac{\partial^2 u}{\partial x^2}$. Verify
that $\dfrac{\partial u}{\partial t} = 2\dfrac{\partial^2 u}{\partial x^2}$: that is, verify that u satisfies
this partial differential equation, called the **heat equation**.

2 Verify that $u = x^2 + 4t^2$ satisfies the **wave
equation** $\dfrac{\partial^2 u}{\partial t^2} = 4\dfrac{\partial^2 u}{\partial x^2}$.

Stationary values of a function of two variables

4.1 Introduction

When studying functions of a single variable we used differentiation to locate the positions of maximum and minimum points. We now extend this technique to functions of two variables.

4.2 Maxima, minima and saddle points

Figure 4.1 shows a graph of the function $z = f(x, y) = -x^2 - y^2$. The point labelled A is called a **maximum point**. If we were to move away from the maximum point in any direction but remain on the surface, the value of the function would decrease.

Figure 4.1
Point A is a maximum point. It is the highest point in its locality.

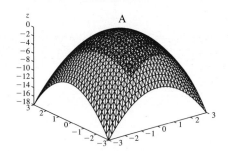

Figure 4.2 shows a graph of the function $z = f(x, y) = x^2 + y^2$. The point labelled B is called a **minimum point**. If we were to move away from the minimum point in any direction but remain on the surface, the value of the function would increase.

Figure 4.2
Point B is a minimum point. It is the lowest point in its locality.

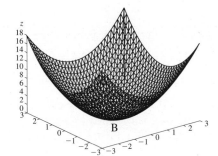

Figure 4.3 shows a graph of the function $z = f(x, y) = 3xy + x + y$. The point labelled C is called a **saddle point**. When leaving a saddle point in one direction, the value of the function increases; when leaving in the perpendicular direction, the value of the function decreases.

Figure 4.3
Point C is a saddle point.

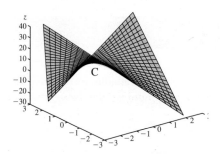

Collectively, maximum points, minimum points and saddle points are known as **stationary points** of the function f.

4.3 Finding the location of stationary points

To locate stationary points both first partial derivatives are equated to zero: that is, we solve

$$\frac{\partial f}{\partial x} = 0 \quad \text{and} \quad \frac{\partial f}{\partial y} = 0$$

Key point

Stationary points are located by solving

$$\frac{\partial f}{\partial x} = 0 \quad \text{and} \quad \frac{\partial f}{\partial y} = 0$$

Example 4.1
Find the coordinates of any stationary points of the function $f(x, y) = x^2 + y^2$.

Solution
We begin by noting that stationary points are located by solving $\frac{\partial f}{\partial x} = 0$ and $\frac{\partial f}{\partial y} = 0$ and so we find the first partial derivatives of f. If $f = x^2 + y^2$ then $\frac{\partial f}{\partial x} = 2x$ and $\frac{\partial f}{\partial y} = 2y$. Equating these to zero produces two simultaneous equations: $2x = 0$ and $2y = 0$. Solving these we find $x = 0$ and $y = 0$. These are the x and y coordinates of the stationary point. The z coordinate is obtained using $z = x^2 + y^2$ and so $z = 0$ also. The graph of this function is shown in Figure 4.2, from which we note that the stationary point is a minimum.

Example 4.2

Find the coordinates of any stationary points of the function $f(x, y) = x^2 + xy - 7y$.

Solution

First of all find the partial derivatives of f and equate both of these derivatives to zero:

$$\frac{\partial f}{\partial x} = \boxed{} \hspace{4cm} 2x + y = 0$$

$$\frac{\partial f}{\partial y} = \boxed{} \hspace{4cm} x - 7 = 0$$

Solve the resulting simultaneous equations. From the second equation

$$x = \boxed{} \hspace{5cm} 7$$

Substitution in the first equation gives $y = -14$.

Therefore a stationary point occurs at $(7, -14)$. The z coordinate is found from $z = x^2 + xy - 7y$.

$$z = \boxed{} \hspace{5cm} 49$$

Example 4.3

Find the coordinates of any stationary points of $f(x, y) = 2x^2 + 3xy - y^2 - 2x$.

Solution

First of all find the partial derivatives of f and equate these to zero.

$$\frac{\partial f}{\partial x} = 0 \boxed{} \hspace{3cm} 4x + 3y - 2 = 0$$

$$\frac{\partial f}{\partial y} = 0 \boxed{} \hspace{3cm} 3x - 2y = 0$$

Solving the simultaneous equations gives

$$x = \boxed{}, y = \boxed{} \hspace{3cm} \frac{4}{17}, \frac{6}{17}$$

Therefore a stationary point occurs at $\left(\frac{4}{17}, \frac{6}{17}\right)$. The z coordinate is found from $z = 2x^2 + 3xy - y^2 - 2x$.

$$z = \boxed{} \hspace{5cm} -4/17$$

Exercises

Locate the position of any stationary points of the following functions:

1   $f(x, y) = x^2 + y^3 - 3y$

2   $f(x, y) = 4xy - 2x^2y$

3   $f(x, y) = -x - y^3$

4   $f(x, y) = \dfrac{x^3}{3} + 3x^2 + xy + \dfrac{y^2}{2} + 6y$

Solutions to exercises

1   (0, 1) and (0, −1)

2   (0, 0) and (2, 0)

3   none

4   (1, −7) and (−6, 0)

4.4   The nature of stationary points

Equating the first partial derivatives to zero locates the stationary points but does not identify them as maximum points, minimum points or points of inflexion. To distinguish between such points we can make use of a test that involves second partial derivatives.

We consider the expression

$$\frac{\partial^2 f}{\partial x^2}\frac{\partial^2 f}{\partial y^2} - \left(\frac{\partial^2 f}{\partial x \partial y}\right)^2$$

We evaluate this expression at each of the stationary points.

- If the result is negative the stationary point is a saddle point.
- If the result is positive and if $\dfrac{\partial^2 f}{\partial x^2}$ is positive we have a minimum point.
- If the result is positive and if $\dfrac{\partial^2 f}{\partial x^2}$ is negative we have a maximum point.
- If the result is zero, this test fails and further tests are required that are beyond the scope of this book.

In summary:

Key point

The nature of a stationary point is determined using the following test:

$$\frac{\partial^2 f}{\partial x^2}\frac{\partial^2 f}{\partial y^2} - \left(\frac{\partial^2 f}{\partial x \partial y}\right)^2 < 0 \quad \text{saddle point}$$

$$\frac{\partial^2 f}{\partial x^2}\frac{\partial^2 f}{\partial y^2} - \left(\frac{\partial^2 f}{\partial x \partial y}\right)^2 > 0 \quad \text{and} \quad \frac{\partial^2 f}{\partial x^2} > 0 \quad \text{minimum point}$$

$$\frac{\partial^2 f}{\partial x^2}\frac{\partial^2 f}{\partial y^2} - \left(\frac{\partial^2 f}{\partial x \partial y}\right)^2 > 0 \quad \text{and} \quad \frac{\partial^2 f}{\partial x^2} < 0 \quad \text{maximum point}$$

Example 4.4

Locate and identify the nature of the stationary values of

$$f(x, y) = x^2 + xy - 7y$$

Solution

This function has been discussed already in Example 4.2. We found that a stationary point was located at (7, −14).

We now determine its nature. For this we need to calculate the second derivatives.

Recall that $\dfrac{\partial f}{\partial x} = 2x + y$ and $\dfrac{\partial f}{\partial y} = x - 7$.

$$\frac{\partial^2 f}{\partial x^2} = \boxed{}$$ 2

$$\frac{\partial^2 f}{\partial y^2} = \boxed{}$$ 0

$$\frac{\partial^2 f}{\partial x \partial y} = \boxed{}$$ 1

We then consider the quantity $\dfrac{\partial^2 f}{\partial x^2}\dfrac{\partial^2 f}{\partial y^2} - \left(\dfrac{\partial^2 f}{\partial x \partial y}\right)^2$. In this case we have

$$\frac{\partial^2 f}{\partial x^2}\frac{\partial^2 f}{\partial y^2} - \left(\frac{\partial^2 f}{\partial x \partial y}\right)^2 = (2)(0) - (1)^2 = -1$$

Then, from the key point immediately above we deduce that this stationary point is a

$\boxed{}$ saddle point

Exercises

Determine the position and nature of the stationary points of the following functions:

1 $f(x, y) = 3xy + x + y$

2 $f(x, y) = x^2 + y^2 - 3y$

3 $f(x, y) = x^2 + y^2 - 3xy$

4 $f(x, y) = \dfrac{1}{x} + \dfrac{1}{y} - \dfrac{3}{xy}$

5 $f(x, y) = e^{xy}$

Solutions to exercises

1 Saddle at $(-1/3, -1/3)$

2 Minimum at $(0, 3/2)$

3 Saddle at $(0, 0)$

4 Saddle at $(3, 3)$

5 Saddle at $(0, 0)$

End of block exercises

Determine the position and nature of the stationary points of the following functions:

1 $f(x, y) = xy + x + y^2$

2 $f(x, y) = -x^2 - y^2 - 3y$

3   $f(x, y) = 7x - 2y^2 - xy$

4   $f(x, y) = \dfrac{1}{x} + \dfrac{1}{y} + \dfrac{1}{xy}$

5   $f(x, y) = e^{-x(y-1)}$

Solutions to exercises

1   Saddle at $(2, -1)$

2   Maximum at $(0, -3/2)$

3   Saddle at $(-28, 7)$

4   Saddle at $(-1, -1)$

5   Saddle at $(0, 1)$

End of chapter exercises

1   If $z = 14x - 13y$ state $\dfrac{\partial z}{\partial x}$ and $\dfrac{\partial z}{\partial y}$.

2   If $w = 5y - 2x$ state $\dfrac{\partial^2 w}{\partial x^2}$ and $\dfrac{\partial^2 w}{\partial y^2}$.

3   If $z = 3x^2 + 7xy - y^2$ find $\dfrac{\partial^2 z}{\partial y \partial x}$ and $\dfrac{\partial^2 z}{\partial x \partial y}$.

4   If $z = 14 - 4xy$ evaluate $\dfrac{\partial z}{\partial x}$ and $\dfrac{\partial z}{\partial y}$ at the point $(1, 2)$.

5   If $z = 4e^{5xy}$ find $\dfrac{\partial z}{\partial x}$ and $\dfrac{\partial z}{\partial y}$.

6   If $y = x \cos t$ find $\dfrac{\partial y}{\partial x}$ and $\dfrac{\partial y}{\partial t}$.

7   If $w = 3xy^2 + 2yz^2$ find all first partial derivatives of w at the point with coordinates $(1, 2, 3)$.

8   If $V = D^{1/4} T^{-5/6}$ find $\dfrac{\partial V}{\partial T}$ and $\dfrac{\partial V}{\partial D}$.

9   If $p = \dfrac{RT}{V}$ where R is a constant, find $\dfrac{\partial p}{\partial V}$ and $\dfrac{\partial p}{\partial T}$.

10   Find $\dfrac{\partial f}{\partial x}, \dfrac{\partial f}{\partial y}, \dfrac{\partial^2 f}{\partial x^2}, \dfrac{\partial^2 f}{\partial y^2}$ and $\dfrac{\partial^2 f}{\partial x \partial y}$ if $f = (x - y)^2$.

11   Calculate $\dfrac{\partial z}{\partial x}$ when (a) $z = \dfrac{y}{x^2} - \dfrac{x}{y^2}$,

(b) $z = e^{x^2 - 4xy}$.

12   Calculate $\dfrac{\partial z}{\partial y}$ when $z = \dfrac{x^2 - 3y^2}{x^2 + y^2}$.

13   If $T = \dfrac{\pi \mu \Omega h D^3}{4c}$ find $\dfrac{\partial T}{\partial D}$ and $\dfrac{\partial T}{\partial c}$.

14   Given $f(x, y) = \sin 4x \cos 3y$ find $\dfrac{\partial^2 f}{\partial x^2}, \dfrac{\partial^2 f}{\partial y^2}$ and $\dfrac{\partial^2 f}{\partial x \partial y}$.

15   Determine the stationary points of $f(x, y) = 2x^2 + 3y^2 + 5x + 12y + 19$.

16   Consider the function $f(x, y) = 5x^2 y$.
(a) Evaluate this function and its first partial derivatives at the point $A(2, 3)$.
(b) Suppose we consider point A. Suppose small changes, $\delta x, \delta y$, are made in the values of x and y so that we move to a nearby point B. It is possible to show that the corresponding change in f is given approximately by $\delta f \approx \dfrac{\partial f}{\partial x} \delta x + \dfrac{\partial f}{\partial y} \delta y$, where the partial derivatives are evaluated at the original point A. Use this result to find the approximate change in the value of f if x is increased to 2.1 and y is increased to 3.2.
(c) Compare your answer in (b) to the value of f at $(2.1, 3.2)$.

Solutions to exercises

1   14, −13

2   0, 0

3   7, 7

4   −8, −4

5   $20ye^{5xy}, 20xe^{5xy}$

6   $\cos t, -x \sin t$

7   12, 30, 24

8   $-\frac{5}{6}D^{1/4}T^{-11/6}, \frac{1}{4}D^{-3/4}T^{-5/6}$

9   $-\dfrac{RT}{V^2}, \dfrac{R}{V}$

10 $2(x - y), -2(x - y), 2, 2, -2$

11 (a) $-\dfrac{2y}{x^3} - \dfrac{1}{y^2}$ (b) $(2x - 4y)e^{x^2-4xy}$

12 $-\dfrac{8yx^2}{(x^2 + y^2)^2}$

13 $\dfrac{3\pi\mu\Omega h D^2}{4c}, -\dfrac{\pi\mu\Omega h D^3}{4c^2}$

14 $-16 \sin 4x \cos 3y, -9 \sin 4x \cos 3y,$
    $-12 \cos 4x \sin 3y$

15 Minimum at $(-5/4, -2)$

16 (a) 60, 60, 20
    (b) $\delta f = (60)(0.1) + (20)(0.2) = 10$
    (c) From part (b) the approximate value of f at
        the new point is $60 + 10 = 70$. The exact
        value is $f(2.1, 3.2) = 70.56$.

Computer and calculator exercises

1   Investigate how to find the first partial
    derivatives of the function $f(x, y) = \tan^{-1}xy$
    using a computer algebra package.

2   Use a computer algebra package to find
    $\dfrac{\partial^2 u}{\partial x^2}, \dfrac{\partial^2 u}{\partial y^2}, \dfrac{\partial u}{\partial z}$ when $u(x, y, z) = \dfrac{1}{z}e^{\left(-\frac{x^2+y^2}{4z}\right)}$

From your results show that u is a solution of
the equation

$$\frac{\partial^2 u}{\partial x^2} + \frac{\partial^2 u}{\partial y^2} = \frac{\partial u}{\partial z}$$

Solutions to exercises

1   $\dfrac{y}{1 + x^2y^2}, \dfrac{x}{1 + x^2y^2}$

2   $\dfrac{\partial^2 u}{\partial x^2} = -\dfrac{e^{\left(-\frac{x^2+y^2}{4z}\right)}}{2z^2} + \dfrac{x^2e^{\left(-\frac{x^2+y^2}{4z}\right)}}{4z^3}$,

    $\dfrac{\partial^2 u}{\partial y^2} = -\dfrac{e^{\left(-\frac{x^2+y^2}{4z}\right)}}{2z^2} + \dfrac{y^2e^{\left(-\frac{x^2+y^2}{4z}\right)}}{4z^3}$,

    $\dfrac{\partial u}{\partial z} = -\dfrac{e^{\left(-\frac{x^2+y^2}{4z}\right)}}{z^2} + \dfrac{(x^2 + y^2)e^{\left(-\frac{x^2+y^2}{4z}\right)}}{4z^3}$

The Laplace transform

The **Laplace transform** is an integral transform: that is, it is defined by an integral. Given a function, $f(t)$, the Laplace transform of $f(t)$ is a function $F(s)$ defined by

$$F(s) = \int_0^\infty e^{-st} f(t) \, dt$$

The original function, $f(t)$, has been transformed into a new function, $F(s)$.

As with differentiation and integration, most Laplace transforms are calculated from a table of standard functions. This is covered in Block 1. Block 2 deals with the **inverse Laplace transform**. Here we know the transformed function, $F(s)$, and we seek the original function, $f(t)$.

The chapter closes with Block 3, which illustrates the application of the transform to the solution of **differential equations**. Both first- and second-order linear constant-coefficient equations can be solved using the Laplace transform. This method has the advantage that initial conditions are automatically incorporated into the solution.

Chapter 22 contents

The Laplace transform

1.1 Introduction

In mathematics the word 'transform' has the same meaning as in everyday language. The Laplace transform is one of a number of integral transforms used by engineers.

The Laplace transform can be used to solve a linear constant-coefficient differential equation by transforming it into an algebraic equation. The resulting algebraic equation is solved and then the transform is reversed to find the solution of the differential equation in terms of the original variable.

The Laplace transform can also be used to calculate transfer functions. These functions describe the elements of an engineering system and are particularly important in the design of control systems.

1.2 Definition of the Laplace transform

Let $f(t)$ be a function of time, t. In many engineering problems t has only non-negative values, that is $t \geq 0$. Hence $f(t)$ will be given for $t \geq 0$, and for $t < 0, f(t)$ is assumed to be zero.

Key point

The **Laplace transform** of $f(t)$ is $F(s)$ and is defined by

$$F(s) = \int_0^\infty e^{-st} f(t) \, dt$$

So, to find the Laplace transform of a function $f(t)$, we multiply it by e^{-st} and then integrate between the limits 0 and ∞. Because of the infinite limit of integration, restrictions sometimes need to be placed on s to ensure that the integral exists. Throughout this chapter it is assumed that s has a value such that this condition is satisfied.

Note that although f is a function of t, the Laplace transform is a function of a new variable, s. It is conventional to use a lower case letter for the function of t, as in $f(t)$, but an upper case letter for the function of s, as in $F(s)$. In addition, we often write $\mathcal{L}\{f(t)\}$ to denote the Laplace transform of the function $f(t)$.

Example 1.1

Calculate the Laplace transform of $f(t) = 3$.

Solution

$$F(s) = \int_0^\infty e^{-st}\, 3\, dt$$

$$= 3 \int_0^\infty e^{-st}\, dt$$

$$= 3 \left[\frac{e^{-st}}{-s} \right]_0^\infty$$

$$= 3 \left[0 + \frac{1}{s} \right]$$

$$= \frac{3}{s}$$

Hence we may write $\mathcal{L}\{3\} = \dfrac{3}{s}$.

Exercises

1 Write down the definition of the Laplace transform of a function $f(t)$.

2 Given a function $f(t)$, write two ways in which its Laplace transform might be expressed.

3 If x is a function of t, that is $x(t)$, how would you write its Laplace transform?

4 If y is a function of t, that is $y(t)$, how would you write its Laplace transform?

5 Use integration by parts to show that the Laplace transform of $f(t) = t$ is $\dfrac{1}{s^2}$.

6 Show that $\mathcal{L}\{e^t\} = \dfrac{1}{s-1}$.

(Hint: write $e^t e^{-st}$ as $e^{(1-s)t}$.)

Solutions to exercises

2 $\mathcal{L}\{f(t)\}$ or $F(s)$

3 $\mathcal{L}\{x(t)\}$ or $X(s)$

4 $\mathcal{L}\{y(t)\}$ or $Y(s)$

1.3   Using a table of Laplace transforms

Although the Laplace transforms of most functions can be calculated from the definition, most engineers use a table. Table 1.1 lists some common functions and their Laplace transforms.

Table 1.1

Function, $f(t)$	Laplace transform, $\mathcal{L}\{f(t)\} = F(s)$	Function, $f(t)$	Laplace transform, $\mathcal{L}\{f(t)\} = F(s)$
1	$\dfrac{1}{s}$	$\sinh bt$	$\dfrac{b}{s^2 - b^2}$
t	$\dfrac{1}{s^2}$	$\cosh bt$	$\dfrac{s}{s^2 - b^2}$
t^2	$\dfrac{2}{s^3}$	$e^{-at}\sinh bt$	$\dfrac{b}{(s + a)^2 - b^2}$
t^n	$\dfrac{n!}{s^{n+1}}$	$e^{-at}\cosh bt$	$\dfrac{s + a}{(s + a)^2 - b^2}$
e^{at}	$\dfrac{1}{s - a}$	$t\sin bt$	$\dfrac{2bs}{(s^2 + b^2)^2}$
e^{-at}	$\dfrac{1}{s + a}$	$t\cos bt$	$\dfrac{s^2 - b^2}{(s^2 + b^2)^2}$
$t^n e^{-at}$	$\dfrac{n!}{(s + a)^{n+1}}$	$u(t)$, unit step	$\dfrac{1}{s}$
$\sin bt$	$\dfrac{b}{s^2 + b^2}$	$u(t - d)$	$\dfrac{e^{-sd}}{s}$
$\cos bt$	$\dfrac{s}{s^2 + b^2}$	$\delta(t)$	1
$e^{-at}\sin bt$	$\dfrac{b}{(s + a)^2 + b^2}$	$\delta(t - d)$	e^{-sd}
$e^{-at}\cos bt$	$\dfrac{s + a}{(s + a)^2 + b^2}$		

Example 1.2
Use Table 1.1 to find the Laplace transform of each of the following functions:
(a) t^4 (b) e^{2t} (c) e^{-3t} (d) $t^2 e^{3t}$ (e) $\sin 2t$ (f) $\cos 5t$ (g) $e^{-t}\sin t$

Solution
(a) We use the result for $f(t) = t^n$ and put $n = 4$. The required Laplace transform is then

$$F(s) = \frac{4!}{s^{4+1}}$$

$$= \frac{4!}{s^5}$$

$$= \frac{24}{s^5}$$

We may write

$$\mathcal{L}\{t^4\} = \frac{24}{s^5}$$

(b) Using the result for e^{at} with $a = 2$ we find

$$\mathcal{L}\{e^{2t}\} = $$

$$\frac{1}{s-2}$$

(c) Using the result for e^{-at} with $a = 3$ we have

$$\mathcal{L}\{e^{-3t}\} = \frac{1}{s+3}$$

(d) We use the result for $t^n e^{-at}$ when $n = 2$ and $a = -3$ to obtain

$$\mathcal{L}\{t^2 e^{3t}\} = $$

$$\frac{2!}{(s-3)^3} = \frac{2}{(s-3)^3}$$

(e) $\mathcal{L}\{\sin 2t\} = \dfrac{2}{s^2 + 4}$

(f) $\mathcal{L}\{\cos 5t\} = $

$$\frac{s}{s^2 + 25}$$

(g) $\mathcal{L}\{e^{-t} \sin t\} = \dfrac{1}{(s+1)^2 + 1}$

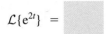

Example 1.3
Find the Laplace transform of each of the following functions:
(a) $e^{t/2}$ (b) $e^{2t} \cos t$ (c) $t^3 e^{2t}$

Solution

(a) We use the result for e^{at} and take $a = \frac{1}{2}$.

$$\mathcal{L}\{e^{t/2}\} = $$

$$\frac{1}{s - \frac{1}{2}}$$

(b) We use the result for $e^{-at} \cos bt$. We take

$$a = , \quad b = $$

$$-2 , 1$$

so

$$\mathcal{L}\{e^{2t} \cos t\} = \boxed{}$$

$$\frac{s-2}{(s-2)^2+1}$$

(c) We use the result for $\boxed{}$

$$t^n e^{-at}$$

We take

$$n = \boxed{}, \quad a = \boxed{}$$

$$3,\ -2$$

Hence

$$\mathcal{L}\{t^3 e^{2t}\} = \boxed{}$$

$$\frac{6}{(s-2)^4}$$

Exercises

1   Use Table 1.1 to find the Laplace transform of the following:
(a) t^5 (b) $\cos 4t$ (c) e^{9t} (d) e^{-6t} (e) $\sin 5t$

2   Find the Laplace transform of each of the following:
(a) $e^{-3t} \sin 5t$ (b) $e^{3t} \cos 2t$

(c) $t \sin 4t$ (d) $t \cos \dfrac{t}{2}$ (e) $t^2 e^{-5t}$

3   Find the Laplace transform of each of the following:
(a) $e^{2t/3}$ (b) $\dfrac{t^2}{e^{4t}}$ (c) $\dfrac{\sin 5t}{e^{6t}}$ (d) $e^{t/2} \cos \dfrac{t}{2}$

(e) $e^{4t} \sin \dfrac{2t}{3}$

Solutions to exercises

1   (a) $\dfrac{5!}{s^6}$ (b) $\dfrac{s}{s^2+16}$ (c) $\dfrac{1}{s-9}$ (d) $\dfrac{1}{s+6}$

(e) $\dfrac{5}{s^2+25}$

2   (a) $\dfrac{5}{(s+3)^2+25}$ (b) $\dfrac{s-3}{(s-3)^2+4}$

(c) $\dfrac{8s}{(s^2+16)^2}$ (d) $\dfrac{s^2-0.25}{(s^2+0.25)^2}$

(e) $\dfrac{2!}{(s+5)^3}$

3   (a) $\dfrac{3}{3s-2}$ (b) $\dfrac{2}{(s+4)^3}$ (c) $\dfrac{5}{(s+6)^2+25}$

(d) $\dfrac{s-0.5}{(s-0.5)^2+0.25}$ (e) $\dfrac{6}{9(s-4)^2+4}$

Since the Laplace transform is defined in terms of an integral it comes as no surprise that some of the properties of integrals are also properties of the Laplace transform. We note two of these properties.

Key point

$$\mathcal{L}\{f + g\} = \mathcal{L}\{f\} + \mathcal{L}\{g\}$$

$$\mathcal{L}\{kf\} = k\mathcal{L}\{f\} \quad \text{if } k \text{ is a constant}$$

The first property states that to find the Laplace transform of two functions added together we simply find the transform of each individual function and then add together these transforms.

The second property states that when a function, f, is multiplied by a constant, k, the corresponding transform is also multiplied by the same constant.

Together, these properties mean that the Laplace transform is what is known as a **linear transform**.

Example 1.4

Find the Laplace transform of each of the following functions:
(a) 3 (b) $3 + 2t$ (c) $3 + 2t - 5t^2$

Solution

(a) We think of 3 as 3(1). From Table 1.1

$$\mathcal{L}\{1\} = \frac{1}{s}$$

and so, by the second property,

$$\mathcal{L}\{3\} = 3\left(\frac{1}{s}\right)$$

$$= \frac{3}{s}$$

(b) We note that

$$\mathcal{L}\{t\} = \frac{1}{s^2}$$

and so, using the second property,

$$\mathcal{L}\{2t\} = \boxed{} \qquad \frac{2}{s^2}$$

Finally, using the first property,

$$\mathcal{L}\{3 + 2t\} = \mathcal{L}\{3\} + \mathcal{L}\{2t\}$$

$$= \frac{3}{s} + \frac{2}{s^2}$$

(c) $\mathcal{L}\{3 + 2t - 5t^2\} = \mathcal{L}\{3\} + \mathcal{L}\{2t\} - \mathcal{L}\{5t^2\}$

$$= \boxed{} \qquad\qquad \frac{3}{s} + \frac{2}{s^2} - \frac{10}{s^3}$$

Example 1.5

Find the Laplace transform of (a) $\dfrac{e^{3t}}{2} - 6 \sin 2t$, (b) $\dfrac{1}{3}\cos 6t - 2e^{-t} \sin t$.

Solution

(a) From Table 1.1

$$L\{e^{3t}\} = \frac{1}{s - 3}$$

and so

$$\mathcal{L}\left\{\frac{e^{3t}}{2}\right\} = \boxed{} \qquad\qquad \frac{1}{2(s - 3)}$$

Similarly

$$\mathcal{L}\{\sin 2t\} = \boxed{} \qquad\qquad \frac{2}{s^2 + 4}$$

and so

$$\mathcal{L}\{6\sin 2t\} = \boxed{} \qquad\qquad \frac{12}{s^2 + 4}$$

Finally

$$\mathcal{L}\left\{\frac{e^{3t}}{2} - 6 \sin 2t\right\} = \frac{1}{2(s - 3)} - \frac{12}{s^2 + 4}$$

(b) We know from Table 1.1 that

$$\mathcal{L}\{\cos 6t\} = \boxed{} \qquad\qquad \frac{s}{s^2 + 36}$$

and hence

$$\mathcal{L}\left\{\frac{1}{3}\cos 6t\right\} = \boxed{} \qquad\qquad \frac{s}{3(s^2 + 36)}$$

Similarly

$$\mathcal{L}\{e^{-t} \sin t\} = \boxed{} \qquad\qquad \frac{1}{(s + 1)^2 + 1}$$

and so

$$\mathcal{L}\{2e^{-t}\sin t\} = \boxed{} \qquad \frac{2}{(s+1)^2+1}$$

So

$$\mathcal{L}\left\{\frac{1}{3}\cos 6t - 2e^{-t}\sin t\right\} = \frac{s}{3(s^2+36)} - \frac{2}{(s+1)^2+1}$$

Exercises

1 Explain what is meant by stating that the Laplace transform is a linear transform. Write down the two linearity properties.

2 Find the Laplace transform of

(a) $2 + 3t$ (b) $2t^2 - 3$ (c) $\dfrac{t^2}{2} - 6$

(d) $2t^3 - 3t^2 + 5t - 6$ (e) $2 - 4t^3 + 5t^5$

3 Find the Laplace transform of

(a) $4e^{2t} + 3e^{-2t}$ (b) $\dfrac{e^{3t}}{2} + 4e^t - 5e^{-t}$

(c) $1 - 3e^{-t} + \dfrac{2e^t}{3}$ (d) $2te^t - 3t^2e^{-t}$

(e) $2t - 3te^{-t} + 4$

4 Find the Laplace transform of each of the following:

(a) $3\sin 2t - 4t\sin 3t$

(b) $5\cos 3t + 3\sin 3t$

(c) $2 - 3\cos\dfrac{2t}{3} + 2t\cos 2t$

(d) $2e^t\cos 2t - 4e^t\sin 2t$

(e) $3e^{-2t}\sin 3t - 2e^{-2t}\cos 3t$

Solutions to exercises

1 $\mathcal{L}\{f + g\} = \mathcal{L}\{f\} + \mathcal{L}\{g\}, \mathcal{L}\{kf\} = k\mathcal{L}\{f\}$ if k is a constant.

2 (a) $\dfrac{2}{s} + \dfrac{3}{s^2}$ (b) $\dfrac{4}{s^3} - \dfrac{3}{s}$ (c) $\dfrac{1}{s^3} - \dfrac{6}{s}$

(d) $\dfrac{12}{s^4} - \dfrac{6}{s^3} + \dfrac{5}{s^2} - \dfrac{6}{s}$ (e) $\dfrac{2}{s} - \dfrac{24}{s^4} + \dfrac{600}{s^6}$

3 (a) $\dfrac{4}{s-2} + \dfrac{3}{s+2}$

(b) $\dfrac{1}{2(s-3)} + \dfrac{4}{s-1} - \dfrac{5}{s+1}$

(c) $\dfrac{1}{s} - \dfrac{3}{s+1} + \dfrac{2}{3(s-1)}$

(d) $\dfrac{2}{(s-1)^2} - \dfrac{6}{(s+1)^3}$

(e) $\dfrac{2}{s^2} - \dfrac{3}{(s+1)^2} + \dfrac{4}{s}$

4 (a) $\dfrac{6}{s^2+4} - \dfrac{24s}{(s^2+9)^2}$ (b) $\dfrac{5s}{s^2+9} + \dfrac{9}{s^2+9}$

(c) $\dfrac{2}{s} - \dfrac{3s}{s^2+\frac{4}{9}} + \dfrac{2(s^2-4)}{(s^2+4)^2}$

(d) $\dfrac{2(s-1)}{(s-1)^2+4} - \dfrac{8}{(s-1)^2+4}$

(e) $\dfrac{9}{(s+2)^2+9} - \dfrac{2(s+2)}{(s+2)^2+9}$

End of block exercises

1 Find the Laplace transform of each of the following:

(a) 7 (b) $\dfrac{2}{3}$ (c) -0.25 (d) $t - 2$ (e) $\dfrac{2t - 1}{3}$

2 Find the Laplace transform of the following:
(a) $0.6t$ (b) $6 - 6t$ (c) $2(t + 1)$
(d) $(t + 1)(t - 1)$ (e) $3t^4 - 2$

3 Find the Laplace transform of the following:
(a) $2 - 3t^2e^{-3t}$ (b) $t(2 - 3t + 4e^t)$
(c) $2t(\sin 3t - 2\cos 3t)$ (d) $3e^{2t}(1 - t + e^t)$
(e) $\cos 2t(e^{-t} + t - 2)$

Solutions to exercises

1 (a) $\dfrac{7}{s}$ (b) $\dfrac{2}{3s}$ (c) $-\dfrac{1}{4s}$ (d) $\dfrac{1}{s^2} - \dfrac{2}{s}$

(e) $\dfrac{2}{3s^2} - \dfrac{1}{3s}$

2 (a) $\dfrac{0.6}{s^2}$ (b) $\dfrac{6}{s} - \dfrac{6}{s^2}$ (c) $2\left(\dfrac{1}{s^2} + \dfrac{1}{s}\right)$

(d) $\dfrac{2}{s^3} - \dfrac{1}{s}$ (e) $\dfrac{72}{s^5} - \dfrac{2}{s}$

3 (a) $\dfrac{2}{s} - \dfrac{6}{(s + 3)^3}$ (b) $\dfrac{2}{s^2} - \dfrac{6}{s^3} + \dfrac{4}{(s - 1)^2}$

(c) $2\left(\dfrac{6s}{(s^2 + 9)^2} - 2\dfrac{(s^2 - 9)}{(s^2 + 9)^2}\right)$

(d) $3\left(\dfrac{1}{s - 2} - \dfrac{1}{(s - 2)^2} + \dfrac{1}{s - 3}\right)$

(e) $\dfrac{s + 1}{(s + 1)^2 + 4} + \dfrac{s^2 - 4}{(s^2 + 4)^2} - \dfrac{2s}{s^2 + 4}$

The inverse Laplace transform

2.1 Introduction

In Block 1 we saw how to use Table 1.1 to find the Laplace transform of a variety of functions. Often, however, we know the Laplace transform of a function without knowing the actual function. In such cases we need to calculate the original function from knowledge of its Laplace transform: that is, we seek the **inverse Laplace transform**. This is the subject of this block.

2.2 Using Table 1.1 to find inverse Laplace transforms

To find the inverse Laplace transform we use Table 1.1 in Block 1. We write any given Laplace transform as the sum of standard expressions given in Table 1.1 and then find the inverse from the table. We use the notation $\mathcal{L}^{-1}$ to denote an inverse Laplace transform, so that

$$\text{if } \mathcal{L}\{f(t)\} = F(s), \quad \text{then } \mathcal{L}^{-1}\{F(s)\} = f(t)$$

In addition we shall make use of the following linearity properties:

Key point

$$\mathcal{L}^{-1}\{F + G\} = \mathcal{L}^{-1}\{F\} + \mathcal{L}^{-1}\{G\}$$
$$\mathcal{L}^{-1}\{kF\} = k\mathcal{L}^{-1}\{F\} \text{ if } k \text{ is a constant}$$

The first property states that to find the inverse Laplace transform of two functions added together we simply find the inverse transform of each individual function and then add together these inverse transforms.

The second property states that when a function, $F(s)$, is multiplied by a constant, k, the corresponding inverse transform is also multiplied by the same constant.

The technique is illustrated by examples.

Example 2.1

Find the inverse Laplace transform of

(a) $\dfrac{1}{s}$ (b) $\dfrac{4}{s}$ (c) $\dfrac{1}{2s}$

Solution

(a) From Table 1.1 we note that

$$\mathcal{L}\{1\} = \frac{1}{s}$$

Hence the inverse Laplace transform of $\dfrac{1}{s}$ is 1. We may write this compactly as

$$\mathcal{L}^{-1}\left\{\frac{1}{s}\right\} = 1$$

Thus the function, $f(t)$, whose Laplace transform is $\dfrac{1}{s}$, is $f(t) = 1$.

(b) We write $\dfrac{4}{s}$ as $4\left(\dfrac{1}{s}\right)$. From (a) we know $\mathcal{L}^{-1}\left\{\dfrac{1}{s}\right\} = 1$ and so

$$\mathcal{L}^{-1}\left\{\frac{4}{s}\right\} = 4\mathcal{L}^{-1}\left\{\frac{1}{s}\right\}$$

$$= 4(1)$$

$$= 4$$

Here we have made use of the second linearity property.

(c) We write $\dfrac{1}{2s}$ as $\dfrac{1}{2}\left(\dfrac{1}{s}\right)$ and so

$$\mathcal{L}^{-1}\left\{\frac{1}{2s}\right\} = \frac{1}{2}$$

Example 2.2
Find the inverse Laplace transform of

(a) $\dfrac{1}{s^2}$ (b) $\dfrac{3}{s^2}$ (c) $\dfrac{3}{s^2} - \dfrac{7}{s}$

Solution
(a) From Table 1.1 we note that $\mathcal{L}\{t\} = \dfrac{1}{s^2}$ and so

$$\mathcal{L}^{-1}\left\{\frac{1}{s^2}\right\} = t$$

The inverse Laplace transform of $\dfrac{1}{s^2}$ is t.

(b)

$$\mathcal{L}^{-1}\left\{\frac{3}{s^2}\right\} = 3\mathcal{L}^{-1}\left\{\frac{1}{s^2}\right\}$$

$$= 3t$$

(c) First note that $\mathcal{L}^{-1}\left\{\dfrac{7}{s}\right\} = 7$. Then using the first linearity property we can write

$$\mathcal{L}^{-1}\left\{\frac{3}{s^2} - \frac{7}{s}\right\} = \mathcal{L}^{-1}\left\{\frac{3}{s^2}\right\} - \mathcal{L}^{-1}\left\{\frac{7}{s}\right\}$$

$$= 3t - 7$$

Example 2.3

Find the inverse Laplace transform of

(a) $\dfrac{2}{s+1} - \dfrac{4}{s-2}$ (b) $\dfrac{3}{(s+1)^6}$

Solution

(a) Using Table 1.1 we see that

$$\mathcal{L}^{-1}\left\{\frac{1}{s+1}\right\} = e^{-t}$$

and so

$$\mathcal{L}^{-1}\left\{\frac{2}{s+1}\right\} = \boxed{}$$

$$2e^{-t}$$

Similarly

$$\mathcal{L}^{-1}\left\{\frac{4}{s-2}\right\} = \boxed{}$$

$$4e^{2t}$$

Hence

$$\mathcal{L}^{-1}\left\{\frac{2}{s+1} - \frac{4}{s-2}\right\} = \boxed{}$$

$$2e^{-t} - 4e^{2t}$$

(b) We note that

$$\mathcal{L}^{-1}\left\{\frac{5!}{(s+1)^6}\right\} = \boxed{}$$

$$t^5e^{-t}$$

Noting that $5! = 120$, then

$$\mathcal{L}^{-1}\left\{\frac{3}{(s+1)^6}\right\} = \boxed{}$$

$$\frac{t^5e^{-t}}{40}$$

Sometimes it is necessary to carry out some preliminary rewriting of a transform in order to be able to use Table 1.1. The best way to proceed is not always obvious, but doing plenty of practice exercises will help. Consider Example 2.4

Example 2.4

Find the inverse Laplace transform of

(a) $\dfrac{5}{(s+2)^2+9}$ (b) $\dfrac{s+2}{(s+2)^2+9}$ (c) $\dfrac{s+7}{(s+2)^2+9}$

Solution

(a) From Table 1.1 we see that

$$\mathcal{L}\{e^{-2t}\sin 3t\} = \boxed{}$$

$$\frac{3}{(s+2)^2+9}$$

so that

$$\mathcal{L}^{-1}\left\{\frac{3}{(s+2)^2+9}\right\} = e^{-2t}\sin 3t$$

Hence

$$\mathcal{L}^{-1}\left\{\frac{5}{(s+2)^2+9}\right\} = \boxed{}$$

$$\frac{5}{3}e^{-2t}\sin 3t$$

(b) $\mathcal{L}^{-1}\left\{\dfrac{s+2}{(s+2)^2+9}\right\} = \boxed{}$

$$e^{-2t}\cos 3t$$

(c) We think of $\dfrac{s+7}{(s+2)^2+9}$ as

$$\frac{5}{(s+2)^2+9} + \frac{s+2}{(s+2)^2+9}$$

So

$$\mathcal{L}^{-1}\left\{\frac{s+7}{(s+2)^2+9}\right\} = \boxed{}$$

$$\frac{5}{3}e^{-2t}\sin 3t + e^{-2t}\cos 3t$$

Exercises

1 Find the inverse Laplace transform of each of the following:

(a) $\dfrac{4}{s^2}$ (b) $-\dfrac{3}{s}$ (c) $\dfrac{5}{s^3}$ (d) $\dfrac{3}{s^5} - \dfrac{2}{s^4}$

(e) $\dfrac{1}{3s^5} - \dfrac{2}{3s^2}$

2 Find the inverse Laplace transform of

(a) $\dfrac{3}{s+2}$ (b) $\dfrac{-2}{s-1}$ (c) $\dfrac{2}{(s+2)^3}$

(d) $\dfrac{1}{s} + \dfrac{1}{s+1}$ (e) $\dfrac{1}{(s-3)^2} + \dfrac{1}{(s-3)^3}$

3 Find the inverse Laplace transform of

(a) $\dfrac{1}{2(s+3)}$ (b) $\dfrac{1}{2s+3}$ (c) $\dfrac{3s}{s^2+1}$

(d) $\dfrac{-6}{s^2+9}$ (e) $\dfrac{s+2}{s^2+4}$

4 Find the inverse Laplace transform of

(a) $\dfrac{s-1}{(s-1)^2+36}$ (b) $\dfrac{3}{(s+2)^2+9}$

(c) $\dfrac{-1}{(s-1)^2+4}$ (d) $\dfrac{s+5}{(s+2)^2+9}$

(e) $\dfrac{s+5}{(s+3)^2+4}$

Solutions to exercises

1 (a) $4t$ (b) -3 (c) $\dfrac{5t^2}{2}$ (d) $\dfrac{t^4}{8} - \dfrac{t^3}{3}$

(e) $\dfrac{t^4}{72} - \dfrac{2t}{3}$

2 (a) $3e^{-2t}$ (b) $-2e^t$ (c) t^2e^{-2t} (d) $1 + e^{-t}$

(e) $te^{3t} + \dfrac{t^2e^{3t}}{2}$

3 (a) $\dfrac{e^{-3t}}{2}$ (b) $\dfrac{e^{-1.5t}}{2}$ (c) $3\cos t$ (d) $-2\sin 3t$

(e) $\cos 2t + \sin 2t$

4 (a) $e^t\cos 6t$ (b) $e^{-2t}\sin 3t$ (c) $-\dfrac{e^t\sin 2t}{2}$

(d) $e^{-2t}\cos 3t + e^{-2t}\sin 3t$

(e) $e^{-3t}\cos 2t + e^{-3t}\sin 2t$

2.3 Inversions requiring completing the square

The technique of completing the square, which was described in Chapter 7 Block 2, is often used to rewrite Laplace transforms in alternative forms so that they can be inverted. Consider the following examples.

Example 2.5

Find the inverse Laplace transform of $\dfrac{1}{s^2 - 2s + 5}$.

Solution

By completing the square we write $s^2 - 2s + 5$ as $(s - 1)^2 + 4$, which is the same as $(s - 1)^2 + 2^2$. From Table 1.1

$$\mathcal{L}\{e^t \sin 2t\} = \frac{2}{(s - 1)^2 + 2^2}$$

and so

$$\mathcal{L}^{-1}\left\{\frac{1}{(s - 1)^2 + 2^2}\right\} = \frac{1}{2}\, e^t \sin 2t$$

Example 2.6

Find the inverse Laplace transform of $\dfrac{2s - 1}{s^2 + 6s + 10}$.

Solution

By completing the square we see that

$$s^2 + 6s + 10 = (s + 3)^2 + 1^2$$

From Table 1.1, the results with $(s + 3)^2 + 1^2$ in the denominator are

$$\mathcal{L}\{e^{-3t} \sin t\} = \frac{1}{(s + 3)^2 + 1^2}, \quad \mathcal{L}\{e^{-3t} \cos t\} = \frac{s + 3}{(s + 3)^2 + 1^2}$$

We need to rewrite the given transform in terms of these two results.

$$\frac{2s - 1}{s^2 + 6s + 10} = \frac{2s - 1}{(s + 3)^2 + 1^2}$$

$$= \frac{2s + 6 - 7}{(s + 3)^2 + 1^2}$$

$$= \frac{2s + 6}{(s + 3)^2 + 1^2} - \frac{7}{(s + 3)^2 + 1^2}$$

$$= 2\frac{s + 3}{(s + 3)^2 + 1^2} - 7\frac{1}{(s + 3)^2 + 1^2}$$

Now

$$\mathcal{L}^{-1}\left\{2\frac{s + 3}{(s + 3)^2 + 1^2}\right\} = 2e^{-3t} \cos t, \quad \mathcal{L}^{-1}\left\{7\frac{1}{(s + 3)^2 + 1^2}\right\} = 7e^{-3t} \sin t$$

and so

$$\mathcal{L}^{-1}\left\{\frac{2s - 1}{s^2 + 6s + 10}\right\} = 2e^{-3t}\cos t - 7e^{-3t}\sin t$$

Example 2.7

Find the inverse Laplace transform of $\dfrac{9 - 4s}{2s^2 + 8s + 40}$.

Solution

We write $2s^2 + 8s + 40$ as $2(s^2 + 4s + 20)$ and then complete the square to obtain

$$2(s^2 + 4s + 20) = \boxed{} \qquad\qquad 2[(s + 2)^2 + 4^2]$$

The relevant parts of Table 1.1 are

$$\mathcal{L}\{e^{-2t}\sin 4t\} = \boxed{} \qquad\qquad \frac{4}{(s + 2)^2 + 4^2}$$

$$\mathcal{L}\{e^{-2t}\cos 4t\} = \boxed{} \qquad\qquad \frac{s + 2}{(s + 2)^2 + 4^2}$$

The given transform is rewritten so that these results can be used.

$$\frac{9 - 4s}{2s^2 + 8s + 40} = \frac{9 - 4s}{2[(s + 2)^2 + 4^2]}$$

$$= \frac{17 - 4(s + 2)}{2[(s + 2)^2 + 4^2]}$$

$$= \frac{8.5 - 2(s + 2)}{(s + 2)^2 + 4^2}$$

$$= \frac{8.5}{(s + 2)^2 + 4^2} - 2\frac{s + 2}{(s + 2)^2 + 4^2}$$

Now

$$\mathcal{L}^{-1}\left\{\frac{8.5}{(s + 2)^2 + 4^2}\right\} = \boxed{} \qquad\qquad 2.125e^{-2t}\sin 4t$$

and

$$\mathcal{L}^{-1}\left\{2\frac{s + 2}{(s + 2)^2 + 4^2}\right\} = \boxed{} \qquad\qquad 2e^{-2t}\cos 4t$$

Hence

$$\mathcal{L}^{-1}\left\{\frac{9 - 4s}{2s^2 + 8s + 40}\right\} = 2.125e^{-2t}\sin 4t - 2e^{-2t}\cos 4t$$

Exercises

1   Find the inverse Laplace transform of each of the following:

(a) $\dfrac{1}{s^2 + 4s + 5}$ (b) $\dfrac{3s - 5}{s^2 + 1}$ (c) $\dfrac{3 - 2s}{s^2 - 4s + 8}$

(d) $\dfrac{1 - 2s}{s^2 + 4}$ (e) $\dfrac{1}{s^2 - 6s + 9}$

2   Find the inverse Laplace transform of

(a) $\dfrac{6 + s}{2s^2 - 4s + 4}$ (b) $\dfrac{s}{2s^2 + 0.5}$ (c) $\dfrac{1 + s}{s^2}$

(d) $\dfrac{3s + 7}{9 + 3s^2}$ (e) $\dfrac{2s - 5}{s^2 + 4s + 13}$

Solutions to exercises

1   (a) $e^{-2t} \sin t$ (b) $3 \cos t - 5 \sin t$

(c) $-2e^{2t} \cos 2t - \dfrac{1}{2} e^{2t} \sin 2t$

(d) $\dfrac{1}{2} \sin 2t - 2 \cos 2t$ (e) te^{3t}

2   (a) $\dfrac{1}{2} e^t \cos t + \dfrac{7}{2} e^t \sin t$ (b) $\dfrac{1}{2} \cos \dfrac{t}{2}$ (c) $t + 1$

(d) $\cos \sqrt{3}t + \dfrac{7}{3\sqrt{3}} \sin \sqrt{3}t$

(e) $2e^{-2t} \cos 3t - 3e^{-2t} \sin 3t$

2.4   Using partial fractions to find an inverse Laplace transform

The method of partial fractions was met in Chapter 7. It is a technique for expressing an algebraic fraction as a sum of simpler fractions. It is also used as an intermediate step when finding the inverse Laplace transform.

Example 2.8

Find the inverse Laplace transform of $\dfrac{5s + 4}{s^2 + 2s}$.

Solution

The algebraic fraction is expressed as a sum of partial fractions.

$$\frac{5s + 4}{s^2 + 2s} = \frac{2}{s} + \frac{3}{s + 2}$$

The inverse Laplace transform of each partial fraction is found.

$$\mathcal{L}^{-1}\left\{\frac{2}{s}\right\} = 2, \quad \mathcal{L}^{-1}\left\{\frac{3}{s + 2}\right\} = 3e^{-2t}$$

and so

$$\mathcal{L}^{-1}\left\{\frac{5s + 4}{s^2 + 2s}\right\} = 2 + 3e^{-2t}$$

Example 2.9
Find the inverse Laplace transform of

$$\frac{s + 7}{s^2 + 4s + 3}$$

Solution
The algebraic fraction is expressed as its partial fractions.

$$\frac{s + 7}{s^2 + 4s + 3} = \frac{3}{s + 1} - \frac{2}{s + 3}$$

The inverse Laplace transform of each partial fraction can now be found.

$$\mathcal{L}^{-1}\left\{\frac{3}{s + 1}\right\} = 3e^{-t}, \quad \mathcal{L}^{-1}\left\{\frac{2}{s + 3}\right\} = 2e^{-3t}$$

and so

$$\mathcal{L}^{-1}\left\{\frac{s + 7}{s^2 + 4s + 3}\right\} = 3e^{-t} - 2e^{-3t}$$

Example 2.10
Find the inverse Laplace transform of

$$\frac{s^2 + 6s + 1}{s^3 + 2s^2 + s}$$

Solution
The fraction is expressed as its partial fractions.

$$\frac{s^2 + 6s + 1}{s^3 + 2s^2 + s} = \boxed{} \qquad \frac{1}{s} + \frac{4}{(s + 1)^2}$$

The inverse Laplace transform of each partial fraction is found.

$$\mathcal{L}^{-1}\left\{\frac{1}{s}\right\} = \boxed{}, \quad \mathcal{L}^{-1}\left\{\frac{4}{(s + 1)^2}\right\} = \boxed{}, \qquad 1, 4te^{-t}$$

and so

$$\mathcal{L}^{-1}\left\{\frac{s^2 + 6s + 1}{s^3 + 2s^2 + s}\right\} = \boxed{} \qquad 1 + 4te^{-t}$$

Exercises

1 Find the inverse Laplace transform of the following:

(a) $\dfrac{3s + 1}{s^2 - 1}$ (b) $\dfrac{7s + 11}{s^2 + s - 6}$ (c) $\dfrac{3s + 2}{s^2}$

(d) $\dfrac{s + 4}{s^2 + 4s + 4}$ (e) $\dfrac{3s^2 + 2s + 5}{(s + 1)(s^2 + 1)}$

2 Find the inverse Laplace transform of each of the following:

(a) $\dfrac{2s^2 - 3s + 1}{(s - 3)(s^2 + 1)}$ (b) $\dfrac{-(s + 6)}{2(s^2 + 6s + 8)}$

(c) $\dfrac{-8s - 2}{s^3 - s^2 - 2s}$ (d) $\dfrac{s^2 + 3s + 2}{s^3 + 2s^2 + 2s}$

(e) $\dfrac{5s^2 - 1}{s^4 - 1}$

Solutions to exercises

1   (a) $2e^t + e^{-t}$ (b) $5e^{2t} + 2e^{-3t}$ (c) $3 + 2t$
    (d) $e^{-2t} + 2te^{-2t}$ (e) $2 \sin t + 3e^{-t}$

2   (a) $e^{3t} + \cos t$ (b) $\frac{1}{2}e^{-4t} - e^{-2t}$
    (c) $1 + 2e^{-t} - 3e^{2t}$ (d) $1 + e^{-t} \sin t$
    (e) $2 \sinh t + 3 \sin t$

End of block exercises

1   Find the inverse Laplace transform of the
    following:

(a) $\dfrac{3}{s^2} - \dfrac{4}{s}$ (b) $\dfrac{2}{s^3} + \dfrac{1}{s^2} - \dfrac{2}{s}$ (c) $\dfrac{12}{s^4} - \dfrac{6}{s^3}$

(d) $\dfrac{1}{5}\left(\dfrac{18}{s^4} - \dfrac{8}{s^3} + \dfrac{2}{s}\right)$

(e) $\dfrac{6\alpha}{s^4} - \dfrac{2\beta}{s^3} + \dfrac{\gamma}{s^2}$, α, β, γ constants

2   Find the inverse Laplace transform of the
    following:

(a) $\dfrac{3}{s+1} + \dfrac{5}{s+2}$

(b) $\dfrac{1}{s-1} + \dfrac{1}{s-2} - \dfrac{1}{s-3}$

(c) $\dfrac{4}{(s+1)^3} + \dfrac{1}{(s+2)^2}$

(d) $\dfrac{2}{(s-2)^4} + \dfrac{1}{(s+2)^3}$

(e) $\dfrac{1}{s} + \dfrac{1}{s+1} - \dfrac{2}{s+2}$

3   Find the inverse Laplace transform of the
    following:

(a) $\dfrac{6}{s^2+9} - \dfrac{s}{2(s^2+9)}$ (b) $\dfrac{2}{(s+4)^2+4}$

(c) $\dfrac{3}{(s+1)^2+9} - \dfrac{2(s+1)}{(s+1)^2+9}$

(d) $\dfrac{1}{2[(s+2)^2+0.25]} + \dfrac{2}{s^3}$

(e) $\dfrac{s-2}{(s-2)^2+16}$

4   Find the inverse Laplace transform of the
    following:

(a) $\dfrac{3s-3}{s^2-s-2}$ (b) $\dfrac{5s+6}{s^2+9}$ (c) $\dfrac{3s+5}{s^2+2s+5}$

(d) $\dfrac{4s^2+7s+6}{2s^3+4s^2}$ (e) $\dfrac{4s+5}{2s^2+5s+2}$

Solutions to exercises

1   (a) $3t - 4$ (b) $t^2 + t - 2$ (c) $2t^3 - 3t^2$

    (d) $\dfrac{3t^3 - 4t^2 + 2}{5}$ (e) $\alpha t^3 - \beta t^2 + \gamma t$

2   (a) $3e^{-t} + 5e^{-2t}$ (b) $e^t + e^{2t} - e^{3t}$

    (c) $2t^2e^{-t} + te^{-2t}$ (d) $\dfrac{t^3e^{2t}}{3} + \dfrac{t^2e^{-2t}}{2}$

    (e) $1 + e^{-t} - 2e^{-2t}$

3   (a) $2 \sin 3t - \dfrac{\cos 3t}{2}$ (b) $e^{-4t} \sin 2t$

(c) $e^{-t} \sin 3t - 2e^{-t} \cos 3t$ (d) $e^{-2t} \sin \dfrac{t}{2} + t^2$

(e) $e^{2t} \cos 4t$

4   (a) $e^{2t} + 2e^{-t}$ (b) $2 \sin 3t + 5 \cos 3t$

    (c) $e^{-t}(\sin 2t + 3 \cos 2t)$

    (d) $1 + \dfrac{3t}{2} + e^{-2t}$

    (e) $e^{-t/2} + e^{-2t}$

Solving differential equations using the Laplace transform

3.1 Introduction

In Blocks 1 and 2 we have seen how to find the Laplace transform and its inverse. These techniques are now applied to finding the solution to differential equations.

Linear constant-coefficient differential equations may be solved using the Laplace transform. By applying the transform, the differential equation is converted into an algebraic equation. This algebraic equation is solved and then the inverse Laplace transform is applied to yield the solution to the differential equation. One advantage of using the Laplace transform is that initial conditions are automatically incorporated into the solution.

3.2 The Laplace transform of a derivative

Let $f(t)$ be a function of t and let $F(s)$ be the Laplace transform of f. The value of f and its derivatives when $t = 0$ are denoted by $f(0), f'(0), f''(0)$ and so on. The nth derivative of f is denoted by $f^{(n)}(t)$. Then it can be shown that the Laplace transform of $f^{(n)}(t)$ is given by

Key point

$$\mathcal{L}\{f^{(n)}(t)\} = s^n F(s) - s^{n-1} f(0) - s^{n-2} f'(0) - \cdots - f^{(n-1)}(0)$$

Two common cases are given when $n = 1$ and $n = 2$. In these cases

Key point

$$\mathcal{L}\{f'(t)\} = sF(s) - f(0)$$
$$\mathcal{L}\{f''(t)\} = s^2 F(s) - sf(0) - f'(0)$$

Example 3.1

The Laplace transform of $f(t)$ is $F(s)$. Given $f(0) = 3$ and $f'(0) = -2$ write expressions for the Laplace transform of
(a) $f'(t)$ (b) $f''(t)$ (c) $2f'' - 3f' + f$

Solution

(a)
$$\mathcal{L}\{f'(t)\} = sF(s) - f(0)$$
$$= sF(s) - 3$$

(b)
$$\mathcal{L}\{f''(t)\} = s^2 F(s) - sf(0) - f'(0)$$
$$= s^2 F(s) - 3s + 2$$

(c) $\mathcal{L}\{2f'' - 3f' + f\} = 2\mathcal{L}\{f''\} - 3\mathcal{L}\{f'\} + \mathcal{L}\{f\}$
$= 2[s^2F(s) - 3s + 2] - 3[sF(s) - 3] + F(s)$
$= (2s^2 - 3s + 1)\, F(s) - 6s + 13$

Example 3.2

Obtain an expression for the Laplace transform of $3x''(t) + 2x'(t) - 2x(t)$ where $\mathcal{L}\{x(t)\} = X(s)$, $x(0) = -4$ and $x'(0) = 3$.

Solution

$\mathcal{L}\{x'(t)\} = $ [blank] $sX(s) + 4$

$\mathcal{L}\{x''(t)\} = $ [blank] $s^2X(s) + 4s - 3$

so

$\mathcal{L}\{3x'' + 2x' - 2x\} = 3\mathcal{L}\{x''\} + 2\mathcal{L}\{x'\} - 2\mathcal{L}\{x\}$
$= 3(s^2X + 4s - 3) + 2(sX + 4) - 2X$

$= $ [blank]

$(3s^2 + 2s - 2)X + 12s - 1$

Exercises

1 The Laplace transform of $y(t)$ is $Y(s)$, $y(0) = 2$ and $y'(0) = -1$. Find the Laplace transform of the following expressions:
(a) y' (b) y'' (c) $2y'' - y' + 3y$
(d) $-y'' + 4y' - 6y$
(e) $\dfrac{y''}{2} + 3y' - y$

2 The Laplace transform of $x(t)$ is $X(s)$, $x(0) = 2$, $x'(0) = 3$, $x''(0) = -1$. Find the Laplace transform of
(a) $2x' - 3x$ (b) $x'' - 2x' + 3x$ (c) x'''

(d) $x''' + 2x'' + 3x' - 4x$
(e) $2x''' - 3x'' - 7x' + 6x$

3 From the definition of the Laplace transform, and using integration by parts, show that
$$\mathcal{L}\{f'(t)\} = sF(s) - f(0)$$

4 From the definition of the Laplace transform, and using integration by parts, show that
$$\mathcal{L}\{f''(t)\} = s^2F(s) - sf(0) - f'(0)$$

Solutions to exercises

1 (a) $sY - 2$

(b) $s^2Y - 2s + 1$

(c) $(2s^2 - s + 3)Y - 4s + 4$

(d) $(-s^2 + 4s - 6)Y + 2s - 9$

(e) $\left(\dfrac{s^2}{2} + 3s - 1\right)Y - s - \dfrac{11}{2}$

2 (a) $(2s - 3)X - 4$

(b) $(s^2 - 2s + 3)X - 2s + 1$

(c) $s^3X - 2s^2 - 3s + 1$

(d) $(s^3 + 2s^2 + 3s - 4)X - 2s^2 - 7s - 11$

(e) $(2s^3 - 3s^2 - 7s + 6)X - 4s^2 + 25$

3.3 Solving differential equations

We are now in a position to solve linear constant-coefficient differential equations using the Laplace transform method. The method consists of three distinct phases:

1 Take the Laplace transform of the given differential equation.
2 Make the transformed variable (usually X or Y) the subject.
3 Apply the inverse Laplace transform to find $x(t)$ (or $y(t)$).

Example 3.3
Solve

$$\frac{dx}{dt} - 2x = 2e^{3t}, \quad x(0) = 2$$

Solution
The Laplace transform of every term is found, noting that the transform of $x(t)$ is $X(s)$. In what follows we write $X(s)$ as simply X in the knowledge that X is a function of s.

$$\mathcal{L}\left\{\frac{dx}{dt}\right\} = sX - x(0)$$
$$= sX - 2$$
$$\mathcal{L}\{2x\} = 2X$$
$$\mathcal{L}\{2e^{3t}\} = \frac{2}{s-3}$$

So, by taking the Laplace transform of each term, the differential equation becomes

$$sX - 2 - 2X = \frac{2}{s-3}$$

Note that this is no longer a differential equation but an algebraic equation for X. This equation is rearranged so that X is made the subject.

$$(s-2)X - 2 = \frac{2}{s-3}$$

$$(s-2)X = \frac{2}{s-3} + 2$$

$$X = \frac{2}{(s-3)(s-2)} + \frac{2}{s-2}$$

$$= \frac{2 + 2(s-3)}{(s-3)(s-2)}$$

$$= \frac{2s-4}{(s-3)(s-2)}$$

$$= \frac{2}{s-3}$$

The inverse Laplace transform is now found.

$$\mathcal{L}^{-1}\left\{\frac{2}{s-3}\right\} = 2e^{3t}$$

so

$$x(t) = 2e^{3t}$$

Thus $x(t) = 2e^{3t}$ is the solution to the given differential equation. Note that the initial condition, $x(0) = 2$, is automatically incorporated into the solution.

Example 3.4 Dynamics – Change in pressure

The pressure, p, of a gas varies with altitude, x, according to

$$\frac{\mathrm{d}p}{\mathrm{d}x} = -Kp$$

where K is a constant. The pressure at ground level (i.e. $x = 0$) is known to be p_0. Solve the equation to find p in terms of x.

Note that in this example, the independent variable is x; the dependent variable is p.

Solution

We are given

$$\frac{\mathrm{d}p}{\mathrm{d}x} = -Kp, \qquad p(0) = p_0$$

Let the Laplace transform of p be $P(s)$.

Taking the Laplace transform of $\dfrac{\mathrm{d}p}{\mathrm{d}x}$:

$$\mathcal{L}\left(\frac{\mathrm{d}p}{\mathrm{d}x}\right) = sP - p(0)$$

$$= sP - p_0$$

Taking the Laplace transform of Kp:

$$\mathcal{L}(Kp) = K\mathcal{L}(p) = KP$$

So taking the Laplace transform of the entire equation gives

$$sP - p_0 = -KP$$

This equation is rearranged to make P the subject.

$$sP + KP = p_0$$
$$P(s + K) = p_0$$
$$P = \frac{p_0}{s + K}$$

We now apply the inverse Laplace transform to find $p(x)$.

$$\mathcal{L}^{-1}(P) = \mathcal{L}^{-1}\left(\frac{p_0}{s + K}\right)$$

$$p(x) = p_0\mathcal{L}\left(\frac{1}{s + K}\right)$$

$$= p_0 e^{-Kx}$$

The pressure exhibits an exponential decay as height, x, is increased.

Note that the differential equation has the same form as those in Examples 2.6 and 2.7 in Chapter 20 (Differential equations) which were solved using a separation of variables method. They could equally well have been solved using the Laplace transform method.

Example 3.5

Solve

$$\frac{d^2y}{dt^2} - y = 2, \quad y(0) = y'(0) = 0$$

Solution

Write $\mathcal{L}\{y\} = Y(s)$ and take the Laplace transform of each term. We write $Y(s)$ concisely as Y.

$$\mathcal{L}\left\{\frac{d^2y}{dt^2}\right\} = \boxed{} \qquad\qquad s^2 Y$$

$$\mathcal{L}\{y\} = \boxed{} \qquad\qquad Y$$

$$\mathcal{L}\{2\} = \boxed{} \qquad\qquad \frac{2}{s}$$

The differential equation then becomes

$$(s^2 - 1)\,Y = \frac{2}{s}$$

from which

$$Y = \boxed{} \qquad\qquad \frac{2}{s(s^2-1)} = \frac{2}{s(s+1)(s-1)}$$

Expressing Y as partial fractions yields

$$Y = \boxed{} \qquad\qquad \frac{1}{s-1} + \frac{1}{s+1} - \frac{2}{s}$$

Taking the inverse Laplace transform produces

$$y(t) = \boxed{} \qquad\qquad e^t + e^{-t} - 2$$

Example 3.6

Solve

$$y'' - y' + y = t, \quad y(0) = y'(0) = 1$$

Solution

We take the Laplace transform of all the terms.

$$\mathcal{L}\{y\} = Y$$

$$\mathcal{L}\{y'\} = \boxed{} \qquad\qquad sY - 1$$

$$\mathcal{L}\{y''\} = \boxed{} \qquad\qquad s^2Y - s - 1$$

$$\mathcal{L}\{t\} = \boxed{} \qquad\qquad \frac{1}{s^2}$$

Hence the equation becomes

$$s^2Y - s - 1 - (sY - 1) + Y = \frac{1}{s^2}$$

$$Y(s^2 - s + 1) = s + \frac{1}{s^2}$$

$$= \frac{s^3 + 1}{s^2}$$

$$Y(s^2 - s + 1) = \frac{(s + 1)(s^2 - s + 1)}{s^2}$$

So, cancelling $(s^2 - s + 1)$ from both sides we obtain

$$Y = \frac{s + 1}{s^2}$$

Written as partial fractions we have

$$Y = \boxed{} \qquad\qquad \frac{1}{s} + \frac{1}{s^2}$$

from which

$$y(t) = \boxed{} \qquad\qquad 1 + t$$

Example 3.7 Electrical Engineering – Current in a circuit

The current, $i(t)$, in a circuit is given by

$$L\frac{di}{dt} + Ri = E \sin t$$

where L, R and E are constants and the initial current is 0 (i.e. $i(0) = 0$). Find $i(t)$.

Solution

Taking the Laplace transform of each component of the equation gives

$$\mathcal{L}\left\{L\frac{di}{dt}\right\} = L\mathcal{L}\left\{\frac{di}{dt}\right\} = L(sI - i(0)) = LsI$$

$$\mathcal{L}\{Ri\} = RI$$

$$\mathcal{L}\{E \sin t\} = E\mathcal{L}\{\sin t\} = E\left(\frac{1}{s^2 + 1}\right) = \frac{E}{s^2 + 1}$$

So

$$LsI + RI = \frac{E}{s^2 + 1}$$

Rearranging for I yields

$$I(Ls + R) = \frac{E}{s^2 + 1}$$

$$I = \frac{E}{(s^2 + 1)(Ls + R)}$$

The right-hand side of this equation must now be written, using partial fractions, in a form suitable for taking the inverse Laplace transform.

$$\frac{E}{(s^2 + 1)(Ls + R)} = \frac{E}{L}\left(\frac{1}{(s^2 + 1)(s + R/L)}\right)$$

$$= \frac{E}{L}\left(\frac{As + B}{s^2 + 1} + \frac{C}{s + R/L}\right)$$

The constants A, B and C can be found to be

$$A = \frac{-L^2}{R^2 + L^2}, \quad B = \frac{RL}{R^2 + L^2}, \quad C = \frac{L^2}{R^2 + L^2}$$

The expression is further written in a form suitable for inversion.

$$I = \frac{E}{L}\left(\frac{As + B}{s^2 + 1} + \frac{C}{s + R/L}\right)$$

$$= \frac{E}{L}\left(\frac{As}{s^2 + 1} + \frac{B}{s^2 + 1} + \frac{C}{s + R/L}\right)$$

$$= \frac{EA}{L}\left(\frac{s}{s^2 + 1}\right) + \frac{EB}{L}\left(\frac{1}{s^2 + 1}\right) + \frac{EC}{L}\left(\frac{1}{s + R/L}\right)$$

Note that the inverse Laplace transform of each expression in brackets can be found using Table 1.1 in Block 1. Taking the inverse Laplace transform yields

$$i(t) = \frac{EA}{L}\cos t + \frac{EB}{L}\sin t + \frac{EC}{L}e^{-Rt/L}$$

With the previously calculated values of A, B and C, $i(t)$ could also be written as

$$i(t) = \frac{E}{R^2 + L^2}\left(R\sin t - L\cos t + Le^{-Rt/L}\right)$$

Example 3.8

Solve

$$\frac{d^2x}{dt^2} - \frac{dx}{dt} + 2x = \cos 3t - 17\sin 3t, \quad x(0) = -1, \quad x'(0) = 6$$

Solution

$$\mathcal{L}\{x\} = X$$

$$\mathcal{L}\left\{\frac{dx}{dt}\right\} = \qquad\qquad sX + 1$$

$$\mathcal{L}\left\{\frac{d^2x}{dt^2}\right\} = \qquad\qquad s^2X + s - 6$$

$$\mathcal{L}\left\{\frac{d^2x}{dt^2} - \frac{dx}{dt} + 2x\right\} = \qquad (s^2 - s + 2)X + s - 7$$

$$\mathcal{L}\{\cos 3t - 17\sin 3t\} = \frac{s}{s^2 + 9} - \frac{51}{s^2 + 9}$$

$$= \frac{s - 51}{s^2 + 9}$$

Hence on taking the Laplace transform of the differential equation we have

$$(s^2 - s + 2)X + s - 7 = \frac{s - 51}{s^2 + 9}$$

$$(s^2 - s + 2)X = \frac{s - 51}{s^2 + 9} - s + 7$$

$$= \frac{}{s^2 + 9} \qquad\qquad -s^3 + 7s^2 - 8s + 12$$

$$= \frac{(s^2 - s + 2)(6 - s)}{s^2 + 9}$$

So, on cancelling $(s^2 - s + 2)$ we obtain

$$X = \frac{6 - s}{s^2 + 9}$$

$$= \frac{6}{s^2 + 9} - \frac{s}{s^2 + 9}$$

Hence

$$x(t) = \qquad\qquad 2\sin 3t - \cos 3t$$

Example 3.9 Mechanical vibration – Mass–spring–damper system.

In this example we establish a model of a mass–spring–damper system. Such systems are common in machines where vibration occurs.

A mass, M, is attached to a solid wall by a spring and damper. Figure 3.1 illustrates this.

Figure 3.1
A mass–spring–damper system.

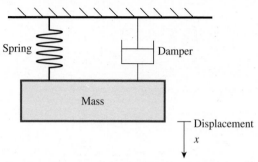

The force F_S in the spring obeys Hooke's Law: that is, the force is proportional to the spring extension. Note that this is an assumption about the spring. For many springs which operate within their limits of elasticity this assumption is accepted as being reasonable.

When the mass is at rest let the spring extension be e. This position is known as the **equilibrium position**. Then the force in the spring is ke, (where k is a constant of proportionality known as the spring stiffness) and this exactly equals the force on the mass due to gravity, Mg, that is,

$$ke = Mg$$

In subsequent motion we measure all distances, x, from the equilibrium position.

$$F_S \propto (x + e)$$
$$F_S = k(x + e)$$

The force, F_d, due to the damper is proportional to the velocity, $\dfrac{dx}{dt}$, of the mass.

$$F_d \propto \frac{dx}{dt}$$
$$F_d = B\frac{dx}{dt}$$

where B is a constant of proportionality known as the damping coefficient. Again, we are making another assumption: that the restraining force in the damper is proportional to the velocity of the mass.

If the mass is pulled from its equilibrium position and then released, the movement of the mass is governed by Newton's second law of motion: that is, force = mass × acceleration. This is expressed mathematically as

$$Mg - ke - kx - B\frac{dx}{dt} = M\frac{d^2x}{dt^2}$$

from which

$$M\frac{d^2x}{dt^2} + B\frac{dx}{dt} + kx = 0$$

As assumptions have been made, this differential equation will provide an approximate description of the actual situation.

Given that $M = 1$, $B = 4$, $k = 8$, $x(0) = 1$ and $\dfrac{dx}{dt}(0) = 0$, find the displacement of the mass in terms of time t.

Solution

The equation and initial conditions are

$$\frac{d^2x}{dt^2} + 4\frac{dx}{dt} + 8x = 0, \quad x(0) = 1, \quad \frac{dx(0)}{dt} = 0$$

Taking the Laplace transform of the equation produces

$$s^2X - s + 4(sX - 1) + 8X = 0$$

from which

$$X = \frac{s + 4}{s^2 + 4s + 8}$$

Completing the square and expressing X in a form suitable for taking the inverse Laplace transform yields

$$X = \frac{s + 2}{(s + 2)^2 + 2^2} + \frac{2}{(s + 2)^2 + 2^2}$$

Taking the inverse Laplace transform produces

$$x(t) = e^{-2t}\cos 2t + e^{-2t}\sin 2t = e^{-2t}(\cos 2t + \sin 2t)$$

Note that $\cos 2t + \sin 2t$ may be expressed in the form $\sqrt{2}\sin\left(2t + \frac{\pi}{4}\right)$ (see Chapter 9 Block 7.8) and so

$$x(t) = \sqrt{2}e^{-2t}\sin\left(2t + \frac{\pi}{4}\right)$$

This is a sine wave with an amplitude of $\sqrt{2}e^{-2t}$ that is decaying exponentially with time. A graph of $x(t)$ is shown in Figure 3.2.

Figure 3.2
Sine wave with exponentially decaying amplitude and exponential envelope.

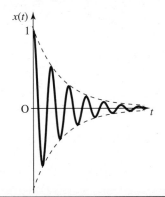

Exercises

1 Solve the following differential equations using the Laplace transform technique:
   (a) $y' - y = 0, \quad y(0) = 3$
   (b) $y' + y = 2e^t, \quad y(0) = 2$
   (c) $y' + 4y = 3e^{2t}, \quad y(0) = 0.5$
   (d) $y' + y = 2\cos t, \quad y(0) = 1$
   (e) $y' - y = 2 - t^2, \quad y(0) = 0$

2 Solve the following equations using the Laplace transform method:
   (a) $x'' + 3x' - 4x = 5e^t, x(0) = 0, x'(0) = 1$
   (b) $x'' - 4x' + 5x = 0, x(0) = 0, x'(0) = 1$
   (c) $y'' + y' + 4y = 4t^2 + 2t + 2 + 2\cos 2t,$
       $y(0) = 0, y'(0) = 2$
   (d) $y'' + 2y' + 9y = 2(3\cos 3t + \sin 3t$
       $+ 3t\cos 3t), y(0) = y'(0) = 0$
   (e) $x'' + 2x' + x = 2e^{-t}, x(0) = x'(0) = 0$

Solutions to exercises

1 (a) $y = 3e^t$ (b) $y = e^t + e^{-t}$ (c) $y = 0.5e^{2t}$
   (d) $y = \sin t + \cos t$ (e) $y = 2t + t^2$

2 (a) $x = te^t$ (b) $x = e^{2t} \sin t$ (c) $y = t^2 + \sin 2t$
   (d) $y = t \sin 3t$ (e) $x = t^2 e^{-t}$

End of block exercises

1 Given $\mathcal{L}\{y\} = Y$, $y(0) = 2$, $y'(0) = -3$ write expressions for the Laplace transform of the following:
   (a) y' (b) y'' (c) $y'' + y'$ (d) $y'' + 3y' - 2y$
   (e) $2y'' - y' + 3y$

   (b) $x' + x = 1 + t + 2\cos t$, $x(0) = 1$
   (c) $4x'' + x = 3$, $x(0) = 3$, $x'(0) = -0.5$
   (d) $x'' + 4x = 0$, $x(0) = 2$, $x'(0) = 0$
   (e) $x'' + x' - 2x = \cos t - 3\sin t$, $x(0) = 2$, $x'(0) = -3$

2 Given $\mathcal{L}\{x\} = X$, $x(0) = x'(0) = 2$, $x''(0) = -1$ write down the Laplace transform of the following expressions:
   (a) x' (b) x'' (c) x''' (d) $2x''' + 3x'' - x' + 2x$
   (e) $-x''' - 2x'' + 3x' + 7x$

4 Solve using the Laplace transform method:
   (a) $y'' + y = 2\cos t$, $y(0) = 0$, $y'(0) = 1$
   (b) $y'' - 2y' = 2e^{2t}$, $y(0) = 0$, $y'(0) = 1$
   (c) $y''' + y'' + y' + y = 1 + t$, $y(0) = 1$, $y'(0) = 0$, $y''(0) = 1$
   (d) $y'' + y = 2(\cos t - \sin t)$, $y(0) = 0$, $y'(0) = 1$
   (e) $y'' + 2y' + 5y = 0$, $y(0) = 1$, $y'(0) = -1$

3 Use the Laplace transform technique to solve
   (a) $x' - 2x = 1 - 2t^2$, $x(0) = (0)$

Solutions to exercises

1 (a) $sY - 2$
   (b) $s^2Y - 2s + 3$
   (c) $(s^2 + s)Y - 2s + 1$
   (d) $(s^2 + 3s - 2)Y - 2s - 3$
   (e) $(2s^2 - s + 3)Y - 4s + 8$

   (d) $(2s^3 + 3s^2 - s + 2)X - 4s^2 - 10s - 2$
   (e) $(-s^3 - 2s^2 + 3s + 7)X + 2s^2 + 6s - 3$

3 (a) $t^2 + t$ (b) $t + \sin t + \cos t$ (c) $3 - \sin\dfrac{t}{2}$
   (d) $2\cos 2t$ (e) $2e^{-2t} + \sin t$

2 (a) $sX - 2$ (b) $s^2X - 2s - 2$
   (c) $s^3X - 2s^2 - 2s + 1$

4 (a) $\sin t + t\sin t$ (b) te^{2t} (c) $e^{-t} + t$
   (d) $t\cos t + t\sin t$ (e) $e^{-t}\cos 2t$

End of chapter exercises

1 Find the Laplace transform of each of the following expressions:
   (a) $t - 3$ (b) $2t^3 + 5t$ (c) $7 - 3t^4$
   (d) $\sin 2t + 2\sin t$ (e) $\cos t + t$

2 Find the Laplace transform of each of the following expressions:
   (a) $2te^{2t}$ (b) $1 - t^3e^{3t}$
   (c) $e^t(1 + \sin t)$ (d) $e^{-2t}(\sin 3t + 2\cos 3t)$
   (e) $t(\cos 2t - 3\sin 2t)$

3   Find the Laplace transform of each of the following expressions:

(a) $(t + 1)^2$ (b) $(e^t + t)^2$ (c) $\dfrac{\sin 4t}{2e^{3t}}$

(d) $2 \sin t \cos t$ (e) $\dfrac{2t^2}{3e^{2t}}$

4   The **first shift theorem** states that if $\mathcal{L}\{f(t)\} = F(s)$, then

$$\mathcal{L}\{e^{-at} f(t)\} = F(s + a)$$

where a is a constant.
(a) From the definition of the Laplace transform show that

$$\mathcal{L}\{e^{-at} f(t)\} = \int_0^\infty e^{-(s+a)t} f(t) \, dt$$

and hence prove the first shift theorem.
(b) Use Table 1.1 in Block 1 and the first shift theorem to find $\mathcal{L}\{u(t - 3)e^{-7t}\}$ where $u(t)$ is the unit step function.

5   Given

$$\mathcal{L}\{g(t)\} = \frac{s + 6}{s^2 + 3}$$

use the result from question 4 to find the Laplace transform of
(a) $e^{-t}g(t)$ (b) $2e^{-2t}g(t)$ (c) $e^{2t}g(t)$

6   The **second shift theorem** states that if $\mathcal{L}\{f(t)\} = F(s)$ then

$$\mathcal{L}\{u(t - d)f(t - d)\} = e^{-sd} F(s), \quad d > 0$$

where $u(t)$ is the unit step function.
(a) Prove this theorem.
(b) Find the Laplace transform of $u(t - 3)(t - 3)^5$.
(c) Find the inverse Laplace transform of

$$\frac{e^{-4s} \, 4!}{s^5}$$

7   Find the inverse Laplace transform of each of the following expressions:

(a) $\dfrac{3}{s} - \dfrac{2}{s^2}$ (b) $\dfrac{16}{s^4}$ (c) $\dfrac{3s}{s^2 + 9}$

(d) $\dfrac{s + 3}{(s + 3)^2 + 1}$ (e) $\dfrac{10}{(s + 2)^3}$

8   Find the inverse Laplace transform of each of the following expressions:

(a) $\dfrac{4}{s^2 + 4s + 5}$ (b) $\dfrac{s + 5}{s^2 + 10s + 29}$

(c) $\dfrac{2s - 3}{s^2 + 6s + 10}$ (d) $\dfrac{s^2 + 6s - 4}{(s^2 + 4)^2}$

(e) $\dfrac{1}{4s^2 + 4s + 1}$

9   Find the inverse Laplace transform of the following expressions:

(a) $\dfrac{3s + 5}{s^2 + 4s + 3}$ (b) $\dfrac{s + 2}{s^2 + s}$

(c) $\dfrac{s^2 - s + 4}{(s - 1)(s^2 - 2s + 5)}$

(d) $\dfrac{2s^3 - 3s^2 + 8s - 3}{(s^2 + 1)(s^2 + 4)}$

(e) $\dfrac{6s^2 - 12s + 2}{(s^2 - 4s + 13)(s^2 + 1)}$

10 Given $\mathcal{L}\{x\} = X, x(0) = 2, x'(0) = 3$ find the Laplace transform of each of the following expressions:
(a) x' (b) x'' (c) $2x' - x$ (d) $3x'' + 2x' - x$
(e) $-2x'' + x' + 3x$

11 Given $\mathcal{L}\{y\} = Y, y(0) = 1, y'(0) = 0$, $y''(0) = -1, y'''(0) = 2$ write an expression for $\mathcal{L}\{y^{(4)}(t)\}$.

12 Solve the following differential equations using the Laplace transform method:
(a) $x' - x = e^t, x(0) = 0$
(b) $x'' + x' + 2x = 4 \cos 2t, x(0) = -1$, $x'(0) = 2$
(c) $x'' + x' - 6x = 1 + 7t - 6t^3$, $x(0) = x'(0) = 0$
(d) $x'' - 4x = 4t, x(0) = x'(0) = 1$
(e) $x'' + x = 2 \sin t, x(0) = x'(0) = 0$

Solutions to exercises

1 (a) $\dfrac{1}{s^2} - \dfrac{3}{s}$ (b) $\dfrac{12}{s^4} + \dfrac{5}{s^2}$ (c) $\dfrac{7}{s} - \dfrac{72}{s^5}$

(d) $\dfrac{2}{s^2 + 4} + \dfrac{2}{s^2 + 1}$ (e) $\dfrac{s}{s^2 + 1} + \dfrac{1}{s^2}$

2 (a) $\dfrac{2}{(s - 2)^2}$ (b) $\dfrac{1}{s} - \dfrac{6}{(s - 3)^4}$

(c) $\dfrac{1}{s - 1} + \dfrac{1}{(s - 1)^2 + 1}$ (d) $\dfrac{2s + 7}{(s + 2)^2 + 9}$

(e) $\dfrac{s^2 - 12s - 4}{(s^2 + 4)^2}$

3 (a) $\dfrac{2}{s^3} + \dfrac{2}{s^2} + \dfrac{1}{s}$ (b) $\dfrac{1}{s - 2} + \dfrac{2}{(s - 1)^2} + \dfrac{2}{s^3}$

(c) $\dfrac{2}{(s + 3)^2 + 16}$ (d) $\dfrac{2}{s^2 + 4}$ (e) $\dfrac{4}{3(s + 2)^3}$

4 (b) $\dfrac{e^{-3(s+7)}}{s + 7}$

5 (a) $\dfrac{s + 7}{(s + 1)^2 + 3}$ (b) $\dfrac{2(s + 8)}{(s + 2)^2 + 3}$

(c) $\dfrac{s + 4}{(s - 2)^2 + 3}$

6 (b) $\dfrac{e^{-3s} 5!}{s^6}$ (c) $u(t - 4)(t - 4)^4$

7 (a) $3 - 2t$ (b) $\dfrac{8t^3}{3}$ (c) $3 \cos 3t$ (d) $e^{-3t} \cos t$

(e) $5t^2 e^{-2t}$

8 (a) $4e^{-2t} \sin t$ (b) $e^{-5t} \cos 2t$

(c) $2e^{-3t} \cos t - 9e^{-3t} \sin t$

(d) $t \cos 2t + 1.5t \sin 2t$ (e) $0.25te^{-0.5t}$

9 (a) $e^{-t} + 2e^{-3t}$ (b) $2 - e^{-t}$ (c) $0.5e^t \sin 2t + e^t$

(d) $2 \cos t - 1.5 \sin 2t$

(e) $e^{2t} \cos 3t + \dfrac{4}{3}e^{2t} \sin 3t - \cos t$

10 (a) $sX - 2$ (b) $s^2X - 2s - 3$

(c) $(2s - 1)X - 4$

(d) $(3s^2 + 2s - 1)X - 6s - 13$

(e) $(-2s^2 + s + 3)X + 4s + 4$

11 $s^4X - s^3 + s - 2$

12 (a) $x = te^t$ (b) $x = \sin 2t - \cos 2t$

(c) $x = \dfrac{t^2}{2} + t^3$ (d) $x = e^{2t} - t$

(e) $x = \sin t - t \cos t$

Statistics and probability

Chapter **23**

Most experimental work in engineering involves the use of statistics and probability. When an experiment is carried out, **data** are collected and analysed. Data may be classified as either **discrete** or **continuous**. This distinction is explained in Block 1. Two key parameters, the centre and the variation, are useful in describing a set of data, and comparing it with other sets. Various ways in which these parameters can be calculated are covered in Blocks 2 and 3.

The second part of the chapter focuses on **probability**. The probability of an event is the likelihood of it occurring, measured on a scale from 0 to 1, where 0 represents impossibility and 1 represents certainty. In any particular trial or experiment there are usually several possible outcomes, each with a different probability. It is useful to think of the total probability, that is 1, being spread out or distributed amongst the various possibilities. This idea gives rise to probability distributions for discrete data and probability density functions for continuous data. The common probability distributions and density functions – binomial, Poisson and normal – are covered in Blocks 7, 8 and 9.

Chapter 23 contents

Data

1.1 Introduction

When a new engineering component is developed, there will be extensive testing at each stage. Much data will be gathered and analysed. It is useful to classify data as either **discrete** or **continuous**.

1.2 Discrete data

Sometimes a variable must take on a value from a set of individually specified values and no other values are possible. Such a variable is then called a **discrete variable**. Consider a quality control engineer who selects five silicon chips at random and tests each one to see whether it works or not. The engineer is interested in the number of chips that work. Let this variable be n. Then n must be a number from the set $\{0, 1, 2, 3, 4, 5\}$. It is impossible for n to have any other values.

As another example, consider the number of people living in a household. This could be 0, 1, 2, 3, . . . and so on. It is impossible to have 2.3 or 1.7 people living in a household.

These are just two examples of variables that are discrete. When a discrete variable is measured several times, the data so generated are called **discrete data**. Discrete data can have only a limited number of values. Other examples of discrete data are

- the number of cars produced in a factory in a week
- the shoe sizes of people in an office block
- the number of times a machine breaks down in 1 year.

1.3 Continuous data

Sometimes a variable can take on any value within a specified range. Such a variable is called **continuous**. For example, consider the weight of a pack of butter produced in a factory. A pack could have any weight between, say, 230 g and 270 g. The weight will be recorded to a particular accuracy, which depends upon the measuring device and the use to which the data will be put. However, the actual weight could be any value in the given range. When a continuous variable is measured several times the data generated are **continuous data**.

Examples of continuous data are

- the diameters of pistons produced in a factory
- the volume of gas produced in a chemical reaction
- the current in a branch of a circuit.

Exercises

1   Explain what is meant by (a) discrete data, (b) continuous data.

2   Give two examples of (a) discrete data, (b) continuous data.

3   Classify the following variables as discrete or continuous:
(a) the number of times a machine breaks down in 12 months

(b) the time between breakdowns of a machine
(c) the capacitance of a capacitor
(d) the amount of money in your pocket
(e) the number of hairs on your head.

Solutions to exercises

2   (a) number of employees of a firm, the number of cars passing a given point in 12 hours
(b) the length of a metal bar, the volume of petrol used on a journey.

3   (a) discrete (b) continuous (c) continuous
(d) discrete (e) discrete

End of block exercises

1   Classify the following variables as discrete or continuous:
(a) the distance travelled before a set of tyres needs replacing
(b) the intensity of light in a room
(c) the force needed to extend a spring by a set amount
(d) the number of bearings in a machine
(e) the percentage mark obtained in an examination.

2   Classify each of the following as discrete or continuous:
(a) the number of pages in a book
(b) the weight of a book
(c) the area of paper needed to make the pages of a book
(d) the price of a book
(e) the number of copies of a book that are sold.

Solutions to exercises

1   (a) continuous (b) continuous
(c) continuous (d) discrete (e) discrete

2   (a) discrete (b) continuous (c) continuous
(d) discrete (e) discrete

Data averages

Introduction

We are often presented with a large amount of data. It may be useful if we can look at a single number that typifies the data. For example, we measure the force at which a certain gauge of wire breaks. If this experiment is repeated many times the force required for breaking will vary: some values will be low, some will be high, and of course there will be some in between. We aim to find a single force that in some way summarises or typifies the measurements that we have made.

A value that typifies a set of data is called an **average**. In statistics there are three important averages: the arithmetic mean, the median and the mode.

2.2 The arithmetic mean

The arithmetic mean is also referred to simply as the **mean**. The mean is found by adding up all the data values and then dividing this total by the number of values:

Key point

$$\text{mean} = \frac{\text{sum of values}}{\text{number of values}}$$

Example 2.1 Mechanical Engineering – Breaking force
The force, in newtons, needed to break a wire was measured and the experiment repeated 10 times. The breaking forces are

$$26 \; 19 \; 17 \; 23 \; 25 \; 20 \; 23 \; 18 \; 20 \; 21$$

Find the mean breaking force.

Solution
The sum of the values is 212. The number of values is 10. Thus

$$\text{mean} = \frac{\text{sum of values}}{\text{number of values}}$$

$$= \frac{212}{10}$$

$$= 21.2$$

To the nearest whole number, the mean breaking force is 21 newtons.

Example 2.2

The temperature, in °C, at which a liquid boils is measured several times. The results are

$$93.7\ 91.4\ 95.3\ 94.9\ 92.3\ 95.8$$

Find the mean temperature.

Solution

Sum of values =	563.4
Number of values =	6
Mean temperature =	$\dfrac{563.4}{6} = 93.9$

A special notation is often used when calculating a mean. Suppose we have n values and we label these $x_1, x_2, x_3, \ldots, x_n$. The sum of these values is denoted by $\sum_{i=1}^{n} x_i$, or more simply as $\sum x$. The mean is denoted by $\bar{x}$, pronounced 'x bar'. Hence

Key point

$$\text{mean} = \bar{x} = \frac{\sum_{i=1}^{n} x_i}{n}$$

Exercises

1 The diameters in mm of some ball bearings were measured; the results are

$$5.1\ 4.9\ 5.0\ 5.2\ 5.1\ 4.8\ 5.2$$

Calculate the mean diameter, giving your answer to 1 d.p.

2 The current, in amps, in a wire was measured several times and the results noted as follows:

$$13.1\ 12.9\ 13.1\ 12.8\ 12.7\ 12.6\ 13.2\ 13.1$$

Calculate the mean current, giving your answer to 1 d.p.

3 The temperature, in °C, at which a liquid froze was recorded several times. The results are

$$-7\ -4\ -1\ -6\ -3\ -2\ -3\ -4$$

Calculate the mean temperature at which the liquid freezes. Give your answer to the nearest integer.

Solutions to exercises

1 5.0 mm

2 12.9 amps

3 −4 °C

The **frequency** of a value is the number of times it occurs. A set of numbers, together with their frequency, is called a **frequency distribution**. For example, suppose the force, in newtons, needed to break a wire is measured and the experiment repeated several times. The results are recorded in Table 2.1.

Table 2.1

Force (N)	Frequency, f
17	3
18	4
19	9
20	11
21	6
22	2

Thus a force of 17 N was recorded three times, a force of 18 N was recorded four times, and so on. Table 2.1 is an example of a frequency distribution. Note that the sum of the frequencies gives the total number of measurements made.

When data are presented in the form of a frequency distribution, the mean can still be calculated. Example 2.3 illustrates the method.

Example 2.3
Table 2.2 is a frequency distribution for the variable x.

Table 2.2

x	Frequency, f
5	2
6	3
7	4
8	2
9	1

Calculate the mean of x: that is $\bar{x}$.

Solution
The value 5 occurs twice and so contributes 5×2 to the sum of the x values. The value 6 occurs three times and so this contributes 6×3 to the sum of the x values. The remaining contributions are 7×4, 8×2 and 9×1. Hence

$$
\begin{aligned}
\text{sum of values} &= (5 \times 2) + (6 \times 3) + (7 \times 4) + (8 \times 2) + (9 \times 1) \\
&= 10 + 18 + 28 + 16 + 9 \\
&= 81
\end{aligned}
$$

The number of values is $2 + 3 + 4 + 2 + 1 = 12$, that is the sum of the frequencies. Finally

$$\text{mean} = \bar{x} = \frac{\text{sum of values}}{\text{number of values}}$$

$$= \frac{81}{12}$$

$$= 6.75$$

The mean of the frequency distribution is 6.75.

By referring to Example 2.3 we see that the sum of the values is found by adding the products $x \times f$. The number of values is found by adding up all the frequencies. If the values are denoted by $x_1, x_2, \ldots, x_n$ and the corresponding frequencies by $f_1, f_2, \ldots, f_n$ then we see that

$$\text{Sum of values} = \sum_{i=1}^{n} x_i f_i$$

$$\text{Number of values} = \sum_{i=1}^{n} f_i$$

These are often written respectively in the more compact form as $\sum xf$ and $\sum f$, where the limits are assumed.

The mean can now be stated:

Key point

$$\text{mean} = \frac{\sum_{i=1}^{n} x_i f_i}{\sum_{i=1}^{n} f_i}$$

$$= \frac{\sum xf}{\sum f}$$

When calculating the mean of the frequency distribution as given in Table 2.2 usually we would extend the table as follows:

x	f	xf
5	2	10
6	3	18
7	4	28
8	2	16
9	1	9
	$\sum f = 12$	$\sum xf = 81$

Example 2.4
Find the mean of the data given in Table 2.1.

Solution

The table is extended.

Force, x (N)	Frequency, f	xf
17	3	51
18	4	72
19	9	171
20	11	220
21	6	126
22	2	44
	$\Sigma f = 35$	$\Sigma xf = 684$

$$\text{mean} = \bar{x} = \frac{\Sigma xf}{\Sigma f}$$

$$= \frac{684}{35}$$

$$= 19.54$$

Exercises

1   Calculate the mean of the frequency distribution

x	f
17	4
18	11
19	14
20	9
21	5

2   The resistance, in ohms, of a certain type of resistor is measured many times. The results are

Resistance (Ω)	Frequency
4.7	6
4.8	11
4.9	4
5.0	8
5.1	3
5.2	7

Calculate the mean value of the resistance, giving your answer to 1 d.p.

3   The lifetimes of a set of components are measured to the nearest 100 hours. The results are

Lifetime (h)	Frequency
0	1
100	1
200	4
300	10
400	17
500	3
600	2
700	10

Calculate the mean lifetime.

4   The temperature of ignition of a certain mixture of gases is recorded to the nearest 10 °C as follows:

Temperature (°C)	Frequency
170	2
180	11
190	10
200	6
210	1
220	2
230	1

Calculate the mean temperature of ignition.

Solutions to exercises

1   19

2   4.9 ohms

3   425 hours

4   190 °C to the nearest 10 °C

2.4   The median

The **median** of a set of numbers is found by listing all the numbers in ascending order and selecting the number that is half-way along the list.

Example 2.5
Find the median of the numbers

$$6\ 8\ 3\ 11\ 10\ 12\ 7\ 9\ 7$$

Solution
The numbers are arranged in ascending order.

$$3\ 6\ 7\ 7\ 8\ 9\ 10\ 11\ 12$$

There are nine numbers in the list. The middle number is the fifth one, that is 8. Therefore, the median is 8.

When there is an even number of values in the list, the median is the mean of the two middle values.

Example 2.6
Find the median of the following temperatures:

$$49\ 50\ 40\ 38\ 41\ 57\ 56\ 52$$

Solution
The numbers are arranged in ascending order:

$$38\ 40\ 41\ 49\ 50\ 52\ 56\ 57$$

There are eight numbers in the list. The middle two values are 49 and 50. The mean of these is $\frac{49 + 50}{2} = 49.5$ and so the median temperature is 49.5.

Exercises

1   Calculate the median of each of the following sets of numbers:
(a) 63 47 51 59 56 50 61 55 58 61
(b) 10 12 16 11 14 11 10 7 13
(c) −4 0 −1 2 3 2 −3 −1
(d) 76 79 81 70 64 62 75 70 71 63 69

Solutions to exercises

1   (a) 57 (b) 11 (c) −0.5 (d) 70

2.5 The mode

The **mode** of a set of values is the value that occurs most often.

Example 2.7
Find the mode of the set of numbers

$$2\ 3\ 3\ 4\ 5\ 5\ 6\ 6\ 6\ 8\ 11$$

Solution
The number 6 occurs three times; this is more than any other number. Hence the mode is 6.

Example 2.8
Find the mode of the set of numbers

$$2\ 2\ 2\ 3\ 3\ 5\ 6\ 11\ 14\ 14\ 14\ 16$$

Solution
In this example there is no single number that occurs most frequently. The numbers 2 and 14 both occur three times. There are two modes. The data are said to be **bimodal**.

Exercises

1 State the mode of each of the following sets of data:
 (a) 5 5 5 6 6 7 7 7 7 9 9 9 11 11
 (b) 4 1 3 6 4 5 1 4 5 3 2
 (c) 4 0 1 3 2 0 2 0 1 3 0 2

Solutions to exercises

1 (a) 7 (b) 4 (c) 0

End of block exercises

1 Find the mean, median and mode of each of the following sets of values:
 (a) 6 9 5 7 6 8 10
 (b) 0 −3 0 −2 1 2 −2 −1 −2 0
 (c) 2 3 1 5 2 5 1 4 3 1 6

2 The resistance of several resistors was measured. The results are

Resistance (Ω)	Frequency
5.0	3
5.5	4
6.0	9
6.5	17
7.0	11
7.5	1

 (a) Calculate the mean resistance.
 (b) Calculate the median.
 (c) Calculate the mode.

3 The mean of the set of values

 5 3 X 10 2 11

 is 8.3. Find X.

4 Find the mean of the data set

 2 2 2 2 2 2 20

 Explain why the mean does not represent the data adequately. Which average would have been more appropriate to use?

5 The force needed to buckle a support beam is measured several times. The results are tabulated thus

Force (×10³ N)	Frequency
2.00	4
2.25	1
2.50	2
2.75	5
3.00	2
3.25	2

 Calculate the mean buckling force.

Solutions to exercises

1 (a) 7.29, 7, 6 (b) −0.7, −0.5, −2 and 0 (bimodal) (c) 3, 3, 1

2 (a) 6.36 Ω (b) 6.5 Ω (c) 6.5 Ω

3 18.8

4 Mean = 4.57. Mode would be more appropriate.

5 2.59 × 10³ N

Variation of data

Block 2 illustrated different ways of describing the central location of a data set. We now look at the variation in a data set.

Consider the mean of the two sets of data: 4 4 4 and 1 3 8. The means are

$$\frac{4 + 4 + 4}{3} = 4 \ \text{ and } \ \frac{1 + 3 + 8}{3} = 4$$

Both data sets have the same mean although the values in the first data set are all the same whereas those in the second set are spread widely. Clearly the mean value does not reflect the variability of the values in a data set. We need additional parameters to describe variability of data. These additional parameters are the **variance** and the **standard deviation**.

3.2 Variance and standard deviation

Both the variance and the standard deviation quantify the variation of a set of data values.

Suppose we have a set of n values: $x_1, x_2, x_3, \ldots, x_n$. The mean of these values is found and labelled as $\bar{x}$. Then the variance is given by

Key point

$$\text{variance} = \frac{\sum_{i=1}^{n} (x_i - \bar{x})^2}{n}$$

Example 3.1
Find the variance of
(a) 4 4 4
(b) 1 3 8

Solution
(a) We have $x_1 = 4$, $x_2 = 4$, $x_3 = 4$. The mean, $\bar{x}$, is calculated to be 4. So

$$x_1 - \bar{x} = 0, \quad x_2 - \bar{x} = 0, \quad x_3 - \bar{x} = 0$$

and so

$$\text{variance} = \frac{\sum_{i=1}^{3} (x_i - \bar{x})^2}{3}$$

$$= \frac{0^2 + 0^2 + 0^2}{3}$$

$$= 0$$

The variance is zero. Since there is no variation in the data values it is not surprising that the variance is zero.

(b) Here we have $x_1 = 1$, $x_2 = 3$, $x_3 = 8$ and $\bar{x} = 4$. Then

$$x_1 - \bar{x} = -3, \quad x_2 - \bar{x} = -1, \quad x_3 - \bar{x} = 4$$

and so

$$\text{variance} = \frac{\sum_{i=1}^{3}(x_i - \bar{x})^2}{3}$$

$$= \frac{(-3)^2 + (-1)^2 + 4^2}{3}$$

$$= \frac{26}{3}$$

$$= 8.67$$

The variance is 8.67.

The **standard deviation** is found from

$$\text{standard deviation} = \sqrt{\text{variance}}$$

Note that the standard deviation has the same units as the given data, and as such is a useful measure of variation.

From Example 3.1 we see that the standard deviation of 4, 4, 4 is $\sqrt{0} = 0$. The standard deviation of 1, 3, 8 is $\sqrt{8.67} = 2.94$.

Example 3.2

Calculate the variance and standard deviation of

$$11 \ 6 \ 9.6 \ 10.2$$

Solution

$$\text{mean} = \frac{11 + 6 + 9.6 + 10.2}{4}$$

$$= 9.2$$

The calculation of the variance is given in Table 3.1.

Table 3.1

x_i	$x_i - \bar{x}$	$(x_i - \bar{x})^2$
11	$11 - 9.2 = 1.8$	$(1.8)^2 = 3.24$
6	$6 - 9.2 = -3.2$	$(-3.2)^2 = 10.24$
9.6	$9.6 - 9.2 = 0.4$	$(0.4)^2 = 0.16$
10.2	$10.2 - 9.2 = 1.0$	$1.0^2 = 1.00$
		$\Sigma(x_i - \bar{x})^2 = 14.64$

$$\text{variance} = \frac{\sum (x_i - \bar{x})^2}{n}$$

$$= \frac{14.64}{4}$$

$$= 3.66$$

$$\text{standard deviation} = \sqrt{\text{variance}}$$

$$= \sqrt{3.66}$$

$$= 1.91$$

Example 3.3

Calculate the variance and standard deviation of

$$0\ 2\ 3\ 6\ 6\ 10$$

Solution

mean =   4.5

Complete the calculation started in Table 3.2.

Table 3.2

x_i	$x_i - \bar{x}$	$(x_i - \bar{x})^2$		
0			−4.5,	20.25
2			−2.5,	6.25
3			−1.5,	2.25
6			1.5,	2.25
6			1.5,	2.25
10			5.5,	30.25

$$\sum (x_i - \bar{x})^2 = \qquad\qquad 63.5$$

$$\text{variance} = \qquad\qquad\qquad \frac{63.5}{6} = 10.58$$

$$\text{standard deviation} = \qquad\qquad \sqrt{10.58} = 3.25$$

When data are presented in the form of a frequency distribution the variance and standard deviation can still be found. Example 3.4 illustrates this.

Example 3.4

The number of hardware faults for each computer in a laboratory containing 30 computers is recorded over a 12-month period. The results, in the form of a frequency distribution, are given in Table 3.3.

Table 3.3

Number of faults, x	Frequency, f
0	11
1	4
2	3
3	7
4	5

So 11 of the computers had no faults during the 12-month period, 4 had one fault, 3 had two faults and so on. Note that the sum of the frequencies gives the total number of computers, that is 30.

Calculate the variance and standard deviation.

Solution

The mean, $\bar{x}$, is found to be 1.7. The remainder of the calculation is set out in Table 3.4.

Table 3.4

x_i	f_i	$x_i - \bar{x}$	$(x_i - \bar{x})^2$	$f_i(x_i - \bar{x})^2$
0	11	−1.7	2.89	31.79
1	4	−0.7	0.49	1.96
2	3	0.3	0.09	0.27
3	7	1.3	1.69	11.83
4	5	2.3	5.29	26.45
	$\sum f_i = 30$			$\sum f_i(x_i - \bar{x})^2 = 72.3$

Then

$$\text{variance} = \frac{\sum f_i (x_i - \bar{x})^2}{n}$$

$$= \frac{72.3}{30}$$

$$= 2.41$$

and

$$\text{standard deviation} = \sqrt{\text{variance}}$$

$$= 1.55$$

Example 3.4 illustrates the formula for calculating the variance of a frequency distribution.

Key point

$$\text{variance} = \frac{\sum f_i (x_i - \bar{x})^2}{n}$$

End of block exercises

1   Find the variance and standard deviation of the following sets of data:
(a) 6 11 10 9 7 8 9
(b) 5.3 7.2 9.1 8.6 5.9 7.3
(c) −6 −6 −5 −1 0 2 1 0 −2
Which set has the greatest variation?

2   Find the variance and standard deviation of the following frequency distribution:

x	f
6	7
7	3
8	2
9	4
10	2

3   The resistances of 50 resistors are measured and the results recorded as follows:

Resistance (Ω)	Frequency
5.0	17
5.5	12
6.0	10
6.5	6
7.0	5

Calculate the standard deviation of the measurements.

4   The standard deviation of the values $x_1, x_2, x_3, \ldots, x_n$ is σ. Calculate the standard deviation of the values $kx_1, kx_2, kx_3, \ldots, kx_n$ where k is a constant.

5   Calculate the variance and standard deviation of the values

$$-6 \; -11 \; 4 \; 0 \; 1 \; -5 \; 6 \; 3$$

Solutions to exercises

1   (a) variance = 2.53, standard deviation = 1.59
(b) 1.81, 1.35 (c) 8.32, 2.88. Set (c) has the greatest variation.

2   variance = 2.14, standard deviation = 1.46

3   variance = 0.44, standard deviation = 0.66

4   $k\sigma$

5   variance = 29.50, standard deviation = 5.43

Elementary probability

4.1 Introduction

When an event is impossible we say the probability of its happening is 0. When an event is certain, we say the probability of its happening is 1. For example, it is impossible to live without oxygen and so the probability of doing this is 0. It is certain that a metal bar will sink when placed in water and so the probability of this happening is 1.

Most events are neither impossible nor certain. They have varying degrees of likelihood. The probability of such events lies between 0 and 1. Events that are likely to happen have probabilities close to 1; events that are unlikely to happen have probabilities close to 0. An event that is as likely to happen as not has a probability of 0.5. For example, the probability of throwing a head with a fair coin is 0.5.

4.2 Experimental and theoretical probabilities

We introduce the notation used in probability theory and calculations.

We let E denote an event and $P(E)$ denote the probability of the event E happening. For example, suppose E is the event 'A 4 is obtained on throwing a die'. The probability of this is $\frac{1}{6}$. We would write this as $P(E) = \frac{1}{6}$.

Since all probabilities lie between 0 and 1 we know that for any event, $E, 0 \leq P(E) \leq 1$.

Key point

For any event, E

$$0 \leq P(E) \leq 1$$

There are two ways in which we ascertain the probability of a particular event: theoretically and experimentally.

To calculate a theoretical probability we need to have knowledge of the event. For example, suppose we toss a fair coin and let H be the event that it lands with the head facing uppermost. Clearly $P(H) = 0.5$. Similarly suppose we roll a fair die and E is the event that a 4 is obtained; then $P(E) = \frac{1}{6}$. These probabilities have been calculated from knowledge of the physical situation. When any experiment is carried out there are usually several possible outcomes, or **events** as we call them. For example, when throwing a coin there are two possible events: the coin lands with the tail uppermost or lands with the head uppermost. Often the possible events have equal probabilities. In the case of throwing a coin, there are two equally likely events. So if

H: the coin lands with the head uppermost
T: the coin lands with the tail uppermost

then clearly

$$P(H) = P(T) = 0.5$$

Example 4.1

A die is rolled. Calculate the probability that a 4 is uppermost.

Solution

When a die is rolled there are six possible events: a 1 is uppermost, a 2 is uppermost and so on. Each of the six events is equally likely to happen and so the probability of each event is $\frac{1}{6}$. In particular, the probability that a 4 is uppermost is $\frac{1}{6}$.

In some circumstances we do not have sufficient information to calculate a theoretical probability. We know that if a coin is unbiased the probability of obtaining a head is $\frac{1}{2}$. However, suppose the coin is biased so that it is more likely to land with the head uppermost than with its tail uppermost. We can experiment by tossing the coin a large number of times and counting the number of heads obtained. Suppose we toss the coin 1000 times and obtain 650 heads. We can estimate the probability of obtaining a head as $\frac{650}{1000} = 0.65$. Such a probability is known as an **experimental probability**. It is accurate only if a large number of experiments have been performed.

Example 4.2

A biased die is thrown 1000 times and a 6 is obtained on 330 occasions. Calculate the probability of obtaining a 6 on a single throw of the die.

Solution

$$\text{probability of throwing a 6} = \frac{330}{1000}$$

$$= 0.33$$

Exercises

1 A company manufactures precision bearings. On an inspection 3 batches out of 500 were rejected. Calculate the probability that a batch is rejected.

2 A pack of 52 cards is shuffled and a single card is drawn. Calculate the probability that it is the queen of hearts.

3 A television manufacturer sold 36000 TV sets of which 297 were returned within 12 months with faults.
   (a) Calculate the probability that a TV set, chosen at random, is returned within 12 months.

(b) A store buys 500 TV sets from the manufacturer. How many can be expected to develop faults within 12 months?

Solutions to exercises

1 0.006

2 $\dfrac{1}{52}$

3 (a) 0.00825 (b) 4.125, that is four sets.

4.3 Compound events

Suppose we roll a fair die and we wish to calculate the probability that the number showing is 3 or higher. To obtain a score of 3 or more, we could throw a 3, 4, 5 or 6: that is, there are four ways of obtaining such a score. When the die is rolled there are six possible outcomes, of which four result in a score of 3 or more. So

$$P(\text{obtaining a score of 3 or more}) = \frac{4}{6}$$
$$= \frac{2}{3}$$

An event such as 'score 3 or higher' is an example of a **compound event**. When all the outcomes of an experiment are equally likely then we can calculate the probability of a compound event, E, using

Key point

$$P(E) = \frac{\text{number of ways that } E \text{ can happen}}{\text{total number of possible outcomes}}$$

Example 4.3
A fair die is rolled. Calculate the probability of obtaining an odd score.

Solution
The chosen event is throwing an odd score: that is, a 1, 3 or 5. Thus there are three ways in which the chosen event can occur out of a total of six equally likely outcomes. So

$$P(\text{odd score}) = \frac{3}{6}$$
$$= \frac{1}{2}$$

Example 4.4
Two fair coins are tossed.
(a) Write down all the possible outcomes.
(b) Calculate the probability of obtaining one or more heads.

Solution
(a) Letting H stand for head and T for tail, the possible outcomes are

$$HH \ HT \ TH \ TT$$

All outcomes are equally likely.

(b)

number of ways of obtaining one or more heads =   3

total number of possible outcomes =   4

probability of obtaining one or more heads = $\dfrac{3}{4}$

Exercises

1   A pack of 52 cards is shuffled and a card is selected. Calculate the probability that the card is
(a) black
(b) red
(c) a club
(d) a jack
(e) a red queen

2   Three fair coins are tossed.
(a) List the eight possible outcomes.
(b) Calculate the probability of obtaining
(i) exactly two heads, (ii) at least two heads, (iii) no heads.

3   A fair die is rolled. Calculate the probability that the number showing is
(a) odd
(b) 2 or more
(c) less than 4

4   Two fair dice are thrown. Calculate the probability that the total is
(a) 6 (b) 8 (c) more than 10

Solutions to exercises

1   (a) $\dfrac{1}{2}$ (b) $\dfrac{1}{2}$ (c) $\dfrac{1}{4}$ (d) $\dfrac{1}{13}$ (e) $\dfrac{1}{26}$

2   (a) HHH, HHT, HTH, HTT, THH, THT, TTH, TTT

(b) (i) $\dfrac{3}{8}$ (ii) $\dfrac{1}{2}$ (iii) $\dfrac{1}{8}$

3   (a) $\dfrac{1}{2}$ (b) $\dfrac{5}{6}$ (c) $\dfrac{1}{2}$

4   (a) $\dfrac{5}{36}$ (b) $\dfrac{5}{36}$ (c) $\dfrac{1}{12}$

4.4   Complementary events

Consider the following situation. A component is tested. Either it works or it does not work, so there are two possible events. When the component is tested one of these events must happen. In addition, each event excludes the other. We say that the two events are **complementary**.

In general, two events are complementary if one of them must happen and, when it does, the other event cannot happen. If A is an event, then the corresponding complementary event is denoted by $\overline{A}$. The sum of probabilities of two complementary events is always one. For example, if the probability that the component works is 0.92, then the probability that it does not work is $1 - 0.92 = 0.08$.

Example 4.5

The events A and B are defined by

A: a score of 5 is rolled with a fair die
B: the box contains more than 10 components

State the complementary events, $\overline{A}$ and $\overline{B}$.

Solution

$\overline{A}$ is the event: a score other than 5 is obtained when the die is rolled.
$\overline{B}$ is the event: the box contains 10 or fewer components.

Exercises

1   State the complement of the following events:
   (a) the component is reliable
   (b) the digit is greater than 7
   (c) the volume is less than or equal to 1 litre
   (d) the machine is not working
   (e) all components have been tested
   (f) at least four components from the batch are unreliable.

Solutions to exercises

1   (a) the component is not reliable
   (b) the digit is 7 or less
   (c) the volume is more than 1 litre
   (d) the machine is working
   (e) some components have not been tested
   (f) three or fewer components from the batch are unreliable.

When a problem involving probabilities is complex, it is sometimes useful to introduce a **tree diagram**. Tree diagrams enable information to be presented in a clear way and aid understanding.

Example 4.6 Reliability Engineering

Machines A and B make components. Of those made by machine A, 95% are reliable; of those made by machine B, 92% are reliable. Machine A makes 70% of the components with machine B making the rest. Calculate the probability that a component picked at random is
(a) made by machine B
(b) made by machine A and is reliable
(c) made by machine B and is unreliable
(d) reliable.

Solution

(a) Since machine A makes 70% of the components then machine B makes the remaining 30%. Hence the probability that a component is made by machine B is 0.3.

To answer (b), (c) and (d) we introduce the tree diagram.

Consider 100 components: 70 are made by machine A and 30 are made by machine B. This is represented as in Figure 4.1.

Consider the 70 components made by machine A: 95% of these are reliable and so 5% are unreliable. Now 95% of 70 = 66.5, 5% of 70 = 3.5, and this information is represented by Figure 4.2.

Similarly, of the 30 components made by machine B, 92% are reliable and so 8% are unreliable. We know that 92% of 30 = 27.6 and 8% of 30 = 2.4. This information is represented by Figure 4.3.

Putting together Figures 4.1, 4.2 and 4.3 we obtain the tree diagram shown in Figure 4.4.

We use Figure 4.4 to answer (b), (c) and (d).

Figure 4.1

Figure 4.2

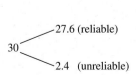

Figure 4.3

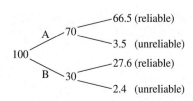

Figure 4.4

(b) We see that there are 66.5 components made by machine A that are reliable from the original 100 components. So

$$P(\text{component is made by A and is reliable}) = \frac{66.5}{100}$$
$$= 0.665$$

(c) We see from Figure 4.4 that there are 2.4 components made by machine B that are unreliable.

$$P(\text{component made by B and is unreliable}) = \frac{2.4}{100}$$
$$= 0.024$$

(d) There are $66.5 + 27.6 = 94.1$ reliable components from the original 100. So

$$P(\text{component is reliable}) = \frac{94.1}{100}$$
$$= 0.941$$

Example 4.7 Reliability Engineering

Machines A, B and C make components. Machine A makes 30% of the components, machine B makes 50% of the components, and machine C makes the remainder. Of those components made by machine A, 93% are reliable, of those made by machine B, 89% are reliable, and of those made by machine C, 96% are reliable.

A component is picked at random. Calculate the probability that it is

(a) made by machine C
(b) made by machine B and is unreliable
(c) made by either machine A or machine B
(d) reliable.

Solution

Consider 100 components. A tree diagram that illustrates the information is shown in Figure 4.5.

Figure 4.5

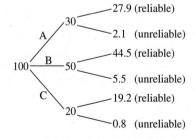

(a) Machine A makes 30% of the components, machine B makes 50% of the components and so machine C makes $(100 - 30 - 50)\% = 20\%$ of the components. So

$$P(\text{component is made by machine C}) = 0.20$$

(b) From Figure 4.5, the number of components made by machine B that are unreliable is

5.5

So

$$P(\text{component is made by B and is unreliable}) = \frac{5.5}{100} = 0.055$$

(c) Out of 100 components, machine A makes 30 and machine B makes 50. So

$$P(\text{component is made by either A or B}) = \frac{30 + 50}{100} = 0.8$$

(d) Out of 100 components there are   reliable ones.   91.6
So

$$P(\text{component is reliable}) = \frac{91.6}{100} = 0.916$$

Exercises

1 Components are made by machines A and B. Machine A makes twice as many components as machine B. When made by machine A, 3% of the components are faulty; when made by machine B, 5% are faulty. Calculate the probability that a component picked at random is
(a) made by machine B
(b) made by machine A and is faulty
(c) made by machine B and is not faulty
(d) faulty.

2 Silicon chips are manufactured by four machines, A, B, C and D. Machines A, B, C and D manufacture 20%, 25%, 35% and 20% of the components respectively. Of those silicon chips manufactured by machine A, 2.1% are faulty. The respective figures for machines B, C and D are 3%, 1.6% and 2.5%.

A silicon chip is selected at random. Calculate the probability that it is
(a) made by machine C and is faulty
(b) made by machine A and is not faulty
(c) faulty.

3 Precision components are made by machines A, B and C. Machines A and C each make 30% of the components with machine B making the rest. The probability that a component is acceptable is 0.91 when made by machine A, 0.95 when made by machine B and 0.88 when made by machine C.
(a) Calculate the probability that a component selected at random is acceptable.
(b) A batch of 2000 components is examined. Calculate the number of components you expect are not acceptable.

Solutions to exercises

1 (a) 0.3333 (b) 0.02 (c) 0.3167 (d) 0.0367

2 (a) 0.0056 (b) 0.1958 (c) 0.0223

3 (a) 0.917 (b) 166

End of block exercises

1 Which of the following numbers cannot represent probabilities?

$$\frac{2}{3}, \frac{1}{11}, -\frac{2}{3}, \frac{3}{7}, \frac{9}{7}, \frac{7}{9},$$
$$0.000, 1.010, 1$$

2 A company manufactures resistors. During a quality control check, 36 out of 2500 resistors failed to perform to the required standard.
  (a) Calculate the probability that a resistor picked at random will fail to perform to the required standard.
  (b) In a batch of 700 resistors how many would you expect to fail?

3 Resistors are manufactured by machines A and B. Machine A makes 60% of the resistors with machine B making the rest. When made by machine A, 3% of the resistors are faulty; when made by machine B, 7% are faulty.
   A resistor is picked at random. Calculate the probability it is
  (a) made by machine B
  (b) made by machine A and is not faulty
  (c) made by machine B and is faulty
  (d) faulty.

4 The probability of throwing a '6' with a fair die is $\frac{1}{6}$. A student makes the following argument.
   If I throw the die once, the probability of obtaining a '6' is $\frac{1}{6}$. So if I throw the die twice, then the probability of throwing a '6' must be $2 \times \frac{1}{6} = \frac{1}{3}$. If I throw the die three times then the probability of throwing a '6' must be $3 \times \frac{1}{6} = \frac{1}{2}$ and so on.
   Is the argument sound? If not, why not?

5 Components are made by machines A, B and C. Machines A and B each make 36% of the components with machine C making the rest. For machine A, 2% of the components made are faulty, for machine B, 6% are faulty, and for machine C, 7% are faulty.
   A component is picked at random. Calculate the probability that it is
  (a) faulty and made by machine C
  (b) not faulty.

6 State the complement of the events:
  (a) the machine is ready
  (b) the assignment was delivered on time
  (c) both printers are working
  (d) at least one person is absent today.

Solutions to exercises

1 $-\frac{2}{3}, \frac{9}{7}, 1.010$

2 (a) 0.0144 (b) 10

3 (a) 0.4 (b) 0.582 (c) 0.028 (d) 0.046

4 The argument is unsound. Using this argument, when the die is thrown seven times, the probability of throwing a '6' is $7 \times \frac{1}{6} = \frac{7}{6}$, which is clearly nonsense.

5 (a) 0.0196 (b) 0.9516

6 (a) the machine is not ready
  (b) the assignment was not delivered on time
  (c) at least one of the printers is not working
  (d) no one is absent today.

Laws of probability

In this block we extend the knowledge of probability gained in Block 4. The concept of **mutually exclusive events** leads naturally to the **addition law of probability**. The probability of an event happening depends upon the conditions that prevail at the time. This gives rise to the idea of **conditional probability**. Finally, once we understand the meaning of **independent events** we can see when to apply the **multiplication law of probability**.

The quality control department of a manufacturing unit examines components and places them into one of three categories:

A: acceptable quality
B: substandard quality
C: reject quality

A component can be placed in only one category. If it is placed in A then this excludes the possibility of its being placed anywhere else. Similarly, if it is placed in B, it cannot possibly belong to A or C. We say the categories are **mutually exclusive**. The events 'The component is placed in category A', 'The component is placed in category B' and 'The component is placed in category C' are **mutually exclusive events**.

After examining many components, the quality control manager calculates the experimental probabilities of a component belonging in category A, B or C as 0.85, 0.10 and 0.05 respectively. Since a component must be placed in one of the three categories, then the three probabilities must add to 1.

If $E_1, E_2, \ldots, E_n$ are n mutually exclusive events then the occurrence of any one of these events excludes the occurrence of all other events.

Example 5.1
A fair die is rolled. Events E_1, E_2, E_3, E_4, E_5 are defined by

E_1: the score is even
E_2: the score is more than 3
E_3: the score is odd
E_4: the score is 1
E_5: the score is 3

Which events are mutually exclusive?

Solution

E_1 and E_3 are mutually exclusive.

E_1, E_4 and E_5 are mutually exclusive.

E_2, E_4 and E_5 are mutually exclusive.

Exercises

1 Two fair coins are tossed. State the events that are mutually exclusive.

E_1: both show heads
E_2: both show tails
E_3: there is one head and one tail
E_4: there is at least one tail

2 A fair die is rolled. The events E_1, E_2, E_3 and E_4 are given by

E_1: the score is less than 2
E_2: the score is divisible by 3
E_3: the score is even
E_4: the score is 5

State which events are mutually exclusive.

Solutions to exercises

1 E_1, E_2 and E_3 are mutually exclusive. E_1 and E_4 are mutually exclusive.

2 E_1, E_2 and E_4 are mutually exclusive. E_1, E_3 and E_4 are also mutually exclusive.

5.3 Addition law of probability

Suppose E_1 and E_2 are mutually exclusive events with probabilities $P(E_1)$ and $P(E_2)$. We consider the event 'E_1 or E_2'.

Key point

The addition law of probability states

$$P(E_1 \text{ or } E_2) = P(E_1) + P(E_2)$$

This idea can be generalised.

Key point

Suppose $E_1, E_2, \ldots, E_n$ are n mutually exclusive events. Then the addition law of probability states

$$P(E_1 \text{ or } E_2 \text{ or} \ldots \text{or } E_n) = P(E_1) + P(E_2) + \cdots + P(E_n)$$

Example 5.2 Reliability Engineering

A machine makes chips that are then classified as one of the following: top quality, standard quality or substandard. An examination of 3000 chips showed 2700 were top quality, 240 were standard quality and 60 were substandard. A chip is selected at random. Calculate the experimental probability that it is

(a) (i) top quality, (ii) standard quality, (iii) substandard
(b) top quality or standard quality
(c) standard quality or substandard

Solution

Let the events E_1, E_2 and E_3 be given by

E_1: the chip selected is top quality
E_2: the chip selected is standard quality
E_3: the chip selected is substandard

These events are mutually exclusive.

(a) (i) probability chip is top quality $= P(E_1) = \dfrac{2700}{3000}$
$$= 0.9$$

(ii) probability chip is standard quality $= P(E_2) = \dfrac{240}{3000}$
$$= 0.08$$

(iii) probability chip is substandard $= P(E_3) = \dfrac{60}{3000}$
$$= 0.02$$

(b) The events are mutually exclusive and so the addition law of probability can be applied.

$$\text{probability chip is top or standard quality } = P(E_1 \text{ or } E_2)$$
$$= P(E_1) + P(E_2)$$
$$= 0.90 + 0.08$$
$$= 0.98$$

(c) probability chip is standard quality or substandard $= P(E_2 \text{ or } E_3)$
$$= P(E_2) + P(E_3)$$
$$= 0.08 + 0.02$$
$$= 0.10$$

Example 5.3 Reliability Engineering

The lifespan, L, of each of 4000 motors was measured and is given in Table 5.1. A motor is randomly selected. Calculate the probability that it had a lifespan of

(a) between 1000 and 2000 hours
(b) between 2000 and 3000 hours
(c) between 1000 and 3000 hours
(d) less than or equal to 4000 hours.

Table 5.1
The lifespan of 4000 motors.

Lifespan of motor (h)	Number
$L \leq 1000$	20
$1000 < L \leq 2000$	75
$2000 < L \leq 3000$	100
$3000 < L \leq 4000$	420
$4000 < L \leq 5000$	585
$L > 5000$	2800

Solution

Let E_1 and E_2 be the events:

E_1: the lifespan is between 1000 and 2000 hours
E_2: the lifespan is between 2000 and 3000 hours

(a) $P(E_1) = \qquad\qquad\qquad \dfrac{75}{4000} = 0.01875$

(b) $P(E_2) = \qquad\qquad\qquad \dfrac{100}{4000} = 0.025$

(c) The events E_1 and E_2 are mutually exclusive and so the addition law of probability can be applied.

$P(1000 < L \leq 3000) = P(E_1 \text{ or } E_2)$

$= \qquad\qquad\qquad$

$= P(E_1) + P(E_2) = 0.04375$

(d) If the lifespan, L, is less than or equal to 4000 hours then L lies between 3000 and 4000 hours, or between 2000 and 3000 hours, or between 1000 and 2000 hours, or is less than 1000 hours. So

$P(L \leq 4000) = P(3000 < L \leq 4000) + P(2000 < L \leq 3000)$
$\qquad\qquad + P(1000 < L \leq 2000) + P(L \leq 1000)$

$= \qquad\qquad\qquad$

$\dfrac{420}{4000} + \dfrac{100}{4000} + \dfrac{75}{4000} + \dfrac{20}{4000}$

$= 0.1538$

Exercises

1 Components are made by machines A, B, C and D. Machine A makes 30% of the components, machine B makes 17%, machine C makes 21% and machine D makes the rest. Calculate the probability that a component is made by
   (a) machine A or machine B
   (b) machine B or machine D
   (c) machine A or machine B or machine D.

2 Components are made by machines A and B. Machine A makes 75% of the components

with machine B making the rest. When made by machine A, 6% of the components are substandard, 2% are standard and the rest are superior. When made by machine B, 3% are substandard, 1% are standard and the rest are superior. A component is picked at random. Calculate the probability it is
   (a) substandard
   (b) standard
   (c) substandard or standard
   (d) standard or superior.

Solutions to exercises

1 (a) 0.47 (b) 0.49 (c) 0.79

2 (a) 0.0525 (b) 0.0175 (c) 0.07 (d) 0.9475

5.4 Conditional probability

Consider components that are manufactured by two machines X and Y. When made by machine X, 93% of the components are acceptable; when made by machine Y, 89% are acceptable. Let the event E be

E: a component is acceptable

If all the components are made by machine X then $P(E) = 0.93$. If all the components are made by machine Y then $P(E) = 0.89$. Clearly the probability of the event depends upon the prevailing conditions. This is intuitive and leads to the idea of **conditional probability**.

Let the events A and B be

A: the component is manufactured by machine X
B: the component is manufactured by machine Y

There is a notation for writing conditional probabilities. The probability that the component is acceptable given it is manufactured by machine X is written $P(E|A)$. Similarly, $P(E|B)$ is the probability that the component is acceptable given it is manufactured by machine Y: that is, the probability of event E happening given that event B has already happened.

In general

Key point

$P(R|S)$ is the probability of event R happening given that event S has already happened.
$P(R|S)$ is the conditional probability of R given S.

Example 5.4 Reliability Engineering

Machines X and Y manufacture components. Machine X makes 65% of the components and machine Y makes the rest. When made by machine X, 95% of the components are acceptable; when made by machine Y, 91% are acceptable. A component is picked at random. Calculate the probability that it is
(a) acceptable, given that it is made by machine Y
(b) acceptable
(c) not acceptable, given that it is made by machine X
(d) made by machine X, given that it is acceptable.

Solution

Consider 100 components. The tree diagram in Figure 5.1 illustrates the given information.

Let the events A, B, E and $\bar{E}$ be defined by

A: the component is made by machine X
B: the component is made by machine Y
E: the component is acceptable
$\bar{E}$: the component is not acceptable

Figure 5.1

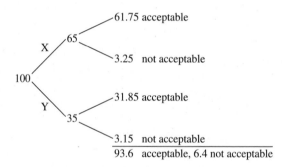

(a) When made by machine Y, 91% of the components are acceptable. So

$$P(E|B) = 0.91$$

(b) From Figure 5.1 we see that there are 93.6 acceptable components from the original 100. So

$$P(E) = 0.936$$

(c) When made by machine X, 95% of the components are acceptable and so 5% are not acceptable. Hence

$$P(\bar{E}|A) = 0.05$$

(d) From Figure 5.1 there are 93.6 acceptable components, of which 61.75 are made by machine X. So

$$P(\text{component is made by machine X given that it is acceptable}) = P(A|E)$$

$$= \frac{61.75}{93.6}$$

$$= 0.6597$$

Example 5.5 Reliability Engineering
A quality control manager records the lifespan, L, of 1000 components of a particular kind. The results are shown in Table 5.2.

Table 5.2
Lifespan of 1000 components (hours).

Lifespan (h)	Number of components
$L < 1000$	69
$1000 \leq L < 2000$	113
$2000 \leq L < 3000$	516
$3000 \leq L < 4000$	201
$L \geq 4000$	101

A component is picked at random. Calculate the probability that it has
(a) a lifespan between 2000 and 3000 hours
(b) a lifespan of 3000 to 4000 hours, given that it is still working after 2000 hours.

Solution
A tree diagram representing the information is shown in Figure 5.2.

Figure 5.2
Tree diagram representing the information of Example 5.5.

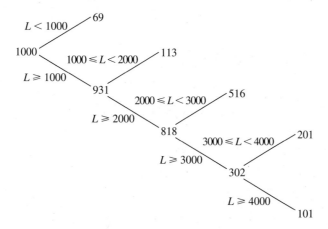

(a) Out of 1000 components, 516 have a lifespan of between 2000 and 3000 hours. So

$$P(2000 \leq L < 3000) = \qquad\qquad \frac{516}{1000} = 0.516$$

(b) There are _____ components still working after 2000 hours.

$$516 + 201 + 101 = 818$$

Of these, _____ have a lifespan of between 3000 and 4000 hours. 201

The probability that the lifespan is between 3000 and 4000 hours, given that the

component is still working after 2000 hours, is $\dfrac{201}{818} = 0.2457$

Exercises

1 Table 5.3 shows the lifespan, L, of 3958 car batteries.

Table 5.3

Lifespan (months)	Number of batteries
$L < 12$	56
$12 \leq L < 24$	219
$24 \leq L < 36$	436
$36 \leq L < 48$	1621
$48 \leq L < 60$	1319
$L \geq 60$	307

Calculate the probability that a battery picked at random
(a) has a lifespan of between 24 and 36 months
(b) has a lifespan of between 12 and 48 months
(c) fails to work more than 48 months
(d) has a lifespan of between 36 and 48 months given it is working after 24 months
(e) has a lifespan greater than 60 months given it is working after 48 months
(f) fails to last more than 48 months given it has lasted 36 months.

2 Resistors are manufactured by machines A, B and C. Machine A manufactures 32% of the production, machine B manufactures 28% of the production and machine C makes the rest. When made by machine A, 3% of the resistors are faulty, when made by machine B, 4.5% are faulty, and when made by machine C, 2.7% are faulty. A resistor is picked at random. Calculate the probability that it is
(a) made by machine C
(b) made by machine A or machine C
(c) faulty

(d) faulty, given it is made by machine B
(e) made by machine B, given it is faulty
(f) made by machine A, given it is not faulty
(g) not made by machine B given it is not faulty.

3 The lifespan, L, of 3000 components is shown in Table 5.4.

Table 5.4

Lifespan (weeks)	Frequency
$0 \leq L < 50$	93
$50 \leq L < 100$	317
$100 \leq L < 150$	1102
$150 \leq L < 200$	1376
$200 \leq L < 250$	112

A component is chosen at random. Calculate the probability that its lifespan is
(a) between 50 and 200 weeks
(b) more than 100 weeks
(c) less than 150 weeks
(d) between 150 and 200 weeks given it is still working after 50 weeks
(e) more than 200 weeks given it is still working after 100 weeks.

4 Components are manufactured by machines A, B, C and D in equal numbers. When made by machine A, 2% of the components are faulty. The figures for machines B, C and D are 3%, 2.5% and 3.5%, respectively. A component is picked at random. Calculate the probability that it is
(a) faulty and made by machine C or faulty and made by machine D
(b) faulty
(c) made by machine A given it is faulty
(d) faulty given it is made by machine C
(e) made by machine B given it is not faulty.

Solutions to exercises

1   (a) 0.1102 (b) 0.5750 (c) 0.5892 (d) 0.4401
    (e) 0.1888 (f) 0.4992

3   (a) 0.9317 (b) 0.8633 (c) 0.504 (d) 0.4733
    (e) 0.0432

2   (a) 0.4 (b) 0.72 (c) 0.033 (d) 0.045
    (e) 0.3818 (f) 0.3210 (g) 0.7235

4   (a) 0.015 (b) 0.0275 (c) 0.1818 (d) 0.025
    (e) 0.2494

5.5   Independent events

Key point

> Two events are independent if the occurrence of either event does not change the probability of the other event occurring.

For independent events we can apply the **multiplication law of probability**.

Key point

> **The multiplication law of probability**
> If E_1 and E_2 are two independent events then
> $$P(E_1 \text{ and } E_2) = P(E_1) \, P(E_2)$$

The multiplication law states that when E_1 and E_2 are independent, then the probability of both events happening is the product of the individual probabilities.

The law can be extended to three or more independent events. For example, if E_1, E_2 and E_3 are independent events then

$$P(E_1 \text{ and } E_2 \text{ and } E_3) = P(E_1) \, P(E_2) \, P(E_3)$$

Example 5.6 Reliability Engineering

A machine manufactures components, 90% of which are not faulty and 10% of which are faulty. Two components are picked at random. Calculate the probability that
(a) both components are not faulty
(b) both components are faulty.

Solution

Let the events E_1 and E_2 be

E_1: the first component is not faulty
E_2: the second component is not faulty

Note that $P(E_1) = P(E_2) = 0.9$. The events are independent.
(a) For both components to be not faulty we require E_1 and E_2 to happen. These events are independent: that is, whether or not E_1 happens does not change the

probability of E_2 happening and vice versa. Hence the multiplication law can be applied.

$$P(\text{both components are not faulty}) = P(E_1 \text{ and } E_2)$$
$$= P(E_1)\, P(E_2)$$
$$= (0.9)(0.9)$$
$$= 0.81$$

(b) The events that are complementary to E_1 and E_2 are

$$\overline{E}_1\text{: the first component is faulty}$$
$$\overline{E}_2\text{: the second component is faulty}$$

Note that $P(\overline{E}_1) = P(\overline{E}_2) = 0.1$.

$$P(\text{both components are faulty}) = P(\overline{E}_1 \text{ and } \overline{E}_2)$$
$$= P(\overline{E}_1)\, P(\overline{E}_2)$$
$$= (0.1)(0.1)$$
$$= 0.01$$

Example 5.7 Reliability Engineering

The probability that a component is faulty is 0.05. Two components are picked at random. Calculate the probability that
(a) both are faulty
(b) both are not faulty
(c) one of the components is not faulty
(d) at least one component is faulty.

Solution
We define events E_1 and E_2 to be

$$E_1\text{: the first component is faulty}$$
$$E_2\text{: the second component is faulty}$$

The events are independent and $P(E_1) = P(E_2) = 0.05$.
   The complementary events, $\overline{E}_1$ and $\overline{E}_2$, are given by

$$\overline{E}_1\text{: the first component is not faulty}$$
$$\overline{E}_2\text{: the second component is not faulty}$$

Then $P(\overline{E}_1) = P(\overline{E}_2) = $ ⬚ $1 - 0.05 = 0.95$

(a) $P(E_1 \text{ and } E_2) = $ ⬚ $(0.05)(0.05) = 0.0025$

(b) $P(\overline{E}_1 \text{ and } \overline{E}_2) = $ ⬚ $(0.95)(0.95) = 0.9025$

(c) If one component is not faulty, then the other component must be faulty. So we have either 'E_1 and $\overline{E}_2$' or '$\overline{E}_1$ and E_2'.

$P(E_1 \text{ and } \overline{E}_2)$ ⬚ $(0.05)(0.95) = 0.0475$

$P(\overline{E}_1 \text{ and } E_2)$ ⬚ $(0.95)(0.05) = 0.0475$

Now the events 'E_1 and $\overline{E}_2$' and '$\overline{E}_1$ and E_2' are mutually exclusive and so the addition law of probability can be used.

$$P(\text{'}E_1 \text{ and } \overline{E}_2\text{' or '}\overline{E}_1 \text{ and } E_2\text{'}) = P(\text{'}E_1 \text{ and } \overline{E}_2\text{'}) + P(\text{'}\overline{E}_1 \text{ and } E_2\text{'})$$
$$= 0.0475 + 0.0475$$
$$= 0.095$$

An alternative way to view this is as follows. We can consider three possible outcomes, X_1, X_2 and X_3, as follows:

X_1: two components are not faulty
X_2: one component is not faulty
X_3: there are no components that are not faulty

The probability of X_1 has been calculated in (b), that is $P(X_1) = 0.9025$.
     The event X_3 may be reworded to

X_3: both components are faulty

and so, by (a), $P(X_3) = 0.0025$.
     Now, since X_1, X_2 and X_3 together cover every possible outcome, we know that

$$P(X_1) + P(X_2) + P(X_3) = 1$$

and so

$$P(X_2) = 1 - P(X_1) - P(X_3)$$
$$= 1 - 0.9025 - 0.0025$$
$$= 0.095$$

(d) 'At least one component is faulty' means that either 'one component is faulty' or 'both components are faulty'. Now

$$P(\text{one component is faulty}) = 0.095$$
$$P(\text{both components are faulty}) = 0.0025$$

and these events are mutually exclusive. So

$$P(\text{at least one component is faulty}) = $$

$$0.095 + 0.0025 = 0.0975$$

Example 5.8
Three coins are tossed. Calculate the probability that all the coins show heads.

Solution
Let the events E_1, E_2 and E_3 be

E_1: first coin shows a head
E_2: second coin shows a head
E_3: third coin shows a head

Clearly, $P(E_1) = P(E_2) = P(E_3) = 0.5$ and all the events are independent. So

$$P(E_1 \text{ and } E_2 \text{ and } E_3) = P(E_1) \, P(E_2) \, P(E_3)$$
$$= (0.5) \, (0.5) \, (0.5)$$
$$= 0.125$$

Exercises

1   A and B are independent events with $P(A) = 0.75$ and $P(B) = 0.9$. The compound event 'A occurs, then A occurs, then B occurs' is denoted by AAB. Other compound events are denoted in a similar way. Calculate the probability of the following compound events occurring:
    (a) AAB (b) BBA (c) $AABB$

2   A fair die is thrown three times. Calculate the probability of obtaining
    (a) three 6s
    (b) three 1s
    (c) two 6s given the first number is a 1
    (d) two 6s given the first number is a 6.

3   Components are made by machines A, B and C. Machine A makes 35% of the components,

    machine B makes 25% and machine C makes the rest. Two components are picked at random. Calculate the probability that
    (a) both are made by machine C
    (b) one is made by machine A and one is made by machine B
    (c) exactly one is made by machine A
    (d) at least one is made by machine B
    (e) both are made by the same machine.

4   A machine makes resistors of which 96% are acceptable and 4% are unacceptable. Three resistors are picked at random. Calculate the probability that
    (a) all are acceptable
    (b) all are unacceptable
    (c) at least one is unacceptable.

Solutions to exercises

1   (a) 0.5063 (b) 0.6075 (c) 0.4556

2   (a) 0.0046 (b) 0.0046 (c) 0.0278
    (d) 0.2778

3   (a) 0.16 (b) 0.175 (c) 0.455 (d) 0.4375
    (e) 0.345

4   (a) 0.8847 (b) 0.000064 (c) 0.1153

End of block exercises

1   The probability that a component is reliable is 0.89. Four components are picked at random. Calculate the probability that
    (a) all are reliable
    (b) all are unreliable
    (c) at least one is unreliable
    (d) at least one is reliable.

2   A machine makes three components, A, B and C. For every 1000 components made, 400 are component A, 250 are component B and the rest are component C. Three components are picked at random. Calculate the probability that
    (a) all are component B
    (b) all are of the same type

(c) all are of a different type

(d) two of the components are of type A.

3 Three machines A, B and C manufacture a component. Machine A makes 30% of the components, machine B makes 35% of the components and machine C makes the rest. Components are classified as either acceptable or faulty: 3% of those made by machine A are faulty, 4% of those made by machine B are faulty, and 3.5% of those made by machine C are faulty.

(a) Find the probability that a component is made either by machine A or by machine C.

(b) Find the probability that it is made by machine B given that it is faulty.

(c) Find the probability that it is not made by machine B given that it is not faulty.

(d) Two components are picked at random. Calculate the probability that exactly one of them is made by machine A.

(e) Three components are picked at random. Calculate the probability that they are made by different machines.

(f) Three components are picked at random. Calculate the probability that they are all acceptable.

4 The lifespans, L, of 807 components were measured and the results recorded in Table 5.5.

Table 5.5
Lifespans of 807 components.

Lifespan (days)	Number
$0 \leq L < 200$	19
$200 \leq L < 300$	86
$300 \leq L < 400$	417
$400 \leq L < 500$	210
$L \geq 500$	75

(a) Calculate the probability that a component has a lifespan of at least 400 days given that it is still working after 200 days.

(b) Calculate the probability of a component failing before reaching 500 days given that it is still working after 300 days.

(c) Calculate the probability of a component failing between 300 and 500 days given that it is still working after 200 days.

5 A box contains 10 red, 12 blue and 8 yellow blocks.

(a) A block is picked at random. Calculate the probability that it is blue.

(b) A block is picked at random, its colour is noted and then it is replaced. This is repeated. Calculate the probability that
(i) both blocks are yellow
(ii) one block is yellow and one is red
(iii) both blocks are the same colour
(iv) the blocks have different colours.

6 A simple circuit is shown in Figure 5.3.

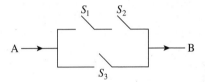

Figure 5.3

S_1, S_2 and S_3 are switches. The probabilities that S_1, S_2 and S_3 are closed are 0.86, 0.91 and 0.70 respectively. The switches open and close independently.

(a) Calculate the probability that S_1 and S_2 are both closed.

(b) Calculate the probability that a current can flow from A to B.

7 Two machines A and B manufacture components, with machine A making 55% and machine B making the rest. Of those made by machine A, 7% are defective; of those made by machine B, 5% are defective.

(a) A component is picked at random. Calculate the probability that it is made by machine A given that it is defective.

(b) Two components are picked at random. Calculate the probability that neither is defective.

(c) Two components are picked at random. Calculate the probability that they are made by different machines.

8 An aeroplane has four engines that all work independently of each other. The probability of an engine failing during a flight is 3×10^{-6}. If the plane can fly on two engines, calculate the probability that the plane is unable to fly.

Solutions to exercises

1   (a) 0.6274 (b) 0.0001 (c) 0.3726 (d) 0.9999

2   (a) 0.0156 (b) 0.1225 (c) 0.21 (d) 0.288

3   (a) 0.65 (b) 0.3972 (c) 0.6517 (d) 0.42
    (e) 0.2205 (f) 0.8979

4   (a) 0.3617 (b) 0.8932 (c) 0.7957

5   (a) 0.4 (b) (i) 0.0711 (ii) 0.1778 (iii) 0.3422
    (iv) 0.6578

6   (a) 0.7826 (b) 0.9348

7   (a) 0.6311 (b) 0.8817 (c) 0.495

8   1.08×10^{-16}

Probability distributions

There are usually several possible outcomes for any given experiment. For example, when a component is tested it may be classified as substandard, satisfactory or first class. Here there are three possible outcomes. The outcome is a discrete variable. As another example, consider measuring the diameter of a piston. It may be 10.101 cm, 10.093 cm, 10.102 cm and so on. Here the outcome is a continuous variable, which can have any value within a specified range.

Probability distributions are used to assign probabilities to the various possible outcomes of an experiment. If the variable is discrete we use a probability distribution; if the variable is continuous we use a modified form of a distribution, called a **probability density function** (p.d.f.).

Key point

The probability distribution of a discrete variable, X, gives the probabilities of all the possible values of X.

Example 6.1

Components produced in a factory are graded as substandard, satisfactory or first class. The percentage of each is 10%, 85% and 5% respectively. Thus the probability distribution is as follows:

Component description	Probability
Substandard	0.10
Satisfactory	0.85
First class	0.05

Example 6.2

A service engineer records the number of calls received in an 8-hour period for 300 similar periods. The results are given as:

Number of calls	Frequency
0	2
1	7
2	26
3	97
4	101
5	53
6	14

Construct a probability distribution for the number of calls.

Solution

Let X be the number of calls received. Note that X is a discrete variable that takes on values 0, 1, 2, . . . , 6. The probability of each different value of X occurring is calculated.

$$P(X = 0) = \frac{2}{300}$$
$$= 0.0067$$

$$P(X = 1) = \frac{7}{300}$$
$$= 0.0233$$

$P(X = 2) = $ ⬜ $\frac{26}{300} = 0.0867$

$P(X = 3) = $ ⬜ $\frac{97}{300} = 0.3233$

$P(X = 4) = $ ⬜ $\frac{101}{300} = 0.3367$

$P(X = 5) = $ ⬜ $\frac{53}{300} = 0.1767$

$P(X = 6) = $ ⬜ $\frac{14}{300} = 0.0467$

The probability distribution is

Number of calls, X	Probability
0	0.0067
1	0.0233
2	0.0867
3	0.3233
4	0.3367
5	0.1767
6	0.0467

A probability distribution may be represented graphically in the form of a **bar chart**. Figure 6.1 is a bar chart that represents the probability distribution of Example 6.2.

Figure 6.1
Bar chart for probability distribution of Example 6.2.

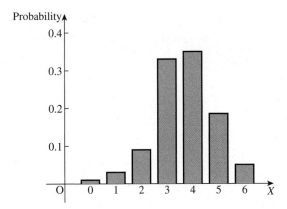

Note that when all the possible values of a variable have been considered, then the sum of the probabilities is 1.

To find the probability of several outcomes, the individual probabilities are added. Example 6.3 illustrates this.

Example 6.3
Using the data from Example 6.2, calculate the probability that the service engineer receives five or six calls in an 8-hour period.

Solution
The individual probabilities are noted: $P(X = 5) = 0.1767$, $P(X = 6) = 0.0467$.

So adding the probabilities we find the probability of receiving five or six calls.

$$P(X = 5 \text{ or } 6) = P(X = 5) + P(X = 6)$$
$$= 0.1767 + 0.0467$$
$$= 0.2234$$

Exercises

1 Explain what is meant by a probability distribution.

2 The number of employees absent each day in a particular firm is recorded over a 200-day period. The results are as follows:

Number absent	Frequency
2	12
3	36
4	81
5	10
6	15
7	46

(a) Calculate the corresponding probability distribution.

(b) Calculate the probability there are five or more absent employees on a day chosen at random.

Solutions to exercises

2 (a)

Number absent	Probability
2	0.06
3	0.18
4	0.405
5	0.05
6	0.075
7	0.23

(b) 0.355.

6.3 Probability density functions

Consider a continuous variable, X, which can take on any value between some prescribed limits. The probability that X lies somewhere between values, say, a and b is found from a **probability density function (p.d.f.)**, $f(x)$.

Key point

The probability that the continuous variable, X, lies between a and b, that is $P(a < X < b)$, is given by

$$P(a < X < b) = \int_a^b f(x)\, \mathrm{d}x$$

where $f(x)$ is the probability density function of X.

Thus the area under the p.d.f. between $x = a$ and $x = b$ gives the probability that the variable X lies between a and b. Figure 6.2 illustrates this.

Figure 6.2

$P(a < X < b) = \int_a^b f(x)\,dx.$

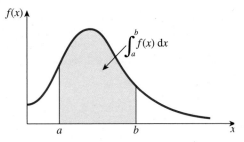

For a function to be a valid p.d.f. it must

1 never be negative
2 have a total area under the curve of 1, representing total probability.

Example 6.4
Given that X is a continuous variable, interpret the following expressions.
(a) $P(15 < X < 20)$
(b) $P(X < 1.7)$
(c) $P(X > 9.6)$
(d) $P(X = 2.6)$

Solution
(a) This represents the probability that X lies between 15 and 20.
(b) This is the probability that X is less than 1.7.
(c) This is the probability that X is greater that 9.6.
(d) $P(X = 2.6)$ is not a valid statement for a continuous variable. Probability statements for continuous variables must always be in terms of intervals. We could calculate $P(2.59 < X < 2.61)$ for example, but not $P(X = 2.6)$.

Example 6.5
A p.d.f. for the continuous variable X is given by

$$f(x) = 2x, \quad 0 < x < 1$$

(a) Check that $f(x)$ is a valid p.d.f.
(b) Calculate $P(0.1 < X < 0.7)$.

Solution
(a) The function, $f(x)$, is defined only on the interval $0 < x < 1$ and on this interval it is never negative. Also

$$\int_0^1 f(x)\,dx = \int_0^1 2x\,dx$$
$$= [x^2]_0^1$$
$$= 1$$

and so the total area under $f(x)$ is 1. Hence $f(x)$ is a valid p.d.f.

(b) $P(0.1 < X < 0.7) = \int_{0.1}^{0.7} 2x\, dx$

$$= [x^2]_{0.1}^{0.7}$$

$$= 0.48$$

The probability that X lies between 0.1 and 0.7 is 0.48.

Example 6.6

A p.d.f., $f(x)$, for a continuous variable X is given by

$$f(x) = \frac{3}{16}(4 - x^2), \quad 0 < x < 2$$

Calculate
(a) $P(0.5 < X < 1.5)$ (b) $P(X < 1)$ (c) $P(X = 1.6)$

Solution

(a) $P(0.5 < X < 1.5) = $

$$\frac{3}{16}\int_{0.5}^{1.5} 4 - x^2\, dx = 0.5469$$

(b) $P(X < 1) = $

$$\frac{3}{16}\int_{0}^{1} 4 - x^2\, dx = 0.6875$$

(c) $P(X = 1.6)$ is not a meaningful statement for a continuous variable.

Exercises

1 Explain what is meant by a probability density function.

2 State two properties that a function must have in order to be a probability density function.

3 A p.d.f., $f(x)$, for a continuous variable X is given by

$$f(x) = \frac{3}{10}(x^2 + 1), \quad 1 < x < 2$$

(a) Verify that $f(x)$ can be a p.d.f.
(b) Find $P(1.7 < X < 2)$.
(c) Find $P(X < 1.5)$.
(d) Find $P(X > 1.25)$.

4 A p.d.f. for a continuous variable X is given by

$$f(x) = e^x, \quad x < 0$$

Calculate
(a) $P(-1 < X < 0)$
(b) $P(-3.5 < X < -3)$
(c) $P(X < -1)$
(d) $P(X > -2)$

5 A p.d.f. for the continuous variable X is given by

$$f(x) = e^{-x}, \quad x > 0$$

If $P(0 < X < \alpha) = 0.5$, find α.

Solutions to exercises

3 (b) 0.3987 (c) 0.3875 (d) 0.8297

4 (a) 0.6321 (b) 0.0196 (c) 0.3679 (d) 0.8647

5 0.6931

End of block exercises

1 A p.d.f., $f(x)$, has the form

$$f(x) = \alpha(x + 1), \quad 0 < x < 1$$

Calculate α.

2 Table 6.1 is a probability distribution.

Table 6.1

x	$P(x)$
6.0	0.13
6.5	0.10
7.0	0.09
7.5	0.02
8.0	0.23
8.5	0.43

Find
(a) $P(x \geq 7.0)$ (b) $P(x > 7.5)$
(c) $P(x \leq 8.0)$ (d) $P(x < 7.5)$
(e) $P(6.5 < x \leq 8.0)$

3 The number of reject components in each of 700 boxes is counted and recorded in Table 6.2.
(a) Construct a frequency distribution from the given data.
(b) Calculate the probability that the number of reject components is between 1 and 4 inclusive.

Table 6.2

Number of rejects per box	Frequency
0	117
1	123
2	236
3	147
4	63
5	14

4 A p.d.f. for the continuous variable X is given by

$$f(x) = \frac{x^2}{21}, \quad 1 < x < 4$$

Calculate
(a) $P(2 < X < 3)$ (b) $P(X > 2.5)$
(c) $P(X < 3.5)$

5 A p.d.f. for the continuous variable X is given by

$$f(x) = 2e^{-2x}, \quad x > 0$$

Calculate
(a) $P(1 < X < 5)$
(b) $P(X > 1)$
(c) $P(X < 2)$
(d) the value of α such that 90% of the X values are less than α.

Solutions to exercises

1 $\frac{2}{3}$

2 (a) 0.77 (b) 0.66 (c) 0.57 (d) 0.32 (e) 0.34

3 (a)

Number of rejects per box	$P(x)$
0	0.1671
1	0.1757
2	0.3371
3	0.2100
4	0.0900
5	0.0200

(b) 0.813

4 (a) 0.3016 (b) 0.7679 (c) 0.6647

5 (a) 0.1353 (b) 0.1353 (c) 0.9871 (d) 1.1513

The binomial distribution

Consider an experiment in which a particular result may or may not occur. For example, a component is tested and it may be acceptable or not be acceptable, a circuit may work or not work. Such outcomes are complementary, and so their probabilities sum to 1. In general, we consider an experiment with possible outcomes of A and its complement $\overline{A}$. We let $P(A) = p$ and $P(\overline{A}) = q$ where clearly $p + q = 1$.

Suppose the experiment is repeated n times. We wish to calculate the probability of A occurring k times and $\overline{A}$ occurring $n - k$ times. For example, in testing a sample of 10 components, a quality control engineer may wish to know the probability of finding eight working components and two non-working components. The calculation of probabilities such as these uses the binomial distribution.

We consider an experiment, or **trial** as it is sometimes known, that is repeated n times. On each occasion the event A may occur with probability p, or its complement, $\overline{A}$, may occur with probability $q = 1 - p$. We assume that p and q are constants. If A occurs k times then $\overline{A}$ must occur $n - k$ times.

We can calculate the probability of no occurrences of A and n occurrences of $\overline{A}$, of one occurrence of A and $n - 1$ occurrences of $\overline{A}$, of two occurrences of A and $n - 2$ occurrences of $\overline{A}$ and so on. Thus we can calculate a probability corresponding to every possible outcome. These probabilities are known as a **binomial distribution**. The number of occurrences of A is known as a **binomial random variable**.

Example 7.1 Reliability Engineering

A machine manufactures components, 90% of which are acceptable. Three components are picked. Calculate the probability that the number of acceptable components is
(a) 0 (b) 1 (c) 2 (d) 3

Solution
We define A and $\overline{A}$ by

A: the component is acceptable

$\overline{A}$: the component is not acceptable

Then clearly, $P(A) = 0.9$ and $P(\overline{A}) = 0.1$.

(a) If there are no acceptable components then all three components are not acceptable. We denote this by $\overline{A}\,\overline{A}\,\overline{A}$. Since the events are independent the multiplication law can be applied.

$$P(\overline{A}\,\overline{A}\,\overline{A}) = P(\overline{A})\,P(\overline{A})\,P(\overline{A})$$
$$= (0.1)(0.1)(0.1)$$
$$= 0.001$$

(b) Here one of the three components is acceptable. We denote by $A\overline{A}\,\overline{A}$ the event: the first component is acceptable, the second is not acceptable, the third is not acceptable. Events $\overline{A}A\overline{A}$ and $\overline{A}\,\overline{A}A$ have obvious meanings.

If one component is acceptable, then $A\overline{A}\,\overline{A}$ or $\overline{A}A\overline{A}$ or $\overline{A}\,\overline{A}A$ occurs. These compound events are mutually exclusive and so the addition law can be applied.

$$P(\text{one component is acceptable}) = P(A\overline{A}\,\overline{A} \text{ or } \overline{A}A\overline{A} \text{ or } \overline{A}\,\overline{A}A)$$
$$= P(A\overline{A}\,\overline{A}) + P(\overline{A}A\overline{A}) + P(\overline{A}\,\overline{A}A)$$
$$= (0.9)(0.1)(0.1) + (0.1)(0.9)(0.1)$$
$$+ (0.1)(0.1)(0.9)$$
$$= 0.027$$

(c) Two of the three components are acceptable. The various possibilities are

$$AA\overline{A}, \quad A\overline{A}A, \quad \overline{A}AA$$

These compound events are mutually exclusive.

$$P(AA\overline{A} \text{ or } A\overline{A}A \text{ or } \overline{A}AA) = P(AA\overline{A}) + P(A\overline{A}A) + P(\overline{A}AA)$$
$$= (0.9)(0.9)(0.1) + (0.9)(0.1)(0.9)$$
$$+ (0.1)(0.9)(0.9)$$
$$= 0.243$$

(d) All three components are acceptable. In this case there is only one possibility, AAA.

$$P(AAA) = (0.9)^3$$
$$= 0.729$$

The results of Example 7.1 are summarised in Table 7.1.

Table 7.1

Number of acceptable components	Probability
0	0.001
1	0.027
2	0.243
3	0.729
	1.000

Since all possibilities have been covered the probabilities sum to 1. Let the number of acceptable components be x. Clearly x is a variable that can have values 0, 1, 2 and 3 with varying likelihood. Table 7.1 shows how the total probability, 1, is distributed between the various possible values of x. It is known as a **probability distribution** of the variable x. Since there are two possible outcomes for each event the distribution is referred to as a binomial distribution.

Exercises

1 The probability that a component is acceptable is 0.95. Four components are picked at random. Calculate the probability that
(a) all four are acceptable
(b) exactly three are acceptable
(c) all four are unacceptable.

2 The probability that a silicon chip works is 0.97. Five such chips are tested. Calculate the probability that
(a) all five chips work
(b) one chip does not work.

Solutions to exercises

1 (a) 0.8145 (b) 0.1715 (c) 6.25×10^{-6}

2 (a) 0.8587 (b) 0.1328

7.3 Combination notation

Before we can make progress and study a general binomial distribution we need to introduce combination notation.

Recall the factorial notation: $2! = 2 \times 1, 3! = 3 \times 2 \times 1, 4! = 4 \times 3 \times 2 \times 1$ and so on. We define the combination notation as follows:

Key point

$$\binom{n}{r} = \frac{n!}{r!(n-r)!}$$

For example, with $n = 5, r = 3$ we have

$$\binom{5}{3} = \frac{5!}{3!(5-3)!}$$

$$= \frac{5!}{3!2!}$$

$$= 10$$

The quantity $\binom{n}{r}$ is the number of ways of selecting r objects from n distinct objects, with the order of selection being unimportant.

Example 7.2
Calculate the number of ways of selecting 3 digits from 10 digits.

Solution

There are $\binom{10}{3}$ ways of making the selection.

$$\binom{10}{3} = \frac{10!}{3!7!}$$

$$= 120$$

Exercises

1   Evaluate

(a) $\binom{10}{5}$ (b) $\binom{8}{7}$ (c) $\binom{9}{9}$

2   Calculate the number of ways of choosing 7 objects from 26 distinct objects.

3   A nuclear power station is to be built. There are 15 possible sites. A team of engineers is commissioned to select a shortlist of three sites. Calculate the number of ways this may be done.

4   Prove that

$$\binom{n}{r} = \binom{n}{n-r}$$

Solutions to exercises

1   (a) 252 (b) 8 (c) 1

2   657800

3   455

7.4 Probability of k occurrences from n trials

We are now able to consider the problem posed in Section 7.1.

An experiment, or trial, has two possible outcomes, A and $\overline{A}$, with respective probabilities of p and q where $p + q = 1$. The trial is repeated n times.

Key point

The binomial distribution

The probability of k occurrences of A and $n - k$ occurrences of $\overline{A}$ in a sequence of n independent trials is

$$\binom{n}{k}p^k q^{n-k}$$

where $P(A) = p$, $P(\overline{A}) = q = 1 - p$.

Example 7.3 Reliability Engineering

The probability that a component is acceptable is 0.97. Ten components are picked at random. Calculate the probability that

(a) nine are acceptable

(b) three are unacceptable.

Solution

Let A and $\overline{A}$ be defined:

$$A\text{: the component is acceptable}$$
$$\overline{A}\text{: the component is unacceptable}$$

Then $P(A) = p = 0.97$, $P(\overline{A}) = q = 1 - 0.97 = 0.03$. Ten components are examined, that is $n = 10$.

(a)

$$P(\text{nine acceptable components}) = \binom{10}{9}p^9q^{10-9}$$

$$= \binom{10}{9}(0.97)^9(0.03)^1$$

$$= 10(0.97)^9(0.03)$$

$$= 0.2281$$

(b) If three components are unacceptable, then seven are acceptable.

$$P(\text{three unacceptable}) = P(\text{seven acceptable}) = \binom{10}{7}p^7q^3$$

$$= 120(0.97)^7(0.03)^3$$

$$= 0.0026$$

Example 7.4 Reliability Engineering

The probability that a component is acceptable is 0.91. Twelve components are selected at random. Calculate the probability that at least 10 are acceptable.

Solution

At least 10 components being acceptable means that 10 or 11 or 12 components are acceptable. These three compound events are mutually exclusive. We calculate the probability of each.

We have $p = \boxed{}$, $q = \boxed{}$, $n = \boxed{}$ 0.91, 0.09, 12

$$P(\text{10 acceptable components}) = \boxed{}\qquad \binom{12}{10}(0.91)^{10}(0.09)^2$$

$$= 0.2082$$

$$P(\text{11 acceptable components}) = \boxed{}\qquad \binom{12}{11}(0.91)^{11}(0.09)$$

$$= 0.3827$$

$$P(12 \text{ acceptable components}) = \binom{12}{12}(0.91)^{12}(0.09)^0$$

$$= 0.3225$$

Since the events are mutually exclusive, the addition law is used.

$$P(10 \text{ or more acceptable components}) = $$

$$0.2082 + 0.3827 + 0.3225$$

$$= 0.9134$$

There is a 91.34% chance that the 12 components contain 10 or more acceptable ones.

Exercises

1 The probability that a car will not develop a major fault within the first 3 years of its life is 0.997. Calculate the probability that of 20 cars selected at random
    (a) 19 will not develop any major faults in the first 3 years
    (b) 19 or more will not develop any major faults in the first 3 years.

2 The probability that a machine has a lifespan of more than 7 years is 0.85. Twelve machines are chosen at random. Calculate the probability that
    (a) 10 have a lifespan of more than 7 years
    (b) 11 have a lifespan of more than 7 years
    (c) 10 or more have a lifespan of more than 7 years.

3 A machine needs all five of its micro-chips to be functional in order to work correctly. The probability that a micro-chip works is 0.99. Calculate the probability that the machine works.

4 The probability that a bearing meets a specification is 0.92. Six bearings are picked at random. Calculate the probability that
    (a) all six meet the specification
    (b) more than four meet the specification
    (c) one or none meets the specification
    (d) exactly four meet the specification.

5 The probability of passing a module on the first attempt is 0.9. A student takes six modules. Calculate the probability that the student
    (a) passes five modules
    (b) passes all modules
    (c) is required to take two or more resits.

Solutions to exercises

1 (a) 0.0567 (b) 0.9984

2 (a) 0.2924 (b) 0.3012 (c) 0.7358

3 0.9510

4 (a) 0.6064 (b) 0.9227 (c) 1.835×10^{-5}
    (d) 6.877×10^{-2}

5 (a) 0.3543 (b) 0.5314 (c) 0.1143

7.5   Mean and variance

Key point

The mean and variance of a binomial random variable, X, are given by

$$\text{mean} = \mu = np$$

$$\text{variance} = \sigma^2 = np(1 - p)$$

where n is the number of trials and p is the probability of an occurrence of X in a single trial.

Example 7.5   Reliability Engineering

The probability that a component is acceptable is 0.96. Fifty components are tested. The number of acceptable components is X. Calculate (a) the mean, (b) the variance of X.

Solution

Here $n = 50$ and $p = 0.96$.
(a) Mean $= np = 50(0.96) = 48$.
(b) Variance $= np(1 - p) = 1.92$.

End of block exercises

1   Evaluate

(a) $\binom{10}{7}$ (b) $\binom{10}{1}$ (c) $\binom{12}{9}$ (d) $\binom{100}{99}$

2   Write a simple expression for

(a) $\binom{n}{1}$ (b) $\binom{n}{n}$ (c) $\binom{n}{n-1}$

3   The probability that a component is acceptable is 0.92. Ten components are selected at random. Calculate the probability that
(a) all 10 are acceptable
(b) eight are acceptable
(c) one is unacceptable
(d) eight or more are acceptable.

4   The probability that a component has a lifespan of more than 6 years is 0.85. Out of 10 components calculate the probability that
(a) nine have a lifespan of more than 6 years
(b) two have a lifespan of less than 6 years.

5   A fair die is thrown eight times. Calculate the probability that
(a) a 6 occurs six times
(b) a 6 never occurs
(c) an odd number of 6s is thrown.

6   A fair coin is tossed five times. Calculate the probability that
(a) five heads are obtained
(b) four heads are obtained
(c) three heads are obtained
(d) two heads are obtained
(e) one head is obtained.

7   A batch of 1500 components is examined and 1411 are found to be acceptable. Five components are picked at random from the production line. Calculate the probability that
(a) all are acceptable
(b) four are acceptable
(c) two are unacceptable.

8   The probability that a component is reliable is 0.92. An engineer wants to collect a number of

components to be at least 95% certain that there are at least seven components that are reliable. Calculate the number of components the engineer should collect.

9 The probability that a component is acceptable is 0.93. An engineer picks 200 components at random. If X is the number of acceptable components calculate (a) the mean, (b) the variance of X.

Solutions to exercises

1 (a) 120 (b) 10 (c) 220 (d) 100

2 (a) n (b) 1 (c) n

3 (a) 0.4344 (b) 0.1478 (c) 0.3777 (d) 0.9599

4 (a) 0.3474 (b) 0.2759

5 (a) 4.1676×10^{-4} (b) 0.2326 (c) 0.4805

6 (a) 0.03125 (b) 0.15625 (c) 0.3125 (d) 0.3125 (e) 0.15625

7 (a) 0.7365 (b) 0.2323 (c) 0.0293

8 9

9 (a) 186 (b) 13.02

The Poisson distribution

8.1 Introduction

We consider a fixed time interval, during which an event may happen 0, 1, 2, 3, . . . times. For example, during a 12-hour period, a fire department may receive 0, 1, 2, 3 or more calls. The department will know from experience that it is likely to receive, say, two or three calls, but on some days it may receive only one, or even none. On other days it will receive four or more calls. It would be useful to the fire department if it knew how likely it was to receive various numbers of calls. The Poisson distribution serves this purpose.

8.2 The Poisson distribution

The Poisson distribution models the number of occurrences of an event in a given time interval. It allows us to calculate the probability of the number of occurrences being 0, 1, 2, 3 and so on. Note that in the Poisson distribution the number of occurrences can be infinite. In contrast, the number of occurrences in a binomial distribution is limited to the number of trials.

The number of occurrences of an event, E, in a given time interval is a discrete variable, which we denote by X. If E does not occur then $X = 0$, if E occurs once then $X = 1$, if E occurs twice then $X = 2$ and so on. If the occurrence of E in any time interval is not affected by its occurrence in any preceding interval, then the probability of X having a particular value can be modelled using the Poisson distribution.

Suppose the number of occurrences, X, in the given time interval is measured many times. The mean of X can then be found; we denote this mean by λ. The probability that X has a value r is given by the Poisson distribution.

Key point

Poisson distribution

$$P(X = r) = \frac{e^{-\lambda} \lambda^r}{r!}, \quad r = 0, 1, 2, 3, \ldots$$

where λ is the mean of the distribution.

Example 8.1

In a particular Poisson distribution $\lambda = 2$. Calculate the probability that (a) $X = 1$, (b) $X = 2$, (c) $X = 3$.

Solution

We are given $\lambda = 2$ and so the distribution is

$$P(X = r) = \frac{e^{-2}2^r}{r!}$$

(a) We evaluate the probability expression with $r = 1$.

$$P(X = 1) = \frac{e^{-2}2^1}{1!}$$
$$= 0.2707$$

(b)
$$P(X = 2) = \frac{e^{-2}2^2}{2!}$$
$$= 0.2707$$

(c)
$$P(X = 3) = \frac{e^{-2}2^3}{3!}$$
$$= 0.1804$$

Example 8.2 Reliability Engineering

A service engineer receives four emergency calls per day on average. Calculate the probability that on a particular day the engineer will receive (a) no calls, (b) one call, (c) two calls, (d) two or fewer calls, (e) more than four calls.

Solution

The average number of calls per day is four, that is $\lambda = 4$. Let X be the number of emergency calls received in 1 day. Then X is a variable with a Poisson distribution given by

$$P(X = r) = \frac{e^{-4}4^r}{r!}$$

(a)
$$P(X = 0) = \frac{e^{-4}4^0}{0!}$$
$$= 0.0183$$

Thus there will be no calls on less than 2 days in 100.

(b)
$$P(X = 1) = \frac{e^{-4}4^1}{1!}$$
$$= 0.0733$$

There will be one call per day in just over 7 days in a 100.

(c)
$$P(X = 2) = \frac{e^{-4}4^2}{2!}$$
$$= 0.1465$$

(d) 'Two or fewer calls' means that the engineer receives two calls, one call or no calls. These events are mutually exclusive and so the addition law can be used.

$$P(X = 0 \text{ or } X = 1 \text{ or } X = 2) = P(X = 0) + P(X = 1) + P(X = 2)$$
$$= 0.0183 + 0.0733 + 0.1465$$
$$= 0.2381$$

(e) Let E be the event

E: the engineer receives more than four calls in a day

Then the complement of E is $\overline{E}$ given by

$\overline{E}$: the engineer receives four or fewer calls in a day

We calculate $P(\overline{E})$.

$$P(\overline{E}) = P(X \leq 4)$$
$$= P(X = 0) + P(X = 1) + P(X = 2) + P(X = 3) + P(X = 4)$$
$$= \frac{e^{-4}4^0}{0!} + \frac{e^{-4}4^1}{1!} + \frac{e^{-4}4^2}{2!} + \frac{e^{-4}4^3}{3!} + \frac{e^{-4}4^4}{4!}$$
$$= 0.6288$$

Since E and $\overline{E}$ are complementary events then

$$P(E) + P(\overline{E}) = 1$$

and so

$$P(E) = 1 - P(\overline{E})$$
$$= 1 - 0.6288$$
$$= 0.3712$$

Example 8.3 Reliability Engineering
A service engineer mends washing machines. In a typical week, five machines will break down. Calculate the probability that in a week
(a) three machines break down
(b) six machines break down
(c) fewer than three machines break down
(d) more than three machines break down.

Solution
On average, five machines break down in a week, that is $\lambda = 5$. Let X be the number of machines that break down in a week. Then X follows a Poisson distribution given by

$$P(X = r) = \boxed{} \qquad\qquad \frac{e^{-5}5^r}{r!}$$

(a) $P(X = 3) = \boxed{} \qquad\qquad \frac{e^{-5}5^3}{3!}$

$$= 0.1404$$

(b) $P(X = 6) =$ ⬚

$$\frac{e^{-5}5^6}{6!}$$

$= 0.1462$

(c) 'Fewer than three breakdowns' means that $X = 0$ or $X = 1$ or $X = 2$. Since these events are mutually exclusive we can use the addition law of probability. Hence

$$P(X < 3) = P(X = 0 \text{ or } X = 1 \text{ or } X = 2)$$

$= $ ⬚

$$P(X = 0) + P(X = 1) + P(X = 2)$$

$= $ ⬚

$$\frac{e^{-5}5^0}{0!} + \frac{e^{-5}5^1}{1!} + \frac{e^{-5}5^2}{2!}$$

$= 0.1247$

(d) Let E be the event

$$E: \text{more than three machines break down}$$

The complementary event, $\overline{E}$, is

$\overline{E} = $ ⬚

three or fewer machines break down

We calculate $P(\overline{E})$.

$$\begin{aligned}
P(\overline{E}) &= P(X \leq 3) \\
&= P(X = 0) + P(X = 1) + P(X = 2) + P(X = 3) \\
&= 0.1247 + P(X = 3) \quad \text{using(c)} \\
&= \text{⬚}
\end{aligned}$$

0.2651

Now

$$P(E) = 1 - P(\overline{E})$$
$$= 0.7349$$

So far we have looked at the probability of a number of occurrences during a time period. Sometimes, however, we consider the probability of a number of occurrences over a length as Example 8.4 illustrates.

Example 8.4 Reliability Engineering

A machine manufactures high-quality pipes. Experience has shown that on average there are three faults per 100 m of pipe. Calculate the probability that in 100 m of pipe there will be (a) no faults, (b) three faults.

Solution
The average number of faults per 100 m is three, that is $\lambda = 3$. Let X be the number of faults per 100 m of pipe. Then X has a Poisson distribution given by

$$P(X = r) = \frac{e^{-3}3^r}{r!}$$

(a)
$$P(X = 0) = \frac{e^{-3}3^0}{0!}$$
$$= 0.0498$$

(b)
$$P(X = 3) = \frac{e^{-3}3^3}{3!}$$
$$= 0.2240$$

Exercises

1   In a computer laboratory, on average three drives malfunction every month. Calculate the probability that in a particular month
(a) two drives malfunction
(b) one drive malfunctions
(c) more than three drives malfunction.

2   A firm has on average seven people absent due to illness per day. Calculate the probability that on a particular day there are
(a) five people absent
(b) eight people absent
(c) more than three people absent.

3   A Poisson distribution is given by

$$P(X = r) = \frac{e^{-3}3^r}{r!}$$

Calculate
(a) $P(X = 2)$ (b) $P(X < 2)$ (c) $P(X \geq 2)$

4   An engineer receives, on average, five emergency calls in an 8-hour period. Calculate the probability that in an 8-hour period the engineer receives
(a) exactly five calls
(b) fewer than three calls
(c) more than four calls.

Solutions to exercises

1   (a) 0.2240 (b) 0.1494 (c) 0.3528

2   (a) 0.1277 (b) 0.1304 (c) 0.9182

3   (a) 0.2240 (b) 0.1991 (c) 0.8009

4   (a) 0.1755 (b) 0.1247 (c) 0.5595

8.3   Poisson approximation to the binomial distribution

The Poisson and binomial distributions are related. Consider a binomial distribution in which n trials take place. Note that n is fixed. Suppose the probability of an event A occurring on a single trial is p. Let X be the number of occurrences of A in n trials. Also consider a Poisson distribution with mean λ.

If n is large and p is small and $\lambda = np$ then the Poisson and binomial distributions are very similar. Table 8.1 lists the probabilities for both distributions where $n = 20$, $p = 0.05$ and $\lambda = 20(0.05) = 1$.

Table 8.1
Probabilities for binomial and Poisson distributions.

	Binomial $P(X = r); n = 20, p = 0.05$	Poisson $P(X = r), \lambda = 1$
$r = 0$	0.3585	0.3679
$r = 1$	0.3774	0.3679
$r = 2$	0.1887	0.1839
$r = 3$	0.0596	0.0613
$r = 4$	0.0133	0.0153
$r = 5$	0.0022	0.0031

Example 8.5 Reliability Engineering

A computer room contains 100 PCs. The probability that a PC breaks down in a week is 0.005. Calculate the probability that in a week
(a) one machine breaks down
(b) no machine breaks down.

Solution

We solve the problem with both the binomial and the Poisson distribution.
Binomial distribution

E: the machine breaks down in a week
n = number of trials = 100
p = probability that E occurs in a single trial = 0.005
X = number of occurrences of E

The binomial distribution is then

$$P(X = r) = \binom{100}{r}(0.005)^r (0.995)^{n-r}$$

(a)
$$P(X = 1) = \binom{100}{1}(0.005)^1 (0.995)^{99}$$
$$= 0.3044$$

(b)
$$P(X = 0) = \binom{100}{0}(0.005)^0 (0.995)^{100}$$
$$= 0.6058$$

Poisson distribution
Strictly speaking, to calculate the probabilities we should use the binomial distribution because the number of occurrences is finite. However, since n, the number of trials, is large, 100, and the probability of breakdown small, 0.005, then a Poisson

distribution can approximate closely to the binomial distribution. The mean of the Poisson distribution is λ where $\lambda = np = 0.5$. The Poisson distribution is then

$$P(X = r) = \frac{e^{-0.5}(0.5)^r}{r!}$$

(a)
$$P(X = 1) = \frac{e^{-0.5}(0.5)^1}{1!}$$
$$= 0.3033$$

(b)
$$P(X = 0) = \frac{e^{-0.5}(0.5)^0}{0!}$$
$$= 0.6065$$

Example 8.6 Reliability Engineering

The probability that a disk drive fails in any month is 0.004. An engineer is responsible for 500 drives. Calculate the probability that in a month the number of drives failing is (a) none, (b) one, (c) more than two.

Use the Poisson approximation to the binomial distribution.

Solution

Number of drives $= n = 500$
Probability of failure $= p = 0.004$

$\lambda = \quad\rule{3em}{1em}$ $\qquad\qquad np = 2$

The Poisson distribution we use is

$$P(X = r) = \quad\rule{3em}{2em} \qquad\qquad \frac{e^{-2}2^r}{r!}$$

(a) $P(X = 0) = \quad\rule{3em}{2em} \qquad\qquad \frac{e^{-2}2^0}{0!}$
$$\qquad\qquad = 0.1353$$

(b) $P(X = 1) = \quad\rule{3em}{2em} \qquad\qquad \frac{e^{-2}2^1}{1!}$
$$\qquad\qquad = 0.2707$$

(c) $P(X > 2) = 1 - [P(X = 0) + P(X = 1) + P(X = 2)]$

$$= \quad\rule{3em}{1em} \qquad\qquad 0.3233$$

Exercises

1 On any day, the probability that a person is absent due to illness is 0.001. In a workforce of 600 people, calculate the probability that on any day the number of people absent is (a) none, (b) one, (c) more than one, (d) less than three.

2 The probability that a component fails within a month is 0.009. If 800 components are examined calculate the probability that the number failing within a month is (a) nine, (b) five, (c) less than three, (d) four or more.

3 A machine manufactures 350 micro-chips per hour. The probability that a chip is faulty is 0.012. Calculate the probability that in a particular hour there are (a) one, (b) three, (c) more than three faulty chips manufactured.

Solutions to exercises

1 (a) 0.5488 (b) 0.3293 (c) 0.1219
(d) 0.9769

2 (a) 0.1070 (b) 0.1204 (c) 0.0255
(d) 0.9281

3 (a) 0.0630 (b) 0.1852 (c) 0.6046

End of block exercises

1 A Poisson distribution is given by

$$P(X = r) = \frac{e^{-4} 4^r}{r!}$$

(a) State the mean value of X.
(b) Calculate the probability that $X = 5$.
(c) Calculate the probability that $X = 2$.
(d) Calculate the probability that $X \geq 2$.

2 In a computer laboratory, a technician replaces, on average, four drives per month. Calculate the probability that in a particular month the technician replaces
(a) four drives
(b) more than four drives
(c) less than four drives.

3 A service engineer receives on average six call-outs in an 8-hour period. Calculate the probability that in an 8-hour period the engineer receives

(a) four call-outs
(b) seven call-outs
(c) between three and five call-outs inclusive.

4 A medium-sized firm has, on average, five people absent on any day. Calculate the probability that on a particular day there are
(a) four people absent
(b) more than four people absent.

5 A Poisson distribution is given by

$$P(X = r) = \frac{e^{-4} 4^r}{r!}$$

Calculate
(a) $P(X < 3)$ (b) $P(X \leq 3)$ (c) $P(X \geq 3)$

6   A workforce comprises 750 people. During any week the probability that a person is absent is 0.01. Use the Poisson approximation to the binomial distribution to calculate the probability that there are
(a) seven absent people
(b) five absent people
(c) more than four absent people.

7   The probability that a component fails within 2 weeks is 0.0075. If 500 components are examined calculate the probability that the number failing within 2 weeks is (a) three, (b) four, (c) less than three.

Solutions to exercises

1   (a) 4 (b) 0.1563 (c) 0.1465 (d) 0.9084

2   (a) 0.1954 (b) 0.3712 (c) 0.4335

3   (a) 0.1339 (b) 0.1377 (c) 0.3837

4   (a) 0.1755 (b) 0.5595

5   (a) 0.2381 (b) 0.4335 (c) 0.7619

6   (a) 0.1465 (b) 0.1094 (c) 0.8679

7   (a) 0.2067 (b) 0.1938 (c) 0.2771

The normal distribution

9.1 Introduction

A variable that can take on any value within a given range is called a **continuous variable**. Examples of continuous variables include the weights of packets of sugar produced in a factory, the diameter of ball bearings and the heights of people. The **normal distribution** allows us to calculate the probability of a continuous variable falling within a particular range of values. Many variables such as weight, volume and density have been found to be modelled by a normal distribution, which is the reason why it has become one of the most important distributions needed by engineers.

In this block we describe the normal distribution and show how it is used to calculate probabilities. An understanding of probability density functions is essential; these were introduced in Block 6 of this chapter.

9.2 The normal probability density function

Suppose a factory produces pistons for a particular model of car. The diameters of the pistons must lie within a given range of values to be acceptable. The quality control manager measures the diameter of several hundred pistons. From these measurements the mean and standard deviation can be calculated.

For a normal distribution we denote the mean by μ and the standard deviation by σ. It is these two values that characterise a normal distribution. Suppose x is a variable, such as length or weight. Then the **normal probability density function**, $N(x)$, is a bell-shaped curve, illustrated in Figure 9.1. Note that $N(x)$ is symmetrical about

Figure 9.1
A typical normal probability density function.

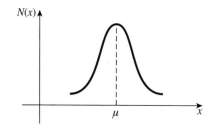

a vertical line placed at $x = \mu$, the mean of the distribution. This means that a variable that is normally distributed is as likely to lie above as below the mean. Furthermore, it is more likely to have a value close to the mean than further from the mean. The symmetry of the normal distribution will be very important when we come to calculate probabilities.

A formula for the normal probability density function $N(x)$ is as follows.

Key point

Normal probability density function

$$N(x) = \frac{1}{\sigma\sqrt{2\pi}} e^{-(x-\mu)^2/2\sigma^2}, \quad -\infty < x < \infty$$

Note that the expression for $N(x)$ involves both μ and σ.

As the values of μ and σ change, the curve, $N(x)$, changes but still retains the bell-shaped feature and the symmetry about $x = \mu$. Figure 9.2 shows two graphs of $N(x)$, both with the same value of σ but with different values of μ.

As μ increases the effect on $N(x)$ is to 'move' it to the right, along the x axis.

Figure 9.3 shows two graphs of $N(x)$, with the same value of μ, but different values of σ.

In Figure 9.3, σ_2 is greater than σ_1. Clearly the spread of the distribution increases as σ increases.

In summary, the value of μ, the mean, determines where the centre of the distribution lies. The value of σ, the standard deviation, determines how far spread out the distribution is.

Figure 9.2
As μ increases, the curve moves to the right.

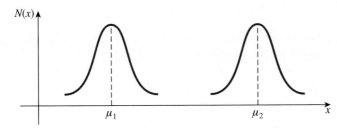

Figure 9.3
As σ increases, the spread of the normal distribution increases.

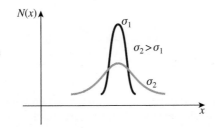

Exercises

1 Figure 9.4 shows two normal distributions, A and B.

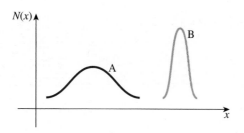

Figure 9.4

(a) Which distribution has the higher mean?
(b) Which distribution has the higher standard deviation?

2 Figure 9.5 shows two normal distributions, A and B.

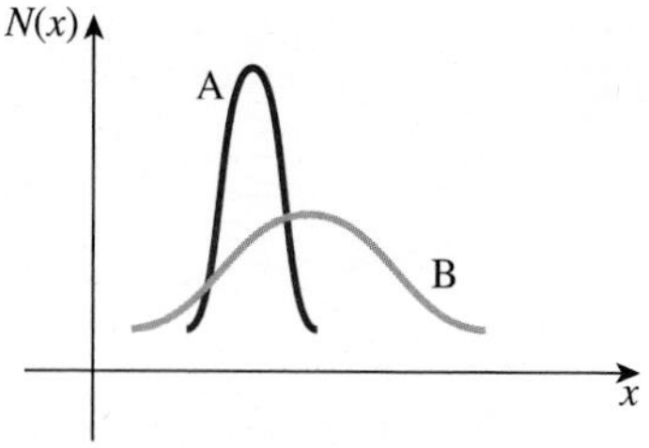

Figure 9.5

(a) Which distribution has the higher mean?
(b) Which distribution has the higher standard deviation?

Solutions to exercises

1 (a) B (b) A

2 (a) B (b) B

9.3 The standard normal distribution

We have seen how a normal distribution, $N(x)$, depends upon the mean, μ, and the standard deviation, σ. We now look at a particular normal distribution with mean 0 and standard deviation 1. Such a distribution is called a **standard normal distribution**. Its probability density function is $N(x) = \dfrac{1}{\sqrt{2\pi}}e^{-x^2/2}$.

Key point

The **standard normal distribution** has a mean of 0 and a standard deviation of 1.

Suppose we have a continuous variable, x, which has a standard normal distribution. Since x is continuous, it can take on any value. We can calculate the probability that x will lie within a given interval, say (a, b).

Key point	If x has a standard normal distribution, $N(x)$, then the probability that x lies within the interval (a, b) is

$$P(a < x < b) = \int_a^b N(x)\, dx$$

The quantity $\int_a^b N(x)\, dx$ is illustrated in Figure 9.6.

Figure 9.6
The shaded area represents the probability that x lies in the interval (a, b).

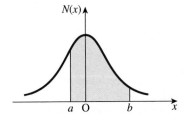

Because $N(x)$ is a probability density function it can be shown that the total area under $N(x)$ is 1, that is

$$\int_{-\infty}^{\infty} N(x)\, dx = 1$$

This represents total probability.

However, we are immediately presented with a problem because $N(x)$ cannot be integrated exactly.

To make numerical calculations easier we introduce a new function $A(z)$ defined by

$$A(z) = P(x < z)$$

Thus $A(z)$ is the probability that x is less than z. Being a probability, $A(z)$ must lie between 0 and 1. In terms of an integral we have

$$A(z) = P(x < z) = \int_{-\infty}^{z} N(x)\, dx$$

The shaded area in Figure 9.7 represents $A(z)$, so it will be helpful to think of $A(z)$ as the area under the graph of $N(x)$ up to the point where $x = z$.

Figure 9.7
The shaded area represents the value of $A(z)$.

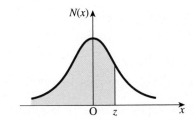

Consider the case where $z = 0$. Then $A(z = 0)$ is illustrated in Figure 9.8.

Recall that $N(x)$ is symmetrical about the mean, which for a standard normal is 0. The total area under the curve is 1. Hence the shaded area in Figure 9.8 is 0.5, that is $A(z = 0) = 0.5$.

Figure 9.8
Owing to
symmetry
$A(z = 0) = 0.5$.

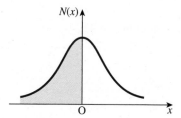

Values of $A(z)$ for other values of z cannot be found by simply inspecting the graph. $A(z)$ must be found by evaluating the integral $\int_{-\infty}^{z} N(x) \, dx$. This is difficult and must be done numerically. Fortunately tables of $A(z)$ are commonly available so that integration is not necessary. Values of $A(z)$ for various values of z are given in Table 9.1 below.

Using Table 9.1 we are able to calculate the probability that x lies within a given interval.

9.4 Calculating probabilities using Table 9.1

Table 9.1 can be used to calculate the probability that x lies within a given interval. Recall that x follows a standard normal distribution: that is, it has a mean of 0 and a standard deviation of 1. The values of $A(z)$ in Table 9.1 give the probability that x is less than z, that is

$$A(z) = P(x < z)$$

Example 9.1
The continuous variable, x, has a standard normal distribution. Calculate the probability that

(a) $x < 0.86$ (b) $x > 0.86$ (c) $x < -0.86$ (d) $x > -0.86$

Solution
(a) This case is illustrated in Figure 9.9.

$$P(x < 0.86) = A(0.86)$$

From Table 9.1, $A(0.86) = 0.8051$. Hence

$$P(x < 0.86) = 0.8051$$

There is a probability of 0.8051 that the variable x is less than 0.86.

Figure 9.9
$P(x < 0.86) = 0.8051.$

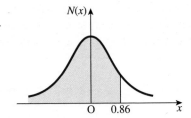

(b) This case is illustrated in Figure 9.10. Since the total area under the curve is 1, we have

$$P(x > 0.86) = 1 - P(x < 0.86)$$
$$= 1 - 0.8051$$
$$= 0.1949$$

Figure 9.10
$P(x > 0.86) = 1 - 0.8051 = 0.1949.$

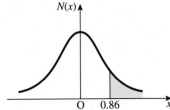

(c) Figure 9.11 illustrates the required area.

Figure 9.11
By symmetry $P(x < -0.86) = P(x > 0.86).$

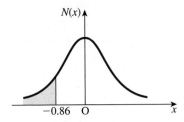

Using the symmetry of the normal distribution we see that

$$P(x < -0.86) = P(x > 0.86)$$
$$= 1 - P(x < 0.86)$$
$$= 0.1949$$

(d) Figure 9.12 illustrates the area.
Again by the symmetry we see that

$$P(x > -0.86) = P(x < 0.86)$$

and so

$$P(x > -0.86) = 0.8051$$

Table 9.1 Cumulative normal probabilities.

z	A(z)	z	A(z)	z	A(z)	z	A(z)	z	A(z)	z	A(z)
0.00	0.5000000	0.30	0.6179114	0.60	0.7257469	0.90	0.8159399	1.20	0.8849303	1.50	0.9331928
0.01	0.5039894	0.31	0.6217195	0.61	0.7290691	0.91	0.8185887	1.21	0.8868606	1.51	0.9344783
0.02	0.5079783	0.32	0.6255158	0.62	0.7323711	0.92	0.8212136	1.22	0.8887676	1.52	0.9357445
0.03	0.5119665	0.33	0.6293000	0.63	0.7356527	0.93	0.8238145	1.23	0.8906514	1.53	0.9369916
0.04	0.5159534	0.34	0.6330717	0.64	0.7389137	0.94	0.8263912	1.24	0.8925123	1.54	0.9382198
0.05	0.5199388	0.35	0.6368307	0.65	0.7421539	0.95	0.8289439	1.25	0.8943502	1.55	0.9394292
0.06	0.5239222	0.36	0.6405764	0.66	0.7453731	0.96	0.8314724	1.26	0.8961653	1.56	0.9406201
0.07	0.5279032	0.37	0.6443088	0.67	0.7485711	0.97	0.8339768	1.27	0.8979577	1.57	0.9417924
0.08	0.5318814	0.38	0.6480273	0.68	0.7517478	0.98	0.8364569	1.28	0.8997274	1.58	0.9429466
0.09	0.5358564	0.39	0.6517317	0.69	0.7549029	0.99	0.8389129	1.29	0.9014747	1.59	0.9440826
0.10	0.5398278	0.40	0.6554217	0.70	0.7580363	1.00	0.8413447	1.30	0.9031995	1.60	0.9452007
0.11	0.5437953	0.41	0.6590970	0.71	0.7611479	1.01	0.8437524	1.31	0.9049021	1.61	0.9463011
0.12	0.5477584	0.42	0.6627573	0.72	0.7642375	1.02	0.8461358	1.32	0.9065825	1.62	0.9473839
0.13	0.5517168	0.43	0.6664022	0.73	0.7673049	1.03	0.8484950	1.33	0.9082409	1.63	0.9484493
0.14	0.5556700	0.44	0.6700314	0.74	0.7703500	1.04	0.8508300	1.34	0.9098773	1.64	0.9494974
0.15	0.5596177	0.45	0.6736448	0.75	0.7733726	1.05	0.8531409	1.35	0.9114920	1.65	0.9505285
0.16	0.5635595	0.46	0.6772419	0.76	0.7763727	1.06	0.8554277	1.36	0.9130850	1.66	0.9515428
0.17	0.5674949	0.47	0.6808225	0.77	0.7793501	1.07	0.8576903	1.37	0.9146565	1.67	0.9525403
0.18	0.5714237	0.48	0.6843863	0.78	0.7823046	1.08	0.8599289	1.38	0.9162067	1.68	0.9535213
0.19	0.5753454	0.49	0.6879331	0.79	0.7852361	1.09	0.8621434	1.39	0.9177356	1.69	0.9544860
0.20	0.5792597	0.50	0.6914625	0.80	0.7881446	1.10	0.8643339	1.40	0.9192433	1.70	0.9554345
0.21	0.5831662	0.51	0.6949743	0.81	0.7910299	1.11	0.8665005	1.41	0.9207302	1.71	0.9563671
0.22	0.5870604	0.52	0.6984682	0.82	0.7938919	1.12	0.8686431	1.42	0.9221962	1.72	0.9572838
0.23	0.5909541	0.53	0.7019440	0.83	0.7967306	1.13	0.8707619	1.43	0.9236415	1.73	0.9581849
0.24	0.5948349	0.54	0.7054015	0.84	0.7995458	1.14	0.8728568	1.44	0.9250663	1.74	0.9590705
0.25	0.5987063	0.55	0.7088403	0.85	0.8023375	1.15	0.8749281	1.45	0.9264707	1.75	0.9599408
0.26	0.6025681	0.56	0.7122603	0.86	0.8051055	1.16	0.8769756	1.46	0.9278550	1.76	0.9607961
0.27	0.6064199	0.57	0.7156612	0.87	0.8078498	1.17	0.8789995	1.47	0.9292191	1.77	0.9616364
0.28	0.6102612	0.58	0.7190427	0.88	0.8105703	1.18	0.8809999	1.48	0.9305634	1.78	0.9624620
0.29	0.6140919	0.59	0.7224047	0.89	0.8132671	1.19	0.8829768	1.49	0.9318879	1.79	0.9632730

z	Prob.	z	Prob.	z	Prob.	z	Prob.	z	Prob.	z	Prob.
1.80	0.9640697	1.95	0.9744119	2.10	0.9821356	2.25	0.9877755	2.40	0.9918025	2.55	0.9946139
1.81	0.9648521	1.96	0.9750021	2.11	0.9825708	2.26	0.9880894	2.41	0.9920237	2.56	0.9947664
1.82	0.9656205	1.97	0.9755808	2.12	0.9829970	2.27	0.9883962	2.42	0.9922397	2.57	0.9949151
1.83	0.9663750	1.98	0.9761482	2.13	0.9834142	2.28	0.9886962	2.43	0.9924506	2.58	0.9950600
1.84	0.9671159	1.99	0.9767045	2.14	0.9838226	2.29	0.9889893	2.44	0.9926564	2.59	0.9952012
1.85	0.9678432	2.00	0.9772499	2.15	0.9842224	2.30	0.9892759	2.45	0.9928572	2.60	0.9953383
1.86	0.9685572	2.01	0.9777844	2.16	0.9846137	2.31	0.9895559	2.46	0.9930531	2.70	0.9965330
1.87	0.9692581	2.02	0.9783083	2.17	0.9849966	2.32	0.9898296	2.47	0.9932443	2.80	0.9974449
1.88	0.9699460	2.03	0.9788217	2.18	0.9853713	2.33	0.9900969	2.48	0.9934309	2.90	0.9981342
1.89	0.9706210	2.04	0.9793248	2.19	0.9857379	2.34	0.9903581	2.49	0.9936128	3.00	0.9986501
1.90	0.9712834	2.05	0.9798178	2.20	0.9860966	2.35	0.9906133	2.50	0.9937903	3.20	0.9993129
1.91	0.9719334	2.06	0.9803007	2.21	0.9864474	2.36	0.9908625	2.51	0.9939634	3.40	0.9996631
1.92	0.9725711	2.07	0.9807738	2.22	0.9867906	2.37	0.9911060	2.52	0.9941323	3.60	0.9998409
1.93	0.9731966	2.08	0.9812372	2.23	0.9871263	2.38	0.9913437	2.53	0.9942966	3.80	0.9999277
1.94	0.9738102	2.09	0.9816911	2.24	0.9874545	2.39	0.9915758	2.54	0.9944574	4.00	0.9999683
										4.50	0.9999966
										5.00	0.9999997
										5.50	0.9999999

Source: This table was originally printed as Table 1 in *Biometrika Tables for Statisticians*, Vol. 1 (1st ed.), edited by E. S. Pearson and H. O. Hartley, 1970. Shown here is the condensed version which was printed in *Statistics: Probability, Inference and Decision*, Vol. 1 & 2 (1st ed.) by Hays, W. L. and Winkler, R. L., 1970. The table is reproduced with kind permission from Oxford University Press and Brooke/Cole, a division of Thomson Learning.

Figure 9.12
$P(x > -0.86) = P(x < 0.86)$.

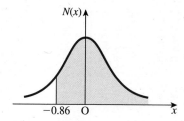

Example 9.2
The variable x has a standard normal distribution. Calculate the probability that
(a) $x < 2.06$ (b) $x < -1.17$ (c) $x > 0.5$ (d) $x > -0.75$

Solution
(a) Figure 9.13 illustrates the area.

Figure 9.13
The shaded area
represents
$P(x < 2.06)$.

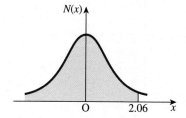

Using Table 9.1 we see

$$P(x < 2.06) = \boxed{}$$

0.9803

(b) Figure 9.14 illustrates the area.

Figure 9.14
The shaded area
represents
$P(x < -1.17)$.

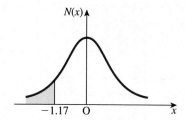

Using the symmetry of the distribution we have

$$P(x < -1.17) = \boxed{}$$

$P(x > 1.17)$

$$= 1 - \boxed{}$$

$P(x < 1.17)$

$$= \boxed{}$$

$1 - 0.8790 = 0.1210$

(c) Figure 9.15 illustrates the area.

Figure 9.15
The shaded area represents $P(x > 0.5)$.

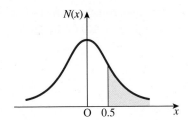

$$P(x > 0.5) = \boxed{} \qquad\qquad \boxed{1 - P(x < 0.5)}$$

$$= \boxed{} \qquad\qquad \boxed{1 - 0.6915 = 0.3085}$$

(d) Figure 9.16 illustrates the required area.

Figure 9.16
The shaded area represents $P(x > -0.75)$.

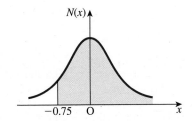

$$P(x > -0.75) = \boxed{} \qquad\qquad \boxed{P(x < 0.75)}$$

$$= \boxed{} \qquad\qquad \boxed{0.7734}$$

Example 9.3

A variable, x, has a normal distribution with a mean of 0 and a standard deviation of 1: that is, it has a standard normal distribution. Calculate
(a) $P(0.6 < x < 1.9)$ (b) $P(-0.86 < x < 0.3)$ (c) $P(-1.5 < x < -0.7)$

Solution

(a) Figure 9.17 illustrates the required area.
Recall that Table 9.1 gives values of $A(z)$ where

$$A(z) = P(x < z)$$

Figure 9.17
The shaded area represents $P(0.6 < x < 1.9)$.

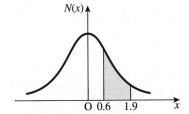

that is, the probability that x is less than some specified value, z. We can view the shaded area in Figure 9.17 as $A(1.9) - A(0.6)$. Figure 9.18 illustrates this.

Figure 9.18
$A(1.9) - A(0.6)$
gives the area
between $x = 0.6$
and $x = 1.9$.

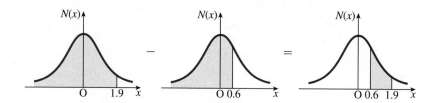

So

$$P(0.6 < x < 1.9) = P(x < 1.9) - P(x < 0.6)$$

$$= A(1.9) - A(0.6)$$

$$= 0.9713 - 0.7257$$

$$= 0.2456$$

There is a probability of 0.2456 that x lies between 0.6 and 1.9.

(b) Using the same reasoning as in (a) we have

$$P(-0.86 < x < 0.3) = P(x < 0.3) - P(x < -0.86)$$

Now

$$P(x < -0.86) = P(x > 0.86)$$

$$= 1 - P(x < 0.86)$$

$$= 0.1949$$

So

$$P(-0.86 < x < 0.3) = P(x < 0.3) - P(x < -0.86)$$

$$= 0.6179 - 0.1949$$

$$= 0.4230$$

(c) $$P(-1.5 < x < -0.7) = P(x < -0.7) - P(x < -1.5)$$

$$= [1 - P(x < 0.7)] - [1 - P(x < 1.5)]$$

$$= P(x < 1.5) - P(x < 0.7)$$

$$= 0.9332 - 0.7580$$

$$= 0.1752$$

Example 9.4

Given that x has a standard normal distribution calculate

(a) $P(1 < x < 2)$ (b) $P(-1.5 < x < 1)$ (c) $P(-1.2 < x < -0.5)$

Solution

(a) $P(1 < x < 2) = P(x < 2) - \boxed{}$	$P(x < 1)$
$= \boxed{} - \boxed{}$	0.9772, 0.8413
$= 0.1359$	
(b) $P(-1.5 < x < 1) = P(x < 1) - P(x < -1.5)$	
$= P(x < 1) - \left(1 - \boxed{}\right)$	$P(x < 1.5)$
$= 0.8413 - \boxed{}$	0.0668
$= 0.7745$	
(c) $P(-1.2 < x < -0.5) = P(x < -0.5) - P(x < -1.2)$	
$= \left(1 - \boxed{}\right) - \left(1 - \boxed{}\right)$	$P(x < 0.5), P(x < 1.2)$
$= \boxed{} - \boxed{}$	0.3085, 0.1151
$= 0.1934$	

Exercises

1 Given that y has a standard normal distribution calculate
(a) $P(y < 1.36)$ (b) $P(y > 1.36)$
(c) $P(y < -0.9)$ (d) $P(y > -0.57)$
(e) $P(0.56 < y < 1.82)$
(f) $P(-0.15 < y < 0)$
(g) $P(-2.1 < y < -1.7)$

2 Given that x has a standard normal distribution calculate
(a) $P(1 < x < 2)$ (b) $P(-1.2 < x < 1.2)$
(c) $P(x < 0)$ (d) $P(-1.75 < x < -0.75)$
(e) $P(x > 0)$ (f) $P(|x| < 1)$
(g) $P(|x| > 1.5)$

3 Given that z has a standard normal distribution calculate the probability that
(a) z lies within two standard deviations of the mean
(b) z lies more than one standard deviation from the mean.

4 The temperature in a fridge follows a normal distribution with a mean of 0° and a standard deviation of 1°. Calculate the probability that the temperature is (a) more than 2°, (b) less than 1°, (c) between −2° and 1°.

Solutions to exercises

1   (a) 0.9131 (b) 0.0869 (c) 0.1841 (d) 0.7157
    (e) 0.2534 (f) 0.0596 (g) 0.0267

2   (a) 0.1359 (b) 0.7699 (c) 0.5 (d) 0.1866
    (e) 0.5 (f) 0.6827 (g) 0.1336

3   (a) 0.9545 (b) 0.3173

4   (a) 0.02275 (b) 0.8413 (c) 0.8186

9.5   Non-standard normal distribution

So far we have calculated probabilities for a standard normal distribution. We now consider non-standard normal distributions. In a non-standard normal distribution, the mean is non-zero and/or the standard deviation is not 1.

Recall that Table 9.1 lists probabilities for a standard normal distribution. In order to use Table 9.1 for problems involving non-standard normal distributions we introduce a way of 'standardising' a variable.

Key point

Suppose a variable, x, has a normal distribution with mean μ and standard deviation σ. We standardise x to Z using the formula

$$Z = \frac{x - \mu}{\sigma}$$

The variable $Z = \dfrac{x - \mu}{\sigma}$ now has a standard normal distribution and so Table 9.1 can be used.

Example 9.5
A variable, x, has a normal distribution with mean 5 and standard deviation 2. Calculate the probability that
(a) $x < 7$ (b) $x > 4$ (c) $2 < x < 6$

Solution
(a) We standardise 7 to

$$Z = \frac{7 - \mu}{\sigma}$$

$$= \frac{7 - 5}{2}$$

$$= 1$$

So $x < 7$ has the same probability as $Z < 7$ where Z follows a standard normal distribution.

$$P(x < 7) = P(Z < 1)$$
$$= 0.8413$$

(b) Applying the standardisation to 4 gives

$$Z = \frac{4 - 5}{2}$$
$$= -0.5$$

So $x > 4$ has the same probability as $Z > -0.5$:

$$P(x > 4) = P(Z > -0.5)$$
$$= P(Z < 0.5)$$
$$= 0.6915$$

(c) Both 2 and 6 are standardised.

$$Z = \frac{2 - 5}{2} = -1.5, \quad Z = \frac{6 - 5}{2} = 0.5$$

So

$$P(2 < x < 6) = P(-1.5 < Z < 0.5)$$
$$= P(Z < 0.5) - P(Z < -1.5)$$
$$= P(Z < 0.5) - [1 - P(Z < 1.5)]$$
$$= P(Z < 0.5) + P(Z < 1.5) - 1$$
$$= 0.6247$$

Example 9.6
The number of miles travelled on 5 litres of petrol by a particular model of car follows a normal distribution with mean 47 and standard deviation 4.3. Calculate the probability that on 5 litres of petrol a car can travel
(a) more than 50 miles
(b) between 45 and 55 miles.

Solution
We let M be the number of miles travelled on 5 litres of petrol. Then M has a normal distribution with mean 47 miles and standard deviation 4.3 miles.
(a) We require the probability that $M > 50$. Using the standardisation we have

$$Z = \boxed{} \qquad\qquad \frac{50 - 47}{4.3} = 0.70$$

So

$$P(M > 50) = P(Z > 0.70) = \boxed{}$$

$$1 - P(Z < 0.70) = 0.2420$$

(b) We standardise 45 and 55.

$$Z = \boxed{}, Z = \boxed{} \qquad\qquad -0.47, 1.86$$

So

$$P(45 < M < 55) = P(-0.47 < Z < 1.86)$$

$$= \qquad P(Z < 1.86) + P(Z < 0.47) - 1$$

$$= 0.6494$$

Exercises

1 A variable, x, has a normal distribution with mean 6 and standard deviation 1.5. Calculate the probability that
   (a) $x < 7.5$ (b) $x < 5$ (c) $x > 7$ (d) $x > 5.5$
   (e) $5 < x < 8$
   (f) x lies within one standard deviation of the mean
   (g) x lies more than two standard deviations from the mean.

2 The temperature, T °C, of a freezer follows a normal distribution with mean -6 °C and standard deviation of 2 °C. Calculate the probability that
   (a) $T > -5$ (b) $T < -7$ (c) $-6 < T < -3$

3 Studies of a particular type of car tyre show that the mileage for which it can be used legally follows a normal distribution with mean 38000 miles and standard deviation 2500 miles. The manufacturers claim that '9 out of 10 of our tyres last more than 35000 miles'. Is the claim justified?

4 The diameters of ball bearings produced in a factory follow a normal distribution with mean 6 mm and standard deviation 0.04 mm. Calculate the probability that a diameter is
   (a) more than 6.05 mm, (b) less than 5.96 mm, (c) between 5.98 and 6.01 mm.

5 A normally distributed variable has mean 2 and standard deviation 2.5. Calculate the probability that the variable is negative.

Solutions to exercises

1 (a) 0.8413 (b) 0.2514 (c) 0.2514 (d) 0.6293
   (e) 0.6568 (f) 0.6829 (g) 0.0455

2 (a) 0.3085 (b) 0.3085 (c) 0.4332

3 Only 88.49% of tyres last more than 35000 miles and so the claim is unjustified.

4 (a) 0.1056 (b) 0.1587 (c) 0.2902

5 0.2119

End of block exercises

1 The variable, z, has a standard normal distribution. Calculate the probability that
   (a) $z < 1.42$ (b) $z < -0.75$ (c) $z > 0.55$
   (d) $z > -1.36$ (e) $0.52 < z < 1.42$
   (f) $-1.38 < z < 0.75$
   (g) $-1.06 < z < 0.08$

2 The variable, x, has a standard normal distribution. Calculate the probability that
   (a) x is within 1.5 standard deviations of the mean
   (b) x is more than 1.75 standard deviations from the mean.

3   The variable, X, has a normal distribution with a mean of 5 and a standard deviation of 3. Calculate the probability that
(a) $X > 10$ (b) $X > 0$ (c) $X < 2$
(d) $X < 1.75$ (e) $1 < X < 9$

4   The lifespan of bulbs has a normal distribution with mean of 10000 hours and a standard deviation of 1200 hours. Calculate
(a) the percentage of bulbs that last more than 11000 hours
(b) the probability that the lifespan of a bulb is between 8000 and 11000 hours
(c) the percentage of bulbs that fail before 8000 hours.

5   The lifespan of car batteries follows a normal distribution with a mean of 5 years and a standard deviation of 10 months. Calculate
(a) the probability that a car battery lasts more than 6 years
(b) the percentage of batteries that fail before they are 3 years old.

6   The scores from IQ tests have a normal distribution with a mean of 100 and a standard deviation of 15.
(a) Calculate the probability that a person has an IQ of more than 120.
(b) Calculate the probability that a person has an IQ of more than 130.
(c) What should a person score in order to be described as being in the top 10% of the population?

7   A factory produces metal bars used in the steering columns of cars. The car manufacturer requires that the bars are 60 ± 0.2 cm long. The lengths of the bars produced in the factory follow a normal distribution with mean 60.2 cm and standard deviation 0.2 cm. What percentage of bars produced by the factory are acceptable to the car manufacturer?

8   A normally distributed variable, Y, has a mean of 30 and a standard deviation of 4. Calculate
(a) $P(Y > 31.6)$ (b) $P(Y < 28.2)$
(c) $P(31 < Y < 35)$ (d) $P(25 < Y < 29)$
(e) $P(Y$ is more than 1.25 standard deviations from the mean).

9   The random variable, x, has a normal distribution. How many standard deviations above the mean must the point P be placed if the tail-end is to represent (a) 10%, (b) 5%, (c) 1% of the total area? (See Figure 9.19.)

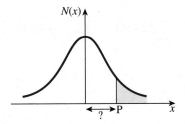

Figure 9.19

10   Consider Figure 9.20.

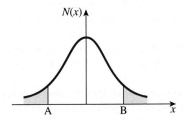

Figure 9.20

The two tail-ends have equal area. How many standard deviations from the mean must A and B be placed if the tail-ends are (a) 10%, (b) 5%, (c) 1% of the total area?

Solutions to exercises

1. (a) 0.9222 (b) 0.2266 (c) 0.2912 (d) 0.9131
   (e) 0.2237 (f) 0.6896 (g) 0.3873

2. (a) 0.8664 (b) 0.0801

3. (a) 0.0475 (b) 0.9525 (c) 0.1587 (d) 0.1401
   (e) 0.8165

4. (a) 20.33% (b) 0.7493 (c) 4.75%

5. (a) 11.51% (b) 0.82%

6. (a) 0.0918 (b) 0.0228 (c) 119

7. 47.72%

8. (a) 0.3446 (b) 0.3264 (c) 0.2956 (d) 0.2956
   (e) 0.2113

9. (a) 1.28 (b) 1.64 (c) 2.33

10. (a) 1.64 (b) 1.96 (c) 2.57

End of chapter exercises

1. Calculate (a) the mean and (b) the standard deviation of the data set

   $$17, 26, 31, 19, 25, 20, 19, 29, 11, 27$$

2. State (a) the median and (b) the mode of the data in question 1.

3. Two identical data points are added to the data set of question 1. The mean of the new data set is 29. Calculate the value of the additional data points.

4. Table 1 shows the results of measuring the petrol consumption of a car over 90 trials.

   Table 1

Miles per gallon	Frequency
42	17
43	18
44	12
45	20
46	23

   (a) Calculate the mean consumption.
   (b) Calculate the standard deviation.

5. State (a) the median and (b) the mode of the distribution of question 4.

6. The data set $A = \{x_1, x_2, x_3, \ldots, x_n\}$ has a mean of $\bar{x}$ and a standard deviation of σ. The data set B is $\{kx_1, kx_2, kx_3, \ldots, kx_n\}$, the data set C is $\{x_1 + k, x_2 + k, x_3 + k, \ldots, x_n + k\}$ where k is a constant.
   (a) State the mean of set B.
   (b) State the mean of set C.
   (c) State the standard deviation of set B.
   (d) State the standard deviation of set C.

7. Out of 6000 components, 39 fail within 12 months of manufacture.
   (a) Calculate the probability that a component picked at random fails within 12 months of manufacture.
   (b) A batch contains 2000 components. How many of these would you expect to fail within 12 months?

8. A fair die is rolled. Calculate the probability that the score is
   (a) 4 (b) 4 or more (c) more than 4 (d) not 4

9. Two fair dice are thrown. Calculate the probability that the total is
   (a) 9 (b) 13 (c) more than 9

10. State the complement of the following statements.
    (a) the score is more than 4
    (b) all of the components are working
    (c) at least two of the machines are not working
    (d) none of the machines are switched on.

11 Components are manufactured by machines A and B. Machine A makes 55% of the components. Of those components made by machine A, 7% are unacceptable; of those made by machine B, 5% are unacceptable. A component is picked at random. Calculate the probability that it is
(a) made by machine B
(b) acceptable
(c) acceptable and made by machine A
(d) unacceptable given it is made by machine B
(e) made by machine A given it is unacceptable.

12 Components are manufactured by three machines, A, B and C. Machine A makes 30% of the components, machine B makes 25% of the components and machine C makes the rest. Of those components made by machine A, 6% are substandard; when made by machine B, 3% are substandard; and when made by machine C, 5% are substandard. A component is picked at random. Calculate the probability that it is
(a) substandard
(b) made by machine B given it is substandard
(c) made by either machine A or machine B
(d) substandard and made by machine B
(e) substandard, given it is made by machine A
(f) made by machine A or is substandard.

13 The lifespans, L, of 2500 components were monitored and recorded in Table 2.

Table 2

Lifespan, L (hours)	Frequency
$0 \leq L \leq 5000$	16
$5000 < L \leq 8000$	132
$8000 < L \leq 9000$	219
$9000 < L \leq 10000$	496
$10000 < L \leq 11000$	1012
$11000 < L \leq 12000$	480
$L > 12000$	145

Calculate the probability that a component picked at random has a lifespan
(a) of between 8000 and 11000 hours
(b) of more than 11000 hours

(c) of more than 10000 hours given that it has already lasted for at least 8000 hours
(d) of between 11000 and 12000 hours given that it has already lasted for at least 9000 hours.

14 The probability that a component is acceptable is 0.91. Ten components are picked at random. Calculate the probability that
(a) eight are acceptable
(b) more than eight are acceptable
(c) three are not acceptable.

15 The probability that a motor will malfunction within 5 years of manufacture is 0.03. Out of eight motors calculate the probability that within 5 years of manufacture
(a) all eight will malfunction
(b) six will malfunction
(c) none will malfunction.

16 The probability that a component works is 0.92. An engineer wants to be at least 99% certain of carrying six working components. Calculate the minimum number of components that the engineer needs to carry.

17 A service engineer receives on average seven calls in a 24-hour period. Calculate the probability that in a 24-hour period the engineer receives
(a) seven calls (b) eight calls (c) six calls
(d) fewer than three calls

18 A firm has 1400 employees. The probability that an employee is absent on any day is 0.006. Use the Poisson approximation to the binomial distribution to calculate the probability that the number of absent employees is
(a) eight (b) nine

19 The lengths of components have a normal distribution, with a mean of 7 cm and a standard deviation of 0.03 cm. Calculate the probability that a component chosen at random has a length
(a) between 6.95 cm and 7.02 cm
(b) more than 7.05 cm
(c) less than 6.96 cm
(d) between 6.95 cm and 6.99 cm.

20 The diameters of bearings have a normal distribution with a mean of 8 mm and a standard deviation of 0.04 mm. In a batch of 6000 bearings how many would you expect to have a diameter of
(a) more than 8.03 mm
(b) less than 7.95 mm
(c) between 8.01 mm and 8.06 mm
(d) more than 2.5 standard deviations from the mean?

Solutions to exercises

1 (a) 22.4 (b) 5.886

2 (a) 22.5 (b) 19

3 62

4 (a) 44.2 (b) 1.475

5 (a) 44 (b) 46

6 (a) $k\bar{x}$ (b) $\bar{x} + k$ (c) $k\sigma$ (d) σ

7 (a) 0.0065 (b) 13

8 (a) $\dfrac{1}{6}$ (b) $\dfrac{1}{2}$ (c) $\dfrac{1}{3}$ (d) $\dfrac{5}{6}$

9 (a) $\dfrac{1}{9}$ (b) 0 (c) $\dfrac{1}{6}$

10 (a) the score is 4 or less
   (b) at least one of the components is not working
   (c) fewer than two of the machines are not working
   (d) some of the machines are switched on.

11 (a) 0.45 (b) 0.939 (c) 0.5115 (d) 0.05 (e) 0.6311

12 (a) 0.048 (b) 0.1563 (c) 0.55 (d) 0.0075 (e) 0.06 (f) 0.33

13 (a) 0.6908 (b) 0.25 (c) 0.6960 (d) 0.2250

14 (a) 0.1714 (b) 0.7746 (c) 0.0452

15 (a) 6.561×10^{-13} (b) 1.921×10^{-8} (c) 0.7837

16 12

17 (a) 0.1490 (b) 0.1304 (c) 0.1490 (d) 0.0296

18 (a) 0.1382 (b) 0.1290

19 (a) 0.7011 (b) 0.0475 (c) 0.0918 (d) 0.3232

20 (a) 1360 (b) 634 (c) 2007 (d) 75

An introduction to Fourier series and the Fourier transform

Chapter **24**

The ability to analyse waveforms of various types is an important engineering skill. Fourier analysis provides a set of mathematical tools that enable the engineer to break down a wave into its various frequency components. It is then possible to predict the effect that a particular waveform may have from knowledge of the effects of its individual frequency components. Often an engineer finds it useful to think of a signal in terms of its frequency components rather than in terms of its time-domain representation. This alternative view is called a frequency-domain representation. It is particularly useful when trying to understand the effect of a filter on a signal. Filters are used extensively in many areas of engineering. In particular, communication engineers use them in signal reception equipment for filtering out unwanted frequencies in the received signal.

When a signal or waveform is periodic, it is possible to represent it as a sum of sine and/or cosine functions of different frequencies and amplitudes. Such a frequency-domain representation is called a **Fourier series**. Calculating a Fourier series is essentially an exercise in integration, in particular the integration of trigonometrical functions. In Block 1 of this chapter you will have the opportunity to calculate several Fourier series. When a signal is not periodic the equivalent frequency-domain representation is found by using a **Fourier transform**. In Block 2, this important transform will be introduced.

This chapter is intended to provide an introduction to these topics rather than give a thorough and rigorous treatment.

Chapter 24 contents

Periodic waveforms and their Fourier representation

Suppose we have a periodic function $f(t)$. Under certain conditions, which will be given at the end of this block, it can be expressed as the sum of an infinite number of sine and/or cosine functions. This infinite sum is known as a **Fourier series**.

For example, Figure 1.1(a) shows a graph of the function $2 \sin t$. Figure 1.1(b) shows a graph of $2 \sin t - \sin 2t$. Notice that we have now included a second sine function that has twice the frequency of the first and a different amplitude. Figures 1.1(c) and 1.1(d) are generated by adding still more sine functions, again with decreasing amplitude and increasing frequency.

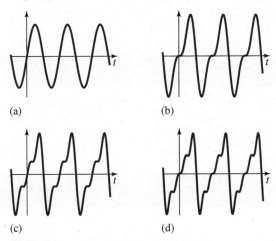

(a)   (b)

(c)   (d)

Figure 1.1

Figures 1.1(c) and 1.1(d) show graphs of $2 \sin t - \sin 2t + \frac{2}{3} \sin 3t$

and $2 \sin t - \sin 2t + \frac{2}{3} \sin 3t - \frac{1}{2} \sin 4t$.

As more and more sine terms are included the graph appears to resemble more and more closely the periodic saw tooth waveform shown in Figure 1.2.

By using an infinite series of sine functions it can be shown that the saw tooth waveform of Figure 1.2 can be represented exactly.

Figure 1.2
A periodic saw
tooth waveform.

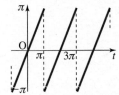

The infinite series of sine functions is the Fourier series of the saw tooth waveform. You will notice that the Fourier series is built up of sine functions of increasing frequencies, and with decreasing amplitudes. In this block you will learn how to calculate the appropriate frequencies and amplitudes for yourself, so that, given any periodic function, you will be able to calculate its corresponding Fourier series.

In the example above it was necessary only to include sine functions to construct the Fourier series, but in general you will need both sine and cosine functions, unless the periodic function under consideration takes rather special forms.

Exercises

1 The Fourier series of the saw tooth waveform can be expressed concisely in the form

$$f(t) = \sum_{n=1}^{\infty} \left(\frac{-2 \cos n\pi}{n} \right) \sin nt$$

Noting that $\cos \pi = -1$, $\cos 2\pi = 1$, $\cos 3\pi = -1$, and so on, use this form to write out the first few terms in the infinite series explicitly.

2 A Fourier series is given by

$$f(t) = \frac{1}{2} + \sum_{n=1}^{\infty} \frac{2}{(2n - 1)\pi} \sin (2n - 1)t$$

Write out the first four terms of this infinite series.

Solutions to exercises

1 $f(t) = 2 \sin t - \sin 2t$
$\qquad + \frac{2}{3} \sin 3t - \frac{1}{2} \sin 4t + \cdots$

2 $\frac{1}{2} + \frac{2}{\pi} \sin t + \frac{2}{3\pi} \sin 3t + \frac{2}{5\pi} \sin 5t + \cdots$

1.2 Calculating a Fourier series

To find a Fourier series for yourself you will make use of several formulae that involve the integration of trigonometrical functions. You will also need the technique of integration by parts. You may wish to revise these topics before proceeding.

Given a periodic function $f(t)$, with period T, calculate the quantities a_0, a_n and b_n from the following formulae. These quantities are known as **Fourier coefficients**:

Key point

Fourier coefficients

$$a_0 = \frac{2}{T} \int_0^T f(t) \, dt$$

$$a_n = \frac{2}{T} \int_0^T f(t) \cos \frac{2n\pi t}{T} \, dt, \quad n = 1, 2, 3, \ldots$$

$$b_n = \frac{2}{T} \int_0^T f(t) \sin \frac{2n\pi t}{T} \, dt, \quad n = 1, 2, 3, \ldots$$

When these quantities have been calculated the Fourier series is given by the following:

Key point

> **Fourier series**
>
> $$f(t) = \frac{a_0}{2} + \sum_{n=1}^{\infty}\left(a_n \cos\frac{2n\pi t}{T} + b_n \sin\frac{2n\pi t}{T}\right)$$

If we write out the first few terms of the infinite series explicitly we find

$$f(t) = \frac{a_0}{2} + a_1 \cos\frac{2\pi t}{T} + b_1 \sin\frac{2\pi t}{T} + a_2 \cos\frac{4\pi t}{T} + b_2 \sin\frac{4\pi t}{T} + \cdots$$

In this form we see that the Fourier coefficients a_n are the amplitudes of the cosine terms in the series, and the b_n are the amplitudes of the sine terms in the series.

These formulae can look quite intimidating when first met, but we shall lead you carefully through the various stages of the calculation in order that you develop confidence in their use. However, you should always sketch a graph of the periodic function before doing anything else.

Example 1.1
Sketch a graph of the periodic function

$$f(t) = \begin{cases} 0 & -5 < t < 0 \\ 1 & 0 < t < 5 \end{cases}$$

of period $T = 10$, from $t = -10$ to $t = 15$.

Solution
Sketch the function for yourself.

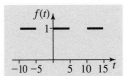

Example 1.2
For the function

$$f(t) = \begin{cases} 0 & -5 < t < 0 \\ 1 & 0 < t < 5 \end{cases}$$

of period $T = 10$, in Example 1.1, evaluate $\frac{2}{T}\int_0^T f(t)\,dt$: that is, calculate the Fourier coefficient a_0.

Solution
First identify the period, T.

$T =$ [ ]   10

We need to find $\frac{2}{10}\int_0^{10} f(t)\,dt$. We are required to integrate $f(t)$ between $t = 0$ and $t = 10$. However, by studying the graph of $f(t)$ you will see that $f(t)$ is zero for t between 5 and 10, and so we only need integrate between 0 and 5, and over this interval the function takes the value 1. Now write down the required integral:

$$a_0 = \qquad\qquad\qquad\qquad\qquad\qquad\qquad \frac{2}{10}\int_0^5 1\,dt$$

Perform the integration to find a_0.

$$\frac{1}{5}\left[t\right]_0^5 = 1$$

We conclude that the Fourier coefficient, a_0, is equal to 1.

Example 1.3

For the function

$$f(t) = \begin{cases} 0 & -5 < t < 0 \\ 1 & 0 < t < 5 \end{cases}$$

of period $T = 10$, in Example 1.1, evaluate $\dfrac{2}{T}\displaystyle\int_0^T f(t)\cos\dfrac{2n\pi t}{T}\,dt$: that is, calculate the Fourier coefficients a_n.

Solution

As in Example 1.2, because $f(t)$ is zero between $t = 5$ and $t = 10$ we need only perform the integration over the interval $t = 0$ to $t = 5$.

Write down the required integral:

$$a_n = \qquad\qquad\qquad\qquad\qquad\qquad \frac{2}{10}\int_0^5 1\cos\frac{2n\pi t}{10}\,dt$$

Perform the integration and complete the calculation:

$$a_n = \qquad\qquad\qquad\qquad\qquad\qquad \frac{1}{5}\left[\frac{\sin\dfrac{n\pi t}{5}}{\dfrac{n\pi}{5}}\right]_0^5$$

If you substitute the limits of integration you will find that all terms are zero since $\sin n\pi = 0$ for all integers n. Hence $a_n = 0$. You have found that all the Fourier coefficients, a_n, are zero. This means that there will be no cosine terms in the Fourier series.

Example 1.4

For the function

$$f(t) = \begin{cases} 0 & -5 < t < 0 \\ 1 & 0 < t < 5 \end{cases}$$

of period $T = 10$, in Example 1.1, evaluate $\dfrac{2}{T}\displaystyle\int_0^T f(t)\sin\dfrac{2n\pi t}{T}\,dt$: that is, calculate the Fourier coefficients b_n.

Solution

As in Examples 1.2 and 1.3, because $f(t)$ is zero between $t = 5$ and $t = 10$ we need only perform the integration over the interval $t = 0$ to $t = 5$.

$$b_n = \boxed{} \qquad \frac{2}{10} \int_0^5 1 \sin \frac{2n\pi t}{10} \, dt$$

Perform the integration:

$$b_n = \boxed{} \qquad \frac{1}{5} \left[-\frac{\cos \dfrac{n\pi t}{5}}{\dfrac{n\pi}{5}} \right]_0^5$$

Insert the limits of integration and so complete the calculation of b_n:

$$\boxed{} \qquad -\frac{1}{n\pi}(\cos n\pi - 1)$$

This result gives us a formula for calculating each Fourier coefficient, b_n. Evaluate the formula when $n = 1$, $n = 2$, and so on:

When $n = 1$, $b_1 = \boxed{} \qquad -\dfrac{1}{\pi}(\cos \pi - 1) = \dfrac{2}{\pi}$

When $n = 2$, $b_2 = \boxed{} \qquad -\dfrac{1}{2\pi}(\cos 2\pi - 1) = 0$

When $n = 3$, $b_3 = \boxed{} \qquad \dfrac{2}{3\pi}$

Continuing in this way you will find $b_4 = 0$, $b_5 = \dfrac{2}{5\pi}$ and so on. These are the Fourier coefficients, b_n.

Example 1.5

Combine the results of Examples 1.2, 1.3 and 1.4 to write down the Fourier series for

$$f(t) = \begin{cases} 0 & -5 < t < 0 \\ 1 & 0 < t < 5 \end{cases}$$

of period 10.

Solution

Using the key point above for the Fourier series,

$$f(t) = \frac{a_0}{2} + \sum_{n=1}^{\infty} \left(a_n \cos \frac{2n\pi t}{T} + b_n \sin \frac{2n\pi t}{T} \right)$$

In this example $a_n = 0$ and $T = 10$ so the series reduces to

$$f(t) = \frac{a_0}{2} + \sum_{n=1}^{\infty} b_n \sin \frac{n\pi t}{5}$$

where the numbers a_0 and b_n are those obtained in Examples 1.2 and 1.4. If we write the series out explicitly we find

$$f(t) = \frac{1}{2} + \frac{2}{\pi} \sin \frac{\pi t}{5} + \frac{2}{3\pi} \sin \frac{3\pi t}{5} + \frac{2}{5\pi} \sin \frac{5\pi t}{5} + \cdots$$

A graph of the first four terms in this series is shown in Figure 1.3. Notice how closely, even with only four terms, it resembles the original graph obtained in Example 1.1.

Figure 1.3
The first four terms in the Fourier series of $f(t)$.

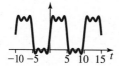

The Fourier series of $f(t)$ has now been found.

Integration over any convenient period

Although the integrals given in the formulae for the Fourier coefficients have limits 0 and T, they can in fact be performed over any complete period. On occasions it may be more convenient to integrate over a different interval, say from $-\frac{T}{2}$ to $\frac{T}{2}$. This is the case in the following example.

Example 1.6
(a) Sketch a graph of the function

$$f(t) = \begin{cases} -t & -\pi < t < 0 \\ 0 & 0 < t < \pi \end{cases}$$

of period 2π, over the interval from -3π to 3π.
(b) Find the Fourier series for $f(t)$.

Solution
(a) First draw the required graph over the interval stated.

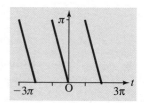

(b) State the value of T:

$$T = 2\pi$$

Because the period is 2π the Fourier series formulae simplify somewhat. For example, terms like $\sin \dfrac{2n\pi t}{T}$ become simply $\sin nt$.

Because the function $f(t)$ is given over the interval from $-\pi$ to π it is convenient to use this complete period in the Fourier coefficient integrals. Noting that $f(t) = 0$ between 0 and π write down the required integrals:

$$a_0 = $$

$$\frac{1}{\pi}\int_{-\pi}^{0} -t\,dt$$

$$a_n = $$

$$\frac{1}{\pi}\int_{-\pi}^{0} -t\cos nt\,dt$$

$$b_n = $$

$$\frac{1}{\pi}\int_{-\pi}^{0} -t\sin nt\,dt$$

Now evaluate a_0:

$$a_0 = $$

$$\frac{1}{\pi}\left[-\frac{t^2}{2}\right]_{-\pi}^{0} = \frac{\pi}{2}$$

To find a_n we note that the integrand is a product and we use integration by parts:

$$a_n = \frac{1}{\pi}\int_{-\pi}^{0} -t\cos nt\,dt = \frac{1}{\pi}\left\{\left[-t\,\frac{\sin nt}{n}\right]_{-\pi}^{0} + \int_{-\pi}^{0}\frac{\sin nt}{n}\,dt\right\}$$

$$= \frac{1}{\pi}\left[-\frac{\cos nt}{n^2}\right]_{-\pi}^{0}$$

$$= \frac{1}{\pi}\left(\frac{\cos n\pi}{n^2} - \frac{1}{n^2}\right)$$

To achieve this result we have made use of the following facts: $\sin 0 = 0$, $\sin(-n\pi) = 0$ for all n, and $\cos(-n\pi) = \cos n\pi$ since the cosine function is even.

Use this result to find a_1, a_2 and a_3 explicitly:

$$a_1 = $$

$$-\frac{2}{\pi}$$

$$a_2 = $$

$$0$$

$$a_3 = $$

$$-\frac{2}{9\pi}$$

To find b_n note that the integrand is a product and use integration by parts in the same way as we did when finding a_n:

$$b_n = $$

$$\frac{1}{\pi}\int_{-\pi}^{0} -t\sin nt\,dt = \frac{1}{\pi}\left\{\left[(-t)\left(-\frac{\cos nt}{n}\right)\right]_{-\pi}^{0} - \int_{-\pi}^{0} -\frac{\cos nt}{n}(-1)\,dt\right\}$$

Simplify this.

$$b_n = \frac{1}{\pi}\left\{\left[\frac{t\cos nt}{n}\right]_{-\pi}^{0} - \int_{-\pi}^{0}\frac{\cos nt}{n}\,dt\right\}$$

The remaining integral is $\displaystyle\int_{-\pi}^{0}\frac{\cos nt}{n}\,dt = \left[\frac{\sin nt}{n^2}\right]_{-\pi}^{0}$, which vanishes.

Complete the calculation:

$$b_n = \frac{1}{\pi}\left\{0 - \frac{-\pi\cos(-n\pi)}{n}\right\} = \frac{1}{n}\cos n\pi$$

Explicitly, $b_1 = -1$, $b_2 = \frac{1}{2}$, $b_3 = -\frac{1}{3}$ and so on.

Putting all these results together we can write the first few terms in the Fourier series as

$$f(t) = \frac{\pi}{4} - \frac{2}{\pi}\cos t - \sin t + \frac{1}{2}\sin 2t - \frac{2}{9\pi}\cos 3t - \frac{1}{3}\sin 3t + \cdots$$

A graph of these first few terms is shown in Figure 1.4.

Figure 1.4

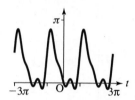

Fourier series of odd functions and even functions

When a function is odd, its Fourier series will contain no cosine or constant terms. Consequently $a_n = 0$ for all n.

When a function is even, its Fourier series will contain no sine terms. Consequently $b_n = 0$ for all n.

These facts will enable you to save considerable effort during the calculation of a Fourier series.

Convergence of the Fourier series

There are certain conditions that must be satisfied before we can be sure that a Fourier series truly represents the function $f(t)$. These conditions, which are called the **Dirichlet conditions**, state that the integral $\int |f(t)|\,dt$ over a complete period must be finite, and $f(t)$ may have no more than a finite number of discontinuities in any finite interval. Fortunately, most functions of interest to engineers satisfy these conditions. At any point where $f(t)$ is continuous, the Fourier series converges to $f(t)$. At any point where $f(t)$ is discontinuous, the Fourier series converges to the average of the two function values at either side of the discontinuity. So, referring to the graph shown in Example 1.1, the Fourier series converges to the value 1 when $x = 2$ say, because $f(t)$ is continuous there and equal to 1, but converges to the value $\frac{1}{2}$ when $x = 5$, the point of discontinuity, and $\frac{1}{2}$ is the average of the function values on either side of $x = 5$.

Exercises

1 (a) Sketch a graph of the function with period 2π defined by

$$f(t) = \begin{cases} 0 & -\pi < t < -\frac{\pi}{2} \\ 4 & -\frac{\pi}{2} \le t \le \frac{\pi}{2} \\ 0 & \frac{\pi}{2} < t < \pi \end{cases}$$

(b) Find the Fourier series for this function.

2 Find the Fourier series representation of the function

$$f(t) = \begin{cases} -4 & -\pi < t \le 0 \\ 4 & 0 < t < \pi \end{cases}$$

of period 2π.

3 Find the Fourier series representation of the function

$$f(t) = \begin{cases} 2(1 + t) & -1 \le t \le 0 \\ 0 & 0 < t < 1 \end{cases}$$

of period 2.

Solutions to exercises

1 (b) $a_0 = 4$, $a_n = \dfrac{8}{n\pi} \sin \dfrac{n\pi}{2}$, $b_n = 0$

$$f(t) = 2 + \frac{8}{\pi} \cos t - \frac{8}{3\pi} \cos 3t$$

$$+ \frac{8}{5\pi} \cos 5t \cdots$$

2 $\dfrac{8}{\pi} \left(2 \sin t + \dfrac{2}{3} \sin 3t + \dfrac{2}{5} \sin 5t + \cdots \right)$

3 $\dfrac{1}{2} + \dfrac{2}{\pi} \left(\dfrac{2}{\pi} \cos \pi t - \sin \pi t - \dfrac{1}{2} \sin 2\pi t \right.$

$$\left. + \frac{2}{9\pi} \cos 3\pi t - \frac{1}{3} \sin 3\pi t \cdots \right)$$

End of block exercises

1 Find the Fourier series representation of the function $f(t) = t^2, 0 < t \le 2\pi$, of period 2π.

2 Find the Fourier series representation of

$$f(t) = \frac{t}{2}, 0 < t < 2\pi, \text{ of period } 2\pi.$$

3 Find the first four non-zero terms in the Fourier series representation of

$$f(t) = \begin{cases} 0 & -\pi < t < -\frac{\pi}{2} \\ 1 & -\frac{\pi}{2} < t < \frac{\pi}{2} \\ 0 & \frac{\pi}{2} < t < \pi \end{cases}$$

of period 2π.

4 Show that the Fourier series representation of $f(x) = x$, $-\pi < x < \pi$, of period 2π, can be written

$$2\sum_{n=1}^{\infty} \frac{(-1)^{n+1}}{n} \sin nx$$

5 (a) Find the Fourier series expansion of

$$f(t) = \begin{cases} 0 & -\pi < t < 0 \\ t^2 & 0 < t < \pi \end{cases}$$

of period 2π.

(b) Use the series to show that

$$\sum_{n=1}^{\infty} \frac{1}{n^2} = \frac{\pi^2}{6}$$

(Hint: let $t = \pi$ and read the comments concerning convergence at the end of this block.)

(c) Use the series to show that

$$\sum_{n=1}^{\infty} (-1)^{n+1} \frac{1}{n^2} = \frac{\pi^2}{12}$$

(Hint: let $t = 0$.)

Solutions to exercises

1 $a_0 = \dfrac{8\pi^2}{3}, a_n = \dfrac{4}{n^2}, b_n = -\dfrac{4\pi}{n}$

2 $a_0 = \pi, a_n = 0, n \geq 1, b_n = \dfrac{-1}{n}$

3 $f(t) = \dfrac{1}{2} + \dfrac{2}{\pi}\left(\cos t - \dfrac{1}{3}\cos 3t\right.$

$$\left. + \dfrac{1}{5}\cos 5t - \cdots \right)$$

5 (a) $a_0 = \dfrac{\pi^2}{3}, a_n = \dfrac{2\cos n\pi}{n^2},$

$$b_n = -\dfrac{\pi}{n}\cos n\pi + \dfrac{2}{\pi n^3}(\cos n\pi - 1)$$

Introducing the Fourier transform

2.1 Introduction

Many practical signals are not periodic. Examples are pulse signals, such as that shown in Figure 2.1, and noise signals. Under certain conditions it can be shown that a non-periodic function $f(t)$ can be expressed not as a sum of sine and cosine waves but as an integral. This idea gives rise to the Fourier transform, which is described in this block.

Figure 2.1
A pulse signal.

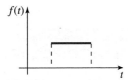

2.2 Calculating a Fourier transform

The Fourier transform of a function $f(t)$ is a function of a new variable ω, which is found from the following formula.

Key point

The Fourier transform of $f(t)$ is a function $F(\omega)$ defined by

$$\mathcal{F}\{f(t)\} = F(\omega) = \int_{-\infty}^{\infty} f(t)e^{-j\omega t}\,dt$$

Some authors use alternative definitions of the Fourier transform, but the one used here is widely accepted. However, when consulting other texts be aware of possible variations.

It is frequently the case that when a Fourier transform is calculated the result is a complex function, as you will see in the examples below.

Example 2.1
Find the Fourier transform of the function defined by

$$f(t) = \begin{cases} 3 & -1 < t < 1 \\ 0 & \text{otherwise} \end{cases}$$

Solution
A graph of $f(t)$ is shown in Figure 2.2.

Figure 2.2
A pulse signal.

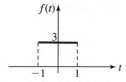

We apply the formula for finding the Fourier transform:

$$\mathcal{F}\{f(t)\} = F(\omega) = \int_{-\infty}^{\infty} f(t)e^{-j\omega t}\, dt$$

Note that in this example $f(t)$ is defined to be zero outside the interval $-1 < t < 1$, and so the integral reduces to

$$\mathcal{F}\{f(t)\} = F(\omega) = \int_{-1}^{1} 3e^{-j\omega t}\, dt$$

$$= \left[\frac{3e^{-j\omega t}}{-j\omega}\right]_{-1}^{1}$$

$$= \frac{3e^{-j\omega}}{-j\omega} - \frac{3e^{j\omega}}{-j\omega}$$

$$= \frac{3e^{j\omega} - 3e^{-j\omega}}{j\omega}$$

This can be written as $F(\omega) = 6\dfrac{\sin \omega}{\omega}$, by making use of Euler's relations (Chapter 11 Block 3). Its graph is shown in Figure 2.3. The function $\dfrac{\sin \omega}{\omega}$ occurs frequently and is often referred to as the **sinc function**.

Figure 2.3
A sinc function.

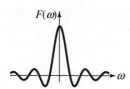

Example 2.2
Find the Fourier transform of the function $f(t) = u(t)e^{-t}$ where $u(t)$ is the unit step function.

Solution
A graph of this function is shown in Figure 2.4. Notice that it is zero when t is negative.

Figure 2.4

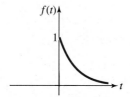

The Fourier transform is

$$F(\omega) = \int_{-\infty}^{\infty} u(t)e^{-t} e^{-j\omega t}\, dt$$

Because $u(t)e^{-t}$ is zero when t is negative, the limits of integration can be altered as follows:

$$F(\omega) = \int_{0}^{\infty} u(t)e^{-t} e^{-j\omega t}\, dt$$

Further, since $u(t) = 1$ when $t \geq 0$ the integral becomes

$$F(\omega) = \int_{0}^{\infty} e^{-t} e^{-j\omega t}\, dt$$

Carry out this integration by combining the two exponential terms into a single term:

$$F(\omega) = \boxed{} \qquad \int_{0}^{\infty} e^{-(1+j\omega)t}\, dt = \left[\frac{e^{-(1+j\omega)t}}{-(1 + j\omega)} \right]_{0}^{\infty}$$

Complete the integration by noting that the contribution from the upper limit is zero, because e^{-t} tends to zero as t tends to infinity.

$$F(\omega) = \boxed{} \qquad\qquad\qquad\qquad \frac{1}{1 + j\omega}$$

This is the required Fourier transform. Unlike the function in Example 2.1 this is complex.

Because the Fourier transform in Example 2.2 is a complex function we cannot immediately plot its graph. However, it is possible to find its modulus and argument, and plot graphs of these against ω. Such plots are called **amplitude spectra** and **phase spectra** respectively.

The amplitude spectrum of $f(t) = u(t)e^{-t}$ is the modulus of $\dfrac{1}{1 + j\omega}$ which equals $\dfrac{1}{\sqrt{1 + \omega^2}}$. A graph of this function is shown in Figure 2.5.

Figure 2.5
The amplitude spectrum of $f(t) = u(t)e^{-t}$.

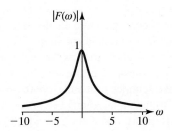

It is usual practice to make use of tables of transforms such as those shown in Table 2.1.

Table 2.1
Table of common Fourier transforms.

$f(t)$	$F(\omega)$
$Au(t)e^{-\alpha t}, \alpha > 0$	$\dfrac{A}{\alpha + j\omega}$
$\begin{cases} 1 & -\alpha \leq t \leq \alpha \\ 0 & \text{otherwise} \end{cases}$	$\dfrac{2\sin\omega\alpha}{\omega}$
A, constant	$2\pi A\delta(\omega)$
$Au(t)$	$A\left(\pi\delta(\omega) - \dfrac{j}{\omega}\right)$
$\delta(t)$	1
$\delta(t - a)$	$e^{-j\omega a}$
$\cos at$	$\pi[\delta(\omega + a) + \delta(\omega - a)]$
$\sin at$	$\dfrac{\pi}{j}[\delta(\omega - a) - \delta(\omega + a)]$
$e^{-\alpha|t|}, \alpha > 0$	$\dfrac{2\alpha}{\alpha^2 + \omega^2}$

Example 2.3 Electronic Engineering – Fourier analysis and digital audio

The mechanism by which sounds can be heard is as follows. A source of sound, for example a singer or a musical instrument, causes the air pressure in its vicinity to rise and fall rapidly. This in turn causes a sound wave to travel through the air until it reaches the ears of a listener, or maybe a microphone. The number of times that the air pressure rises and falls each second is the **frequency** of the sound wave, measured in cycles per second, or hertz.

If a sound wave has a frequency of 440 Hz then a listener will hear a single note, A. If the frequency is higher than this, the pitch of the note will be higher too. If the frequency is lower, then a lower-pitched note will be produced. Musical pieces and songs will be made up from sounds at many different frequencies. Most human beings can hear frequencies only in the range 20 Hz up to 20000 Hz. Sounds with frequencies outside this range will not be heard by humans, although they may be heard by other creatures.

When a sound wave arrives at a microphone it causes a thin diaphragm to vibrate at the same frequency. This motion is converted into an electrical voltage within the microphone. This electrical voltage is a time-domain signal. The voltage will vary with time just as the frequency of the sound varies as the musical pitch is changed during the course of a song or piece of music. To store this continuous voltage signal on a digital computer or CD we take samples of it and store these instead. To obtain a high-quality recording we need to take a large number of samples every second. Then, by taking the Fourier transform we can determine which sound frequencies are being sung or played. A form of Fourier transform especially adapted to deal with sampled signals is known as the **discrete Fourier transform**, or DFT.

For a good introduction to the mathematics underlying the DFT the reader is referred to *Engineering Mathematics: A Foundation for Electronic, Electrical, Communications and Systems Engineers*, 4th edn by A. Croft, R. Davison, M. Hargreaves and J. Flint (Pearson Education).

One of the technical difficulties associated with storing digital audio is the enormous volume of data that has to be stored. If some of these data can be discarded then this problem can be reduced. One way this can be achieved is to ignore any sounds having frequencies greater than 22000 Hz. An important result in digital audio theory is called the **Nyquist sampling theorem**. This states that a signal needs to be sampled at a rate that is twice the highest signal we want to record. Hence if we want to record a signal as high as 22 kHz we need to sample at 44 kHz. It is for this reason that, for CD purposes, an audio signal is sampled at 44100 times each second. We have already shown in Chapter 5, Section 2.8, that with 16-bit technology, and at this sampling rate, we need around 10 megabytes to store a 1-minute stereo recording.

An extremely important modern application of mathematics is to find ways in which these large files can be reduced in size or **compressed**. PC users will know that there are techniques for compressing binary data files. Some of these techniques are known as **loss-less**. This means that none of the original information is lost in the compression process. Audio .wav files are produced by loss-less compression. Unfortunately such files are still enormous. The key to achieving substantial savings in file size is to introduce so-called **lossy** compression. The most widely known lossy compression technique is MP3. This technique works by dividing the frequency range into a large number of different bands. Within each band some of the frequency components are discarded. This is possible because of the way the human ear works. For example, suppose two very close frequencies are being sounded, one played loudly and one played quietly. The quieter sound is masked by the louder one and so its corresponding frequency component can be discarded without a noticeable effect on the sound heard.

To understand compression techniques such as MP3 thoroughly requires not only a knowledge of the physiology of the human ear but also a good understanding of the Fourier transform and related techniques.

Exercises

1 Find, using Table 2.1, the Fourier transform of
$$f(t) = \begin{cases} 1 & -1 \leq t \leq 1 \\ 0 & \text{otherwise} \end{cases}$$

2 Find, by direct integration, the Fourier transform of
$$f(t) = \begin{cases} e^{-t} & t \geq 0 \\ e^{t} & t \leq 0 \end{cases}$$

3 Find, using Table 2.1, the Fourier transform of $f(t) = 4$.

4 Find, using Table 2.1, the Fourier transform of $f(t) = e^{-3t}$ if $t > 0$ and 0 otherwise.

Solutions to exercises

1 $\dfrac{2 \sin \omega}{\omega}$

3 $8\pi\delta(\omega)$

2 $\dfrac{2}{1 + \omega^2}$

4 $\dfrac{1}{3 + j\omega}$

2.3 The frequency-domain representation of a non-periodic function

Under certain conditions a non-periodic function $f(t)$ can be expressed in the following form:

$$f(t) = \frac{1}{2\pi} \int_{-\infty}^{\infty} F(\omega)\, e^{j\omega t}\, d\omega$$

where $F(\omega)$ is the Fourier transform of $f(t)$. Such a representation is called a **Fourier integral representation**.

Example 2.4
Find the Fourier integral representation of the function $f(t)$ given in Example 2.1.

Solution
In Example 2.1 the pulse function was shown to have Fourier transform $F(\omega) = \dfrac{6 \sin \omega}{\omega}$. Hence the Fourier integral representation of the pulse is

$$f(t) = \frac{1}{2\pi} \int_{-\infty}^{\infty} \frac{6 \sin \omega}{\omega}\, e^{j\omega t}\, d\omega$$

2.4 Linearity of the Fourier transform

The Fourier transform possesses a property known as **linearity**, which can be used to find transforms of a wider range of functions than those already discussed.

Key point

The Fourier transform is a linear transform. This means that if k is a constant:

$$\mathcal{F}\{f(t) + g(t)\} = \mathcal{F}\{f(t)\} + \mathcal{F}\{g(t)\}$$

$$\mathcal{F}\{kf(t)\} = k\mathcal{F}\{f(t)\}$$

Example 2.5
Use the linearity properties and the table of transforms to find the Fourier transform of $u(t)e^{-t} + u(t)e^{-2t}$.

Solution
The transforms of $u(t)e^{-t}$ and $u(t)e^{-2t}$ can be obtained separately from the table.

$$\mathcal{F}\{u(t)e^{-t}\} = \boxed{} \qquad\qquad \frac{1}{1 + j\omega}$$

$$\mathcal{F}\{u(t)e^{-2t}\} = \boxed{} \qquad\qquad \frac{1}{2 + j\omega}$$

Now write down the required Fourier transform:

$$\boxed{} \qquad\qquad \frac{1}{1 + j\omega} + \frac{1}{2 + j\omega}$$

Exercise

1   Find the Fourier transform of
$3u(t) - 7u(t)e^{-3t}$.

Solution to exercise

1   $3\left(\pi\delta(\omega) - \dfrac{j}{\omega}\right) - \dfrac{7}{3 + j\omega}$.

End of block exercises

1   Find the Fourier transform of $6u(t)e^{-3t}$.

2   Find the Fourier transform of $f(t) = 5$ for $-2 \le t \le 2$, and 0 otherwise.

3   Find the function $f(t)$ that has Fourier transform

$$F(\omega) = \frac{3}{2 + j\omega} - \frac{7}{1 + j\omega}$$

4   The **first shift theorem** states that if $F(\omega)$ is the Fourier transform of $f(t)$ then

$$\mathcal{F}\{e^{jat} f(t)\} = F(\omega - a)$$

where a is a constant. Use this theorem to find the Fourier transform of

$$g(t) = \begin{cases} e^{2jt} & -3 \le t \le 3 \\ 0 & \text{otherwise} \end{cases}$$

5   The **second shift theorem** states that if $F(\omega)$ is the Fourier transform of $f(t)$ then

$$\mathcal{F}\{f(t - a)\} = e^{-ja\omega} F(\omega)$$

(a) Prove this theorem from the definition of the Fourier transform.
(b) Given that

$$\mathcal{F}\{u(t)e^{-t}\} = \frac{1}{1 + j\omega}$$

use the second shift theorem to find

$$\mathcal{F}\{u(t + 4)e^{-(t+4)}\}$$

(c) Verify the result in (b) by direct integration.

6   Find the Fourier transform of $f(t) = 7$ for $-3 \le t \le 3$, and 0 otherwise.

Solutions to exercises

1   $\dfrac{6}{3 + j\omega}$

2   $\dfrac{10 \sin 2\omega}{\omega}$

3   $3u(t)e^{-2t} - 7u(t)e^{-t}$

4   $\dfrac{2 \sin 3(\omega - 2)}{\omega - 2}$

5   (b) $\dfrac{e^{4j\omega}}{1 + j\omega}$

6   $\dfrac{14 \sin 3\omega}{\omega}$

End of chapter exercises

1   (a) Sketch a graph of the function $f(t) = e^{-|t|}$.
    (b) Show from the definition of the Fourier transform that

$$\mathcal{F}\{f(t)\} = F(\omega) = \frac{2}{1 + \omega^2}$$

    (c) Show that the Fourier transform of $f(t)$ is an even function of ω.

2   Sketch a graph of $f(t) = e^{-|3t|}$ and find its Fourier transform.

3   (a) Sketch a graph of

$$f(t) = \begin{cases} e^{-2t} & t > 0 \\ -e^{2t} & t < 0 \end{cases}$$

    (b) Find the Fourier transform of $f(t)$.
    (c) Show that this Fourier transform is purely imaginary.

4   (a) Sketch a graph of $u(-t)$ where $u(t)$ is the unit step function.
    (b) Sketch a graph of $f(t) = 7e^{2t}u(-t)$.
    (c) Find the Fourier transform of $f(t)$.

5   (a) Sketch a graph of three cycles of the function with period 2π given by

$$f(t) = 1 - \frac{|t|}{\pi}, \quad -\pi \le t < \pi$$

    (b) Find its Fourier series representation.

6   Find the Fourier series representation of the function with period 2 given by

$$f(t) = \begin{cases} 3t & 0 < t < 1 \\ 3 & 1 < t < 2 \end{cases}$$

7   Find the Fourier series representation of the function with period 0.02 defined by

$$f(t) = \begin{cases} 1 & 0 \le t < 0.01 \\ 0 & 0.01 \le t < 0.02 \end{cases}$$

8   Find the Fourier series representation of the function with period 1 given by

$$f(t) = \begin{cases} t & 0 < t < \frac{1}{2} \\ 0 & \frac{1}{2} < t < 1 \end{cases}$$

9   From the definition of the Fourier transform find $F(\omega)$ when $f(t) = u(t)te^{-3t}$.

Solutions to exercises

2 $\dfrac{6}{9 + \omega^2}$

3 (b) $\dfrac{-2j\omega}{4 + \omega^2}$

4 (c) $\dfrac{7}{2 - j\omega}$

5 (b) $b_n = 0, a_0 = 1, a_n = \dfrac{4}{n^2\pi^2}$ if n is odd and zero if n is even.

6 $a_0 = \dfrac{9}{2}, a_n = \dfrac{3}{n^2\pi^2}(\cos n\pi - 1), b_n = -\dfrac{3}{n\pi}$

7 $a_0 = 1, a_n = 0,$ for $n \geq 1, b_n = \dfrac{1}{n\pi}[1 - (-1)^n]$

8 $a_0 = \dfrac{1}{4}, a_n = \dfrac{1}{2\pi^2 n^2}[(-1)^n - 1], b_n = \dfrac{(-1)^{n+1}}{2\pi n}$

9 $\dfrac{1}{(3 + j\omega)^2}$

Typical examination
papers

This section contains two typical examination papers, with questions
drawn from a wide range of topics within the book.

Full solutions are available on the Companion Website at
www.pearsoned.co.uk/croft.

Paper 1

1 (a) Express as partial fractions

$$\frac{x - 8}{2x^2 + 3x - 2}$$

(b) Solve the equation

$$2v^2 - v + 6 = 0$$

(c) Simplify the following expression as much as possible:

$$\sin 2A(\tan A + \cot A)$$

2 (a) Consider the function

$$f(x) = \frac{2x}{x^2 - x - 2}$$

   (i) State any poles of f.
   (ii) Sketch f.
   (iii) State the equations of any asymptotes of f.
(b) The function g is defined by

$$g(t) = 5(3 - 2e^{-t}), \quad t \geq 0$$

   (i) State the domain of g.
   (ii) State the range of g.
   (iii) Calculate the maximum possible value of g.
   (iv) Calculate the value of t at which $g = 10$.

3 (a) Given

$$A = \begin{pmatrix} 1 & -2 \\ 4 & 3 \end{pmatrix}, \quad B = \begin{pmatrix} 0 & 4 \\ -3 & 2 \end{pmatrix}, \quad C = \begin{pmatrix} 2 & 3 & 1 \\ -1 & 0 & 4 \end{pmatrix}$$

find, if possible,
(i) AB (ii) CB (iii) $|B|$ (iv) B^{-1} (v) the transpose of C
If a calculation is not possible then state this clearly.
(b) Consider the system

$$4x - 3y = 7$$
$$x + 6y = -32$$

   (i) Write the system in the form

$$DX = E$$

where D and E are matrices and $X = \begin{pmatrix} x \\ y \end{pmatrix}$.

   (ii) Calculate D^{-1}.
   (iii) Use matrix multiplication to find X and hence state the values of x and y.

4 (a) The vectors a and b are given by

$$a = (4, 2, -1) \quad b = (2, -3, 1)$$

Calculate
   (i) $a \cdot b$
   (ii) $|a|$ and $|b|$
   (iii) the angle between a and b.

(b) Vectors c and d are given by

$$c = (5, 3, -2) \quad d = (-1, 4, 1)$$

Calculate
   (i) a unit vector that is perpendicular to c
   (ii) a unit vector that is perpendicular to d
   (iii) a unit vector that is perpendicular to both c and d.

(c) Given vectors u, v and w the triple scalar product is $u \cdot (v \times w)$.
Show that

$$u \cdot (v \times w) = (u \times v) \cdot w$$

5 The function y is defined by

$$y(x) = 2x^3 + 3x^2 - 36x + 9$$

(a) Determine $\frac{dy}{dx}$.

(b) Determine $\frac{d^2y}{dx^2}$.

(c) Locate and identify all maximum and minimum points of y.

(d) Locate all the points of inflexion of y.

6 (a) Evaluate

$$\text{(i) } \int_0^2 (\cos 3x - 1)\, dx \quad \text{(ii) } \int_1^3 \left(e^{-2x} + \frac{2}{x} \right) dx$$

(b) Use a suitable substitution to evaluate

$$\int_0^2 x^2 \sqrt{x^3 + 9}\, dx$$

(c) Evaluate

$$\int_1^2 3x e^{2x}\, dx$$

7 (a) The complex numbers z_1, z_2 and z_3 are defined by

$$z_1 = 3 + 2j \quad z_2 = 1 - 3j \quad z_3 = -2 + j$$

   (i) Calculate $|z_2|$.
   (ii) Calculate $z_1 z_2$ in Cartesian form.
   (iii) Calculate $\frac{z_2}{z_3}$ in Cartesian form.
   (iv) Express z_1 in polar form.

(b) Form the quadratic equation whose roots are $z = -3 + 2j$ and $z = -3 - 2j$.

Paper 2

1 (a) Solve the following trigonometrical equation, stating all the solutions between 0° and 360°:

$$4 \sin \theta = 3 \cos \theta$$

(b) Express $4 \sin 2t - 3 \cos 2t$ in the form $R \cos(2t + \alpha)$, $\alpha \geq 0°$. Hence find the smallest positive value of t for which

$$4 \sin 2t - 3 \cos 2t = 4$$

(c) Figure Q1 shows three forces acting at the origin. Find the resultant force.

Figure Q1

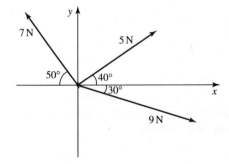

2 (a) If

$$(7A)^{-1} = \begin{pmatrix} 1 & 0 & 1 \\ 1 & 1 & 0 \\ 1 & 0 & 0 \end{pmatrix}$$

find A.

(b) Determine the eigenvalues and corresponding eigenvectors of the system

$$9x + 4y = \lambda x$$
$$-2x + 3y = \lambda y$$

3 (a) Find all values of z such that

$$z^3 = j$$

State your solutions in polar form.

(b) An LCR circuit has a voltage source of 2 V applied with a frequency of 10^3 Hz, a capacitor of 1.5×10^{-4} F, an inductor of 3×10^{-3} H and a 6 Ω resistor. Calculate the complex impedance in cartesian form.

(c) Describe the path traced out by a point represented by the complex number

$$z = 2e^{j\theta}$$

as θ varies from 0 to π.

4 A uniform lamina is enclosed by the curve $y = 2x^2 + 1$, $x = 1$, $x = 2$ and the x axis.

(a) Sketch the lamina.

(b) Calculate the volume generated when the lamina is rotated about the x axis.

(c) Show that the moment of inertia of the lamina about the y axis is $\frac{13M}{5}$ where M is the mass of the lamina.

5 (a) The sequence $x[k]$ is defined by

$$x[k] = \frac{2k + 1}{3k + 1}, \quad k = 2, 3, 4, \ldots$$

   (i) State the third term.

   (ii) State the limit of $x[k]$ as k tends to infinity.

   (iii) State a sequence $g[k]$ that has identical terms to $x[k]$ but which starts at $k = 0$.

(b) A function $f(x)$ is such that $f(0) = 3$, $f'(0) = -2$, $f''(0) = 1$ and $f'''(0) = 4$. Using the Maclaurin series state a cubic approximation to $f(x)$. Hence find an approximation to $f(0.5)$.

(c) Find the first three non-zero terms in the binomial expansion of $f(x) = \sqrt{1 + x^2}$. State the range of validity of your approximation.

6 (a) Consider the differential equation

$$x\frac{dy}{dx} = x + y, \quad y(1) = 2$$

   (i) Write the equation in standard form

$$\frac{dy}{dx} + Py = Q$$

   (ii) Determine the integrating factor.

   (iii) Find the general solution.

   (iv) Find the particular solution satisfying $y(1) = 2$.

(b) Solve

$$\frac{d^2y}{dx^2} - \frac{dy}{dx} - 2y = x + e^{2x}, \quad y(0) = 0, y'(0) = 0$$

7 (a) State the Laplace transform of (i) $t \sin 3t$, (ii) $e^{-2t} t \sin 3t$, [Hint: Use the first shift theorem.]

(b) Calculate the inverse Laplace transform of (i) $\dfrac{1}{s^2 + 4s + 5}$, (ii) $\dfrac{s}{s^2 + 4s + 5}$.

(c) Solve the following differential equation using Laplace transforms:

$$\frac{dx}{dt} + 3x = 1 + t, \quad x(0) = 0$$